Schaller

Synthesis of Organometallic Compounds

Inorganic Chemistry
A Textbook Series

Synthesis of Organometallic Compounds

A Practical Guide

Edited by

Sanshiro Komiya

Tokyo University of Agriculture and Technology

JOHN WILEY & SONS

Chichester · New York · Weinheim · Brisbane · Singapore · Toronto

National 01243 779777
International (+44) 1243 776777
e-mail (for orders and customer service enquiries): cs-books@wiley.co.uk
Visit our Home Page on http://www.wiley.co.uk
or http://www.wiley.com

Reprinted October 1999

Other Wiley Editorial Offices

John Wiley & Sons, Inc., 605 Third Avenue,
New York, NY 10158-0012, USA

VCH Verlagsgesellschaft mbH, Pappelallee 3,
D-69469 Weinheim, Germany

Jacaranda Wiley Ltd, 33 Park Road, Milton,
Queensland 4064, Australia

John Wiley & Sons (Asia) Pte Ltd, 2 Clementi Loop #02-01,
Jin Xing Distripark, Singapore 129809

John Wiley & Sons (Canada) Ltd, 22 Worcester Road,
Rexdale, Ontario M9W 1L1, Canada

British Library Cataloguing in Publication Data

A catalogue record for this book is available from the British Library

ISBN 0 471 97070 0; 0 471 97195 2 (pbk.)

Typeset in 10/12pt Times by Thomson Press (India) Ltd, New Delhi
Printed and bound in Great Britain by Bookcraft (Bath) Ltd.
This book is printed on acid-free paper responsibly manufactured from sustainable forestation, for which at least two trees are planted for each one used for paper production.

Contents

Preface

This textbook is intended for undergraduate students starting organometallic chemistry and researchers who want to use organometallic compounds, but are not professionals in organometallic chemistry. Although there are already many textbooks of organometallic chemistry that are relatively self-contained, the lack of practical guidance in organometallic chemistry is a deterrent to the use of organometallic compounds by beginners and nonprofessionals. Organometallic compounds are formed by nearly all metals and show a variety of structures and reactivities. Thus large compilations such as *Comprehensive Organometallic Chemistry*, *Dictionary of Organometallic Compounds*, *Organometallic Synthesis*, and *Inorganic Synthesis* have to be read to understand the chemistry of organometallic compounds as well as to learn synthetic methods; such books are inconvenient for these outsiders. A book that provides the most important references to organometallic compounds including practical preparation and chemistry would be useful not only in undergraduate or graduate school courses but also in the research laboratory. This book describes briefly the concepts of organometallic chemistry and provides an overview of the chemistry of each metal including the synthesis and handling of its important organometallic compounds. The idea of publishing this type of book in English originated from the Japanese book planned and published by Professors H. Suzuki and S. Komiya and editorially supervised by A. Yamamoto, which has been well received in our country. However this version is completely revised.

Parts of the book were edited and written during my stay at Indiana University and at the Australian National University. Particular thanks are due to Professor Kenneth G. Caulton and Professor Martin A. Bennett for brushing up our English and giving advice. I also thank Professor Akira Nakamura at Osaka University for giving me the opportunity to edit this book which has been entirely written by young Japanese organometallic chemists. Acknowledgment is also made to the excellent services of John Wiley & Sons Ltd for publishing the book. I also express my hearty gratitude to my good friend Professor Akira Miyashita at Saitama University for his great contribution to this book during the early stages. Unfortunately, owing to serious illness, he was unable to complete his contribution. Thus the enormous efforts, due to urgent preparation of manuscripts, by Professor Ito at Yokohama University and Professor Mashima at Osaka University are greatly acknowledged.

Finally, I would like to express my hearty thanks to my wife and family for their continuous help and encouragement.

Sanshiro Komiya
Tokyo, 1996

1 Introduction

S. Komiya, *Tokyo University of Agriculture and Technology*

Organometallic chemistry needs no special ideas if general chemical concepts are accepted. Many organic chemists feel that metals, especially those of the transition series, have various types of bonding schemes with molecules, atoms and ligands, and that the valency of the metal may change arbitrarily. In, fact, molecular compounds of transition metals have well-defined structures, such as octahedral, square planar, trigonal bipyramidal, etc, depending on the electronic state of the metal. On the other hand, organometallic compounds are generally believed to be very air- and-moisture sensitive, since very well known organometallic compounds such as alkyl lithiums and Grignard reagents are vigorously hydrolyzed in solution and organoaluminums are even flammable on exposing to air. Furthermore organotransition metal complexes are active intermediates in many catalyses. These facts probably make many researchers hesitate to use apparently unstable organometallic species in the laboratory. However in recent years, by virtue of the versatility of organometallic compounds in organic synthesis under mild conditions, many organic chemists are now using organometallic compounds as catalysts as well as reagents for creating new highly regio- and stereoselective reactions. Significant developments in these fields are now considered to be highly dependent on the organometallic reagents.

Organometallic chemistry is essentially based on coordination chemistry and organic chemistry. It is not too much to say that Werner's concept of coordination compounds began the development of coordination chemistry in the last 100 years, since it provided the basis for understanding complex inorganic compounds at a molecular level. However, inorganic and organic chemistries unfortunately tended to develop quite independently. Coordination chemists have concentrated on structure and bonding in relation to spectroscopy both experimentally and theoretically, whereas organic groups have used compounds containing metal–carbon bonds as a tool of organic synthesis based on organic chemistry. As a result, inorganic chemists have provided very important structural and theoretical concepts relating to coordination compounds, though they still had resistance to handling air-sensitive organometallic compounds. Coordination chemists are now attempting to resolve problems both in solid state materials by building clusters and on the roles of metals in biology at a molecular level. On the other hand, many highly selective and efficient organic synthetic reactions and catalyses using transition and main-group metals are still developing and attracting growing interest. Selectivities in metal mediated organic reactions are now competing with those of enzymes. It is generally considered that, after the discovery of ferrocene

Synthesis of Organometallic Compounds: A Practical Guide. Edited by S. Komiya

in 1951, organometallic chemistry has achieved explosive development. Organometallic chemists have helped to eliminate the barrier between organic and inorganic chemistry by dealing with all inorganic and organic compounds at a molecular level. As a result, the important concepts such as π-back bonding, agostic interaction, β-hydrogen elimination, reductive elimination, insertion, etc., have been introduced into the field of chemistry.

In recent years, scientists and chemists in fields other than organometallic chemistry have been frequently required, for their own purposes, to handle organometallic compounds which are believed to be very unstable and toxic. However, the problem is really not so difficult if one knows the general techniques for handling under inert gases and in vacuum. General concepts in organometallic chemistry are also not unusual, if both organic and inorganic chemistries are treated together. The purpose of this textbook is to serve as a practical guide to understand the general concepts of organometallic chemistry and methods of handling unstable compounds for graduate and undergraduate students and scientists who are not specialists in organometallic chemistry.

This book is divided in two parts: general concepts and the chemistry of individual metals, including practical synthetic methods for representative organometallic compounds. Chapters 2 and 3 summarize important fundamentals in organometallic chemistry. Chapter 4 describes experimental techniques, where the simplest ways to manipulate air-sensitive compounds are also included. Specialized techniques requring expensive facilities are not mentioned in detail, since they have already been described in references. In Chapters 5–17, the general chemistry of individual metals is summarized with references. Half of each chapter includes practical methods for the synthesis of organometallic compounds, including experimental tricks, which are usually not found in books, although some of them are referred to the original references.

2 Fundamentals of Organometallic Compounds

S. Komiya, *Tokyo University of Agriculture and Technology*

Organometallic compounds are generally defined as compounds having at least one metal–carbon bond. However, some compounds that do not contain any metal–carbon bonds, such as zerovalent metal complexes, hydrides, and dinitrogen complexes, are also admitted as members of this class because of their close relation to organometallic compounds. Though silicon is the element just below carbon in the Periodic Table and may be difficult to classify as a metal, organosilicon compounds are usually regarded as organometallic compounds. It is really not important to define organometallic compounds separately from other inorganic compounds, since some inorganic compounds are very important in this field. Thus the strict definition of organometallic compounds is avoided here. For convenience, organometallic compounds mentioned in this book are classified by the group of the element in the Periodic Table: main group elements and d-block metals (transition metals). Compounds of some elements of non-metallic groups 15–18 such as sulfur, selenium, and phosphorus are not included in this book.

2.1 Synthesis of organometallic compounds

Organometallic compounds are generally prepared by two general methods; (1) reactions using elemental metals and (2) reactions of already formed chemical compounds.

2.1.1 Synthesis from Elemental Metals

Organometallic compounds are synthesized from elemental metals by the following general methods: (a) preparation of Grignard reagents, alkyl lithiums, (b) metal-hydrocarbon reactions such as the synthesis of cyclopentadienyl sodium, NaCp, (c) the direct reaction of metals with CO producing metal carbonyls, and (d) metal vapor synthesis using high vacuum and high temperature techniques.

(a) $M + RX \rightarrow RMX$
(b) $2M + 2RH \rightarrow 2RM + H_2$
(c) $M + CO \rightarrow M(CO)_n$
(d) $M(vapor) + Substrate \rightarrow RM$

Synthesis of Organometallic Compounds: A Practical Guide. Edited by S. Komiya

The synthesis of organosilicon compounds by the direct reaction of Si with methyl chloride is a commercially important process.

2.1.2 Synthesis by Reactions

These procedures are more commonly used to prepare organometallic compounds. The following are representative methods (see each chapter): (a) exchange reactions such as the metathesis of metal halides with alkylating reagents, (b) metal–halogen or –hydrogen exchange processes, (c) addition reactions such as insertion reactions (hydrometallation, carbometallation, oxymetallation, carbonylation), (d) oxidative additions, (e) decarbonylation, (f) β-hydrogen elimination, (g) anionic *ate* complex formation with carbanions, (h) reductive carbonylation of metal oxides with CO, and (i) electrochemical methods for preparing organometallic compounds in unusual oxidation states.

(a) $MX + RM' \rightarrow RM + M'X$
(b) $RM + R'H$ (or $R'X$) $\rightarrow R'M + RH$ (or RX)
(c) $RM + A$ (= olefin, CO etc) $\rightarrow R\text{–}A\text{–}M$
(d) $ML_n + AB$ (= RX, ester etc) $\rightarrow M(A)(B)L_m$
(e) $RCOM \rightarrow RM + CO$
(f) $CH_3CH_2M \rightarrow HM(C_2H_4)$
(g) $MR_n + M'R' \rightarrow M'+[MR_nR']^-$
(h) $MO_n + mCO \rightarrow M(CO)_x + nCO_2$
(i) $RM^{n+} + e^-$ (or $-e^-$) $\rightarrow RM^{(n-1)+}$ (or $RM^{(n+1)+}$)

2.2 Properties of Metal–Carbon Bonds

One of the most important factors describing main group organometallic compounds is electronegativity and electronic structure. Valence bond theory generally predicts the structure of monomeric main group compounds: silicon and tin with sp^3 hybridization usually favor compounds of tetrahedral structure. However, hydrides or alkyl compounds of boron and aluminum tend to form dimers with electron-deficient bonding. Since the electronegativity of alkaline metals is small in comparison with carbon, the bonds in these organometallic compounds show considerable ionic character. Carboanionic character of alkyl lithiums is more markedly than that of Grignard reagents or alkyl aluminums. The metal–carbon bond of elements further to the right in the Periodic Table becomes more covalent, as observed, for example, in organosilicon compounds. Thus, organometallic compounds such as alkyl lithiums, Grignard reagents, and alkyl aluminums are very reactive towards hydrolysis, whereas organosilicon compounds are generally very stable to hydrolysis, in part because of the covalency of the bond.

Metal–carbon bonds in organotransition metal compounds are generally more covalent than those of main group metals. Though organometallic compounds were believed to be unstable because of the intrinsic instability of transition metal–carbon σ-bonds, it is now clear that the dissociation energies of transition metal–carbon bonds are not essentially low and are comparable to those of main group metal–carbon bonds. The apparent instability usually arises from low energy pathways for decomposition via β-hydrogen elimination, reductive elimination, oxidation, hydrolysis, etc. The bond dissociation energy of main group metal–carbon bond generally decreases on descending the same periodic group. Although there are only limited data, those of transition metal–alkyl bonds increase. In both cases bond dissociation enthalpy parallels the bond dissociation energy. The formation of stable metal– carbon multiple bonds is also a noteworthy feature of transition metal organometallic chemistry. Compounds having double and triple bonds are called carbene and carbyne complexes, respectively. These transition metal-stabilized reactive species are intermediates in various catalytic reactions.

M—C: alkyl complex
M=C: carbene complex
M≡C: carbyne complex

Since the electronegativities of transition metals and carbon are relatively close to each other, these bonds are considered more covalent than those in main group metal alkyls. Alkyl complexes are generally considered as the simplest intermediates in various transition metal promoted catalyses and organic reactions. Carbene complexes can be regarded as stabilized derivatives of free carbenes showing very interesting chemical reactivities and are also known to act as active intermediates in olefin metathesis and perhaps polymerization. Carbyne complexes can be regarded as analogues of surface organometallic species in heterogeneous catalysis. It is interesting that these organic ligands on the transition metal would provide more sophisticated reaction conditions, because the selectivity and activity of the reactions can be controlled by designing the electronic and steric properties of ligands. Besides these bonds, coordination bond, π-back bond, electron-deficient bond, etc, also play an important roles in organometallic chemistry. Coordination bond includes not only electron donation from Lewis bases such as amines and phosphines, but also those by C=C double bond, aromatic ring, cyclopentadienyl, η^3-allyl, and even C—H single bond (agostic interaction) or hydrogen (H_2), if they have appropriate orbital overlapping with transition metal orbitals. π-Back bonding from filled metal orbital to ligand empty π^* or σ^*-orbital will strengthen these bonds, but weakens the bond in ligands. σ-Donation from a ligand to a metal and π-back donation from a metal to ligand strengthen the metal–ligand bond and modify the chemical character of ligands. Details of structure and chemical reactivities of these bonds are found in books by Yamamoto, Crabtree, etc, listed in the references to this chapter.

Much attention is also paid to the hypervalency in the organometallic compounds of main group elements (Si, P, S) because of their high reactivities and structural interest. They are not described in this chapter, but will be discussed in Chapter 17.

In recent years, metal–metal bonds in clusters have also attracted much attention in relation to catalysis and solid state chemistry. They show more complicated bonding schemes in a sense of classical bonding. There are increasing numbers of reports that show not only their novel structure but also unusual and surprising chemical reactivities of complexes having more than two transition metals. Mechanisms and applications of many of these reactions are not well understood and still remain unresolved. These will be serendipitous research areas to be developed in the field of organometallic chemistry. Chemistry of clusters and metal–metal bonds are not described here in detail, but readers are strongly recommended to refer to the literature.

2.2.1 π Bonding

Coordination of the ligand to the metal usually increases the electron density of the central metal, if only electron donation is considered. This apparently contradicts the electroneutrality principle of atoms in molecules, when ligands coordinate to the electron-rich low valent metals. Typical examples are metal carbonyls in which the metal atom is frequently zero-valent.

When CO coordinates to a metal by its σ-HOMO at carbon, filled metal dπ orbitals of the transition metal will also overlap with the low lying π-LUMO of CO to give a π-back-bonding interaction, as shown in Figure 2.1. This implies that CO not only donates two electrons by coordination, but also receives two electrons from the metal at the same time, thus stabilizing the M—CO bonds and weakening the C≡O triple bond. The bond can also be described in terms of the following resonance scheme (canonical form) (Figure 2.2).

Thus the M—C bond possesses partial double bond character and becomes stronger than the single coordination bond. Back bonding usually increases the bonding energy

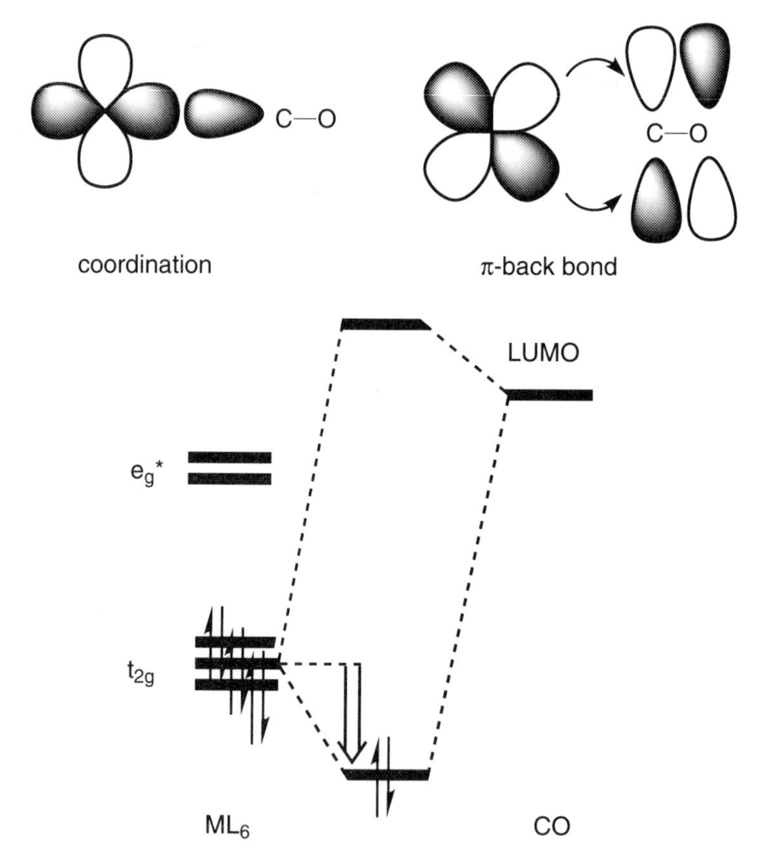

Figure 2.1
Schematic Drawing of Back Bonding of CO

$\ddot{M}^{-}\leftarrow C\equiv \ddot{O}^{+} \longleftrightarrow M=C=\ddot{O}$

Figure 2.2
Resonance Structure of M(CO) Bond

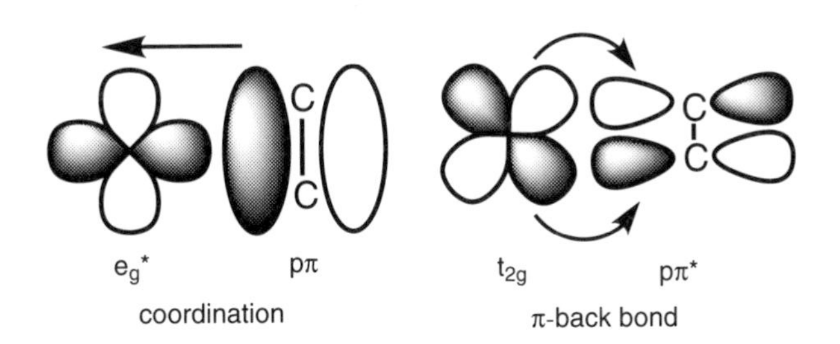

Figure 2.3
Coordination and π-Back Bonds of Ethylene Complex

between metal and ligand, when the metal is electron-rich having high energy HOMO orbitals and the ligand has low lying empty orbitals with proper symmetry. The magnitude of the back bonding is highly dependent on the oxidation state of the metal and ligands employed. For example, the ν(CO) frequency of metal carbonyls of the same structure decreases in the order of $[Fe(CO)_4]^{2-}$ (1790 cm^{-1}) < $[Co(CO)_4]^-$ (1890 cm^{-1}) < $[Ni(CO)_4]$ (2060 cm^{-1}), reflecting the decreasing extent of back bonding as the metal oxidation number becomes less negative.

The side-on type bonding of an olefin to a metal shown in Figure 2.3 is another typical example of π-back bonding. The olefin coördinates to the metal by using the HOMO π-orbital of the olefin to reduce the π-electron density at the C=C double bond. Since the pπ* orbital of the olefin has the proper symmetry to overlap with a metal t_{2g} orbital, an additional bond is also formed. Low valent transition metals are more capable of releasing electrons to the pπ* orbital and thus make stronger chemical bonds with olefins. C=C bond distances are generally elongated to 1.35–1.55 Å on coordination. Other ligands such as dinitrogen, isonitrile, and tertiary phosphines are also capable of participating in this π-back bonding. It should be noted that when the organic molecule coordinates to the metal, its electronic and steric properties change, leading to different chemical reactivity. This is one of the fundamental ways that transition metals can promote selective organic chemical transformations and catalyses.

2.2.2 3-Center-2-Electron Bond (3c-2e Bond)

$AlMe_3$ and BH_3 usually exist as dimers (Figure 2.4). The Al—Me—Al and B—H—B bonds involve only two electrons, which are usually not sufficient to make two single bonds. The bonding can, however, be easily understood by molecular orbital considerations. Considering the B—H—B bonding, combination of two sp^3 orbitals of B and one unique H s orbital gives three orbitals, i.e. bonding, antibonding and nonbonding

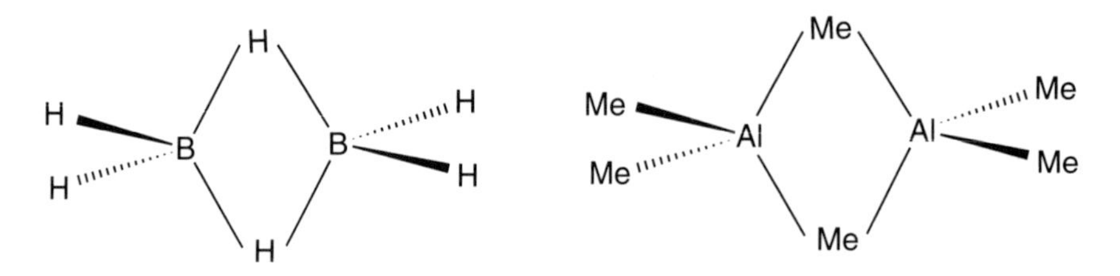

Figure 2.4
3c-2e Electron Deficient Bond

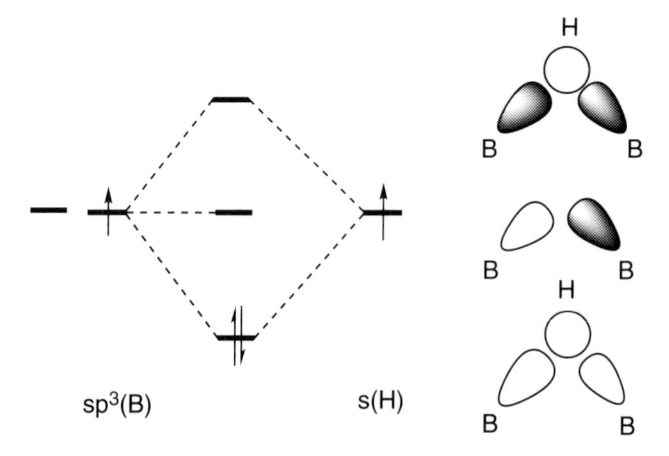

Figure 2.5
Molecular Orbital of B_2H_6 (only B—H—B linkage is shown for clarity)

orbitals as shown in Figure 2.5. The two electrons are placed in the low lying bonding orbital to stabilize the bond. This implies that the bond order of bridging B—H is half and in fact the bond distance of bridging B—H is significantly longer than those of terminal ones.

2.2.3 18 electron Rule

Electron counting in transition metal complexes (18 e rule) is an useful tool to understand their stability and structure, although it does not apply to all transition metal complexes—only for a majority of compounds containing π-acceptor ligands. The 18 electron rule is an extension of the idea of the octet rule, which applies to atoms having only s and p orbitals. The idea is that the molecule will be stable when the central atom has the same electronic structure as noble gases of the same row. A similar concept can be applied to transition metal complexes having d electrons. The compound is considered most stable when the total number of electrons around the atom becomes the same as that of noble gases in the same row (Effective Atomic Number rule). However, this counting method turns out to be more simple and easy if one count only valence electrons; namely only the sum of d-electrons of the metal and donating electrons from the ligands is counted. Thus metals in the same triad are expected to have the same number of d electrons and similar structural and chemical features. A compound that satisfies the 18 e rule is said to be coordinatively saturated, whereas one with less than 18 electrons is coordinatively unsaturated. Coordinatively unsaturated complexes are often reactive intermediates in various reactions. The systematic counting method for the general formula of $[MX_aL_b]^{c+}$ is well documented in the books by Crabtree and Yamamoto.

$$\text{e count} = N + a + 2b - c \tag{1}$$

where N represents the group number of the metal (which corresponds to the number of d electrons of the zero valent metal for the counting except for groups higher than 11), a is the number of one-electron donors, b is the number of two-electron donors, and c is the formal charge on the metal. The group numbers do not necessarily correspond to the number of d electrons in zero-valent metal atoms, since the total number of valence electrons is always considered in the calculation. The formal valency (oxidation state) of the metal is $a + c$, the coordination number is $a + b$. X and L indicate one- and two-electron ligands, respectively. Examples of a one-electron ligand are H, Me, Ph, Cl, and Br, and those of a two-electron ligand are NH_3, PPh_3, CO and $CH_2{=}CH_2$. These can be combined to produce a ligand that donates more than three electrons. For example η^3-C_3H_5, η^4-butadiene, η^5-Cp (cyclopentadienyl), and η^6-benzene are considered 3, 4, 5, and 6 electron donors, respectively. Anionic X^- ligands such as Cl^- can also be treated as two-electron ligands. In this case the metal loses one electron to form a M^+ cation, though the total electron around the metal does not change. Some ligands such as NO and OR are non-innocent, because the number of electrons donated to the metal varies depending on the structure and formalizm of the bonding. The M—NO bond is considered as a single bond (one-electron donor) when it shows bent structure. However, if a pair of electrons at the N atom also join the M—N bond, the NO ligand acts as a three electron donor giving linear bonding scheme to the metal. A similar phenomenon can be applied for M—OR bonds containing a π-donation from the OR ligand.

Although olefins generally donate two electrons by the pπ electrons, the bond can be regarded as as two M—C bonds forming metallacyclopropane. Olefins with electron withdrawing substituents are more likely to form bonds of this type. The two-electron donation does not change the formal oxidation state of the metal, but the formation of a metallacyclopropane ring would increase the oxidation state of the metal by two. Though efficient back bonding leads to the formation of metallacyclopropane structure and lengthens the carbon–carbon bond, most olefin complexes reported have an intermediate character.

Examples of compounds that obey the 18 e rule are shown in Figure 2.6. One method of counting (A) assumes a covalent bond between M and X, whereas the other (B) assumes an ionic bond where an electron is removed from M to X to give an anionic ligand X^-. It is a matter of taste which method is used; the same result is obtained.

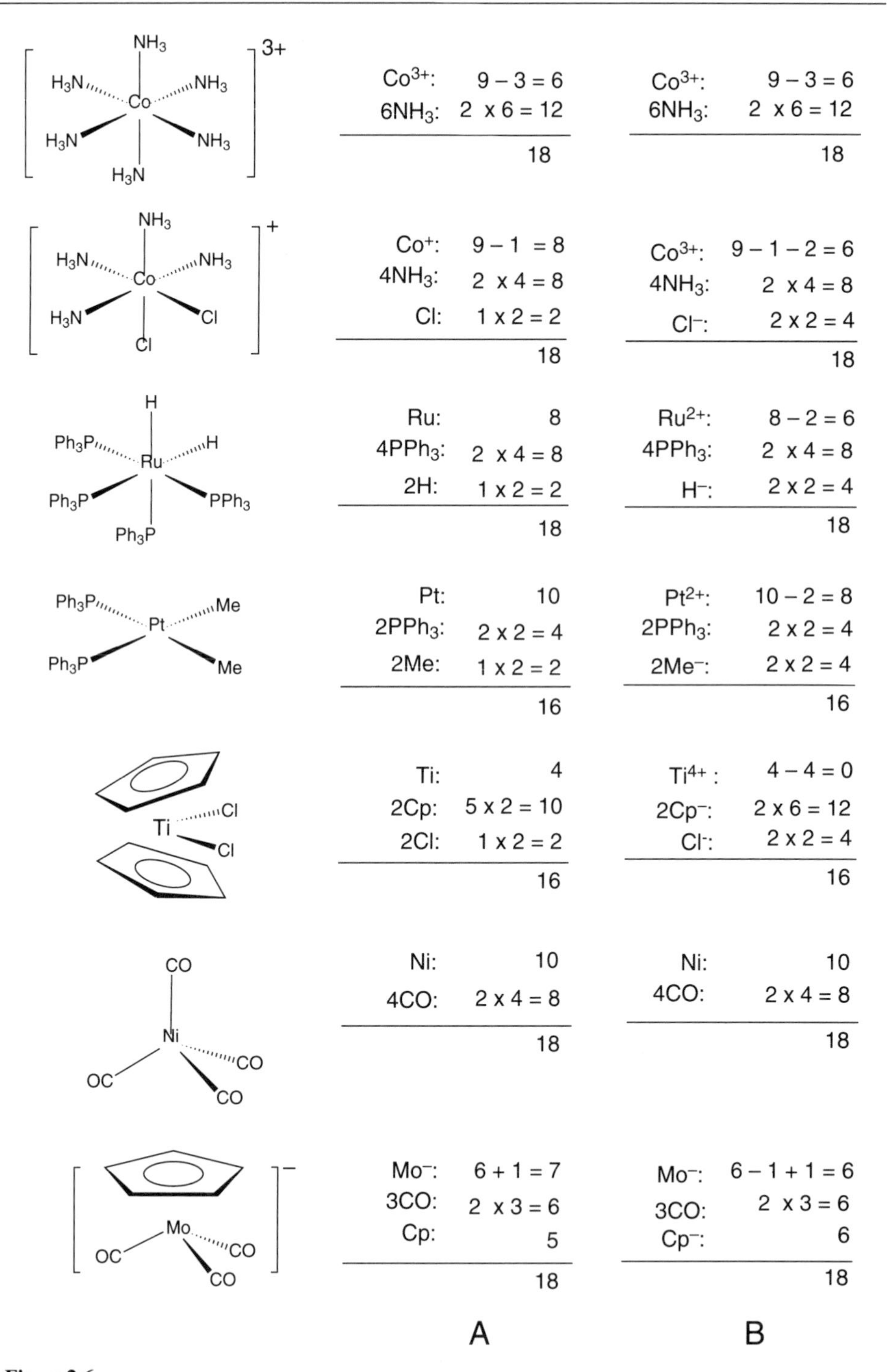

Figure 2.6
A Few Examples of Electron Counting of Transition Metal Complexes. A (Covalent Bond of M–X), B (Ionic Bond of M^{+}–X^{-})

2.2.4 Dihydrogen Complexes and Agostic Interactions

As noted in 2.1.1, chemical bonds can be created if an appropriate overlap of orbitals is present and the consequent energy advantage is established. Thus even dihydrogen, which has only bonding and antibonding orbitals, can also coordinate to transition metals under suitable circumstances. This bond is generally weak and the coordinated dihydrogen can be easily replaced by other ligands. This is an interesting feature that is related to the adsorption of dihydrogen at a molecular level on a metal surface in heterogeneous catalysis and is regarded as a first step of the interaction (Figure 2.7). When the H—H separation is not short enough to make a bond, the resulting compound is considered as a metal dihydride. The relaxation time (T_1) in NMR is believed to provide an available index of distinction between dihydrogen (< ~20 ms) and dihydride (> ~300 ms).

Another interesting bonding phenomenon is the so-called "agostic interaction." The word "agostic", coined by M. S. Brookhart and M. L. H. Green, means "hold to oneself" in Greek and the effect is illustrated by the following example. The ethyl ligand in $TiEtCl_3L_2$ is highly bent towards Ti and one of the terminal hydrogen atoms binds with Ti, making a cyclic structure (Figure 2.8). This bond results from the overlap between a

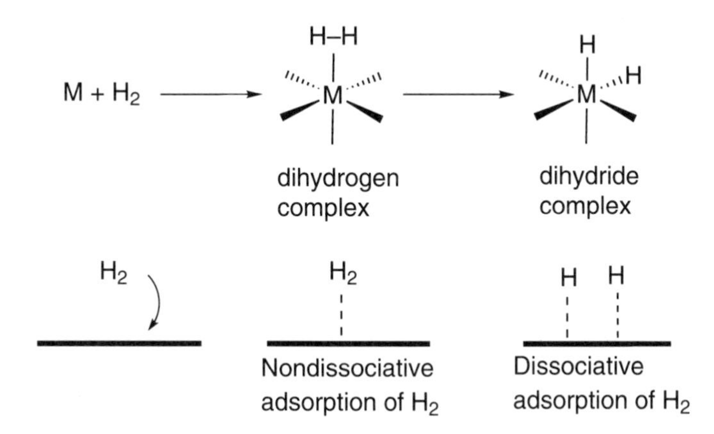

Figure 2.7
Dihydrogen and Dihydride Complexes in relation to Chemical Adsorption on the Heterogeneous Catalyst Surface

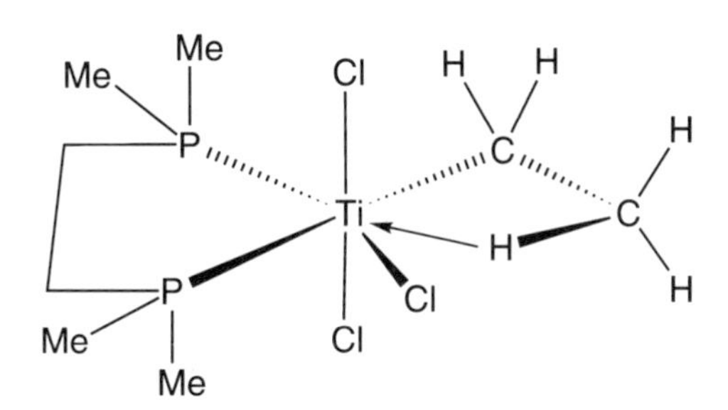

Figure 2.8
Agostic Interaction of β-CH with Ti

filled C—H σ-bond orbital and an empty d orbital of Ti. The interaction is considered as a prior step for bond rupture (oxidative addition), but in this case further oxidation of metal is unreasonable, since Ti(IV) has no available d-electrons.

2.2.5 Trans Effect and Trans Influence

The ligand trans to a certain ligand L receives a strong effect or influence from L. In ligand substitution reactions of square planar complexes, the rate of substitution varies significantly depending upon the ligand employed. One typical example is the synthesis of cis- and trans-dichlorodiammineplatinum(II) (Figure 2.9). The cis isomer (cisplatin is an effective anticancer drug) is prepared by the stepwise reactions of tetrachloroplatinate(II) with ammonia, whereas the trans isomer can be synthesized by the ligand substitution reaction of tetra(ammine)platinum(II) cation with chloride anion.

The reactions are best interpreted by the selective displacement of ligands trans to the chloride ligand in both cases. The rate of ligand displacement trans to chloride is much higher than that trans to NH_3, thus resulting in selective reaction. This kinetic effect is called the trans effect. The magnitude of this effect is known to be in the following order: H_2O, OH, NH_3, py < Cl, Br < SCN, I, NO_2, Ph < Me, $SC(NH_2)_2$ < H, PR_3 < C_2H_4, CN, CO. One of the origin of the trans effect is stabilization of the 5-coordinate transition state by π back bonding, which facilitates the reaction (Figure 2.10). σ-Interaction at the trans ligand also weakens the M—X bond to lower the activation energy. These phenomena were widely accepted and are very sensitive to the ligands and metals.

In addition, the physical properties of certain metal-to-ligand bonds, such as bond distances, chemical shifts and coupling costants in NMR and stretching bands in IR, also vary very much according to the trans ligand present. For example, the Pt—P bond distance of cis-$PtCl_2(PEt_3)_2$ is longer than that of the trans isomer, whereas the Pt—Cl distance of the cis isomer is shorter than that of the trans isomer, as shown in Figure 2.11. The results indicate that the bond trans to P is much weaker than that to Cl. Thus static (i.e. ground state) influence is called the trans influence. NMR and IR also show similar dependence. The order of magnitude of this influence is NO_3 < MeCN < Cl < OAc < I < py < SCN < $SbPh_3$ < SPh < $AsPh_3$ < CO < CN < PPh_3 < Me.

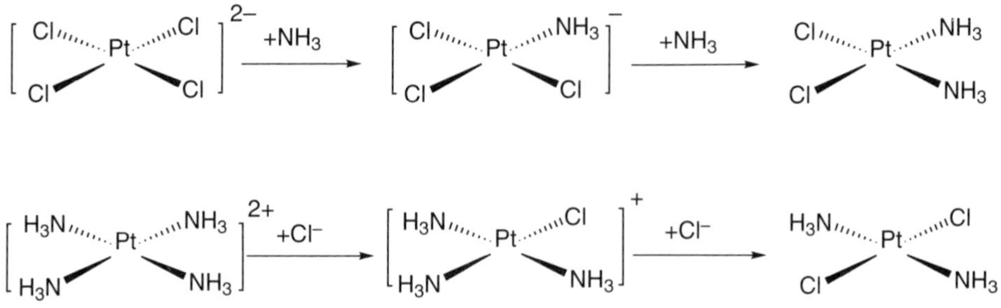

$$[PtCl_4]^{2-} \xrightarrow{+NH_3} [PtCl_3(NH_3)]^{-} \xrightarrow{+NH_3} cis\text{-}PtCl_2(NH_3)_2$$

$$[Pt(NH_3)_4]^{2+} \xrightarrow{+Cl^-} [PtCl(NH_3)_3]^{+} \xrightarrow{+Cl^-} trans\text{-}PtCl_2(NH_3)_2$$

Figure 2.9
Trans Effect in Selective Synthesis of cis and trans Dichlorodiammineplatinum(II)

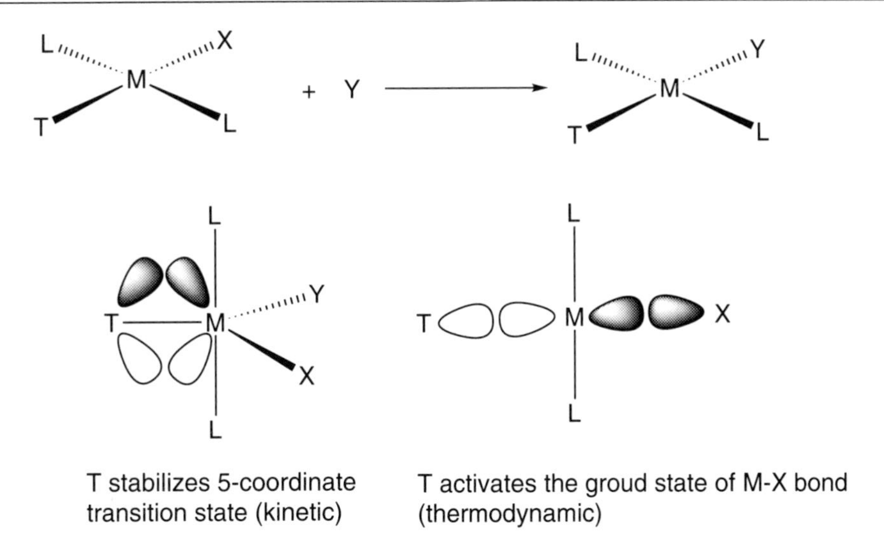

Figure 2.10
Origin of trans Effect and trans Influence of T in Selective Displacement of X by Y

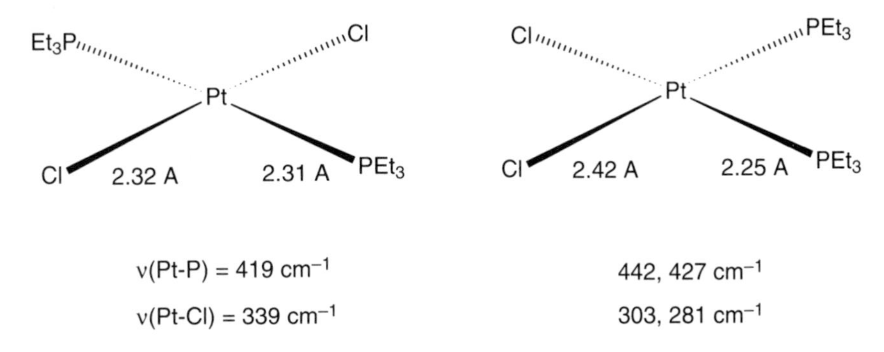

Figure 2.11
Trans Influence in $PtCl_2(PEt_3)_2$

Trans effect and trans influence are strongly related but they are not the same, since the trans-influence is a thermodynamic (bond-weakening) phenomenon, whereas the trans-effect is a kinetic phenomenon.

2.2.6 Fluxionality

Organometallic compounds are frequently stereochemically non-rigid in solution, an effect which is called fluxionality. NMR is one of the best tools to observe this phenomenon, because the rates are generally in the range of 10^2 to 10^5 s^{-1}. The ^{13}C NMR spectrum of trigonal bypyramidal $Fe(CO)_5$ (Figure 12) shows only one signal, because of fast exchange between the equatorial and axial CO ligands on the NMR time scale.

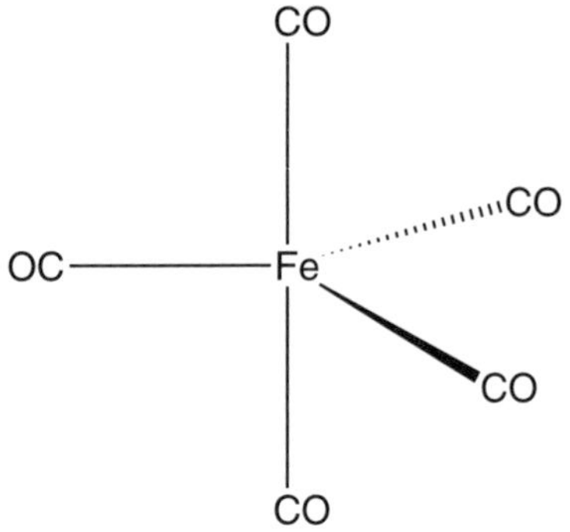

CO
OC—Fe—CO
CO
CO

Figure 2.12
Structure of 5-coordinate $Fe(CO)_5$

The exchange does not involve a dissociation of bonds, but is an intramolecular process which occurs without bond rupture.

The η^3-allyl ligands in $Zr(C_3H_5)_4$ also show fluxionality. The anti and syn protons usually appear in ^1H NMR as two doublets at different chemical shift with a small coupling to each other at –66 °C. These signals collapse to give only one doublet on heating to –20 °C. The fast proton exchange occurs between syn and anti protons through rotation about the C—C bond of an η^1-allyl intermediate (Figure 2.13). (η^n-L denotes that the ligand L bonds to metal by n atoms in L. η^1- and η^3-allyls are also called σ- and π-allyl, respectively.) The ethylene ligands of $RhCp(C_2H_4)_2$ also rotate about the Rh-ethylene π-bond at high temperature with activation energy of 68 kJ/mol. The chemical bond between Rh and ethylene is maintained during the rotation, since the coupling between Rh and ethylene remains intact. The fluxionalities involving facile intramolecular exchange processes of the coordination sites of ligands such as tertiary phosphine ligands are also known. These phenomena should always be taken into account in organometallic reactions.

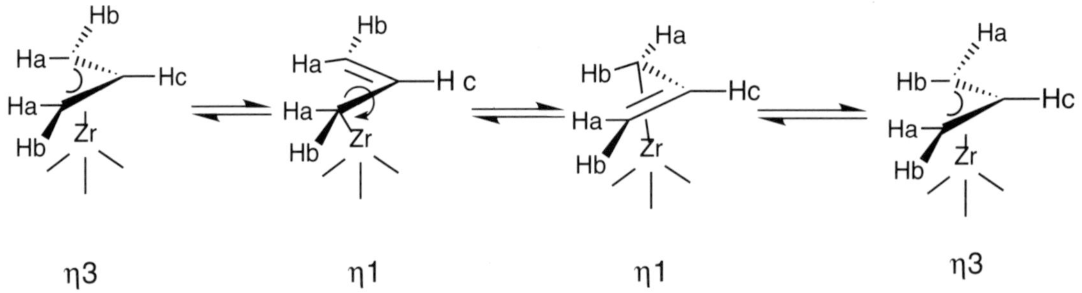

Hb Ha Hc Ha Hb Zr η3 ⇌ Hb Ha H c Ha Hb Zr η1 ⇌ Ha Hb Hc Ha Hb Zr η1 ⇌ Ha Hb Hc Ha Hb Zr η3

Figure 2.13
Facile Intramolecular Rearrangement of $Zr(C_3H_5)_4$

2.3 Reactions of Organometallic Compounds

Organometallic reactions are constituted of the following fundamental types of reactions: (1) coordination and dissociation (2) oxidative addition and reductive elimination, (3) insertion and deinsertion, (4) reaction at the coordinated ligand, (5) electron transfer. A brief summary of these reactions is given below. For details, the reader should refer to advanced books of organometallic chemistry.

2.3.1 Coordination and Dissociation

The coordination bond is a bond in which two bonding electrons formally originate from the ligand. Metal complexes showing facile ligand exchange are called substitution-labile (reactions being complete within 1 min at room temperature at 0.1 M solution); these in which ligand exchange is slow are called substitution-inert (reactions being too slow to measure or slow enough to follow at ordinary conditions by conventional techniques). In the former case two electron ligands can simply dissociate to give coordinatively unsaturated species and multistep equilibria of ligand dissociation are established. The coordinatively unsaturated species may be associated with weakly coordinating solvents.

$$ML_n \underset{}{\overset{K_1}{\rightleftharpoons}} ML_{n-1} + L \tag{2a}$$

$$ML_{n-1} \underset{}{\overset{K_2}{\rightleftharpoons}} ML_{n-2} + L \tag{2b}$$

$$ML_{n-2} \underset{}{\overset{K_3}{\rightleftharpoons}} ML_{n-3} + L \tag{2c}$$

$$\vdots$$

In contrast, one-electron ligands, such as alkyl and hydride, are usually not susceptible to facile dissociation; this would cause either homolytic fission of the bond or liberation as a free carbanion, which are both unlikely since the electronegativity difference between the atoms is not large enough. If the ligand is more polarizable as $M^+ X^-$, it may be better to represent the bond as coordination by the two-electron donor X^- to M^+.

$$M \overset{\bullet}{\underset{\bullet}{—}} X \qquad M^+ \leftarrow\!\!—\, :X^- \tag{3}$$

There are in general two mechanisms for the ligand exchange process, associative and dissociative. Six-coordinate octahedral d^6 complexes tend to dissociate to induce ligand exchange; they are unlikely to bind one more ligand in an associative process, because this would form an unstable 20 electron species. Thus, the rate for the ligand exchange process must be smaller than that of ligand dissociation. Five-coordinate d^8 and four-coordinate d^{10} metal complexes also dissociate in a similar manner. The 18 electron rule implies that coordinatively saturated complexes must dissociate ligands to induce ligand exchange. In contrast, association of one more ligand with four-coordinate square planar d^8 complexes (16 e) is possible, because of the coordinative unsaturation at the metal. Selective formation of a trigonal bipyramidal intermediate accounts for the selective ligand displacement process as shown in Figure 2.14.

Dissociative Ligand Exchange

$$ML_n \rightleftharpoons ML_{n-1} + L \quad (4)$$

$$ML_{n-1} + L' \rightleftharpoons ML_{n-1}L' \quad (5)$$

Associative Ligand Exchange

$$ML_n + L' \rightleftharpoons ML_nL' \quad (6)$$

$$ML_nL' \rightleftharpoons ML_{n-1}L' + L \quad (7)$$

However, it should be noted that fluxionality of the 5-coordinate intermediate both in associative and dissociative processes may reduce the stereoselectivity. Dissociation from square planar 16e complexes is also known to occur in the thermolyses of organotransition metal complexes such as $AuMe_3L$ and PtR_2L_2. The 3-coordinate 14 electron species thus formed has T-shape structure and not the one with C_3 symmetry. It is interesting to note that Au favors reductive elimination (Figure 2.15), whereas Pt

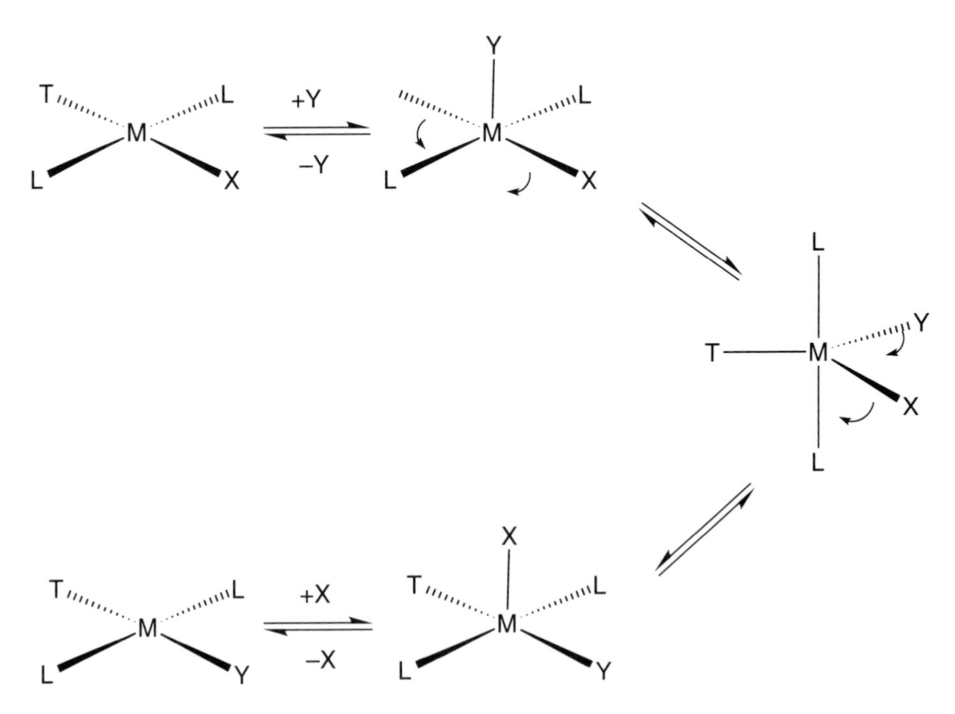

Figure 2.14
Stereoselective Associative Ligand Exchange of 4-coordinate Square Planar Complex

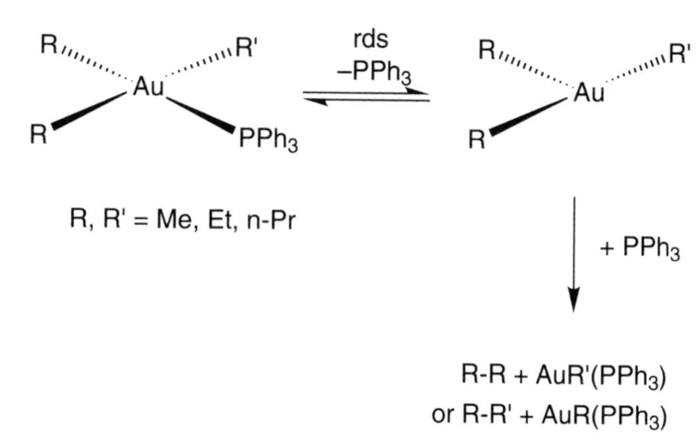

Figure 2.15
Dissociative Ligand Exchange Prior to Reductive Elimination of $AuR_3(PPh_3)$

favors β–hydrogen elimination (Figure 2.16), even though both have the same d^8 square planar structure. Other mechanisms involving intermediates or transition states with different coordination numbers, such as two for group 11 metals and seven for early transition metals, lanthanides and actinides, are also known.

The exchange rate and stability of complexes are also highly dependent on the ligands and metals employed. Steric factors are often important in these ligand exchange processes. Steric and electronic properties of ligands are defined later.

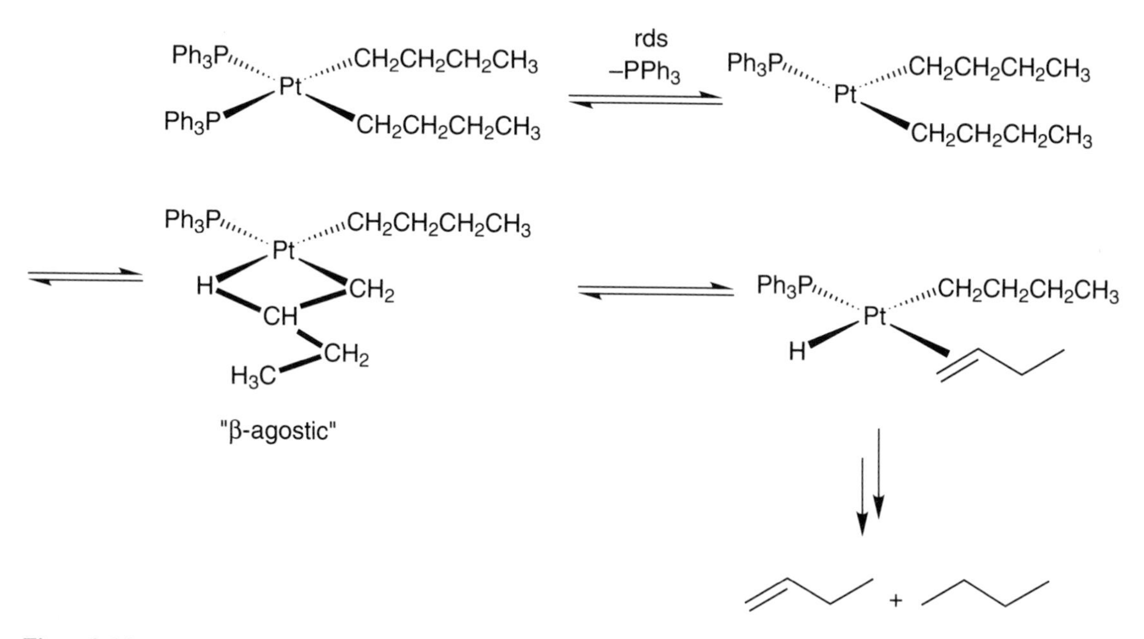

Figure 2.16
β-Hydrogen Elimination in Dialkylplatinum(II) complex

2.3.2 Oxidative Addition and Reductive Elimination

Oxidative addition and reductive elimination are the formal chemical processes involving oxidation and reduction of metal atoms accompanied by bond cleavage and formation between ligands A and B, respectively, as shown below. Thus, since A and B are one-electron ligands, the oxidation statte, electron count, and coordination number increase by two units in the oxidative addition. Oxidative addition to dinuclear or 17-electron complexes results in change of the oxidation state, electron count, and coordination number increase by one unit. The reductive elimination is the inverse process of oxidative addition and vice versa.

$$L_nM^{(n)+} + A\text{–}B \underset{\text{reductive elimination}}{\overset{\text{oxidative addition}}{\rightleftharpoons}} L_nM^{(n+2)+}\langle^{A}_{B}$$

$$L_nM^{(n)+}\text{–}M^{(n)+}L_n \text{ (or } L_nM^{(n)+}\text{)} + A\text{–}B \underset{\text{reductive elimination}}{\overset{\text{oxidative addition}}{\rightleftharpoons}} L_nM^{(n+1)+}\text{– }A + L_nM^{(n+1)+}\text{–}B$$

In many stoichiometric and catalytic chemical transformations using organometallics, oxidative addition is believed to be the first step which introduces and activates substrates to the metal leading to organometallic species, and reductive elimination is the last step in which the products are released. Although little is known yet about the mechanisms of these processes, the most important point is that both bond fission and formation take place by virtue of metal oxidation and reduction under mild conditions and reactive organometallic species can be formed in the reactions. A typical example of oxidative addition has been demonstrated for Vaska's complex $IrCl(CO)L_2$ as shown in Figure 2.17.

Reaction of trans-$IrCl(CO)_2L$ with hydrogen gas gives a cis-dihydride with increase in oxidation state of Ir from +1 to +3. A three-centered neutral transition state has

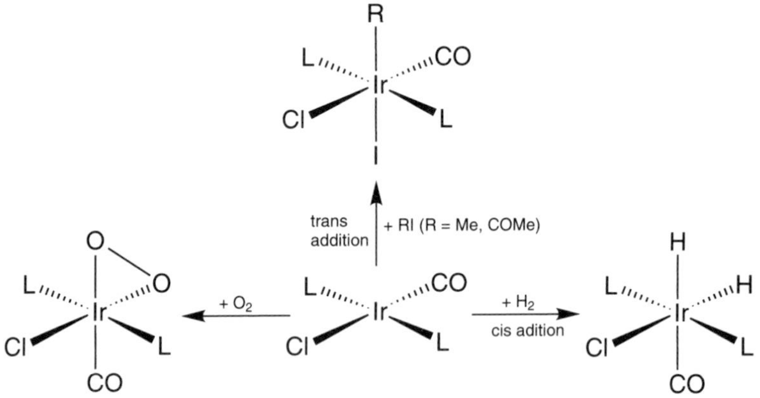

Figure 2.17
Reactions of Vaska's Complex Including cis and trans Oxidative addition and Coordination

been proposed for this reaction. It should be noted that the H—H bond (450 kJ/mol) has been cleaved on the metal under ambient conditions, thus providing a model for the first step of catalytic hydrogenation. Side-on coordination of dihydrogen forces the interaction of the σ^* orbital of H_2 with a filled $d\pi$ orbital to induce smooth bond fission (Figure 2.18). This is somewhat similar to the concept of back-bonding.

In contrast, organic halides such as methyl iodide and acetyl iodide oxidatively add to Ir(I) to give trans-organo(iodo)iridium(III) complexes exclusively. The trans product is considered to be formed by an S_N2 type mechanism in which inversion at the carbon center of the alkyl ligand is involved (Figure 2.19).

Vinyl and aryl halides also add to the metals such as Ni(0), Pd(0) and Pt(0) regioselectively. In oxidative addition of aryl halide, a nucleophilic substitution mechanism is also proposed. Sometimes electron transfer processes are also considered to be involved in oxidative addition, which gives anion radicals or free radicals. However, these processes sometimes lose stereoselectivity. In general, oxidative additions take place by a variety of mechanisms.

Oxidative additions of other bonds such as C—O, C—S, C—N, C—H and even C—C are also known. The reactions can be used to develop new methodologies in organic synthesis. Allyl esters and ethers are frequently used in selective allylations of

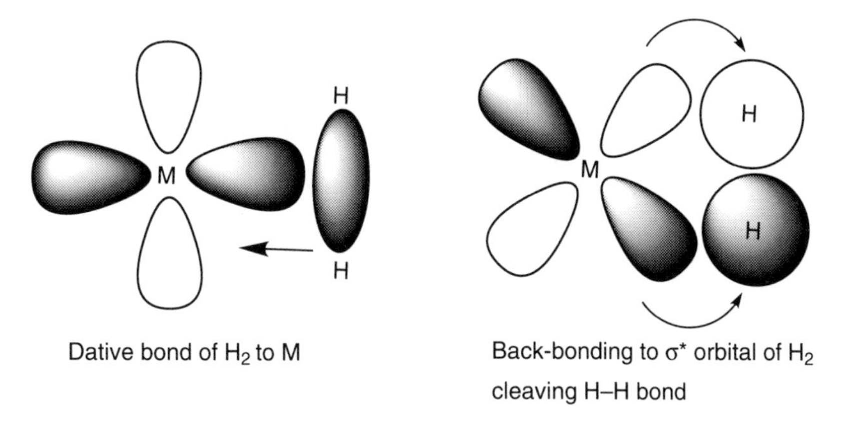

Figure 2.18
Side-on Interaction of H_2 with Transition Metal

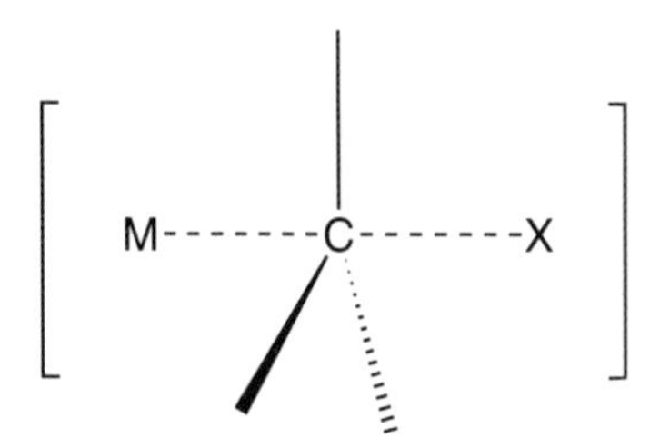

Figure 2.19
S_N2 Type Transition State

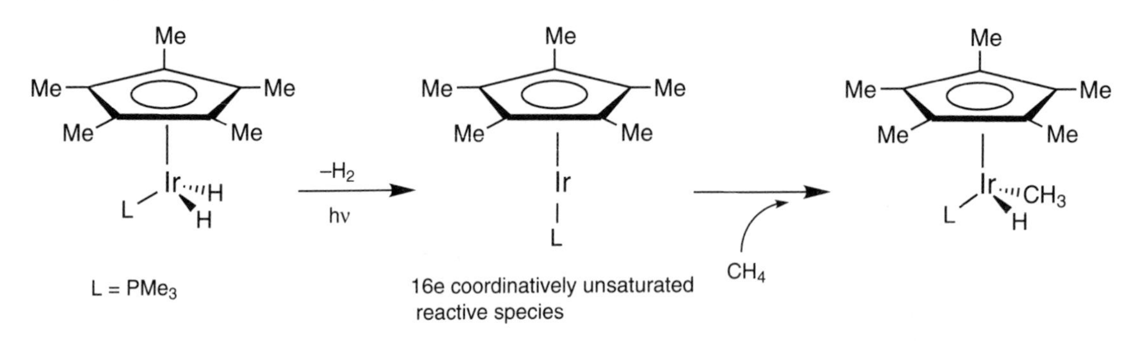

Figure 2.20
Oxidative Addition of Methane to Iridium(I)

nucleophiles and electrophiles catalyzed by Pd and Ru complexes. In these reactions C—O bond oxidative addition is considered to be a crucial first step. C—H and C—C bond oxidative addition by transition metal complexes are recent intriguing research developments and C—H bonds in hydrocarbons such as methane and benzene are now known to oxidatively add to certain low-valent transition metal complexes of Ru, Rh, Ir, Fe and Re, giving hydrido(alkyl)metal complexes (Figure 2.20). Recently, it has been proved possible to oxidatively add an unactivated C(sp^2)—C(sp^3) bond in arene derivatives to an iridium(I) center. However, applications of these fundamental reactions are still not well developed so far and are being sought. Some known examples are carbonylation of arenes and alkanes and dehydrogenation of alkane with Rh catalysts, carboxylation of alkanes with Pd complex, and aldol and Michael additions with Ru complexes.

Reductive elimination is also an important step in organometallic reactions. In general the process forms a new C—C bond with high regio- and stereoselectivity and is considered to proceed in a concerted fashion, the two organic ligands being eliminated from cis positions. A trans elimination process is geometrically impossible and usually symmetry forbidden (Figure 2.21).

It is interesting to note that bond forming reactions by reductive elimination between atoms of similar or the same electronegativities such as C—C and C—H smoothly take place by virtue of metal reduction, which are usually difficult processes in the sense of organic chemistry. In addition the processes also proceed under neutral mild conditions, where undesired side reactions may be prevented. The mechanism of reductive elimination has also been studied extensively on Ni, Pd, and Au complexes as well as theoretically. The processes sometimes require prior ligand dissociation to give unstable T-shaped intermediates from which facile cis reductive elimination takes

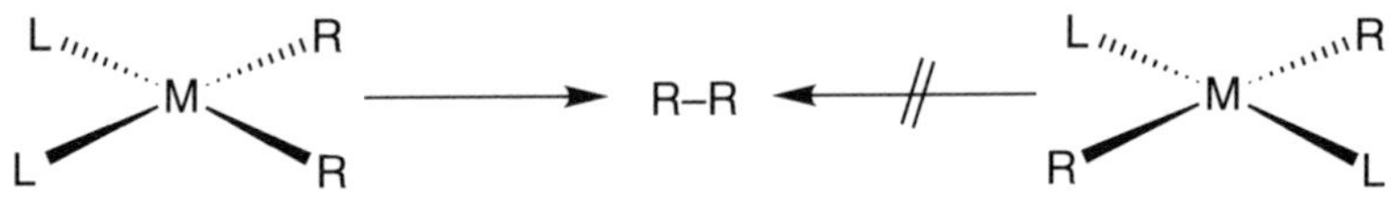

Figure 2.21
Reductive Elimination from Transition Metal Dialkyls

place, especially when a large contribution of ligand field stabilization is involved. Non-dissociative and associative processes are also known for reductive elimination pathways. The more electron-donating are the leaving ligands, the more are they reductively eliminated. For example, CH_3 ligands are more susceptible to reductive elimination than CF_3 groups. Aryl or vinyl ligands are also much better leaving groups than CH_3 due to effective overlapping of pπ-orbitals with neighboring M—C bond orbitals. However, acetylide ligands are stable to reductive elimination probably because of their intrinsically strong M—C bond energy. Electron withdrawal from the central metal by coordination of olefins with electron-withdrawing substituents or electron removal by oxidants generally accelerates the reductive elimination, partly because of the effective stabilization of the low valent inorganic products. Highly selective reductive elimination of two organic groups (or acyl and aryloxo) on $Ni(II)R_2(bpy)$ can be achieved on interaction with acrylonitrile or aryl halides (Figure 2.22). The inorganic product is $Ni(0)(olefin)_n(bpy)$ or Ni(II)ArX(bpy), respectively.

The following nickel catalyzed Grignard coupling reaction with aryl halide is a good example utilizing oxidative addition and reductive elimination as key steps as shown in Figure 2.23. Aryl halide oxidatively adds to Ni(0) to give an arylhalonickel(II) complex

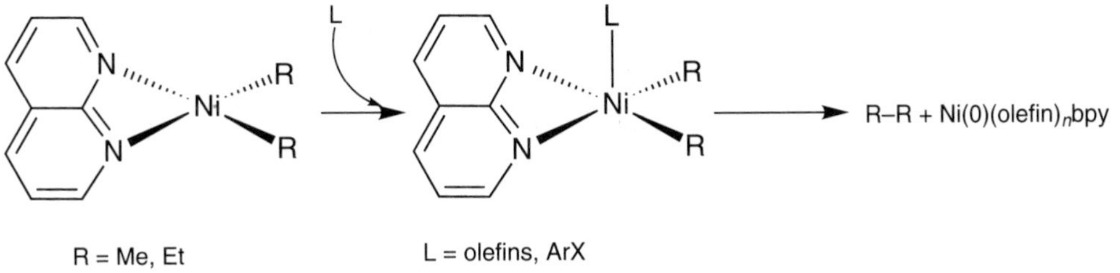

Figure 2.22
Reductive Elimination from NiR_2bpy Induced by Olefins Having Electron-Withdrawing Group

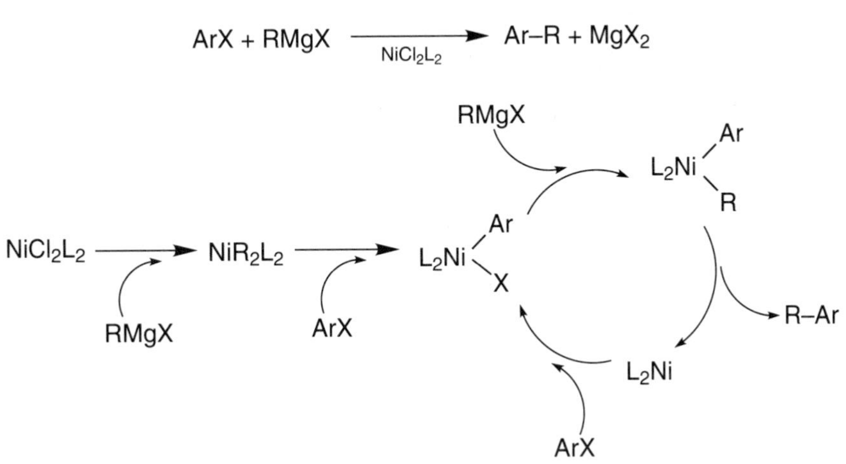

Figure 2.23
Mechanism of Ni-Catalyzed Cross-Coupling of Aryl Halide with Grignard Reagent

which is alkylated with a Grignard reagent, followed by reductive elimination of the product from dialkylnickel(II) species. Oxidative addition of aryl halide reforms the arylhalonickel(II) species. The mechanism may also involve Ni(I) intermediates via an electron transfer process.

2.3.3 Insertion and Deinsertion

The insertion of unsaturated organic molecules such as olefins, acetylenes, CO and RNC into metal–carbon bonds is a characteristic and important reaction in organometallic chemistry. The reaction is also termed carbometallation of the unsaturated molecule (i.e., multiply bonded reagents):

$$\text{M–R} + \text{Y} \longrightarrow \text{M–Y–R}$$

Typical examples of insertion are shown in Figure 2.24. The reactions lead to new C—C bond formation. Successive insertion of olefin frequently gives highly stereoregularly controlled polymers that are important materials. Catalytic hydrogenation of olefin also includes insertion of olefin into M—H bond followed by hydrogenolysis with hydrogen or reductive elimination of alkane. Isomerization of terminal to internal olefins also contains reversible insertion processes. Many of transition metal-catalyzed carbonylations, such as the Oxo and Monsanto processes that include insertion as a key step are major industrial processes using homogeneous catalyses. Combination of the insertion with other organometallic processes, has been widely explored and developed in organic and catalytic reactions (see those chapters on individual group metals). Insertion can be described as the 1,2-addition of an organometallic species across an olefin or acetylene and 1,1-addition to CO or RNC. The insertion reactions frequently require prior coordination of substrates, since a

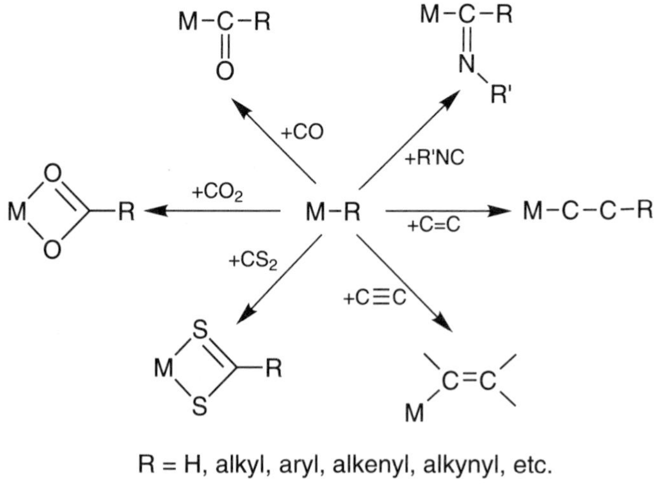

Figure 2.24
Insertions Into an M–R Bond

strong retarding rate effect of added stabilizing auxiliary ligand is sometimes observed. The stereochemistry around the multiple bond is usually cis. The prior coordination or interaction is considered to be a very important step in determining regio- and enantioselectivity of the reaction and is usually controlled by the ligand employed. Highly selective asymmetric hydrogenation of prochiral olefins such as N-acetoamidocinnamate by Rh complexes having a bidentate chiral ligand (BINAP) proceeds through the facile oxidative addition of H_2 followed by insertion of the olefin into the hydride in the thermodynamically unstable hydrido(olefin)rhodium(I) intermediate. A concerted processes including 3- or 4-center transition state for the insertion is the most typical, which gives only cis addition product. The alkyl group has been proved to migrate to the unsaturated ligand both in CO and olefin insertions.

The reverse process of insertion is called as deinsertion. β-Hydrogen in the alkyl chain is generally the most susceptible to this reaction, though α, γ and δ hydrogens can also be eliminated under certain reaction conditions, where low energy pathways are chemically blocked. The reactions involve intramolecular activation of C—H bond by the transition metal to give a hydrido(olefin)metal complex and the stereochemistry is usually cis. This is one of the reason why many transition metal alkyl intermediates are apt to decompose to release olefins in spite of their large intrinsic M—C bond dissociation energy. An unfavorable β-hydrogen elimination can be suppressed by protecting vacant sites for coordination of β-C—H bond by adding excess of innocuous ligands such as tertiary phosphines. Thus the thermolysis of transition metal alkyls leading to β-elimination usually involves prior dissociation of ligand.

2.3.4 Reactions of Coordinated Ligands

The chemical reactivities of the organic ligands (molecules) vary very much when coordinated to the transition metal. For example, olefins, which are usually unreactive to nucleophiles, do react with various nucleophiles such as malonates, acetates, hydroxides, etc, leading to new carbon–carbon bond formation. Transition metals in high oxidation states such as Pd(II) and Ni(II) are most suitable for this reaction, since they tend to withdraw electron density from the olefins. Similar reactivity is also observed in π-allyl metal species. The stereoselectivity of these reactions is usually trans, since the incoming reagents usually approach from the side opposite the metal. However, this is not always true and the incoming reagents can also attack the electrophilic metal; subsequent insertion gives a cis adduct (Figure 2.25).

Thus the selectivity and reactivities of coordinated olefins are highly dependent on the mechanisms and steric and electronic environments of the transition metal.

The reactions of coordinated dienes, acetylenes, allyls, arenes, CO, alkyls, etc, are well documented in many organometallic textbooks.

Though transition metal-to-carbon or -hydrogen bonds are more covalent than those of alkali and alkaline earth metals, they are capable of reacting with electrophiles, possibly by a concerted mechanism. One interesting reaction pattern is the σ-bond metathesis of group 4 organotransition metal and organoactinide (or lanthanide) complexes with alkanes. Alkane coordination increases C—H polarization to set the stage for CH activation and is regarded as another possible way in addtion to the oxidative addition method of C—H bond activation using metals.

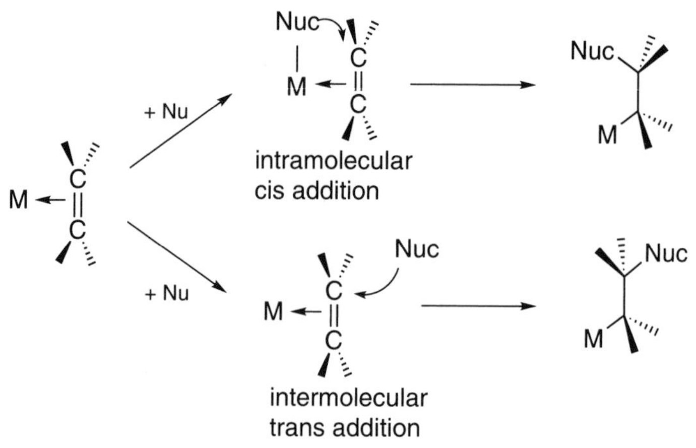

+ Nu
intramolecular
cis addition

+ Nu
intermolecular
trans addition

Figure 2.25
Reaction Pathways of Coordinated Olefin

$$M{-}CH_3 + CH_4 \rightleftharpoons \begin{array}{ccc} M & \cdots & CH_3 \\ \vdots & & \vdots \\ H_3C & \cdots & H \end{array}$$

2.3.5 Electron Transfer

Electron transfer reactions are frequently encountered in the reactions of transition metal complexes, especially with strong oxidants such as $[IrCl_6]^{3-}$ and $CuCl_2$. Electrochemical oxidation proceeds without structural change of the transition metal complexes. As noted previously, the chemical reactivity of organotransition metal complexes depends on their oxidation states. Without going into detail, we give one example in organoiron chemistry, where thermolysis products alter dramatically depending on the metal oxidation state (Figure 2.31). Thermolysis of $Fe(II)Et_2(bpy)_2$ is known to liberate exclusively ethylene and ethane by β-hydrogen elimination from one ethyl ligand followed by reductive elimination of ethyl and hydrido ligands, whereas $[Fe(III)Et_2(bpy)_2]^+$, obtained by electrochemical oxidation, releases an ethyl radical. Further oxidation to $[Fe(IV)Et_2(bpy)_2]^{2+}$ leads to the formation of only a reductive elimination product, butane (Figure 2.26).

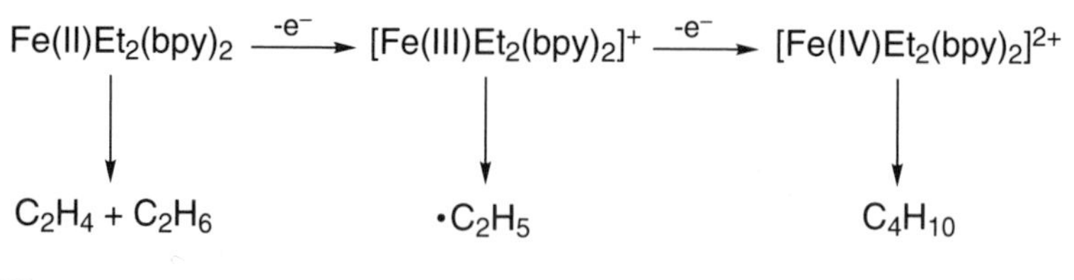

$$\begin{array}{ccccc} Fe(II)Et_2(bpy)_2 & \xrightarrow{-e^-} & [Fe(III)Et_2(bpy)_2]^+ & \xrightarrow{-e^-} & [Fe(IV)Et_2(bpy)_2]^{2+} \\ \downarrow & & \downarrow & & \downarrow \\ C_2H_4 + C_2H_6 & & \bullet C_2H_5 & & C_4H_{10} \end{array}$$

Figure 2.26
Thermolysis of $[FeEt_2(bpy)_2]^{n+}$ ($n = 0, 1, 2$)

The results indicate that Fe(II) favors β-hydrogen elimination, but Fe(III) and Fe(IV) induce radical formation and reductive elimination, respectively. Preferential reductive elimination from Fe(IV) is consistent with the fact that many reductive eliminations are accelerated by chemical oxidation or electron removal from metal as mentioned above. Electron transfers involving transition metals are more frequently observed in the organometallic reactions. They are sometimes difficult to differentiate from nucleophilic reactions of transition metal complexes unless the free radicals can be trapped, since the trends in electron transfer and nucleophilic reactions usually parallel each other.

Only a limited number of organotransition metal complexes having odd numbers of electrons are known, partly because of some difficulties in treatment and characterization, and partly because of the stability of complexes having an 18 e and 16 e count. They are usually prepared by chemical electron transfer reactions or electrochemical oxidation/reduction and are likely to play an important role in fields such as catalysis and new materials.

References

The following are reference books recommended to read for details on organometallic chemistry, which can be referred to for convenience.

Organometallic Chemistry

(a) R. H. Crabtree, *The Organometallic Chemistry of the Transition Metals*, 2nd. ed., John Wiley, New York (1994), (b) C. Elschenbroich, A. Salzer, *Organometallics*, 2nd. ed., VCH, Weinheim (1992), (c) A. Yamamoto, *Organotransition Metal Chemistry*, John Wiley, New York (1990), (d) J. P. Collman, L. S. Hegedus, J. R. Norton, R. G. Finke, *Principles and Applications of Organometallic Chemistry*, 2nd. ed., University Science Books, Mill Valley, CA. (1987), (e) C. M. Lukehart, *Fundamental Organometallic Chemistry*, Brooks, Cole, Monterey, CA. (1985), (f) M. Bochmann, *Organometallics 1 and 2*, Oxford Science Publications, Oxford (1994), (g) J. K. Kochi, *Organometallic Mechanisms and Catalyses*, Academic Press, New York (1979), (h) G. Wilkinson, F. G. A. Stone, E. Abel Eds, *Comprehensive Organometallic Chemistry*, Pergamon Press, Oxford (1982), (i) E. Abel, F. G. A. Stone, G. Wilkinson Eds, *Comprehensive Organometallic Chemistry*, 2nd. ed., Pergamon Press, Oxford (1995), (j) J. Buckingham, Ed., *Dictionary of Organometallic Compounds*, Chapman and Hall, London, (k) J. S. Thayler, *Organometallic Chemistry: An Overview*, VCH, New York (1987), (l) J. J. Eisch, R. B. King, *Organometallic Synthesis* vol. 2, Academic Press, New York (1981), (m) M. Dub, Ed., *Organometallic Compounds, Methods of Synthesis, Physical Constants and Chemical Reactions*, 2nd ed., Springer-Verlag, Berlin (1966), (n) A. N. Nesmeyanov and K. A. Kocheshkov, eds., *Methods of Elemento-Organic Chemistry*, North Holland, Amsterdam (1967), (o) W. A. Herrmann, Ed., *Synthetic Methods of Organometallic and Inorganic Chemistry*, Thieme, Stuttgart (1996).

Inorganic Chemistry

(a) F. A. Cotton, G. Wilkinson, *Advanced Inorganic Chemistry*, 5th. ed., John Wiley, New York (1988), (b) J. E. Huheey, *Inorganic Chemistry Principles of Structure and Reactivities*, 4th. ed., Harper & Row, New York (1993), (c) D. F. Shriver, P. W. Atkins, C. H. Langford, *Inorganic Chemistry*, 2nd. ed., Freeman, New York (1994), (d) F. Basolo, R. G. Pearson, *Mechanisms of Inorganic Reactions*, 2nd. ed., John Wiley, New York (1967), (e) G. Wilkinson, R. D. Gillard, J. E. McCleverty, Eds., *Comprehensive Coordination Chemistry*, Pergamon Press, Oxford (1987),

(f) A. F. Trotman-Dickenson, Ed., *Comprehensive Inorganic Chemistry*, Pergamon Press, Oxford (1973), (g) M. Chisholm, *Early Transition Metal Clusters with π-Donor Ligands*, VCH, New York (1995), (h) B. F. G. Johnson, Ed., *Transition Metal Clusters*, Wiley-Interscience, New York (1980), (i) F. A. Cotton, R. A. Walton, *Multiple Bonds between Metal Atoms*, John Wiley, New York (1982).

Organic Synthesis

(a) E. Negishi, *Organometallics in Organic Synthesis*, John Wiley, New York (1980), (b) S. G. Davies, *Organo-Transition Metal Chemistry: Applications to Organic Synthesis*, Pergamon, Oxford (1982), (c) H. Alper, *Transition Metal Organometallics in Organic Synthesis*, Vols I and II, Academic Press, New York (1978), (d) A. J. Peason, *Metallo-organic Chemistry*, Wiley-Interscience, Chichester (1985), (e) R. F. Heck, *Organotransition Metal Chemistry, A Mechanistic Approach*, Academic Press, New York (1974).

Electron Transfer

(a) J. K. Kochi, *Organometallic Mechanisms and Catalysis*, Academic Press, New York (1979), (b) D. Astruc, *Electron Transfer and Radical Process in Transition-Metal Chemistry*, VCH, New York (1995).

Homogeneous Catalysis

(a) P. A. Chaloner, *Handbook of Coordination Catalysis in Organic Chemistry*, Butterworth, London (1986), (b) G. W. Parshall and S. D. Ittel, Homogeneous Catalysis, Wiley, New York (1992), (c) A. Nakamura and M. Tsutsui, *Principles and Applications of Homogeneous Catalysis*, Wiley, New York (1980), (d) M. M. Taqui Khan, A. E. Martell, *Homogeneous Catalysis by Transition Metal Complexes*, Academic Press, New York (1974), (e) C. Masters, *Homogeneous Transition-Metal Catalysis: A Gentle Art*, Chapman and Hall, London (1981) (f) G. Henrici-Olive, S. Olive, *Coordination and Catalysis*, Verlag Chemie, Weinheim, (1976), (g) L. H. Pignolet, Ed., *Homogeneous Catalysis with Metal Phosphine Complexes*, Prenum, New York (1983), (h) J. Falbe, Ed., *New Synthesis with Carbon Monoxide*, Springer-Verlag, Berlin (1980), (i) I. Wender, P. Pino, *Organic Synthesis via Metal Carbonyls*, Wiley-Interscience, New York, Vol. 1 (1968), Vol. 2 (1977), (j) *Carbon Monoxide in Organic Synthesis*, Springer-Verlag, Berlin (1970), (k) W. Keim, Ed., *Catalysis in C1 Chemistry*, D. Reidel, Dordrecht (1983), (l) J. Boor, *Ziegler–Natta Catalysts and Polymerizations*, Academic Press, New York (1978), (m) C. C. Price, E. J. Vandenberg, Eds., *Coordination Polymerization*, Plenum Press, New York (1983), (n) J. C. W. Chien, Ed., *Coordination Polymerization*, Academic Press, New York (1975).

3 Ligands

S. Komiya, *Tokyo University of Agriculture and Technology*

Ligands play very important roles in organometallic and coordination chemistry, since they can bring about drastic changes in the chemical and physical properties of transition metal complexes [1,3]. Thus, the products in many transition metal catalyzed reactions depend on the ligand employed. In other words, some reactions can be controlled simply by ligand selection [1]. For example, $RhCl(PPh_3)_3$ (Wilkinson's catalyst) having bulky triphenylphosphine ligands is a good catalyst for hydrogenation of terminal olefins, but not internal olefins. In Rh catalyzed hydroformylation, addition of tertiary phosphine increases the ratio of commercially important linear aldehydes to branched ones. This may be because of the instability of branched alkyl intermediates due to the coordination of the bulky phosphine ligand. Recently many transition metal-mediated asymmetric reductions and oxidations have been achieved by employing appropriate chiral ligands [2]. Coordination of the chiral ligand to the metal generates an asymmetric environment at the metal, which discriminates between the faces of prochiral substances such as substituted olefins on coordination. Thermodynamic and kinetic energy differences of the two intermediates allow highly selective asymmetric reactions to be carried out [2]. On the other hand, addition of excess ligand to the reaction system often retards the reactions. This may be due to the blocking of coordination sites of substrates for the following reactions. This suggests that it is quite important to create a coordinatively unsaturated species in transition metal mediated reactions, though the supporting ligands also play crucial roles in selectivity.

Transition metal hydrides are considered as containing hydride ligands (H^-), since the metal is usually less electronegative than hydrogen [3]. However the hydride in hydridotetracarbonylcobalt acts as a strong acid (proton), similar to sulfuric acid, even though it is formally referred to a hydride (H^-). Strong back-bonding from CO ligands assists in electron removal from the hydride to the metal. Coordinated olefins become more susceptible to electrophilic reactions when they are coordinated to high valent transition metal complexes, although free olefins generally react only with electrophiles [4]. The susceptibility to electrophilic attack arises from decreasing electron density at the C=C double bond of the olefin on coordination. Thus, ligands have significant influence not only on the chemical and physical properties of the ligands but also on those of the ligand itself on coordination. Molecular mechanics and physical modeling consideration can sometimes help to predict the steric influence on structure and reactivity.

The hard and soft acid and base (HSAB) principle is also an important concept that allows qualitative prediction of the nature of the metal–ligand bonds [5]. Soft metals

Synthesis of Organometallic Compounds: A Practical Guide. Edited by S. Komiya

favor binding to soft ligands and hard metals favor binding to hard ligands. Hard metal ions include alkali metals, alkaline earth metals and hydrogen ion (proton) which are usually non-polarizable, whereas soft metal ions include the heavier transition metals and those in lower oxidation states (polarizable). The tendency to complex with soft metals is as follows: N << P, O < S < Se ~ Te, F < Cl < Br < I. Soft metals tend to prefer ligands that have available empty orbitals for back-bonding and that can make covalent bonds. In other words, soft metals have a low energy LUMO and soft ligands (bases) have a high energy HOMO, thus causing the bond to be more covalent. Thus the soft platinum metals in low oxidation states favor unsaturated or polarizable ligands such as ethylene, PPh_3, Br^-, and I^-.

In this chapter, a comprehensive description of various ligands including N, O, S, P, As ligands, asymmetric ligands, etc, is not given; only the steric and electronic effects of commonly encountered phosphorus ligands are briefly summarized. These provide the basis for the selection of ligands. The design and development of new types of ligands for specific purposes is a growth area in organometallic and coordination chemistry.

3.1 Electronic Effect

Ligands having electron-donating substituents are generally strong donors. Thus, trialkylphosphines donate more strongly than triarylphosphines to metals. As a result, metal complexes with trialkylphosphines are more electron-rich than their triarylphosphine analogues and therefore such compounds are considered to be more susceptible to nucleophilic reactions and also to have stronger basicity. The pK_a values of the conjugate acids of phosphines ($[HPR_3]^+$) are frequently used as an index of this property and are proportional to the sum of Taft's σ^* value of three substituents (electron-donating ability of organic substituents) on the P atom, as expressed in the following equations (Figure 3.1) [6].

Tertiary phosphines	$pK_a = 7.85 - 2.67\Sigma\sigma^*$
Secondary phosphines	$pK_a = 5.13 - 2.61\Sigma\sigma^*$
Primary phosphines	$pK_a = 2.46 - 2.64\Sigma\sigma^*$
Tertiary amines	$pK_a = 9.61 - 3.30\Sigma\sigma^*$
Secondary amines	$pK_a = 12.13 - 3.23\Sigma\sigma^*$
Primary amines	$pK_a = 13.23 - 3.14\Sigma\sigma^*$

It is interesting that all slopes of the lines for phosphines or amines are almost the same, no matter whether they are primary, secondary or tertiary bases. This suggests that the substituent effect on the P or N atom has a linear relationship to the basicity by virtue of the linear free energy relationship (LFER). They are often used as an index of donor strength of the ligand. Selected values of pK_a are listed in Table 3.1. The electron-donating ability decreases in the following order; tertiary, secondary and primary in PR_3 or NR_3 series. Phosphites are weaker electron donors than phosphines, whereas amines are stronger electron donors than phosphines. The higher the basicity of the ligand, the stronger is the donor ability in general.

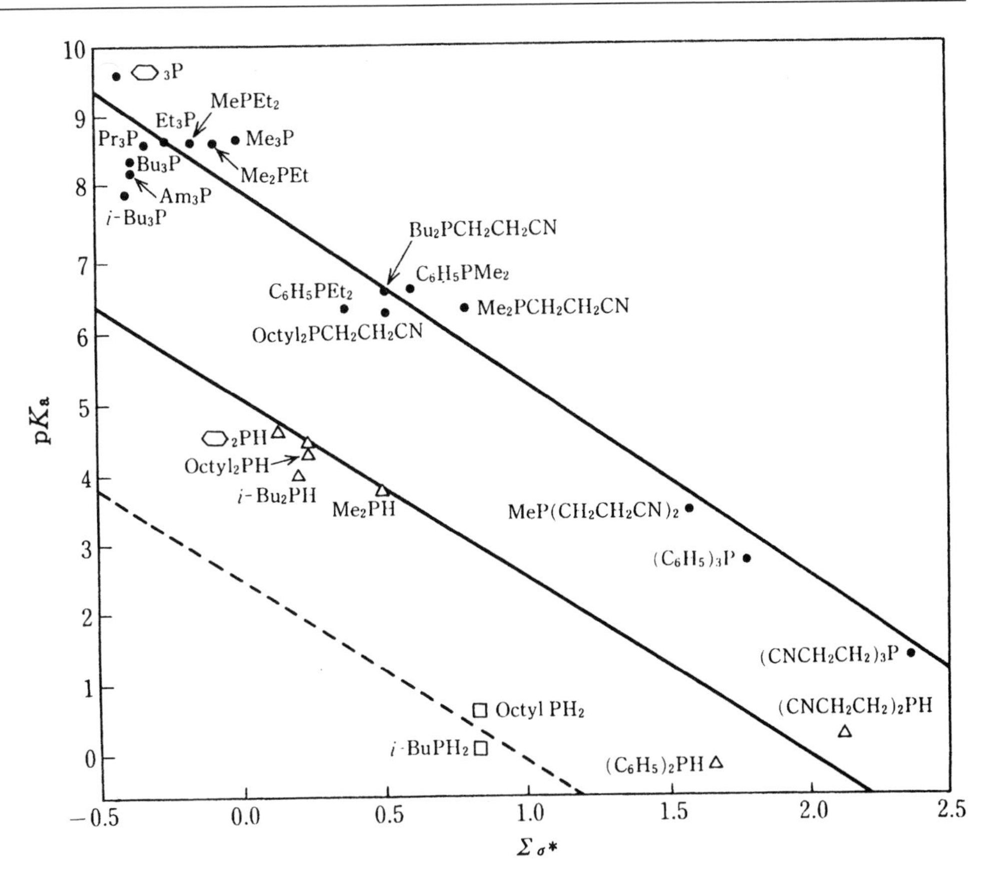

Figure 3.1
Relation between pK_a of the Conjugate Acids of PR_3 and the Sum of Taft's σ* Values. This figure is reproduced from ref. [6a] with permission.

Table 3.1 Some pK_a Values of the Conjugate Acids of Tertiary Phosphine Ligands [7].

Ligand	pK_a	Ligand	pK_a
PCy_3	9.70	PEt_2Ph	6.25
PEt_3	8.69	PCy_2H	4.55
PMe_3	9.65	PBu_2H	4.51
PPr_3	8.64	$P(p\text{-}C_6H_4OMe)_3$	4.46
PMe_2Et	8.61	$P(octyl)_2H$	4.41
$PMeEt_2$	8.61	$P(CH_2CH_2OBu)_2H$	4.15
PBu_3	8.43	PPh_3	2.73
$P(amyl)_3$	8.33	$P(CH_2CH_2CH)_3$	1.37
$P(CH_2CH_2OBu)_3$	8.03	$P(CH_2CH_2CN)_2H$	0.41
$P(i\text{-}Bu)_3$	7.97	PPh_2H	0.03
$PCy_2(CH_2CH_2CN)$	7.13	$PBuH_2$	–0.03
$P(CH_2CH_2Ph)_3$	6.60	$P(i\text{-}Bu)H_2$	–0.02
PMe_2Ph	6.50		

Electronic effects in tertiary phosphine ligands have been estimated by using ν(CO) IR frequencies of $Ni(CO)_3(PR_3)$ in CD_2Cl_2 [7]. The following is the empirical equation for the trend:

$$\nu(CO) = 2056.1 + \Sigma X_i\,(cm^{-1})$$

when X_i denotes a substituent parameter of R on the P atom: R = t-Bu (0), n-Bu (1.4), Et (1.8), Me (2.6), Ph (4.3), H (8.3), OPh (9.7), Cl (14.8), F (18.2), and CF_3 (19.6). Table 3.2 summarizes the ν(CO) values for various tertiary phosphine ligands. The larger the X_i value, the stronger the PR_3 donor. This idea is based on the ability of back-donation

Table 3.2 Carbonyl Stretching Bands of $Ni(CO)L_3$ for Various Tertiary Phosphine Ligands [7].

Ligand	$\nu(cm^{-1})$	Ligand	$\nu(cm^{-1})$
P(*t*-Bu)$_3$	2056.1	P(OEt)$_2$Ph	2074.2
PCy$_3$	2056.4	P(OPh)Ph$_2$	2074.6
P(*o*-C$_6$H$_4$OMe)$_3$	2058.3	PPh$_2$C$_6$F$_5$	2074.8
P(*i*-Pr)$_3$	2059.2	P(O-*i*-Pr)$_3$	2075.9
PBu$_3$	2060.3	PPh$_2$(O-*o*-C$_6$H$_4$Cl)	2076.1
PEt$_3$	2061.7	P(OEt)$_3$	2076.3
P(NMe$_2$)$_3$	2061.9	PH$_2$Ph	2077.0
PEt$_2$Ph	2063.7	P(CH$_2$CH$_2$CN)$_3$	2077.9
PMe$_3$	2064.1	P(OCH$_2$)$_2$Ph	2078.7
PMe$_2$Ph	2065.3	P(OCH$_2$CH$_2$OMe)$_3$	2079.3
PPh$_2$(*o*-C$_6$H$_4$OMe)	2066.1	P(OMe)$_3$	2079.5
P(*p*-C$_6$H$_4$OMe)$_3$	2066.1	P(OPh)$_2$Ph	2079.8
PBz$_3$	2066.4	PClPh$_2$	2080.7
P(*o*-Tol)$_3$	2066.6	PMe$_2$CF$_3$	2080.9
P(*p*-Tol)$_3$	2066.7	P(O-2,4-C$_6$H$_3$Me$_2$)$_3$	2083.2
PEtPh$_2$	2066.7	P(OCH$_2$CH$_2$Cl)$_3$	2084.0
PMePh$_2$	2067.0	P(O-*p*-Tol)$_3$	2084.1
P(*m*-Tol)$_2$	2067.2	P(O-*p*-C$_6$H$_4$OMe)$_3$	2084.1
PPh$_2$NMe$_2$	2067.3	P(O-*o*-Tol)$_3$	2084.1
PPh$_3$(2,4,6-C$_6$H$_2$Me$_3$)	2067.4	P(O-*o*-C$_6$H$_4$-*i*-Pr)$_3$	2084.6
PPhBz$_2$	2067.6	P(O-*o*-C$_6$H$_4$Ph)$_3$	2085.0
PPh$_2$(*p*-C$_6$H$_4$OMe)	2068.2	P(OPh)$_3$	2085.3
PPh$_2$Bz	2068.4	P(O-*o*-C$_6$H$_4$-*t*-Bu)$_3$	2086.1
PPh$_3$	2068.9	P(OCH$_2$)$_2$OPh	2086.5
PPh$_2$(CH=CH$_2$)	2069.3	P(OCH$_2$)$_3$CPr	2086.8
P(CH=CH$_2$)$_2$	2069.5	P(OCH$_2$)$_3$CEt	2086.8
PPh$_2$(*p*-C$_6$H$_4$F)	2069.5	P(OCH$_2$)$_3$CMe	2087.3
PPh$_2$(*m*-C$_6$H$_4$F)	2070.0	P(OCH$_2$CH$_2$CN)$_3$	2087.6
PPh$_2$(CH$_2$CH$_2$Cl)	2070.8	P(O-*o*-Tol-*p*-Cl)$_3$	2088.2
P(*p*-C$_6$H$_4$F)$_3$	2071.3	P(O-*p*-C$_6$H$_4$Cl)$_3$	2089.3
P(OEt)Ph$_2$	2071.6	P(C$_6$F$_5$)$_3$	2090.9
P(OMe)Ph$_2$	2072.0	(P(OCH$_2$CCl$_3$)$_3$	2091.7
P(O-*i*-Pr)$_2$Ph	2072.2	PCl$_2$Ph	2092.1
P(*p*-C$_6$H$_4$Cl)$_2$	2072.8	P(O-*p*-C$_6$H$_4$CN)$_3$	2092.8
PHPh$_2$	2073.3	PCl$_3$	2097.0
P(OBu)$_2$Ph	2073.4	PF$_3$	2110.8
P(*m*-C$_6$H$_4$F)$_3$	2074.1		

from nickel to CO (see Chapter 2). When the P ligand is a stronger donor, nickel becomes more electron rich (the metal HOMO is destabilized), thus allowing more effective back-bonding to the pπ* orbital of the CO ligands, resulting in a decrease in the CO stretching frequency.

3.2 Steric Effect

The steric bulk of the ligand is also a highly important factor in determining the influence of the ligand on the selectivity and reactivity of organometallic compounds. The cone angle, shown in Figure 3.2, is frequently used as an index of the steric bulk of phosphorus ligands [7]. The cone angles θ are estimated by assuming an average M—P bond distance of 2.28 Å and maximum occupation of van der Waals spheres of the substituents on the P atom. When three substituents on the P atom are different, these values are simply averaged. The estimated cone angles for various ligands are listed in Table 3.3.

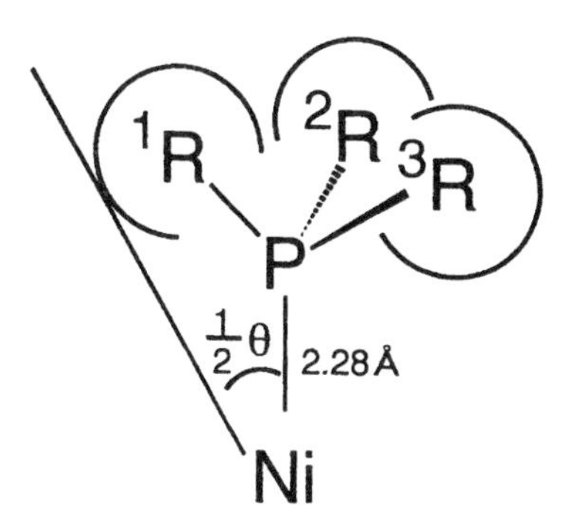

Figure 3.2
Estimation of Cone Angle

Table 3.3 Cone Angles of Tertiary Phosphine Ligands (Cone Angles of Bidentate Ligands [7]. Cone Angles of $Ph_2P(CH_2)_nPPh_2$ are Estimated by Assuming PMP Angles of 74, 85, 90° for $n = 1, 2, 3$, Respectively)

Ligand	Θ(deg)	Ligand	Θ(deg)
PH_3	87	$PMePh_2$	136
PH_2Ph	101	PEt_2Ph	136
$P(OCH_2)_3CR$	101	$P(CF_3)_3$	137
PF_3	104	$PEtPh_2$	140
$P(OCH_3)(CH_3)_2$	106	P(O-o-Tol)$_3$	141
$P(OMe)_3$	107	$Cy_2PCH_2CH_2PCy_2$	142
$Me_2PCH_2CH_2PMe_2$	107	P(*i*-Bu)$_3$	143
$P(OEt)_3$	109	PPh_3	145
$P(OCH_2CH_2Cl)_3$	110	P(*p*-Tol)$_3$	145
$P(CH_2O)_3CR$	114	P(*m*-C_6H_4F)$_3$	145

(*Continued*)

Table 3.3 (*Contd.*)

Ligand	Θ(deg)	Ligand	Θ(deg)
$Et_2PCH_2CH_2PEt_2$	115	P(O-o-C_6H_4-*i*-Pr)$_3$	148
$P(OMe)_2Ph$	115	PPh_2(*i*-Pr)	150
$P(OMe)_2Et$	115	P(O-o-C_6H_4Ph)$_3$	152
$P(OEt)_2Ph$	116	PPh_2(*t*-Bu)	157
PMe_3	118	$P(NMe_2)_3$	157
$Ph_2PCH_2PPh_2$	121	$PPh_2C_6F_6$	158
PMe_2Ph	122	P(*i*-Pr)$_3$	160
PMe_2CF_3	124	P(*s*-Bu)$_3$	160
$Ph_2PCH_2CH_2PPh_2$	125	PBz_3	165
$Ph_2P(CH_2)_3PPh_2$	127	PPh(*t*-Bu)$_2$	170
$PHPh_2$	128	P(O-*t*-Bu)$_3$	172
P(O-*i*-Pr)$_3$	130	P(O-o-C_6H_4-*t*-Bu)$_3$	175
PBr_3	131	P(neopentyl)$_3$	~180
$P(OMe)Ph_2$	132	P(*t*-Bu)$_3$	182
PEt_3	132	$P(C_5F_5)_3$	184
PBu_3	132	P(O-2,6-$C_6H_3Me_2$)$_3$	190
PPr_3	132	P(o-Tol)$_3$	194
$P(CH_2CH_2CN)_3$	132	P(mesityl)$_3$	212
$P(OEt)Ph_2$	133		

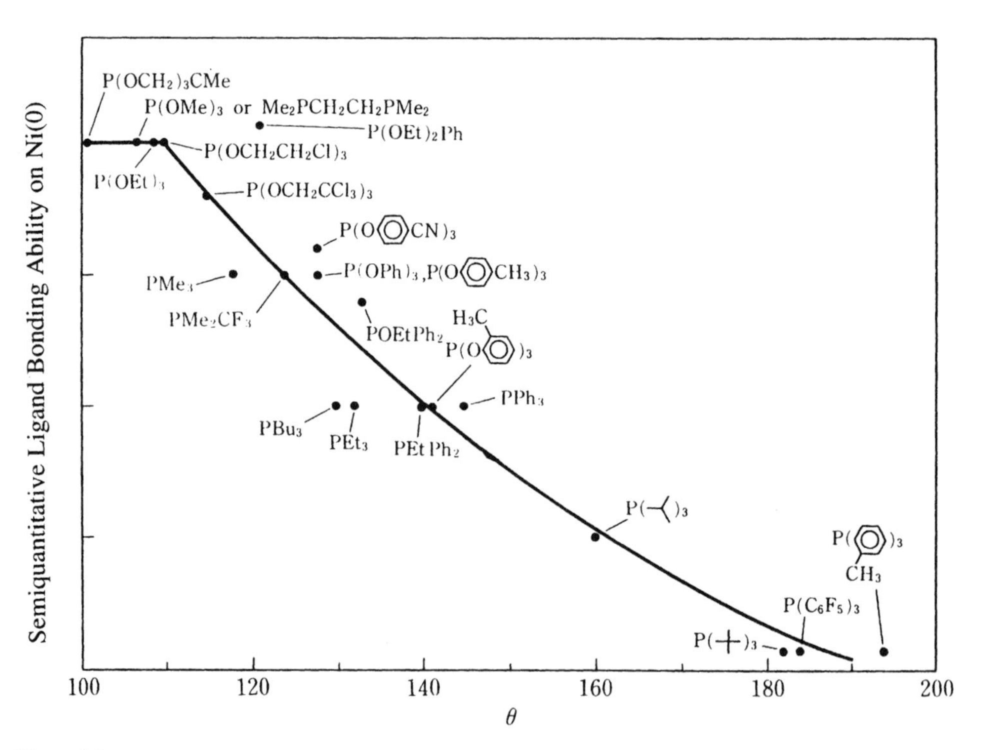

Figure 3.3
Relation between Cone Angles θ of L and Stability Constants of NiL_4 Complexes. This Figure is Reproduced from Ref.[7] with Permission

There is a good correlation between the logarithms of the ligand dissociation constants K (reverse of stability constant) of NiL_4 and the cone angles of L. As the cone angle increases, so does the *K* value:

$$NiL_4 \underset{K}{\rightleftharpoons} NiL_3 + L$$

In this reaction, the log *K* values do not correlate with electronic factors as measuered by ν(CO) as described above. The results indicate that the steric factor is more important than electronic in determining thermodynamic stability of ML_4 type complexes. The environment around the metal is more congested than imagined for the coordination number 4. The cone angle is now widely accepted as an index of steric factors for tertiary phosphine ligands in large areas of transition metal chemistry including organometallics.

The bulk of bidentate ligands are more difficult to quantify, since cone angles of chelating ligands are defined by assuming their PMP angles. The bite angle β as shown in Figure 3.4 can be regarded as an index of the degree of steric crowding around the metal [8]. Natural bite angles β_n are defined by Casey as the preferred chelation angle determined only by ligand backbone constraint by molecular mechanics calculation (M—P = 2.30 Å). They parallel the X-ray PMP angles reasonably well. For example, the natural bite angles of α,ω-(diphenylphosphino)alkane increase with increase in the number of methylene groups of $Ph_2P(CH_2)_nPPh_2$ (M—P = 2.33 Å): $\beta_n = 78°$ ($n = 2$), 87° ($n = 3$), 98° ($n = 4$), and the actual bite angles of $PtX_2(Ph_2P(CH_2)_nPPh_2)$ are 86° ($n = 2$), 92° ($n = 3$), 95° ($n = 4$), respectively. Though bite angles are relatively variable on coordination, chelating ligands with large βn (120°) are capable of coordinating in the equatorial plane of the tbp structure, whereas those with β_n of *ca.* 90° favor coordination at apical and equatorial sites [8]. Rh and Ni catalysts, having certain chelating ligands with a range of natural bite angles of 100–110°, are known to give linear products selectively in hydroformylation and hydrocyanation. In ansa-metallocenes, the angle between two lines connecting the metal and the centers of two Cp rings is employed for estimating the bite angles of the bis(cyclopentadienyl) ligands.

Bulky ligands are especially useful for synthesizing highly reactive, coordinatively unsaturated complexes such as PtL_2 and $Ru(CO)_2L_2$ [9], since they prevent further coordination of ligands and/or oligomerization which would normally occur to satisfy the 18 electron rule. However, one should not ignore the electronic factors for the structural consideration, since they essentially determine the structure of transition metal complexes.

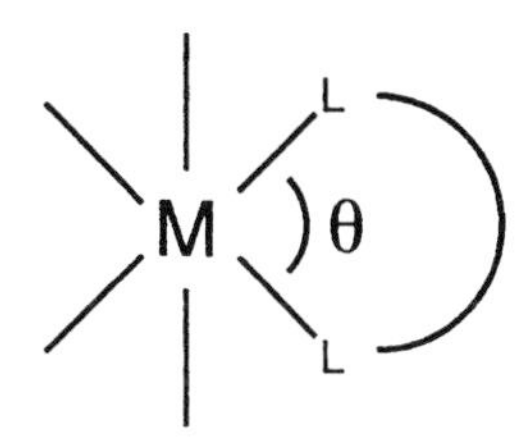

Figure 3.4
Bite angle of bidentate ligand

Delicate control of the steric environment of transition metal complexes is extremely important in the highly selective organic transformations using transition metal catalyses. Recent highly selective Rh or Ru catalyzed asymmetric hydrogenation achieving more than 99 %ee selectivity is considered to be one of the remarkable successes resulted from highly elaborate steric control of the ligand. Lists of specific chiral ligands should be consulted in specialized books [2].

References

[1] G. W. Parshall and S. D. Ittel, *Homogeneous Catalysis*, John Wiley, New York (1992).

[2] [a] R. Noyori, *Asymmetric Catalysis in Organic Synthesis*, John Wiley, New York (1994).
[b] I. Ojima, *Catalytic Asymmetric Synthesis*, VCH, New York (1993).
[c] H. B. Kagan, *Comprehensive Organometallic Chemistry*, Eds. G. WIlkinson, F. G. A. Stone, E. W. Abel, vol. 8, Pergamon, New York (1982); H. B. Kagan, *Asymmetric Synthesis*, vol. 5, Academic Press, New York (1985).
[d] H. Brunner, *Topics in Stereochemistry*, vol. 18 (1988).
[e] M. Sawamura, Y. Ito, *Chem. Rev.*, **92**, 857 (1992).
[f] J. Seyden-Penne, *Chiral Auxiliaries and Ligands in Asymmetric Synthesis*, John Wiley, New York (1995).

[3] A. Dedieu (ed.), *Transition Metal Hydrides*, VCH, New York (1992).

[4] [a] R. H. Crabtree, *The Organometallic Chemistry of the Transition Metals*, 2nd. ed., John Wiley, New York (1994).
[b] A. Yamamoto, *Organotransition Metal Chemistry*, John Wiley, New York (1990).
[c] P. Collman, L. S. Hegedus, J. R. Norton, R. G. Finke, *Principles and Applications of Organometallic Chemistry*, 2nd. ed., University Science Books, Mill Valley, CA (1987).

[5] [a] F. A. Cotton, G. Wilkinson, *Advanced Inorganic Chemistry*, 5th. ed., John Wiley, New York (1988).
[b] J. E. Huheey, *Inorganic Chemistry-Principles of Structure and Reactivities*, 4th. ed., Harper & Row, New York (1993).
[c] D. F. Shriver, P. W. Atkins, C. H. Langford, *Inorganic Chemistry*, 2nd. ed., Freeman, New York (1994).
[d) F. Basolo, R. G. Pearson, *Mechanisms of Inorganic Reactions*, 2nd. ed., John Wiley, New York (1967).

[6] [a] W. A. Henderson, Jr., C. A. Streuli, *J. Am. Chem. Soc.*, **82**, 5791 (1960).
[b] C. A. Streuli, *Anal. Chem.*, **32**, 985 (1960).

[7] C. A. Tolman, *Chem. Rev.*, **77**, 313 (1977).

[8] [a] C. P. Casey, G. T. Whiteker, *Isr. J. Chem.*, **30**, 299 (1992).
[b] M. Kranenburg, P. C. Kamer, P. W. N. M. van Leeuwen, D. Vogt, W. Keim, *J. Chem. Soc., Chem. Commun.*, 2177 (1995).
[c] M. Kranenburg, Y. E. M. van der Burgt, P. C. Kamer, P. W. N. M. van Leeuwen, K. Goubitz, J. Fraanje, *Organometallics*, **14**, 3081 (1995).

[9] [a] S. Otsuka, T. Yoshida, M. Matsumoto, K. Nakatsu, *J. Am. Chem. Soc.*, **98**, 5850 (1976).
[b] T. Yoshida, S. Otsuka, *J. Am. Chem. Soc.*, **99**, 2134 (1977).
[c] J. Fornies, M. Green, J. L. Spencer, F. G. A. Stone, *J. Chem. Soc. Dalton Trans.*, 1006 (1977).
[d] A. Immerzi, A. Musco, *J. Chem. Soc., Chem. Commun.*, 400 (1974).
[e] M. Ogasawara, S. A. Macgregor, W. E. Streib, K. Folting, O. Eisenstein, K. G. Caulton, *J. Am. Chem. Soc.*, **117**, 8869 (1995).

4 Manipulation of Air-sensitive Compounds

S. Komiya, *Tokyo University of Agriculture and Technology*

4.1 Introduction

There are many techniques for handling air- and moisture-sensitive compounds which are fairly well documented in books. Occasionally very expensive glove boxes or high vacuum systems have to be set up. If these facilities are available, most air-sensitive compounds can be handled without problems. However, some of them are difficult to manage for beginners and people who are not familiar with air-sensitive compounds and who may even hesitate to buy such materials. For example, organic chemists usually make an inert gas atmosphere by bubbling Ar or N_2 gas into the solution. For some organometallic compounds, this method is not sufficient and a small amount of air contamination leads to decomposition of these compounds. In *Organic Synthesis* it is described that "Under the best conditions, NaCp gives pale yellow or orange solution. Traces of air lead to red or purple solutions, lowering the reaction yield appreciably". This can be easily overcome if the simple apparatus described here to handle air-sensitive compounds is assembled in the laboratory. Most of the techniques described in this chapter are so-called Schlenk techniques, which make use of flasks equipped with a three-way stopcock. They were originally introduced from Germany and independently developed in Yamamoto's group in Japan. In contrast, chemists in the USA favor the use of dry-box techniques in combination with double manifold vacuum and nitrogen lines. In this book simple and inexpensive methods are described that are especially designed for beginners. Other literature dealing with the manipulation of air-sensitive compounds should also be read, since these techniques are often a matter of individual taste. Readers are also highly recommended to develop their own skill in manipulating their specific compounds or reactions.

4.2 Basic Apparatus (Vacuum and Nitrogen Lines)

Apparatus for handling air-sensitive compounds varies, depending on how stable the compounds are in air, how much purity is required, the scale, and the physical state of the compounds: solid, liquid or gas. If expensive high-quality apparatus is used for handling all chemicals and reactions, it may soon degrade and break down. Some

Synthesis of Organometallic Compounds: A Practical Guide. Edited by S. Komiya

apparatus is very inconvenient for specific purposes. In this section, the most basic apparatus, necessary at least on a lab bench, is described. Individual handling methods are described later.

Vacuum and inert gas lines are the most fundamental pieces of equipment to be set up. Figures 4.1, 4.2 and 4.3 depict their general structure. To obtain a good vacuum, at least one liquid nitrogen trap should be connected between the pump and vacuum system, which usually provides 10^{-3} mmHg vacuum. Liquids collected in the liquid nitrogen trap during an experiment should be removed when turning the vacuum line off. This will extend the life of the vacuum pump and oil, and avoid contamination of the vacuum oil by liquids or gases. When high vacuum is required, an oil diffusion pump, which should give 10^{-6} mmHg, can be installed between the vacuum pump and liquid nitrogen trap. For the high vacuum experiment, it is desirable to keep the vacuum line under vacuum as much as possible, even when it is not used. The vacuum line can be connected to a manometer to measure the gas pressure, so that gas volume can be measured accurately. Several two-way stopcocks having standard high quality male glass joints are fitted to the vacuum line. Stopcocks whose interior can be kept under vacuum to avoid gas leaks by loosening are also obtainable. Two types of vacuum grease are generally used for stopcocks: silicone and Apiezon. The latter is of much higher quality, though it is expensive. When putting grease on the ground glass, air contamination between ground joints should be avoided by first putting a few spots of grease on the male stopcock, and then installing the stopcock to spread the grease uniformly. Various greaseless teflon stopcocks are also commercially available.

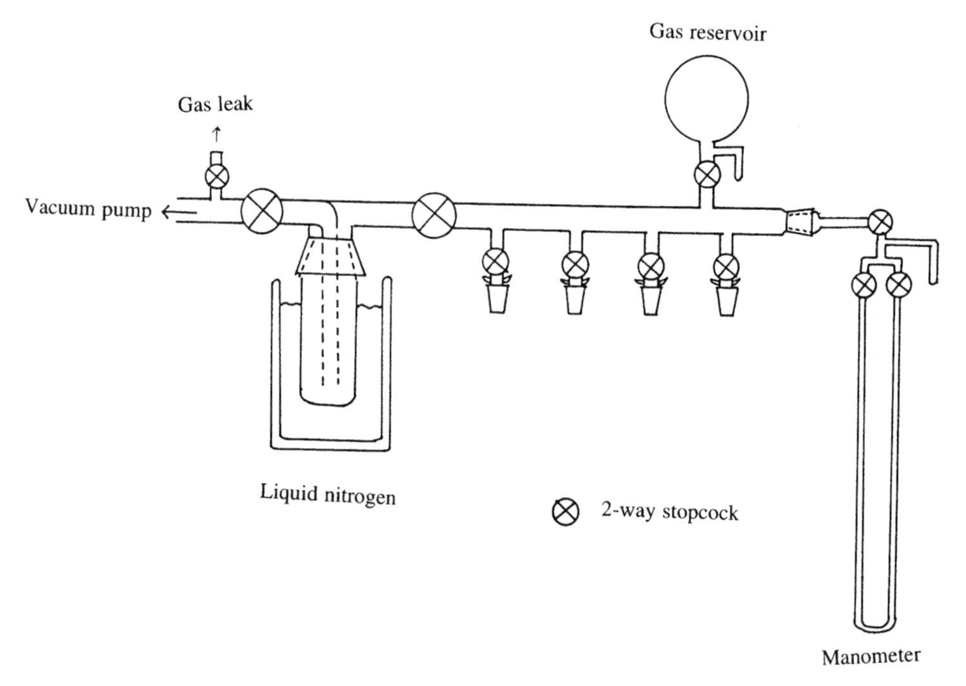

Figure 4.1
Standard Vacuum Line with Gas Reservoir and Mercury Manometer.

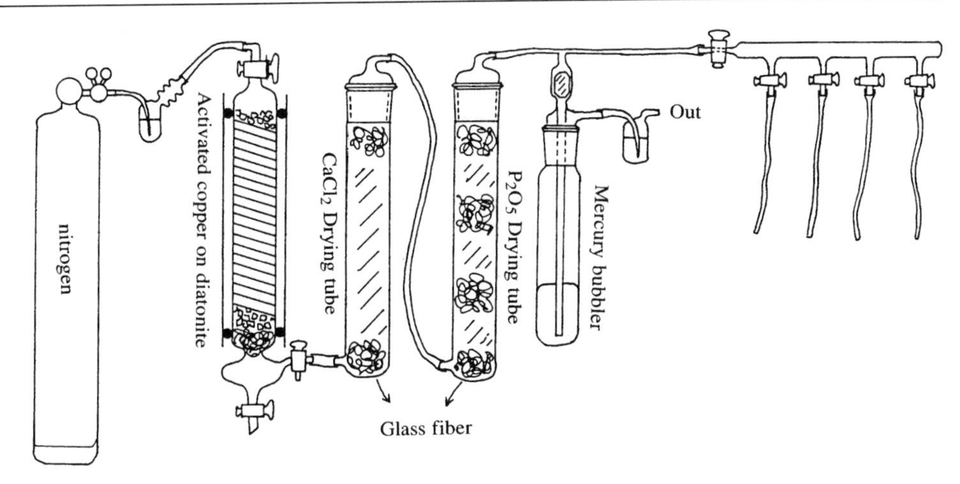

Figure 4.2
Standard Nitrogen Line. Nitrogen gas of high purity is further purified by passage first through the active copper filler then through $CaCl_2$ and P_2O_5 towers to remove traces of moisture. A mercury bubbler can also be placed in the line to obtain a pressure slightly above atmospheric.

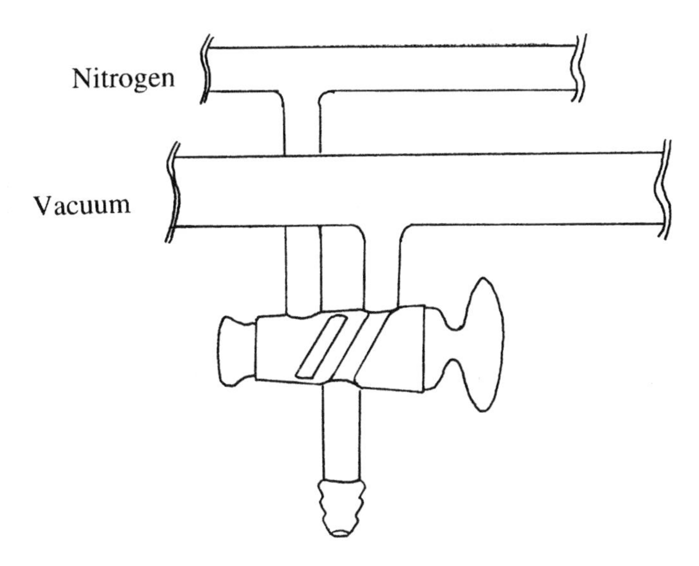

Figure 4.3
Double Manifold Vacuum Line for Nitrogen and Vacuum. Turning the stopcock 180° switches the line from nitrogen to vacuum and vice versa.

Pure nitrogen or argon is usually used as an inert gas. To remove any trace of air in these gases, it is desirable to use a deoxygenation column and drying towers, although many synthetic experiments do not require such purification. The deoxygenation column consists of pellets of activated copper supported on diatomite, which operate at 180 °C. This is reusable and can be reduced periodically at the same temperature with hydrogen. For drying these gases, $CaCl_2$, P_2O_5 and molecular sieves 5A are used as fillers. The commercially available drying agent SICAPENT is useful, because its color change indicates whether the drying reagents are still active. It is desirable to put a mercury bubbler (or an excess pressure bulb which is free of mercury) before gas outlet stopcocks, to obtain slightly higher pressure of the inert gas and to avoid unnecessary and dangerous high pressure inside. An excess of gas should be always exhausted from this mercury bubbler. To make the mercury bubbler more compact, a back current stopper can be put only 10 cm above the mercury level, though more than 76 cm height inlet glass tube can also be used as a substitute. This excess of pressure is necessary when a solution is transferred by cannula or filtration. It is also convenient to place two simple glass bubblers containing mineral oil both at the inlet and the outlet of the gas line, since the gas flow can be easily verified at a glance. The Tygon or rubber tube may be furnished at the end of a two-way stopcock to introduce the gas into the Schlenk tube. In addition a small vacuum pump should be provided near the bench to evacuate the flask for replacing the atmosphere by inert gas.

If a double manifold vacuum and nitrogen stopcocks are preferred, thick tygon tubing should be used for connection because it withstands evacuation. In this case, the additional small vacuum pump is not necessary, since evacuation and filling with nitrogen are carried out by simply rotating the double manifold stopcock.

4.3 Handling of Air-sensitive Materials

4.3.1 The Simplest Method

The methods described here are applicable to compounds that are only moderately air-sensitive, but do not decompose in an hour or for reactions that are not affected by a small amount of air or moisture. However, these methods are still very convenient in the laboratory where no equipment is provided.

4.3.1.1 Glove Bag

A glove bag is a commercially available plastic bag with a nitrogen inlet and gloves as shown in Figure 4.4. After placing the bottles in the plastic bag, air is forced out by flattening the bag. After sealing the end of the bag by wrapping around the stick, nitrogen gas is introduced to inflate the bag. The gas is released from the bag and refilled with nitrogen. After three repetitions of this operation, air-sensitive compounds can be manipulated from the side gloves while keeping the inside under slightly increased pressure by nitrogen gas. Since this method can be operated without any other special equipment, it is sometimes very useful to open bottles containing air-sensitive and hygroscopic materials.

4.3.1.2 Balloon

Many organic chemists use balloons for carrying out reactions under inert atmosphere. Methods vary, but some typical arrangements are shown in Figure 4.5. Although any type of balloon can be used, high quality rubber balloons made for medical use are the best. Usually, a needle or two-way stopcock is connected tightly with the balloon. Nitrogen gas is introduced into the balloon from the gas cylinder and then is discarded.

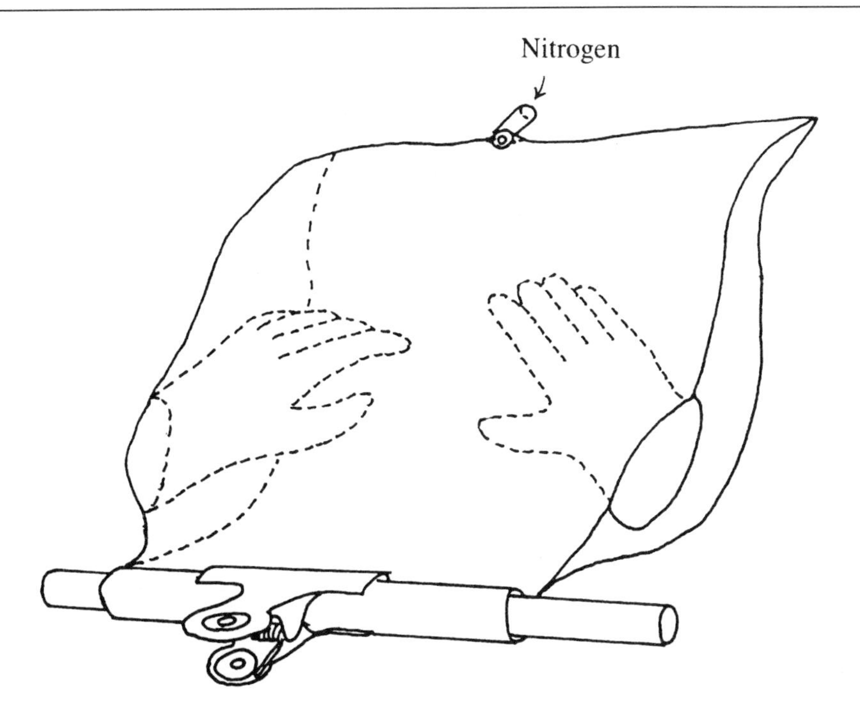

Figure 4.4
Glove Bag

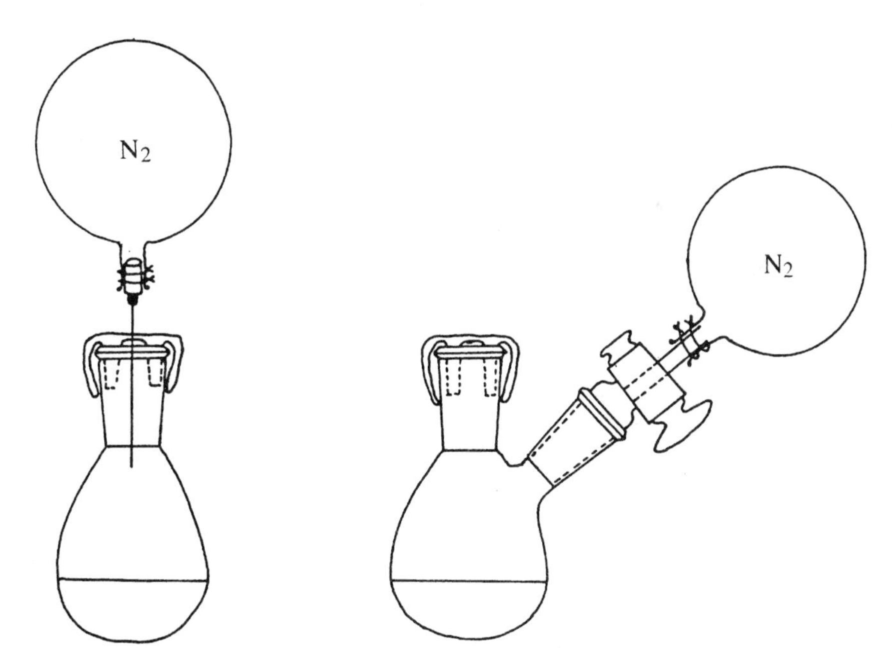

Figure 4.5
Balloon Technique

After three repetitions of this operation, the balloon is inflated as much as desired. The balloon thus prepared can be equipped on the flask with septum rubber or glass joint. It is better to replace the atmosphere in the flask by nitrogen before connection. As long as the balloon is swelling, the inside of the flask is maintained under some nitrogen pressure. It is very useful if there happens to be a pressure increase in the flask as a result of the reaction, which might cause an explosion in a closed system. However, it should be noted that the method is not good enough for many organometallic compounds and it is difficult to keep the nitrogen atmosphere for a long time. Nevertheless this method is still widely used, especially in organic syntheses, since it is simple, easy and inexpensive and works nicely for most preparations and reactions of organic compounds, even in the presence of transition metal catalysts. This is probably because the final products are stable in air and usually workup is carried out in air.

4.3.2 Simple But Effective Methods

4.3.2.1 Three Needles Technique

A round bottom flask containing the solid reactants and a magnetic stirring bar is fitted with a rubber septum cap secured with a rubber band. The flask is evacuated by vacuum pump and then filled with nitrogen with a needle adaptor. This is repeated three times. The flask will then be maintained under sufficient nitrogen atmosphere. When transferring the solution or liquid by a syringe through the septum, two needles for inlet and outlet of inert gas should be attached to the septum rubber. These equalize the pressure inside and outside the flask and prevent spilling the liquid. A cannula can also be used to transfer liquid to another flask or to remove liquid by application of inert gas pressure. The outlet needle at the flask should be removed in order to increase slightly the pressure inside the flask. It must be remembered that the rubber septum is not a complete seal and over a period, it absorbs gases and volatile liquids in the flask, and, once used, it is more likely to have small leaks especially when the flask is evacuated. Therefore it is not recommended to keep the flask under vacuum with a rubber septum, but to keep it under nitrogen atmosphere for a longer period.

4.3.2.2 Schlenk Technique

A Schlenk tube is a flask that has at least one arm where inert gas can be introduced. Usually a three- or two-way stopcock is fitted at the end of the arm. Some typical Schlenk flasks are shown in Figure 4.7. A combination of a commercially available round bottom flask with a two- or three-way stopcock or T-shape stopcock can also be used as a Schlenk flask. The shape of Schlenk-ware varies depends on individual taste. The tube shown in Figure 4.7, which was developed originally in the Max-Planck-Institute for Coal Research at Mulheim, Ruhr in Germany, can be rested against the pipes on the lab bench and can be stored conveniently. Tubes with a straight arm are often favored, since they are easier to clean. The bottom part of the flask is rather flat which allows the stirring bar to rotate more efficiently. When vacuum and nitrogen lines are assembled separately, it is very convenient to have a three-way stopcock for the Schlenk tube, whereas a two-way stopcock is sufficient when one uses a combined double manifold vacuum and nitrogen line.

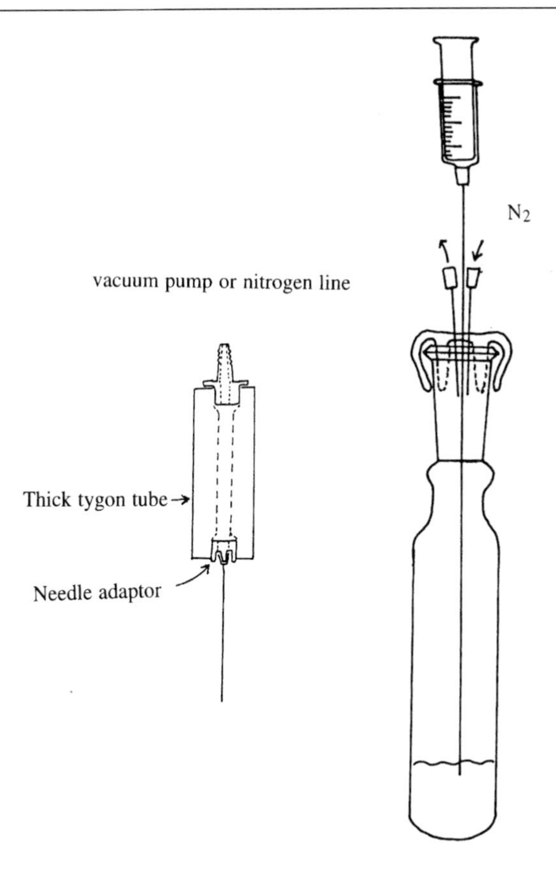

Figure 4.6
Three Needle Technique. Liquids are easily transferred by this method. It is useful to prepare the needle adapter (left) from a broken syringe.

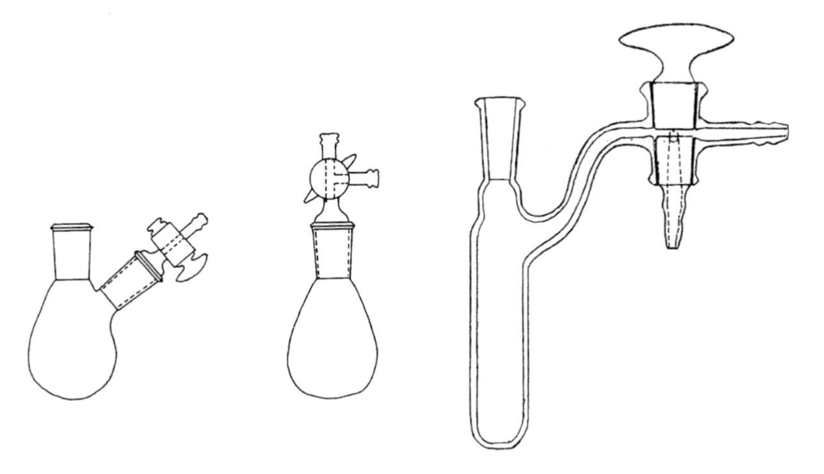

Figure 4.7
Schlenk Tubes. The shape of Schlenk flasks is down to individual preference, but it is highly recommended to use ground glass joints of high quality.

In use, the air in a Schlenk flask should be replaced at least three times by inert gas using a pump and fill procedure and all subsequent handling should be carried out under nitrogen flow. Nitrogen gas flow is always monitored easily by looking at the bubblers equipped in the nitrogen line. Solid is transferred by a spatula or by connecting two Schlenk tubes with a joint (Figure 4.8). To avoid air contamination, protection by glass pants (or trousers) can also be provided, if necessary. However, most air-sensitive compounds, even if they are stable only for a few seconds in air, can be transferred without the glass pants if one has sufficient experimental skill. If this method is not adequate, the transfer should be done only under high vacuum or in a well--maintained glove box (see literature methods). When liquid is transferred by a syringe, great care should be taken, since solutions are generally much more susceptible to oxidation or hydrolysis. A syringe should be filled with pure nitrogen and emptied three times before use. The syringe will then be ready to use without contamination by air. It should be noted that the rate of drawing nitrogen gas should not be greater than that of nitrogen flow. A cannula is also very useful for transferring liquids through a rubber septum. Furthermore if a fine glass fiber is fitted at the tip, a heterogeneous solution can be filtered (Figure 4.9). When a septum is placed on the Schlenk tube, it is impor-

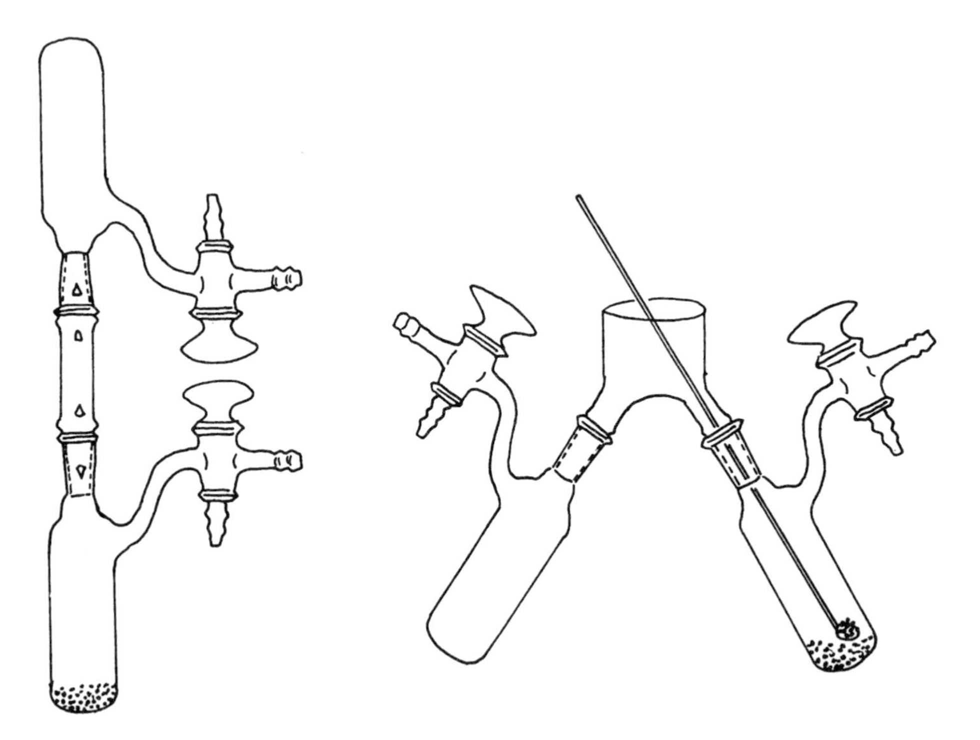

Figure 4.8.
Transfer of Solid Materials. Glass pants are used when highly air-sensitive compounds are employed. Covering the pants with a thin plastic sheet help avoids further decomposition. Solids can also be transferred by direct connection of two Schlenk flasks, but great care should be taken when disconnecting the flasks, since powder adhering to the surface near the inlet may be blown off.

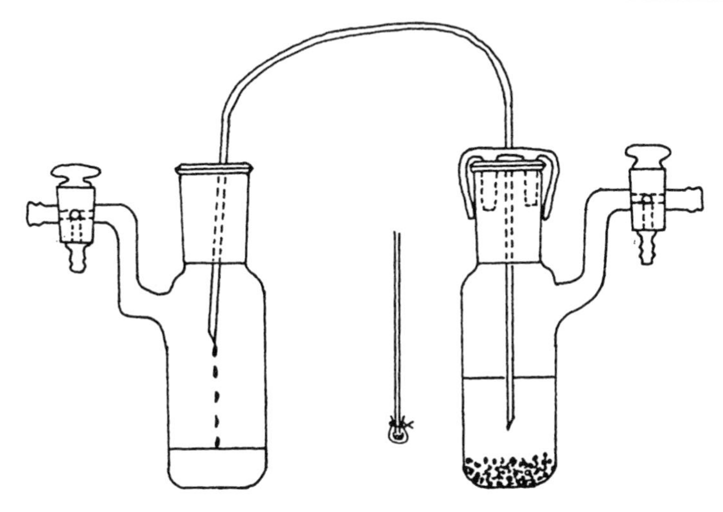

Figure 4.9
Filtration by Cannula. The cannula method is a simple and effective means for transferring liquid. When a rubber septum is fitted at the inlet of the Schlenk flask under nitrogen, extra air trapped inside should be removed promptly by inserting the needle into the septum. The tip of the cannula can be covered with fine glass fiber for filtration.

tant to remove the extra air inside the septum by puncturing with a needle as quickly as possible. The cannula method is more reliable than syringe transfer and more suitable for beginners who are manipulating a small amount of liquid, although syringe transfer is quicker. These techniques permit all experimental work such as transfer, filtration, reflux, distillation, recrystallization, concentration, mixing, chromatography, autoclave reactions, IR and NMR sampling, preparation of a single crystal for X-ray analysis, etc, to be done under an inert atmosphere. Figure 4.10 shows some useful

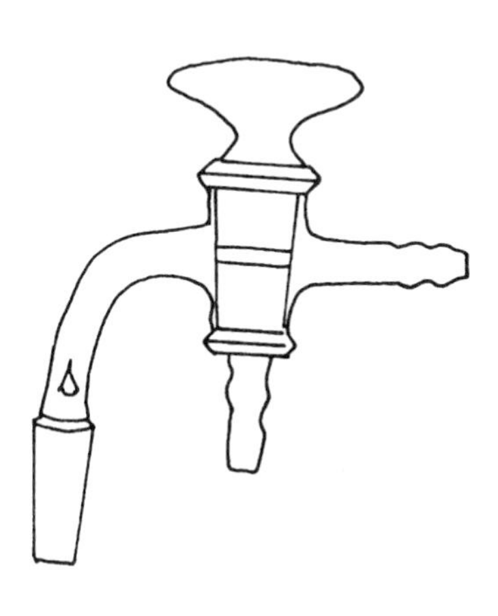

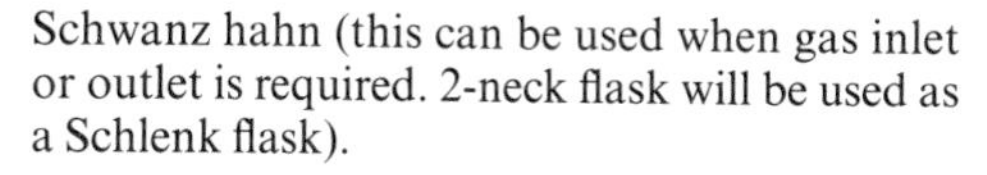

Schwanz hahn (this can be used when gas inlet or outlet is required. 2-neck flask will be used as a Schlenk flask).

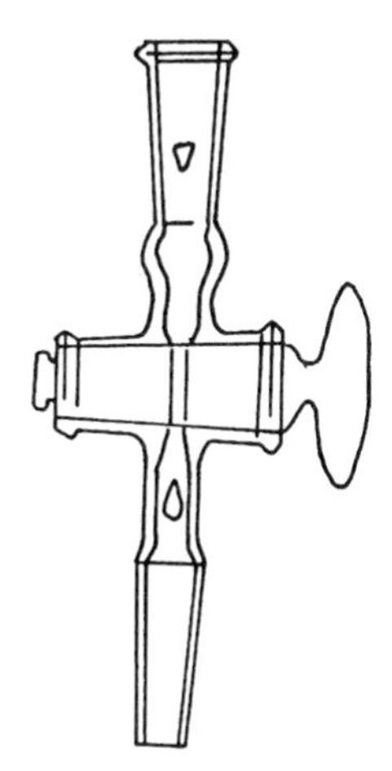

Vacuum stopcock (use for connection of Schlenk flask and vacuum line).

Figure 4.10 *(continued)*

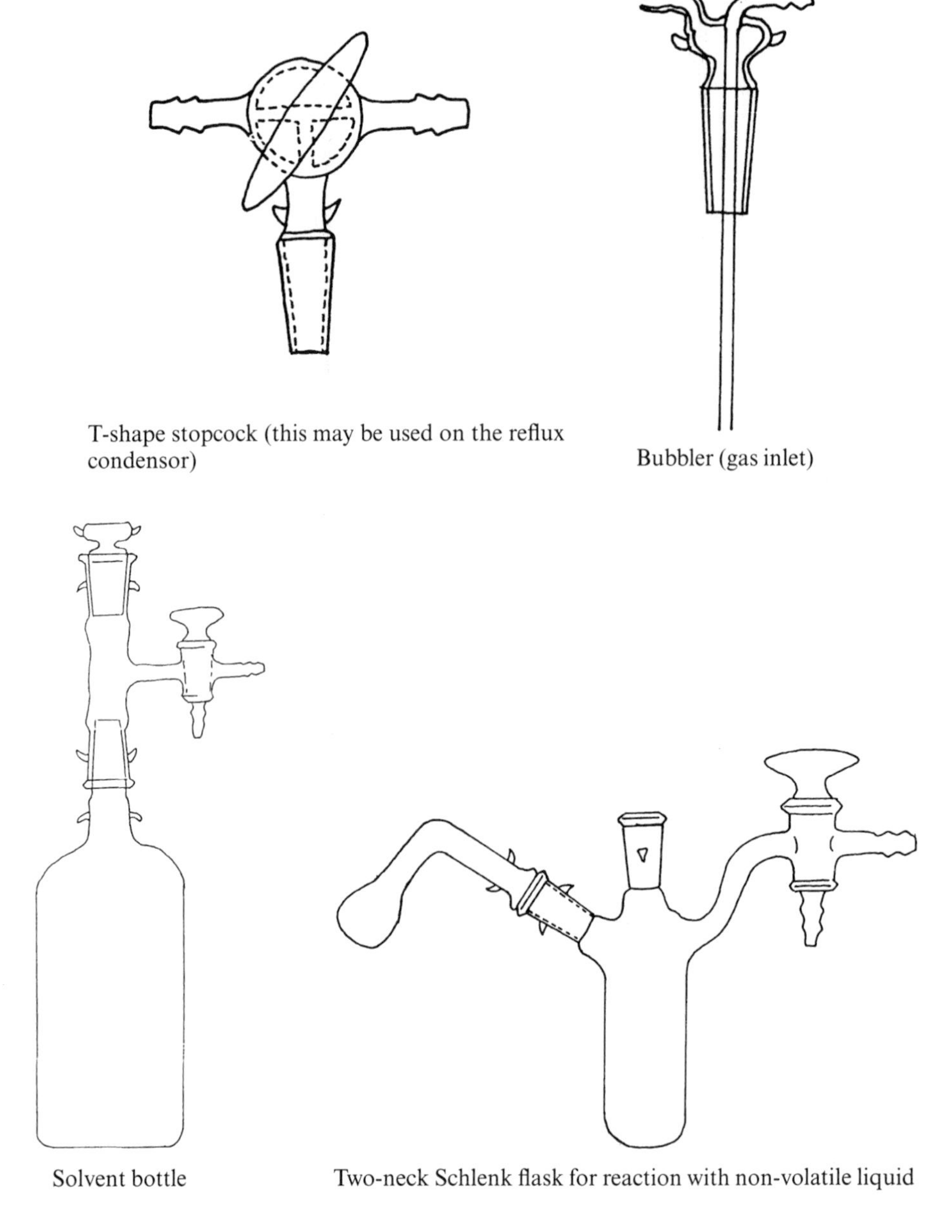

Figure 4.10
Useful glassware for handling air-sensitive materials

glassware for such experiments. When one becomes accustomed to these operations, one will find specific methods that will be more suitable for specific compounds and reactions.

4.3.3 More Advanced Methods

4.3.3.1 Glove Boxes

Glove boxes are now commercially available, but are still very expensive. If the apparatus is operated properly, chemicals and equipments can be handled as if they were in air on the lab bench. Details of manipulation should be obtained from the authorized books or manuals. Glove boxes are very useful for the preparation of samples for physical measurements, such as NMR, IR, single crystals for X-ray, etc. Small items such as a balance and cyclic voltammetry can be placed in the glove box. They are also useful for storage of chemicals. Continual care must be taken to keep the inside completely free of air, by using deoxygenation and drying towers with highly efficient gas circulation. Despite their advantages, glove boxes are not suitable for handling volatile materials, since these remain in the glove box. Handling with large gloves is sometimes clumsier than with manipulation on the lab bench. However, there is much less chance of contamination by air or moisture. The methods used are a matter of taste, once the ideas of handling are well understood.

4.3.3.2 Vacuum Technique

Commercially available nitrogen and argon gases today are of very high purity and can usually be used directly without further purification for most organometallic compounds. However, a trace of oxygen or moisture sometimes induces autocatalytic decomposition of compounds even under inert gas. This is sometimes observed because of continuous accumulation of oxygen and moisture by the gas flow.

All the experimental methods can also be achieved under vacuum, but detailed procedures of manipulation are beyond the scope of this book. Appropriate books in the references should be consulted. The glassware becomes more complex and specialized, and a vacuum pump of high performance is required. Therefore only when the compounds cannot be treated by the methods so far described should vacuum techniques be considered. However, when gases are to be employed, the vacuum technique is indispensable. The volumes of gases, which can be condensed at liquid nitrogen temperature, are easily estimated by releasing the gases in the closed system and measuring the pressure. The dead volume of the closed part in a vacuum line should be calibrated by filling with water or mercury before use. Condensable gases are easily transferred by a trap to trap (bulb to bulb) method (see also Section 4.2.4.2). The gases can be purified by vacuum transfer through a drying tube. The desired amount of gases can be easily introduced into a reaction vessel by a trap-to-trap distillation. However, noncondensable gases at liquid nitrogen temperature such as hydrogen, nitrogen, carbon monoxide and methane should be handled with great care. These gases should be introduced into the reaction vessel with prior purification, if necessary. For gas–solid reactions in a closed system over heterogeneous catalyst, the gas is circulated by a magnetic power pump in order to obtain efficient mixing of gas and solid.

4.3.4 Some Examples of Handling

The methods described below mainly employ Shlenk flasks. Some of the operations can be done in the glove box, and are not mentioned here.

4.3.4.1 Filtration

There are several methods of filtration, and one example of solution transfer by cannula has been described earlier. The cannula is inserted into the rubber septum at the top of the Schlenk tube. If the inlet of teflon or stainless steel cannula is stuffed with a fine glass fiber (such as that used for GC columns) and/or wrapped with paper tissue (Kimwipe) tied by thin wire, the solution can be transferred with filtration. The rate of filtration varies very much depending on how tight the glass fiber is stuffed in and how fine the precipitates are. In a similar manner, a glass bridge filter (Figure 4.11) having a fritted glass filter at the end is also used for filtration. #G3 fritted glass is most commonly used for filtration, though others can also be used, depending on the size of precipitates. Sometimes filtration becomes very slow because the precipitates clog the filter. In this case nitrogen pressure must be released to remove the solid on the filter and filtration started again. Thus, filtration should be carried out by touching only the surface of the solution with the tube and slowly lowering as the surface goes down. This is the most important trick for fast filtration.

Normal filter paper can also be used for filtration under nitrogen. A glass filter tube (Figure 4.11) containing a folded filter paper is connected with the Schlenk flask. Then

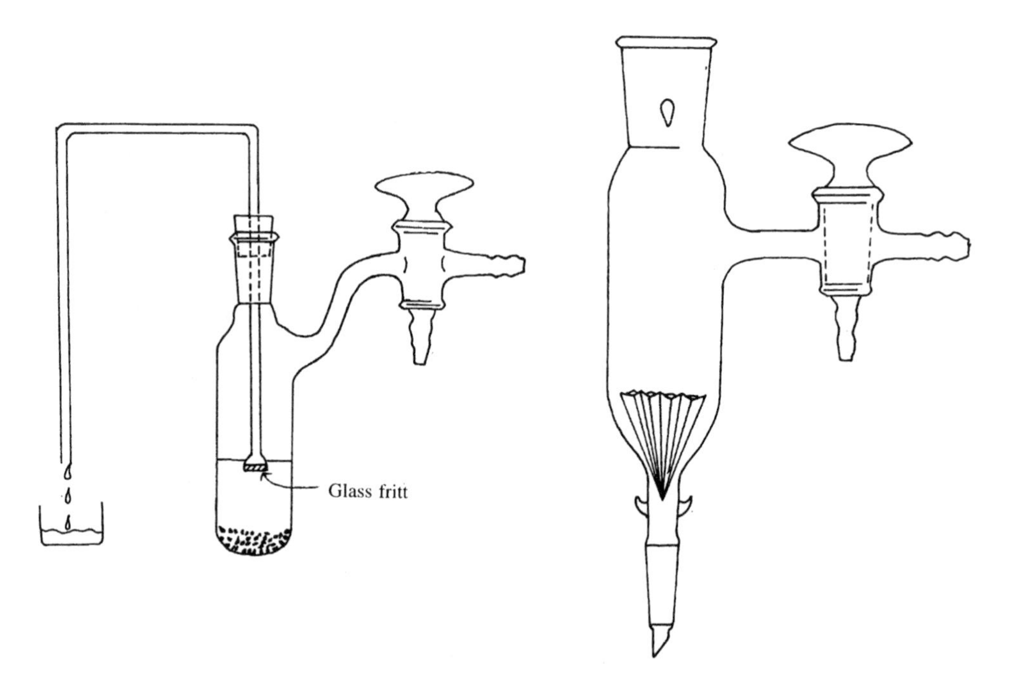

Figure 4.11
Filtration Using Normal Filter Paper and Bridge Filter. A folded filter paper is placed in the Schlenk filter flask (right) and the whole system is filled with nitrogen. Liquid is transferred by the syringe or cannula to filter. Inert gas flow should not pass from the Schlenk flask but filtration glassware should be used to prevent the evaporation. When filtering, the glass frit of the bridge filter should always touch the surface of the solution and then be slowly lowered for smooth filtration. If the solid sticks to the glass filter, the inside pressure is reduced to normal by lifting the rubber holder, which removes the solid easily.

air in the whole system is replaced by nitrogen by pump and re-filling three times. The heterogeneous solution is transferred by a syringe or cannula onto the filter paper. Instead of the filter paper a glass frit can be put in the glass filter tube, which can also be surrounded by a glass jacket that is cooled by dry ice/alcohol to handle thermally unstable materials.

The most important trick in filtration is always to wait until the solids have settled as much as possible and to filter the clear solution first. Then the remaining turbid part can be filtered. This often saves time, especially when precipitates are very fine.

A column packed with filler such as celite or neutral alumina can also be used for filtration or removal of undesired materials (Figure 4.12). After the whole system is filled with nitrogen, the solution is transferred slowly by a syringe or cannula to the top of the filler. The clear solution will be obtained from the bottom. In order to handle a small amount of solution such as for NMR samples, a disposable pipet with filler filters very efficiently. A small amount of fine glass fiber (glass wool) is placed at the bottom of the pipet to avoid leakage of solid. As expected, the pressure difference before and after the filter affects very much the speed of filtration. Thus the evacuation of the receiver flask during filtration effectively accelerates the filtration, though one must always look out for an unexpected air leak.

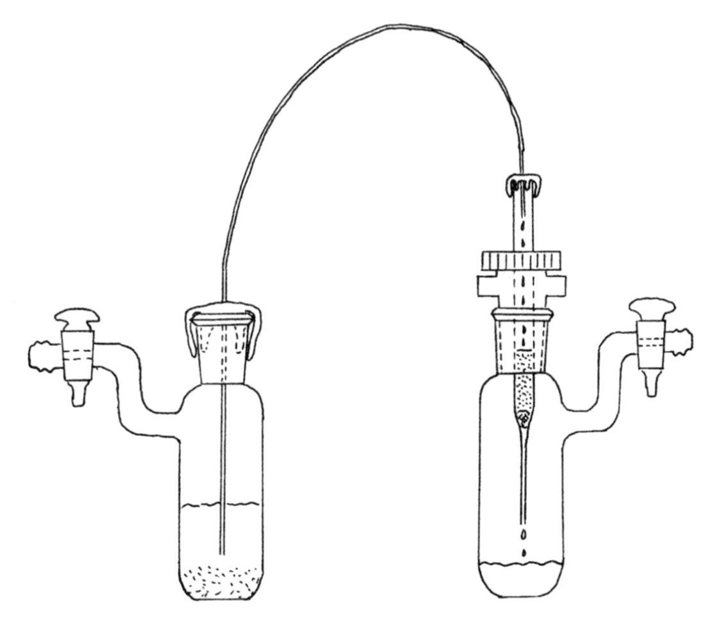

Figure 4.12
Simple Filtration by Celite. A disposable pipet filled with celite is fitted with a rubber septum and is placed at the inlet of a Schlenk flask by means of a thermometer holder. Then the whole system is evacuated and filled with inert gas. Transfer of liquid is carried out by the cannula method. The Schlenk receiver should be at ambient pressure, although careful evacuation can accelerate the filtration.

4.3.4.2 *Distillation*

Distillation is carried out essentially as usual. The whole system is usually filled with nitrogen by slow flow of nitrogen into the side arm of a Schlenk flask containing the liquid and/or by bubbling nitrogen into the liquid to be distilled. It is better to have a slight nitrogen flow in the receiving Schlenk flask to avoid air contamination during the distillation, but nitrogen gas bubbling may be stopped, since it always lowers the boiling point. Distillation under reduced pressure is carried out similarly. In this case the inert gas should be introduced through the pressure controller such as a capillary fitted to the distillation flask or the needle valve of a Bunsen burner that is placed between receiver and vacuum pump.

In a vacuum system, liquids can be transferred by a trap-to-trap (bulb-to-bulb) distillation. It should be noted that small amounts of non-condensable gases at liquid nitrogen temperature severely lower the rate of transfer. Therefore all the air or inert gases should be completely evacuated at liquid nitrogen temperature before vacuum transfer. This operation should be repeated three times to completely remove non-condensable gases (freeze–pump–thaw method). When the receiver flask that is connected with the flask containing a volatile material in a closed system is cooled by liquid nitrogen, the volatiles are condensed or solidified into the flask. Great care must be taken when the flask containing the volatiles is opened, since quick connection induces abrupt boiling. Cold ethanol at –20 ~ –40 °C, a temperature that can be

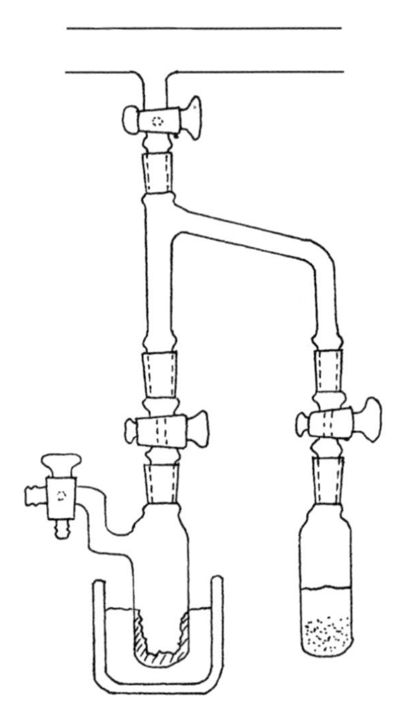

Figure 4.13
Trap-to-Trap (Bulb-to-Bulb) Distillation. Before the operation is carried out, all the noncondensable gases at liquid nitrogen temperature should be evacuated by at least three pump-and-thaw treatments.

achieved with an electric cold bath, will suffice in practice for transferring relatively volatile solvents such as ether, hexane and THF, thus saving expensive dry ice and liquid nitrogen. Solvents such as ether, THF, benzene, toluene, dioxane, hexane and pentane that has been dried over benzophenone ketyl and many organic unsaturated compounds dried over calcium hydride can easily be distilled into the Schlenk flask by this method without contacting air and moisture. In these cases any evolved hydrogen gas should be removed periodically by the pump-and-thaw method.

4.3.4.3 Reflux or heating solution

When the solution is subjected to reflux or to heat under inert gas, a cooling tower such as a reflux condenser should be attached to the Schlenk tube. After the whole system is filled with an inert gas, the solution is heated with stirring by means of a temperature-controlled oil bath. The inert gas should not flow from the Schlenk flask toward the condenser, because this will drive evaporated solvent from the flask and decrease its amount quickly. A steady slow nitrogen flow should pass above the reflux condenser to protect the solution from an unexpected back current flow of gas. When the condenser is to be attached to a Schlenk flask that already contains a solution, a slow nitrogen gas flow from the Schlenk flask may be made to fill the condenser with nitrogen at low temperature. Heating the solution without a reflux condenser may contaminate the nitrogen line with undesirable volatiles.

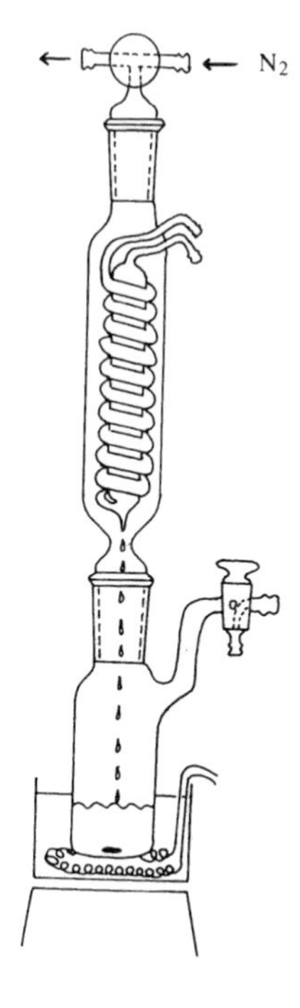

Figure 4.14
Refluxing Solution

4.3.4.4 *Measuring Liquid by Pipet*

For quantitative measurements requiring more accurate measurement of a volume of liquid than is possible with a syringe, pipettes and commercially available pipette bulbs are used under nitrogen. First, the side gas inlet of the pipette bulb is connected to a nitrogen gas line to introduce nitrogen gas. A few minutes gas flow is enough to replace the air in the connecting tube to nitrogen. Then only nitrogen gas is taken very slowly from the Schlenk tube that is already connected to a nitrogen line and discarded from the top of the pipette bulb. After three repetitions of this procedure, the pipette is ready to take the solution under nitrogen. Since the tip of the pipette might be exposed to air for a short time during transferring, the operation must be done promptly, or the inlets of the Schlenk flask should be protected by appropriate guards such as glass pants.

4.3.4.5 *Toepler Pump (Quantitative measurement of gas volume)*

The volume of gases formed in a reaction can be easily measured without exposure to air by means of a Toepler pump. Although automatic versions can be purchased commercially, a hand-operated one is described here for convenience. Figure 4.15 shows a schematic drawing of the Toepler pump. A flask containing an unknown amount of gases in solution is connected at stopcock A of the Toepler pump and gas sampling tubes are attached at stopcock D. The system is completely evacuated before use. Then

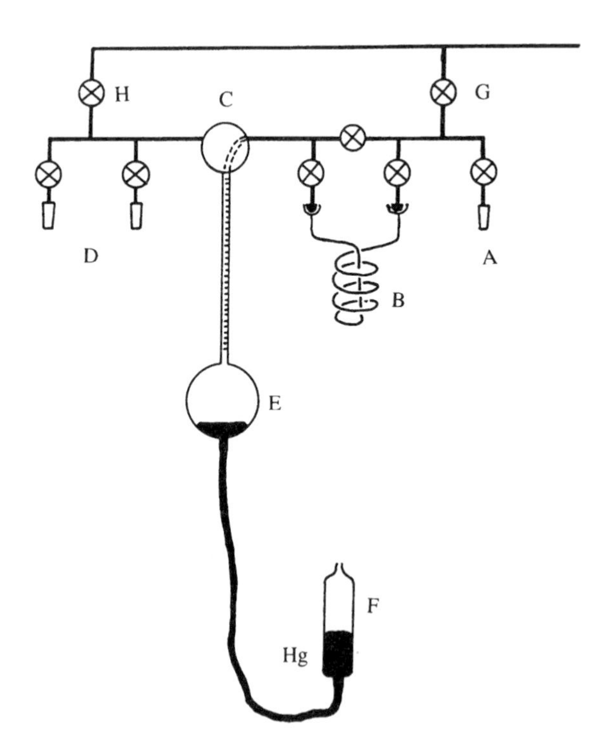

Figure 4.15
Toepler Pump

the flask and trap B are cooled with liquid nitrogen. When the two-way stopcock (120 °) C is turned to connect to the sample side (right side), non-condensable gases at this temperature, such as nitrogen, hydrogen, methane and carbon monoxide, diffuse into a collecting glass ball E. After closing the stopcock C, the mercury container F is slowly raised until the mercury levels in the scale and container become equal, when the gas volume is estimated by reading the scale at room temperature and pressure. The scale on the glass tube below the stopcock C should be calibrated in advance. Then, mercury container F is lowered once and the stopcock C is turned to connect to the left sampling tube side. Mercury container F is slowly raised till the mercury reaches just to the bottom level of stopcock C, so that all the gases in the receiving glass ball are transferred to the left receiver side. After closing stopcock C, the level of mercury container F is lowered again. Then stopcock C is turned to connect A and E to diffuse the rest of gases into E. The procedure is repeated until no significant amount of gas is left. After redissolving the sample solution, the same procedure is repeated until the confined gases in solution are expelled completely. The sum of these measurements is the volume of those gases which are noncondensable at liquid nitrogen temperature. The contents of gases collected in the sample tube in the left side are analyzed independently by GC, from which the amounts of all gases are estimated quantitatively. Then if liquid nitrogen baths of the trap B and the sample flask are replaced by a dry ice/ethanol bath, gases such as ethylene, ethane, acetylene, and carbon dioxide can be quantitatively analyzed analogously.

More simply, gases in the closed system are quantitatively analyzed only by GC using the internal standard method. It should be noted that some skill may be required to introduce an accurate amount of standard gases into the flask. It is convenient to have a septum rubber cap in one outlet of the flask from which gases are introduced and taken. Two digits of accuracy can be obtained by use of a hypodermic syringe that has been calibrated by filling with water. A certain amount of standard gas is taken from the gas cylinder. A soft rubber tube is attached to the outlet of the gas cylinder and the other side is closed with a pinch cock. When the gas cylinder is opened very slightly, the pressure in the rubber tube increases and the gas is easily taken from this tube by the syringe. The gas taken in the syringe is discarded and the gas is taken again. After a few repetitions of these operations, an accurately measured amount of pure gas is taken and injected into the flask via septum as quickly as possible. After a few minutes allowance for equilibration of gases and liquids at a constant temperature (preferably in a thermostatted bath), the composition of the gas phase is analyzed by GC to give the exact amount of the gases in the flask. Very careful calibration using standard samples under the same conditions is required, especially when samples contain both gases and liquids, since the solubility of gases in solvents varies considerably depending on the temperature and the amount of solvent used. Several trial runs using known samples may be necessary in order to develop skill. Nevertheless, since this is such a simple and inexpensive method to obtain reasonably quantitative data for gas analysis, it is still useful and valuable.

4.3.4.6 NMR Sample

A NMR sample tube is connected with a female glass joint by a teflon connecting joint (Figure 4.16) or welding. If this is attached to a glass joint having a three-way stop-

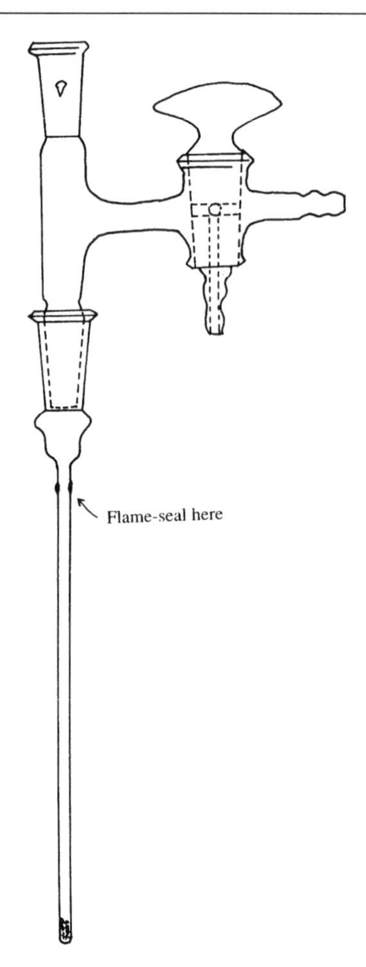

Figure 4.16
Preparation of NMR Sample

cock, it can be used like Schlenk-type glassware. Solid and liquid samples are transferred by the methods described previously. When deuterated solvent is introduced into the NMR tube by vacuum transfer (trap-to-trap method), the cooling container (Dewar flask) of liquid nitrogen should first be set level with the bottom of the NMR tube and then raised gradually. This frequently prevents wasting of excess solvent as well as time. A commercially available rubber septum for an NMR tube can be used for sealing the NMR tube. Alternatively, the tube can be closed by sealing with a small torch. In order to avoid air contamination, extra nitrogen flow to the connection part from both inside and outside should be applied when installing the septum cap. Though this method seems unreliable because of its apparent simplicity, it actually works very well in most cases. Another merit of this method is that NMR tubes are recyclable. The preparation of NMR samples is also done easily in a drybox.

Gases can also be introduced in the NMR tube (commercially available) using the joint shown in Figure 4.16. However, great care must be taken to avoid high pressure in

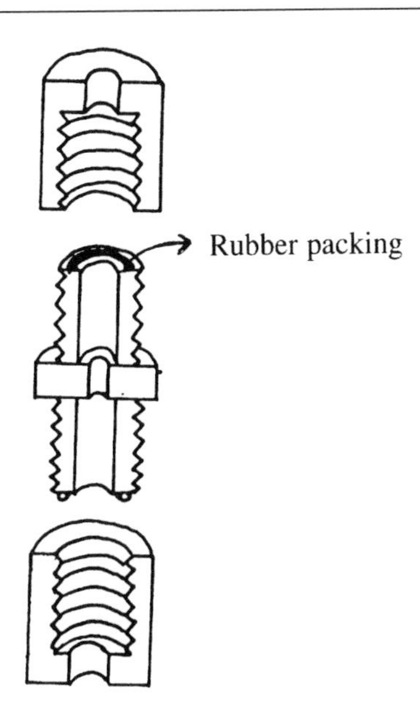

Figure 4.17
Joint for NMR Tube

the tube if gases are introduced by a trap-to-trap distillation. Usually a pressure of less than 1 atm should be employed. A NMR tube which withstands 5 atm is also commercially available. However, if high pressure NMR work is contemplated, a specialist should be consulted.

4.3.4.7 IR Sample

KBr disks for IR can also be prepared easily without use of a glove box. As shown in Figure 4.18, nitrogen gas is introduced into a molding apparatus from an outlet for evacuation. After a few minutes nitrogen gas flow, a certain amount of KBr powder ground beforehand is placed onto a clean face of the press by spatula. Then an air-sensitive sample is taken and mixed with KBr promptly. After putting a cap on firmly with a forceps, pressing the molding apparatus for a few minutes gives a disk pellet. Evacuation during pressing is not usually necessary to give a transparent disk.

A liquid or solution sample can also be transferred into a solution cell by using appropriate glassware and joints, since this is a simple application of the above techniques, though the drybox technique may be easier for this purpose.

4.3.4.8. UV/VIS Sample

A 1 cm UV/Vis cell is sealed on to one of the side-arms of a Schlenk flask as shown in Figure 4.19. A solution or solid sample is transferred into the flask. Then reagents or solvents are introduced by vacuum distillation or by use of a syringe. The solution is transferred into a cell by tilting the glassware.

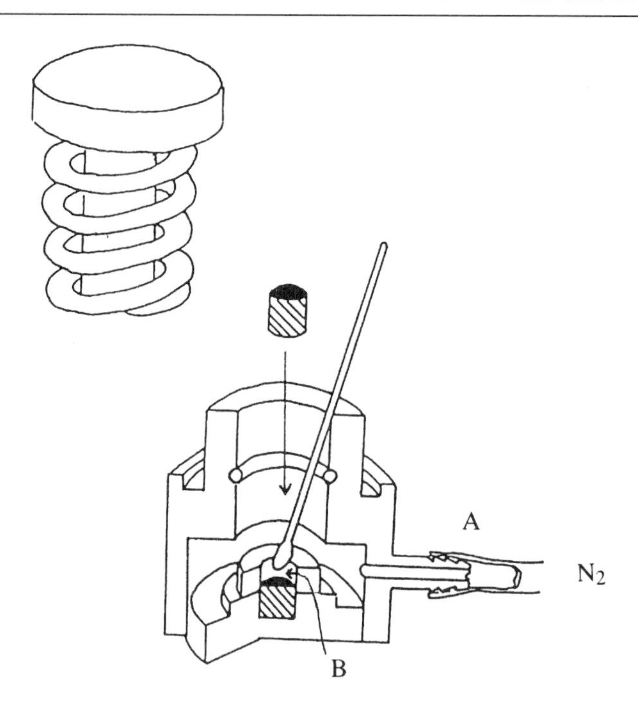

Figure 4.18
Preparation of IR Sample

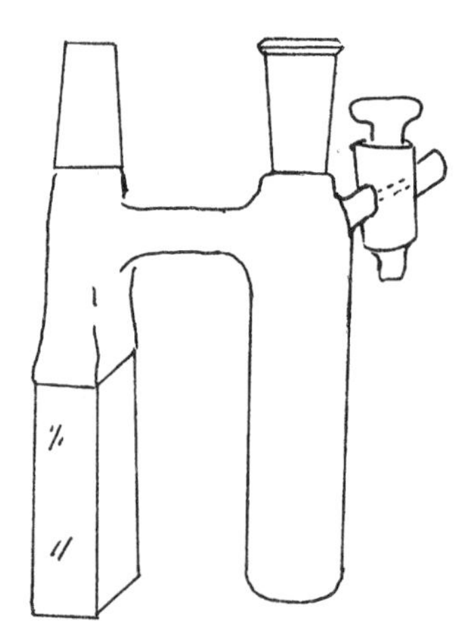

Figure 4.19
Preparation of Sample of UV/Vis Measurement

4.3.4.9 A Crystal for X-Ray

The piece of apparatus for mounting an air-sensitive crystal in a capillary tube for X-ray diffraction is drawn in Figure 4.20. A tiny drop of glue such as silicone grease is

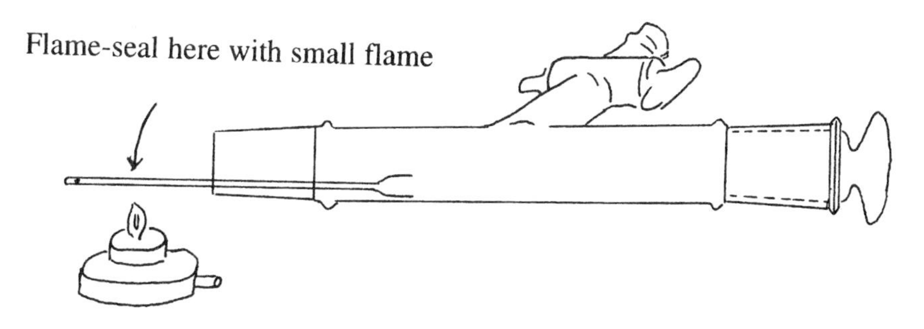

Figure 4. 20
Preparation of a Single Crystal for X-Ray Crystallography

stuck inside the capillary beforehand. Then the capillary and a selected crystal are placed in the flask shown in Figure 4.20 under nitrogen. The crystal is picked up with a thin glass rod and is mounted on the glue by pushing it with a thin glass capillary. A crystal that easily loses included solvent can be mounted without glue while it is still wet. The capillary sample is partly moved out from the bottom side of the flask and is sealed by a very thin flame that can be easily prepared by using only the bottom part of Bunsen burner. Another procedure is to quickly mount a crystal on the tip of the capillary and to cool it immediately to –196 °C by an efficient flow of liquid nitrogen to prevent oxidation by air, but this method requires considerable skill.

References

[1] (a) D. F. Shriver, M. A. Drezdzon, *Manipulation of Air-Sensitive Compounds*, 2nd ed., John Wiley, New York (1986).
(b) M. Y.Wayda, M. Y. Darensbourg, Ed., *Experimental Organometallic Chemistry, ACS Symp. Ser. 357*, American Chemical Society, Washington DC. (1987).
(c) R. B. King, *Organometallic Syntheses*, vol. 1, *Transition Metal Compounds*, Academic Press, New York (1965).
(d) J. J. Eisch, *Organometallic Syntheses*, vol. 2, *Nontransition Metal Compounds*, Academic Press, New York (1981).
(e) J. D. Woollins, Ed., *Inorganic Experiments*, VCH, Weinheim (1994).
(f) J. Leonard, B. Lygo, G. Procter, *Advanced Practical Organic Chemistry*, 2nd ed., Blackie, London (1995).
(g) A. Yamamoto, *Organotransition Metal Chemistry Fundamental Concepts and Applications*, John Wiley, New York (1986).
(h) H. C. Brown, *Organic Synthesis via Boranes*, John Wiley, New York (1975).
(i) Chemical Society of Japan, Ed., *Shin-Jikkenkagaku Koza*, vol. 12, Maruzen, Tokyo (1976); *Jikkenkagaku Koza*, 4th. Ed., vols, 1, 2 (1991).

5 Group 3 (Sc, Y, Lanthanide) Metal Compounds

K. Tatsumi, *Nagoya University*

5.1 Introduction

This chapter deals with organometallic compounds of scandium, yttrium, lanthanum, and lanthanoids. The normal oxidation state of these group 3 metals is +3, and the large ionic radii and presence of vacant outer d, s, and p orbitals make them strong Lewis acids. Thus the coordination number is usually large, which may reach 9 or even 12, and the metal–ligand bonds are mostly ionic in nature. For instance, the electron flow from the cyclopentadienyl anions to metal cation in MCp_3 was reported not to exceed 0.3e [1]. This indicates the degree of bond covalency to be small, and the M—Cp bond ionicity increases slightly from Sc, Y, La, and to heavier lanthanoids. However, according to luminescence spectra of a series of Ce(III) complexes, the presence of a significant degree of covalency has been noticed [2].

The lanthanide series is characterized by consecutive filling of electrons in the 4f shell. However, the radial extension of the f orbitals is not great, so that their participation in bonding appears to be limited, and the outer 5d, 6s, and 6p orbitals are responsible for the interactions with ligands. High spin states are favored due to the small f-orbital size and the concomitant large electron–electron repulsive integrals, and the organolanthanide complexes are paramagnetic except for those of f^0 and f^{14} metal elements such as La(III) and Lu(III). Since the complete and half-complete shells with f^0, f^7, and f^{14} electronic configurations tend to exhibit enhanced stability, the unusual oxidation states having these electronic states emerge, e.g. Ce(IV), Eu(II), Yb(II), while Sm(II), Dy(II), and Tm(II) are also known.

5.2 Organometallic Compounds of Group 3 and Lanthanide Metals

The most fundamental precursors to organometallic complexes of group 3 elements are the trichlorides, MCl_3. The commercially available trichlorides are hydrated, and even for those claimed dehydrated it is advisable to take the appropriate steps necessary to dry them further before use. The divalent iodide complexes MI_2 are available for certain group 3 elements such as samarium and ytterbium, and they serve as conve-

Synthesis of Organometallic Compounds: A Practical Guide. Edited by S. Komiya

nient precursors for the preparation of divalent organometallic complexes. In some cases, the metals themselves can be used as starting materials.

The organometallic chemistry of the group 3 metals is dominated by the complexes having cyclopentadienyl and/or its derivatives. It has been reported that approximately 90% of previously described organolanthanides are those containing such anionic π ligands [3]. The first member of this class is the tris(cyclopentadienyl) complexes, MCp_3, which are now known for all of the group 3 metals. The tris(cyclopentadienyl) lanthanides were synthesized by Birmingham and Wilkinson in 1954, and they are actually the first unambiguously characterized organometallic complexes of lanthanides [4]. The two remaining members of this family, $[MCp_2Cl]_2$ and $MCpCl_2(THF)_3$, were prepared several years later by Dubeck *et al.* [5]. The yttrium analogues, $[YCp_2Cl]_2$ and $YCpCl_2(THF)_3$, appeared more recently [6,7].

The cyclopentadienyl complexes of group 3 metals are air- and/or moisture sensitive, and solvent-free MCp_3 were prepared by the reactions of the corresponding trichlorides with sodium and potassium salts of cyclopentadienyl in THF followed by the sublimation of the crude products (eq (1) and (2)) [4,8]. The europium complex is vulnerable to sublimation, probably due to reduction to the divalent state, and it is necessary to carry out the reaction in THF under carefully controlled conditions [9]. An intriguing route to tris(cyclopentadienyl) lanthanides has been reported, which consists of reactions between lanthanide metals and thallium cyclopentadienyl giving rise to MCp_3 or $MCp_3(THF)$ depending on the metals (eq (3)) [10]. This may serve as a general method for preparation of the wide range of MCp_3 complexes and their derivatives.

$$MCl_3 + 3\,NaCp \xrightarrow{THF} MCp_3 + 3\,NaCl \qquad (1)$$

M = Sc, Y, La, Ce, Pr, Nd, Sm, Gd, Dy, Er, Yb

$$MCl_3 + 3\,KCp \xrightarrow{C_6H_6 \text{ or } Et_2O} MCp_3 + 3\,KCl \qquad (2)$$

M = Tb, Ho, Tm, Lu, Yb

$$\text{“M”} + 3\,TlCp \xrightarrow{THF} MCp_3 \;(M = Er, Yb) \quad \text{or} \quad MCp_3(THF) \;(M = Ce, Nd, Sm, Gd, Er) \qquad (3)$$

Neither of the MCp_3 structures is monomeric, but they consist of infinite chains, in which two pentahapto Cp ligands are bound to a single metal center and the other Cp bridges two metals [11]. Although the cyclopentadienyl bridge occur in a μ–η^1, η^2 (or μ–η^3, η^4) manner, the ring remains nearly planar and the Cp—M bonding is strongly ionic. The size of the lanthanide ions are too large to settle with three pentahapto cyclopentadienyl ligands. Thus MCp_3 complexes, except for the scandium complex, tend to accommodate a donor ligand such as THF, phosphine, isocyanide, etc, forming formally ten-coordinate base adducts, Mcp3L [11d,12]. In spite of the fact that the Cp complexes are extremely hydrolytic, the aquo adducts have been isolated in small quantities, $YCp_3(OH_2)$ and $Ho(C_5MeH_4)_3(OH_2)$ [13]. For the early lanthanides (La, Ce and Pr), the MCp_3 complexes can form 1:2 adducts with acetonitrile or propionitrile when an excess amount of the nitriles is introduced (eq (4)) [14a]. Isocyanides also form 1:2 adducts with $LaCp_3$ (eq (5)) [14b]. The coordination geometry is trigonal–bipyramid, and two nitrile (or isocyanide) ligands sit at the axial sites.

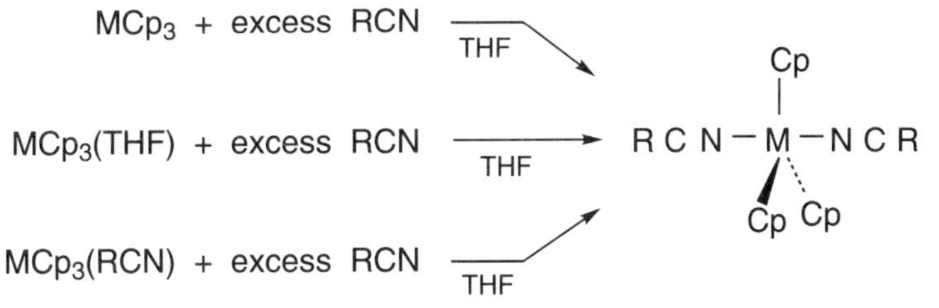

$$MCp_3 + \text{excess } RCN \xrightarrow{THF} ; \quad MCp_3(THF) + \text{excess } RCN \xrightarrow{THF} ; \quad MCp_3(RCN) + \text{excess } RCN \xrightarrow{THF} \quad RCN{-}M(Cp)_3{-}NCR \tag{4}$$

$$LaCp_3 + \text{excess } R'NC \longrightarrow MCp_3(CNR')_2 \qquad R' = \text{cyclo-}C_6H_{11} \tag{5}$$

By using pyrazine as a donor ligand, the dinuclear ytterbium complex, $YbCp_3(\mu\text{–}NC_4H_4N)\,YbCp_3$, was synthesized [15]. The related anionic dinuclear samarium complexes, $[Li(DME)_3]\,[SmCp_3(\mu\text{–}N_3)SmCp_3]$ were prepared by reacting $SmCp_3$ with LiN_3 in DME [16], while a similar reaction between $LuCp_3$ and NaH or NaD in THF generated the hydride (or deuteride) bridge, $[Na(THF)_6][LuCp_3(\mu\text{–}H)LuCp_3]$ [17].

The infinite chain structure of tris(cyclopentadienyl) complexes can be unraveled by increasing the steric hindrance of the Cp ligand. In the case of C_5MeH_4, the lanthanum, cerium and neodymium complexes are tetrameric, while the ytterbium analogue is monomeric [18]. The even bulkier 1,3-bis(trimethylsilyl)cyclopentadienyl) ligand provides monomeric complexes for cerium and samarium [19].

Chlorobis(cyclopentadienyl) complexes of the form MCp_2Cl or $MCp_2Cl(L)$ have been useful organometallic precursors of group 3 elements. In the absence of coordinating solvents, the solvent(L)-free complexes exist as dimers with two chlorides bridging two metal atoms. The dimeric form appears to be yet coordinatively unsaturated for early lanthanides, La–Nd, and chlorobis(cyclopentadienyl) complexes of these

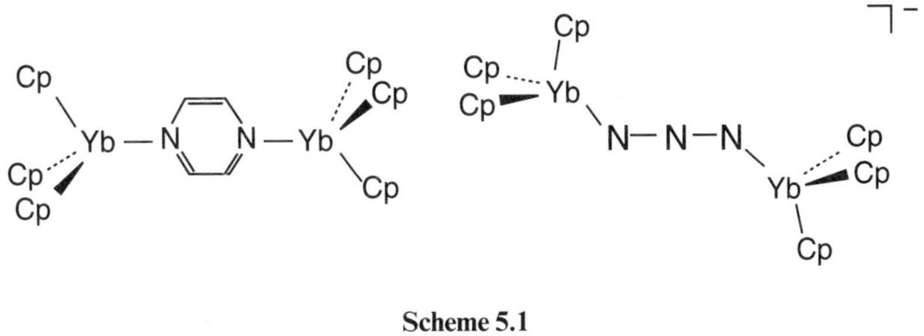

Scheme 5.1

large elements are not available. One obvious method for preparation of MCp_2Cl or $MCp_2Cl(L)$ is to carry out the reaction between MCl_3 and NaCp in a molar ratio of 1:2 (eq (6)) [20]. The sodium salt can be replaced by TlCp or $MgCp_2$ (eq (7)). The other approaches include treatment of MCp_3 with NH_4Cl in benzene (eq (8)), the reaction of MCp_3 with HCl (eq (9)), and comproportionation between MCl_3 and MCp_3 (eq (10)).

$$MCl_3 + 2\ NaCp \xrightarrow[THF]{} MCp_2Cl + 2\ NaCl \tag{6}$$

M = Sm, Gd, Dy, Ho, Er, Yb, Lu

$$ScCl_3 + MgCp_2 \xrightarrow[THF]{} ScCp_2Cl + MgCl_2 \tag{7}$$

$$MCl_3 + NH_4Cl \xrightarrow[C_6H_6]{} MCp_2Cl + NH_3 + C_5H_6 \tag{8}$$

M = Sm, Yb

$$MCp_3 + HCl \longrightarrow MCp_2Cl + C_5H_6 \tag{9}$$

$$MCl_3 + 2\ MCp_3 \longrightarrow 3\ MCp_2Cl \tag{10}$$

M = Sm, Gd, Dy, Ho, Er, Yb

When the lithium salt of cyclopentadienyl is used, anionic complexes of the type $[MCp_2Cl_2]LiL_2$ are formed. The lithium cation is actually incorporated in the complex by bridging two chlorides, where L may be THF or $\frac{1}{2}$ (TMEDA), etc, and the entire molecule is soluble in benzene and toluene. A structurally characterized example is $[Nd(C_5Me_5)_2Cl_2]Li(THF)_2$ (Scheme 5.2(a)) [21].

Another interesting approach to bis(cyclopentadienyl) complexes is the use of lanthanide triflates [22a,b]. The anhydrous triflate complexes $M(O_3SCF_3)_3$ are easily obtained by treatment of lanthanide oxides with a stoichiometric amount of triflic acid and subsequent acetonitrile extraction of the crude products. It was reported that the reaction of $M(O_3SCF_3)_3$ with NaCp produced $[MCp_2(O_3SCF_3)]_2$ [22c,d]. Like many of the chloride analogues, the structure of the triflate complex is dimeric with bridging triflate ligands (Scheme 5.2 (b)). The readily accessible triflate complexes could become a convenient entry to organolanthanide complexes.

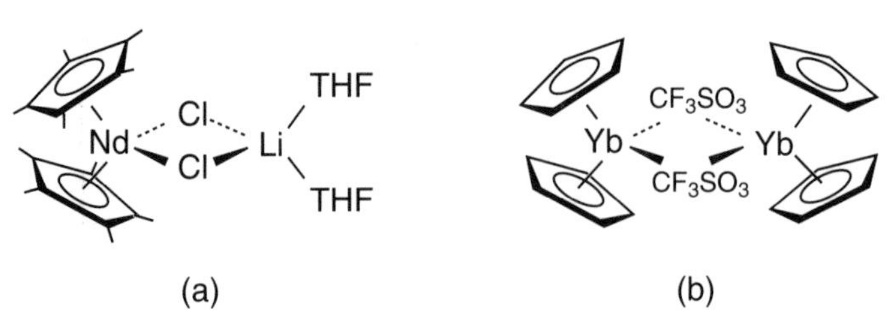

Scheme 5.2

$$M_2O_3 + 3\,CF_3SO_3H \longrightarrow M(O_3SCF_3)_3 + 3\,H_2O$$

$$M(O_3SCF_3)_3 + 2\,NaCp \longrightarrow \tfrac{1}{2}[MCp_2(\mu\text{-}O_3SCF_3)]_2 + 2NaO_3SCF_3 \quad (11)$$

M = Sc, Yb, Lu

Numerous substitution reactions of MCp_2Cl complexes and their derivatives have been reported. A typical example is formation of hydrocarbonyl complexes, which are normally prepared from the reactions of MCp_2Cl with equimolar amounts of the lithium salts of alkyls and aryls (eq (12)). For the synthesis of corresponding allyl complexes, allyl magnesium bromide is used (eq (13)). Because the monomeric MCp_2R molecules are coordinatively unsaturated, the complexes often assume dimeric structures when they are isolated in non-coodinating medium [23a]. As most of the alkylation reactions are performed in THF, many THF adducts have been prepared [20]. In order to obtain monomeric structures, bulky alkyl groups and/or highly substituted cyclopentadienyls have to be employed. An example of this class is $Y(C_5Me_5)_2\{CH(SiMe_3)_2\}$ [23b]. The monomeric structure is stabilized by bulkiness of the $CH(SiMe_3)_2$ ligand, and perhaps by agostic interactions between yttrium and the

$$\text{"}MCp_2Cl\text{"} + LiR \xrightarrow[THF]{} MCp_2R(THF) + LiCl \quad (12)$$

M = Y, Nd, Sm, Dy, Er, Tm, Yb, Lu

R = Me, Et, ^iPr, ^tBu, CH_2TMS, Bz, Ph, C_6H_4Me-*p* etc.

$$\text{"}MCp_2Cl\text{"} + C_3H_5MgBr \longrightarrow MCp_2(\eta^3\text{-}C_3H_5) + MgClBr \quad (13)$$

M = Sc, Sm, Ho, Er

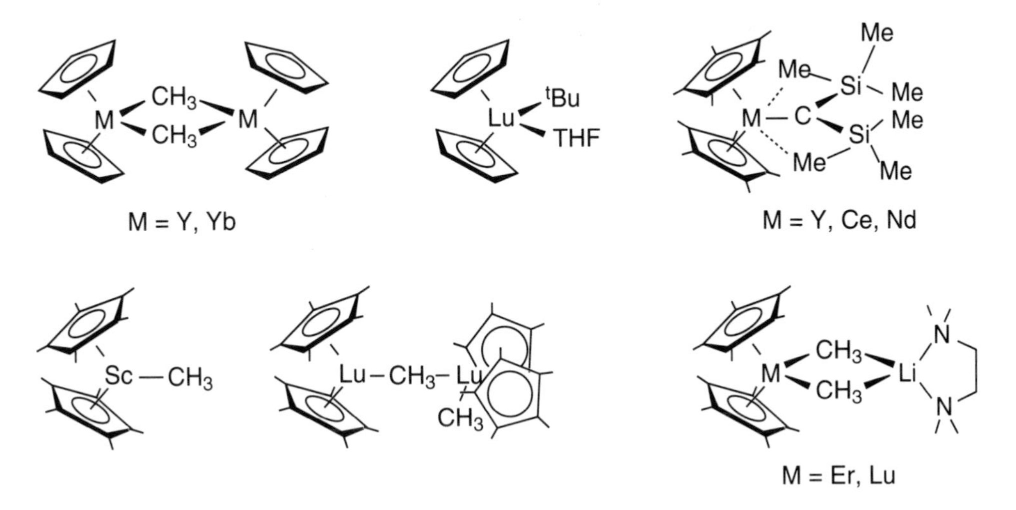

Scheme 5.3

methyl groups of the alkyl ligand as well. In this respect, successful isolation of monomeric $Sc(C_5Me_5)_2(CH_3)$ is noteworthy [23c]. The lutetium analogue $[Lu(C_5Me_5)_2(CH_3)]_2$ provides another interesting structure in solid, which is dimeric with a single methyl bridge [23d]. This compound exists in solution in a monomer–dimer equilibrium. When a MCp_2Cl precursor is treated with two equivalents of methyllithium, a monomeric dimethyl complex anion is obtained, in which the lithium cation is bound to two methyl groups. The lithium is further coordinated by THF or TMEDA giving rise to $[MCp_2Me_2]Li(THF)_2$ or $[MCp_2Me_2]Li(TMEDA)$.

It is known that some of the $M—CH_3$ bonds in "$MCp_2(CH_3)$" are reactive toward terminal alkynes to give $MCp_2(C{\equiv}CR)$ (eq (14)) [24]. The complexes are dimeric in nature with bridging alkynide ligands. Substitution reactions of the methyl group in $M(C_5Me_5)_2(CH_3)$ proceed also with $^{13}CH_3$, ethane, and benzene (eq (15)) [23c, 25]. The rate of C—H activation varies depending on the metal, and $Y(C_5Me_5)_2(CH_3)$ reacts with CH_4 5 times faster than the Lu analogue and 250 times faster than the Sc complex. Bis(cyclopentadienyl)alkyl and aryl complexes are also capable of activating dihydrogen to generate dimeric hydride complexes (eq (16)). The non-solvated hydride dimer was isolated for lutetium, while for larger lanthanides the hydride complexes tend to accommodate THF molecules [14a, 20, 26]. The hydride-bridged dimers are highly reactive. For instance, $[YCp_2(\mu\text{-}H)(THF)]_2$ and its C_5MeH_4 congener react with alkenes, alkynes, allene, nitriles to give corresponding inserted products, e.g. $YCp'_2R(THF)$, $YCp'_2(CR{=}CRH)(THF)$, $[YCp'_2(C{\equiv}CR)]_2$, $YCp'_2(\eta^3\text{-allyl})(THF)$ and $[YCp'_2(\mu\text{-NCHR})(THF)]_2$ ($Cp' = Cp$, C_5MeH_4) [26b].

$$MCp'_2(CH_3) + RC{\equiv}CH \longrightarrow [Cp_2M(\mu\text{-}C{\equiv}CR)_2MCp_2] + CH_4 \quad (14)$$

M = Er; Cp' = Cp

M = Yb; Cp' = C_5MeH_4

R = ^tBu, n-C_6H_{13}

$$M(C_5Me_5)_2(CH_3) + RH \longrightarrow M(C_5Me_5)_2(R) + CH_4 \quad (15)$$

M = Sc, Y, Yb

$$LuCp_2(R) + H_2 \longrightarrow [Cp_2Lu(\mu\text{-}H)_2LuCp_2] + RH$$

R = alkyl, aryl

(16)

$$2\ MCp_2(R)(THF) + H_2 \longrightarrow [Cp_2(THF)M(\mu\text{-}H)_2M(THF)Cp_2] + RH$$

M = Y, Er, Lu

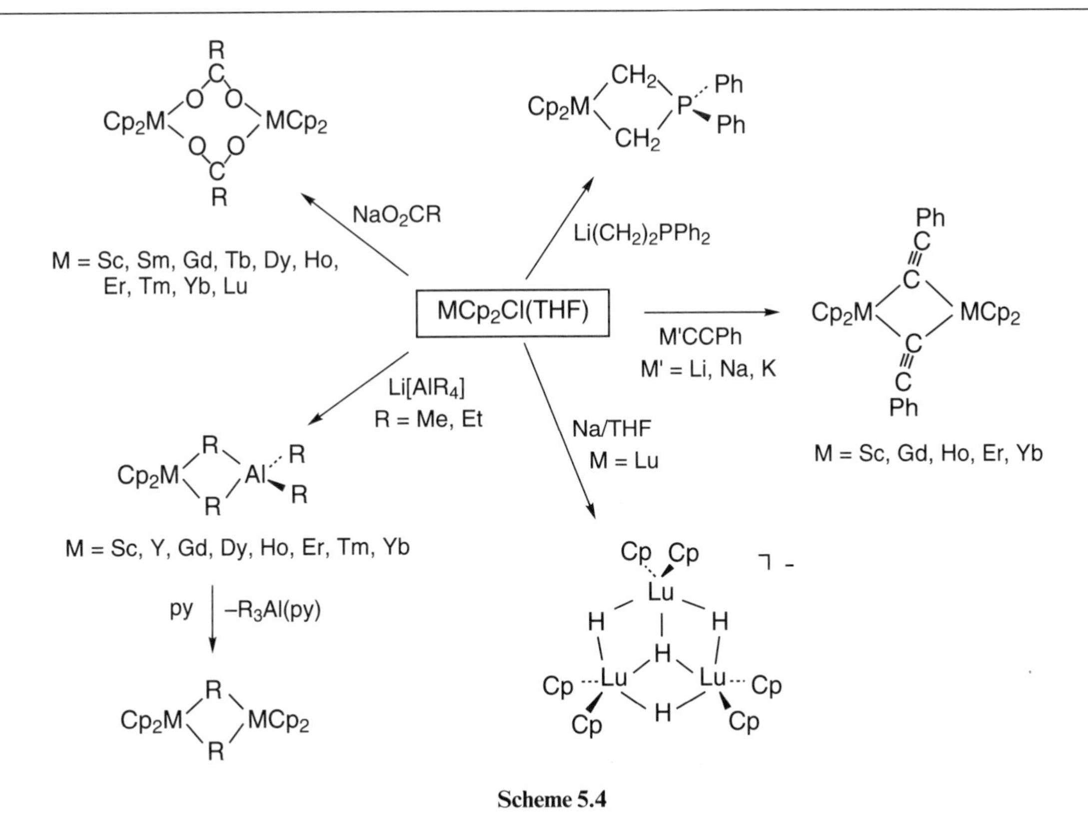

Scheme 5.4

The other representative substitution reactions of MCp_2Cl complexes are summarized in Scheme 5.4. Alkali metal salts of alkynides, carboxylates and phosphoylides react readily with MCp_2Cl producing corresponding bis(cyclopentadienyl) complexes [20, 27]. Likewise, treatment with lithium tetraalkylaluminates gives $MCp_2(AlR_4)$ which in turn lead to alkyl-bridged dimers upon addition of pyridine [23a]. The reaction of $LuCp_2Cl(THF)$ with metallic sodium in THF resulted in formation of the trinuclear hydride complex, $[Na(THF)_6][\{LuCp_2(\mu\text{-}H)\}_3(\mu_3\text{-}H)]$ [28]. The occurrence of triply-bridging hydride is rather common in the structures of lanthanide hydrides. Certain anionic bis(cyclopentadienyl)alkyl complexes decompose via β-hydrogen elimination leading to $[\{MCp_2(\mu\text{-}H)\}_3(\mu_3\text{-}H)]^-$. A more convenient way to synthesize hydride complexes of this class may be the reaction of $[YCp_2(\mu\text{-}H)(THF)]_2$ with LiH, LiMe, or Li^tBu [29].

Chemistry of mono(cyclopentadienyl) complexes of group 3 elements is less developed. The selective preparation of the coodinatively unsaturated $MCpCl_2$ or their solvated complexes is not an easy task. Without solvation monomeric complexes cannot be prepared even if very bulky cyclopentadienyl ligands are employed. The usual strategy to synthesize "$MCpCl_2$" is treatment of anhydrous trichlorides with one equivalent of NaCp (eq (17)) in THF, or an appropriate ligand redistribution reaction between MCl_3 and MCp_3 (eq (18)) [20, 30]. The resulting mono(cyclopentadienyl) complexes carry three THF molecules, or four THF molecules may be incorporated in some cases. A synthetic complication arises from the non-stoichiometric eatiincorporation of salts

$$MCl_3 + NaCp \xrightarrow[THF]{} MCpCl_2(THF)_3 + NaCl \quad (17)$$

$$2\,MCl_3 + MCp_3 \xrightarrow[THF]{} 3\,MCpCl_2(THF)_3 \quad (18)$$

and concomitant difficulty of purifying the product. Another complication comes from the possibility of disproportionation reactions to give MCp_2Cl and MCp_3. The use of lanthanide triflates (see eq (11)) may be a way to avoid the problems associated with incorporation of salts in the products. Thus $LuCp(O_3SCF_3)_2(THF)_3$ was synthesized straightforwardly from $Lu(O_3SCF_3)_3$ according to eq (19), which further reacts with two equivalents of $LiCH_2(SiMe_3)$ yielding $LuCp\{CH_2(SiMe_3)\}_2(THF)_3$ [31].

$$Lu(O_3SCF_3)_3 + NaCp \longrightarrow LuCp(O_3SCF_3)_2(THF)_3 + NaO_3SCF_3$$

$$LuCp(O_3SCF_3)_2(THF)_3 + 2\,LiCH_2(SiMe_3) \longrightarrow LuCp\{CH_2(SiMe_3)\}_2(THF)_3 + 2\,LiO_3SCF_3 \quad (19)$$

The reaction of $LuCpCl_2(THF)_3$ with disodium naphthalide in DME resulted in formation of a novel lanthanide naphthalide complex $LuCp(C_{10}H_8)(DME)$ (Scheme 5.5 (a)) [32]. Another interesting mono(cyclopentadienyl) complex is the sandwich structure of ScCp(OEP) (OEP = octaethylporphyrin) (Scheme 5.5 (b)), which was prepared by reacting ScCl(OEP) with cyclopentadienide [33].

Preparation of mono(pentamethylcyclopentadienyl) dichlorides of group 3 elements is also complicated, the situation of which parallels the synthsis of the mono(cyclopentadienyl) analogues. Reactions between lanthanide trichlorides and alkali metal salts of pentamethylcyclopentadienide in a molar ratio of 1:1 often end up with "ate" complexes of the type $M(C_5Me_5)Cl_3M'(L)_n$ (M′ = Li, Na; L = THF, Et_2O) [14a, 20]. In these complexes alkali metal chloride is incorporated, and the entire molecules are neutral. The iodide complexes have less tendency to take up alkali metal halide, and treatment of $MI_3(THF)_3$ with $K(C_5Me_5)$ produces the desired diiodide complexes $M(C_5Me_5)I_2(THF)_3$ (eq (20)). The THF-solvated triiodide can be prepared by heating lanthanide metals with EtI in THF for 1 day, and a subsequent Soxhlet extraction with THF for several days [34]. The reaction of $La(C_5Me_5)I_2(THF)_3$ with

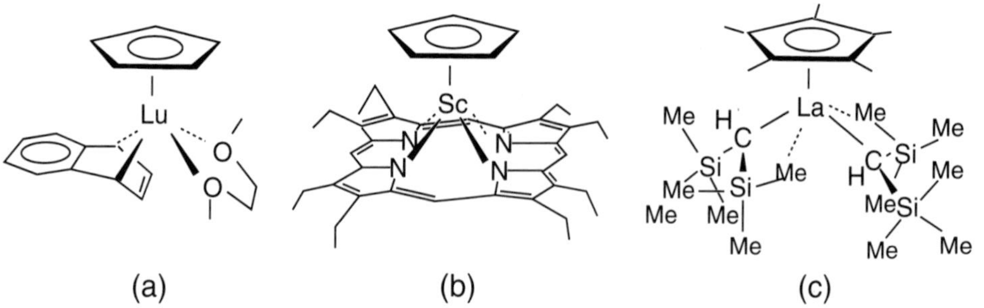

Lu O O (a) | Sc N N N N (b) | La Me H C Si Me Si Me Me Me Me Me C Si H Me MeSi Me Me (c)

Scheme 5.5

$$MI_3(THF)_3 + K(C_5Me_5) \longrightarrow M(C_5Me_5)I_2(THF)_3 + KI \qquad (20)$$
$$M = Ce, La$$

2 equivalents of $KCH(SiMe_3)_2$ affords salt-free $La(C_5Me_5)[CH(SiMe_3)_2]_2(THF)$ [35]. Interestingly, THF-free $[La(C_5Me_5)I_2]_n$ can be obtained upon treating $MI_3(THF)_3$ with Me_3SiI, and its reaction with 2 equivalents of $KCH(SiMe_3)_2$ gives $La(C_5Me_5)[CH(SiMe_3)_2]_2$. The apparently coordinatively unsaturated THF-free complex is stabilized by agostic interactions (Scheme 5.5 (c)). Removal of THF directly from preformed $La(C_5Me_5)[CH(SiMe_3)_2]_2(THF)$ is also achieved by the action of Me_3SiI.

As shown above, organometallic complexes of group 3 elements often incorporate donor solvent molecules and/or alkali metal halides. In order to avoid solvation and incorporation of alkali metal halides, appropriate reaction conditions have to be chosen. One such example is successful synthesis of the three-coordinate lanthanum alkyl complexes, $M\{CH(SiMe_3)_2\}_3$ (M = Y, La, Sm, Lu) (Scheme 5.6 (a)). The highly coordinatively unsaturated alkyl compound was prepared from $M\{OC_6H_2(^tBu)_3\text{-}2,4,6\}_3$ and $LiCH(SiMe_3)_2$ by treating them in pentane. In this condition, $Li\{OC_6H_2(^tBu)_3\text{-}2,4,6\}$ precipitates out and the highly soluble alkyl complexes are easily separated [36a]. Interestingly the solvent-free complexes adopt a pyramidalized coordination geometry. In contrast, when the lutetium analogue was prepared by the reaction between $LuCl_3$ and $KCH(SiMe_3)_2$ in ether, the solvated KCl adduct $M\{CH(SiMe_3)_2\}_3(\mu\text{-}Cl)K(Et_2O)$ resulted, which was also formed by treating $Lu\{CH(SiMe_3)_2\}_3$ with KCl [36b]. The ether molecule may be removed under vacuum to give $Lu\{CH(SiMe_3)_2\}_3(\mu\text{-}$

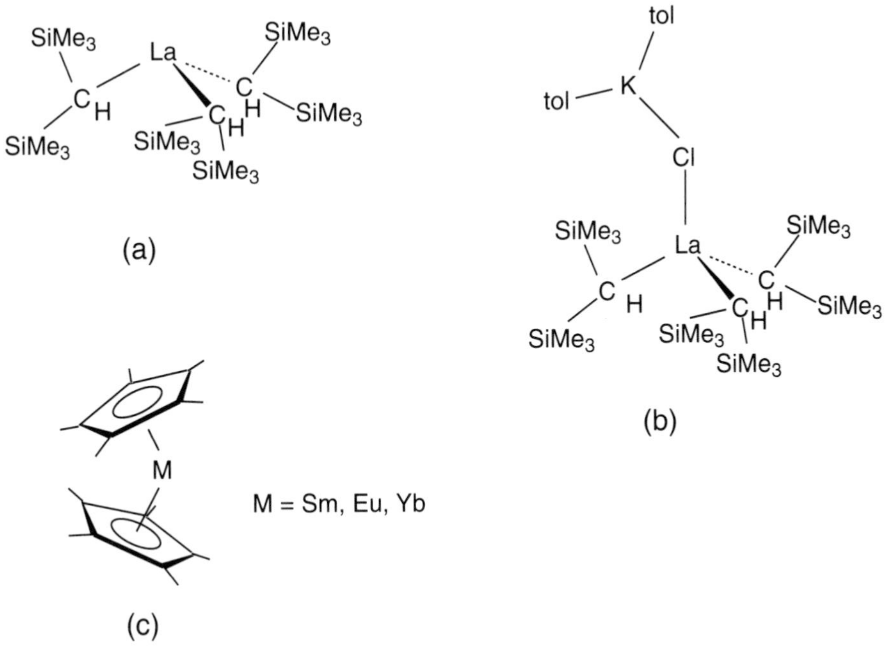

Scheme 5.6

Cl)K, and resolvation by toluene gives Lu{$CH(SiMe_3)_2$}$_3$(μ-Cl)K(toluene) and Lu{$CH(SiMe_3)_2$}$_3$(μ-Cl)K(toluene)$_2$ (Scheme 5.6 (b)).

The most intriguing coordinatively unsaturated complexes are the base-free divalent metallocene complexes, $M(C_5Me_5)_2$ (M = Sm, Eu, Yb) (Scheme 5.6 (c)) [37]. These complexes exhibit bent metallocene structures with the Cp*—M—Cp* angles of 140.1° (Sm), 140.3° (Eu), and 158° (Yb), even though there are no ligands at the equatorial girdle. The THF solvates of the samarium, europeum, and ytterbium complexes, $M(C_5Me_5)_2(THF)_2$, are prepared straightforwardly from the reaction of divalent iodide complexs according to eq (21). Unsolvated decamethylmetallocenes can then be obtained by sublimation of the THF adducts. Many interesting reactions of $Sm(C_5Me_5)_2$ and $Sm(C_5Me_5)_2(THF)_2$ have been reported, some of which are shown in Scheme 5.7 [38]. The divalent samarocene often reduces substrates, and thereby induces their reductive coupling. Very rare π-adducts of lanthanides have also been achieved using unsolvated $M(C_5Me_5)_2$. Dinitrogen molecule is reversively coordinated at Sm in an unusual $\mu\text{-}\eta^2{:}\eta^2$ manner. In the case of ytterbium, the π-bonded alkyne complex $Yb(C_5Me_5)_2(MeC{\equiv}CMe)$ was isolated (eq (22a)) [39]. The ytterbium-to-

$$MI_2(THF)_n + 2\,K(C_5Me_5) \xrightarrow{THF} M(C_5Me_5)_2(THF)_2 + 2\,KI$$
$$M(C_5Me_5)_2(THF)_2 \xrightarrow{\text{sublimation}} M(C_5Me_5)_2 \qquad (21)$$
$$M = Sm, Eu, Yb$$

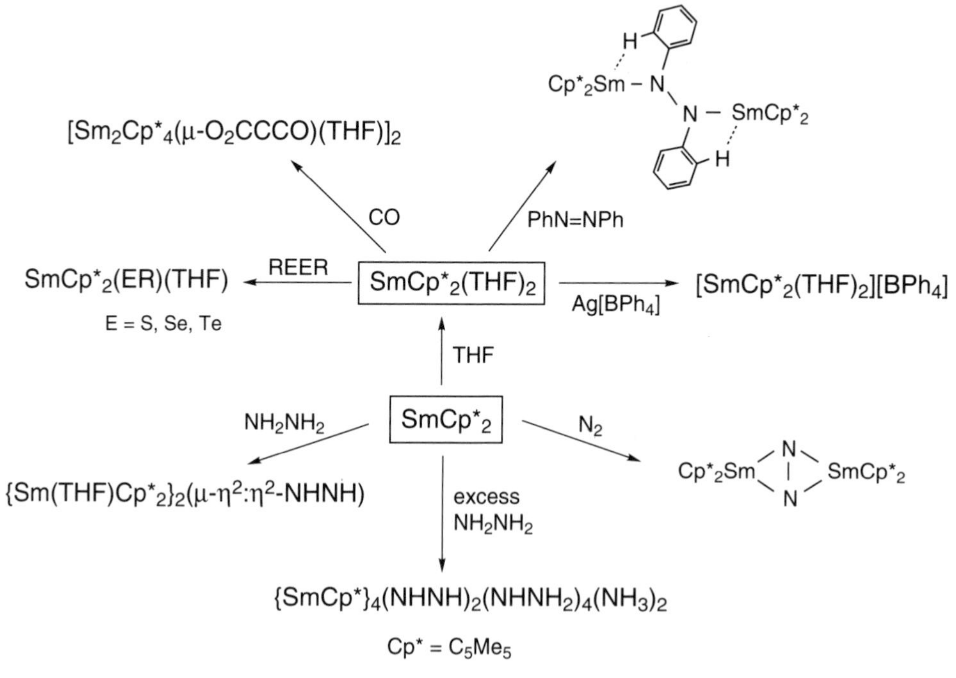

Scheme 5.7

$$Yb(C_5Me_5)_2 + MeC{\equiv}CMe \longrightarrow Yb(C_5Me_5)_2(MeC{\equiv}CMe) \quad (22a)$$

$$Yb(C_5Me_5)_2 + Pt(PPh_3)_2(H_2C{=}CH_2) \longrightarrow Yb(C_5Me_5)_2(\mu\text{-}H_2C{=}CH_2)Pt(PPh_3)_2 \quad (22b)$$

$$Yb(C_5Me_5)_2 + BeMe(C_5Me_5)_2 \longrightarrow Yb(C_5Me_5)_2(\mu\text{-}Me)Be(C_5Me_5) \quad (22c)$$

alkyne bonding is very week, due to lack of back-donation interations. The decamnethylmetallocene was found to form a π-bond even with the platinum-bound ethylene in $Pt(PPh_3)_2(H_2C{=}CH_2)$ when treated with $Yb(C_5Me_5)_2$ (eq (22b)). The reaction of $Yb(C_5Me_5)_2$ with $Be(C_5Me_5)Me$ resulted in $Yb(C_5Me_5)_2(\mu\text{-}Me)Be(C_5Me_5)$ having the nearly linear Yb—Me—Be structure (eq 22c) [40].

5.3 Synthesis of Group 3 and Lanthanide Metal Compounds

All the procedures must be conducted under inert atmosphere using dry solvents.

5.3.1 Chlorobis(η^5-cyclopentadienyl)lanthanides(III), $Ln(\eta^5\text{-}C_5H_5)_2Cl$[5a]

$$LnCl_3 + 2\ Na(C_5H_5) \xrightarrow[THF]{} Ln(C_5H_5)_2Cl + 2\ NaCl \quad (23a)$$

A THF solution (50 mL) of $Na(C_5H_5)$(4.23 g, 48 mmol) is added dropwise to a THF suspension (100 mL) of anhydrous $LuCl_3$ (7.4 g, 26 mmol) at 0°C. After stirring the resulting orange solution for 12 h, the solvents is removed *in vacuo* to give a yellow residue. Sublimation of the residue at 120–240°C under 10–5 Torr produces $Lu(\eta^5\text{-}C_5H_5)_2Cl$ as pale-green crystals. Yield 72%. Air- and moisture-sensitive. The analogous $Ln(\eta^5\text{-}C_5H_5)_2Cl$ complexes of Sm, Gd, Dy, Ho, Er, Yb and Lu can be obtained by a similar procedure in moderate yields.

		$Ln(\eta^5\text{-}C_5H_5)_2Cl$	
Ln	color	Mp(°C)	μ_{eff}
Sm	yellow	200 (dec.)	1.62
Gd	colorless	140(dec.)	8.86
Dy	yellow	345	10.6
Ho	yellow-orange	342	10.3
Er	pink	200 (dec.)	9.79
Yb	orange-red	240(dec.)	4.81
Lu	pale-green	319	0

The above procedure fails to give chlorobis (cyclopentadienyl) complexes for early (and thus large) lanthanides (La–Nd), probably due to the steric reasons. Thus the use of bulkier cyclopentadienyl derivatives would facilitate isolation of such complexes.

An example is the successful synthesis of $Nd(\eta^5\text{-}C_5Me_5)_2Cl_2Li(THF)_2$ as described below (eq (23b)) [21].

$$NdCl_3 + 2\,Li(C_5Me_5) \xrightarrow[\text{pentane/THF}]{} Nd(C_5Me_5)_2Cl_2Li(THF) + LiCl \quad (23b)$$

A white slurry of $Li(C_5Me_5)$ 1.66 g, 11.7 mmol) in pentane is added to a slurry of dehydrated $NdCl_3$ (1.46 g, 5.84 mmol) in THF. Stirring the mixture for 2 days at room temperature results in the formation of a cloudy light-blue solution. Evaporation of the solvent *in vacuo* forms a sticky blue-green residue which is extracted with toluene to remove any unreacted $NdCl_3$ and byproduct LiCl. Removal of toluene *in vacuo* gives a blue powder, and it is dissolved in pentane. Upon cooling the resulting blue solution, blue-violet crystals of $Nd(\eta^5\text{-}C_5M_5)_2Cl_2Li(THF)_2$ are formed. Several crops of crystals can be collected by repeating partial solvent removal and cooling. Yield 10–15%. IR(Nujol mull, KBr, cm^{-1}): several weak bands associated with C_5Me_5, and peaks at 1045, 915 and 895 arising from coordinated THF. μ_{eff}': 3.50 μ_B. ^{1}H-NMR (C_6D_6, δ): 9.18 (C_5Me_5), 1.13(THF), 0.11(THF). Air- and moisture-sensitive.

5.3.2 Dichloro (η^5-cyclopentadienyl) tris(tetrahydrofuran)lanthanides(III), $Ln(\eta^5\text{-}C_5H_5)Cl_2(THF)_3$ (30b,c)

$$LnCl_3 + Na(C_5H_5) \xrightarrow[\text{THF}]{} Ln(C_5H_5)Cl_2(THF)_3 + NaCl \quad (24)$$

The lanthanide(III) dichloride complexes carrying a single cyclopentadienide may in principle be synthesized from the reactions of lanthanide trichlorides with 1 equivalent of $Na(C_5H_5)$.

A THF (10 mL) solution of $Na(C_5H_5)$ (1.7 g, 19 mmol) is added with stirring at room temperature to a suspension of anhydrous $SmCl_3$ (4.9 g, 19 mmol) in THF (125 mL). After stirring the mixture for 12 h–3 days, an extra 100 mL of THF is added and the solution is heated to *ca.* 60 °C until the product is dissolved. The solution is filtered and the filtrate is cooled to –20 °C, producing $Sm(\eta^5\text{-}C_5H_5)Cl_2(THF)_3$ as beige crystalline material. Yield 30–40%. Air- and moisture-sensitive. The analogous $Ln(\eta^5\text{-}C_5Me_5)Cl_2$ complexes of Sm, Eu, Gd, Dy, Ho, Er, Yb, and Lu can be obtained by a similar procedure.

	$Ln(\eta^5\text{-}C_5H_5)_2Cl_2$	
Ln	color	mp(°C)
Sm	beige	50–240 (dec.)
Eu	purple	50–240 (dec.)
Gd	blue-violet	82–86
Dy	colorless	85–90
Ho	yellow	84–92
Er	pink	91–94
Yb	orange	78–81
Lu	colorless	76–78

5.3.3 Bis(η^5-pentamethylcyclopentadienyl) bis(tetrahydrofuran) samarium(II), $Sm(\eta^5\text{-}C_5Me_5)_2(THF)_2$ [37d]

$$SmI_2 + 2\,K(C_5Me_5) \xrightarrow[THF]{} Sm(C_5Me_5)_2(THF)_2 + 2\,KI \qquad (25a)$$

The title complex was first prepared from the reaction between zerovalent samarium vapor and pentamethylcyclopentadiene at –120 °C [37a]. Later it was found that $SmI_2(THF)_2$ was a more convenient entry to the samarocene(II) complex as treated with $K(C_5Me_5)$.

Solutions of $SmI_2(THF)_2$ is prepared from excess Sm metal and predried 1,2-$C_2H_4I_2$ in THF. The reaction mixture is stirred until it becomes a homogeneous dark blue color without a trace of the insoluble yellow SmI_3. The solution is then filtered and removal of the solvent yields $SmI_2(THF)_2$ as a dark blue powder. The reaction proceeds quantitatively in 1,2-$C_2H_4I_2$. The complex KC_5Me_5 is prepared by slow addition of a slight excess (by 3%) of water-free C_5Me_5H to a vigorously stirring suspension of KH in THF. The reaction mixture is stirred until H_2 evolution ceases. The insoluble white $K(C_5Me_5)$ obtained therefrom is washed by hexane and dried.

The potassium salt of pentamethylcyclopentadienide, $K(C_5Me_5)$ (5.43 g, 31.2 mmol) is added to a stirring solution of $SmI_2(THF)_2$ (7.78 g, 14.2 mmol) in THF (75 mL). The color of the solution turns immediately from blue-green to purple, and white solids of KI are formed. After 4 h at room temperature, the THF is removed and 100 mL of toluene is added. The resulting solution of $Sm(\eta^5\text{-}C_5Me_5)_2$ $(THF)_2$ with suspended KI is stirred vigorously for 10 h and filtered. The filtrate is dried to give a greenish-purple solid. The toluene extraction procedure is repeated to insure that all iodide containing species are excluded. Then this solid is redissolved in THF and then the solvent is removed to yield the purple disolvate, $Sm(\eta^5\text{-}C_5Me_5)_2(THF)_2$ (5.95 g). Yield 74%. Recrystallization from THF at –25 °C overnight gives the product as purple crystals in a 69% yield. Air- and moisture-sensitive.

Properties: ^{1}H-NMR (THF-d_8, 25 °C, δ): 1.58 (s, 30 C_5Me_5), 3.95 (m, 8, THF), 1.72 (m, 8, THF). The NMR spectra in THF-d_8 are concentration independent, while those in benzene-d_8 vary depending on concentration and the THF content in the complex. As the THF content is lowered the C_5Me_5 resonance moves downfield, and in some samples, the C_5Me_5 signal is shifted as far downfield as 4.0 ppm and the THF signals as far upfield as –3.8 ppm. μ_{eff}: 3.6 μ_B. IR(KBr, cm^{-1}): 3100-2725(s), 2705(w), 1440(s), 1370(w), 1240(m), 1210(w), 1080(s), 1040(s), 850(w), 895(s), 795(m).

The unusual THF-free samarocene $Sm(\eta^5\text{-}C_5Me_5)_2$ can be obtained by sublimation of $Sm(\eta^5\text{-}C_5Me_5)_2(THF)_2$ according to eq. (25b) [37a]

$$Sm(C_5Me_5)_2(THF)_2 \xrightarrow[85\text{–}125°C]{\text{sublimation}} Sm(C_5Me_5)_2 \qquad (25b)$$

As $Sm(\eta^5\text{-}C_5Me_5)_2(THF)_2$ is heated up to 75 °C, a pressure surge starts to occur. At 85 °C THF molecules are gradually removed, and the THF-free complex $Sm(\eta^5\text{-}C_5Me_5)_2$ sublimes as a dark green solid. The highest yield of $Sm(\eta^5\text{-}C_5Me_5)_2$ is 74%, which is achieved upon sublimation at 125 °C. Air- and moisture-sensitive.

Properties : ^{1}H-NMR (C_6D_6, $\delta(\Delta\nu_{1/2}$ in HZ)): 1.53 (20) (s, C_5Me_5); in toluene-d_8, $\Delta\nu_{1/2}$ = 9 Hz. ^{13}C-NMR (toluene-d_8, δ): –98.18 (s, C_5Me_5), 99.04 (q, J = 130 Hz, $C_5(CH_3)_5$). IR (KBr, cm^{-1}): 2970(s), 2923(s), 2869(s), 1440(s), 1260(m), 1240(m), 1130(s), 1060(s), 1020(w), 805(s), 790(s).

5.3.4 Tris(trifluoromethanesulfonato)lanthanides(III), Ln $(O_3SCF_3)_3$ [22a, b]

All syntheses are carried out in an aqueous solution. Triflic acid, HSO_3CF_3 is distilled by the trap-to-trap method before use, and it is diluted with water (10^{-3}–10^{-2} M). Stoichiometric quantities of this diluted triflic acid is added to the lanthanide chloride or lanthanide oxide and heating at boiling for 30 min to 1 h. The unreacted lanthanide chloride or lanthanide oxide, if any, is removed by filtration, and the water is removed from the filtrate under reduced pressure. The resulting hydrate is then dried by heating under vacuum (1.33 Pa) at 180 °C to 200 °C for several days.

$$\begin{array}{c} LnCl_3 \\ \text{or} \\ 1/2\,Ln_2O_3 \end{array} + (3+n)\,HSO_3CF_3 \xrightarrow[100^\circ C]{H_2O} Ln(O_3SCF_3)\cdot n\,H_3OSO_3CF_3 + \begin{array}{c} 3\,HCl \\ \text{or} \\ 3/2\,H_2O \end{array} \tag{26}$$

$$Ln(O_3SCF_3)\cdot n\,H_3OSO_3CF_3 \xrightarrow[190-200^\circ C,\ -n\,H_3OSO_3CF_3]{\text{in vacuo}} Ln(O_3SCF_3)_3$$

Direct reactions of anhydrous triflic acid with lanthanide chlorides or oxides often resulted in complexes solvated by the acid, which are viscous liquids with a very low vapor pressure. Thus it is advisable to perform reactions with diluted trific acid in an aqueous solution. If anhydrous lanthanide chlorides are used, they are to be hydrated with care in water before use in order to avoid contamination by side products. For instance, anhydrous $ScCl_3$ reacts violently with water, the sample is first cooled at –180 °C and water is added slowly. The mixture is then warmed in 12 h to room temperature. The anhydrous compound, $ScCl_3 \cdot nH_2O$ (n ≌ 3.7) is obtained as a white solid after boiling the solution to dryness.

Reactant		$Ln(O_3SCF_3)_3$	
	color	IR	Raman
$ScCl_3 \cdot nH_2O$	grey	583w, 601w, 634s, 736sh, 1030s, 1140m, 1209s, 1322s, 1350sh	
La_2O_3	white	566w, 643s, 771vw, 1065m, 1167sh, 1195s, 1224s, 1290sh, 1332sh	334s, 335m, 520w, 587m, 653w, 778s, 1070s, 1179m, 1188m, 1239m, 1313 w
$NdCl_3 \cdot nH_2O$	white-pink	541w, 607w, 664m, 772vw, 1052s, 1159sh, 1196s, 1240s, 1280sh, 1337sh	332s, 358s, 582w, 735sbr, 780s, 930sbr, 1085s, 1157w, 1188w, 1210w

Sm_2O_3	white	521w, 590w, 640s, 777vw, 1057m, 1168sh, 1214s, 1290sh, 1339sh	275m, 337s, 353s, 518m, 580m, 658w, 777s, 1077s, 1232sbr, 1413sbr
$GdCl_3{\cdot}6H_2O$	white	545w, 580w, 640m, 775vw, 1054m, 1164sh, 1215s, 1235s, 1301s, 1338sh	327s, 349s, 512m, 580m, 656w, 772s, 1081s, 1158m, 1203m, 1226m, 1274m, 1342w
Er_2O_3 or $ErCl_3{\cdot}6H_2O$	white	583m, 642m, 775vw, 1060m, 1166sh, 1217s, 1304m, 1351sh	340m, 357s, 430w, 522m, 582m, 661w, 779s, 1092s, 1164m, 1205m, 1225m, 1240s, 1288w

5.3.5 η⁵-Cyclopentadienyltris(tetrahydrofuran)bis(trifluoromethanesulfonato)lutetium(III) and Bis(η⁵-Cyclopentadienyl) tetrahydrofuran-trifluoromethanesulfonato-lutetium(III), $Lu(\eta^5\text{-}C_5H_5)(O_3SCF_3)_2(THF)_3$ and $Lu(\eta^5\text{-}C_5H_5)_2(O_3SCF_3)(THF)$ [31]

$$Lu(O_3SCF_3)_3 + Na(C_5H_5) \xrightarrow[-\,NaO_3SCF_3]{THF} Lu(C_5H_5)(O_3SCF_3)_2(THF)_3 \quad (27a)$$

$$Lu(O_3SCF_3)_3 + 2\,Na(C_5H_5) \xrightarrow[-\,2\,NaO_3SCF_3]{THF} Lu(C_5H_5)_2(O_3SCF_3)(THF) \quad (27b)$$

Reactions of $Lu(O_3SCF_3)_3$ with $Na(C_5H_5)$ in THF generate readily the title monocyclopentadienyl- and biscyclopentadienyl–lutetium complexes when the appropriate stoichiometric amount of $Na(C_5H_5)$ is added. The highly reactive monocyclopentadienyl–lutetium unit seems to be stabilized by the bulky triflate anions.

A THF solution of $Na(C_5H_5)$ (19.4 mL, 0.727 M, 14.1 mmol) is added dropwise to a THF solution (80 mL) of $Lu(O_3SCF_3)3$ (8.80 g, 14.1 mmol) in 3 h at room tempperature. The resulting light-yellow solution is dried *in vacuo*, and the sticky yellowish-brown residue is extracted from ether (70 mL). After filtrating off the insoluble NaO_3SCF_3, the complex $Lu(\eta^5\text{-}C_5H_5)(O_3SCF_3)_2(THF)_3$ is isolated from the ether solution as colorless crystals. X-ray quality crystals can be obtained by recrystallization from concentrated THF solution. Two THF molecules are removed from the complex under reduced pressure to give $Lu(\eta^5\text{-}C_5H_5)$ $(O_3SCF_3)_2(THF)$. Yield 68%. Air- and moisture-sensitive.

Properties : ^{1}H-NMR (THF-d_8, 25 °C, δ): 6.26 (s, C_5H_5), 3.60 (m, THF), 1.73 (m, THF). ^{13}C-NMR (THF-d_8, 25 °C, δ): 6.26 (s, C_5H_5), 3.60 (m, THF), 1.73 (m, THF). ^{13}C-NMR (THF-d_8, 25 °C, δ): 120.05 (quart, ^1J(C-F) 318.4 Hz, CF_3), 111.98 (s, C_5H_5), 68.16 (s, THF), 26.01 (s, THF).

A THF solution of $Na(C_5H_5)$ (8.7 mL, 1.488 M, 13.0 mmol) is added to a 80 mL THF solution of $Lu(O_3SCF_3)_3$ (4.04g, 6.5 mmol) at room temperature. The resulting yellowish-orange solution is stirred for 5 h at room temperature, and the solvent is removed under reduced pressure. Then ether (100 mL) is added to the residue. From the ether extract, after filtration, $Lu(\eta^5\text{-}C_5H_5)_2(O_3SCF_3)(THF)$ is obtained as colorless needles. The THF molecule comes off from the complex *in vacuo* to give $Lu(\eta^5\text{-}C_5H_5)_2(O_3SCF_3)$. Yield 61%. Air- and moisture-sensitive.

Properties: ^{1}H-NMR (THF-d_8, 25 °C, δ): 6.15 (s, C_5H_5), 3.58 (m, THF), 1.73 (m, THF). ^{13}C-NMR (THF-d_8, 25 °C, δ): 120.08 (quart, ^1J(C-F) 319.2 Hz, CF_3), 111.68 (s, C_5H_5), 68.20 (s, THF), 26.20(s, THF).

References

[1] R. G. Hayes and J. L. Thomas, *Organomet. Chem. Rev. A.*, *7*, 1 (1971).
[2] P. N. Hazin, J. W. Bruno, H. G. Brittain, *Organometallics*, **6**, 913 (1987).
[3] F. T. Edelmann, *Angew. Chem., Int. Ed. Engl.*, **34**, 2466 (1995).
[4] (a) G. Wilkinson, J. M. Birmingham, *J. Am. Chem. Soc.*, **76**, 6210 (1954).
(b) J. M. Birmingham, G. Wilkinson, *J. Am. Chem. Soc.*, **78**, 42 (1956).
[5] (a) R. E. Maginn, S. Manastyrskyj, M. Dubeck, *J. Am. Chem. Soc.*, **85**, 672 (1963).
(b) S. Manastyrskyj, R. E. Maginn, M. Dubeck, *Inorg. Chem.*, **2**, 904 (1963).
[6] D. R. Rogers, J. L. Atwood, D. F. Foust, M. D. Rausch, *J. Organomet. Chem.*, **265**, 241 (1984).
[7] X. Zhou, Z. Wu, H. Ma, Z. Xu, X. You, *Polyhedron*, **13**, 375 (1984).
[8] (a) E. O. Fischer, H. Fischer, *J. Organomet. Chem.*, **3**, 181 (1965).
(b) F. Calderazzo, R. Pappalardo, S. Losi, *J. Inorg. Nucl. Chem.*, **28**, 987 (1966).
[9] (a) S. Manastyrskyj, M. Dubeck, *Inorg. Chem.*, **3**, 1647 (1964).
(b) M. Tsutsui, T. Takino, D. Lorenz, *Z. Naturforsch*, **B21**, 1 (1966).
[10] (a) G. B. Deacon, A. J. Koplick, T. D. Tuong, *Aust. J. Chem.*, **37**, 517 (1984).
(b) G. B. Deacon, C. M. Forsyth, R. H. Newnham, T. D. Tuong, *Aust. J. Chem.*, **40**, 895 (1987).
[11] (a) C. H. Wong, T. Lee, Y. Lee, *Acta Crystallogr.*, **B25**, 2580 (1969).
(b) J. L. Atwood, K. D. Smith, *J. Am. Chem. Soc.*, **95**, 1488 (1973).
(c) S. H. Eggers, J. Kopf, R. D. Fischer, *Organometallics*, **5**, 383 (1986).
(d) S. H. Eggers, H. Schultze, J. Kopf, R. D. Fischer, *Angew. Chem., Int. Ed. Engl.*, **25**, 656 (1986).
[12] (a) J. M. Birmingham and G. Wilkinson, *J. Am. Chem. Soc.*, **78**, 42 (1956).
(b) E. O. Fischer, H. Fischer, *J. Organomet. Chem.*, **6**, 141 (1966).
(c) J. H. Burns, W. H. Baldwin, *J. Organomet. Chem.*, **120**, 361 (1976).
(d) W. Q. Chen, G. Y. Lin, J. S. Xia, G. C. Wei, Y. Zhang, Z. S. Jin, *J. Organomet. Chem.*, **467**, 75 (1994).
[13] H. Schumann, F. H. Görlitz, F. E. Hahn, J. Pickardt, C. Qian, Z. Xie, *Z. Anorg. Allg. Chem.*, **609**, 131 (1992).
[14] (a) C. J. Schaverien, *Adv. Organomet. Chem.*, **24**, 131 (1985).
(b) S. H. Eggers, R. D. Fischer, *J. Organomet. Chem.*, **315**, C61 (1986).
[15] E. C. Baker, K. N. Raymond, *Inorg. Chem.*, **16**, 2710 (1977).
[16] H. Schumann, C. Janiak, J. Pickardt, *J. Organomet. Chem.*, **349**, 117 (1988).
[17] W. J. Evans, J. H. Meadows, A. L. Wayda, W. E. Hunter, J. L. Atwood, *J. Am. Chem. Soc.*, **104**, 2015 (1982).
[18] (a) S. D. Stults, R. A. Andersen, A. Zalkin, *Organometallics*, **9**, 115 (1990).
(b) M. Booij, N. H. Kiers, A. Meetsma, J. H. Teuben, W. J. J. Smeets, A. L. Spec, *Organometallics*, **8**, 2545 (1989).
[19] (a) S. D. Stults, R. A. Andersen, A. Zalkin, *Organometallics*, **9**, 1623 (1990).
(b) W. J. Evans, R. A. Keyer, J. W. Ziller, *J. Organomet. Chem.*, **394**, 87 (1990).
[20] H. Schumann, *Angew. Chem., Int. Ed. Engl.*, **23**, 474 (1984).
[21] A. L. Wayda, W. J. Evans, *Inorg. Chem.*, **19**, 2190 (1980).
[22] (a) M. E. M. Hamidi, J.-L. Pascal, *Polyhedron*, **13**, 1787 (1994).

(b) J. H. Forsberg, V. T. Spaziano, T. M. Balasubramanian, G. K. Liu, S. A. Kinsley, C. A. Duckworth, J. J. Poteruca, P. S. Brown, J. L. Miller, *J. Org. Chem.*, **52**, 1017 (1987).
(c) J. Stehr, R. D. Fischer, *J. Organomet. Chem.*, **430**, C1 (1992).
(d) H. Schumann, J. A. Meese-Marktscheffel, A. Dietrich, F. H. Görlitz, *J. Organomet. Chem.*, **430**, 299 (1992).
[23] (a) J. Holton, M. F. Lappert, G. H. Ballard, R. Pearce, J. L. Atwood, W. E. Hunter, *J. Chem. Soc., Dalton Trans.*, **54** (1979).
(b) K. H. den Haan, J. L. de Boer, J. H. Teuben, A. L. Spek, B. Kojić-Prodić, G. R. Haya, R. Huis, *Organometallics*, **5**, 1726 (1986).
(c) M. E. Thompson, S. M. Baxter, A. R. Bulls, B. J. Burger, M. C. Nolan, B. D. Santasiero, W. P. Schaefer, J. E. Bercaw, *J. Am. Chem. Soc.*, **109**, 203 (1987).
(d) P. L. Watson, G. W. Parshall, *Acc. Chem. Res.*, **18**, 51 (1985).
[24] (a) W. J. Evans, A. L. Wayda, W. E. Hunter, J. L. Atwood, *J. Chem. Soc., Chem. Commun.*, 292 (1981).
(b) R. D. Fischer, G. Bielang, *J. Organomet. Chem.*, **191**, 61 (1980).
[25] (a) P. L. Watson, *J. Am. Chem. Soc.*, **105**, 6491 (1983).
(b) G. Perkin, E. Bunel, B. J. Burger, M. S. Trimmer, A. van Asselt, J. E. Bercaw, *J. Mol. Cat.*, **41**, 21 (1987).
[26] (a) W. J. Evans, J. H. Meadows, A. L. Wayda, W. E. Hunter, J. L. Atwood, *J. Am. Chem. Soc.*, **104**, 2008 (1982).
(b) W. J. Evans, J. H. Meadows, W. E. Hunter, J. L. Atwood, *J. Am.Chem. Soc.*, **106**, 1291 (1984).
[27] (a) H. Schumann, F. W. Reier, *J. Organomet. Chem.*, **235**, 287 (1982).
(b) W. J. Evans, D. K. Drummond, L. A. Hughes, H. Zhang, J. L. Atwood, *Polyhedron*, **7**, 1693 (1988).
[28] H. Schumann, W. Genthe, E. Hahn, M. B. Hossain, D. van der Helm, *J. Organomet. Chem.*, **299**, 67 (1986).
[29] W. J. Evans, J. H. Meadows, T. P. Hanusa, *J. Am. Chem. Soc.*, **106**, 4454 (1984).
[30] (a) R. D. Rogers, L. M. Rogers, *J. Organomet. Chem.*, **442**, 225 (1992).
(b) C. S. Day, V. W. Day, R. D. Ernst, S. H. Vollmer, *Organometallics*, **1**, 998 (1982).
(c) S. Manastyrskyj, R. E. Maginn, M. Dubeck, *Inorg. Chem.*, **2**, 904 (1963).
[31] H. Schumann, J. A. Meese-Marktscheffel, A. Dietrich, *J. Organomet. Chem.*, **377**, C5 (1989).
[32] A. V. Protchenko, L. N. Zakharov, M. N. Bochkarev, Y. T. Struchkov, *J. Organomet. Chem.*, **447**, 209 (1993).
[33] J. Arnold, C. G. Hoffmann, D. Y. Dawson, F. J. Hollander, *Organometallics*, **12**, 3645 (1993).
[34] P. N. Hazin, J. C. Huffman, J. W. Bruno, *Organometallics*, **6**, 23 (1987).
[35] H. van der Heijden, C. J. Schaverien, A. G. Orpen, *Organometallics*, **8**, 255 (1989).
[36] (a) P. B. Hitchcock, M. F. Lappert, R. G. Smith, R. A. Bartlett, P. P. Power, *J. Chem. Soc., Chem. Commun.*, 1007 (1988).
(b) C. J. Schaverien, J. B. van Mechelen, *Organometallics*, **10**, 1704 (1991).
[37] (a) W. J. Evans, L. A. Hughes, T. P. Hanusa, *J. Am. Chem. Soc.*, **106**, 4270 (1984).
(b) W. J. Evans, L. A. Hughes, T. P. Hanusa, *Organometallics*, **5**, 1285 (1986).
(c) R. A. Andersen, J. M. Boncella, C. J. Burns, R. Blom, A. Haaland, H. V. Volden, *J. Organomet. Chem.*, **312**, C49 (1986).
(d) W. J. Evans, J. W. Grate, H. W. Choi, I. Bloom, W. E. Hunter, J. L. Atwood, *J. Am. Chem. Soc.*, **107**, 941 (1985).
[38] (a) W. J. Evans, J. W. Grate, L. A. Hughes, H. Zhang, J. L. Atwood, *J. Am. Chem. Soc.*, **107**, 3728 (1985).
(b) W. J. Evans, D. K. Drummond, S. G. Bott, J. L. Atwood, *Organometallics*, **5**, 5389 (1986).
(c) W. J. Evans, T. A. Ulibarri, J. W. Ziller, *J. Am. Chem. Soc.*, **110**, 6877 (1988).
(d) W. J. Evans, T. A. Ulibarri, L. R. Chamberlain, J. W. Ziller, D. Alvarez, *Organometallics*, **9**, 2124 (1990).

(e) W. J. Evans, G. Kociok-Köhn, J. W. Ziller, *Angew. Chem., Int. Ed. Engl.*, **31**, 1081 (1992).
(f) A. Recknagel, M. Noltemeyer, D. Stalke, U. Pieper, H.-G. Schmidt, F. T. Edelmann, *J. Organomet. Chem.*, **411**, 347 (1991).
(g) K.-G. Wang, E. D. Stevens, S. P. Nolan, *Organometallics*, **11**, 1011(1992).
[39] C. J. Burns, R. A. Andersen, *J. Am. Chem. Soc.*, **109**, 941 (1987).
[40] C. J. Burns, R. A. Andersen, *J. Am. Chem. Soc.*, **109**, 5853 (1987).

6 Group 4 (Ti, Zr, Hf) Metal Compounds

K. Mashima, *Osaka University*

6.1 Introduction

The nature of the metal–carbon bond of Group 4 metals is partially ionic, while that of late transition metals is generally covalent. The extent of ionicity is strongly dependent on the formal oxidation states and auxiliary ligands of each metal. Because the highest oxidation state, M(IV), is d^0, π-back donation from a metal center to a ligand is not possible. Metal centers in lower oxidation states form π-complexes that are less stable than those of late transition metal complexes. The reactivity of organometallic compounds of Group 4 metals is expected to lie between that of corresponding compounds of main group metals. The discovery of homogeneous metallocene catalysts for α-olefin polymerization has recently promoted the rapid development of organometallic chemistry of Group 4 metals. In this short text, some aspects of organometallic chemistry are only briefly reviewed, and books and review articles are cited for readers interested in more details [1–7].

6.2 Synthesis of Organometallic Complexes of Group 4 Metals

6.2.1 Hydride Complexes

The monohydride complex of zirconocene Cp_2ZrHCl (**2**), the so-called Schwartz reagent, has been prepared by treatment of Cp_2ZrCl_2 (**1**) with one equivalent of $LiAlH(O^tBu)_3$ or Vitride [8], but the product is sometimes contaminated by Cp_2ZrH_2 (**3**). The most convenient and practical method reported so far is a two-step reaction: (i) reduction of **1** with $LiAlH_4$ to give a mixture of **2** and Cp_2ZrH_2 (**3**); (ii) treatment of the reaction mixture with dichloromethane giving **2** in modest yield [9]. The pure dihydride complex **3** is prepared by reaction of $[Cp_2ZrCl]_2O$ with $LiAlH_4$ [10].

$$\underset{1}{Cp_2ZrCl_2} \xrightarrow{LiAlH_4} \underset{2}{Cp_2Zr(H)Cl} + \underset{3}{Cp_2ZrH_2} \quad (\text{3} \xrightarrow{CH_2Cl_2} \text{2}) \tag{1}$$

Scheme 6.1 shows typical reactions of **2** [11]. Hydrozirconation of 1-octene produces predominantly a primary alkyl complex **4**. When 4-octene is used, isomerization

Synthesis of Organometallic Compounds: A Practical Guide. Edited by S. Komiya

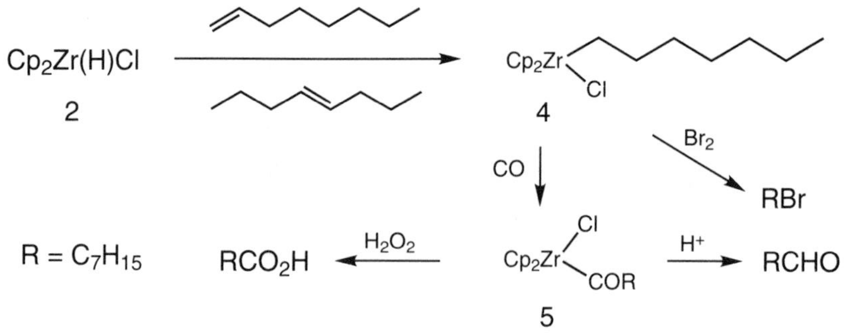

$Cp_2Zr(H)Cl$ (2) → $Cp_2Zr(C_8H_{17})Cl$ (4); 4 + Br_2 → RBr; 4 + CO → $Cp_2Zr(Cl)(COR)$ (5); 5 + H^+ → RCHO; 5 + H_2O_2 → RCO_2H; R = C_7H_{15}

Scheme 6.1

to 1-octene and following hydrozirconation results in exclusive formation of **4**. Carbonylation of **4** gives an acyl complex **5**, which produces an aldehyde with acids and a carboxylic acid with H_2O_2 [11–13]. **5** has a η^2-acyl structure and its dominant reactivity is attributed to a contribution of the oxycarbene-type resonance of acyl moiety [14,15]. Initial coordination of CO to d^0 Cp_2MR_2 occurs along the line perpendicular to the Cp centroid-metal–Cp centroid plane, approaching the LUMO ($1a_1$ orbital) [15,16]. Hydrozirconation of internal alkynes affords *cis*-alkenyl complexes, $Cp_2Zr[(E)\text{-}C(R){=}CRH]Cl$ (**6**),via *cis* addition into the Zr—H bond [17,18].

Complex **2** is polymeric and poorly soluble in common solvents. A soluble version of Schwartz reagent is the dimeric triflate analog, $[Cp_2Zr(OTf)(\mu\text{-}H)]_2$ (**7**), which is soluble and highly reactive towards unsaturated organic molecules [19].

6.2.2 Alkyl Complexes

Alkyl complexes of titanium have been extensively studied both from the point of view of the nature of the metal–carbon bond as well as their application as alkylating reagents in organic reactions [4–6, 20, 21]. The simplest methyl complex, $MeTiCl_3$ (**8**), is readily obtained by the reaction of $TiCl_4$ with half an equivalent of $ZnMe_2$ [22, 23]. An ether adduct of **8** is synthesized by treatment of $TiCl_4$ with an ethereal solution of MeLi [24]. Alkoxy substitution on titanium stabilizes the methyl complexes. Reaction of $TiCl(O^iPr)_3$ (**9**) with MeLi affords $MeTi(O^iPr)_3$ (**10**), which is thermally very stable and can be distilled under reduced pressure (48–53 °C/0.01 mmHg) [25]. The phenyl derivative **11** is prepared by the similar reaction of **9** with PhLi.

$$\underset{9}{Cl{-}Ti(OCHMe_2)_3} + RLi \longrightarrow R{-}Ti(OCHMe_2)_3 \quad (2)$$

10: R = Me
11: R = Ph

These compounds are unique alkylating or arylating reagents and their reactivity has been reviewed [25,26]. Some chemoselective reactions are presented. The addition reaction of aldehydes to **10** proceeds faster than that of ketones [23, 25, 27, 28]. Compound **8** reacts chemoselectively with an acetal, and not with an ester group [29].

The structure of $TiMeCl_3(dmpe)$ (**12**) has been determined both by X-ray and neutron diffraction analysis. The latter confirms that there is a unique agostic interaction of the α-hydrogen atom with the titanium atom [30,31]. This weak interaction stabi-

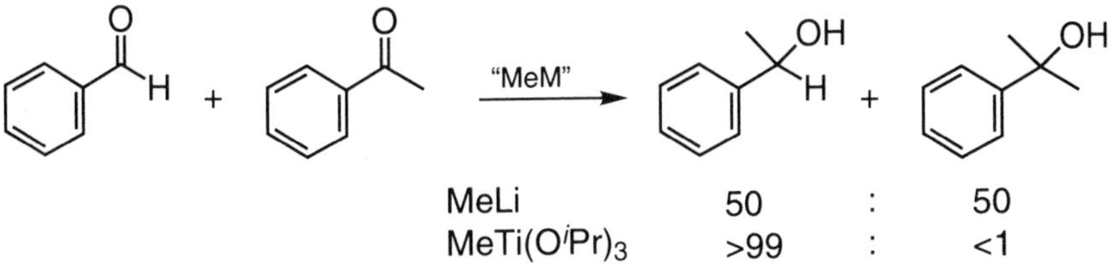

"MeM" (3)

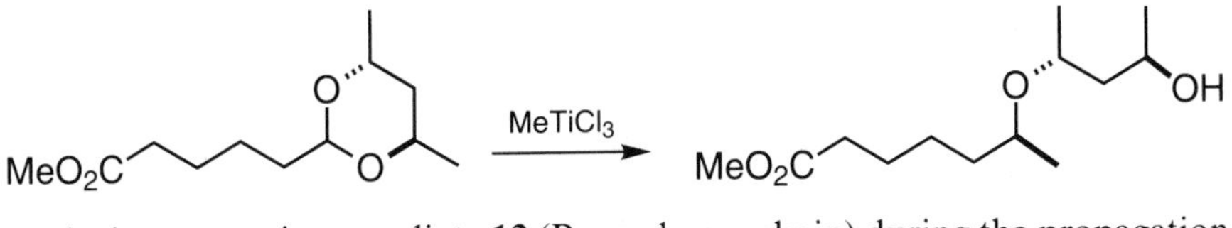

MeLi	50	:	50
$MeTi(O^iPr)_3$	>99	:	<1

MeO_2C … $\xrightarrow{MeTiCl_3}$ MeO_2C … OH (4)

lizes the important intermediate **13** (P = polymer chain) during the propagation process of α-olefin polymerization [32–35]. The existence of this agostic interaction has also been supported by theoretical studies [36–40]. Ethyl and higher alkyl complexes of group 4 metals are thermally unstable and undergo β-hydrogen elimination, producing low valent metal species, whereas a chelating diphosphine ligand stabilizes the ethyl titanium complex, $TiEtCl_3(dmpe)$ (**14**). It is noteworthy that **14** contains a novel β-agostic interaction, which is the transition state of β-hydrogen elimination [31, 41].

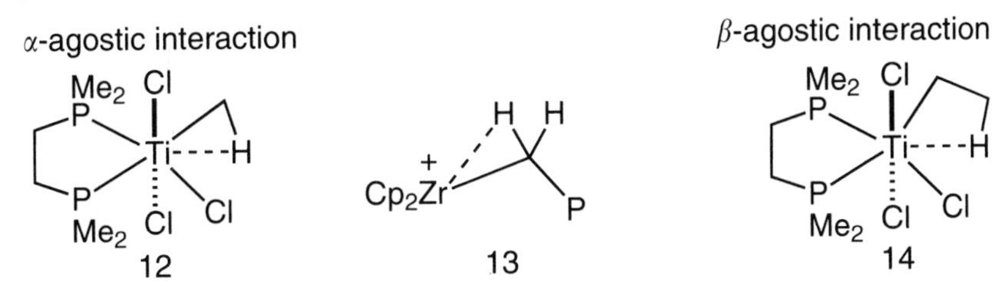

Except for the tetrabenzyls $M(CH_2Ph)_4$ (M = Ti, Zr, Hf), peralkyl complexes, in contrast to mono- and dialkyls, are generally thermally unstable. Introduction of Cp and its derivatives as ligands stabilizes trimethyl complexes of zirconium and hafnium [42]. However, $CpTiMe_3$ (**15**), which is prepared from $CpTiCl_3$ and MeLi, is thermally fragile, decomposing above 0°C [43]. Dimethyltitanocene, Cp_2TiMe_2 (**16**), is also thermally comparatively unstable. The decomposition process of **15** and **16** involves carbene species, and thereby these complexes are used as methylation reagents as well as catalysts for ring opening metathesis polymerization (see next section) [44–46].

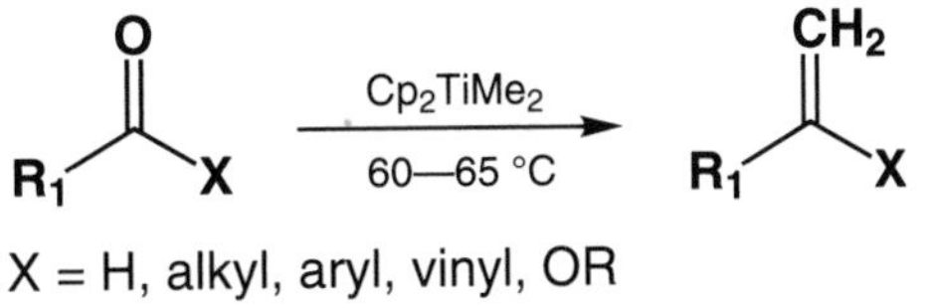

$$R_1C(=O)X \xrightarrow[60\text{—}65\ °C]{Cp_2TiMe_2} R_1C(=CH_2)X \quad (5)$$

X = H, alkyl, aryl, vinyl, OR

Cationic metallocene–alkyl complexes are attracting much interest because of their catalytic activity for α-olefin polymerization [47, 48]. The most convenient method for generating the ion pairs is the reaction of metallocene dialkyl complexes of zirconium and hafnium with $B(C_6F_5)_3$, forming cationic complexes $[Cp_2MR]^+[RB(C_6F_5)_3]^-$ (M = Zr, Hf) [49,50]. *ansa*-Type cationic compounds have been applied in the stereospecific polymerization of α-olefins [51].

6.2.3 Alkylidene Complexes and 4-Membered Metallacycles

Although the simple methylene complex of titanocene is not isolable, this species can be stabilized by Lewis acids such as $AlClMe_2$. In 1978, Tebbe *et al.* In contrast, isolated $Cp_2Ti(\mu\text{-}CH_2)(\mu\text{-}Cl)AlMe_2$ (**17**) by the reaction of Cp_2TiCl_2 with two equivalents of $AlMe_3$ [52]. The reaction of zirconocene dichloride **1** with $AlMe_3$ does not produce a similar methylene complex. This combination acts as a selective reagent for the carboalumination of alkynes, providing (*E*)-2-methyl-1-alkenylalane s in high yield and high stereoselectivity [53–55].

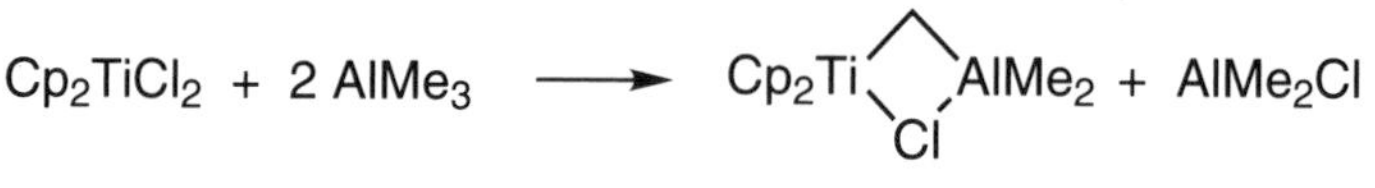

$$Cp_2TiCl_2 + 2\ AlMe_3 \longrightarrow \underset{\mathbf{17}}{Cp_2Ti(\mu\text{-}CH_2)(\mu\text{-}Cl)AlMe_2} + AlMe_2Cl \tag{6}$$

Complex **17** shows versatile reactivity in both catalytic and stoichiometric reactions. It is a catalyst for olefin metathesis, in which a metallacyclobutane is proposed to be a key intermediate [52]. Grubbs *et al.* successfully isolated titanacyclobutanes **18** by treatment of **17** with alkenes in the presence of a Lewis base such as dimethylaminopyridine [56,57]. When cyclic olefins such as norbornene are treated with a catalytic amount of **18**, ring opening metathesis polymerization occurs [58].

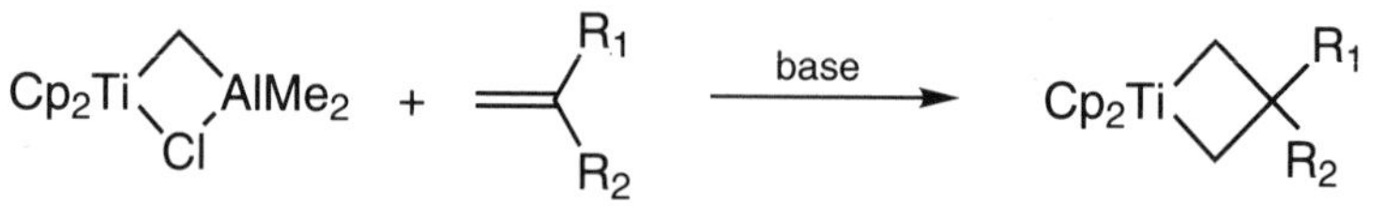

$$\underset{\mathbf{17}}{Cp_2Ti(\mu\text{-}CH_2)(\mu\text{-}Cl)AlMe_2} + CH_2{=}CR_1R_2 \xrightarrow{\text{base}} \underset{\mathbf{18}}{Cp_2Ti\text{-}CH_2CR_1R_2CH_2} \tag{7}$$

Complexes **17** and **18** are superb precursors to a methylene species [59,60], whose reactions are schematically shown in Scheme 6.2. The methylene species can be

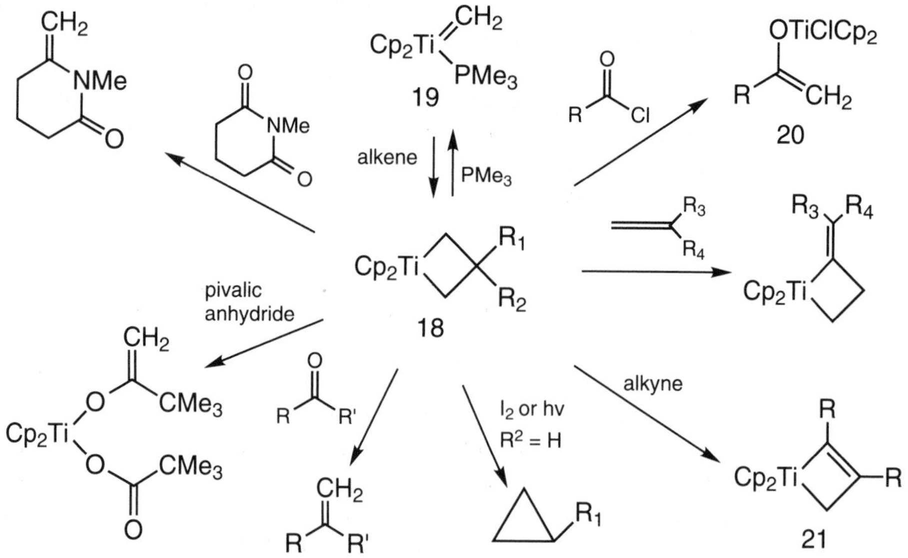

Scheme 6.2

trapped by coordination of trimethylphosphine to give **19** [61]. Remarkably, the methylenation of carbonyl groups in esters, lactones, imides, and acid anhydrides [62–64] proceeds, whereas methylenation of these substrates with Wittig reagents is difficult. Reaction with an acid chloride results in the formation of enolate complex **20** [65,66]. Oxidation of **18** with iodine affords cyclopropanes [67–70].

Treatment of **18** with alkynes results in the formation of metallacyclobutenes **21** [71], which are alternatively synthesized by the reaction of dimethyl titanocene **16** with alkynes [72,73]. The latter reaction involves methane elimination from an initially formed alkenyl–methyl complex **22** giving **21**.

$$Cp_2TiMe_2\ (16) \xrightarrow{\text{alkyne}} Cp_2Ti(C(R)=C(R)Me)(Me)\ (22) \xrightarrow{-CH_4} 21 \quad (8)$$

Recently, Stryker *et al.* have developed a new synthetic method that provides β-substituted metallacyclobutanes. Alkylation of the cationic allyl complex **23** with nucleophiles [74] and thermal reaction of the bis(allyl) complex **25** [75] result in the formation of metallacyclobutanes **24** and **26**, respectively. Radical addition of RX to Ti(III)-allyl complex **23** takes place exclusively at the center carbon atom, providing a metallacyclobutane [76].

$$Cp^*_2Ti(\eta^3\text{-allyl})^+\ (23) \xrightarrow[\text{THF, 9h}]{\text{+ enolate (OK, Ph)}} 24 \quad (9)$$

$$Cp^*_2Zr(allyl)_2\ (25) \xrightarrow{\text{heat}} 26 \quad (10)$$

The metallocene complex **27** containing a M=X double bond undergoes overall [2 + 2] cycloaddition with an internal alkynes to give heterometallacyclobutenes (**28**) [77]. A formal [2 + 2] cycloaddition of Cp_2Zr (=N^tBu)(thf) with imine affords a 2,4-diazametallacyclobutane, whose further reaction with imines results in an imine metathesis reaction [78]; azametallacyclobutene is an intermediate in the $Cp_2Zr(NHR)_2$-assisted hydroamination of alkynes and allene [79].

$$Cp'_2M{=}X\ (27) + {}^1R{-}C{\equiv}C{-}R^2 \xrightarrow{[2+2]} Cp'_2M(C(R^1){=}C(R^2))X\ (28) \quad (11)$$

X = O, S, NR

6.2.4 Five-Membered Metallacycles

Reaction of Cp_2TiCl_2 with 1,4-dilithiobutane at –78 °C leads to the titanacyclopentane (**29**), which is thermally unstable and releases ethylene together with 1-butene [80]. Metallacyclopentane is in equilibrium with bis(olefin) complex and this equilibrium is discussed based on EHMO calculation by Hoffmann *et al.* [16,81].

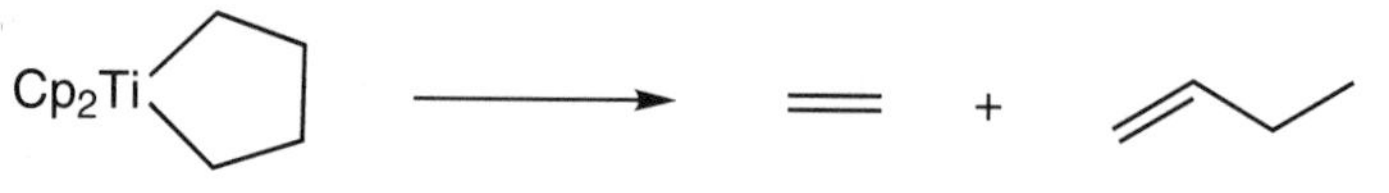

Cp_2Ti (cyclopentane ring) **29** ⟶ ethylene + 1-butene (12)

Metallocene species "Cp_2M" are generated *in situ* from reduction of Cp_2MX_2 [82, 83] or from phosphine or dinitrogen complexes of metallocene. Moreover, the addition of two equivalent of n-BuLi to a solution of **1** affords zirconocene species "Cp_2Zr" (**30**) [84,85], which is used as a precursor for some reactions such as hydrosilylation of alkenes [86,87], carbometalation reaction and so on. Metallocene species of titanium and zirconium have been utilized for selective carbon–carbon bond formation [88–90]. Treatment of **30** with a 1,6- or 1,7-diene results in ring closure with excellent *trans*-fused control to give **31**, whereas the cyclization of 1,6-heptadiene with Cp^*ZrCl_3/Na-Hg gives the dimeric complex **32**, which has *cis* stereochemistry about the carbon–carbon bond (Scheme 6.3) [91].

Alkene complexes of Ti, Zr and Hf have been intensively investigated with regard to the nature of bonding and the close relation to olefin oligomerization and polymerization. Alkene complexes of zirconocene and hafnocene are isolated as the trimethylphosphine adduct, $Cp_2M(\eta\text{-alkene})(PMe_3)$ (**33**) [92–94]. $Cp^*_2Ti(CH_2{=}CH_2)$ (**34**) is a 16 electron ethylene complex with a rich reaction chemistry as summarized in Scheme 6.4 [95–99]. The reaction profile of **34** indicates that the metallacyclopropane canonical form makes an important contribution [100].

Dimeric alkene complexes, in which two formal Zr(III) units are bridged by an alkene, have been prepared. Some ethylene complexes have been characterized by X-ray analysis: $[Cp_2ZrClAlEt_3]_2(\mu\text{-}CH_2CH_2)$ [101], $[Cp_2ZrMe]_2(\mu\text{-}CH_2CH_2)$ [102] and $[MX_3(PEt_3)_2]_2(\mu\text{-}CH_2CH_2)$ (M = Zr, Hf; X = Cl, Br) [103].

The zirconium-catalyzed carbometalation reaction has been developed from the initial observation by Dzhemilev in 1983 that Cp_2ZrCl_2 (**1**) catalyzes the addition of ethylmagnesium halide to unactivated alkenes [104–106]. A plausible mechanism for this carbometalation involves zirconocene and zirconacyclopentane species (Scheme 6.5) [107–109]. Diastereo- and enantioselective carbomagnesation is accomplished by use of chiral *ansa*-zirconocene derivative [110].

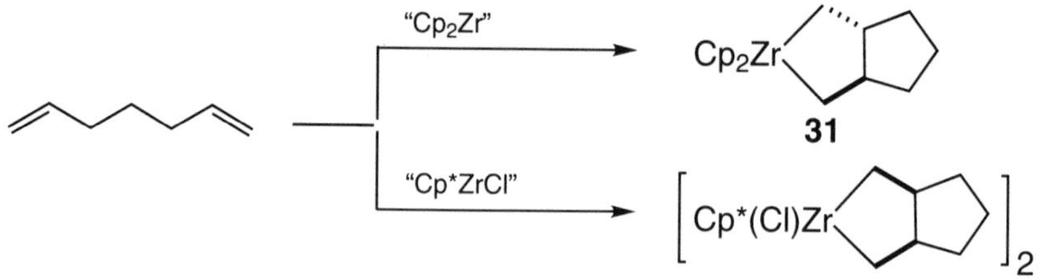

"Cp_2Zr" → Cp_2Zr (bicyclic) **31**

"Cp*ZrCl" → $[Cp^*(Cl)Zr$ (bicyclic)$]_2$ **32**

Scheme 6.3

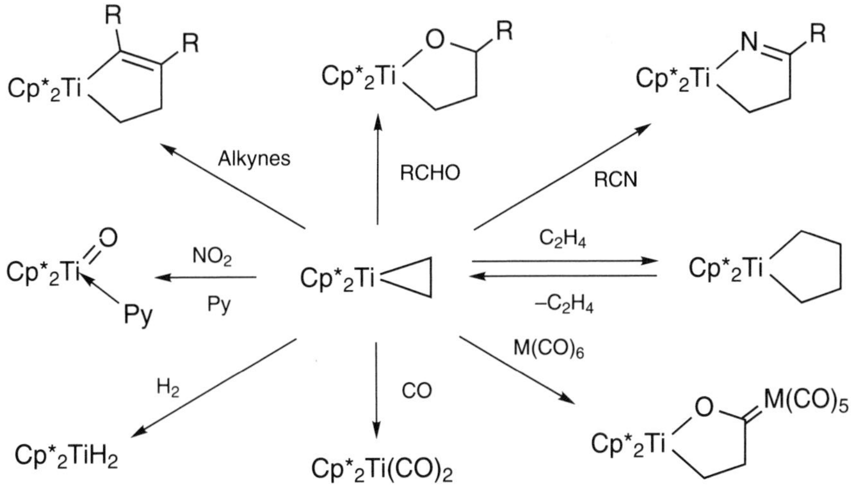

Scheme 6.4

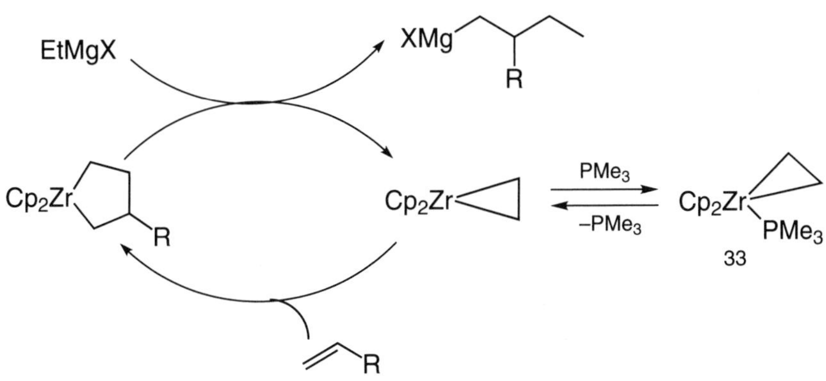

Scheme 6.5

$Cp_2Ti(PMe_3)_2$ (**35**) catalyzes the conversion of enyne and *tert*-$BuMe_2SiCN$ into bicyclic iminocyclopentenone [111] (Scheme 6.6).

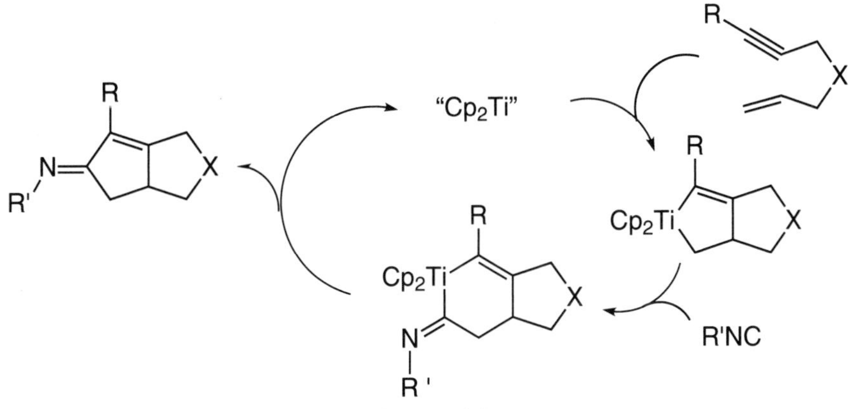

Scheme 6.6

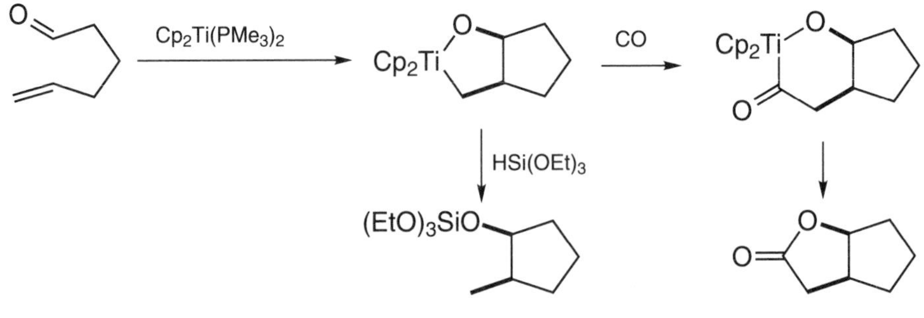
Cp2Ti(PMe3)2
Cp2Ti
CO
Cp2Ti
HSi(OEt)3
(EtO)3SiO

Scheme 6.7

The stoichiometric reaction of 1,6-enone and **35** gives bicyclic oxatitanacycle whose carbonylation resulted in the formation of bicyclic γ-butyrolactone (Scheme 6.7) [112,113]. Addition of H_2SiPh_2 and $HSi(OEt)_3$ to the oxatitanacycle cleaves Ti—O bond to generate titanocene species together with silyl ether product, and thus this process goes catalytically [114, 115].

Alkyne complex **37** is prepared by intramolecular hydrogen abstraction followed by methane elimination from the methyl metallocene compound **36** in the presence of a donor ligand (Scheme 6.8). **37** reacts with alkenes, alkynes, carbonyl compounds, and nitriles to give 5-membered metallacycles [89, 116–117]. In the absence of any donor ligand, metallacyclopentadiene **38** is formed. Complex **37** is also isolated by trapping **30** with an alkyne [89].

Dinitrogen complexes of titanocene and zirconocene are excellent sources for generating metallocene species *in situ*, which are highly active and are readily trapped by organic and inorganic unsaturated compounds. Eq (13) is an example that involves carbonyl coupling in iron-carbonyl to give the metallacyclic compound [118].

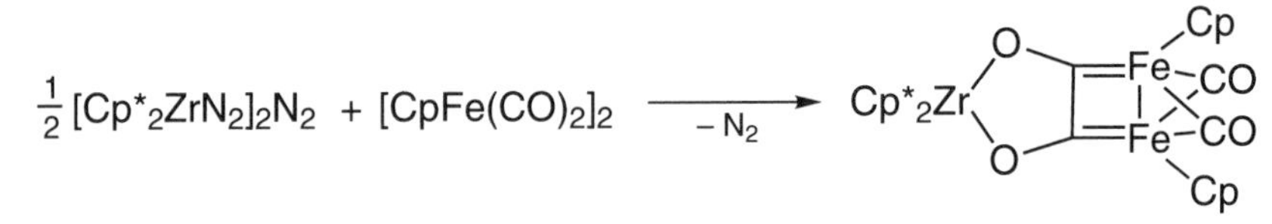
1/2 [Cp*2ZrN2]2N2 + [CpFe(CO)2]2
- N2
Cp*2Zr
Fe
CO
Cp
(13)

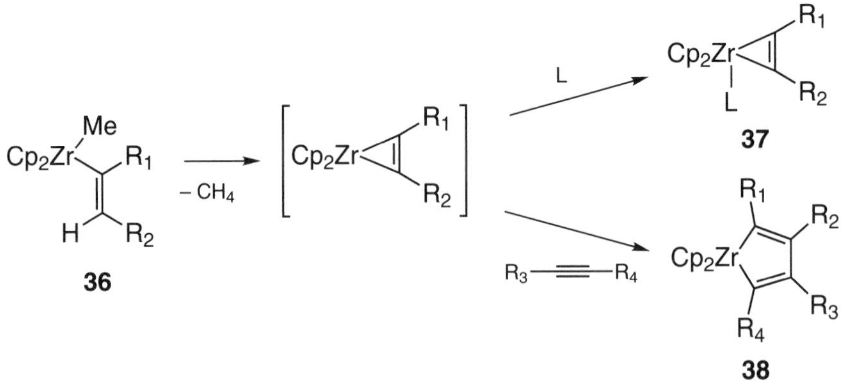
Me
Cp2Zr
R1
R2
H
36
- CH4
L
37
R3
R4
38

Scheme 6.8

Harrod *et al.* discovered the catalytic dehydrogenative oligomerization/polymerization of organosilanes assisted by d^0 metallocene silyl derivatives of Group 4 metals [119–121]. A σ-bond metathesis mechanism has been proposed [122,123]. Recently cation-like zirconocene derivatives were found to be catalysts for catalytic dehydrocoupling of phenylsilane giving poly(phenylsilanes) having long silicon chains [124]. Dehydropolymerization of secondary stannanes to high molecular weight polystannanes is achieved by using $CpCp^*ZrMe[Si(SiMe_3)_3]$ as a catalyst [125].

6.2.5 Allyl Complexes

Allyl complexes of Ti(III) have been prepared by the reaction of Cp_2TiCl_2 with two equivalents of allyl Grignard reagent or with two equivalents of isopropyl Grignard reagent in the presence of 1,3-diene [126,127]. In the first reaction, one equivalent of Grignard reagent acts as reducing reagent, and the resulting Cp_2TiCl reacts further with allyl Grignard reagent to give **39**. In the second reaction, the hydride species Cp_2TiH reacts with 1,3-diene to give the allyl complex **40**. Oxidative addition of allyl halides to **39** and **40** provide Ti(IV)–allyl compounds $Cp_2Ti(allyl)X$ (**41**) [128]. Allyl complexes **39–41** react with carbonyl compounds such as ketones and aldehydes.

$$Cp_2TiCl_2 + 2\ RCH{=}CHCH_2MgBr \longrightarrow Cp_2Ti(\eta^3\text{-}CH_2CHCHR) \quad \mathbf{39} \tag{14}$$

$$Cp_2TiCl_2 + \text{(2-R-1,3-butadiene)} + 2\ Me_2CHMgBr \longrightarrow Cp_2Ti(\eta^3\text{-}CH_2CRCHCH_3) \quad \mathbf{40} \tag{15}$$

6.2.6 Diene and Trimethylenemethane Complexes

Mono–diene complexes of zirconocene and hafnocene have been prepared by two methods [129–131], viz the photochemical reaction of diphenylzirconocene in the presence of diene and the reaction of metallocene dichlorides with diene magnesium adduct. The structures and reactivity of *s-cis*-diene complexes indicate that the metallacyclopentene (**B**) is the preferred canonical form. Complexes of the type Cp_2Zr(*s-trans*-1,3-diene), have been prepared; they were the first examples of this mode of coordination (**C**). Insertion of unsaturated compounds into a diene coordinated to zirconocene results in regioselective C—C bond formation [132–136].

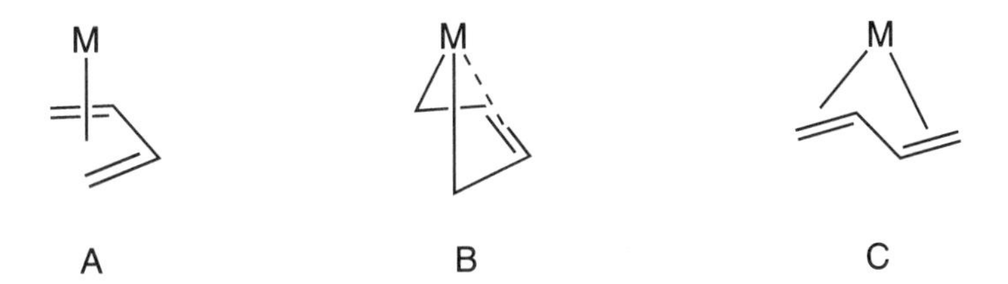

Monocyclopentadienyl–diene complexes such as CpM(dmpe)(butadiene)R (**42**) (M = Zr, Hf; R = Cl, H) [137] and CpM(diene)X (**43**) (M = Ti, Zr, Hf) [138,139] have been prepared. Complex **42** exhibits catalytic activity for the dimerization of ethylene to 1-butene. Insertion reaction of unsaturated organic compounds such as alkene, alkyne, ketone, nitrile, and so on into diene of **43** has also been reported [140,141]. Phosphine–butadiene complexes M(butadiene)R[N(SiMe$_2$CH$_2$PR'$_2$)$_2$] (M = Zr, Hf) have been reported [142]. The trimethylenemethane (TMM) ligand, which is isoelectronic with 1,3-diene, coordinates to the metallocene unit in η^3-fashion to form Cp*$_2$Zr(TMM) [143], whereas the coordination mode of TMM in [Cp*Zr(TMM)(μ-Cl)$_2$Li(tmeda)] is tetrahapto [144].

The simplest known homoleptic bis(diene) complex of a Group 4 metal is the highly electron-deficient, 12-electron species Ti(η-1,4-^{t}Bu$_2$C$_4$H$_4$)$_2$ (**44**), which is made by cocondensation of electron-beam vaporized titanium with an excess of 1,4-di-*tert*-butylbuta-1,3-diene and isolated by sublimation [145]. Bis(naphthalene)titanium is thermally unstable, although anionic derivatives such as $[Ti(C_{10}H_8)_2]^{2-}$ and $[Ti(C_{10}H_8)_2SnMe_3]^-$ are stable [146]. Bis(diene) complexes M(diene)$_2$(dmpe) (Ti, Zr, Hf) are prepared either by the amalgam reduction of MCl$_4$(dmpe) in the presence of diene or by the reaction of MCl$_4$(dmpe) with magnesium-diene adduct [147–149]. The anionic tris(diene)zirconate(2-) complex, $[Zr(\eta^4\text{-}C_{10}H_8)_3]^{2-}$ (**45**), is prepared by the reaction of ZrCl$_4$(THF)$_2$ with KC$_{10}$H$_8$ in the presence of [2.2.2]cryptand [150]. Complex **45** has the same structural features as the isoelectronic tris(η^4-butadiene)molybdenum [151,152].

6.2.7 Organometallic Compounds Bearing Alkoxy and Amide Ligands

The alkoxytitanium propene compound Ti(η^2-propene)(O^{i}Pr)$_2$ (**46**) [153], which is believed to be generated from Ti(O^{i}Pr)$_4$ and two equivalents of *i*-PrMgCl, reacts with internal alkynes to give titanium–alkyne compounds Ti(η-alkyne) (O^{i}Pr)$_2$ (**47**) in quantitative yield (Scheme 6.9) [154,155]. **46** reacts with carboxylic esters to produce cyclopropan-1-ols in modest yields [156,157]. Oxidative addition of allyllic halides or allyllic alcohols to **46** proceeds readily to form allyl titanium compounds **48**, whose reaction with aldehyde provides a stereoselective synthesis of homoallylic alcohols [153].

Rothwell *et al.* prepared some metallacyclic complexes **49–51** bearing bulky aryloxo ligands (ArO = 2,6-diphenylphenoxide or 2,6-diisopropylphenoxide) [158,159]. Some chelating aryloxy complexes such as **52** in combination with MAO are catalysts for

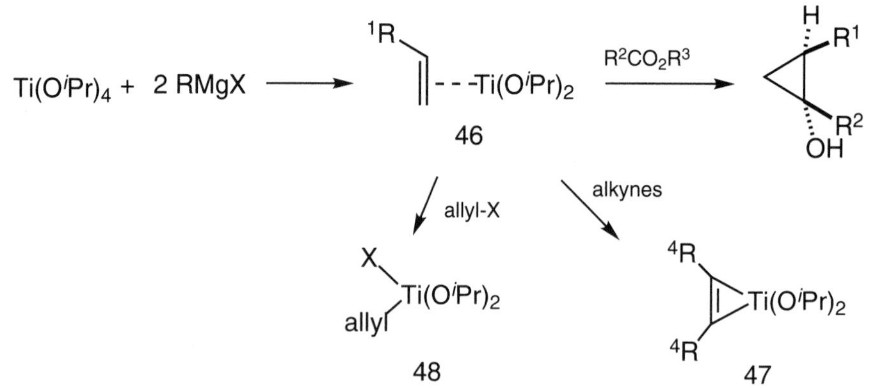

Ti(O^{i}Pr)$_4$ + 2 RMgX → ^{1}R-alkene---Ti(O^{i}Pr)$_2$ (**46**) —R^{2}CO$_2$R^3→ cyclopropanol (H, R^1, R^2, OH)

46 —allyl-X→ X(allyl)Ti(O^{i}Pr)$_2$ (**48**)

46 —alkynes→ ^{4}R-alkyne-Ti(O^{i}Pr)$_2$ (**47**)

Scheme 6.9

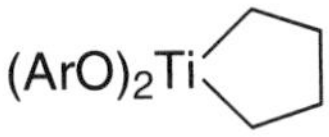

$(ArO)_2Ti$ 49 $(ArO)_2Ti$ R R 50 $(ArO)_2Ti$ R R R R 51 SiR_3 O MX_2 O SiR_3 52

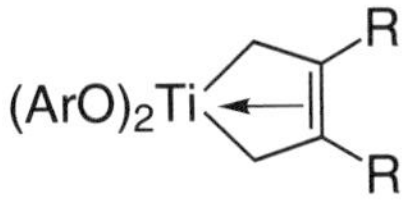

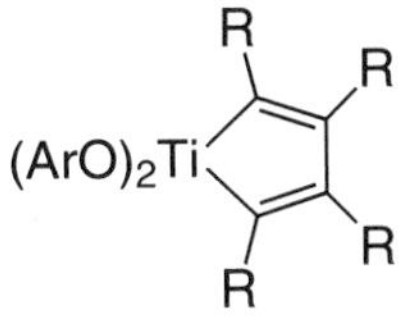

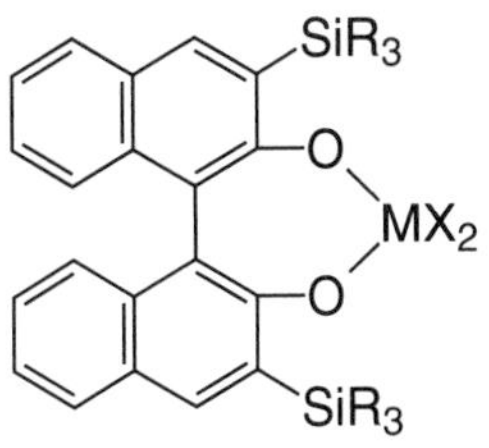

olefin polymerization [160]. Gambarotta *et al.* synthesized various amide complexes of titanium. Dinitrogen titanium(II) amido complexes [161,162] as well as methyl, allyl and methylene complexes [163,164] have been reported recently.

6.2.8 Carbonyl Complexes

Neutral $Ti(CO)_6$ is an extremely unstable compound which decomposed even below −220 °C, as shown by matrix isolation spectroscopy [165]. The much more stable phosphine derivatives $Ti(CO)_3(dmpe)_2$, $Ti(CO)_5(dmpe)$, $Ti(CO)_5(PMe_3)_2$, $Ti(CO)_4(PMe_3)_3$ have been isolated [166–168]. In contrast, the dianionic salt $[Ti(CO)_6]^{2-}$ (**53**) is thermally much more stable and decomposes only above 200 °C. Complex **53** was obtained by reductive carbonylation of $Ti(CO)_3(dmpe)_2$ by alkali metal naphthalenides in the presence of cryptand [169]. Carbonylation of **79** also produces **53** [170]. The naphthalenide-assisted reductive carbonylation of the zirconium tetrachloride afforded the zirconium analog $[Zr(CO)_6]^{2-}$ (**54**) [171], which was also derived by carbonylation of the tris(diene) dianion **45** [150]. One anion $[R_3Sn]^-$ effectively stabilizes $Ti(CO)_6$ as an air stable monoanionic salt, $[R_3SnTi(CO)_6]^-$ [172].

Carbonyl compounds of metallocene and half-metallocene are highly stable. Metallocene carbonyl complexes, $(\eta^5\text{-}C_5R_5)_2M(CO)_2$ (M = Ti, Zr, Hf; R = H, Me) have been characterized by IR spectra and X-ray analysis [173]. Anionic complexes $[(\eta^5\text{-}C_5R_5)M(CO)_4]^-$ (M = Ti, Zr, Hf; R = H, Me) (**55**) were prepared by alkali metal naphthalenide reduction of $(\eta^5\text{-}C_5R_5)MCl_3$ followed by carbonylation [174,175].

6.2.9 Arene Complexes

Well-defined arene complexes of Group 4 metals in various oxidation states have been isolated. The air- and moisture-sensitive complexes $Ti(\eta^6\text{-arene})_2$ (**56**) have a sandwich structure similar to that of the related chromium compounds [176–178]. They have been used for deoxygenation of propylene oxide and coupling reaction of organic carbonyl compounds [179]. The first synthesis of **56** was cocondensation of metal vapor with arene matrix [176]. Two more convenient methods are: reduction of $TiCl_4$ with $K[BEt_3H]$ in arene solvent [180] and reaction of $TiCl_4(THF)_2$ with arene anions followed by treatment with iodine [170,176]. The latter method involves the formation of an anionic titanate complex, $[Ti(\eta^6\text{-arene})_2]^-$ (**57**), which can also be formed from KH and **56** [181].

Titanium(II) arene complexes, $(\eta^6\text{-arene})Ti(AlCl_4)_2$, are readily obtained from the reaction between $TiCl_4$, aluminium powder and aluminium chloride in refluxing aromatic solvent [182]. Metal halides have been used as Lewis acid catalysts for various organic reactions. Recently the interaction of these metal halides with hydrocarbons, especially aromatic hydrocarbons, has been investigated. $ZrCl_4$ is insoluble in noncoordinating solvents, however it is soluble in dichloromethane in the presence of 1,2,4,5-$Me_4C_6H_2$. From the reaction mixture, $(\eta^6\text{-}C_6Me_6)Zr_2Cl_8$ was isolated and characterized by X-ray analysis [183], showing that $ZrCl_4$ had promoted methyl group rearrangement. A similar reaction of $TiCl_4$ and hexamethylbenezene results in the formation of the salt, $[(\eta^6\text{-}C_6Me_6)TiCl_3][Ti_2Cl_9]$, which can also be synthesized by cyclotrimerization of 2-butyne in the presence of $TiCl_4$ [184].

Cationic arene complexes of Group 4 metals have attracted much interest in connection with the polymerization of α-olefins. Reaction of $Zr(CH_2Ph)_4$ with 1 equivalent of $B(C_6F_5)_3$ resulted in the formation of cationic arene complex, $[Zr(CH_2Ph)_3]^+[B(CH_2Ph)(C_6F_5)_3]^-$ (**58**), X-ray analysis of which revealed a η^6-bonding interaction between the metal and the aromatic ring of the B-benzyl [185]. **58** catalyzes polymerization of ethylene and propene [185,186]. Treatment of $Zr(CH_2Ph)_4$ with 1 equivalent of $[PhNMe_2H][BPh_4]$ gave $[Zr(CH_2Ph)_3][(\eta^6\text{-}C_6H_5)BPh_3]$ (**59**) [187]. The titanium analog of **59** has not yet been synthesized. However, the monocyclopentadienyl complex $CpTiMe_3$ (**15**) reacts readily in THF to give the 14-electron zwitterionic arene complex $[CpTiMe_2(THF)][BPh_4]$ (**60**) [187], whose thermal stability is in sharp contrast to the thermal instability of $CpTiMe_3$. The cationic η^6-arene complexes, $[(\eta^5\text{-}C_5R_5)MR_2(\eta^6\text{-arene})]^+$ (**61**) (M = Ti, Zr, Hf; R = Me, CH_2Ph) have been characterized and their catalytic activities for polymerization have been studied [188,189].

6.2.10 Enolate and Homoenolate Complexes

$TiCl_4$ is one of the Lewis acids most often used in organic chemistry. For example, the well-known Mukaiyama aldol reaction is the $TiCl_4$-induced aldol cross-coupling reaction of silyl enol ether with aldehydes and ketones [190,191]. Metallocene–enolate complexes such as $Cp_2M(OCH{=}CR_2)$ and $Cp_2M(Me)(OCH{=}CR_2)$(M = Ti, Zr, R = H, Me), which are known to be *O*-enolate by X-ray crystallography [192], react with aldehydes to produce β-keto alcohols with erythroselectivity, independent of the stereochemistry of the starting enolate.

As shown in Scheme 6.10, treatment of $TiCl_4$ with trimethylsiloxycyclopropane affords the titanium homoenolate compound **62** [193–195], which is a chlorine-bridged dimer as revealed by X-ray analysis [196]. The enolate **62** reacts with aldehydes, not

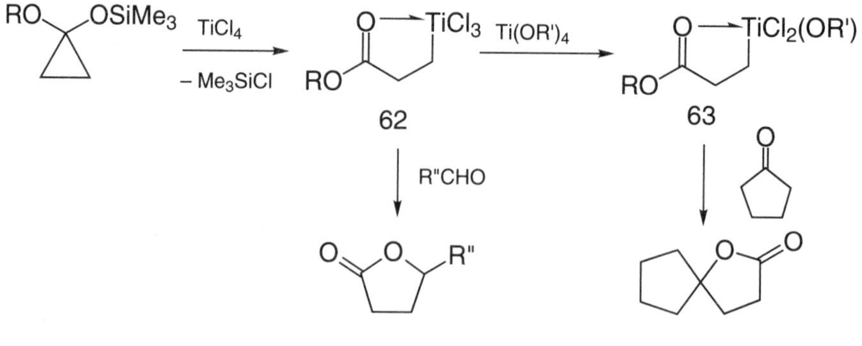

RO OSiMe$_3$ → ($TiCl_4$, $-Me_3SiCl$) → **62** (RO, O→$TiCl_3$) → ($Ti(OR')_4$) → **63** (RO, O→$TiCl_2(OR')$)

62 → (R"CHO) → lactone (O, O, R")

63 → (cyclopentanone) → spirolactone

Scheme 6.10

with ketones. However, enolate **63**, which is readily derived from the ligand exchange reaction of **62** with $Ti(OR')_4$, reacts with both aldehydes and ketones.

6.2.11 Chiral Complexes As Catalysts for Asymmetric Reactions

The asymmetric carbonyl-ene reaction has been developed by using chiral aryloxy complexes, i.e, **64** [197,198] and **65** [199]. Titanium complexes bearing chiral diols such as **66** and **67** have been used as catalysts for various enantioselective reactions such as synthesis of cyanohydrin, aldol reactions [200], [2 + 2] cycloaddition reactions [201], the reaction of diketene with aldehyde [202], hydroboration [203], allylation of aldehydes [204], and Diels–Alder reactions [205].

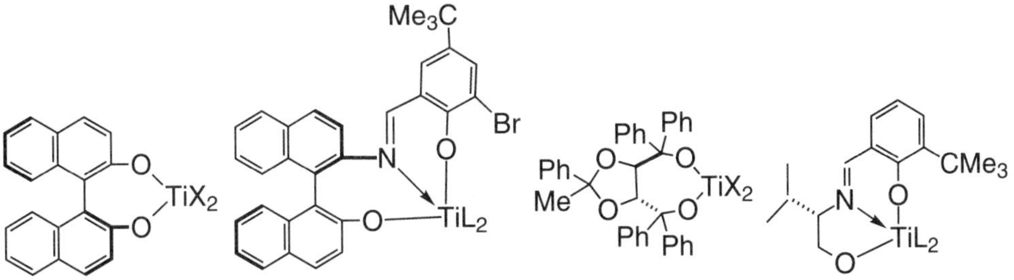

64 65 66 67

A chiral *ansa*-metallocene triflate complex catalyzes the Diels–Alder cycloaddition reaction between an oxazolidinone-based dienophile and cyclopentadiene [206]. Triflate in titanocene and zirconocene complexes is labile [207,208] and thus the polarity of solvent influences the reactivity and enantioselectivity. Asymmetric hydrogenation of imines and enamines catalyzed by chiral *ansa*-titanocene catalyst provides amines with high enantioselectivity [209,210].

6.3 Synthesis of Group 4 Metal Complexes

6.3.1 General Remarks

All reagents, solvents and apparatus must be dry, and moisture and oxygen should be excluded. All operations should be carried out in a glove-box or by standard Schlenk techniques. Chemicals used in the following synthesis are mostly commercially available. Abbreviations: dmpe = 1,2-Bis(dimethylphosphino)ethane; Cp = cyclopentadienyl; Cp* = pentamethylcyclopentadienyl.

(1) Schwartz's Reagent, Cp_2ZrHCl (2) [9]

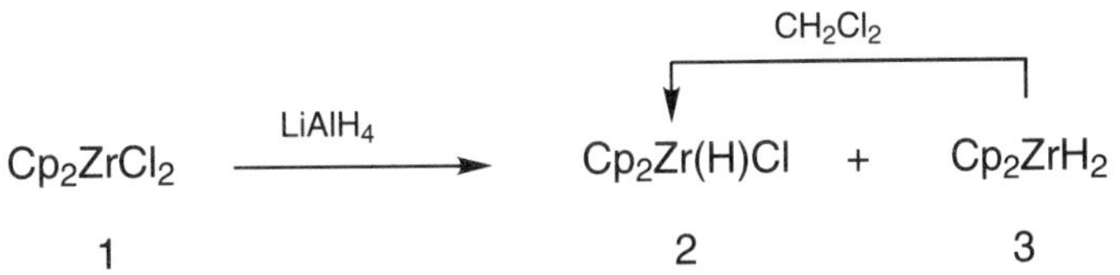

$$Cp_2ZrCl_2 \ (\mathbf{1}) \xrightarrow{LiAlH_4} Cp_2Zr(H)Cl \ (\mathbf{2}) + Cp_2ZrH_2 \ (\mathbf{3}) \quad [\mathbf{3} \xrightarrow{CH_2Cl_2} \mathbf{2}] \qquad (1)$$

Complex **1** (100 g, 0.342 mol) is dissolved in dry THF (650 mL, heating required) in a 1L Schlenk flask under argon. To this solution (at or slightly above room temperature) is added dropwise over a 45 min period a filtered solution of $LiAlH_4$ in diethyl ether (prepared from 3.6 g, 94.9 mmol of 95% $LiAlH_4$ and dry ether (100 mL) followed by filtration using a cannula flitted with a piece of glass fiber filter). The resulting suspension is allowed to stir at room temperature for 90 min. It was then Schlenk filtered under argon through a frit The white solid is washed on the frit with THF (4 × 75 mL), CH_2Cl_2 (2 × 100 mL) and ether (4 × 50 mL) with stirring or agitation of a stirbar immersed in the slurry . The resulting white solid is dried in vacuo to give a white powder: 65–81 g, 77–92% yield (94–96% purity).[1]

(2) Triisopropoxymethyltitanium $TiMe(O^iPr)_3$ (10) [25]

$$Cl{-}Ti(OCHMe_2)_3 + RLi \longrightarrow R{-}Ti(OCHMe_2)_3 \quad (2)$$

9 → 10: R = Me

A 2 L three-necked flask equipped with a dropping funnel and magnetic stirrer is charged with 250 ml of ether and 130.3 g (0.50 mol) of $TiCl(O^iPr)_3$ (**9**) and cooled to −40 °C. The equivalent amount of methyllithium (ether solution) is slowly added and the solution allowed to come to room temperature within 1.5 h. The solvent is removed *in vacuo* and the yellow product distilled directly from the precipitated lithium chloride at 48–53 °C/0.01 Torr: 113 g (94% yield)[2].

Properties: ^1H NMR(CCl_4): δ 0.5 (s, 3H), 1.3 (d, 18H), 4.5 (m, 3H).

(3) Tebbe Reagent, Cp2Ti(μ-Cl)(μ-CH_2)$AlMe_2$ (17) [52,57]

$$Cp_2TiCl_2 + 2\,AlMe_3 \longrightarrow Cp_2Ti(\mu\text{-}CH_2)(\mu\text{-}Cl)AlMe_2 + AlMe_2Cl \quad (6)$$

17

The original preparation method has been reported by Tebbe *et al.* [52]. The slightly modified procedure described by Grubbs et al. [57] is given here. Neat $AlMe_3$ (42 mL, 440 mmol)[3] is added via cannula to a suspension of Cp_2TiCl_2 (50 g, 200 mmol) in toluene (200 mL) to give a homogeneous red solution. Evolution of methane begins immediately. After this solution is stirred for 48 h, all volatiles are removed by vacuum distillation into a cold trap (caution: the $AlMe_2Cl$ evolved reacts violently with protic media and due precautions should be exercised).[4] A small sample of the resulting red crystalline material may be assayed for the presence of the

---

[1] A small sample is assayed in a 5 mm NMR tube in C_6D_6 by treatment with a known excess of acetone, and the relative areas of the signals for the mono and diisopropoxides are determined by ^1H NMR [18].

[2] The compound is air sensitive, but can be kept under nitrogen in a refrigerator for weeks or months.

[3] $AlMe_3$: mp 15 °C, bp 126 °C/760 mmHg. Flammable liquid that is highly sensitive to air and moisture, most conveniently handled as a toluene solution (10–20%).

[4] The toluene solution in the trap contains alkyl aluminium compounds and this solution should be carefully destroyed under inert atmosphere.

$Cp_2(Cl)Ti(\mu\text{-}CH_2)Al(Me)(Cl)$ (the $—CH_2—$ resonance appears at δ 7.68 as an unresolved AB quartet). This material is converted to the titled compound by addition of an equivalent amount of $AlMe_3$ to the reaction mixture, which is redissolved in toluene. The resulting solution is filtered through a pad of Celite supported on a coarse frit and concentrated to the point of saturation (*ca.* 150–170 mL total volume). This saturated solution is carefully layered with an equal volume of hexane or petroleum ether and allowed to stand undisturbed at –20 °C for 2–3 days. The supernantant is removed via cannula and the red crystalline mass is washed with several portions of petroleum ether at –20 °C. The solid is dried under high vacuum. Typical yields are 30–35 g (53–61%). An additional crop of less pure material (10–15 g) is obtained by concentrating the mother liquor as above.

Properties : ^1H NMR (C_6D_6, 22 °C, 220 MHz) δ 8.49 (2H, s, CH_2), 5.84 (10H, s, Cp), –0.06 (6H, s, $AlMe_2$); ^{13}C NMR (C_6D_6) δ 188 (CH_2).

(4) 1,1-Bis(cyclopentadienyl)-1-titana-3-*tert*-butylcyclobutane, $Cp_2Ti(CH_2CH(CMe_3)CH_2)$ [56]

$$Cp_2Ti(\mu\text{-}CH_2)(\mu\text{-}Cl)AlMe_2\ (\mathbf{17}) + CH_2{=}CR_1R_2 \xrightarrow{\text{base}} Cp_2Ti(CH_2CR_1R_2CH_2)\ (\mathbf{18}) \quad R_1 = CMe_3,\ R_2 = H \tag{7}$$

To **17** (1 g, 3.5 mmol) dissolved in toluene (6 mL) is added 3,3-dimethyl-1-butene (500 µL, 3.9 mmol) and then (dimethylamino)pyridine (DMAP, 472 mg, 4 mmol). The resulting red solution is transferred into 50 mL of vigorously stirred, cold (–20 °C) pentane or petroleum ether. The DMAP–$AlMe_2Cl$ adduct precipitates as a yellow-orange mass, which is rapidly filtered to give a clear red solution. This solution is evaporated to dryness by vacuum distillation to yield 770–800 mg of the title compound (80–83% yield)[1].

Properties : ^1H NMR (C_6D_6, 90 MHz) δ 5.61 (5H, s, Cp), 5.43 (5H, s, Cp), 2.24 (2H, d of d, H_{1a} and H_{3a}, J_{AB} = 8.5 Hz), 1.92 (2H, d of d, H_{1b} and H_{3b}), 0.93 (9H, s, CMe_3), –0.02 (^1H, m, H_2); ^{13}C NMR (C_6D_6) δ 110.6, 110.1 (Cp), 67.6 (C_α), 18.4 (C_β).

(5) Bis(pentamethylcyclopentadienyl)(ethylene)titanium(II), $Cp^*_2Ti(CH_2{=}CH_2)$ (34) [95]

$$Cp^*_2TiCl_2 \xrightarrow[\text{Na/Hg}]{C_2H_4} Cp^*_2Ti(C_2H_4)\ (\mathbf{34}) \tag{16}$$

Sodium amalgam (300 g, 0.9% w/w) is added via syringe to an argon-blanketed toluene slurry (150 mL) of $Cp^*_2TiCl_2$ (3.40 g, 8.73 mmol). The argon atmosphere is replaced

[1]Similarly, treatment of **17** with 1-alkenes and alkynes affords the corresponding metallacyclobutane and metallacyclobutene compounds, respectively [57,59]. Heating the dimethyl complex of titanocene generates the methylene species [44–46,71–73].

with ethylene and maintained at *ca.* 700 Torr while the mixture is stirred for 72 h. The resulting yellow-brown solution is filtered through a pad of Celite[1], and the toluene and excess C_2H_4 are removed *in vacuo* to yield the crude product. Recrystallization from petroleum ether affords 2.5 g (80%) of bright green $Cp^*_2Ti(CH_2{=}CH_2)$.

Properties : ^{1}H NMR (C_6D_6): δ 1.68 (s, Cp*), 2.02 (s, $CH_2{=}CH_2$); ^{13}C NMR (C_6D_6) δ 11.9 (q, C_5Me_5, 125 Hz), 105.1 (t, C_2H_4, 143.6), 119.8 (s, C_5Me_5).

(6) Bis(cyclopentadienyl)(1,3-butadiene)zirconium(II), Cp_2Zr(1,3-butadiene) [212]

$$Cp_2ZrCl_2 + 1/n\,[Mg(C_4H_6)(THF)_2]_n \longrightarrow Cp_2Zr(\text{s-cis-}C_4H_6) + Cp_2Zr(\text{s-trans-}C_4H_6) \quad (17)$$

s-cis s-trans

To a stirred slurry of Cp_2ZrCl_2 (2.9 g, 10 mmol) in THF (25 mL) in a 100 mL Schlenk tube is added a suspension of magnesium butadiene adduct (2.2 g, 10 mmol) in THF (150 mL) at −40 °C. The mixture is stirred for 1h at 30 °C and then heated to 60 °C for 10 min The solution turns from pale yellow to deep red. After evaporation of the solution to dryness, a mixture of hexane (50 mL) and benzene (15 mL) is added and the mixture is heated at reflux for 10 min. under an argon atmosphere. The precipitated $MgCl_2{\cdot}2THF$ is separated by centrifugation using a two-necked glass tube fitted with rubber stoppers. Removal of benzene–hexane by trap-to-trap distillation gives Cp_2Zr(1,3-butadiene) (a mixture of s-*cis*- and s-*trans*-butadiene isomers) as deep red crystals in 70% yield. The sample was purified by recrystallization from hexane at −20 °C, mp 73 °C (dec)[2].

Properties: s-cis isomer: ^{1}H NMR (toluene-d_8, −70 °C): δ −0.69 (m, H_{1anti}), 3.45 (m, H_{1syn}), 4.78 (m, H_2), 4.84 (s, Cp), 5.37 (s, Cp); s-trans isomer: ^{1}H NMR (toluene-d_8, 38 °C): δ 1.22 (m, H1), 2.90 (m, H2), 3.22 (m, H1′), 4.92 (s, Cp).

(7) Bis(cyclopentadienyl)dimethylzirconium, Cp_2ZrMe_2 [217,218]

$$Cp_2ZrCl_2 + 2\,MeLi \xrightarrow[\text{ether}]{} Cp_2ZrMe_2 \quad (18)$$

To a stirred suspension of Cp_2ZrCl_2 (18.0 g, 61.6 mmol) in diethyl ether (100 mL) at room temperature is added methyllithium solution (145 mL of 0.85 M solution, 123.2 mmol, in diethyl ether) over 1 h. After stirring overnight the mixture is filtered under

---

[1] This complex is prepared in quantitative yield by treatment of dinitrogen complex, $[TiCp^*_2]_2(\mu\text{-}N_2)$, with ethylene [95]. Alternatively, reaction of $Cp^*_2TiCl_2$ with 2 equivalents of n-BuLi under ethylene affords $Cp^*_2Ti(CH_2{=}CH_2)$ in 43% yield [211].

[2] Similarly, Cp_2M(1,3-diene) (M = Zr, Hf; 1,3-diene = isoprene, 2,3-dimethylbutadiene, 1,4-diphenylbutadiene) are prepared [212]. The ratio of *cis* and *trans* isomers depends on the diene ligand as well as crystallization condition (*cis* isomer turns thermally *trans* one). The title complex can be alternatively synthesized by the photolysis of diphenyl zirconocene in the presence of butadiene [213,214] and treatment of Cp_2MCl_2 (M = Zr, Hf) with 2 equivalents. of vinyl lithium [215,216].

argon and the filtrate is evaporated to dryness under reduced pressure. After two sublimations at 100–110°C/10–4 mmHg, pure Cp_2ZrMe_2 is obtained as a colorless solid (11.4 g, 73% yield), mp 190 °C (dec)[1].

Properties : ^1H NMR ($CDCl_3$): δ 6.11 (s, Cp), –0.39 (s, Me).

(8) Bis(cyclopentadienyl)chloromethylzirconium, $Cp_2Zr(Me)(Cl)$ [220,221]

$$Cp_2ZrCl_2 + Cp_2ZrMe_2 \xrightarrow[\text{toluene}]{\text{heat}} 2\ Cp_2Zr(Me)Cl \qquad (19)$$

A solution of Cp_2ZrMe_2 and Cp_2ZrCl_2, each 0.5 M, in toluene is heated to 130 °C for 35 h. $Cp_2Zr(Me)(Cl)$ can be isolated in >90% yield by recrystallization from toluene/hexane[2].

Properties : ^1H NMR (THF-d_8): δ 6.23 (s, Cp), 0.21 (s, Me).

(9) $[Cp_2ZrMe]^+[BMe(C_6F_5)_3]^-$ [49]

$$Cp_2ZrMe_2 + B(C_6F_5)_3 \longrightarrow [Cp_2ZrMe][MeB(C_6F_5)_3] \qquad (20)$$

Cp_2ZrMe_2 (0.100 g, 0.398 mmol) and $B(C_6F_5)_3$ (0.205 g, 0.400 mmol) are charged into a 25 mL reaction flask in a glove box. On a vacuum line, benzene (15 mL) is then vacuum-transferred into this flask at –78 °C. The mixture is slowly warmed to room temperature and stirred for 1.5 h. Large quantities of solid are observed to precipitate. Pentane (10 mL) is next vacuum-transferred into the flask, and, after stirring, the mixture is filtered. The light yellow solid is collected, washed once with 5 mL of pentane, and dried under vacuum: yield 72%.

Properties : ^1H NMR ($CDCl_3$): δ 5.37 (s, Cp), 0.26 (s, Zr-CH_3), 0.10 (brs, B-CH_3).

(10) Isopropyl 3-(trichlorotitanio)propionate [194]

$$TiCl_4 + \text{1-}^i\text{PrO-1-OSiMe}_3\text{-cyclopropane} \xrightarrow[-\ Me_3SiCl]{} {}^i\text{PrO-C(=O}\rightarrow TiCl_3\text{)-CH}_2\text{CH}_2\text{-} \qquad (21)$$

To a water-cooled solution of $TiCl_4$ (110 μL, 1.0 mmol) in 2.0 mL of hexane is added 1-trimethylsiloxy-1-isopropylcyclopropane (220 mL, 1.0 mmol) at 21 °C over 20 s. The initially formed white milky mixture turns brown in in 10 s, and finally deep purple microcrystals fall out. Heat evolution continues for a few minutes. The mixture is left to stand for 30 min; NMR analysis of the supernatant reveals the absence of the prod-

[1]Reaction of Cp_2ZrI_2 with methyllithium affords dimethyl complex uncontaminated by the monomethyl compound [219]. Reaction of Cp_2ZrCl_2 with 2 equivalents of PhLi gives Cp_2ZrPh_2 [218].

[2]$Cp_2Zr(Me)(Cl)$ has also been prepared by reaction of $[Cp_2Zr]_2O$ with $AlMe_3$ [222], but, the method described above is more convenient for easy handling. Bromo and iodo derivatives are also prepared similarly [221].

uct and the quantitative formation of chlorotrimethylsilane. The procedure can be scaled up to a 10 g scale without modification. This product melts at 90–95 °C and sublimes with some decomposition at 90–110 °C (0.005 mmHg)[1].

Properties : ^{1}H NMR ($CDCl_3$): δ 1.51 (d, $CHMe_2$, J = 6 Hz), 2.40 (t, $TiCH_2CH_2$, J = 7 Hz), 3.38 (t, $TiCH_2CH_2$, J = 7 Hz), 5.65 (qq, $CHMe_2$). IR (KBr): ν 1610 cm^{-1}.

References

[1] E. W. Abel, F. G. A. Stone, G. Wilkinson, Eds., *Comprehensive Organometallic Chemistry II*, Vol. 4 (1995).
[2] P. C. Wailes, R. S. P. Coutts, H. Weigold, *Organometallic Chemistry of Titanium, Zirconium and Hafnium.*, Academic Press, New York (1974).
[3] D. J. Cardin, M. F. Lappert, C. L. Raston, *Chemistry of Organo-Zirconium and -Hafnium Compounds*, John Wiley, Chichester (1986).
[4] B. Weidmann, D. Seebach, *Angew. Chem., Int. Ed. Engl.*, **22**, 31 (1983).
[5] M. T. Reetz, *Organotitanium Reagents in Organic Synthesis*, Springer, Berlin (1986).
[6] M. T. Reetz, *Titanium in Organic Synthesis. Organometallics in Synthesis: A Manual*, Ed. M. Schlosser., John Wiley, Chichester, p. 195 (1994).
[7] E. Negishi, T. Takahashi, *Synthesis*, **1** (1988).
[8] D. W. Hart, J. Schwartz, *J. Am. Chem. Soc.* **96**, 8115 (1974).
[9] S. L. Buchwald, S. J. LaMaire, R. B. Nielsen, B. T. Watson, S. M. King, *Tetrahedron Lett.*, **28**, 3895 (1987).
[10] P. C. Wailes, H. Weigold, *J. Organomet. Chem.*, **24**, 405 (1970).
[11] J. Schwartz, J. A. Labinger, *Angew. Chem., Int. Ed. Engl.*, **15**, 333 (1976).
[12] C. A. Bertelo, J. Schwartz, *J. Am. Chem. Soc.*, **97**, 228 (1975).
[13] C. A. Bertelo, J. Schwartz, *J. Am. Chem. Soc.*, **98**, 262 (1976).
[14] G. Fachinetti, G. Fochi, C. Floriani, *J. Chem. Soc., Dalton Trans.* 1946 (1977).
[15] G. Erker, *Acc. Chem. Res.*, **17**, 103 (1984); K. Tatsumi, A. Nakamura, P. Hofmann, P. Stauffert, R. Hoffmann, *J. Am. Chem. Soc.*, **107**, 4440 (1985) and references cited therein.
[16] J. W. Lauher, R. Hoffmann, *J. Am. Chem. Soc.*, **98**, 1729 (1976).
[17] D. W. Hart, T. F. Blackburn, J. Schwartz, *J. Am. Chem. Soc.*, **97**, 679 (1975).
[18] D. B. Carr, J. Schwartz, *J. Am. Chem. Soc.*, **101**, 3521 (1979).
[19] G. A. Luinstra, U. Rief, M. H. Prosenc, *Organometallics*, **14**, 1551 (1995).
[20] D. Seebach, A. K. Beck, M. Schiess, L. Widler, A. Wonnacott, *Pure Appl. Chem.* **55**, 1807 (1983).
[21] M. T. Reetz, *Pure Appl. Chem.*, **57**, 1781 (1985).
[22] K. H. Thiele, *Pure Appl. Chem.*, **30**, 575 (1972).
[23] J. D. McCowan, *Can. J. Chem.*, **51**, 1083 (1973).
[24] M. T. Reetz, S. H. Kyung, M. Hüllmann, *Tetrahedron*, **42**, 2931 (1986).
[25] M. T. Reetz, J. Westermann, R. Steinbach, B. Wenderoth, R. Peter, R. Ostarek, S. Maus, *Chem. Ber.*, **118**, 1421 (1985).
[26] B. Weidmann, L. Widler, A. G. Olivero, C. D. Maycock, D. Seebach, *Helv. Chim. Acta*, **64**, 357 (1981).
[27] M. T. Reetz, S. Maus, *Tetrahedron*, **43**, 101 (1987).
[28] K. Kostova, M. Hess, *Helv. Chim. Acta*, **67**, 1713 (1984).
[29] A. Mori, K. Maruoka, H. Yamamoto, *Tetrahedron Lett.*, **25**, 4421 (1984).
[30] Z. Dawoodi, M. L. H. Green, V. S. B. Mtetwa, K. Prout, *J. Chem. Soc., Chem. Commun.*, 1410 (1982).
[31] Z. Dawoodi, M. L. H. Green, V. S. B. Mtetwa, K. Prout, A. J. Schultz, J. M. Williams, T. F. Koetzle, *J. Chem. Soc., Dalton Trans.*, 1629 (1986).

[1] The similarly prepared ethyl ester is a dimer, as revealed by X-ray analysis [196].

[32] M. Brookhart, M. L. H. Green, L.-L. Wong, *Prog. Inorg. Chem.*, **36**, 1 (1988) and references cited therein.
[33] W. E. Piers, J. E. Bercaw, *J. Am. Chem. Soc.*, **112**, 9406 (1990).
[34] H. Krauledat, H. H. Brintzinger, *Angew. Chem., Int. Ed. Engl.*, **29**, 1412 (1990).
[35] M. K. Leclerc, H. H. Brintzinger, *J. Am. Chem. Soc.*, **117**, 1651 (1995).
[36] H. Kawamura-Kuribayashi, N. Koga, K. Morokuma, *J. Am. Chem. Soc.*, **114**, 8687 (1992).
[37] L. A. Castonguay, A. K. Rappé, *J. Am. Chem. Soc.*, **114**, 5832 (1992).
[38] M.-H. Prosenc, C. Janiak, H.-H. Brintzinger, *Organometallics*, **11**, 4036 (1992).
[39] T. K. Woo, L. Fan, T. Ziegler, *Organometallics*, **13**, 432 (1994).
[40] H. Weiss, M. Ehrig, R. Ahlrichs, *J. Am. Chem. Soc.*, **116**, 4919 (1994).
[41] Z. Dawoodi, M. L. H. Green, V. S. B. Mtetwa, K. Prout, *J. Chem. Soc., Chem. Commun.*, 802 (1982).
[42] S. J. Lancaster, O. B. Robinson, M. Bochmann, S. J. Coles, M. B. Hursthouse, *Organometallics*, **14**, 2456 (1995).
[43] U. Giannini, S. Cesca, *Tetrahedron Lett.*, **19** (1960).
[44] N. A. Petasis, E. I. Bzowej, *J. Am. Chem. Soc.*, **112**, 6392 (1990).
[45] N. A. Petasis, D.-K. Fu, *J. Am. Chem. Soc.*, **115**, 7208 (1993).
[46] N. A. Petasis, S.-P. Lu, *J. Am. Chem. Soc.*, **117**, 6394 (1995).
[47] R. F. Jordan, *Adv. Organomet. Chem.*, **32**, 325 (1991).
[48] T. J. Marks, *Acc. Chem. Res.*, **25**, 57 (1992).
[49] X. Yang, C. L. Stern, T. J. Marks, *J. Am. Chem. Soc.*, **116**, 10015 (1994) and references cited therein.
[50] P. A. Deck, T. J. Marks, *J. Am. Chem. Soc.*, **117**, 6128 (1995).
[51] H. H. Brintzinger, D. Fischer, R. Mülhaupt, B. Rieger, R. M. Waymouth, *Angew. Chem., Int. Ed. Engl.*, **34**, 1143 (1995).
[52] F. N. Tebbe, G. W. Parshall, G. S. Reddy, *J. Am. Chem. Soc.*, **100**, 3611 (1978).
[53] E. Negishi, *Pure Appl. Chem.*, **53**, 2333 (1981).
[54] E. Negishi, D. E. Van Horn, T. Yoshida, *J. Am. Chem. Soc.*, **107**, 6639 (1985).
[55] E. Negishi, *Acc. Chem. Res.* **20**, 65 (1987).
[56] T. R. Howard, J. B. Lee, R. H. Grubbs, *J. Am. Chem. Soc.*, **102**, 6876 (1980); J. B. Lee, K. C. Ott, R. H. Grubbs, *J. Am. Chem. Soc.*, **104**, 7491 (1982).
[57] T. Ikariya, S. C. H. Ho, R. H. Grubbs, *Organometallics*, **4**, 199 (1985).
[58] R. H. Grubbs, W. Tumas, *Science*, **243**, 907 (1989).
[59] K. A. Brown-Wensley, S. L. Buchwald, L. Cannizzo, L. Clawson, S. Ho, D. Meinhardt, J. R. Stille, D. Straus, R. H. Grubbs, *Pure Appl. Chem.*, **55**, 1733 (1983).
[60] J. M. Hawkins, R. H. Grubbs, *J. Am. Chem. Soc.*, **110**, 2821 (1988).
[61] J. D. Meinhart, E. V. Anslyn, R. H. Grubbs, *Organometallics*, **8**, 583 (1989).
[62] S. H. Pine, R. Zahler, D. A. Evans, R. H. Grubbs, *J. Am. Chem. Soc.*, **102**, 3270 (1980); L. Clawson, S. L. Buchwald, R. H. Grubbs, *Tetrahedron Lett.*, **25**, 5733 (1984).
[63] S. H. Pine, R. J. Pettit, G. D. Geib, S. G. Cruz, C. H. Gallego, T. Tijerina, R. D. Pine, *J. Org. Chem.*, **50**, 1212 (1985).
[64] L. F. Cannizzo, R. H. Grubbs, *J. Org. Chem.*, **50**, 2316 (1985).
[65] J. R. Stille, R. H. Grubbs, *J. Am. Chem. Soc.*, **105**, 1664 (1983).
[66] T.-S. Cho, S.-B. Huang, *Tetrahedron Lett.*, **24**, 2169 (1983).
[67] S. C. H. Ho, D. A. Straus, R. H. Grubbs, *J. Am. Chem. Soc.*, **106**, 1533 (1984).
[68] W. Tumas, D. R. Wheeler, R. H. Grubbs, *J. Am. Chem. Soc.*, **109**, 6182 (1987).
[69] M. J. Burk, D. L. Staley, W. Tumas, *J. Chem. Soc., Chem. Commun.*, 809 (1990).
[70] M. J. Burk, W. Tumas, D. R. Wheeler, *J. Am. Chem. Soc.*, **112**, 6133 (1990).
[71] R. J. McKinney, T. H. Tulip, D. L. Thorn, T. S. Coolbaugh, F. N. Tebbe, *J. Am. Chem. Soc.*, **103**, 5584 (1981).
[72] N. A. Petasis, D.-K. Fu, *Organometallics*, **12**, 3776 (1993).
[73] K. M. Doxsee, J. J. J. Juliette, J. K. M. Mouser, K. Zientara, *Organometallics*, **12**, 4682 (1993); K. M. Doxsee, J. J. J. Juliette, K. Zientara, G. Nieckarz, *J. Am. Chem. Soc.*, **116**, 2147 (1994).
[74] E. B. Tjaden, C. L. Casty, J. M. Stryker, *J. Am. Chem. Soc.*, **115**, 9814 (1993).
[75] E. B. Tjaden, J. M. Stryker, *J. Am. Chem. Soc.*, **115**, 2083 (1993).

[76] G. L. Casty, J. M. Stryker, *J. Am. Chem. Soc.*, **117**, 7814 (1995).
[77] J. L. Polse, R. A. Andersen, R. G. Bergman, *J. Am. Chem. Soc.*, **117**, 5393 (1995) and references cited therein.
[78] K. E. Meyer, P. J. Walsh, R. G. Bergman, *J. Am. Chem. Soc.*, **116**, 2669 (1994).
[79] P. J. Walsh, A. M. Baranger, R. G. Bergman, *J. Am. Chem. Soc.*, **114**, 1708 (1992).
[80] J. X. McDermott, M. E. Wilson, G. M. Whitesides, *J. Am. Chem. Soc.*, **98**, 6529 (1976).
[81] A. Stockis, R. Hoffmann, *J. Am. Chem. Soc.*, **102**, 2952 (1980).
[82] S. Thanedar, M. F. Farona, *J. Organomet. Chem.*, **235**, 65 (1982).
[83] K. I. Gell, J. Schwartz, *J. Am. Chem. Soc.*, **103**, 2687 (1981).
[84] E. Negishi, F. E. Cederbaum, T. Takahashi, *Tetrahedron Lett.*, **27**, 2829 (1986).
[85] E. Negishi, T. Takahashi, *Acc. Chem. Res.*, **27**, 124 (1994); Y. Hanzawa, H. Ito, T. Taguchi, *Synlett*, 299 (1995).
[86] T. Takahashi, M. Hasegawa, N. Suzuki, M. Saburi, C. J. Rousset, P. E. Fanwick, E. Negishi, *J. Am. Chem. Soc.*, **113**, 8564 (1991).
[87] J. Y. Corey, X.-H. Zhu, *Organometallics*, **11**, 672 (1992).
[88] E. Negishi, T. Takahashi, *Synthesis*, **1** (1988).
[89] S. L. Buchwald, R. B. Nielsen, *Chem. Rev.*, **88**, 1047 (1988).
[90] E. Negishi, T. Takahashi, *Ald. Chim. Acta.*, **18**, 31 (1985).
[91] D. F. Taber, J. P. Louey, Y. Wang, W. A. Nugent, D. A. Dixon, R. L. Harlow, *J. Am. Chem. Soc.*, **116**, 9457 (1994) and references cited therein.
[92] H. G. Alt, C. E. Denner, U. Thewalt, M. D. Rausch, *J. Organomet. Chem.*, **356**, C83 (1988).
[93] T. Takahashi, M. Murakami, M. Kunishige, M. Saburi, Y. Uchida, K. Kozawa, T. Uchida, D. R. Swanson, E. Negishi, *Chem. Lett.* 761 (1989); T. Takahashi, M. Tamura, M. Saburi, Y. Uchida, E. Negishi, *J. Chem. Soc., Chem. Commun.* 852 (1989).
[94] S. L. Buchwald, K. A. Kreutzer, R. A. Fisher, *J. Am. Chem. Soc.*, **112**, 4600 (1990).
[95] S. A. Cohen, P. R. Auburn, J. E. Bercaw, *J. Am. Chem. Soc.*, **105**, 1136 (1983).
[96] S. A. Cohen, J. E. Bercaw, *Organometallics*, **4**, 1006 (1985).
[97] K. Mashima, H. Haraguchi, A. Ohyoshi, N. Sakai, H. Takaya, *Organometallics*, **10**, 2731 (1991).
[98] K. Mashima, K. Jyodoi, A. Ohyoshi, H. Takaya, *Bull. Chem. Soc. Jpn.*, **64**, 2065 (1991).
[99] M. R. Smith III, P. T. Matsunaga, R. A. Andersen, *J. Am. Chem. Soc.*, **115**, 7049 (1993).
[100] M. L. Steigerwald, W. A. Goddard III, *J. Am. Chem. Soc.*, **107**, 5027 (1985).
[101] W. Kaminsky, J. Kopf, H. Sinn, H.-J. Vollmer, *Angew. Chem., Int. Ed. Engl.*, **15**, 629 (1976).
[102] T. Takahashi, K. Kasai, N. Suzuki, K. Nakajima, E. Negishi, *Organometallics*, **13**, 3413 (1994).
[103] F. A. Cotton, P. A. Kibala, *Inorg. Chem.*, **29**, 3192 (1990); J. H. Wengrovius, R. R. Schrock, C. S. Day, *Inorg. Chem.*, **20**, 1845 (1981).
[104] U. M. Dzhemilev, O. S. Vostrikova, R. M. Sultanov, *Izv. Akad. Nauk SSSR, Ser. Khim.*, 218 (1983).
[105] U. M. Dzhemilev, O. S. Vostrikova, *J. Organomet. Chem.*, **285**, 43 (1985).
[106] U. M. Dzhemilev, O. S. Vostrikova, G. A. Tolstikov, *J. Organomet. Chem.*, **304**, 17 (1986) and references cited therein.
[107] T. Takahashi, T. Seki, Y. Nitto, M. Saburi, C. J. Rousset, E. Negishi, *J. Am. Chem. Soc.*, **113**, 6266 (1991).
[108] A. H. Hoveyda, Z. Xu, *J. Am. Chem. Soc.*, **113**, 5079 (1991).
[109] K. S. Knight, R. M. Waymouth, *J. Am. Chem. Soc.*, **113**, 6268 (1991).
[110] M. T. Didiuk, C. W. Johannes, J. P. Morken, A. H. Hoveyda, *J. Am. Chem. Soc.*, **117**, 7097 (1995) and references cited therein.
[111] S. C. Berk, R. B. Grossman, S. L. Buchwald, *J. Am. Chem. Soc.*, **115**, 4912 (1993).
[112] D. F. Hewlett, R. J. Whitby, *J. Chem. Soc., Chem. Commun.*, 1684 (1990).
[113] W. E. Crowe, A. T. Vu, *J. Am. Chem. Soc.*, **118**, 1557 (1996).
[114] N. M. Kablaoui, S. L. Buchwald, *J. Am. Chem. Soc.*, **117**, 6785 (1995).
[115] W. E. Crowe, M. J. Rachita, *J. Am. Chem. Soc.*, **117**, 6787 (1995).
[116] M. A. Bennett, H. P. Schwemlein, *Angew. Chem., Int. Ed. Engl.*, **28**, 1296 (1989).
[117] J. Cámpora, S. L. Buchwald, *Organometallics*, **12**, 4182 (1993); U. Rosenthal, A. Ohff, M.

Michalik, H. Görls, V. V. Burlakov, V. B. Shur, *Angew. Chem., Int. Ed. Engl.*, **32**, 1193 (1993).
[118] D. H. Berry, J. E. Bercaw, A. J. Jircitano, K. B. Mertes, *J. Am. Chem. Soc.*, **104**, 4712 (1982).
[119] C. T. Aitken, J. F. Harrod, E. Samuel, *J. Organomet. Chem.*, **279**, C11 (1985).
[120] C. T. Aitken, J. F. Harrod, E. Samuel, *J. Am. Chem. Soc.*, **108**, 4059 (1986).
[121] J. F. Harrod, T. Ziegler, V. Tschinke, *Organometallics*, **9**, 897 (1990).
[122] H.-G. Woo, J. F. Walzer, T. D. Tilley, *J. Am. Chem. Soc.*, **114**, 7047 (1992).
[123] T. D. Tilley, *Acc. Chem. Res.*, **26**, 22 (1993).
[124] V. K. Dioumaev, J. F. Harrod, *Organometallics*, **13**, 1548 (1994).
[125] T. Imori, V. Lu, H. Cai, T. D. Tilley, *J. Am. Chem. Soc.*, **117**, 9931 (1995).
[126] H. A. Martin, F. Jellinek, *J. Organomet. Chem.*, **8**, 115 (1967).
[127] H. A. Martin, F. Jellinek, *J. Organomet. Chem.*, **12**, 149 (1968).
[128] F. Sato, K. Iida, S. Iijima, H. Moriya, M. Sato, *J. Chem. Soc., Chem. Commun.* 1140 (1981).
[129] H. Yasuda, K. Tatsumi, A. Nakamura, *Acc. Chem. Res.*, **18**, 120 (1985).
[130] G. Erker, C. Krüger, G. Müller, *Adv. Organomet. Chem.*, **24**, 1 (1985).
[131] H. Yasuda, A. Nakamura, *Angew. Chem., Int. Ed. Engl.*, **26**, 723 (1987).
[132] H. Yasuda, T. Okamoto, K. Mashima, A. Nakamura, *J. Organomet. Chem.* (1989).
[133] G. Erker, *Angew. Chem., Int. Ed. Engl.*, **28**, 397 (1989).
[134] G. Erker, R. Pfaff, D. Kowalski, E.-U. Würthwein, C. Krüger, R. Goddard, *J. Org. Chem.*, **58**, 6771 (1993).
[135] G. Erker, R. Pfaff, *Organometallics*, **12**, 1921 (1993).
[136] L. López, M. Berlekamp, D. Kowalski, G. Erker, *Angew. Chem., Int. Ed. Engl.*, **33**, 1114 (1994).
[137] Y. Wielstra, A. Meetsma, S. Ganbarotta, *Organometallics*, **8**, 258 (1989).
[138] J. Blenkers, B. Hessen, F. van Bolhuis, A. J. Wagner, J. H. Teuben, *Organometallics*, **6**, 459 (1987).
[139] H. Yamamoto, H. Yasuda, K. Tatsumi, K. Lee, A. Nakamura, J. Chen, Y. Kai, N. Kasai, *Organometallics*, **8**, 105 (1989).
[140] B. Hessen, J. Blenkers, J. H. Teuben, G. Helgesson, S. Jagner, *Organometallics*, **8**, 830 (1989).
[141] B. Hessen, J. Blenkers, J. H. Teuben, G. Helgesson, S. Jagner, *Organometallics*, **8**, 2809 (1989).
[142] M. D. Fryzuk, T. S. Haddad, S. J. Rettig, *Organometallics*, **8**, 1723 (1989).
[143] G. E. Herberich, C. Kreuder, U. Englert, *Angew. Chem., Int. Ed. Engl.*, **33**, 2465 (1994).
[144] G. C. Bazan, G. Rodriguez, B. P. Cleary, *J. Am. Chem. Soc.*, **116**, 2177 (1994).
[145] F. G. N. Cloke, A. McCamley, *J. Chem. Soc., Chem. Commun.*, 1470 (1991).
[146] J. E. Ellis, D. W. Blackburn, P. Yuen, M. Jang, *J. Am. Chem. Soc.*, **115**, 11616 (1993).
[147] S. S. Wreford, M. B. Fischer, J.-S. Lee, E. J. James, S. C. Nyburg, *J. Chem. Soc., Chem. Commun.*, 458 (1981).
[148] S. Datta, S. S. Wreford, R. P. Beatty, T. J. Neese, *J. Am. Chem. Soc.*, **101**, 1053 (1979).
[149] R. P. Beatty, S. Datta, S. S. Wreford, *Inorg. Chem.*, **18**, 3139 (1979).
[150] M. Jang, J. E. Ellis, *Angew. Chem., Int. Ed. Engl.*,. **33**, 1973 (1994).
[151] V. Akhmedov, M. T. Anthony, M. L. H. Green, D. Young, *J. Chem. Soc., Dalton Trans.*, 1412 (1975).
[152] P. S. Skell, M. J. McGlinchey, *Angew. Chem., Int. Ed. Engl.*, **14**, 195 (1975).
[153] A. Kasatkin, T. Nakagawa, S. Okamoto, F. Sato, *J. Am. Chem. Soc.*, **117**, 3881 (1995).
[154] Y. Gao, K. Harada, F. Sato, *Tetrahedron Lett.*, **36**, 5913 (1995).
[155] K. Harada, H. Urabe, F. Sato, *Tetrahedron Lett.*, **36**, 3203 (1995).
[156] E. J. Corey, S. A. Rao, M. C. Noe, *J. Am. Chem. Soc.*, **116**, 9345 (1994).
[157] J. Lee, H. Kim, J. K. Cha, *J. Am. Chem. Soc.*, **117**, 9919 (1995).
[158] J. E. Hill, G. Balaich, P. E. Fanwick, I. P. Rothwell, *Organometallics*, **12**, 2911 (1993).
[159] G. J. Balaich, J. E. Hill, S. A. Waratuke, P. E. Franwick, I. P. Rothwell, *Organometallics*, **14**, 656 (1995) and references cited therein.
[160] A. van der Linden, C. J. Schaverien, N. Meijboom, C. Ganter, A. G. Orpen, *J. Am. Chem. Soc.*, **117**, 3008 (1995) and references cited therein.

[161] R. Duchateau, S. Gambarotta, N. Beydoun, C. Bensimon, *J. Am. Chem. Soc.*, **113**, 8986 (1991).
[162] N. Beydoun, R. Duchateau, S. Gambarotta, *J. Chem. Soc., Chem. Commun.*, 244 (1992).
[163] D. G. Dick, R. Duchateau, J. J. H. Edema, S. Gambarotta, *Inorg. Chem.* **32**, 1959 (1993).
[164] L. Scoles, R. Minhas, R. Duchateau, J. Jubb, S. Gambarotta, *Organometallics*, **13**, 4978 (1994).
[165] R. Busby, W. Klotzbucher, G. A. Ozin, *Inorg. Chem.*, **16**, 822 (1977).
[166] P. J. Domaille, R. L. Harlow, S. S. Wreford, *Organometallics*, **1**, 935 (1982).
[167] K. M. Chi, S. R. Frerichs, B. K. Stein, D. W. Blackburn, J. E. Ellis, *J. Am. Chem. Soc.*, **110**, 163 (1988).
[168] K. M. Chi, S. R. Frerichs, J. E. Ellis, *J. Chem. Soc., Chem. Commun.*, 1013 (1988).
[169] K. M. Chi, S. R. Frerichs, S. B. Philson, J. E. Ellis, *J. Am. Chem. Soc.*, **110**, 303 (1988).
[170] D. W. Blackburn, D. Britton, J. E. Ellis, *Angew. Chem., Int. Ed. Engl.*, **31**, 1495 (1992).
[171] K. M. Chi, S. R. Frerichs, S. B. Philson, J. E. Ellis, *Angew. Chem., Int. Ed. Engl.*, **26**, 1190 (1987).
[172] J. E. Ellis, P. Yuen, *Inorg. Chem.*, **32**, 4998 (1993).
[173] D. J. Shikora, M. D. Rausch, R. D. Rogers, J. L. Atwood, *J. Am. Chem. Soc.*, **103**, 1265 (1981) and reference cited therein.
[174] J. E. Ellis, S. R. Frerichs, B. K. Stein, *Organometallics*, **12**, 1048 (1993).
[175] J. E. Ellis, B. K. Stein, S. R. Frerichs, *J. Am. Chem. Soc.*, **115**, 4066 (1993).
[176] M. T. Anthony, M. L. H. Green, D. Young, *J. Chem. Soc., Dalton Trans.*, 1419 (1975).
[177] P. N. Hawker, P. E. Kündig, P. L. Timms, *J. Chem. Soc., Chem. Commun.*, 730 (1978).
[178] U. Thewalt, F. Österle, *J. Organomet. Chem.*, **172**, 317 (1979).
[179] H. Ledon, I. Tkatchenko, D. Young, *Tetrahedron Lett.*, 173 (1979).
[180] H. Bönnemann, B. Korall, *Angew. Chem., Int. Ed. Engl.*, **31**, 1490 (1992).
[181] J. A. Bandy, A. Berry, M. L. H. Green, R. N. Perutz, K. Prout, J.-N. Verpeaux, *J. Chem. Soc., Chem. Commun.*, 729 (1984).
[182] S. Pasynkiewicz, R. Giezynski, S. Dzierzgowski, *J. Organomet. Chem.*, **54**, 203 (1973).
[183] E. Solari, F. Musso, R. Ferguson, C. Floriani, A. Chiesi-Villa, C. Rizzoli, *Angew. Chem., Int. Ed. Engl.*, **34**, 1510 (1995).
[184] E. Solari, C. Floriani, K. Schenk, A. Chiesi-Villa, C. Rizzoli, M. Rosi, A. Sgamellotti, *Inorg. Chem.*, **33**, 2018 (1994).
[185] C. Pellecchia, A. Grassi, A. Immirzi, *J. Am. Chem. Soc.*, **115**, 1160 (1993).
[186] C. Pellecchia, A. Grassi, A. Zambelli, *Organometallics*, **13**, 298 (1994).
[187] M. Bochmann, G. Karger, A. J. Jaggar, *J. Chem. Soc., Chem. Commun.*, 1038 (1990).
[188] D. J. Gillis, M.-J. Tudoret, M. C. Baird, *J. Am. Chem. Soc.*, **115**, 2543 (1993).
[189] C. Pellecchia, A. Immirzi, A. Grassi, A. Zambelli, *Organometallics*, **12**, 4473 (1993).
[190] T. Mukaiyama, *Angew. Chem., Int. Ed. Engl.*, **16**, 817 (1977).
[191] T. Mukaiyama, *Organic Reactions*, **28**, 203 (1982).
[192] M. D. Curtis, S. Thanedar, W. M. Butler, *Organometallics*, **3**, 1855 (1984).
[193] E. Nakamura, I. Kuwajima, *J. Am. Chem. Soc.*, **105**, 651 (1983).
[194] E. Nakamura, H. Oshino, I. Kuwajima, *J. Am. Chem. Soc.*, **108**, 3745 (1986).
[195] I. Kuwajima, E. Nakamura, *Comprehensive Organic Synthesis*, **2**, 441 (1991).
[196] P. G. Cozzi, T. Carofiglio, C. Floriani, A. Chiesi-Villa, C. Rizzoli, *Organometallics*, **12**, 2845 (1993).
[197] K. Mikami, M. Shimizu, *Chem. Rev.*, **92**, 1021 (1992).
[198] K. Mikami, M. Terada, S. Narisawa, T. Nakai, *Synlett*, **255** (1992).
[199] E. M. Carreira, W. Lee, R. A. Singer, *J. Am. Chem. Soc.*, **117**, 3649 (1995).
[200] K. Narasaka, *Synthesis*, 1 (1991); M. Hayashi, T. Inoue, Y. Miyamoto, N. Oguni, *Tetrahedron*, **50**, 4385 (1994); G. Keck, D. Krishnamurthy, *J. Am. Chem. Soc.*, **117**, 2363 (1995).
[201] K. Narasaka, Y. Hayashi, H. Shimadzu, S. Niihata, *J. Am. Chem. Soc.*, **114**, 8869 (1992).
[202] M. Hayashi, T. Inoue, N. Oguni, *J. Chem. Soc., Chem. Commun.*, 341 (1994).
[203] G. Giffels, C. Dreisbach, U. Kragl, M. Weigerding, H. Waldmann, C. Wandrey, *Angew. Chem., Int. Ed. Engl.*, **34**, 2005 (1995).
[204] S. Aoki, K. Mikami, M. Terada, T. Nakai, *Tetrahedron*, **49**, 1783 (1993); A. L. Costa, M.

G. Piazza, E. Tagliavini, C. Trombini, A. Umani-Ronchi, *J. Am. Chem. Soc.*, **115**, 7001 (1993); G. E. Keck, K. H. Tarbet, L. S. Geraci, *J. Am. Chem. Soc.*, **115**, 8467 (1993).
[205] H. B. Kagan, D. Riant, *Chem. Rev.*, **92**, 1007 (1992).
[206] J. B. Jaquith, J. Guan, S. Wang, S. Collins, *Organometallics*, **14**, 1079 (1995).
[207] T. K. Hollis, N. P. Robinson, B. Bosnich, *Organometallics*, **12**, 2745 (1992).[
[208] T. K. Hollis, N. P. Robinson, B. Bosnich, *J. Am. Chem. Soc.*, **114**, 5464 (1992).
[209] C. A. Willoughby, S. L. Buchwald, *J. Am. Chem. Soc.*, **116**, 11703 (1994).
[210] N. E. Lee, S. L. Buchwald, *J. Am. Chem. Soc.*, **116**, 5985 (1994).
[211] K. Mashima, N. Sakai, H. Takaya, *Bull. Chem. Soc. Jpn.*, **64**, 2475 (1991).
[212] H. Yasuda, Y. Kajiwara, K. Mashima, K. Nagasuna, K. Lee, A. Nakamura, *Organometallics*, **1**, 388 (1982).
[213] G. Erker, J. Wicher, K. Engel, F. Rosenfeldt, W. Dietrich, C. Krüger, *J. Am. Chem. Soc.*, **102**, 6344 (1980).
[214] G. Erker, J. Wicher, K. Engel, C. Krüger, *Chem. Ber.*, **115**, 3300 (1982).
[215] R. Beckhaus, K.-H. Thiele, *J. Organomet. Chem.*, **317**, 23 (1986).
[216] P. Czisch, G. Erker, H.-G. Korth, R. Sustmann, *Organometallics*, **3**, 945 (1984).
[217] P. C. Wailes, H. Weigold, A. P. Bell, *J. Organomet. Chem.*, **34**, 155 (1972).
[218] E. Samuel, M. D. Rausch, *J. Am. Chem. Soc.*, **95**, 6263 (1973).
[219] W. E. Hunter, D. C. Hrncir, R. V. Bynum, R. A. Penttila, J. L. Atwood, *Organometallics*, **2**, 750 (1983).
[220] P. J. Walsh, F. J. Hollander, R. G. Bergman, *J. Am. Chem. Soc.*, **110**, 8729 (1988).
[221] R. F. Jordan, *J. Organomet. Chem.*, **294**, 321 (1985).
[222] P. C. Wailes, H. Weigold, A. P. Bell, *J. Organomet. Chem.*, **33**, 181 (1970).

7 Group 5 (V, Nb, Ta) Metal Compounds

K. Tatsumi, *Nagoya University*

7.1 Introduction

The development of organometallic chemistry of the group 5 elements is a relatively more recent phenomenon than that of the neighboring transition elements. However, the rapid growth of this field has provided a wide variety of complexes having nearly any possible organic ligand, and they have unfolded a rich and beautiful chemistry of their own. Although the application of group 5 complexes to organic synthesis is still limited, certain complexes have been used for carbon–carbon bond forming reactions, e. g. polymerization of unsaturated hydrocarbons, cross-coupling of imines and ketones, and cyclization of dienes. Group 5 organometallics are now thought to be essential to understanding a number of metal-catalyzed processes. Representatives are a series of tantalum alkylidene complexes which played a central role in clarifying mechanisms of olefin metathesis reactions. More recently, some tantalum complexes afforded conceptual models for Fischer–Tropsch catalysis, and for carbon–nitrogen and carbon–sulfur bond breaking processes of hydrodenitrogenation and hydrodesulfurization catalysts, respectively. Like group 4 metal complexes, vanadium, niobium and tantalum complexes are strongly oxophilic, so that they are often unstable towards moisture and air.

7.2 Vanadium Complexes

The coordination chemistry of vanadium is strongly influenced by the redox properties of the metal center. Conversely, in order to selectively isolate desired complexes with a specific oxidation state in high yield, it is important to choose suitable reaction conditions and adequate auxiliary ligands. Low-valency vanadium complexes consist mostly of those containing carbonyl ligands, while the higher oxidation states (+4 and +5) of vanadium are often associated with an oxo ligand in the ubiquitous vanadyl (VO^{2+} and VO^{3+}) units. Thus a large number of vanadyl complexes have been synthesized. Vanadyl occurs in certain petroleums in the form of oxovanadium(IV) porphyrins, which cause problems in refineries. High-valency vanadium is also present in bioinorganic systems, and a high concentration of vanadium has been noticed in the blood of ascidians such as sea squirts and tunicates.

Synthesis of Organometallic Compounds: A Practical Guide. Edited by S. Komiya

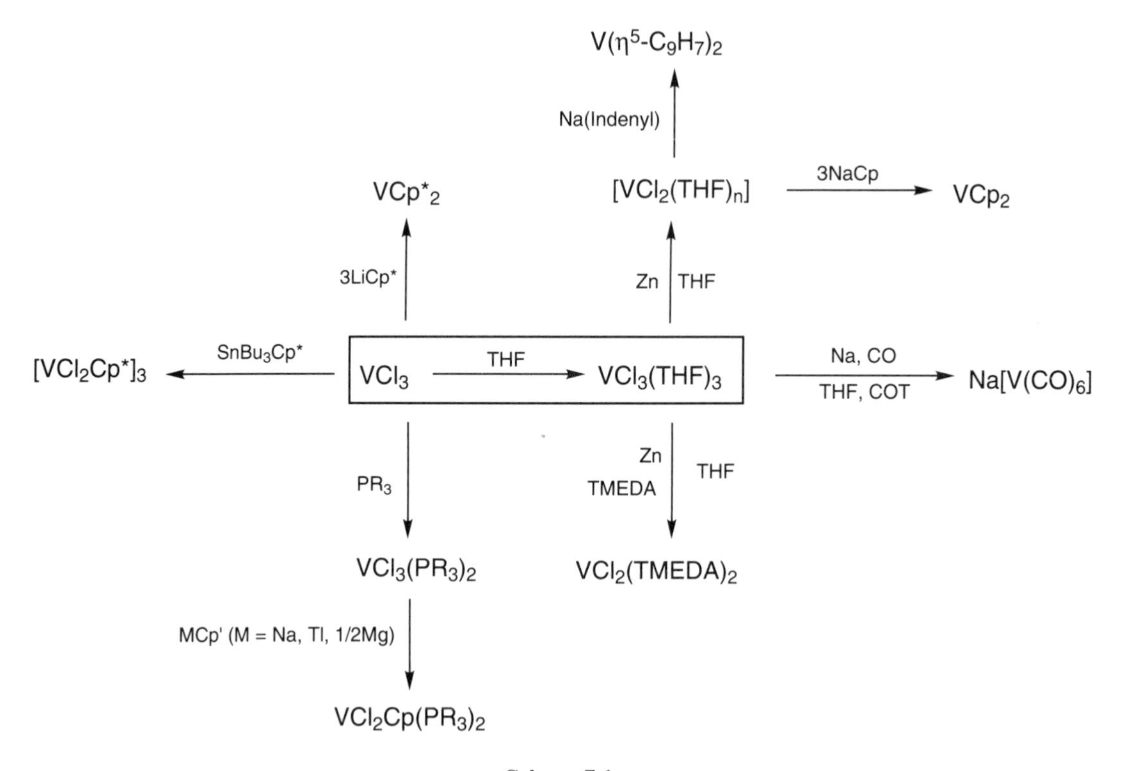

Scheme 7.1

The progress of organovanadium chemistry has been slower than that of the other first transition elements. This is partly because vanadium in most oxidation states tends to form paramagnetic complexes and their characterization by NMR is not very easy. Nevertheless, there are a good number of complexes having cyclopentadienyl and its derivatives. Typical examples include $VCp(CO)_4$, VCp_2, $VClCp_2$, and VCl_2Cp_2, where the oxidation state of vanadium spans +1 to +4. These cyclopentadienyl complexes have been used as convenient entries to other organovanadium complexes.

The basic precursors for organometallic complexes of vanadium, and for coordination complexes as well, are the commercially available V(III) chloride and its THF adduct, VCl_3 and $VCl_3(THF)_3$, as shown in Scheme 7.1. Reductive carbonylation of $VCl_3(THF)_3$ under atmospheric pressure of CO with sodium sand at room temperature leads to $Na[V(CO)_6]$ [1]. This reaction, however, requires a large quantity of cyclooctatetraene (cot) as a catalyst. Without cot, it is necessary to introduce CO at a high pressure and elevated temperature. The electronically "precise" anionic carbonyl complex can then be converted into the neutral 17-electron $V(CO)_6$, by treating $Na[V(CO)_6]$ with an acid, or by sublimation of the tetraalkyl ammonium salts in the presence of H_3PO_4 [2].

When the reduction of $VCl_3(THF)_3$ by zinc or magnesium is carried out in the presence of diphosphines, TMEDA (= N, N, N′, N′-tetramethylethylenediamine), or pyridine, V(II) chlorides of general formula trans-VCl_2L_4 (L_2 = diphos, TMEDA; L = py) are readily formed [3]. Likewise reduction of $VCl_3(THF)_3$ with zinc dust followed by addition of cyclopentadienide gives rise to VCp_2 [4]. The ring-substituted vanadocenes as well as bis(indenyl)vanadium (indenyl=η^5-C_9H_7) are prepared in a similar manner [5]. In the case of decamethylvanadocene, VCp^*_2 (Cp^* = pentamethylcyclopentadienyl), addition of excess $LiCp^*$ to a THF solution of $VCl_3(THF)_3$ gives a high yield of product [6]. The X-ray analysis of VCp^*_2 exhibits a normal sandwich structure.

Oxidation of VCp_2 with AgCl suspended in THF readily gives $VClCp_2$ and then VCl_2Cp_2 [7]. Vanadocene monochloride can also be synthesized by oxidation of VCp_2 with RCl or $CHCl_3$ [8], and the analogous $VClCp^*_2$ by oxidation of VCp^*_2 with Cl_2 or HCl [9]. Alternatively, these complexes are prepared from the reaction of the corresponding vanadocene and $PbCl_2$, where elemental lead is generated as the by-product [10]. It was reported that the direct reaction of VCl_3 with TlCp in THF provided $VClCp_2$ in high yield, if TlCp is sublimed prior to use [11].

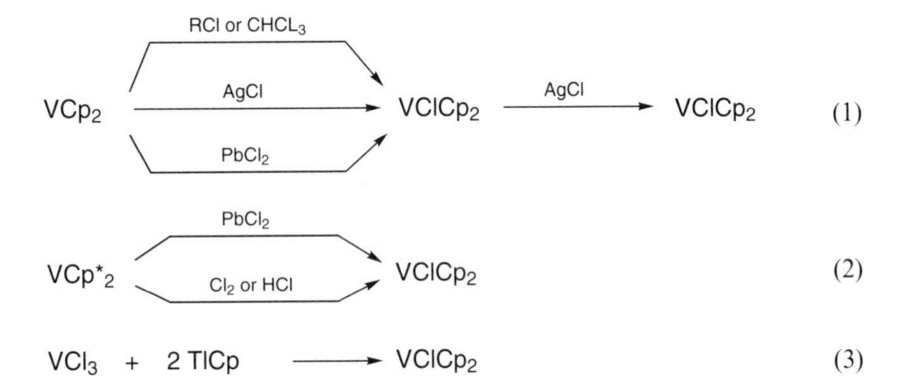

The reaction of $VCl_3(THF)_3$ with $SnBu_3Cp^*$ resulted in the halfsandwich V(III) complex, $[VCl_2Cp^*]_3$ [12]. Aside from $[VCl_2Cp^*]_3$, facile syntheses of monocyclopentadienyl vanadium dichloride, VCl_2Cp' or $VCl_2Cp'L_n$ (L = neutral donor ligand), from $VCl_3(THF)_3$ have not been reported. However, if the preformed alkylphosphine adducts $VCl_3(PR_3)_2$ (R = Me, Et) are used as precursors, the desired complexes of general formula $VCl_2Cp'(PR_3)_2$ (Cp' = C_5H_5, C_5H_4Me) can be prepared upon treating the adduct with NaCp′, TlCp′, or $\frac{1}{2}$ $MgCp'_2$ in THF [13].

Synthesis and isolation of homoleptic alkyl, aryl, and allyl complexes from $VCl_3(THF)_3$ is not well documented, except for a few thermally stable examples of this class (Scheme 7.2). The reaction of $VCl_3(THF)_3$ with excess LiPh in ether generated $Li_4VPh_6(Et_2O)_4$, the sole example of a homoleptic V(II) aryl complex [14]. During this reaction the vanadium center is reduced, and the four lithium cations in the complex cap the trigonal faces of the VPh_6 octahedron. The analogous reaction with $Li[2,6\text{-}(MeO)_2C_6H_3]$ gave rise to $V_2[2,6\text{-}(MeO)_2C_6H_3]_4$, where each vanadium is again reduced to V(II) [15]. In contrast to the mononuclear structure of $Li_4VPh_6(Et_2O)_4$, some of the methoxy groups and the aryl carbons link two metal centers of $V_2[2,6\text{-}(MeO)_2C_6H_3]_4$ resulting in a very short V—V triple bond. Interestingly, addition of three equivalent of MesMgBr (Mes = 2,4,6-$Me_3C_6H_2$) to $VCl_3(THF)_3$ in THF led to the formation of the mononuclear V(III) aryl complex, $V(Mes)_3(THF)$ [16]. When the alkylation was carried out with LiNp (Np = neopentyl) under an N_2 atmosphere, dinitrogen was incorporated, thus forming $(VNp_3)_2(N_2)$ [17]. The dinitrogen

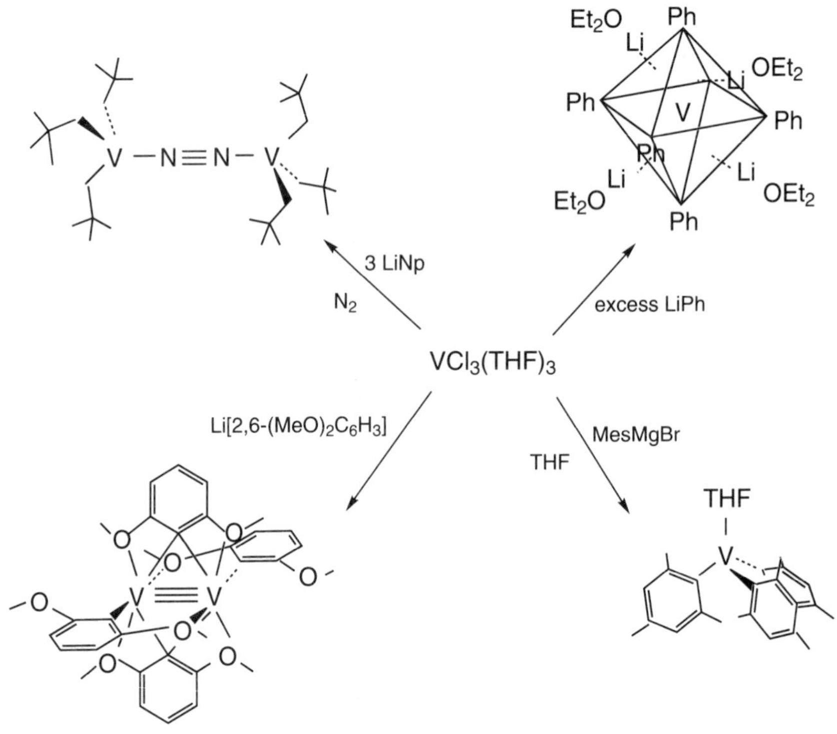

Scheme 7.2

bridges two vanadiums(III) with a linear V—N—N—V spine, and the N_2 molecule is labile.

It was reported that reaction of VCl_3 with $Mg(C_3H_5)X$ produces the homoleptic allyl complex $V(\eta^3\text{-}C_3H_5)_3$, which spontaneously decomposes even at low temperature [18]. The pentenyl analog, $V(\eta^3\text{-}C_5H_9)_3$, is slightly more stable.

The low-valency cyclopentadienyl vanadium complexes are usually stabilized by carbon monoxide, and $VCp(CO)_4$ is the most useful precursor to various low-valency organovanadium complexes. There are a number of synthetic routes to $VCp(CO)_4$. One convenient method is to carry out the reaction between NaCp and VCl_3 in THF *in situ* and then to carbonylate under 60 atm CO pressure at 120 °C [19]. Reduction of preformed vanadocene by potassium, and subsequent carbonylation also gives rise to $VCp(CO)_4$ [20]. These methods, however, cannot be applied to alkyl-substituted cyclopentadienyl derivatives. It is necessary to treat alkyl-substituted cyclopentadiene

$$VCp_2 \xrightarrow{\text{K or Na sand}} [VCp_2]^- \xrightarrow{CO} VCp(CO)_4 \qquad (4)$$

$$VCl_3 + NaCp \xrightarrow[\text{CO 60 atm}]{} VCp(CO)_4$$

$$V(CO)_6 + Cp'H \longrightarrow VCp'(CO)_4 \qquad (5)$$

directly with $V(CO)_6$, which produces the corresponding $VCp'(CO)_4$ in 45–85% yield, where Cp′ is C_5Me_5, C_5RH_4 (R = Me, Et, trityl), or indenyl, etc [21].

The substitution of CO in the 17-electron neutral carbonyl complex $V(CO)_6$ with P-donor ligands proceeds under mild conditions to form $V(CO)_5L$ (L = phosphine, phosphite) [22]. Since $V(CO)_6$ tends to be reduced to $[V(CO)_6]^-$, substitution reactions, particularly those with N - and O-donor Lewis bases, often lead to with disproportionation products formulated as $[VL_n][V(CO)_6]$ or $[VL_n][V(CO)_6]_2$ [23]. The disproportionation reactions occur more slowly for the phosphine derivatives $V(CO)_{6-n}(PR_3)_n$ [24]. On the other hand, photolysis of the 18- electron anionic complex $[V(CO)_6]^-$ in the presence of phosphines gives rise to the corresponding $[V(CO)_{6-n}(PR_3)_n]^-$ derivatives [25].

$$V(CO)_6 + PR_3 \longrightarrow V(CO)_5(PR_3) \tag{6}$$

$$V(CO)_6 \xrightarrow{L} [VL_n][V(CO)_6] \text{ or } [VL_n][V(CO)_6]_2 \tag{7}$$

L = N-donors, O-donors
n = 4 or 6

Certain organometallic complexes of high-valency vanadium can be synthesized from VCl_4 (Scheme 7.3) and $VOCl_3$ (Scheme 7.4). Reaction of VCl_4 with a cyclopentadienyl anion, e.g. MgCpCl, and its derivatives provides a general synthetic route to $VCl_2Cp'_2$ (Cp′ = C_5H_5, $C_5(CR_3)H_4$, $C_5(SiR_3)H_4$) [26]. Ring-bridged vanadocene dichlorides $[Me_3E(\eta^5\text{-}C_5H_5)_2]VCl_2$ (E = Si, Ge) have also been synthesized using analogous procedures [27]. Treatment of VCl_2Cp_2 with $SOCl_2$ leads to formation of the monomeric vanadium(IV) trichloride $VCpCl_3$ [28]. Alternatively, the trichloride derivatives $VCp'Cl_3$ (Cp′ = C_5H_5, C_5MeH_4, C_5Me_5, C_5Me_4Et) can be prepared in high yields by oxidation of $VCp'(CO)_4$ with Cl_2 [29].

$$VCp'(CO)_4 \xrightarrow{Cl_2} VCp'Cl_3 \tag{8}$$

Homoleptic alkyl and aryl complexes of vanadium(IV) can in principle be synthesized from VCl_4. However their thermal instability of makes them difficult to isolate. In some cases, the use of lithium alkyls and alkyl Grignards leads to reduction from V(IV) to V(III). Interestingly, [V(1-norbornyl)$_4$], prepared from VCl_4 and Li(1-norbornyl), is relatively stable up to 100 °C and is only moderately air sensitive [30]. The most air/ thermally stable complex of this class is [V(Mes)$_4$], which was isolated in nearly quantitative yield by air oxidation of Li[V(Mes)$_4$] [31].

$$V(Mes)_3THF \xrightarrow{Li(Mes)} Li[V(Mes)_4] \xrightarrow{air} V(Mes)_4 \tag{9}$$

Reduction of VCl_4 by $AlCl_3$ and Al powder (Friedel–Crafts method) in boiling benzene yields the vanadium(I) bis(benzene) complex salt, $[V(\eta\text{-}C_6H_6)_2][AlCl_4]$. The neutral bis (benzene) complex was prepared by hydrolysis of the salt [32].

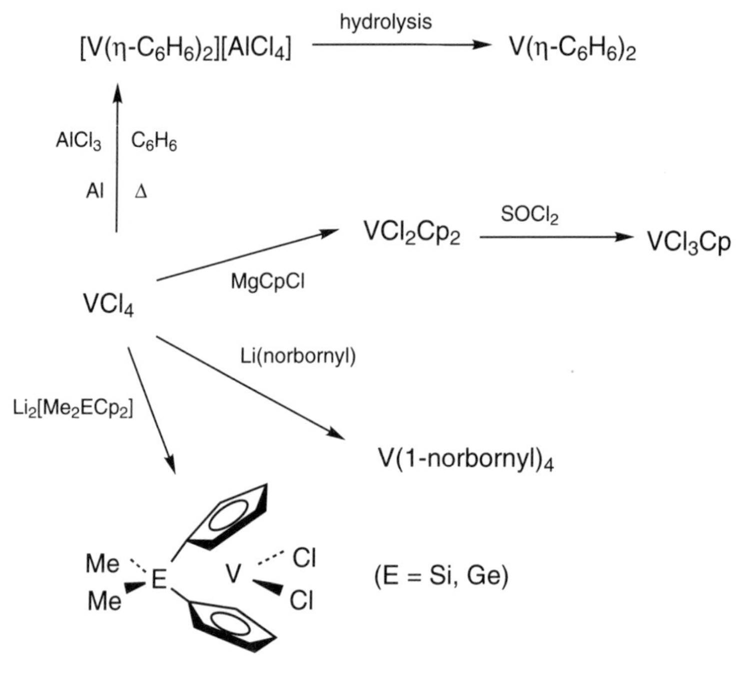

Scheme 7.3

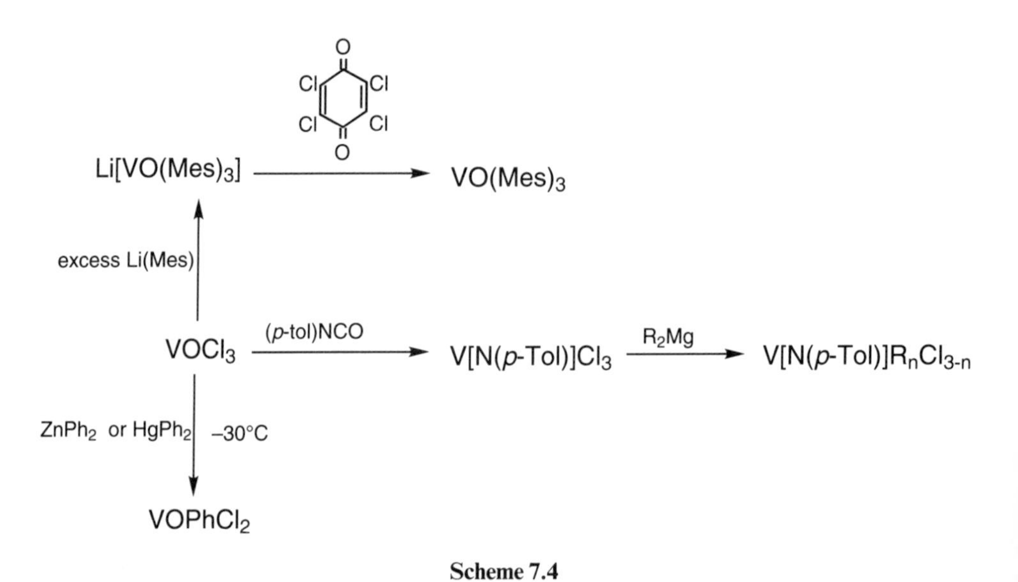

Scheme 7.4

Alkylation of $VOCl_3$ usually give rise to partially alkylated compounds which may decompose explosively. For instance, the low-temperature reaction of $VOCl_3$ with $ZnPh_2$ or $HgPh_2$ in pentane or in liquid propane produces $VOCl_2Ph$ in high yields [33]. In contrast, when the reaction of $VOCl_3$ is carried out with excess Li(Mes), paramagnetic Li[V(Mes)$_3$O] is formed which is in turn converted into neutral pentavalent vanadium species V(Mes)$_3$O by oxidation with chloranil [34]. Replacement of the oxo group of $VOCl_3$ with imido ligands facilitates isolation of alkyl complexes and relevant alkoxy and amide complexes. Introduction of imido ligands is achieved by treating $VOCl_3$ with *p*-substituted arylisocyanates, *p*-XC_6H_4NCO(X = Me, OMe), to give V[N(*p*-XC_6H_4)]Cl_3 [35]. Further alkylation, for instance, with $(Me_3SiCH_2)_2Mg$ leads to V[N(*p*-XC_6H_4)]$(Me_3SiCH_2)_nCl_{3-n}$ the product depending on the stoichiometric ratio.

7.3 Synthesis of Vanadium Compounds

All procedures must be conducted under inert atmosphere using dry solvents.

(1) *trans*-Dichlorobis(N, N, N′, N′-tetramethylethylenediamine)vanadium(II), *trans*-$VCl_2(TMEDA)_2$[3]

$$VCl_3 \xrightarrow[\text{THF}]{\text{Zn}} [V_2Cl_3(THF)_6]_2[Zn_2Cl_6] \xrightarrow[\text{THF}]{\text{TMEDA}} trans\text{-}V_2Cl_2(TMEDA)_2 \quad (10)$$

The original method to prepare the title V(II) complex requires two steps. The first step consists of reduction of $VCl_3(THF)_3$ with zinc powder [36], and then the resulting complex salt $[V_2Cl_3(THF)_6]_2[Zn_2Cl_6]$ is treated with TMEDA in THF. Thus a 200 mL Schlenk tube equipped with a reflux condenser and a magnetic stirring bar is charged with THF(45 mL), pre-dried zinc powder (2.4 g, 37 mmol), and VCl_3 (5.5 g, 35 mmol). The THF suspension is stirred for 18 h at room temperature, and the precipitate becomes green in color. If the reaction does not proceed smoothly, the THF suspension may be refluxed. After the supernatant is decantated off, the green powder is dissolved in CH_2Cl_2 to give a dark green solution. The solution is filtered carefully to remove zinc chloride and unreacted zinc powder, and partial removal of the solvent followed by addition of hexane gives $[V_2Cl_3(THF)_6]_2[Zn_2Cl_6]$ as a green powder (13.1 g, 8.05 mmol, 92%).

To a THF (120 mL) suspension of $[V_2Cl_3(THF)_6]_2[Zn_2Cl_6]$ (12.3 g, 7.6 mmol) in a three-necked round-bottomed 200 mL flask, equipped with a reflux condenser and a magnetic stirring bar, 30 mL of TMEDA is added. The THF suspension is refluxed for 1 h and is filtered. Light-blue micro crystals are formed from the resulting light-blue solution overnight in a refrigerator. Analytically pure large-scale crystals are obtained after recrystallization from THF (150 mL) containing a small amount of TMEDA. Yield 41%.

Propeties: IR(Nujol mull, KBr, cm^{-1}): 1380 (m), 1360(s), 1290(w), 1280(s), 1235(s), 1185(s), 1150(s), 1120(s), 1105(w), 1100(w), 1070(s), 1045(m), 1015(s), 955(s), 920(m), 775(s), 720(w), 595(s), 490(s), 470(s), 460(w). μ_{eff}: 3.68μ_B. Highly air and moisture sensitive.

The title complex can also be prepared from the reaction of $VCl_3(THF)_3$ with Mg metal in THF and in the presence of TMEDA. A THF suspension (10 mL) of

$VCl_3(THF)_3$ (0.16 g, 0.43 mmol) becomes a homogeneous purple solution upon introduction of TMEDA (0.15 mL). Addition of Mg (0.13 g, 5.3 mmol) to this solution results in a color change from purple to light blue in 30 min. Stirring the suspension for 2 h, followed by a workup similar to the above, gives *trans*-$VCl_2(TMEDA)_2$ in 50–60% yields [37].

Analogous V(II) dichlorides, *trans*-VCl_2(N, N,N′-trimethylethylenediamine)$_2$, *trans*-$VCl_2(pyridine)_4$ and *trans*-$VCl_2(pyrrolidine)_2$ can be prepared in a similar manner. In the case of $VCl_2(pyridine)_4$, both reduction of $VCl_3(THF)_3$ by zinc powder and addition of pyridine may be carried out in one pot, so that it is not necessary to isolate $[V_2Cl_3(THF)_6]_2[Zn_2Cl_6]$.

The chemistry of cyclopentadienyl-free and CO-free low-valent organovanadium complexes has been little explored due to synthetic difficulties. Thus readily available *trans*-$VCl_2(TMEDA)_2$ is a convenient entry to rare divalent organovanadiums. Reaction of $VCl_2(TMEDA)_2$ with two equivalents of Li[*o*-$C_6H_4CH_2NMe_2$] and subsequent addition of one equivalent of pyridine under nitrogen atmosphere forms the dinitrogen complex $[V(o\text{-}C_6H_4CH_2NMe_2)_2(py)_2]_2(\mu\text{-}N_2)_2$ [38].

$$trans\text{-}VCl_2(TMEDA)_2 \xrightarrow[\text{ether}]{Li[o\text{-}C_6H_4CH_2NMe_2]} \xrightarrow[N_2]{\text{pyridine}} [V(o\text{-}C_6H_4CH_2NMe_2)_2(py)_2]_2(\mu\text{-}N_2)_2 \quad (11)$$

The V(II) tetrakis- and bis(alkynide) complexes, $[Li(TMEDA)]_2[V(\mu\text{-}C{\equiv}CPh)_4(TMEDA)]$ and $V(C{\equiv}CPh)_2(TMEDA)_2$ have been synthesized from the reaction of $VCl_2(TMEDA)_2$ with four and two equivalents of Li(C≡CPh), respectively [39].

(2) Tetracarbonyl(η^5-pentamethylcyclopentadienyl)vanadium(I), $V(\eta^5\text{-}C_5Me_5)(CO)_4$ [21a]

$$V(CO)_6 + C_5Me_5H \xrightarrow[\text{hexane}]{} V(\eta^5\text{-}C_5Me_5)(CO)_4 + 2\,CO + 1/2\,H_2 \quad (12)$$

Vanadium hexacarbonyl, $V(CO)_6$ (2.42 g, 1L mmol) is dissolved in 250 mL of hexane in a three-necked round-bottomed 500 mL flask equipped with a condenser. Pentamethylcyclopentadiene (2.7 mL, 17 mmol) is added via syringe, and the solution is refluxed for 3 h. After the solution has been evaporated under reduced pressure to *ca.* 30 mL, it is loaded onto a silica gel column (Merck grade 60, 70–230 mesh, 60 × 1.6 cm). The desired complex is eluted with pentane/toluene (10:1) as a red-orange band, and the solvent is removed *in vacuo*. The complex may be recrystallized by redissolving the resulting solid in pentane, and refrigerating the concentrated solution, to give $V(\eta^5\text{-}C_5Me_5)(CO)_4$ as reddish-orange needles. Yield 45%.

Properties: IR(KBr, cm^{-1}): ν(CO); 2001 (m), 1990(s). ^{1}H-NMR (THF-d_8, 28°C, δ): 1.93 (s, CH_3). mp 130–131 °C. Air- and moisture-sensitive.

The analogous vanadocene–carbonyl complex $V(\eta^5\text{-}C_5H_5)(CO)_4$ can be synthesized from reduction of $V(\eta^5\text{-}C_5H_5)_2$ with Na or with K under atmospheric CO [20]. The

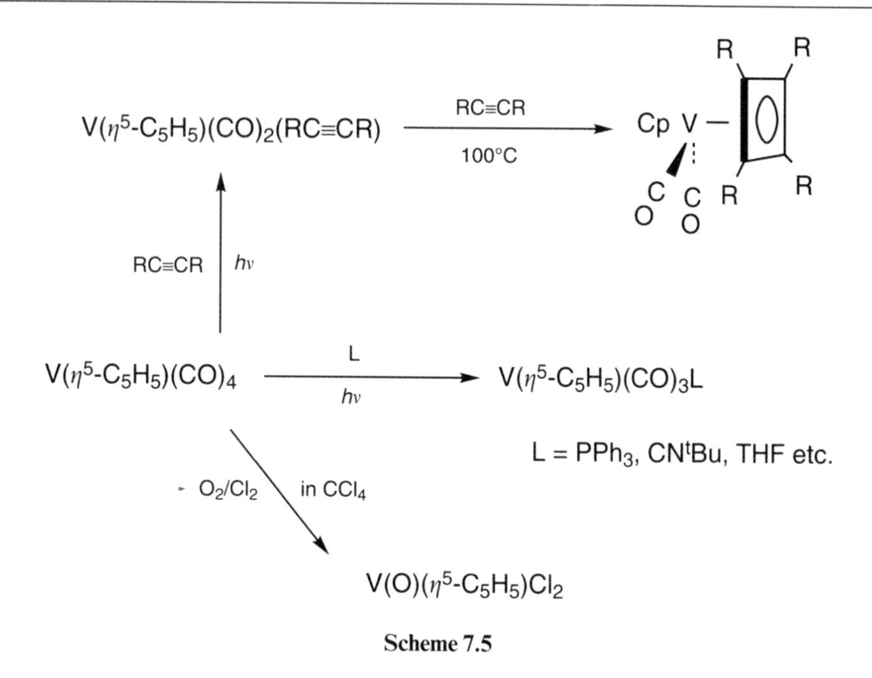

Scheme 7.5

procedure is rather straightforward, and the complex can be purified by sublimation (0.1 – 1 mmHg, 80–100°C). Yield 75%. mp 138 °C.

One or more carbonyl ligands in $V(\eta^5\text{-}C_5Me_5)(CO)_4$ and $V(\eta^5\text{-}C_5H_5)(CO)_4$ can be replaced with a number of nucleophiles upon photolysis at low temperatures. Selected reactions of $V(\eta^5\text{-}C_5H_5)(CO)_4$ are shown in Scheme 7.5 [40].

(3) Bis(η⁵-pentamethylcyclopentadienyl)vanadium(II), $V(\eta^5\text{-}C_5Me_5)_2$ [6]

$$
\begin{aligned}
&VCl_3 + 3\,Li(C_5Me_5) \xrightarrow[THF]{} V(\eta^5\text{-}C_5Me_5)_2 + 3\,LiCl + Me_5C_5\bullet \\
&VCl_2(THF)_n + 2\,Na(C_5Me_5) \xrightarrow[THF]{} V(\eta^5\text{-}C_5Me_5)_2 + 2\,NaCl
\end{aligned}
\qquad (13)
$$

Pentamethylcyclopentadiene (30.0 g, 220.6 mmol) and 50 mL of THF are placed in a 200 mL Schlenk tube equipped with a magnetic stirring bar. To this solution, 240 mmol of LiMe is introduced dropwise under vigorous stirring, and a light yellow solid gradually precipitates. After the addition of LiMe is complete, the suspension is kept stirring for 2 h. The light–yellow solid obtained is separated from the supernatant liquid by filtration or by a centrifuge, and is washed with ether. The resulting white solid is dried *in vacuo* for several hours to give $Li(C_5Me_5)$ (25.5 g), which probably contain THF molecules coordinating to Li. It is essential to use this preformed white solid in order to synthesize $V(\eta^5\text{-}C_5Me_5)_2$ in high yields according to the procedure described below.

A THF (500 mL) suspension of VCl_3 (5.25 g, 33.36 mmol) and $Li(C_5Me_5)$ (14.2 g, 100 mmol) is refluxed for 2 days in a three-necked round-bottomed 1000 mL flask equipped with a condenser. After the solid is filtered off, the mother liquid is dried *in*

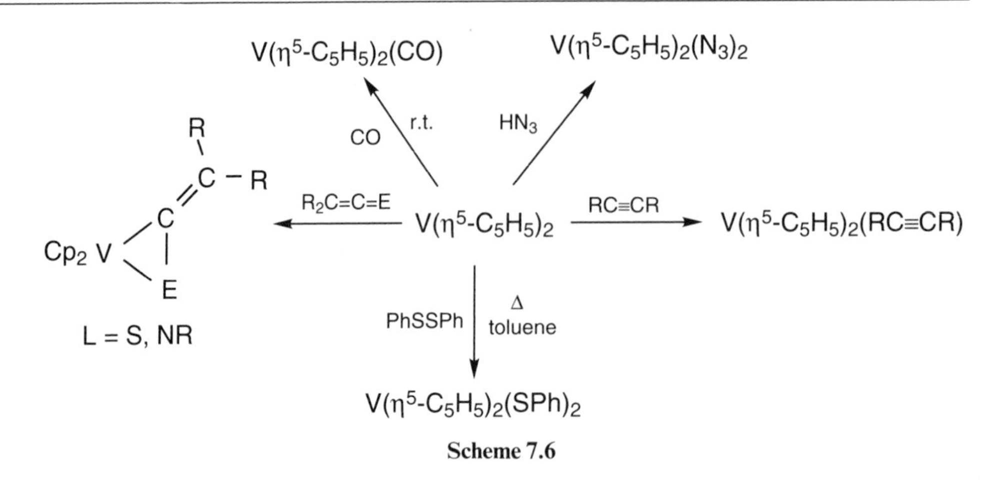

Scheme 7.6

vacuo. The residue is then sublimed at 170–190 °C to give $V(\eta^5\text{-}C_5Me_5)_2$ as a red solid. The compound can be recrystallized from n-hexane or pentane. Yield 70 %.

Alternatively, the title complex may be synthesized from the reaction between $VCl_2(THF)_n$ and $Na(C_5Me_5)$ according to the following procedure [41].

In a 100 mL Schlenk tube a suspension of $VCl_2(THF)_n$ in THF (50 mL) is prepared from VCl_3 (3.11 g, 19.8 mmol) and zinc dust (0.65 g, 9.9 mmol) according to the method described by Kohler and Prössdorf [4]. The stirred suspension is transferred through a cannula to a THF (100 mL) solution of $Na(C_5Me_5)$ (4.00 g, 25.3 mmol) in a 200 mL Schlenk tube equipped with a condenser, and the mixture is refluxed for 7 h. The resulting dark purple solution is dried *in vacuo*, and the residue is heated under vacuum (60 °C, 10^{-3} Torr) for 6 h to remove oily impurities. The solid is then extracted with pentane (50 mL) , washed with pentane until washings become colorless, and the extracts are filtered. Solvent is removed under reduced pressure to give a red microcrystalline solid. Sublimation (100 °C, 10^{-5} Torr) and subsequent recrystallization from pentane give $V(\eta^5\text{-}C_5Me_5)_2$ as dark red prisms. Yield 65 %. $Li(C_5Me_5)$ may be substituted for $Na(C_5Me_5)$, but with a significant reduction in yield.

Properties: IR(Nujol mull, KBr, cm^{-1}): 2989 (m), 2940(m), 2895(s), 2850(m), 2750(w), 1470(m), 1448(m), 1422(m), 1373(m), 1355(w), 1065(m), 1023(w), 722(w), 587(w), 463(m), 422(w), 233(w). μ_{eff}: 3.78μ_B. mp 300 °C. Air- and moisture-sensitive.

Non-substituted vanadocene $V(\eta^5\text{-}C_5H_5)_2$ as well as $V(\eta^5\text{-}C_5MeH_4)_2$ can be prepared by the second route using the Na and Li salts of the corresponding cyclopentadienyl [4]. Scheme 7.6 shows typical reactions of $V(\eta^5\text{-}C_5H_5)_2$, some of which occur also with $V(\eta^5\text{-}C_5Me_5)_2$ [6,8a,42].

(4) Dichloro(η5-cyclopentadienyl)bis(triethylphosphine)vanadium(II), $V(\eta^5\text{-}C_5H_5)Cl_2(PEt_3)_2$ [13]

$$VCl_3(THF)_3 + 2\ PEt_3 \xrightarrow[-3\ THF]{} VCl_3(PEt_3)_2 \xrightarrow[-1/2\ MgCl_2]{1/2\ Mg(C_5H_5)_2} V(\eta^5\text{-}C_5H_5)Cl_2(PEt_3) \quad (14)$$

The title complex can be prepared from the reaction of the preformed phosphine adduct $VCl_3(PEt_3)_2$ with $Mg(C_5H_5)_2$ or from the one-pot reaction of PEt_3, $VCl_3(THF)_3$, and $Mg(C_5H_5)_2$.

THF is syringed into a 200 mL Schlenk tube containing $VCl_3(THF)_3$ (7.78 g, 20.8 mmol) with stirring. Subsequent addition of PEt_3 (6.20 mL, 42.1 mmol) at room temperature results in a dark solution, which is stirred for 3 h and then is cooled slowly to −80 °C. Crude $VCl_3(PEt_3)_2$ (6.86 g, 81% yield) separates as dark crystals. The compound may be recrystallized from pentane to give analytically pure dark red crystals. The title complex is synthesized by treating this phosphine adduct with $Mg(C_5H_5)_2$ in THF.

The general procedure described below can be applied preparation of $V(C_5H_5)Cl_2(PR_3)_2$ (R = Me, Et) on a large scale. Since isolation of $VCl_3(PR_3)_2$ prior to the reaction with $Mg(C_5H_5)_2$ may lower the overall yields, the following procedure uses $VCl_3(PR_3)_2$ which is prepared *in situ* without separation.

Addition of PR_3 (200 mL) to a THF solution (1200 mL) of $VCl_3(THF)_3$ (100 mmol) at room temperature results in a dark solution after stirring for 3 h. The solution is cooled to −80 °C, and $Mg(C_5H_5)_2$ (50 mmol) in 100 mL of THF is added dropwise over 2 h. Upon warming the reaction mixture slowly to room temperature, the color of the solution turns blue. The mixture is stirred for another 2 h at room temperature. Removal of solvent *in vacuo* gives a sticky residue, which is stirred with 200 mL of pentane for 15 min. Solvent is evaporated from the extract to remove the remaining THF. The resulting solid is carefully powdered and extracted with pentane. The blue pentane extract is cooled slowly to −25 °C and the blue $V(C_5H_5)Cl_2(PR_3)_2$ crystallizes out. Yield 71% (R=Me), 88% (R=Et).

Properties: $V(C_5H_5)Cl_2(PMe_3)_2$: ^{1}H-NMR (toluene-d_8, 20°C, $\delta(\Delta\nu_{1/2}$ in Hz)); 17.6 (1085) (PMe_3). mp. 198 °C. $V(C_5H_5)Cl_2(PEt_3)_2$: ^{1}H-NMR (toluene-d_8, 20 °C, δ ($\Delta\nu_{1/2}$ in Hz)); 18.8 (2280) ($P(CH_2CH_3)_3$), −1.49 (310) ($P(CH_2CH_3)_3$). mp 92 °C. Both compounds are air- and moisture-sensitive.

Use of $VBr_3(THF)_3$, instead of its chloride congener, generates $V(C_5H_5)Br_2(PR_3)_2$. However, attempts to prepare analogous complexes with PPh_3, PPh_2Me, $PPhMe_2$, Pcy_3, DMPE, and DPPE failed. On the other hand, although introduction of methyl substituted cyclopentadienyl is possible upon using $Na(C_5Me_5)$, pentamethylcyclopentadienyl derivatives cannot be synthesized in this manner with $Li(C_5Me_5)$, $Na(C_5Me_5)$, or $Mg(C_5Me_5)Cl$.

(5) Tris(2,4,6-trimethylphenyl)tetrahydrofuranvanadium(III), $V(2,4,6\text{-}Me_3C_6H_2)_3(THF)$ [16]

$$VCl_3(THF)_3 + 3\,Mg(2,4,6\text{-}Me_3C_6H_2)Br \xrightarrow[-2\,THF]{-3\,MgBr} V(2,4,6\text{-}Me_3C_6H_2)_3(THF) \qquad (15)$$

A 300 mL THF solution of the Grignard reagent, MesMgBr (Mes = 2,4,6-$Me_3C_6H_2$), is prepared from the reaction of mesitylbromide MesBr(75 g) and Mg (12g) in THF followed by filtration. To the stirred solution, a 200 mL THF solution of $VCl_3(THF)_3$ is

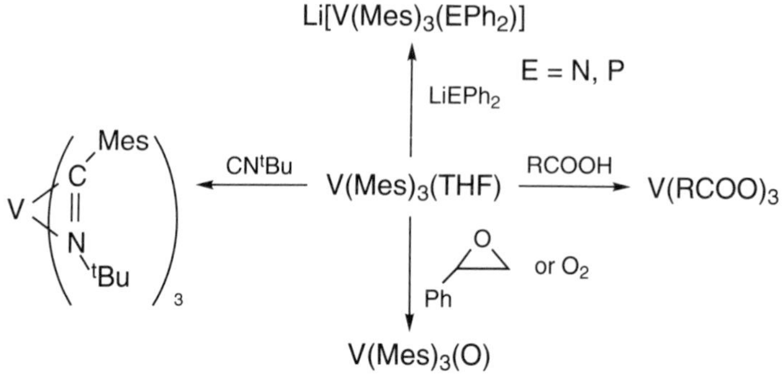

Scheme 7.7

added dropwise at room temperature, and the mixture is stirred for 2 h to give a blue solution. Addition of dioxane leads to precipitation of the magnesium salt which is separated by filtration. When the resulting blue solution is allowed to stand at −20 °C, analytically pure blue crystals of $V(Mes)_3(THF)$ are formed. Yield 50–60%.

Properties: IR(Nujol mull, cm^{-1}): 3010 (m), 2970(m), 2925(s), 2850(m), 1738(m), 1618(m), 1598(m), 1545(m), 1450(s), 1404(m), 1382(m), 1283(m), 1260(w), 1224(m), 1064(m), 1040(m), 1008(s), 946(w), 912(w), 874(s), 860(ss), 712(w), 705(w), 690(w), 583(m), 540(m). μ_{eff}: 2.73μ_B. mp. 300 °C. Air- and moisture-sensitive.

The title complex is a rare example of a fully characterized homoleptic vanadium aryl (and alkyl) complexes which are fully characterized, and its reactivity has been widely investigated. Some reactions are shown in Scheme 7.7 [43].

7.4 Niobium and Tantalum Complexes

Compared to vanadium, the higher oxidation states are much more favored in niobium and tantalum, and the majority of their organometallic complexes occur with oxidation state of +5. Strong oxophilicity and resistance to reduction are properties shared with the neighboring zirconium and hafnium. In the lower oxidation states, the heavy group 5 transition metals are prone to form metal–metal bonds, a trend which resembles that of the group 6 elements. These group 5 and group 6 metals also share an ability to form multiple bonds with non-metal ligands, such as carbenes (or alkylidenes), nitrenes (imides), oxides and sulfides. Roughly speaking, in their high oxidation states niobium and tantalum exhibit chemistries similar to those of the group 4 transition metals, while in the lower oxidation states their chemistries are comparable to those of the group 6 transition metals.

The most important starting materials for organometallic complexes of niobium and tantalum are their pentachlorides, $NbCl_5$ and $TaCl_5$. The commercially available pentachlorides can be sublimed if further purification is necessary. They are commonly used for access to both high-valency and low-valency compounds. The crystal structure of $NbCl_5$ consists of dimeric Nb_2Cl_{10} with two chlorides bridging as shown in Scheme 7.8, and the Ta analogue is presumed to have a similar structure. In non-

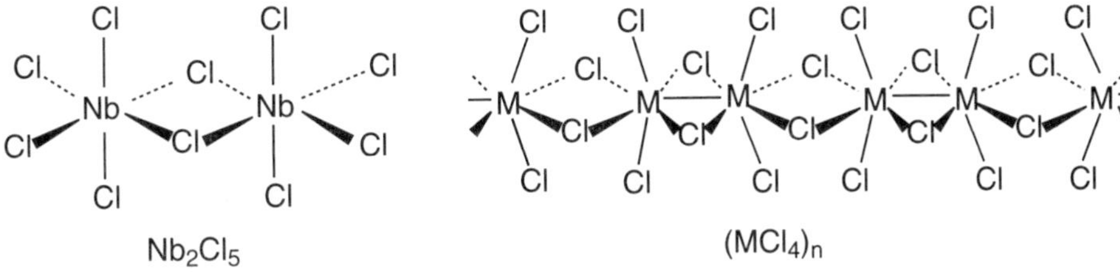

Nb_2Cl_{10} $(MCl_4)_n$

Scheme 7.8

complexing solvents, the dimeric structure could be retained, while in donor solvents monomeric adducts are formed. In the gas phase, MCl_5 is monomeric with a trigonal bipyramid structure. The pentachlorides are strongly electrophilic, and catalyze Friedel–Crafts type reactions. Since the MCl_5 compounds promotes polymerization of THF, albeit slowly, it is advisable to avoid using THF as a solvent for their reactions. Although neutral adducts of the pentachlorides MCl_5L (L = nitriles, ethers, phosphine oxides, amines, organic sulfides, etc) are available [44] reactions with O-donors occasionally produce oxo trichloro complexes as well, particularly in the case of niobium.

The tetrachlorides of niobium and tantalum offer attractive entries into chemistry of oxidation state +4 or lower. They form one-dimensional polymers consisting of edge-sharing polyoctahedra, which exhibit bond alternation with metal–metal bonding pairs (Scheme 7.8). Because of the chain structure, the relatively air stable tetrachlorides are diamagnetic and poorly soluble in most organic solvents. Their more soluble molecular adducts such as $NbCl_4(THF)_2$ and $NbCl_4(MeCN)_2$ are available, while the phosphine adducts often occur as dinuclear complexes formulated as $Nb_2Cl_8(PR_3)_2$ or $Nb_2Cl_8(PR_3)_4$. The dinuclear structure contains a metal–metal single bond, and may be regarded as the remnant of the polymeric MCl_4 structure. Tantalum appears to be less prone to adopt dinuclear structures, and forms the seven-coordinate $TaCl_4(PR_3)_3$ or six-coordinate $TaCl_4(PR_3)_2$ depending on the size of the phosphines.

The majority of low-valency complexes of niobium and tantalum contain carbonyl ligand(s). Carbonyl complexes of group 5 metals are usually less stable than those of the corresponding group 6 carbonyls, particularly so for heavier group 5 elements. While the neutral hexacarbonyl complex of vanadium, $V(CO)_6$, is accessible and constitutes a convenient entry to low-valency organovanadiums, its niobium and tantalum congeners have not been isolated. On the other hand, anionic 18-electron complexes $[M(CO)_6]^-$ (M=Nb, Ta) are stable, and can be prepared on a large scale according to eq (16). Reduction of the pentahalides by sodium naphthalenide or

$$MCl_5 \xrightarrow[\text{CO 1 atm}]{M'[C_{10}H_8]} M'[M(CO)_6] \quad (\text{M = Nb, Ta; M}' = \text{Na, K}) \tag{16}$$

potassium naphthalenide is followed by addition of an atmospheric pressure of CO [45]. Previously reported route to $[M(CO)_6]^-$ consists of a similar reduction of the pentachlorides with Na/K alloy in the presence of diglyme. However this method requires a high pressure of CO (5000–6000 psi) and yields are low (*ca* 15%) [46].

Typical reactions of $[M(CO)_6]^-$ are shown in Scheme 9. Photochemical replacement of CO by phosphines gives a series of phosphine-substituted derivatives formulated as $[M(CO)_5PR_3]^-$, and the analogous reactions with diphosphines lead to the tetracar-

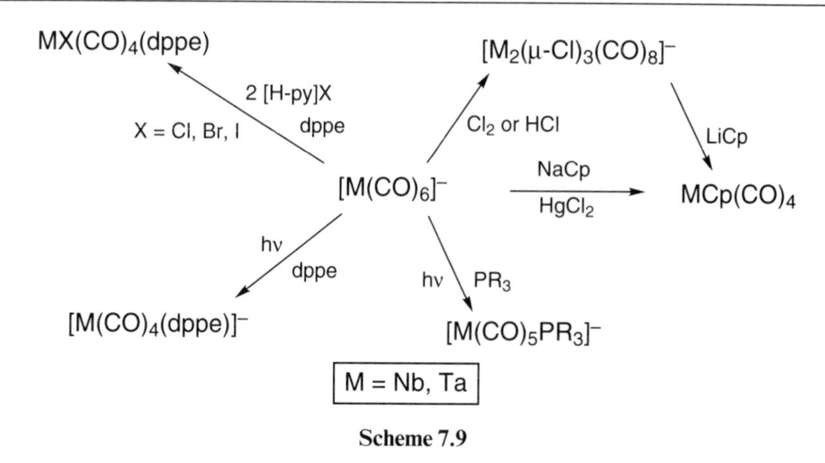

Scheme 7.9

bonyl complexes $[M(CO)_4(P\text{–}P)]^-$ [47]. Treatment of $[M(CO)_6]^-$ with Cl_2 or HCl produces the binuclear chloride/carbonyl complexes $[M_2(\mu\text{-}Cl)_3(CO)_8]^-$, where the metal is oxidized to the +1 state. When similar oxidation reactions with I_2 are carried out under the presence of diphosphines, mononuclear 7-coordinate complexes $MI(CO)_4(P\text{–}P)$ are obtained, which can also be synthesized by addition of diphosphines to a THF solution of $[M_2(\mu\text{-}I)_3(CO)_8]^-$ at low temperatures. Use of 2 equivalents of pyridinium halides [H-py]X and 1 equivalent of diphosphines provides a more general synthetic route to $MX(CO)_4(P\text{–}P)$ (X = Cl, Br, I) [48].

Another important entry to group 5 metal carbonyls is provided by the neutral 18-electron complexes of the type $MCp(CO)_4$ (M=Nb, Ta) and their ring-substituted derivatives. The cyclopentadienyl complexes are synthesized in moderate yields from the reaction of $[M(CO)_6]^-$ with NaCp in the presence of $HgCl_2$ as shown in Scheme 7.9 [49]. Alternatively high yields of $MCp(CO)_4$ can be obtained by oxidation of $[M(CO)_6]^-$ with Cl_2 or HCl, which generates the aforementioned binuclear anionic complex $[M_2(\mu\text{-}Cl)_3(CO)_8]^-$, and subsequent reaction with LiCp [50]. The ring-substituted derivatives are generally prepared by Zn-promoted reductive carbonylation of $MCp'Cl_4$ under 300–400 atm CO pressure (eq (17)) [51a]. Successive introduction of methyl groups

$$MCp'Cl_4 \xrightarrow[\text{CO 300–400 atm}]{\text{Zn}} MCp'(CO)_4 + 2\,ZnCl_2 \qquad (17)$$

in the ring seems to facilitate the reductive carbonylation. For tantalum, an improved route to $TaCp^*(CO)_4$ (Cp* = pentamethylcyclopentadienyl) has been developed as shown in eq (18), where Na/Hg reduction of $TaCp^*Cl_4$ is attained under 1 atm CO in the presence of PMe_3 [51b].

$$TaCp^*Cl_4 \xrightarrow[\text{CO 1 atm}]{\text{Na/Hg, PMe}_3} \underset{47\%}{TaCp^*(CO)_4} + \underset{7\%}{TaCp^*(CO)_3(PMe_3)}$$

$$TaCp^*Cl_4 \xrightarrow{\text{PMe}_3,\ \text{Mg}} \underset{66\%}{TaCp^*Cl_2(PMe_3)_2} \xrightarrow{\text{CO 1 atm}} \underset{94\%}{TaCp^*Cl_2(CO)_2(PMe_3)_2} \xrightarrow[\text{CO 1 atm}]{\text{Na/Hg}} \underset{65\%}{TaCp^*(CO)_4} \qquad (18)$$

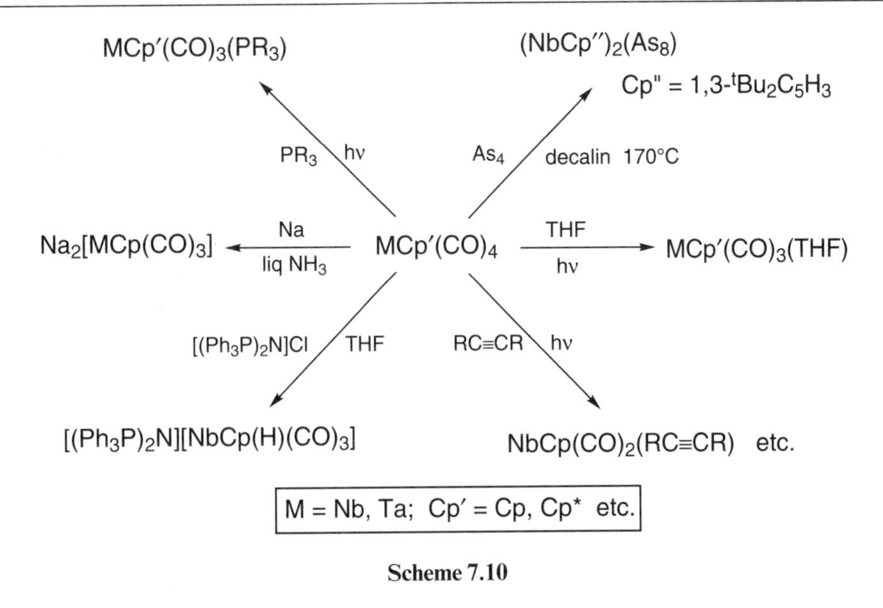

Scheme 7.10

Scheme 7.10 shows representative reactions of $MCp'(CO)_4$ (Cp′ = Cp, Cp*). Treatment of $MCp(CO)_4$ with Na in liquid NH_3 generates $Na_2[MCp(CO)_3]$, while a hydride complex of niobium $[(Ph_3P)_2N][NbCp(H)(CO)_3]$ is formed upon reduction of $NbCp(CO)_4$ with Na/Hg in THF followed by a cation exchange with $[(Ph_3P)_2N]Cl$ [52]. Upon refluxing a decalin solution of $MCp'(CO)_4$, the carbonyls are replaced by polypnictogens such as As_n [53]. Photosubstitution reactions of $MCp(CO)_4$ readily occur with various phosphines and diphosphines to give $MCp(CO)_3PR_3$ and $MCp(CO)_2(P\text{–}P)$, respectively. Likewise, photolysis of $NbCp(CO)_4$ in the presence of alkynes leads to a series of alkyne complexes, $NbCp(CO)_2(RC{\equiv}CR)$, $NbCp(CO)(RC{\equiv}CR)_2$, and $[NbCp(CO)(\mu\text{-}RC{\equiv}CR)]_2$ [54]. The analogously prepared THF adduct, $NbCp(CO)_3(THF)$, serves as another useful precursor which, for instance, reacts readily with H_2S and CH_3SH to give the dinuclear sulfide and thiolate complexes (eq (19)) [55]. Interestingly,

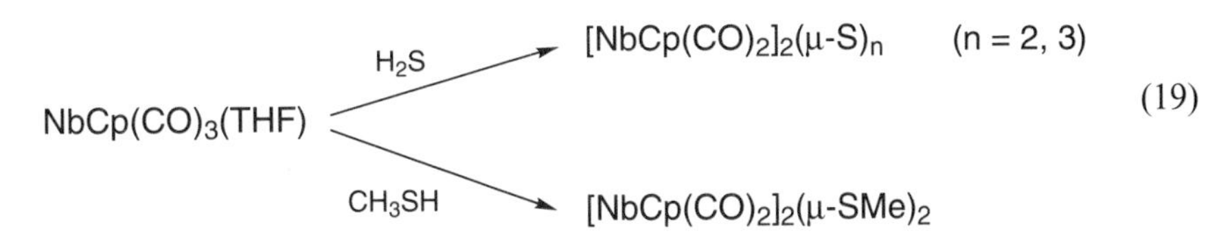

(19)

$NbCp(CO)_3(THF)$ can be handled at ambient temperatures, although the vanadium analog $VCp(CO)_3(THF)$ is unstable and decomposes above –20 °C,

Cyclopentadienyl/chloride complexes of the types $MCp'Cl_4$ and MCp'_2Cl_2 are extremely versatile starting materials for organoniobium and organotantalum chemistry with the oxidation states of +5 and +4. A large number of interesting derivatives have been prepared from these chloride complexes. The first convenient syntheses of the tetrachlorides, which appeared in 1977, employed 1 equivalent of tin reagents, as shown in eq (20) [56a,b,c]. It was later found that corresponding trimethylsilyl reagents

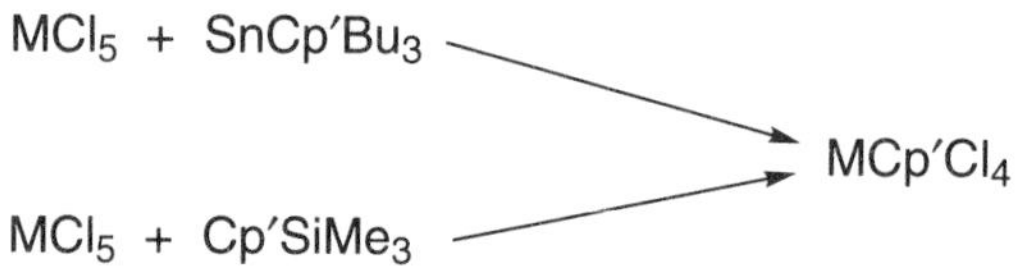

$$MCl_5 + SnCp'Bu_3 \longrightarrow MCp'Cl_4 \longleftarrow MCl_5 + Cp'SiMe_3 \qquad (20)$$

M = Nb, Ta; Cp′ = Cp, Cp* etc.

(1 equivalent) react readily with the pentachlorides to give $Mcp'Cl_4$ [56d,e]. If the analogous reactions are carried out between $NbCl_5$ and an excess of sodium or lithium salts of cyclopentadienyl, biscyclopentadienyl chloride complexes $NbCp_2Cl_2$ tend to be formed (eq (21)) [56a,b,f,g]. These sodium and lithium salts promote reduction of the metals. Alternatively, niobocenedichloride can be prepared from the reaction of $NbCl_5$ with two equivalents of the tin reagent, or by use of the pre-reduced Nb(IV) chloride, $NbCl_4(THF)_2$ [56h]. The tantalum complex, $TaCp_2Cl_2$, is available by basically similar preparative routes.

$$NbCl_5 + \text{excess NaCp} \longrightarrow NbCp_2Cl_2$$
$$NbCl_5 + 2\,Sn(MeCp)Bu_3 \longrightarrow Nb(MeCp)_2Cl_2 \qquad (21)$$
$$NbCl_5 \xrightarrow{Al} \xrightarrow{THF} NbCl_4(THF)_2 \xrightarrow{2\,LiCp} NbCp_2Cl_2$$

Addition of $NaBH_4$ to the $NbCl_5$/LiCp* reaction system generates $NbCp^*_2BH_4$, which provides a convenient entry to $NbCp^*_2Cl_2$ and $NbCp^*_2H_3$ as shown in Scheme 7.11 [57a]. This strategy does not apply to the preparation of the Ta congeners. However, reduction of $TaCp^*Cl_4$ with magnesium in the presence of PMe_3 gives rise to $TaCp^*Cl_3(PMe_3)$, and the subsequent replacement of one chloride with Cp* leads to $TaCp^*_2Cl_2$. This method allows us to synthesize the mixed-ring complex $TaCp^*Cp'Cl_2$ by substituting an appropriate cyclopentadienyl derivative for KCp* [57b].

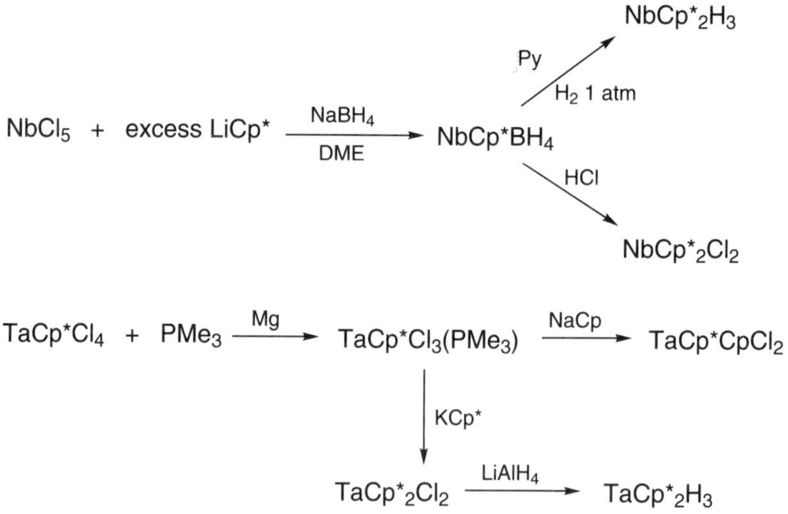

$$NbCl_5 + \text{excess LiCp}^* \xrightarrow[DME]{NaBH_4} NbCp^*BH_4$$
$$NbCp^*BH_4 \xrightarrow[Py]{H_2\ 1\ atm} NbCp^*_2H_3$$
$$NbCp^*BH_4 \xrightarrow{HCl} NbCp^*_2Cl_2$$

$$TaCp^*Cl_4 + PMe_3 \xrightarrow{Mg} TaCp^*Cl_3(PMe_3) \xrightarrow{NaCp} TaCp^*CpCl_2$$
$$TaCp^*Cl_3(PMe_3) \xrightarrow{KCp^*} TaCp^*_2Cl_2 \xrightarrow{LiAlH_4} TaCp^*_2H_3$$

Scheme 7.11

The trihydride structure of MCp'_2H_3 is unique to niobium and tantalum. The bent-metallocene fragment of these metals appears to be ideal for accommodating three hydrides because of the favorable oxidation state of +5 and perhaps because the void created at the equatorial girdle has the right size for three hydrides. The compounds can be easily handled and yet show interesting reactivity. The trihydrides continue to serve as important starting compounds for a wealth of chemistry. Scheme 7.12 shows their transformation to some useful precursors and to heterometallic clusters [56b,57,58].

As was mentioned earlier, the monocyclopentadienyl and biscyclopentadienyl complexes of niobium and tantalum constitute an extensive array of organometallic derivatives. An epoch-making example of this class is a series of tantalum alkylidene complexes which attracted attentions as prototypes for the formation of metal–carbon double bonds by metal atoms in high oxidation states. Many synthetic routes to the alkylidene complexes have been developed, α-hydrogen abstraction from an alkyl ligand being a common strategy. For instance, dehydochlorination of $TaCp^*(CH_2Ph)_2Cl_2$ and $TaCp'(CH_2Ph)_3Cl$ occurs by thermolysis (or photolysis) and by addition of $Ph_3P{=}CH_2$, respectively, to afford corresponding alkylidene complexes (eq (22)) [59].

$$\begin{aligned} &TaCp'(CH_2Ph)_3Cl \xrightarrow[-\,Ph_3MePCl]{Ph_3P=CH_2} TaCp'(=CHPh)(CH_2Ph)_2 \\ &TaCp^*(CH_2Ph)_2Cl_2 \xrightarrow[-\,toluene]{\Delta\ or\ h\nu} TaCp^*(=CHPh)Cl_2 \end{aligned} \tag{22}$$

The bulkiness of the cyclopentadienyl group is probably a factor which promotes activation of a C—H bond at the α position. Thus, addition of $LiCp^*$ to a solution of $Ta(CH_2Ph)_3Cl_2$ results in formation of $TaCp^*(=CHPh)(CH_2Ph)Cl$, while the reaction of $TaCp^*Cl_4$ with 4 equivalents of $Li(CH_2SiMe_3)$ gives $TaCp^*(=CHSiMe_3)(CH_2SiMe_3)_2$ [60]. Another interesting route is decarbonylation of $TaCp^*_2(CO)Me$ by photolysis, which produces $TaCp^*_2(=CH_2)H$ (eq (23)) [61]. The immediate product of

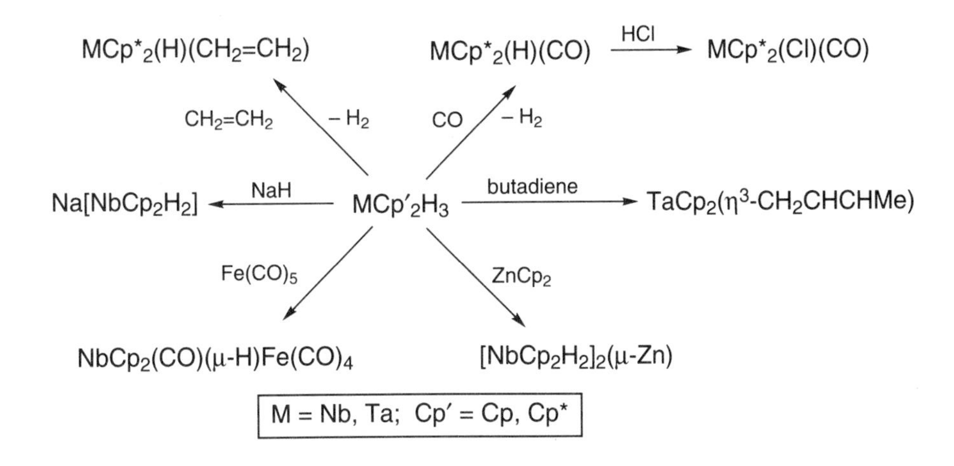

Scheme 7.12

this reaction could be the coordinatively unsaturated methyl complex, and it is in equilibrium with the methylene complex.

$$TaCp^*_2(CO)Me \xrightarrow[-CO]{h\nu} [TaCp^*_2Me] \rightleftharpoons TaCp^*_2(=CH_2)H \quad (23)$$

Reactivity characteristic of alkylidene complexes of tantalum is that the α-carbon is susceptible to electrophilic attack, in contrast to the electron-deficient α-carbon of Fischer-type carbene complexes of group 6 transition metals [62]. Based on this unique property of the alkylidene metal–carbon double bond, a range of new types of reactions has been developed. The discovery of the alkylidene complexes of tantalum was a key to understanding the mechanism of olefin metathesis, and they continue to play important roles in C—H bond activation, alkyne polymerization, and ring-opening metathesis polymerization.

Introduction of bulky aryloxides and alkoxides as auxiliary ligands added a new flavor to the chemistry of niobium and tantalum. Reactions of MCl_5 with hydroxide and small alkoxides usually lead to oligomeric or polymeric structures which are difficult to characterize. However, use of bulky OAr (or OR) groups allows access to monomeric chloride/ aryloxide (alkoxide) complexes $M(OAr)_nCl_{5-n}$. For instance, treatment of $TaCl_5$ with an excess of 2,6-di-tert-butylphenoxide leads to $Ta\{O(2,6\text{-}C_6H_3{}^tBu_2\}_2Cl_3$, while a similar reaction with less bulky 3,5-dimethylphenoxide gives $Ta\{O(3,5\text{-}C_6H_3Me_2\}_5$ [63]. On the other hand, $TaCl_5$ reacts with 2 and 3 equivalents of

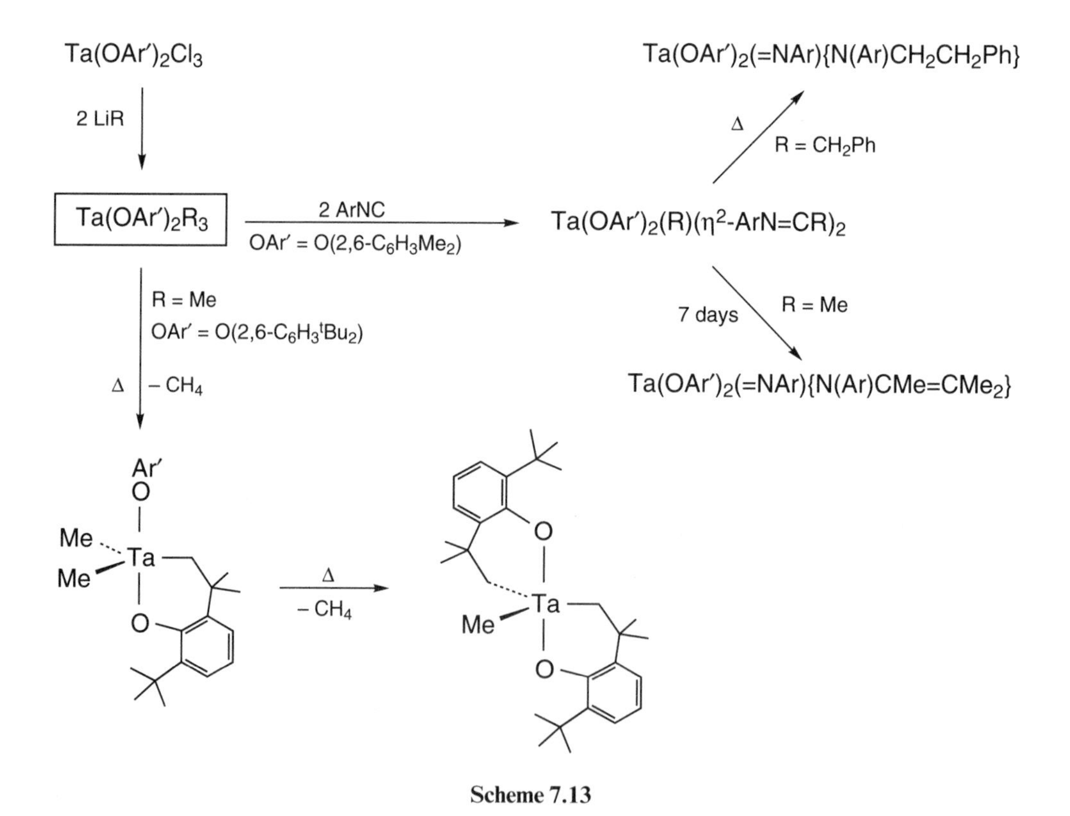

Scheme 7.13

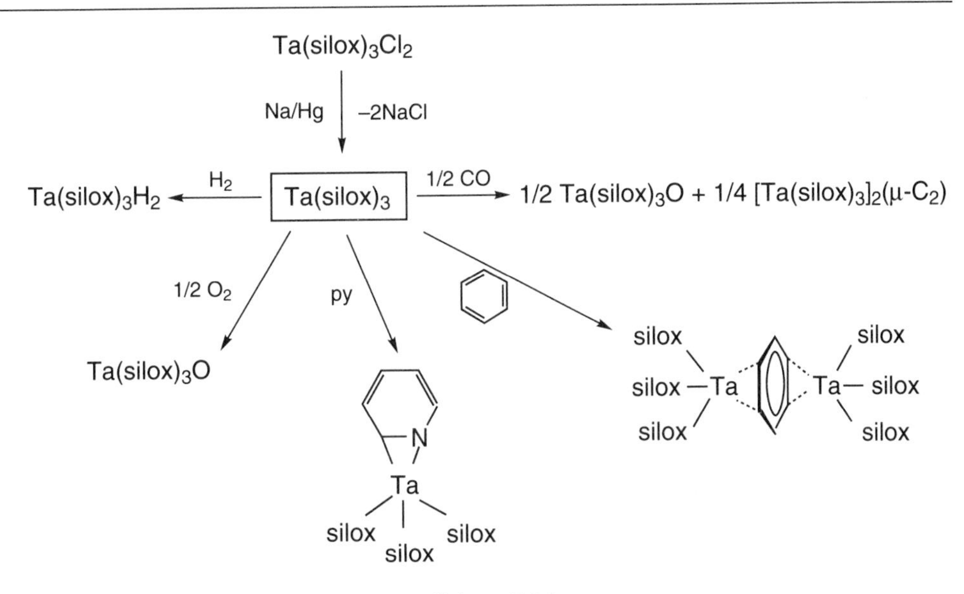

Scheme 7.14

Na–silox ($NaOSi^tBu_3$) to yield $Ta(silox)_2Cl_3$ and $Ta(silox)_3Cl_2$, respectively [64]. Some intriguing reactions of these complexes are of note.

The alkyl/aryloxide complexes $Ta\{O(2,6\text{-}C_6H_3Me_2)\}_2R_3$ (R = Me, CH_2Ph), prepared by reactions of the corresponding trichlorides with LiR, undergo migratory insertion of 2 equivalents of arylisocyanides (CNAr) to give the bis(η^2-iminoacyl) complexes. The benzyl complex thermally generates the mixed imido/amido complex $Ta\{O(2,6\text{-}C_6H_3Me_2)\}_2({=}NAr)\{N(Ar)CH_2CH_2Ph\}$, presumably via η^2-imine/η^2-iminoacyl intermediate and subsequent transfer of α-hydrogens of the η^2-imine to the η^2-iminoacyl. When a solution of the methyl complex was allowed to stand for 7 days, $Ta\{O(2,6\text{-}C_6H_3Me_2)\}_2({=}NAr)\{N(Ar)CMe{=}CMe_2\}$ was isolated, which was thought to be formed via iminoacyl-imine coupling followed by an electrocyclic rearrangement. Thermolysis of a similar trimethyl complex $Ta\{O(2,6\text{-}C_6H_3{}^tBu_2)\}_2Me_3$ resulted in interesting two sequential C—H bond activation processes of the ^{t}Bu groups that generated methane and mono- and bis(cyclometallated) products.

Reduction of $Ta(silox)_3Cl_2$ with Na/Hg leads to a three-coordinate alkoxide complex $Ta(silox)_3$. The coordinatively unsaturated tantalum complex is capable of cleaving H_2 and O_2 bonds resulting in the hydride and oxo complexes as illustrated in Scheme 7.14. Carbon monoxide is also split upon carbonylation of $Ta(silox)_3$ generating the oxo and μ-dicarbide complexes. This reaction models the C—O bond cleavage and C—C bond formation believed to occur in the Fischer–Tropsch reaction, and the ketenylidene complex $Ta(silox)_3({=}C{=}C{=}O)$ was postulated as the key intermediate. On the other hand, when $Ta(silox)_3$ was treated with pyridine and benzene, remarkable η^2-coordinated complexes were formed.

7.5 Synthesis of Niobium and Tantalum Compounds

All the procedures must be conducted under inert atmosphere using dry solvents.

(1) Hexacarbonylniobate and hexacarbonyltantalate(-I), $[M(CO)_6]^-$ (M = Nb, Ta) [45]

$$MCl_5 + Li(MeNp) \text{ or } LiNp \xrightarrow[DME,\ -50 \sim -70°C]{CO\ /\ {}^nBu_4NBr} [{}^nBu_4N][M(CO)_6] \quad (24)$$

M = Nb, Ta; Np = naphthalenide

The traditional methods for preparation of the title carbonyl complexes involved very high-pressure reductive carbonylations of the pentahalides. On the other hand, the present method, developed by J. E. Ellis *et al*, provides a series of hexacarbonylmetalates efficiently using atmospheric carbon monoxide.

Freshly sublimed $TaCl_5$ (20.1 g, 56.1 mmol) is slowly added to 300 mL of vigorously stirred DME in a three-necked round-bottomed flask held at –50 °C. Rapid addition of $TaCl_5$ must be avoided, because the exothermic reaction causes strong local heating. In another reaction vessel, 12 mL of 30% lithium dispersion (6.5 equivalents of Li) in mineral oil is transferred by syringe to a DME (425 mL) solution of 1-methylnaphthalene (55 mL) at 0 °C. The deep green (or black) Li–MeNp mixture is stirred for 2.5 h , during which time the solution gradually warms to room temperature. Then the solution is cooled to –60 °C with stirring. To this cold Li-MeNp solution, the aforementioned DME solution of $TaCl_5$ at –60 °C is added via a large diameter cannula. Immediately after the addition of the $TaCl_5$, color of the solution becomes deep-red brown. Carbon monoxide is bubbled into the solution through a Nujol bubbler vent, first vigorously for 1h and then relatively slowly for 14 h. The temperature must be maintained low at –60 °C during the carbonylation reaction. The resulting very dark solution is filtered into a flask containing $^nBu_4NBr$ (18.2 g, 56.1 mmol) to give a dark solution. Removal of DME from the solution by a rotary evaporator at 30–50 °C leaves nearly black tar, which is then triturated and washed with pentane (5 × 200 mL) until it solidifies to a brown yellow solid. The solid is dried *in vacuo* and recrystallized from 95% ethanol/acetone (100 mL each). Addition of water (300 mL) facilitates precipitation of the product, which is filtered and washed with water and dried *in vacuo* to give $[^nBu_4N][Ta(CO)_6]$ (18.0 g, 54%).

If lithium naphthalenide is used in place of Li–MeNp, a bright orange solution is obtained after a cation exchange with $^nBu_4NBr$. Addition of excess pentane to this solution often gives the product as a homogeneous yellow solid.

A series of complex salts of $[Ta(CO)_6]^-$ and $[Nb(CO)_6]^-$ are synthesized by similar procedures described above. For the details, see [45]. The Nujol mull IR spectra in the ν(CO) region exhibit a very broad asymmetric absorption band, while the spectra in polar solvents show one relatively sharp intense band. All the complexes are very oxygen sensitive in solution, while the solid samples can be handled in air for brief periods of time.

Complex	ν(CO)(cm^{-1})	mp, (°C)
$[^nBu_4N][Ta(CO)_6]$	~1835 (s, br), 1863 sh (Nujol mull) 1860(s) (in CH_3CN)	117–118
$[Et_4N][Ta(CO)_6]$	~1835(s, br), 1875 sh, 1851 sh (Nujol mull)	142 (darkened) >190 (dec.)

(*continued*)

Complex	ν(CO)(cm^{-1})	mp, (°C)
$Na[Ta(CO)_6]$	1874(m), 1835(s, br) (Nujol mull) 1860(s) (in CH_3CN)	pyrophoric
$[Na(diglyme)][Ta(CO)_6]$	2015(w), 1835(s, br) (Nujol mull)	170–172 (dec.)
$[PPh_4][Ta(CO)_6]$	1859(s, br) 1837(s, br) (Nujol mull)	>170
$[^nBu_4N][Nb(CO)_6]$	~1830(s, br) (Nujol mull)	116–117
$[Et_4N][Nb(CO)_6]$	~1837(s, br) (Nujol mull) 1860(s) (in CH_3CN)	>140 (darkened)
$[Na(diglyme)][Nb(CO)_6]$	2015(w), 1834(s, br) (Nujol mull)	146–149 (dec.)

(2) Tetrachloro(η^5-pentamethylcyclopentadienyl)niobium(V) and Tetrachloro(η^5-pentamethylcyclopentadienyl)tantalum(V), $M(\eta^5\text{-}C_5Me_5)Cl_4$ (M = Nb, Ta) [56]

(Method 1)

$$MCl_5 + {}^nBu_3Sn(C_5Me_5) \xrightarrow{CH_2Cl_2} M(\eta^5\text{-}C_5Me_5)Cl_4$$

(Method 2)

$$MCl_5 + Me_3Si(C_5Me_5) \xrightarrow{toluene} M(\eta^5\text{-}C_5Me_5)Cl_4 \qquad (25)$$

M = Nb, Ta

Tetrachloro(cyclopentadienyl) complexes of niobium and tantalum, and their ring-substituted derivatives can be prepared by two general methods. One method uses tin reagents of cyclopentadienes and the other uses trimethylsilyl reagents, which are reacted with niobium or tantulum pentachloride. Here the former method (Method 1) is described for the synthesis of $Ta(\eta^5\text{-}C_5Me_5)Cl_4$, and the latter (Method 2) for the synthesis of $Nb(\eta^5\text{-}C_5Me_5)Cl_4$.

(Method 1) The mixture of preformed sodium pentamethylcyclopentadienide (7.9 g, 50 mmol) and n-Bu_3SnCl (16.3 g, 50 mmol) in 400 mL of toluene is stirred at 20 °C for 4 days followed by filtration to give a light yellow-orange filtrate. The solid remaining on the filter is washed with three 50 mL portions of toluene. All filtrates are combined and the solvent is removed *in vacuo* resulted in yellow-orange liquid. This liquid is distilled at 0.005 torr and the fraction between 115 and 120 °C is collected to yield (n-Bu_3)$Sn(C_5Me_5)$ as a light yellow-orange liquid (16.7 g, 78%).

Properties: ^{1}H-NMR (δ): 1.86 (s, $C_5(CH_3)_5$), satellite peaks with *J*(Sn-H) 17 Hz, 0.92 (m, C_4H_9), 1.38 (m, C_4H_9). Air- and moisture-sensitive.

A $TaCl_5$ (3.2 g, 8.9 mmol) suspension in CH_2Cl_2 (100 mL) is maintained at –78 °C. n-$Bu_3Sn(C_5Me_5)$ (4.3 g, 10.1 mmol) is then added to the suspension with stirring, and the mixture is allowed to warm to room temperature and stirred for 24 h. During this time the color changed from light orange-yellow to deep red brown. Removal of the solvent under reduced pressure resulted in the red-brown solid, which is washed several times with light petroleum ether and is recrystalized from chloroform and petroleum ether to yield Ta(η^5-C_5Me_5)Cl_4 as bright yellow-orange crystalline solid (3.9 g, 95 %).

(Method 2) By use of a pressure-equalized dropping funnel, $SiMe_3(C_5Me_5)$ (0.9 g, 4.3 mmol) is added to a stirred solution of $NbCl_5$ (1.1 g, 4.1 mmol) in toluene at 80 °C. The mixture is stirred at 100 °C for 2h and the solvent is evaporated to dryness. The resulting solid is washed with two portions of hexane (40 ml) and dried in vacuo to give analytically pure Nb(η^5-C_5Me_5)Cl_4 as dark-red microcrystals. Yield 70 %. Mp 145 °C.

In the case of the synthesis of Ta(η^5-C_5Me_5)Cl_4, the reaction of $SiMe_3(C_5Me_5)$ with $TaCl_5$ in hot toluene gives rise to the product as an orange-yellow precipitate, which can be readily separated by filtration followed by washing with toluene.

(3) Tetracarbonyl(η^5-pentamethylcyclopentadienyl)tantalum(I), Ta(η^5-C_5Me_5)$(CO)_4$ [51b]

$$\mathrm{Ta(\eta^5\text{-}C_5Me_5)Cl_4 + PMe_3 + Na/Hg \xrightarrow[\text{THF, }-94^\circ C]{CO} Ta(\eta^5\text{-}C_5Me_5)(CO)_4} \quad (26)$$

Trimethylphosphine (0.084 g, 1.1 mmol) is vacuum condensed from a graduated cold finger onto a mixture of Ta(η^5-C_5Me_5)Cl_4 (0.5 g, 1.1 mmol) and sodium amalgam (22.5 g, 0.5% w/w, 4.95 mmol) in THF (40 ml) cooled to –94 °C in a thick-walled 150 ml Rotoflo glass ampoule. Then 1 atm CO is introduced into the vessel and the mixture is allowed to warm to room temperature with stirring. In 1h, the color of the solution turns red-brown. After refilling with 1 atm CO at –94 °C, the mixture is stirred at room temperature for a further 20 h. The color of the solution turns further from purple to brown and finally to orange-brown. The volatile components are removed *in vacuo* and the residue is extracted into toluene (80 mL). Filtration of the resulting orange solution, concentration of it to 10 mL, and subsequent cooling to –78 °C, generates Ta(η^5-C_5Me_5)$(CO)_4$ as orange crystals. Yield 47 % (0.22 g). The remaining filtrate after this recrystallization contains a side product Ta(η^5-C_5Me_5)$(CO)_3(PMe_3)$ in ca. 7 % yield.

Properties: IR(THF solution, CsI, cm^{-1}): 2020 (s), 1905(s, br). ^{1}H-NMR ($CDCl_3$, 25 °C, δ): 2.14 (s, $C_5(CH_3)_5$).

(4) Bis(η^5-pentamethylcyclopentadienyl)niobium(V) tetrahydroborate, Nb(η^5-C_5Me_5)$_2$(BH_4) [57a]

$$\mathrm{NbCl_5 + Li(C_5Me_5) + NaBH_4 \xrightarrow[\text{DME, }80^\circ C]{} Nb(\eta^5\text{-}C_5Me_5)_2(BH_4)} \quad (27)$$

To a dimethoxyethane slurry (250 ml) of Li(C_5Me_5) (31.4 g, 0.22 mol) is added $NaBH_4$ (22.7 g, 0.68 mol). The slurry is cooled to –80 °C, and $NbCl_5$ (27.0 g, 0.10 mol) is slowly added over a 30 min period with stirring under an argon flow. The resulting brown mixture is allowed to warm to room temperature and then heated to reflux for 3 days. The volatile components are removed under reduced pressure and the purple residue is sublimed at 120 °C, 10^{-3} Torr. The green sublimate is separated and washed with petroleum ether, and then sublimed again under the same condition to give Nb(η^5-C_5Me_5)$_2$(BH_4) as a green solid. Yield 12.1 g (32 %). mp. 239–241 °C.

Properties: IR(Nujol mull, cm^{-1}): 2716(w), 2452(s), 2428(s), 2312(m), 1728(w), 1620(m), 1516(m), 1483(s), 1417(m), 1390(w), 1171(s), 1028(s), 864(s), 798(w), 420(m). ^{1}H-NMR (benzene-d_6, δ): 1.67 (s, $C_5(CH_3)_5$), 5.2 (s, br, BH_4), –18.2 (s, br, BH_4).

(5) Tris(hydrido)bis(η^5-pentamethylcyclopentadienyl)niobium(V), Nb(η^5-C_5Me_5)$_2$$H_3$ [57a]

$$Nb(\eta^5\text{-}C_5Me_5)_2(BH_4) + C_5H_5N \xrightarrow[\text{toluene}]{H_2} Nb(\eta^5\text{-}C_5Me_5)_2H_3 + C_5H_5NBH_3 \qquad (28)$$

To a toluene solution (10 mL) of Nb(η^5-C_5Me_5)$_2$(BH_4) (5.61 g, 14.8 mmol) is added pyridine (15 mL) at –196 °C. The solution is allowed to warm to room temperature, to which 1 atm of H_2 is added. After stirring for 8 h, the resulting red solution is filtered and cooled to prduce a light pink solid. Washing the solid with cold petroleum ether, and subsequent recrystallization from octane and petroleum ether under 1 atm of H_2 gives the white product Nb(η^5-C_5Me_5)$_2$$H_3$. Yield 78 %.

Properties: IR(Nujol mull, cm^{-1}): 1752, 1697, 1027, 773. ^{1}H-NMR (benzene-d_6, δ): 1.81 (s, $C_5(CH_3)_5$), –2.31 (d, br, $^2J_{HH}$ = 4 Hz, NbH_2H), –1.23 (t, br, NbH_2*H*).

References

[1] F. Calderazzo, U. Englert, G. Pampaloni, *J. Organomet. Chem.*, **250**, C33 (1983).

[2] (a) F. Calderazzo, G. Pampaloni, *Organometallic Syntheses*, Academic Press, New York, Vol. 4, p. 4 (1988).
(b) J. E. Ellis, R. A. Faltyneck, G. L. Rochfort, R. E. Stevens, G. A. Zank, *Inorg. Chem.*, **19**, 1082 (1980).

[3] (a) G. S. Giralami, G. Wilkinson, A. M. R. Galas, M. Thornton-Pett, M. B. Hursthouse, *J. Chem. Soc., Dalton Trans.*, 1339 (1985).
(b) J. J. E. Edema, W. Stauthamer, F. van Bolhuis, S. Gambarotta, W. J. J. Smeets, A. L. Spek, *Inorg. Chem.*, **29**, 1302 (1990).

[4] F. H. Köhler, W. Prossdorf, *Z. Naturforsch*, **B32**, 1026 (1977).

[5] (a) F. H. Köhler, W. A. Geike, *J. Organomet. Chem.*, **328**, 35 (1987).
(b) J. R. Bocarsby, C. Floriani, A. Chiesi-Villa, C. Guastini, *Inorg. Chem.*, **26**, 1871 (1987).

[6] S. Gambarotta, C. Floriani, A. Chiesi-Villa, C. Guastini, *Inorg. Chem.*, **23**, 1739 (1984).

[7] E. N. Gladyshev, P. Ya. Bayushikin, V. K. Cherkasov, V. S. Sokolov, *Bull. Acad. Sci. USSR, Div. Chem. Sci.*, **28**, 1070 (1979).

[8] (a) H. J. de Liefde Meijer, F. Jellinek, *Inorg. Chim. Acta*, **4**, 651 (1970).
(b) H. J. de Liefde Meijer, M. H. Jansen, G. J. M. van der Kerk, *Recl. Trav. Chim. Pays-Bas*, **80**, 831 (1961).
(c) R. Sustmann, G. Kopp, *J. Organomet. Chem.*, **347**, 325 (1988).

[9] (a) J. E. Ellis, R. A. Faltyneck, G. L. Rochfort, R. E. Stevens, G. A. Zank, *Inorg. Chem.*, **19**, 1082 (1980).
(b C. J. Curtis, J. C. Smart, J. L. Robbins, *Organometallics*, **4**, 1283 (1985).

[10] G. A. Luinstra, J. H. Teuben, *J. Chem. Soc., Chem. Commun.*, 1470 (1990).

[11] L. E. Manzer, *Inorg. Synth.*, **21**, 84 (1982).

[12] R. Poli, *Chem. Rev.*, **91**, 509 (1991).

[13] J. Nieman, J. H. Teuben, J. C. Huffman, K. G. Caulton, *J. Organomet. Chem.*, **255**, 193 (1983).

[14] (a) R. Kirmse, J. Stach, G. Kreisel, *Z. Chemistry*, **20**, 420 (1980).
(b) M. M. Olmstead, P. P. Power, S. C. Shoner, *Organometallics*, **7**, 1380 (1988).

[15] (a) W. Seidel, G. Kreisel, H. Z. Mennega, *Z. Chemistry*, **16**, 492 (1976).

(b) F. A. Cotton, G. E. Lewis, G. N. Mott, *Inorg. Chem.*, **22**, 560 (1983).
[16] (a) W. Seidel, G. Kreisel, *Z. Anorg. Allg. Chem.*, **435**, 146 (1977).
(b) S. Gambarotta, C. Floriani, A. Chiesi-Villa, C. Guastini, *J. Chem. Soc. Chem. Commun.*, 886 (1984).
[17] J. K. Buijink, A. Meetsma, J. H. Teuben, *Organometallics*, **6**, 2004 (1993).
[18] G. Wilke, B. Bogdanovic, P. Hardt, P. Heinbach, W. Keim, M. Kröner, W. Oberkirch, K. Tanaka, E. Steinrücke, D. Walter, H. Zimmermann, *Angew. Chem., Int. Ed. Engl.*, **5**, 151 (1966), and references therein.
[19] R. B. King, *Organometallic Syntheses*, Academic Press, New York, Vol. 1, p. 105 (1965).
[20] K. Jonas and V. Wiskamp, *Z. Naturforschung*, **B38**, 1113 (1983).
[21] (a) W. A. Hermann, W. Kalcher, *Chem. Ber.*, **115**, 3886 (1982).
(b) M. Hoch, A. Duch, D. Rehder, *Inorg. Chem.*, **25**, 2907 (1986).
[22] (a) J. E. Ellis, R. S. Faltynek, G. L. Rochfort, R. E. Stevens, G. A. Zank, *Inorg. Chem.*, **19**, 1082 (1980).
(b) Q. Z. Shi, T. G. Richmond, W. C. Trogler, F. Basolo, *J. Am. Chem. Soc.*, **104**, 4032 (1982).
(c) K. Ihmels, D. Rehder, *J. Organomet. Chem.*, **232**, 151 (1983).
[23] T. G. Richmond, Q. Z. Shi, W. C. Trogler, F. Basolo, *J. Am. Chem. Soc.*, **106**, 76 (1984).
[24] Q. Z. Shi, T. G. Richmond, W. C. Trogler, F. Basolo, *J. Am. Chem. Soc.*, **106**, 71 (1984).
[25] D. Rehder, U. Puttfarcken, *Z. Naturforsch.*, **B37**, 348 (1982).
[26] H. Köpf, N. Klouras, *Chem. Scripta*, **19**, 122 (1982).
[27] N. Klouras, *Z. Naturforschung*, **B46**, 650 (1991).
[28] (a) K. H. Thiele, L. Oswald, *Z. Anorg. Allg. Chem.*, **423**, 231 (1976).
(b) D. B.Morse, D. N. Hendrickson, T. B. Rauchfuss, S. R. Wilson, *Organometallics*, **7**, 496 (1988).
[29] (a) M. Herberhold, W. Kremnitz, M. Kuhnlein, M. L. Ziegler, K. Brunn, *Z. Naturforschung*, **B42**, 1520 (1987).
(b) M. S. Hammer, L. Messerle, *Inorg. Chem.*, **29**, 1780 (1990).
[30] B. K. Bower, H. G. Tennent, *J. Am. Chem. Soc.*, **94**, 2512 (1972).
[31] W. Seidel, G. Kreisel, *Z. Chem.*, **17**, 115 (1976).
[32] (a) E. O. Fischer, H. Y. P. Kögler, *Chem. Ber.*, **90**, 250 (1957).
(b) G. Braur, *Handbook of Preparative Inorganic Chemistry*, Academic Press New York, Vol. 2, p. 1289 (1965).
[33] (a) K.-H. Thiele, *Pure Appl. Chem.*, **30**, 575 (1972).
(b) K.-H. Thiele, W. Schumann, S. Wagner, W. Brüser, *Z. Anorg. Allg. Chem.*, **390**, 280 (1972).
[34] (a) W. Seidel, G. Kreisel, *Z. Chem.*, **21**, 295 (1981).
(b) W. Seidel, G. Kreisel, *Z. Chemistry*, **22**, 113 (1982).
[35] D. D. Devore, J. D. Lichtenhan, F. Takusagawa, E. A. Maatta, *J. Am. Chem. Soc.*, **109**, 7408 (1987).
[36] (a) R. J. Bouma, J. H. Teuben, W. R. Beukema, R. L. Bansemer, J. C. Huffman, K. G. Caulton, *Inorg. Chem.*, **23**, 2715 (1984).
(b) J. A. M. Canich, F. A. Cotton, S. A. Duraj, W. J. Roth, *Polyhedron*, **6**, 1433 (1987).
[37] A. Morita, K. Tatsumi, unpublished result.
[38] J. J. H. Edema, A. Meetsma, S. Gambarotta, *J. Am. Chem. Soc.*, **111**, 6878 (1989).
[39] H. Kawaguchi, K. Tatsumi, *Organometallics*, **14**, 4294 (1995).
[40] (a) R. Tsumura, N. Hagihara, *Bull. Chem. Soc. Jpn.*, **38**, 1901 (1965).
(b) D. F. Foust, M. D. Rausch, *J. Organomet. Chem.*, **239**, 321 (1982).
(c) H. G. Alt, H. E. Engelhardt, A. Razavi, M. D. Rausch, R. D. Rogers, *Z. Naturforsch*, **B43**, 438 (1988).
[41] J. L. Robbins, N. Edelstein, B. Spencer, J. C. Smart, *J. Am. Chem. Soc.*, **104**, 1882 (1982).
[42] (a) R. Drews, D. Wormsbächer, U. Behrens, *J. Organomet. Chem.*, **272**, C40 (1984).
(b) J. Benecke, R. Drews, U. Behrens, F. Edelmann, K. Keller, H. W. Roesky, *J. Organomet. Chem.*, **320**, C31 (1987).
(c) T. Sielisch, U. Behrens, *J. Organomet. Chem.*, **310**, 179 (1986).
(d) J. L. Petersen, L. Griffith, *Inorg. Chem.*, **19**, 1852 (1980).
(e) G. Fachinetti, C. Floriani, A. Chiesi-Villa, C. Guastini, *Inorg. Chem.*, **18**, 2282 (1979).

(f) G. Fachinetti, C. Floriani, *J. Chem. Soc., Dalton Trans.*, 2433 (1974).
(g) M. Morán, M. Gayoso, *J. Organomet. Chem.*, **243**, 423 (1983).
[43] (a) J. Ruiz, M. Vivanco, C. Floriani, A. Chiesi-Villa, C. Guastini, *J. Chem. Soc., Chem. Commun.*, 762 (1991).
(b) M. Vivanco, J. Ruiz, C. Floriani, A. Chiesi-Villa, C. Guastini, *Organometallics*, **9**, 2185 (1990).
(c) G. Kresel, W. Seidel, *Z. Anorg. Allg. Chem.*, **478**, 106 (1981).
(d) G. Kresel, W. Seidel, *Z. Chemistry*, **26**, 260 (1986).
[44] (a) G. A. Ozin, R. Walton, *J. Chem. Soc.* **(A)**, 2236 (1970).
(b) F. Fairbrother, *The Chemistry of Niobium and Tantalum*, Elsevier, New York (1967).
(c) S. R. Wade, G. R. Willey, *Inorg. Chim. Acta*, **72**, 201 (1983).
(d) J. C. Bünzli, A. E. Merbach, *Helv. Chim. Acta*, **55**, 2867 (1972).
(e) C. M. P. Fravez, H. Rollier, A. E. Merbach, *Helv. Chim. Acta*, **59**, 2383 (1976).
(f) L. G. Hubert-Pfalzgraf, M. Tsunoda, *Inorg. Chim. Acta*, **38**, 43 (1980).
[45] C. G. Dewey, J. E. Ellis, K. L. Fjare, K. M. Pfahl, G. F. P. Warnock, *Organometallics*, **2**, 388 (1983).
[46] J. E. Ellis, A. Davison, *Inorg. Synth.*, **16**, 68 (1976).
[47] (a) A. Davison, J. E. Ellis, *J. Organomet. Chem.*, **31**, 239 (1971).
(b) J. E. Ellis and R. A. Faltynek, *Inorg. Chem.*, **15**, 3168 (1976).
(c) H. -C. Bechthold, D. Rehder, *J. Organomet. Chem.*, **233**, 215 (1976).
[48] (a) F. Calderazzo, G. Pampaloni, U. Englert, J. Strähle, *J. Organomet. Chem.*, **383**, 45 (1990).
(b) F. Calderazzo, G. Pampaloni, G. Pelizzi, F. Vitali, *Organometallics*, **7**, 1083 (1988).
[49] R. P. M. Werner, A. H. Filbey, S. A. Manastyrskyj, *Inorg. Chem.*, **3**, 298 (1964).
[50] F. Calderazzo, M. Castellani, G. Pampaloni, P. F. Zanazzi, *J. Chem. Soc., Dalton Trans.*, 1989 (1985).
[51] (a) W. A. Herrmann, W. Kalcher, H. Biersack, I. Bernal, M. Creswick, *Chem. Ber.*, **114**, 3558 (1981).
(b) V. C. Gibson, T. P. Kee, W. Clegg, *J. Chem. Soc., Dalton Trans.*, 3199 (1990).
[52] (a) K. M. Pfahl, J. E Ellis, *Organometallics*, **3**, 230 (1984).
(b) P. Oltmanns, D. Rehder, *J. Organomet. Chem.*, **345**, 87 (1988).
[53] O. J. Scherer, R. Winter, G. Heckmann, G. Wolmershäuser, *Angew. Chem. Int. Ed. Engl.*, **30**, 850 (1991).
[54] (a) J. W. Freeman, F. Basolo, *Organometallics*, **10**, 256 (1991).
(b) A. N. Nesmeyanov, A. I. Gusev, A. A. Pasynskii, K. N. Anisimov, N. E. Kolobova, Yu. T. Struchkov, *J. Chem. Soc., Chem. Commun.*, 277 (1969).
(c) H. Alt, H. E. Engelhardt, *Z. Naturforsch.*, **B42**, 711 (1987).
[55] W. A. Herrmann, H. B. Biersack, M. L. Ziegler, B. Balbach, *J. Organomet. Chem.*, **206**, C33 (1981).
[56] (a) M. J. Bunker, A. deCian, M. L. H. Green, *J. Chem. Soc., Chem. Commun.*, 59 (1977).
(b) M. J. Bunker, A. deCian, M. L. H. Green, J. J. E. Moreau, N. Siganporia, *J. Chem. Soc., Dalton Trans.*, 2155 (1988).
(c) R. J. Burt, J. Chatt, G. L. Leigh, J. H. Teuben, A. Westerhof, *J. Organomet. Chem.*, **129**, C33 (1977).
(d) M. A. Cardoso, R. J. H. Clark, S. Moorhouse, *J. Chem. Soc., Dalton Trans.*, 1156 (1980).
(e) H. Yasuda, T. Arai, A. Nakamura, *Organometallic Syntheses*, Vol. 4, Academic Press, New York, p. 20 (1988).
(f) C. R. Lucas, *Inorg. Synth.*, **16**, 107 (1976).
(g) M. J. Curtis, L. G. Bell, W. H. Butler, *Organometallics*, **4**, 701 (1985).
(h) P. B. Hitchcock, M. F. Lappert, C. R. C. Milne, *J. Chem. Soc., Dalton Trans.*, 180 (1981).
[57] (a) R. A. Bell, S. A. Cohen, N. M. Doherty, R. S. Threlkel, J. E. Bercaw, *Organometallics*, **5**, 972 (1986).
(b) V. C. Gibson, J. E. Bercaw, W. J. Bruton, Jr., R. D. Sanner, *Organometallics*, **5**, 976 (1986).
[58] (a) D. A. Lemenovskii, I. E. Nifant'ev, I. F. Urazowski, E. G. Perevalova, T. V. Timofeeva, Yu. L. Slovokhotov, Yu. T. Struchkov, *J. Organomet. Chem.*, **342**, 31 (1988).
(b) P. H. M. Budzelaar, K. H. den Haan, J. Boersma, G. J. M. van der Kerk, A. L. Spek, *Organometallics*, **3**, 156 (1984).

(c) K. S. Wong, J. A. Labinger, *J. Am. Chem. Soc.*, **102**, 3652 (1980).
[59] L. W. Messerle, P. Jennische, R. R. Schrock, G. Stucky, *J. Am. Chem. Soc.*, **102**, 6744 (1980).
[60] I. de Castro, J. de la Mata, P. Gomez-Sal, P. Royo, J. M. Selas, *Polyhedron*, **11**, 1023 (1992).
[61] A. van Asselt, B. J. Burger, V. C. Gibson, J. E. Bercaw, *J. Am. Chem. Soc.*, **108**, 5347 (1986).
[62] R. R. Schrock, *Acc. Chem. Res.*, **12**, 98 (1979).
[63] (a) L. Chamberlain, J. Keddington, I. P. Rothwell, J. C. Huffman, *Organometallics*, **1**, 1538 (1982).
(b) L. Chamberlain, I. P. Rothwell, K. Folting, J. C. Huffman, *J. Chem. Soc., Dalton Trans.*, 155 (1987).
(c) L. Chamberlain, I. P. Rothwell, J. C. Huffman, *J. Chem. Soc., Chem. Commun.*, 1203 (1986).
[64] (a) D. R. Neithamer, R. E. LaPointe, R. A. Wheeler, D. S. Richerson, G. D. van Duyne and P. T. Wolczanski, *J. Am. Chem. Soc.*, **111**, 9056 (1989).
(b) D. R. Neithamer, L. Párkányl, J. F. Mitchell, P. T. Wolczanski, *J. Am. Chem. Soc.*, **110**, 4421 (1988).
(c) K. J. Covert, D. R. Neithamer, M. C. Zonnevylle, R. E. LaPonte, C. P. Schaller, P. T. Wolczanski, *Inorg. Chem.*, **30**, 2494 (1991).

8 Group 6 (Cr, Mo, W) Metal Compounds

T. Ito, *Yokohama National University*

8.1 Introduction

All three metals belonging to the group 6 triad display a wide variety of oxidation states ranging from −2 to +6 although the first row element chromium differs somewhat in its chemistry from the other members of the triad, molybdenum and tungsten. For instance, chromium compounds with the +2 valency state are ubiquitous, whereas such compounds of molybdenum and tungsten structurally differ from Cr compounds. Furthermore, although paramagnetic d^3 complexes of Cr(III) are stable, very few such complexes of Mo and W have been reported. Chromocene, Cp_2Cr (Cp = η^5-C_5H_5), can be isolated as a stable solid at room temperature, whereas the corresponding compounds molybdenocene and tungstenocene can only be detected as highly reactive intermediates. It is intriguing also that the low-valency complexes with the phosphine ligands, e.g. $[ML_2(PR_3)_4]$ (L = alkenes, alkynes), which are fairly ubiquitous for molybdenum and tungsten, has not been so popular for chromium.

Organometallic compounds of group 6 transition metals may be divided into the following six categories.

8.2 The Homoleptic Alkyl and Aryl Complexes and the Related Compounds

Rather limited numbers of homoleptic alkyl or aryl complexes of the group 6 metals have been reported since it is usually difficult to attain an 18-electron configuration by summing d^6 electrons of the central metal and the ligand electrons of alkyl or aryl ligands which donate one electron each to the metal center. Coordinatively unsaturated homoleptic alkyls (or aryls) of the type MR_4 (M = Cr, Mo, W) are known for alkyl ligands lacking β-hydrogens such as R = Me, $CH_2{}^tBu$, or CH_2SiMe_3 [1]. X-ray structure analysis of the tolyl complex Mo(*o*-tolyl)$_4$ revealed that the complex possesses a slightly distorted tetrahedral geometry [2]. In the case of tungsten, octahedral hexamethyltungsten(VI), $W(CH_3)_6$, has been synthesized by the transmetallation reaction using trimethylaluminum (see Section 9.8 (1))[3].

Synthesis of Organometallic Compounds: A Practical Guide. Edited by S. Komiya

8.3 Complexes with One or More Carbonyl Ligands

The octahedral complexes of the type $M(CO)_6$ (M = Cr, Mo, W) are air-stable, hydrophobic white crystalline solids which are readily sublimable under vacuum. They are very slightly soluble in non-polar solvents and are slightly soluble in polar solvents such as THF and $CHCl_3$. Some physical and structural properties of $M(CO)_6$ are summarized in Table 8.1 [4].

The hexacarbonyls, $M(CO)_6$, are convenient starting materials for a variety of syntheses. One or more CO ligands can be displaced by Lewis bases such as tertiary phosphines, isonitriles, and amines, and the products undergo oxidative addition to give Mo(II) species. Representative reactions of molybdenum hexacarbonyl are summarized in Scheme 8.1. Since CO is a good π-accepting ligand, good σ-donors such as tertiary phosphines, amines or ethers usually displace at most three CO groups from $M(CO)_6$.

Table 8.1 Representative Data for the Physical and Structural Properties of $M(CO)_6$ [4]

	$Cr(CO)_6$	$Mo(CO)_6$	$W(CO)_6$
Molecular weight	220.06	264.00	351.91
Vapor pressure, Torr/°C	1.63/50	0.27/30	0.35/50
	58.9/100	42.8/100	14.1/100
Melting point, °C (under air)	130 (dec.)	150 (dec.)	*ca.* 150 (dec.)
(under vacuum)	150(2)	146(2)	166
M—C bond energy, kJ mol^{-1}	107	151	179
Bond length, Å (M—C)	1.915(1)	2.063(3)	2.058(3)
(C—O)	1.140(1)	1.145(2)	1.148(3)
IR (gas) ν(CO), cm^{-1}	2000.4	2004	1998

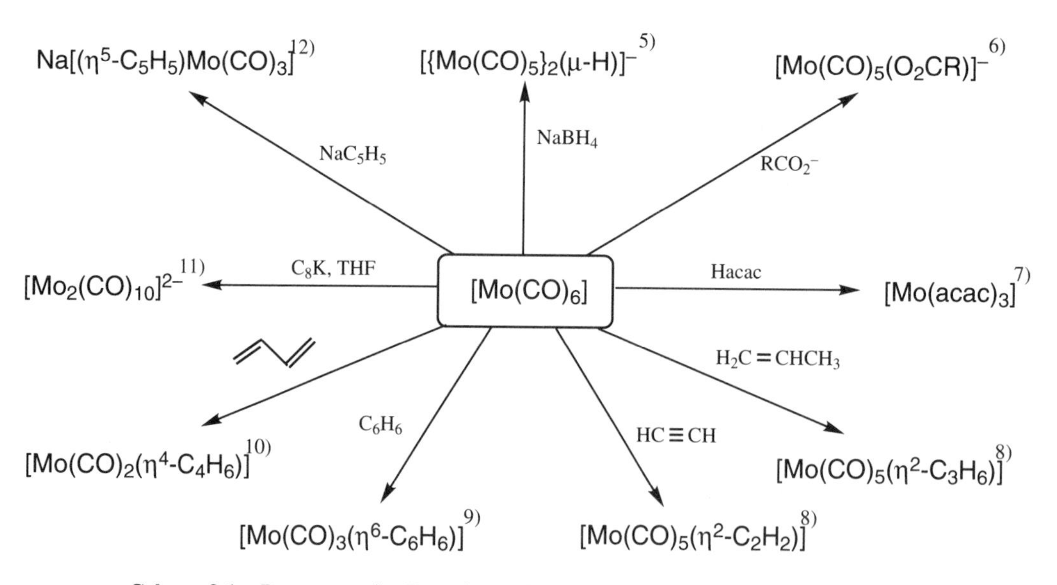

Scheme 8.1 Representative Reactions of $[Mo(CO)_6$ (Hacac = 2,4-pentanedione).

8.4 Carbene and Carbyne Complexes

Compounds with metal–carbon double bonds, which are now known as carbene complexes, were first recognized in 1964 when E. O. Fischer and A. Massbol reported the syntheses of $(OC)_5M{=}CR(OR')$ ($M{=}Cr$, Mo, and W) [13]. This type of carbene complex, which features a low oxidation number of the central metal and heteroatom(s) on the α-carbon, is called a Fischer-type carbene complex and is prepared typically by the following reaction starting from the hexacarbonyls (eq. (1)).

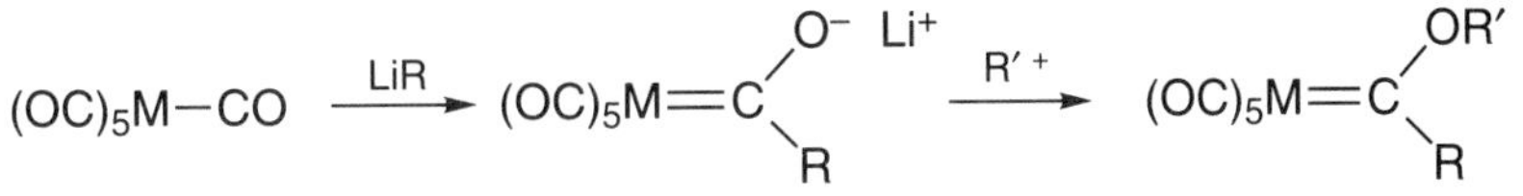

$$(OC)_5M{-}CO \xrightarrow{LiR} (OC)_5M{=}C(O^- Li^+)(R) \xrightarrow{R'^+} (OC)_5M{=}C(OR')(R) \quad (1)$$

Another type of carbene complex is characterized by a high oxidation state of the central metal and an α-carbon atom that does not usually bear a hetero atom. It is called an alkylidene complex or Schrock-type complex, since R. R. Schrock first synthesized a tantalum complex of this type [14]. Formation of the tantalum carbon double bond is based on the α-elimination reaction of a neopentyl ligand as shown in eq.(2).

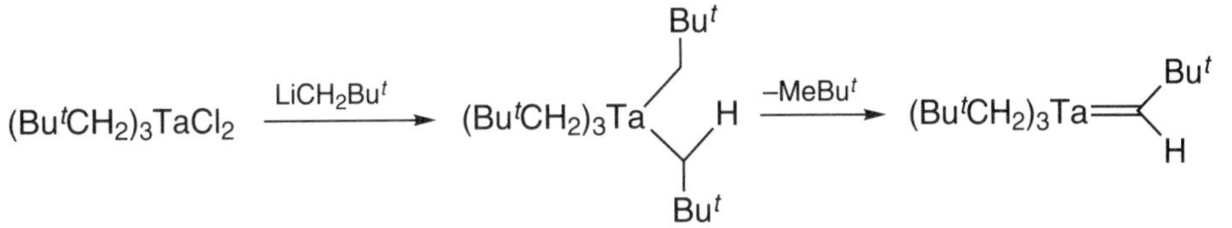

$$(Bu^tCH_2)_3TaCl_2 \xrightarrow{LiCH_2Bu^t} (Bu^tCH_2)_3Ta(CH_2Bu^t)(CHBu^t)\cdots H \xrightarrow{-MeBu^t} (Bu^tCH_2)_3Ta{=}C(Bu^t)(H) \quad (2)$$

A tungsten complex which possesses all three types of metal–carbon bonds, single (alkyl), double (alkylidene), and triple bonds (alkylidyne) has been prepared as shown in eq. (3) and structurally characterized [15].

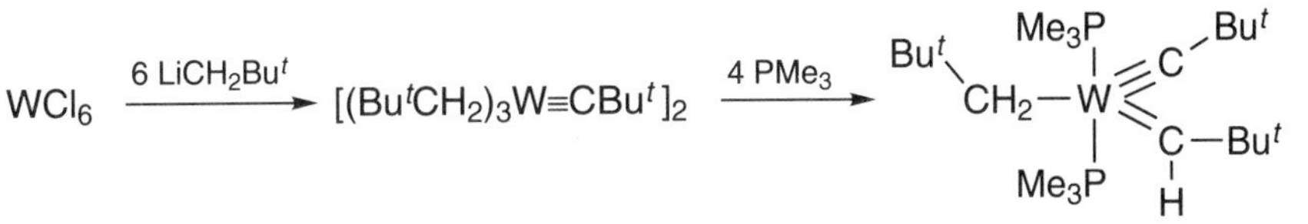

$$WCl_6 \xrightarrow{6\ LiCH_2Bu^t} [(Bu^tCH_2)_3W{\equiv}CBu^t]_2 \xrightarrow{4\ PMe_3} Bu^tCH_2{-}W(PMe_3)_2({\equiv}C{-}Bu^t)({=}C(H){-}Bu^t) \quad (3)$$

The stability of Fischer-type carbene complexes is explained in terms of the resonance hybrids shown in Scheme 8.2 in which the unshared electron pairs on the

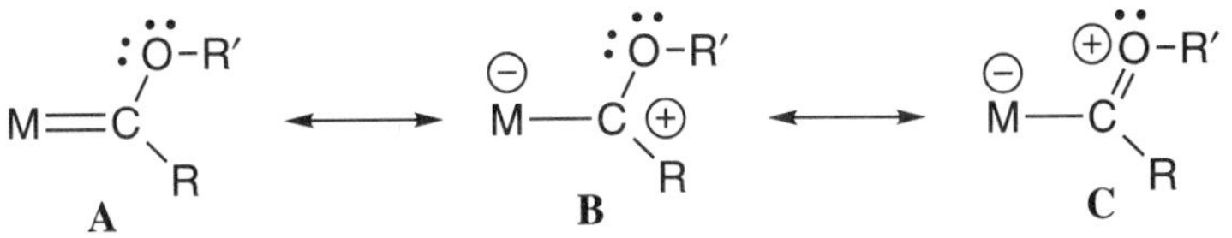

$$\underset{\mathbf{A}}{M{=}C(\ddot{O}{-}R')(R)} \longleftrightarrow \underset{\mathbf{B}}{\overset{\ominus}{M}{-}\overset{\oplus}{C}(\ddot{O}{-}R')(R)} \longleftrightarrow \underset{\mathbf{C}}{\overset{\ominus}{M}{-}C({=}\overset{\oplus}{\ddot{O}}{-}R')(R)}$$

Scheme 8.2 The Canonical Forms of Fischer Carbene Complex.

heteroatom play an important role. The presence of the canonical form **B** suggests that the carbene carbon atom should be electrophilic.

A Schrock-type diphenylmethylidene complex has been prepared by nucleophilic attack of phenyl lithium at the carbene carbon of the Fischer-type methoxyphenylcarbene complex (eq. (4)), which demonstrates the electrophilic nature of the carbene carbon in Fischer–carbene complexes [16, 17].

$$(OC)_5W{=}C(OMe)(Ph) \xrightarrow[-78\,^{\circ}C]{PhLi} \left[(OC)_5\overset{\ominus}{W}{-}C(OMe)(Ph){-}Ph\right] Li^{\oplus} \xrightarrow[-78\,^{\circ}C]{HCl} (OC)_5W{=}C(Ph)(Ph) \quad (4)$$

The carbene carbon atom in high oxidation state complexes of Schrock-type carbenes, by contrast, shows nucleophilic behavior. The bonding mode of these two types of carbene complexes may be schematically represented as shown in Scheme 8.3.

Studies on these carbene complexes, especially those of the Schrock type, have attracted special interest in connection with the mechanism of catalytic olefin metathesis reactions. The formation of metallacyclobutane intermediate from the oxidative cycloaddition reaction between carbene complex and olefin was found to be an important key step in the catalytic cycle (eq. (5)).

$$M{=}CHR \; + \; R'CH{=}CHR' \longrightarrow \begin{matrix} M & — & CH{\cdot}R \\ | & & | \\ 'R{-}CH & — & CH{-}R' \end{matrix} \longrightarrow M{=}CHR' \; + \; RCH{=}CHR' \quad (5)$$

Recently special interest in the Schrock carbenes, especially those of Ta, Mo, W, Re, and Ru, have been focused on the stereospecific polymerization of the cyclic olefins or dienes which is called a ring opening metathesis polymerization (ROMP). A typical example of ROMP is shown in the following equation [18].

$$\text{(norbornadiene, X, X)} \xrightarrow{\text{“Mo=C(H)R”}} Mo{=}\!\left[\!{=}CH{-}C_5H_4X_2{-}CH{=}\right]_n\!{=}C(H)Bu^t \quad (6)$$

“Mo=C(H)R” = $(Bu^tCH_2O)_2Mo(=N{-}2,6\text{-}^iPr_2C_6H_3)(=C(H)Bu^t)$ X = CF_3, CO_2Me

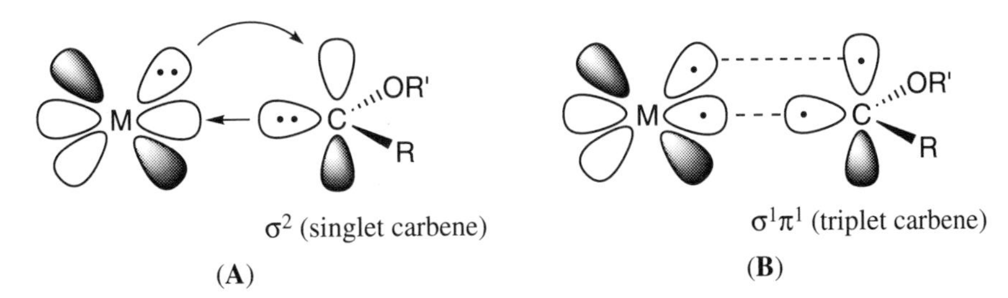

Scheme 8.3 Bonding Mode in the Fischer Carbenes (A) and the Schrock Carbenes (B).

8.5 Complexes with η^5-cyclopentadienyl Ligand(s)

Complexes of the group 6 transition metals containing η^5-cyclopentadienyl ligand, η^5-C_5H_5 (hereafter abbreviated as Cp), may be classified roughly into two categories: one is those which contain only one Cp ligand of the type $CpML_n$, which are often called "half sandwich" or "piano stool" complexes, the other is those which possess two Cp ligands of the type Cp_2ML_n, which generally have a wedge-shaped bent metallocene structure.

Typical examples of the half sandwich type Cp complex are anionic zerovalent metal complexes of the type $[CpM(CO)_3]^-$ and their binuclear derivatives $[Cp_2M_2(CO)_6]$ [12, 19]. Starting from the anion $[CpM(CO)_3]^-$, a variety of complexes have been derived, including the neutral hydrides, $CpMo(H)(CO)_3$, alkyls, $CpM(R)(CO)_3$, [12] and the cationic allyl or diene complexes, $[CpMo(\eta^3$-$C_3H_5)(NO)(CO)]^+$ [20, 21] or $[CpMo(\eta^4$-$C_4H_6)(CO)_2]^+$ [21, 22]. The stereochemistries of nucleophilic attack on the last two complexes have been extensively studied [23]. Permethylcyclopentadienyl of these complexes, e.g. $(\eta^5$-$C_5Me_5)Mo(H)(CO)_3$ [24], have also been extensively reported.

As mentioned above, the only simple ferrocene analogue of the group 6 metals that can be isolated is chromocene, Cp_2Cr, which is prepared by the reaction of Na^+Cp

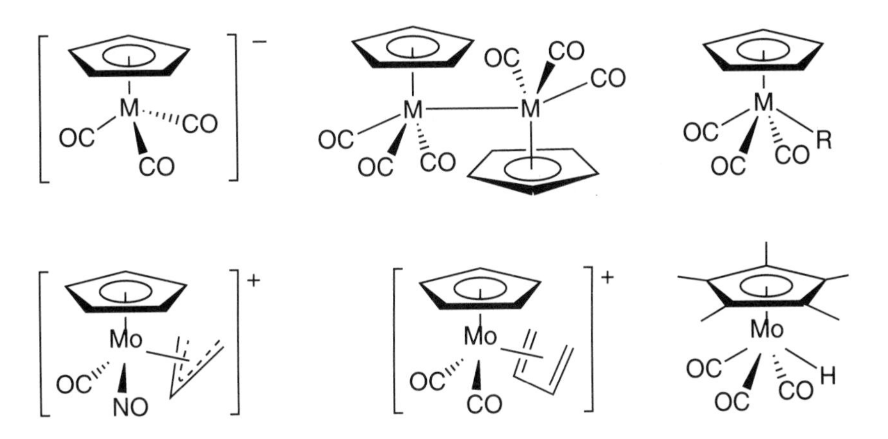

Representative half sandwich complexes of Cr, Mo, and W.

with $CrCl_2$ [19, 25]. The chromocene thus obtained, the electron count of which is 16e, forms red, air-sensitive, paramagnetic needles; mp 173 °C, μ_{eff} = 3.20 B.M. (the calculated spin only value for two unpaired electrons is 2.83 B.M.). In the case of molybdenum or tungsten, this type of metallocene is not isolable and the dihydrido derivatives of the metallocene, Cp_2MH_2 (M = Mo, W), are most commonly used as starting materials for the synthesis of a wide variety of molybdenocene and tungstenocene derivatives (see below) [26, 27].

Since the bent metallocene derivatives Cp_2MH_2 (M = Mo or W) have three orbitals as shown below, the d^2 electron pair of Mo(IV) occupy one of three orbitals; thus these compounds behave as Lewis bases and are easily protonated with acids such as acetic acid, HCl, [26] or *p*-toluenesulfonic acid [28] to give the cationic trihydrides, $[Cp_2MH_3]^+$.

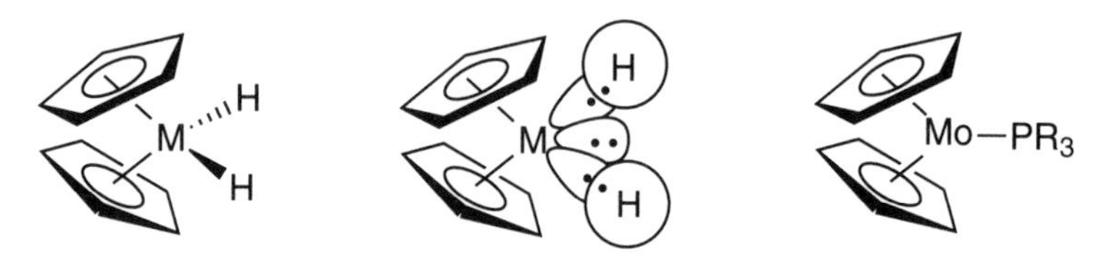

Neutral molybdenocene derivatives coordinated with tertiary phosphines, $Cp_2Mo(PR_3)$ (R = Ph, Et, *cyclo*-C_6H_{11}, *n*-Bu, OEt) have been prepared from Cp_2MoH_2 by the reaction steps shown below (eq. (7)–(9)) [29].

$$Cp_2MoH_2 + TsOH{\cdot}H_2O \xrightarrow[80\ °C]{EtOH} Cp_2MoH(OTs) + H_2 + H_2O \qquad (7)$$

$$Cp_2MoH(OTs) + PR_3 \overset{r.t.}{\rightleftharpoons} [Cp_2MoH(PR_3)]^+TsO^- \qquad (8)$$

$$[Cp_2MoH(PR_3)]^+TsO^- + NaOH \xrightarrow[EtOH]{r.t.} Cp_2Mo(PR_3) + NaOTs + H_2O \qquad (9)$$

R = Ph. Et, *cyclo*-C_6H_{11}, *n*-Bu, OEt

Green *et al.* have reported that dihydridotungstenocene, Cp_2WH_2, on irradiation with a medium pressure lamp in benzene, afforded the hydridophenyl derivative as a result of the aromatic intermolecular C—H activation by the photochemically generated intermediate tungstenocene (eq. (10)) [30].

$$Cp_2WH_2 \xrightarrow[-H_2]{h\nu} \{Cp_2W\} \xrightarrow{C_6H_6} Cp_2W(H)(C_6H_5) \qquad (10)$$

8.6 Arene Complexes and Other π-complexes

Sandwich complexes of the type (η^6-arene)$_2$M and the half sandwich complexes (η^6-arene)M(CO)$_3$ are the representative of this class. As an arene ligand, benzene as well as biphenyl, naphthalene, and their derivatives are often employed. The preparation of

$(\eta^6\text{-}C_6H_6)_2Mo$ and $(\eta^6\text{-}C_6H_5OCH_3)Cr(CO)_3$ will be described as typical examples in the later section. Mixed cyclopentadienyl-arene complexes have also been synthesized as shown in eq. (11) [31].

$$[(C_6H_6)Mo(C_3H_5)Cl]_2 + C_5H_6 \xrightarrow[C_6H_6]{EtAlCl_2} \xrightarrow[KOH\ aq]{Na_2S_2O_4} (C_6H_6)\text{-}Mo\text{-}(C_5H_5) \qquad (11)$$

Complexes of the type (η^6-arene)Cr(CO)3 are very important from the viewpoint of selective organic synthesis involving aromatic substitution reactions since the tricarbonylchromium moiety attached to the arene ring can modify its reactivity. Interesting stereo- and regioselective organic syntheses have been developed based on the specific properties of (η^6-arene)$Cr(CO)_3$ [32]. A typical example of nucleophilic substitution on the aromatic ring by a carbanion is shown in eq (12).

$$(C_6H_6)Cr(CO)_3 \overset{R^-}{\rightleftharpoons} [(C_6H_6R)Cr(CO)_3]^- \xrightarrow{\text{oxidation}} C_6H_5R \qquad (12)$$

The following homoleptic η-allyl complexes of the group 6 metals are known, the paramagnetic $Cr(\eta^3\text{-}C_3H_5)_3$, diamagnetic $M(\eta^3\text{-}C_3H_5)_4$ (M = Mo, W), and dimeric $M_2(\eta^3\text{-}C_3H_5)_4$ (M = Cr, Mo) [33]. They are highly air-sensitive solids; X-ray analysis of $Mo_2(\eta^3\text{-}C_3H_5)_4$ has shown that it has two terminal and two bridging allyl ligands [33b]. $Mo(\eta^4\text{-}C_4H_6)_3$ is an example of a homoleptic diene complex which has been prepared by metal–vapor synthesis directly from metallic molybdenum and butadiene [34].

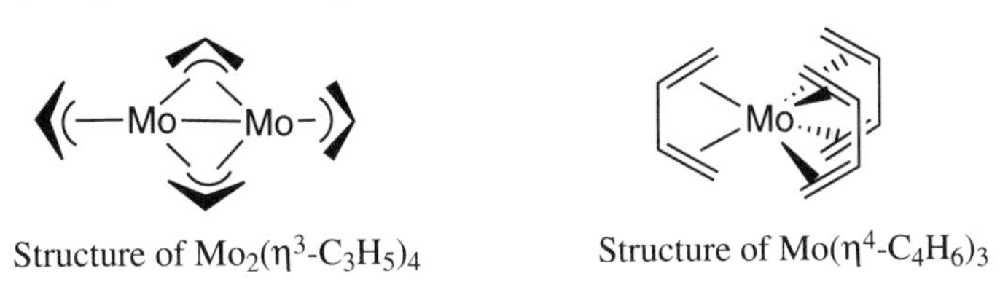

Structure of $Mo_2(\eta^3\text{-}C_3H_5)_4$ Structure of $Mo(\eta^4\text{-}C_4H_6)_3$

8.7 Organometallic Complexes with Tertiary Phosphine Ligands

In contrast to the late transition metals, especially group 10 metals, which form a large number of alkyl or aryl complexes stabilized with tertiary phosphine ligands, few complexes of this type are known for group 6 metals. Instead, various kinds of polyhydrido derivatives or zero-valency metal complexes of molybdenum and tungsten stabilized by tertiary phosphines with or without N_2 or CO have been synthesized. Examples are $MH_4(PR_3)_4$, $MH_2(CO)(PR_3)_4$, $M(N_2)_2(PR_3)_4$, $M(CO)_n(PR_3)_{6-n}$, $M(alkene)_2(PR_3)_4$, $M(PMe_3)_6$ (M = Mo or W); some will be described in detail in the following section. Interestingly, very few analogous chromium complexes containing tertiary phosphine ligands have been prepared.

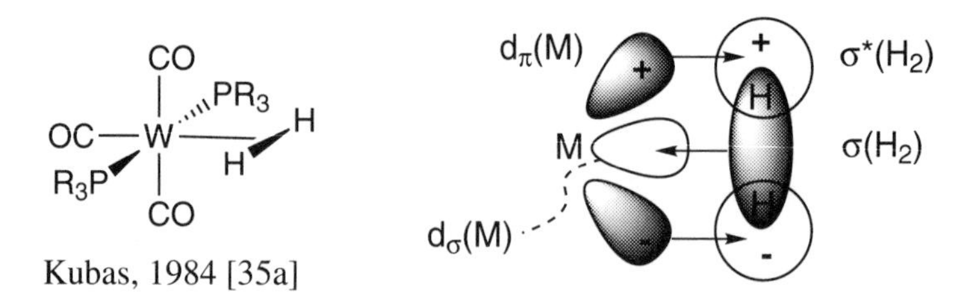

Figure 8.1 Bonding mode in the η^2-dihydrogen complex [37].

The first known η^2-dihydrogen complexes, so-called "non-classical" hydrido complexes, were the molybdenum and tungsten compounds, $M(CO)_3(PR_3)_2(\eta^2\text{-}H_2)$ (M = Mo, W; R = *cyclo*-C_6H_{11}, *i*-Pr) isolated for the first time by Kubas [35]. In order for the dihydrogen molecule to be able to coordinate to the metal without cleaving the H—H bond, it is necessary that the back-donation of the electron from metal d orbital to the antibonding σ^* orbital of the H_2 is minimized (Figures 8.1). This was shown by a comparison of the T_1 values in the ^{1}H NMR measurement in the two complexes, $MoH_2(CO)(R_2PCH_2CH_2PR_2)_2$ (R = Et and Ph). The T_1 value (370 ms) for the complex with the more electron donating depe ligand (R = Et) is much longer than the value of 20 ms for the complex containing the more π-accepting dppe ligand [36].

8.8 Synthesis of Group 6 Metal Compounds

(1) Hexamethyltungsten(VI), $W(CH_3)_6$ [38]

$$WCl_6 + 6Al(CH_3)_3 \longrightarrow \xrightarrow{NMe_3} W(CH_3)_6 + 6Al(CH_3)_2Cl\{N(CH_3)_3\} \tag{13}$$

To a suspension of tungsten(VI) hexachloride in isopentane (*ca.* 10 mL per 1 g of WCl_6) cooled to –70 °C, is added dropwise trimethylaluminum (6 mols per mol of WCl_6) during a period of 10–15 min with a vigorous stirring. (Caution: Trimethylaluminum is a highly pyrophoric liquid (bp 126 °C, mp 15 °C) that reacts violently with air, water and alcohols. Special care should be paid so as to handle under a rigorously air- and moisture-free atmosphere.) After the addition is completed, the temperature of the system is raised gradually to room temperature. After stirring the solution at room temperature for 15 min, the system is cooled again to –70 °C and an excess of $N(CH_3)_3$ is added carefully. A slurry of $Al(CH_3)_2Cl\{N(CH_3)_3\}$ is found with evolution of heat. After the exotherm has ceased, the system is filtered at –70 °C and the volume of the deep reddish orange filtrate is reduced to half by evaporation under reduced pressure. After cooling the concentrated solution at –70 °C, precipitated $Al(CH_3)_2Cl\{N(CH_3)_3\}$ is filtered off to leave $W(CH_3)_6$ as a red oil. Yield 60–70%. Although the purity of the oil is high enough for ordinary synthetic purposes, it is possible to purify the oil by recrystallizing it at –100 °C or by subliming it under a vacuum to give red crystals. (Caution: The title complex is explosive at room temperature and reacts violently with the air.)

An alternative synthetic route employing methyllithium has been developed as follows [39].

$$WCl_6 + LiCH_3 \rightarrow W(CH_3)_6 + 6LiCl \qquad (14)$$

To a suspension of tungsten(VI) hexachloride (12 g, 30 mmol) in diethyl ether cooled at –20 °C is added dropwise methyllithium (90 mmol) during a period of 30 min. The resulting greenish brown solution turns to dark brown on raising the temperature to room temperature. The solvent is removed by evaporation under vacuum at about 0 °C and the residue is extracted with petroleum ether. The solvent is evaporated off from the deep red extract under vacuum at –20 °C and the red crystals of $W(CH_3)_6$ are vacuum sublimed onto a cold finger kept at –20 °C. Yield 40%.

Properties: red crystals when pure, which readily form an oil. Soluble in hydrocarbons and can be sublimed onto a –10 °C probe. Upon warming to room temperature, $W(CH_3)_6$ decomposes violently evolving methane and ethane. IR (CCl_4): ν(C—H) = 2980, 2870 cm^{-1}; δ(C—H) = 1395, 1090, 800 cm^{-1}; ν(W—C) = 482 cm^{-1}. ^{1}H NMR (toluene-d_8) : $\delta(CH_3)$ = 1.62 ppm (singlet with W-satellite, J_{WC} = 3.0 Hz).

(2) Pentacarbonyl{(methoxy)methylmethylene}chromium(0), $Cr(CO)_5\{=C(CH_3)(OCH_3)\}$ [40,41]

$$Cr(CO)_6 + CH_3Li \rightarrow Li[Cr(CO)_5(COCH_3)] \qquad (15)$$

$$Li[Cr(CO)_5(COCH_3)] + [(CH_3)_3O][BF_4] \rightarrow Cr(CO)_5\{=C(CH_3)(OCH_3)\} \qquad (16)$$

In a three-necked, round-bottomed flask equipped with a nitrogen gas inlet, a gas outlet with a paraffin bubbler, and a pressure-equalized dropping funnel, are placed hexacarbonylchromium(0) (5.0 g, 22.7 mmol) and dry diethyl ether (100 mL). To the stirred suspension is added dropwise a diethyl ether solution of methyllithum (1.0 M, 25 mL, 25 mmol) from the dropping funnel over a period of 15–20 min. As the reaction proceeds, the color of the solution changes from yellow to dark brown while most of the $Cr(CO)_6$ dissolve. From the reaction mixture solvent is evaporated off under reduced pressure to leave solid $Li[Cr(CO)_5(COCH_3)]$. The solid is dissolved in water (60–80 mL) and the solution is filtered. To the filtrate is added solid trimethyloxonium tetrafluoroborate in small portions until the aqueous solution is neutralized. Care should be taken not to make the solution too acidic, since the by-product HBF_4 may react with the product. After addition of a small amount of Na_2CO_3 to make the solution slightly basic, the carbene complex is quickly extracted with pentane. The extract is dried with sodium sulfate, and the dried solution is concentrated to a volume of 20–30 mL on a rotary evaporator. Cooling to dry ice–acetone temperature causes the carbene complex to crystallize. It is recrystallized from diethyl ether–hexane to give yellow needles of pentacarbonyl{(methoxy)methylmethylene}chromium(0). 2.0 g (35%).

Following an essentially similar procedure, and allowing one of the carbene complex to react with some nucleophilic reagent, a variety of the Fischer-type carbene complexes of the type $(CO)_5Cr{=}C(R)X$ (R = CH_3, C_2H_5, C_6H_5, $CH{=}CH_2$, $C_{10}H_7$, $C_6H_4N(CH_3)_2$, $C_6H_4OCH_3$, CF_3; X = OCH_3, SCH_3, $OCOCH_3$, NH_2, NHR′, NR'_2, etc) have been prepared [42].

Properties: a dull yellow crystalline solid. mp 34 °C. Decomposes slowly in the air in the solid state. Soluble in pentane, hexane, heptane, benzene, 1,4-dioxane, tetrahydrofuran, chloroform, dichloromethane, and methanol. IR (cyclohexane): $\nu(CO) = 2065$, 1985, 1965, and 1950 cm^{-1}. ^1H NMR ($CDCl_3$): $\delta(CH_3O) = 3.85$ ppm, $\delta(CH_3) = 2.30$ ppm. ^{13}C NMR: $\delta(Cr{-}C) = 362.3$ ppm, $\delta(CO_{cis}) = 217.6$ ppm, $\delta(CO_{trans}) = 223.6$ ppm.

(3) {*n*-Butyl(methoxy)methylene}pentacarbonylmolybdenum(0), $Mo(CO)_5\{{=}C(n\text{-}C_4H_9)(OCH_3)\}$ [42]

$$Mo(CO)_6 + n\text{-}C_4H_9Li \longrightarrow \xrightarrow{CH_3OSO_2CF_3} Mo(CO)_5\{{=}C(n\text{-}C_4H_9)(OCH_3)\} \quad (17)$$

To a suspension of molybdenum hexacarbonyl (0.53 g, 2.0 mmol) in diethyl ether (10 mL), cooled to 0 °C, is added a hexane solution (1.6 M) of *n*-butyllithium (1.25 mL, 2.0 mmol). The mixture is stirred until all $Mo(CO)_6$ has disappeared, when the solution becomes reddish orange. Methyl trifluoromethanesulfonate (methyl triflate) (0.45 mL, 4.0 mmol) is slowly added to the solution. After the addition is completed, the solution is stirred at 0 °C for 5 min. The reaction mixture is treated with saturated aqueous $NaHCO_3$, then extracted with hexane. The combined organic layer is dried over $MgSO_4$ and chromatographed on silica gel to give $Mo(CO)_5\{{=}C(n\text{-}C_4H_9)(OCH_3)\}$. 0.38 g, 56%. Methyl fluorosulfonate CH_3OSO_2F can be used in place of $CH_3OSO_2CF_3$ to give the same product in yield of 54%.

When methyllithium and phenyllithium are employed in place of *n*-butyllithium, the corresponding carbene complexes, $Mo(CO)_5\{{=}C(CH_3)(OCH_3)\}$ and $Mo(CO)_5\{{=}C(C_6H_5)(OCH_3)\}$, respectively, are obtained, though they are less stable than the *n*-butyl analog and can not be obtained in a satisfactorily pure state.

The present preparative method involving direct methylation via addition of methyl ester of fluoro or trifluoromethylsulfonate is superior to the conventional method such as shown above in view of simplicity and reproducibility.

Properties: ^1H NMR ($CDCl_3$): $\delta(CH_3O) = 4.68$ ppm (s), $\delta(n\text{-Bu}) = 0.90$ (t), 1.34 (sextet), 1.51 1.43 (m), 3.24 (t) ppm.

(4) Bis(η^5-cyclopentadienyl)dihydridomolybdenum(IV), $Mo(\eta^5\text{-}C_5H_5)_2H_2$ [26,43]

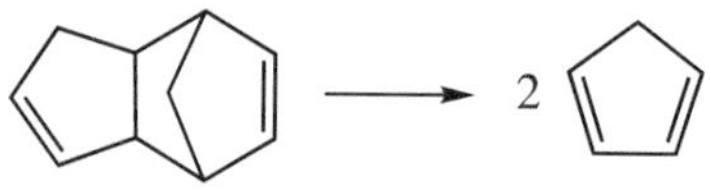

$$\text{(dicyclopentadiene)} \longrightarrow 2\ \text{(cyclopentadiene)} \quad (18)$$

$$\text{(cyclopentadiene)} + Na \xrightarrow{THF} Na^+(C_5H_5^-) + 1/2\ H_2 \quad (19)$$

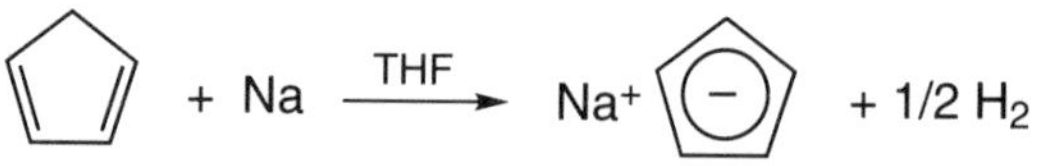

$$MoCl_5 + 2Na^+(C_5H_5)^- + NaBH_4 \xrightarrow{THF} Mo(\eta^5\text{-}C_5H_5)_2H_2 \quad (20)$$

Since the product $Mo(\eta^5\text{-}C_5H_5)_2H_2$ is highly sensitive to the air, the following procedure should be conducted under dry, oxygen-free nitrogen or argon gas.

Just prior to the synthesis of $Mo(\eta^5\text{-}C_5H_5)_2H_2$, cyclopentadiene should be prepared by cracking dicyclopentadiene, the Diels-Alder dimer of cyclopentadiene. In a three-necked, round-bottomed flask equipped with a nitrogen gas inlet, a reflux condenser, and a pressure-equalized dropping funnel, is placed sodium dispersion (50% in paraffin, 52 g, 1.14 mol) and the dispersion is rinsed 10 times with 100 mL of hexane to remove paraffin. After being dried in a stream of nitrogen, the powdery sodium is treated with dry tetrahydrofuran (THF) (350 mL) and the suspension is cooled to –10 °C with an ice–salt bath. At the same temperature, cyclopentadiene (120 mL) is added to the stirred suspension dropwise through a dropping funnel. After the addition is completed, the mixture is stirred at room temperature for 5 h.

To the THF solution of sodium cyclopentadienide (0.5 mol in 250 mL THF) thus prepared is added sodium borohydride (10 g, 0.263 mol). To the stirred mixture, cooled with an ice bath, is added slowly powdery molybdenum pentachloride (27 g, 0.1 mol) with stirring. After addition is completed, the mixture is heated under reflux for 4 h. The solvent is removed by evaporation under vacuum to leave a dark brown solid from which the product dihydride is isolated by employing one of the following two procedures.

(1) The residue is submitted to vacuum sublimation (*ca* 120 °C / 10^{-2} mmHg) to give a yellow solid which is crystallized from diethyl ether–light petroleum at –78 °C.

(2) The residue is dissolved in 3M aqueous HCl and the resulting colorless solution of $[Mo(\eta^5\text{-}C_5H_5)_2H_3]^+Cl^-$ is filtered. Neutralization of the filtrate with 2M aqueous NaOH affords yellow dihydride which is collected by filtration.

The product thus isolated is purified by sublimation under a vacuum at 50 °C to give yellow crystals of $Mo(\eta^5\text{-}C_5H_5)_2H_2$. Yield 30–50%.

An alternative method of preparation of $Mo(\eta^5\text{-}C_5H_5)_2H_2$ employs metal vapor synthesis, in which molybdenum atom vapor and cyclopentadiene react directly under a high vacuum. Yield *ca* 50%. [44].

$$C_5H_6 + Mo \longrightarrow Mo(\eta^5\text{-}C_5H_5)_2H_2 \qquad (21)$$

Properties: Yellow crystals. Highly sensitive to the air both in the solid state and in solution. mp 183–185 °C. IR: $\nu(Mo—H) = 1847\ cm^{-1}$. ^1H NMR(C_6D_6): $\delta(MoH) = -8.76$ ppm (m); $\delta(C_5H_5) = 4.36$ ppm (t).

(5) Bis(η^5-cyclopentadienyl)dihydridotungsten(IV), $W(\eta^5\text{-}C_5H_5)_2H_2$ [26,27,43]

$$WCl_6 + 2Na(C_5H_5) + NaBH_4 \xrightarrow{THF} W(\eta^5\text{-}C_5H_5)_2H_2 \qquad (22)$$

Essentially the same procedure as shown above for the preparation of $Mo(\eta^5\text{-}C_5H_5)_2H_2$ is applicable for $W(\eta^5\text{-}C_5H_5)_2H_2$, using tungsten hexachloride (40 g, 0.1 mol) in place of $MoCl_5$. Since the temperature needed for a vacuum sublimation (120 °C / 10^{-3} mmHg) is higher than for the molybdenum analogue, method (2) above, i.e., extraction with acid, for the isolation of the product is preferred to the sublimation method (1).

Properties: Yellow crystals. Less sensitive to the air than molybdenum analog. mp 163–165 °C. IR (KBr): ν(W—H) = 1905 cm^{-1}. ^{1}H NMR(C_6D_6): δ(WH) = –12.28 ppm (septet, J = 0.75 Hz); δ(C_5H_5) = 4.24 ppm (t, J = 0.75 Hz).

(6) (η^6-Anisole)tricarbonylchromium(0), Cr(η^6-C_6H_5OMe)$(CO)_3$ [45]

$$Cr(CO)_6 + C_6H_5OCH_3 \longrightarrow Cr(\eta^6\text{-}C_6H_5OMe)(CO)_3 + 3CO \quad (23)$$

The reaction should be conducted under inert atmosphere in a well ventilated hood because carbon monoxide evolves during the reaction.

In a three-necked, round-bottomed flask equipped with a nitrogen gas inlet, a Liebig condenser, and a pressure-equalized dropping funnel, are placed hexacarbonylchromium(0) (4.0 g, 18 mmol), anisole (25 mL), di-*n*-butyl ether (120 mL) and tetrahydrofuran (10 mL). The mixture was heated under reflux with stirring for 24 h. The resulting yellow solution is cooled to room temperature and filtered through Celite (or kieselguhr or anhydrous silica gel) to remove the small amount of decomposition products. The filtrate is concentrated by rotary evaporator and finally the solvent is evaporated off under a high vacuum to leave a deep yellow oil. Addition of light petroleum or hexane (20 mL) to the oil yields yellow crystals of Cr(η^6-C_6H_5OMe)$(CO)_3$. 4.1 g (92%). The product can be further purified by recrystallization from benzene–light petroleum or from diethyl ether–light petroleum, or by sublimation *in vacuo*.

Various arenechromium complexes of the type Cr(η^6-arene)$(CO)_3$ (arene = C_6H_6, C_6H_5F, C_6H_5Cl, $C_6H_5(NMe_2)$, C_6H_5COOMe) have been prepared following essentially the same procedure as described above [46].

Properties: Yellow crystals. mp 83–84 °C. It is fairly stable to the air in a solid state but deteriorates slowly in solution. IR (cyclohexane): ν(CO) = 1980, 1908 cm^{-1}. ^{1}H NMR ($CDCl_3$): δ(CH_3O) = 3.6 ppm, δ(C_6H_5) = 4.77 (1H, t), 5.03 (2H, d), and 5.4 (2H, t) ppm.

(7) Bis(η^6-benzene)molybdenum(0), Mo(η^6-$C_6H_6)_2$ [47]

$$3MoCl_5 + 4Al + 6C_6H_6 \xrightarrow{AlCl_3} 3[Mo(\eta^6\text{-}C_6H_6)_2][AlCl_4] + AlCl_3 \quad (24)$$

$$6[Mo(\eta^6\text{-}C_6H_6)_2]^+ + 8OH^- \longrightarrow 5Mo(\eta^6\text{-}C_6H_6)_2 + [MoO_4]^- + 4H_2O + 2C_6H_6 \quad (25)$$

In a 500 mL round-bottomed flask are placed molybdenum pentachloride (65 g, 238 mmol), anhydrous aluminum chloride (140 g, 1.05 mol), and aluminum powder (8.5 g, 315 mmol), and the mixture is mixed well. To the mixture, benzene (250 mL) is added slowly over a period of *ca* 15 min. If the temperature of the contents increases too much on addition of benzene, the flask should be cooled by an ice bath. After the reaction mixture has cooled to room temperature, the flask is shaken well to mix the contents and a reflux condenser is attached. A Teflon sleeve should be put on the flask joint to prevent it from freezing and the condenser should be tightly wired to the neck of the flask. The nitrogen pressure is adjusted as 15–17 psi using mercury bubbler. The flask is immersed in the oil bath and the temperature of the bath is raised gradually to 120 °C. After refluxing for 12 h, the mixture is cooled to room temperature and the flask is shaken to mix the contents well. Then the flask is heated again under reflux for 12 h. After the flask has cooled to room temperature, the benzene layer is removed by decantation and the residue is washed four times with hexane (250 mL). The solid thus obtained is dried well under a vacuum for 14 h, while the flask is shaken periodically to break down the bulk solid into small lumps.

To a 3 L three-necked round-bottomed flask containing an aqueous solution of KOH (750 g KOH / 1750 mL H_2O) cooled to –15 °C, the small lumps of solid obtained as above are added slowly over a period of *ca* 2 h with vigorous stirring under a counter stream of nitrogen. The addition should be at such a rate that the temperature of the contents do not exceed –5 °C. After the addition, the contents of the flask are allowed to warm slowly to room temperature and stirring is continued for another 2 h. The contents are filtered through Celite and the residue is washed with water (250 mL) and dried under vacuum. Care should be taken because the solid thus obtained is highly pyrophoric. The solid is finely pulverized and extracted with portions of hot benzene (total volume: 1 L). The extract is filtered while hot and the green filtrate is concentrated to a volume of *ca* 100 mL. On allowing the solution to stand at 6 °C for 1 h, bright green crystals of $Mo(\eta^6\text{-}C_6H_6)_2$ come out which are filtered off, washed with light petroleum (100 mL), and dried *in vacuo*. Yield: 17.3 g (34.5%).

Use of toluene or 1,3,5-mesitylene in place of benzene in the above reaction has been shown to give the corresponding bis(η^6-arene) complexes [48].

The same reaction in a smaller scale (using 5.0 g of $MoCl_5$) has been reported by using a glass ampoule to give $Mo(\eta^6\text{-}C_6H_6)_2$ in yield of 71%. [49].

Alternatively, $Mo(\eta^6\text{-arene})_2$ complexes (arene = C_6H_6, $C_6H_5CH_3$, C_6H_5F, C_6H_5Cl, $C_6H_5(NMe_2)$, C_6H_5COOMe) have been prepared by metal atom vapor synthesis from molybdenum metal and arene [50,51].

Properties of $Mo(\eta^6\text{-}C_6H_6)_2$: Bright green crystals. mp 115 °C. ^1H NMR (C_6D_6): $\delta(C_6H_6) = 4.60$ ppm (s).

(8) Tris(η^3-allyl)chromium(III), $Cr(\eta^3\text{-}C_3H_5)_3$ [52]

$$CrCl_3 + 3C_3H_5MgCl \longrightarrow Cr(\eta^3\text{-}C_3H_5)_3 \tag{26}$$

To a Schlenk flask containing a suspension of anhydrous chromium(III) trichloride (4.7 g, 29.6 mmol) in dry diethyl ether (30 mL), cooled to –20 – –30 °C, is added a diethyl ether solution of allylmagnesium chloride (0.5 M, 195 mL, 98 mmol) over 2 h with good stirring. Stirring is continued for another 3 h at the same temperature, then the solution is allowed to stand overnight at –78 °C. The dark red solution thus obtained is filtered and the solvent is removed from the filtrate by evaporation under reduced pressure (0.1–0.5 mmHg) at –40 °C. After removal of diethyl ether, the residue is extracted with three 150 mL portions of precooled pentane. The extraction should be conducted at –40 °C to avoid decomposition. Evaporating the solvent from the extract at –40 °C yields the red powdery $Cr(\eta^3\text{-}C_3H_5)_3$. 3.6–4.1 g (69–79%). The product could be further purified to give red crystals by concentrating the pentane solution to two-thirds volume at –40 °C and crystallizing at –78 °C. Special care should be taken to avoid contact with air because the solid is highly pyrophoric.

Properties: a dark red crystalline solid. Monomeric in benzene and in dioxane. Paramagnetic with $\mu_{eff} = 3.78$ B.M. mp 77–79 °C (dec.). $Cr(\eta^3\text{-}C_3H_5)_3$ undergoes transformation to the dimer, $Cr_2(\eta^3\text{-}C_3H_5)_4$, thermally or photochemically, which has been independently prepared from $CrCl_3$ by its reaction with allylmagnesium chloride at –18 °C followed by warming to room temperature [53].

(9) Tetrahydridobis{1,2-bis(diphenylphosphino)ethane}molybdenum(IV), $MoH_4(Ph_2PCH_2CH_2PPh_2)_2$ [54]

$$MoCl_4(CH_3CN)_2 + Ph_2PCH_2CH_2PPh_2 \longrightarrow MoCl_4(Ph_2PCH_2CH_2PPh_2) \quad (27)$$

$$MoCl_4(Ph_2PCH_2CH_2PPh_2) + Ph_2PCH_2CH_2PPh_2 + NaBH_4 \longrightarrow MoH_4(Ph_2PCH_2CH_2PPh_2)_2 \quad (28)$$

To a flask containing $MoCl_5$ (25.0 g, 91.5 mmol) is added acetonitrile (100 mL) to result in gas evolution and exotherm, and the mixture is stirred for 2 ds. The resulting brown precipitate is collected by filtration, washed with benzene and hexane, and dried *in vacuo* to yield $MoCl_4(CH_3CN)_2$ [55]. (20.2 g, 69%; IR (KBr): ν(CN) = 2270, 2300 cm^{-1}). To the dichloromethane (100 mL) solution of $Ph_2PCH_2CH_2PPh_2$ (diphos, 0.021 mol) is added $MoCl_4(CH_3CN)_2$ (0.01 mol) during a period of 1 h with stirring. Addition of light petroleum (350 mL) to the solution separates a solid which is collected and washed with light petroleum (400 mL) and dried *in vacuo* for 15 min. Formation of $MoCl_4(Ph_2PCH_2CH_2PPh_2)$ was ascertained by observing the absence of ν(CN) in the IR spectrum.

The mixture of $MoCl_4(Ph_2PCH_2CH_2PPh_2)$ (6.80 g, 0.01 mmol) thus formed and diphos (4.78 g, 0.12 mol) dispersed in dry ethanol (150 mL) is stirred at 50 °C for 30 min. To the solution, $NaBH_4$ (3.0 g, 0.08 mol) is added with stirring over a period of 5 min. The solution is heated under reflux for 30 min and then cooled slowly to room temperature. The yellow precipitate thus formed is collected by filtration and dissolved in benzene at 50 °C. Addition of ethanol (200 mL) to the solution crystallizes the product. Repetition of this crystallization procedure yields yellow crystals of $MoH_4(Ph_2PCH_2CH_2PPh_2)_2$, which are dried *in vacuo* at 25 °C for 5 h. Yield: 1.5 g.

The analogous complexes MoH_4L_4 with L = unidentate phosphines such as $P(C_2H_5)_3$, $P(CH_3)Ph_2$, $P(C_2H_5)Ph_2$, and $P(OC_2H_5)_2Ph$ are similarly prepared, although in some cases (for L = $P(CH_3)Ph_2$ and $P(C_2H_5)Ph_2$) the reaction should be conducted under an atmosphere of argon or helium instead of nitrogen to obtain satisfactory yield [54]. Tungsten analogs of MoH_4L_4 are also prepared similarly starting from WCl_6.

Properties of $MoH_4(Ph_2PCH_2CH_2PPh_2)_2$: yellow crystalline powder. IR: ν(Mo—H) = 1750, 1830 cm^{-1}. ^1H NMR (C_6D_6): δ(Mo—H) = –3.64 ppm (quintet, $^2J_{P\text{-}H}$ = 30.3 Hz).

(10) Hexahydridotris(trimethylphosphine)tungsten(VI), $WH_6(PMe_3)_3$ [56]

$$WCl_6 \longrightarrow WCl_4(PMe_3)_3 \text{ [56a]} \quad (29)$$

$$WCl_4(PMe_3)_3 + LiAlH_4 \longrightarrow WH_6(PMe_3)_3 \text{ [56b]} \quad (30)$$

To a stirred suspension of tetrachlorotris(trimethylphosphine)tungsten(IV) (10 g, 18.1 mmol), which is prepared by stirring WCl_4 suspended in THF in the presence of trimethylphosphine [55a], in diethyl ether (400 mL) is added lithium aluminum hydride (7 g, 184 mmol). The color changes from red to yellowish green and hydrogen gas is evolved. After stirring for 1 h the mixture is cooled to –78 °C and methanol/diethyl ether (1:1 v/v, 30 ml) is added dropwise over a period of 30 min. The mixture is stirred at –78 °C for another 1 h and then allowed to warm to room temperature. On raising the temperature, further gas evolution and a color change to brown is observed. After having been stirred for 1 h at room temperature, the mixture is centrifuged to collect a fine powdery product suspended in the system and the supernatant layer is removed. The residue is extracted with four portions of hexane (100 mL) and the collected extracts are reduced in volume to *ca* 50 mL by evaporation under reduced pressure. Cooling the solution to –20 °C affords several crops of white crystals. Yield:3.5 g (46%).

Properties: Colorless crystals. mp 115 °C (dec.). IR (Nujol): 1745 s, 1420 s, 1305 m, 1300 s, 1280 s, 950 s, 770 w, 725 s, and 680 s cm^{-1}. ^1H NMR (C_6D_6): δ(W–H) = –2.61 ppm (quintet, $^2J_{PH}$ = 37.6 Hz with ^{183}W satellites, J_{WH} = 27.0 Hz), δ(CH_3) = 1.57 ppm (pseudo-triplet).

(11) *trans*-Bis(dinitrogen)bis{1,2-bis(diphenylphosphino)ethane}molybdenum(0), *trans*-$Mo(N_2)_2(Ph_2PCH_2CH_2PPh_2)_2$ [57]

$$MoCl_4(CH_3CN)_2 + Ph_2PCH_2CH_2PPh_2 \longrightarrow MoCl_4(Ph_2PCH_2CH_2PPh_2) \quad (31)$$

$$MoCl_4(Ph_2PCH_2CH_2PPh_2) + Ph_2PCH_2CH_2PPh_2 + 2N_2 + Na\text{–}Hg \longrightarrow trans\text{-}Mo(N)_2(Ph_2PCH_2CH_2PPh_2)_2 \quad (32)$$

To a mixture of $MoCl_4(Ph_2PCH_2CH_2PPh_2)$ (2,5 g, 3.9 mmol), prepared as described above, and $Ph_2PCH_2CH_2PPh_2$ (diphos) (1.25 g, 3.1 mmol) dissolved in tetrahydrofuran (30 mL), 1% sodium amalgam (40 g, 17 mmol as Na) is added and the mixture is stirred at room temperature for 6 h under an atmosphere of nitrogen. From the resulting orange-brown suspension, mercury is removed by decantation and the suspension is filtered through a No.3 glass filter. From the filtrate solvent is evaporated *in vacuo* to leave oily material which is dissolved in benzene (5 mL). Addition of methanol (50 mL) cooled at 0 °C to the benzene solution precipitates the product. This is collected by filtration, and the orange solid thus obtained is recrystallized from benzene–methanol (1:10) to give orange crystalline *trans*-$Mo(N_2)_2(Ph_2PCH_2CH_2PPh_2)_2$. Yield: 1.3 g (36%).

Although *trans*-$Mo(N_2)_2(Ph_2PCH_2CH_2PPh_2)_2$ was originally prepared by the route employing molybdenum(III) acetylacetonate and triethylaluminum, the procedure described above is preferable because handling of pyrophoric triethylaluminum needs special caution and the yield is not very high (13%) [58].

$$Mo(acac)_3 + Ph_2PCH_2CH_2PPh_2 + Al(C_2H_5)_3 + 2N_2 \longrightarrow trans\text{-}Mo(N_2)_2(Ph_2PCH_2CH_2PPh_2)_2 \quad (33)$$

An alternative route in which $MoCl_3$(tetrahydrofuran)$_3$ is reduced with magnesium under a nitrogen atmosphere to give trans-$Mo(N_2)_2(Ph_2PCH_2CH_2PPh_2)_2$ in fairly high yield has been reported [59, 60].

Properties: orange crystals. mp 165 °C (dec.). Slightly air sensitive in the solid state but rapidly oxidized in solution. Soluble in tetrahydrofuran and toluene. IR (KBr): $\nu(N_2)$ = 2020 vw, 1970 vs cm^{-1}.

References

[1] G. Wilkinson, F. G. A. Stone, E. W. Abel, Ed., *Comprehensive Organometallic Chemistry* , Vol. 3, Sections 26.1.3, 27.1.3, and 28.1.3, Pergamon Press, Oxford (1982).
[2] M. L. Ziegler, K. Weidenhammer, H. Zeiner, P. K. Skell, *Angew. Chem., Int. Ed. Engl.*, **15**, 695 (1976).
[3] R. A. Andersen, E. Carmona-Guzman, J. F. Gibson, G. Wilkinson, *J. Chem. Soc., Dalton Trans.*, 2204 (1976).
[4] G. Wilkinson, F. G. A. Stone, E. W. Abel, Ed., *Comprehensive Organometallic Chemistry*, Vol. 3, Sections 26.1, 27.1, and 28.1, Pergamon Press, Oxford (1982).
[5] R. G. Hayter, *J. Am. Chem. Soc.*, **88**, 4376 (1966).
[6] G. Doyle, *J. Organomet. Chem.*, **84**, 323 (1975).
[7] M. L. Larson, F. W. Moore, *Inorg. Chem.*, **1**, 856 (1962).
[8] I. W. Stolz, G. R. Dobson, R. K. Sheline, *Inorg. Chem.*, **2**, 1264 (1963).
[9] B. Nicholls, M. C. Whitting, *J. Chem. Soc.*, 551 (1959).
[10] E. O. Fischer, H. P. Kogler, P. Kuzel, *Chem. Ber.*, **93**, 3006 (1960).
[11] C. Ungurenasu, M. Palie, *J. Chem. Soc., Chem. Commun.*, 388 (1975).
[12] T. S. Piper, G. Wilkinson, *J. Inorg. Nucl. Chem.*, **3**, 104 (1956).
[13] E. O. Fischer. A., Massböl, *Angew. Chem. Int. Ed. Engl.*, **3**, 580 (1964).
[14] R. R. Schrock, *J. Am. Chem. Soc.*, **96**, 6796 (1974).
[15] D. N. Clark, R. R. Schrock, *J. Am. Chem. Soc.*, **100**, 6774 (1978).
[16] E. O. Fischer, W. Held, F. R. Kreissl, *Chem. Ber.*, **110**, 3842 (1977).
[17] C. P. Casey, S. W. Polichnowski, H. E. Tuinstra, L. D. Albin, J. C. Calabrese, *Inorg. Chem.*, **17**, 3045 (1978).
[18] G. C. Bazan, E. Khosravi, R. R. Schrock, *J. Am. Chem. Soc.*, **112**, 8378 (1990).
[19] E. O. Fischer, W. Hafner, H. O. Stahl, *Z. Anorg. Allg. Chem.*, **282**, 47 (1955).
[20] J. A. McCleverty, A. J. Murray, *Transition Met. Chem.*, **4**, 273 (1979).
[21] J. W. Faller, A. M. Rosan, *Ann. N. Y. Acad. Sci.*, **295**, 186 (1977).
[22] J. W. Faller, A. M. Rosan, *J. Am. Chem. Soc.*, **99**, 4858 (1977).
[23] A. J. Pearson, *Advances Metal-Organic Chemistry*, L. S. Liebeskind (ed.), Vol. 1, p.1, JAI Press Inc., Greenwich (1988).
[24] S. P. Nolan, C. D. Hoff, J. T. Landrum, *J. Organomet. Chem.*, **282**, 357 (1985).
[25] G. Wilkinson, F. A. Cotton, J. M. Birmingham, *J. Inorg. Nucl. Chem.*, **2**, 96 (1956); F. H. Köhler, *J. Organomet. Chem.*, **110**, 235 (1976).
[26] M. L. H. Green, J. A. McCleverty, L. Pratt, G. Wilkinson, *J. Chem. Soc.*, 4854 (1961).
[27] M. L. H. Green, P. J. Knowles, *J. Chem. Soc., Perkin Trans.*, 989 (1973).
[28] T. Igarashi, T. Ito, *Chem. Lett.*, 1699 (1985).
[29] T. Ito, T. Tokunaga, M. Minato, T. Nakamura, *Chem. Lett.*, 1893 (1991); M. Minato, H. Tomita, T. Igarashi, J.-G. Ren, T. Tokunaga, and T. Ito, *J. Organometallic Chem.*, **473**, 149 (1994).
[30] L. Farrugia, M. L. H. Green, *J. Chem. Soc., Chem. Commun.*, 416 (1975); M. L. H. Green, *Pure App. Chem.*, **50**, 27 (1978).
[31] M. L. H. Green, J. Knight, J. A. Segal, *J. Chem. Soc., Chem. Commun.*, 283 (1975).
[32] (a) S. G. Davies, *J. Organomet. Chem.*, **400**, 223 (1990); *Organotransition Metal Chemistry. Application to Organic Synthesis*, Pergamon Press, Oxford, 1982.
(b) M. Uemura, *Advances in Metal-Organic Chemistry*, **2**, 195 (1991).
[33] (a) G. Wilke, B. Bogdanovic, P. Hardt, P. Heimbach, W. Keim, M. Kröner, W. Oberkirch, K. Tanaka, E. Steinrücke, D. Walter, H. Zimmerman, *Angew. Chem., Int. Ed. Engl.*, **5**, 151 (1966).
(b) F. A. Cotton, J. R. Pipal, *J. Am. Chem. Soc.*, **93**, 5441 (1971)

(c) S. O'Brien, M. Fishwick, B. McDermott, M. G. H. Wallbridge, G. A. Wright, *Inorg. Synth.*, **13**, 73 (1972).
[34] P. S. Skell, M. J. McGlinchey, *Angew. Chem., Int. Ed. Engl.*, **14**, 195 (1975).
[35] (a) G. J. Kubas, R. R. Ryan, B. I. Swanson, P. J. Vergamini, H. J. Wasserman, *J. Am. Chem. Soc.*, **106**, 451 (1984).
(b) G. J. Kubas, C. J. Unkefer, B. I. Swanson, E. Fukushima, *J. Am. Chem. Soc.*, **108**, 7000 (1986).
(c) G. J. Kubas, *Accts. Chem. Res.* **23**, 120 (1988).
[36] G. J. Kubas, R. R. Ryan, C. J. Unkefer, *J. Am. Chem. Soc.*, **109**, 8113 (1987).
[37] R. H. Crabtree, X. Luo, D. Michos, *Chemtracts, Inorg. Chem.*, **3**, 245 (1991).
[38] L. Gayler, K. Mertis, G. Wilkinson, *J. Organometallic Chem.*, **85**, C37, C65 (1972).
[39] A. J. Shortland, G. Wilkinson, *J. Chem. Soc. Dalton Trans.*, 872 (1973).
[40] R. Aumann, E. O. Fischer, *Chem. Ber.*, **101**, 964 (1968).
[41] C. T. Lam, C. D. Malkiewich, C. V. Senoff, *Inorg. Synth.*, **17**, 95 (1977).
[42] D. F. Harvey, M. F. Brown, *Tetrahedron Lett.*, **31**, 2529 (1990).
[43] R. B. King, *Organometallic Syntheses*, Vol. 1, p. 79, Academic Press, New York (1989).
[44] E. M. van Dam, W. N. Brent, M. P. Silvon, P. S. Skell, *J. Am. Chem. Soc.*, **97**, 465 (1965).
[45] C. A. L. Mahaffy, P. L. Pauson, *Inorg. Synth.*, **19**, 155 (1979).
[46] W. R. Jackson, B. Nicholls, M. C. Whiting, *J. Chem. Soc.*, 469 (1960).
[47] W. E. Silverthorn, *Inorg. Synth.*, **17**, 54 (1977).
[48] M. L. H. Green, W. E. Silverthorn, *J. Chem. Soc., Dalton Trans.*, **1973**, 301.
[49] E. O. Fischer, F. Scherer, H. O. Stahl, *Chem. Ber.*, **93**, 2065 (1961).
[50] M. P. Silvon, E. M. van Dam, P. S. Skell, *J. Am. Chem. Soc.*, **96**, 1945 (1974).
[51] M. L. H. Green, *J. Organomet. Chem.*, **200**, 119 (1980).
[52] J. K. Becconsall, B. E. Job, S. O'Brien, *J. Chem. Soc. (A)*, **1967**, 423; S. O'Brien, M. Fishwick, B. McDerrmott, M. G. H. Wallbridge, G. A. Wright, *Inorg. Synth.*, **13**, 73 (1972).
[53] T. Aoki, A. Furusaki, Y. Tomiie, K. Ono, K. Tanaka, *Bull. Chem. Soc. Jpn.*, **42**, 545 (1969).
[54] P. Meakin, L. J. Guggenberger, W. G. Peet, E. L. Muetterties, J. P. Jesson, *J. Am. Chem. Soc.*, **95**, 1467 (1973).
[55] F. A. Allen, B. J. Brisdon, G. Wilkinson, *J. Chem. Soc.*, 4531 (1964).
[56] (a) P. R. Sharp, R. R. Schrock, *J. Am. Chem. Soc.*, **102**, 1430 (1980).
(b) D. Lyons, G. Wilkinson, *J. Chem. Soc., Dalton Trans.*, 587 (1985).
[57] T. A. George, C. Seibold, *Inorg. Chem.*, **12**, 2544 (1973).
[58] M. Hidai, K. Tominari, Y. Uchida, *J. Am. Chem. Soc.*, **94**, 110 (1972).
[59] J. Chatt, A. G. Wedd, *J. Organomet. Chem.*, **27**, C15 (1971).
[60] J. R. Dilworth, R. L. Richards, *Inorg. Synth.*, **20**, 119 (1980).

9 Group 7 (Mn, Tc, Re) Metal Compounds

T. Ito, *Yokohama National University*

9.1 Introduction

Transition metals of groups 6 and 7 display the characteristics of early and late transition metals because they reside in between those two categories in the Periodic Table. Among the group 7 triad, the chemistry of manganese differs considerably from that of technetium and rhenium. In contrast to manganese, which was isolated as early as 1774 by Gahn, the other two metals were not discovered until 1937 for technetium and 1925 for rhenium. Technetium, which is a radioactive metal, was the first element to be produced artificially; named after the Greek 'technikos' which means 'artificial', it has been produced in tonne quantity from fission products of uranium nuclear fuel [1]. There are 19 isotopes of technetium, with atomic masses ranging from 90 to 108. The half-lives of ^{97}Tc and ^{98}Tc are 2.6×10^6 years and 4.2×10^6 years, respectively. The isomeric ^{95m}Tc, whose half-life is 61 d, is useful as a tracer because it produces gamma rays [2]. Since its organometallic chemistry is still in its infancy, the chemistry of technetium will not be described in detail in this chapter.

9.2 Organometallic Chemistry of Manganese

Manganese compounds with the oxidation state ranging from –3 to 7 are known, those of Mn(II) being most stable. Mn(VII) species such as $KMnO_4$ and Mn_2O_7 are known to be powerful oxidizing agents. As for organometallic complexes, the carbonyl complexes $Mn_2(CO)_{10}$, $Mn(CO)_5X$, and $CpMn(CO)_3$ are probably the most important, and rich chemistries have been reported for these complexes.

Dimanganese decacarbonyl, which is prepared by the reduction of manganese(II) halide in the presence of carbon monoxide under pressure [3], is a yellow crystalline, air-stable solid soluble in organic solvents, which sublimes readily *in vacuo* at *ca* 50 °C. A single-crystal X-ray diffraction study showed that $Mn_2(CO)_{10}$ possesses the D_{5d} structure as shown below.

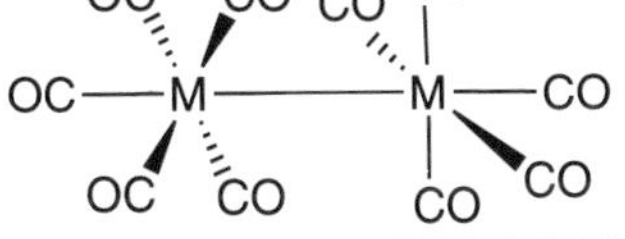

Synthesis of Organometallic Compounds: A Practical Guide. Edited by S. Komiya

Table 9.1 Representative Data for the Physical and Structural Properties of $Mn_2(CO)_{10}$ [4]

	$Mn_2(CO)_{10}$	$Tc_2(CO)_{10}$	$Re_2(CO)_{10}$
Molecular weight	389.98	476.11	652.52
Appearance	Yellow crystals	White crystals	White crystals
Melting point, °C (under air)	110 (dec.)		
(under vacuum)	154–155	160	177
M—M Bond energy, kJ mol^{-1}	154	160	166
Bond length (M—M), Å	2.9038(6)	3.036(6)	3.0413(11)
Raman ν(M—M), cm^{-1}	160	148	122
IR (hexane soln.) ν(CO), cm^{-1}	2045.8 2014.7 1983.8		

In Table 9.1 are compared the physical and structural properties of the decacarbonyls $Mn_2(CO)_{10}$ (M = Mn, Te, Re).

Dimanganese decacarbonyl is an important starting material for manganese carbonyl chemistry: the Mn—Mn bond is cleaved by treatment with various reagents, e.g. cleavage by halogens gives the halides $Mn(CO)_5X$ (X=Cl, Br, and I) and by sodium amalgam in tetrahydrofuran affords the anionic manganese(-I) pentacarbonyl, $Na[Mn(CO)_5]$. In contrast, a cationic Mn(I) hexacarbonyl is formed by its treatment with $NOPF_6$. The reactions of dimanganese decacarbonyl are briefly summarized in Scheme 9.1.

The photochemical reaction of $Mn_2(CO)_{10}$ with PPh_3 has been shown to form disubstituted product, $Mn_2(CO)_8(PPh_3)_2$, as the major product together with a minor amount (*ca* 5%) of monosubstituted $Mn_2(CO)_9(PPh_3)$ (Scheme 9.2). This finding indicates that the reaction proceeds via initial photocleavage of the Mn—Mn bond to yield 17-electron radicals $Mn(CO)_5$, which is known to be substitution-labile, followed by thermal substitution of one of the CO ligands with PPh_3 at the radical stage [5].

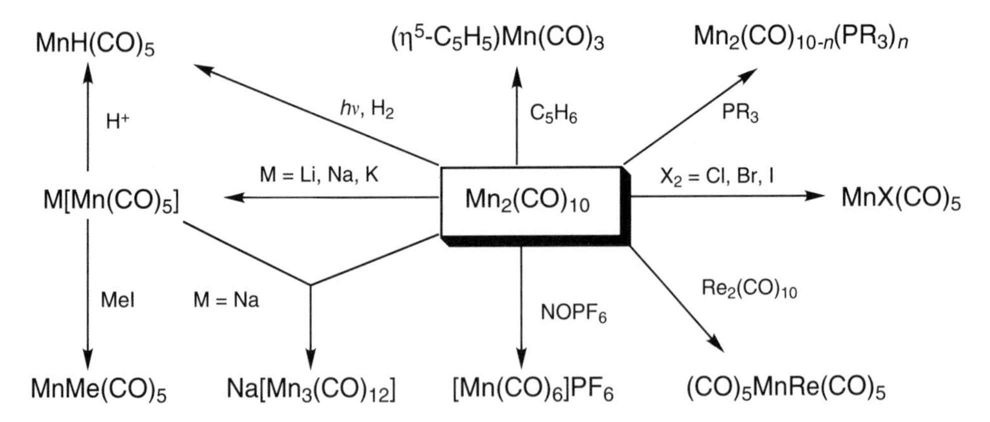

Scheme 9.1 Representative Reactions of $Mn_2(CO)_{10}$ [4b].

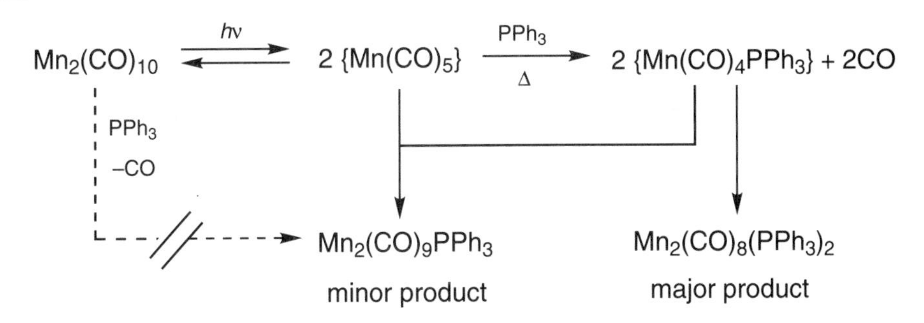

Scheme 9.2 Photochemical Reactions of $Mn_2(CO)_{10}$ with PPh_3.

Manganocene and permethylmanganocene $(\eta^5\text{-}C_5R_5)_2Mn$ (R = H, Me) have been prepared by the reaction between anhydrous MnX_2 (X = Cl, Br, I) and $Na[C_5R_5]$ in tetrahydrofuran. $(\eta^5\text{-}C_5H_5)_2Mn$ is a brown, highly air-sensitive, volatile solid, which is antiferromagnetic below 180 °C. Its color changes to pink between 180 °C and the mp 193 °C and in this range the compound is paramagnetic, and high spin. An X-ray crystallographic study on $(\eta^5\text{-}C_5H_5)_2Mn$ showed that it has a polymeric structure in which chains take zigzag arrangement of $(\eta^5\text{-}C_5H_5)Mn$ units bridged by the second C_5H_5 group.

η^5-Cyclopentadienyltricarbonylmanganese(I), $(\eta^5\text{-}C_5H_5)Mn(CO)_3$, so-called "cymantrene," constitutes another important category of organo-manganese chemistry. In addition to the method involving treatment of $Mn_2(CO)_{10}$ with cyclopentadiene as shown in Scheme 9.1, cymantrene has been prepared either by carbonylation of $(\eta^5\text{-}C_5H_5)_2Mn$ (eq (1)) or by the direct method starting from $MnCl_2(py)_2$ and CO under pressure (eq (2)).

$$(\eta^5\text{-}C_5H_5)_2Mn + 3\,CO \longrightarrow (\eta^5\text{-}C_5H_5)Mn(CO)_3\ [6] \quad (1)$$

$$2\,MnCl_2(C_5H_5N)_2 + Mg + 2\,C_6H_6 + 6\,CO \xrightarrow[H_2]{DMF} (\eta^5\text{-}C_5H_5)Mn(CO)_3\ [7] \quad (2)$$

η^5-Cyclopentadienyltricarbonylmanganese(I) is an air-stable, pale-yellow crystalline solid with mp 76.8–77.1 °C and is readily sublimable at room temperature *in vacuo*. The analogous (η^5-methylcyclopentadienyl)tricarbonylmanganese(I), $(\eta^5\text{-}C_5H_4Me)Mn(CO)_3$, has been commercially produced by the direct carbonylation method similar to that shown above starting from $MnCl_2(py)_2$ and methylcyclopentadiene, since it is used as an antiknock agent in gasoline.

The reactions of $(\eta^5\text{-}C_5H_5)Mn(CO)_3$ have been studied extensively; they include electrophilic ring substitution and photosubstitution of the carbonyl ligand(s). Photolysis of $(\eta^5\text{-}C_5H_5)Mn(CO)_3$ in tetrahydrofuran (THF) gives $(\eta^5\text{-}C_5H_5)Mn(CO)_2(THF)$, which is useful for the preparation of a series of monosubstituted cymantrenes. Treatment of $(\eta^5\text{-}C_5H_5)Mn(CO)_3$ with sodium nitrite and HCl at elevated temperatures affords the cationic η^5-cyclopentadienyl(dicarbonyl)nitrosylmanganese(II). Substitution of one of the two carbonyl ligands with PPh_3 gives the racemic cation, $[(\eta^5\text{-}C_5H_5)Mn(CO)(NO)(PPh_3)]^+$ [8], which can be separated into opti-

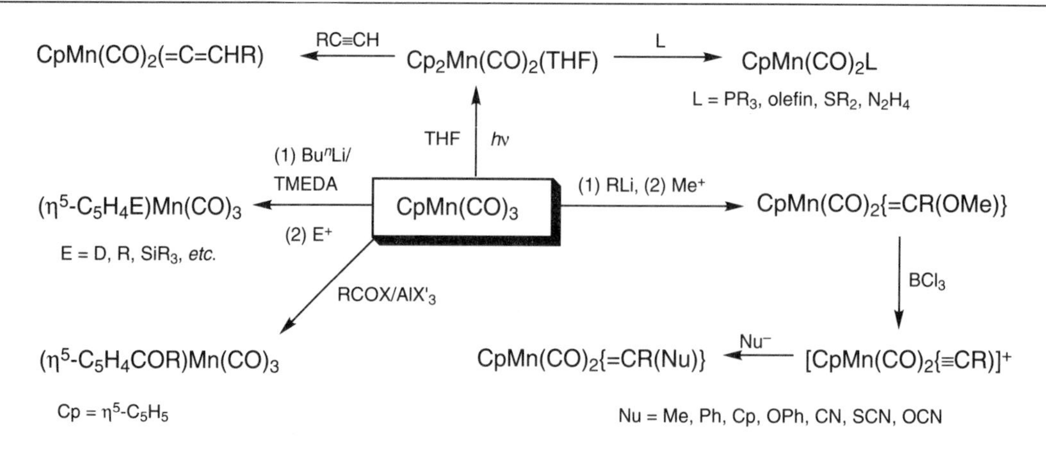

Scheme 9.3 Some Reactions of η^5-cyclopentadienyltricarbonylmanganese(I), η^5 $(C_5H_5)Mn(CO)_3$, (Cymantrene).

cally pure isomers via the diastereomeric isomers formed by the treatment with chiral sodium *l*-mentholate [9]. Some reactions of $(\eta^5\text{-}C_5H_5)Mn(CO)_3$ are summarized in Scheme 9.3.

Oxidation of the hydrazine complex formed by the reaction of $(\eta^5\text{-}C_5H_5)Mn(CO)_2(THF)$ with hydrazine (Scheme 9.2) gives a dinitrogen complex $(\eta^5\text{-}C_5H_5)Mn(CO)_2(N_2)$. Methylation of the N_2 ligand with MeLi followed by treatment with methyl cation converts the N_2 complex into a dimethyldiazene complex, from which the diazene ligand is released by applying N_2 pressure (eq (3)) [10].

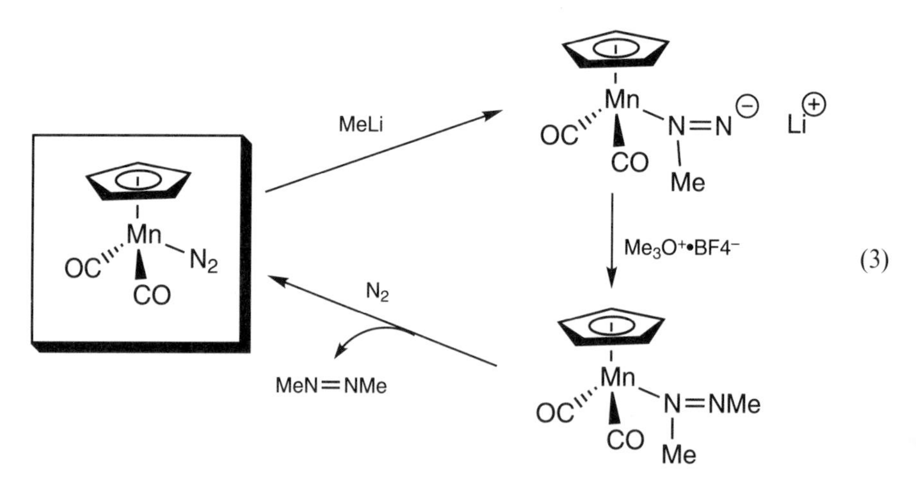

(3)

The first σ-alkyl or σ-acyl complexes of manganese to be characterized were $Mn(CH_3)(CO)_5$ and $Mn(COCH_3)(CO)_5$, which were prepared from $Na[Mn(CO)_5]$ and CH_3I and CH_3COCl, respectively [11]. $Mn(COCH_3)(CO)_5$ is also formed by carbonyl insertion into the Mn—C bond of $Mn(CH_3)(CO)_5$. The investigation of this reaction using CO labeled with ^{13}C and ^{14}C revealed that it is in fact an alkyl migration reaction (eq (4)); the name "migratory insertion reaction" was given to this type of reaction [12].

$$MeMn(CO)_5 + {}^*CO \longrightarrow Me({}^*CO)Mn(CO)_4 \quad (4)$$

$^*CO = {}^{13}CO$ or ${}^{14}CO$

The homoleptic alkyl complexes of the type MnR_2 have been prepared by the reaction of $MnCl_2$ with Grignard reagent. Most of these complexes are found to be dimeric or polymeric; e.g. $[Mn(CH_2CMe_2Ph)_2]_2$, $[Mn(CH_2CMe_3)_2]_4$, and $[Mn(CH_2SiMe_3)_2]_n$ [13]. A phenylmanganese complex of the type $MnPh_2(PCy_3)$ (Cy = cyclohexyl), that is prepared from $Mn(acac)_3$ and $AlPh_3{\cdot}Et_2O$ in the presence of PCy_3, has been shown to undergo Grignard-like reactions such as the insertion of CO_2 or R_2CO into the Mn—C bond [14].

9.3 Organometallic Chemistry of Rhenium

In contrast to manganese, in which oxidation number of +2 is the most stable form, rhenium can take wide range of oxidation states; high oxidation states such as +4 and +5 have an extensive chemistry.

Binuclear carbonyl complexes with M—M bond analogous to $Mn_2(CO)_{10}$ have been prepared for technetium and rhenium. The rhenium analogue has been synthesized by the carbon monoxide reduction of Re_2O_7 or $KReO_4$ (eq (5)) [3] or by reduction of anhydrous Re(III) or Re(V) chloride by sodium under CO pressure (eq (6)). [15]. The physical and structural properties of $Tc_2(CO)_{10}$ and $Re_2(CO)_{10}$ are listed in Table 9.1.

$$Re_2O_7 + 17CO \xrightarrow[200\text{–}250\ \text{atm}]{250\text{–}270\ °C} Re_2(CO)_{10} + 7CO_2\uparrow \quad (5)$$

$$ReCl_3 / ReCl_5 + Na + CO \xrightarrow[250\ \text{atm}]{THF / 130\ °C} Re_2(CO)_{10} \quad (6)$$

$Re_2(CO)_{10}$ is an important starting material in various facets of organorhenium chemistry. Reduction of $Re_2(CO)_{10}$ with $NaBH_4$ followed by protonolysis with H_3PO_4 gives a mixture of hydrides, $ReH(CO)_5$, $Re_3H_3(CO)_{12}$, and $Re_3H(CO)_{14}$ [16]; bubbling H_2 through a decahydronaphthalene solution of $Re_2(CO)_{10}$ heated at 160 °C gives the *tetrahedro*-cluster $Re_4H_4(CO)_{12}$ [17]; and bromination of $Re_2(CO)_{10}$ in cyclohexane affords $ReBr(CO)_5$ [18]. Furthermore, treatment of $Re_2(CO)_{10}$ with dicyclopentadiene forms another important starting complex, $(\eta^5\text{-}C_5H_5)Re(CO)_3$ [19], which is also prepared by the reaction between $ReBr(CO)_5$ and sodium cyclopentadienide (eq (7)) [20].

$$Re_2(CO)_{10} + \text{dicyclopentadiene} \longrightarrow (\eta^5\text{-}C_5H_5)Re(CO)_3 \longleftarrow Na^+ C_5H_5^- + ReBr(CO)_5 \quad (7)$$

The pentamethylcyclopentadienyl analogue of $(\eta^5\text{-}C_5H_5)Re(CO)_3$, $(\eta^5\text{-}C_5Me_5)Re(CO)_3$, has been similarly prepared by the reaction between $Re_2(CO)_{10}$ and pentamethylcyclopentadiene at 150–210 °C (eq (8)) [21]. Pentamethylcyclopentadienyltricarbonylrhenium(I) thus obtained is converted into a cationic (carbonyl)nitrosyl derivative by treatment with $NO^+BF_4^-$. A hydride reduction of the coordinated carbonyl to methyl has been shown (eq (9)) [21, 22].

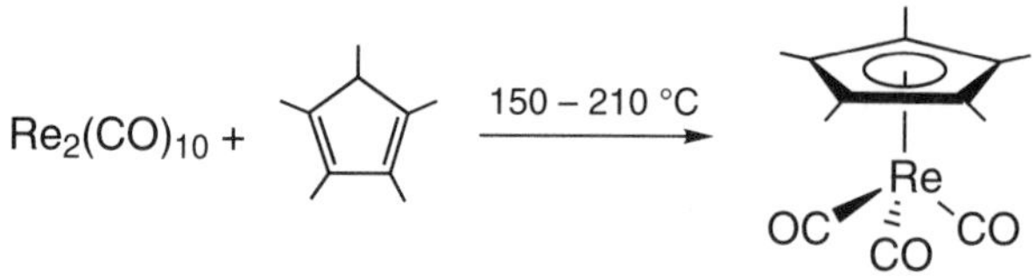

$$Re_2(CO)_{10} + C_5Me_5H \xrightarrow{150-210\ °C} (\eta^5\text{-}C_5Me_5)Re(CO)_3 \qquad (8)$$

$$(\eta^5\text{-}C_5Me_5)Re(CO)_3 \xrightarrow{NO^+BF_4^-} [(\eta^5\text{-}C_5Me_5)Re(CO)_2(NO)]^+BF_4^- \xrightarrow{PPh_3} [(\eta^5\text{-}C_5Me_5)Re(PPh_3)(CO)(NO)]^+BF_4^- \xrightarrow[2)\ BH_3\bullet THF]{1)\ LiEt_3BH} (\eta^5\text{-}C_5Me_5)Re(PPh_3)(CH_3)(NO) \qquad (9)$$

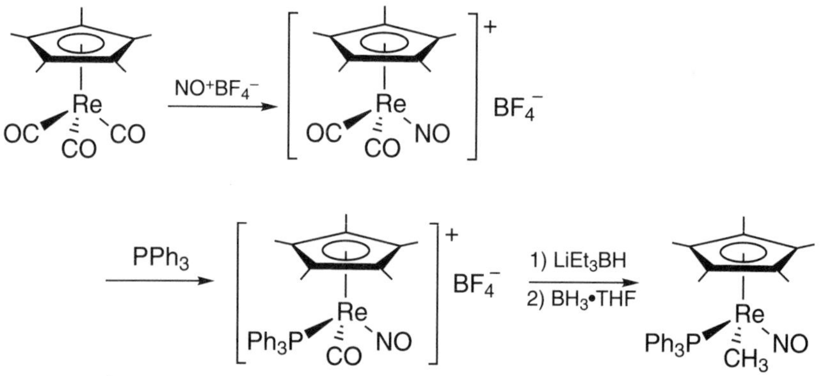

As in the case of the manganese analogue, tetrahydrofuran (THF) displaces one of three carbonyl ligands from $(\eta^5\text{-}C_5H_5)Re(CO)_3$ or $(\eta^5\text{-}C_5Me_5)Re(CO)_3$ on irradiation of their solutions with a high pressure mercury lamp. Taking advantage of the lability of the THF ligand in $(\eta^5\text{-}C_5H_5)Re(CO)_2(THF)$ or $(\eta^5\text{-}C_5Me_5)Re(CO)_2(THF)$ one can prepare various kind of monosubstituted derivatives. The following reaction shown in eq (10) is one such example [23,24].

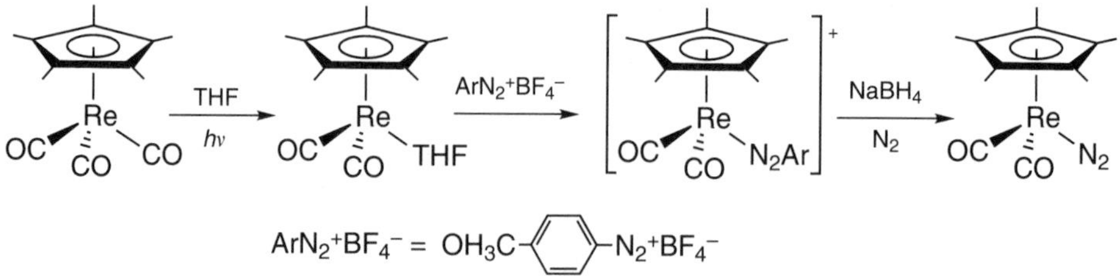

$$(\eta^5\text{-}C_5Me_5)Re(CO)_3 \xrightarrow[h\nu]{THF} (\eta^5\text{-}C_5Me_5)Re(CO)_2(THF) \xrightarrow{ArN_2^+BF_4^-} [(\eta^5\text{-}C_5Me_5)Re(CO)_2(N_2Ar)]^+ \xrightarrow[N_2]{NaBH_4} (\eta^5\text{-}C_5Me_5)Re(CO)_2(N_2) \qquad (10)$$

$$ArN_2^+BF_4^- = OH_3C\text{-}C_6H_4\text{-}N_2^+BF_4^-$$

Bis(cyclopentadienyl)hydridorhenium(III), $(\eta^5\text{-}C_5H_5)_2ReH$, which is prepared by the reduction of Re_2Cl_{10} with $NaBH_4$ in the presence of sodium cyclopentadienide, constitutes another important category of cyclopentadienyl derivatives of rhenium [25]. $(\eta^5\text{-}C_5H_5)_2ReH$ forms yellow crystals (mp 161–162 °C) that are sublimable at *ca* 80 °C / 0.01 mmHg and are highly sensitive to the air. A significant feature of this

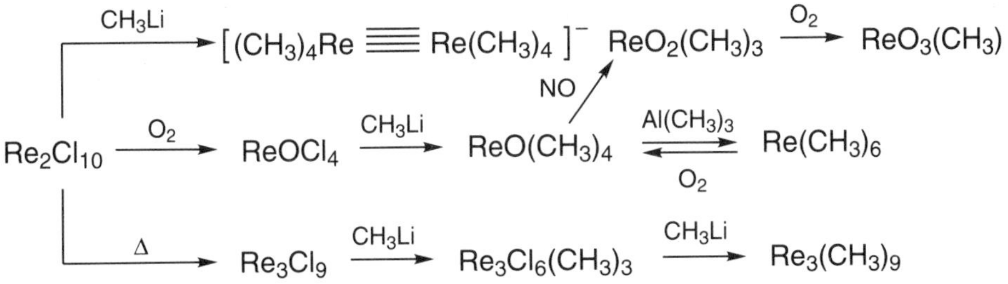

Scheme 9.4 Reactions of Re_2Cl_2.

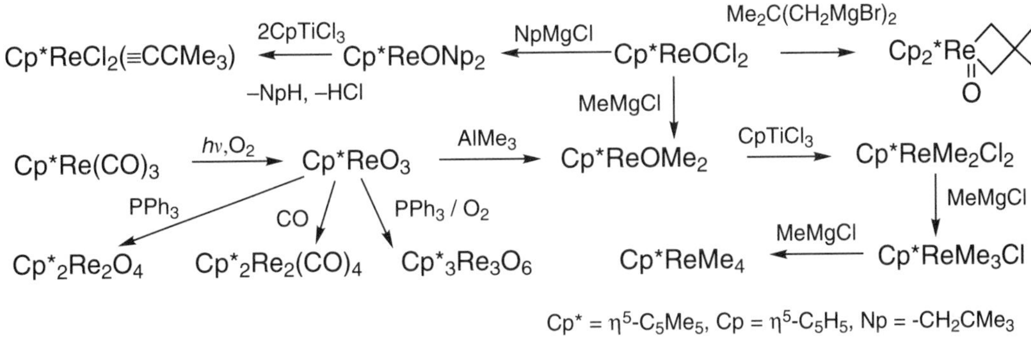

Cp* = η^5-C_5Me_5, Cp = η^5-C_5H_5, Np = -CH_2CMe_3

Scheme 9.5

hydride is the high basicity of the metal center; it is as basic as the nitrogen atom in organic amines. As a consequence, the compound dissolves in dilute acids to form solutions of the protonated cation $[(\eta^5\text{-}C_5H_5)_2ReH_2]^+$ and reacts with methyl iodide to give the cationic dimethyl derivative, $[(\eta^5\text{-}C_5H_5)_2ReMe_2]^+$ [26].

One of the characteristic features of rhenium is its tendency to form stable complexes in a high oxidation state such as Re(VI) and Re(VII) and a series of rhenium oxo-alkyls and a variety of polyhydrides of this type have been extensively studied.

In Scheme 9.4, various oxo-alkyls of Re(VI) and Re(VII) starting from rhenium(V) chloride are shown [27]. The binuclear oxo-alkyl complexes of formula $[ReR_2O_2]_2$ (R = neopentyl), have been prepared by the reaction between $ReOCl_3(PPh_3)_2$ and dineopentylzinc [28].

Herrmann *et al.* have developed an extensive chemistry of the oxorhenium(V) complexes of the type $(\eta^5\text{-}C_5Me_5)ReOX_2$ which are prepared from $(\eta^5\text{-}C_5Me_5)Re(CO)_3$ [29]. An outline of the reactions is given in Scheme 9.5 [30].

The highest number of hydride ligands attached to a single metal atom has been found in the long-known homoleptic, dianionic rhenium nonahydride, $K_2[ReH_9]$ (Scheme 9.6). Its molecular structure has been determined by a neutron diffraction study as tricapped trigonal prismatic with an average Re—H distance of 1.68(1) Å (Figure 9.1a) [31]. Later, a series of rhenium polyhydrides stabilized by tertiary phosphine ligands has been synthesized and characterized spectroscopically and by the use of X-ray or neutron diffraction crystallography [32]. Most of these polyhydrides

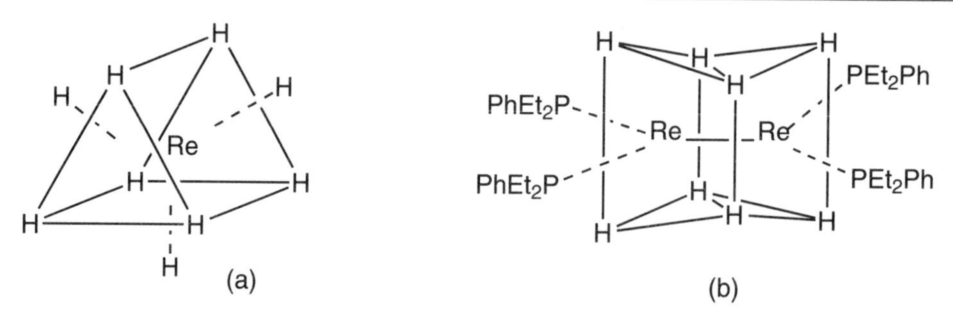

Figure 9.1 The Tricapped Trigonal Prismatic Geometry of the $[ReH_9]^{2-}$ Anion (a) and the Central Core of $H_8Re_2(PEt_2Ph)_4$ (b).

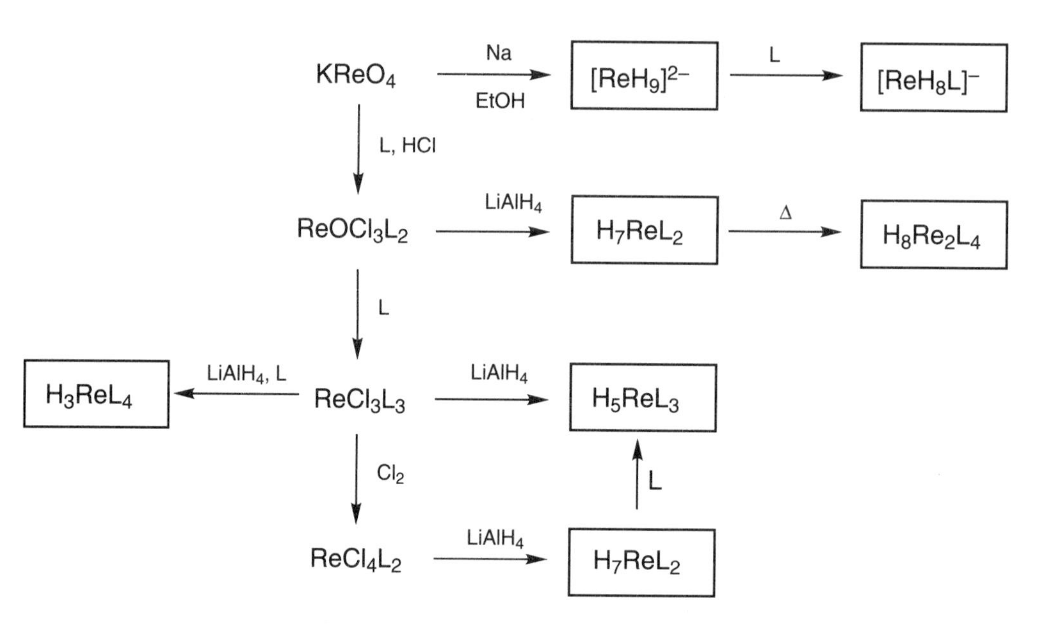

Scheme 9.6 Polyhydrides of Rhenium

shown in Scheme 9.6 are colorless or light yellow, diamagnetic, and air stable in the solid state. The molecular structure of dimeric $H_8Re_2(PEt_2Ph)_4$ determined by neutron diffraction (Figure 9.1b) revealed unambiguously that it has four bridging and four terminal hydrogen atoms (the average Re—H distances are 1.669(7) Å for the terminal hydrides and 1.878(7) Å for the bridging hydrides) [33].

9.4 Synthesis of Group 7 Metal Compounds

(1) Pentacarbonylhydridomanganese(I), $MnH(CO)_5$ [34]

$$Mn_2(CO)_{10} + 2Na\text{-}Hg \longrightarrow 2Na[Mn(CO)_5] \quad (11)$$

$$2Na[Mn(CO)_5] + H_3PO_4 \longrightarrow MnH(CO)_5 + NaH_2PO_4 \quad (12)$$

Sodium amalgam is first prepared in a 300 mL three-necked, round-bottomed flask fitted with a nitrogen inlet, a pressure-equalized dropping funnel, and a reflux condenser. Under an atmosphere of nitrogen, mercury (4 mL, 54 g) is placed in the flask and sodium metal (0.45 g, 19.5 mmol) was carefully added in several portions with stirring. One must ensure that every piece of sodium has reacted with mercury before the next portion is added. After addition of sodium is completed, the liquid amalgam is allowed to cool to room temperature and a tetrahydrofuran (60 mL) solution of dimanganese decacarbonyl (3.0 g, 7.7 mmol) is added from the dropping funnel. The reaction mixture is stirred vigorously for 2 h, during which period the yellow color of the $Mn_2(CO)_{10}$ disappears and becomes a grey suspension. Then the mixture is allowed to stand and the grayish green supernatant solution containing $Na[Mn(CO)_5]$ is transferred by a hypodermic syringe to a 300 mL three-necked, round-bottomed flask. Solvent is removed by evaporation *in vacuo* (*ca* 30 mmHg) to leave a grayish residue. After admitting nitrogen to the flask, a dropping funnel containing 85% aqueous orthophosphoric acid (30 mL) is attached. The apparatus is connected to a vacuum system through a trap cooled to –196 °C that contains phosphorus(V) oxide (25 g). The apparatus is evacuated, and H_3PO_4 is added slowly (1 h), while the product and water distill into the cooled trap. At the end of the addition, the mixture is warmed to 50 °C. The product is distilled from the trap into the vacuum system and fractionated in the vacuum system to remove traces of tetrahydrofuran and finally distilled a second time over P_2O_5 *in vacuo* to give $MnH(CO)_5$ as a slightly yellow liquid [2.5 g (83%)].

$MnH(CO)_5$ can also be synthesized by the reaction of $Mn_2(CO)_{10}$ with H_2 and CO under pressure at 200 °C or by reduction of $Mn_2(CO)_{10}$ with magnesium metal in aqueous methanolic hydrochloric acid [35]. Treatment of $Mn_2(CO)_{10}$ with aqueous KOH at 60 °C followed by acidification with H_3PO_4 gives trinuclear cluster, $[HMn(CO)_4]_3$ (mp 123–125 °C (dec.), IR (cyclohexane): ν(CO) = 2078 vs, 2032 vs, 2008 vs, 1985 vs cm^{-1}) [36].

Properties: colorless, foul-smelling liquid. Usually it appears yellow due to the contamination by a small amount of $Mn_2(CO)_{10}$. mp –24.6 °C. Bp. 111 °C (extrapolated value). Vapor pressure = *ca* 5 mmHg at room temperature. Weak acid ($K = 0.8 \times 10^{-7}$) and very air-sensitive. IR (cyclohexane): ν(CO) = 2118 w, 2016 vs, 2007 s cm^{-1}; ν(Mn—H) = 1783 w cm^{-1}. ^1H NMR: δ(Mn—H) = –7.5 ppm.

(2) Pentacarbonylmethylmanganese(I), $MnMe(CO)_5$ [37]

$$Mn_2(CO)_{10} + 2Na\text{-}Hg \longrightarrow 2Na[Mn(CO)_5] \quad (13)$$

$$2Na[Mn(CO)_5] + CH_3I \longrightarrow MnCH_3(CO)_5 + NaI \quad (14)$$

The first step is the preparation of $Na[Mn(CO)_5]$ described above for the preparation of $MnH(CO)_5$. To a tetrahydrofuran (50 mL) solution of $Na[Mn(CO)_5]$ (prepared as above from 60.6 g of 10% sodium amalgam and $Mn_2(CO)_{10}$ (780 mg, 2 mmol)), a tetrahydrofuran (20 mL) solution of methyl iodide (600 mg, 4.2 mmol) is added dropwise from the dropping funnel and the mixture is stirred at room temperature for 30 min. Solvent is evaporated *in vacuo* by means of a water pump. Since the product is so volatile it is necessary to take care not to evacuate too much. The resulting solid is sublimed at 25 °C / 0.1 mmHg to yield colorless crystals of $MnMe(CO)_5$. Yield *ca* 90%.

When $XMn(CO)_4L$ or $XMn(CO)_3L_3$ (X = Cl, Br; L = PPh_3, $P(OPh)_3$) are employed in place of $Mn_2(CO)_{10}$, the corresponding phosphine substituted methyl(carbonyl) complexes, $Mn(CH_3)(CO)_4L$ or $Mn(CH_3)(CO)_3L_3$, were formed via the intermediates, $Na[Mn(CO)_4L]$ or $Na[Mn(CO)_3L_3]$, respectively [38,39].

A procedure similar to that described above is also applicable for the preparation of $Re(CH_3)(CO)_5$ starting from $Re_2(CO)_{10}$ [40].

Properties: Colorless, air-stable crystals. Special precautions are necessary because of its high volatility. mp 95 °C. IR (hexane)ν: (CO) = 2085 w, 2000 vs, 1960 s cm^{-1}. ^1H NMR: $\delta(CH_3)$ = *ca* 0.05 ppm.

(3) Dicarbonyl(η^5-cyclopentadienyl)(1-methoxyethylidene)manganese(I), $Mn(\eta^5\text{-}C_5H_5)\{=C(OCH_3)CH_3\}(CO)_2$ [41]

$$Mn(\eta^5\text{-}C_5H_5)(CO)_3 + LiCH_3 \longrightarrow Li[Mn(\eta^5\text{-}C_5H_5)(CO)_2(COCH_3)] \quad (15)$$

$$Li[Mn(\eta^5\text{-}C_5H_5)(CO)_2(COCH_3)] + H^+ \longrightarrow [Mn(\eta^5\text{-}C_5H_5)\{=C(OH)CH_3\}(CO)_2] \quad (16)$$

$$\{Mn(\eta^5\text{-}C_5H_5)\{=C(OH)CH_3\}(CO)_2\} + CH_2N_2 \longrightarrow Mn(\eta^5\text{-}C_5H_5)\{=C(OCH_3)CH_3\}(CO)_2 \quad (17)$$

Tricarbonyl(η^5-cyclopentadienyl)manganese (8.16 g, 40 mmol) is allowed to react with methyllithium (40 mmol) in dry diethyl ether (100 mL) to give a yellow, very air- and moisture-sensitive precipitate of an adduct, $Li[Mn(\eta^5\text{-}C_5H_5)(CO)_2(COCH_3)]$, which is filtered and dried *in vacuo*. Yield: 7.80 g (86%).

A suspension of $Li[Mn(\eta^5\text{-}C_5H_5)(CO)_2(COCH_3)]$ (1.72 g, 7.61 mmol) in dry diethyl ether is treated with dilute H_2SO_4 (100 mL). The mixture is stirred vigorously to form the protonated intermediate, which is methylated by the corresponding amount of diazomethane. Chromatography and distillation *in vacuo* at 60 °C give $Mn(\eta^5\text{-}C_5H_5)\{=C(OCH_3)CH_3\}(CO)_2$ as a red-brown liquid. Yield: 0.44 g (18%).

Following similar method, the phenylmethoxycarbene analogue, $Mn(\eta^5\text{-}C_5H_5)\{=C(OCH_3)C_6H_5\}(CO)_2$ is obtained by using phenyllithium in place of methyllithium. (Yield 49%).

Properties: red-brown liquid. IR (Nujol): ν(CO) = 1965 vs, 1894 vs cm^{-1}. ^1H NMR: $\delta(OCH_3)$ = 3.88 ppm; $\delta(C{-}CH_3)$ = 2.33 ppm; $\delta(C_5H_5)$ = 4.37 ppm.

(4) Tricarbonyl(η^5-pentamethylcyclopentadienyl)rhenium(I), $Re\{\eta^5\text{-}C_5(CH_3)_5\}(CO)_3$ [21]

$$Re_2(CO)_{10} + (CH_3)_5C_5H \longrightarrow Re\{\eta^5\text{-}C_5(CH_3)_5\}(CO)_3 \quad (18)$$

Since this reaction is accompanied by evolution of CO gas, the reaction should be conducted in a good fume cupboard. In a 100 mL round-bottomed flask fitted with a reflux condenser are placed $Re_2(CO)_{10}$ (10.0 g, 15.3 mmol) and pentamethylcyclopentadiene (6.88 g, 50.5 mmol) and the mixture is heated under nitrogen at 150 °C with stirring for *ca* 0.5 h, while evolution of CO gas is observed. The temperature is then gradually raised to 210 °C. After 0.5 h at 210 °C, evolution of gas had ceased and heating is continued for 2–5 h in all. The flask is allowed to cool slowly to room temperature. That no $Re_2(CO)_{10}$ remained is confirmed at this stage by silica gel TLC. The flask is chilled in the freezer and the crystals are filtered, washed several times with cold hexane and air-dried to give micro-plates of $Re\{\eta^5\text{-}C_5(CH_3)_5\}(CO)_3$. Yield: 11.7 g (95%).

Properties: colorless plates. mp 149–150 °C. IR (hexane): ν(CO) = 2015 s, 1915 s cm^{-1}. ^1H NMR ($CDCl_3$): δ(CH_3) = 2.21 ppm (s). ^{13}C NMR (acetone-d_6): δ(CO) = 199.1 ppm; δ(C_5Me_5) = 99.6 ppm; δ(C_5Me_5) = 10.7 ppm.

(5) η^5-Cyclopentadienyltris(trimethylphosphine)rhenium(I), $Re(\eta^5\text{-}C_5H_5)(PMe_3)_3$ [42]

$$KReO_4 + PMe_3 + \text{aqueous } HCl \longrightarrow mer\text{-}ReCl_3(PMe_3)_3 \quad (19)$$

$$ReCl_3(PMe_3)_3 + C_5H_6 + K_2CO_3 + Na\text{–}Hg \longrightarrow Re(\eta^5\text{-}C_5H_5)(PMe_3)_3 \quad (20)$$

A slurry of potassium perrhenate (1.50 g, 5.18 mmol) in absolute ethanol (10 mL) is degassed in a glass bomb of size 3 × 12 cm by evacuating while stirring at –78 °C. To the slurry is added PMe_3 (2.6 g, 34 mmol) by a trap-to-trap method in a vacuum. The mixture is brought to –78 °C to which deoxygenated, 36% aqueous HCl (1.2 mL, 14 mmol) is added. The resulting slurry is warmed to 110 °C for 19 h which causes the color of the solution to become bright lime-green. After having been cooled to –78 °C, the solution is treated with a second aliquot of 36% aqueous HCl (5.0 mL, 58 mmol). The resulting black solution is heated at 110 °C for 4 d when the solution becomes dark red. The solution is concentrated *in vacuo* to *ca* one half of its volume and then partitioned between water (100 mL) and benzene (50 mL). The aqueous layer is reextracted with 50 mL of benzene and the combined organic phases are washed with water (100 mL) and brine (100 mL). After being dried over K_2CO_3, the solution is concentrated on a rotary evaporator to give a brown-yellow, crystalline solid which is purified by flash chromatography (silica gel, 12 × 5 cm) in air. Elution with diethyl ether/hexane (2:1) produces an orange band followed by an yellow band. Evaporation of the yellow fraction provides paramagnetic *mer*-$ReCl_3(PMe_3)_3$ as a bright yellow powder. Yield: 2.29 g (85%). mp 174–175 °C. IR (KBr): 950 s cm^{-1}. ^1H NMR (benzene-d_6): δ(CH_3) = 2.26 (s, 18H), –2.59 ppm (s, 9H).

In a 500 mL flask are placed a slurry of *mer*-$ReCl_3(PMe_3)_3$ (3.00 g, 5.77 mmol), cyclopentadiene (25 mL, *ca* 310 mmol), K_2CO_3 (1.94 g, 14.0 mmol), and tetrahydrofuran (80 mL). To the vigorously stirred slurry sodium amalgam (0.787%, 40.34 g, 13.8 mmol Na) is added at once causing a color change to green and then to brown. After having been stirred for 18 h, the dark green slurry is filtered and the filtrate is evaporated to leave a dark green solid. Rapid filtration of the 5% ether/pentane solution of the solid through 4 × 14 cm alumina(III) column at –105 °C provides 150 mL of a pale yellow solution which is evaporated to dryness to give a pale yellow solid. In order to remove small amounts of contaminating dicyclopentadiene, the solid is dissolved in cyclohexane (75 mL) and the solution is frozen. The coolant is removed, and the volatile materials are sublimed off under high vacuum to give $Re(\eta^5\text{-}C_5H_5)(PMe_3)_3$ as an extremely air sensitive, fluffy white powder. Yield: 1.04 g (38%).

Irradiation of a benzene solution of $Re(\eta^5\text{-}C_5H_5)(PMe_3)_3$ in Pyrex vessels with a medium pressure mercury lamp gives $Re(\eta^5\text{-}C_5H_5)Ph(H)(PMe_3)_2$ in 64% yield.

Irradiation of $Re(\eta^5\text{-}C_5H_5)(PMe_3)_3$ in hexane, cyclopentane, or cyclohexane with methane affords the corresponding C—H activation products, $Re(\eta^5\text{-}C_5H_5)(n\text{-}C_6H_{13})(H)(PMe_3)_2$ (15%), $Re(\eta^5\text{-}C_5H_5)(c\text{-}C_5H_9)(H)(PMe_3)_2$ (17%), or $Re(\eta^5\text{-}C_5H_5)(CH_3)(H)(PMe_3)_2$ (43%), respectively [42].

Properties: fluffy white powder. mp decomposed above 250 °C without melting. Extremely air-sensitive. IR (benzene): 927 s cm^{-1}. ^1H NMR (benzene-d_6): $\delta(C_5H_5)$ = 4.19 ppm (quartet, J = 0.8 Hz, 5H); $\delta(CH_3)$ = 1.40 ppm (virtual d, J = 7.0 Hz, 27H). $^{13}C\{^1H\}$NMR (benzene-d_6): $\delta(C_5H_5)$ = 71.93 ppm (quartet, J_p = 1.7 Hz); $\delta(CH_3)$ = 30.10 ppm (m). $^{31}P\{^1H\}$NMR (benzene-d_6): $\delta(P(CH_3)_3)$ = –37.67 ppm (br, s).

(6) Methyltrioxorhenium(VII), $ReO_3(CH_3)$ [43, 44]

$$ReOCl_3(PPh_3)_2 + LiCH_3 \longrightarrow \xrightarrow{H_2O_2} ReO(CH_3)_4 \quad (21)$$

$$ReO(CH_3)_4 + O_2 \longrightarrow ReO_3CH_3 \quad (22)$$

To a suspension of $ReOCl_3(PPh_3)_2$ (8.31 g, 10.0 mmol) in diethyl ether (120 mL) is added slowly a diethyl ether solution (*ca* 1M) of methyllithium (80 mL, 80.0 mmol) at –78 °C. The mixture is allowed to warm slowly to room temperature. After having been stirred for 0.5 h, the dark brown solution is cooled to –78 °C, and water (*ca* 10 mL) is added dropwise to destroy excess of MeLi. After being allowed to warm to room temperature the solution is cooled again to –30 °C and hydrogen peroxide (2.5 g, 30% H_2O_2 diluted in 20 mL of H_2O) is added dropwise very slowly with vigorous stirring. The color of the solution rapidly changes to red and stirring is continued after warming to room temperature for *ca* 0.5 h. The solution is then cooled to –78 °C and the ether layer is filtered to remove ice, and dried first with anhydrous $CaCl_2$ at room temperature, and then with molecular sieves. The solution is cooled again to –78 °C, filtered, and the solvent is carefully removed at –50 °C. From the residue, crystalline $ReO(CH_3)_4$ is sublimed *in vacuo* onto a cold probe at –78 °C (10^{-3} mmHg). 1.26 g (*ca* 48%). Reddish purple crystals mp *ca* 45 °C. IR (CS_2): ν(Re O) = 1002 cm^{-1}. Very volatile. Sensitive to air but thermally very stable.

Tetramethyloxorhenium thus obtained (*ca* 100 mg) is placed in a 1 L round-bottomed flask. Dry air is admitted up to atmospheric pressure. The growth of long needles of ReO_3CH_3 can be seen in a few days, and after 4 weeks the yield exceeds 50%. The product is purified by sublimation *in vacuo*.

Properties: colorless needles. mp 110 °C. Stable in air. Soluble in acetonitrile, benzene, chloroform, ethanol or diethyl ether, and sparingly soluble in CS_2 or hexane. IR (CS_2): ν(Re—O) = 990 w and 960 s cm^{-1}. ^1H NMR ($CDCl_3$): $\delta(CH_3)$ = 2.6 ppm.

(7) (η^5-Pentamethylcyclopentadienyl)trioxorhenium(VII), $ReO_3\{\eta^5\text{-}C_5(CH_3)_5\}$ [45]

$$Re\{\eta^5\text{-}C_5(CH_3)_5\}(CO)_2(THF) + O_2 \longrightarrow ReO_3\{\eta^5\text{-}C_5(CH_3)_5\} \quad (23)$$

A tetrahyrofuran solution of dicarbonyl(η^5-pentamethylcyclopentadienyl)(tetrahydrofuran)rhenium(I) is placed in an autoclave and set aside under O_2 pressure (450 psi) for a day. The solvent is removed by evaporation *in vacuo* and the residue is column chromatographed on silica gel. Organic products are eluted and the organorhenium complex is eluted as a yellow band with diethyl ether. Evaporation of the yellow fraction and recrystallization of the resulting solid give yellow needles of $ReO_3\{\eta^5\text{-}C_5(CH_3)_5\}$. Yield 55%.

Properties: yellow needles. mp 192 °C. Soluble in benzene, diethyl ether, dichloromethane, tetrahydrofuran, and methanol. Air-stable both in solid state and in solution. Sublimable (40 °C / 10^{-2} Torr). IR (KBr): ν(Re—O) = 910 s and 881 vs cm^{-1}. ^1H NMR ($CDCl_3$): $\delta(CH_3) = 2.16$ ppm.

(8) Bis(tricyclohexylphosphine)heptahydridorhenium(VII), $ReH_7\{P(c\text{-}C_6H_{11})_3\}_2$ [46]

$$ReOCl_3(PPh_3)_2 + 2P(c\text{-}C_6H_{11})_3 \longrightarrow ReOCl_3\{P(c\text{-}C_6H_{11})_3\}_2 \quad (24)$$

$$ReOCl_3\{P(c\text{-}C_6H_{11})_3\}_2 + NaBH_4 \longrightarrow ReH_7\{P(c\text{-}C_6H_{11})_3\}_2 \quad (25)$$

To a slurry of $ReOCl_3(PPh_3)_2$ [47] (1.9 g, 2.3 mmol) in benzene (200 mL) is added an excess of $P(c\text{-}C_6H_{11})_3$ (2.9 g, 10 mmol). The yellow slurry turns into a clear green solution on stirring for 10–20 min. Stirring is continued for 1 h until a green slurry appears. The solvent is removed under vacuum and the green solid residue is extracted with 5 × 20 mL portions of diethyl ether to give a green solution and a yellow powder of $ReOCl_3\{P(c\text{-}C_6H_{11})_3\}_2$. Yield: 2.3 g (100%). IR (KBr): ν(Re—O) = 969 cm^{-1}; ν(C—H) = 2850, 2930 cm^{-1}. $^{31}P\{1H\}$NMR ($CDCl_3$): ν($P(c\text{-}C_6H_{11})_3$) = –27.0 ppm.

$ReOCl_3\{P(c\text{-}C_6H_{11})_3\}_2$ prepared as above (1.5 g, 1.7 mmol) and $NaBH_4$ (1.8 g, 48 mmol) are added to absolute ethanol (100 mL) and the mixture is stirred for 20 h at room temperature. After removal of ethanol *in vacuo*, the residue is extracted with 4 × 20 mL benzene. The benzene is removed under vacuum and the residue washed with absolute ethanol, giving $ReH_7\{P(c\text{-}C_6H_{11})_3\}_2$ as a beige solid in 70% yield. Further purification is achieved by recrystallization from benzene–hexane.

Properties: white powder. IR (KBr): ν(Re—H) = 1970 w br, 1935 sh, 1915 s cm^{-1}. ^1H NMR (benzene-d_6): $\delta(C_5H_5)$ = 4.19 ppm (quartet, J = 0.8 Hz, 5H); $\delta(c\text{-}C_6H_{11})$ = 2.22 (d, J = 13 Hz, 12H), 1.80 (18H), 1.64 (18H), 1.20 (18H) ppm; –6.50 ppm (t, $J_{P\text{-}H}$= 20 Hz). $^{13}C\{^1H\}$NMR (benzene-d_6): $\delta(c\text{-}C_6H_{11})$ = 40.4 (d of t, 1C), 30.5 (t, 2C), 28.1 (t of t, 2C), 27.2 (t, 1C) ppm. $^{31}P\{^1H\}$NMR (benzene-d_6): $\delta(P(c\text{-}C_6H_{11})_3)$ = 47.2 ppm.

(9) Bis{1,1-bis(trifluoromethyl)ethoxo}(2,6-diisopropylphenylimido)-(neopentylidyne)rhenium(VII), $Re\{N(2,6\text{-}C_6H_3\text{-}i\text{-}Pr)\}(CCMe_3)\{OCMe(CF_3)_2\}_2$ [48]

$$Re\{N(2,6\text{-}C_6H_3\text{-}i\text{-}Pr)\}_2(CHCMe_3)Cl + HCl\ (gas) \longrightarrow [N(2,6\text{-}C_6H_3\text{-}i\text{-}Pr)H_3][Re\{NH(2,6\text{-}C_6H_3\text{-}i\text{-}Pr)\}(CCMe_3)Cl_4] \quad (26)$$

$$[N(2,6\text{-}C_6H_3\text{-}i\text{-}Pr)H_3][Re\{NH(2,6\text{-}C_6H_3\text{-}i\text{-}Pr)\}(CCMe_3)Cl_4] + NEt_4Cl \longrightarrow [NEt_4][Re\{NH(2,6\text{-}C_6H_3\text{-}i\text{-}Pr)\}(CCMe_3)Cl_4] \quad (27)$$

$$[NEt_4][Re\{NH(2,6\text{-}C_6H_3\text{-}i\text{-}Pr)\}(CCMe_3)Cl_4] + LiOC(CF_3)_2Me \longrightarrow Re\{N(2,6\text{-}C_6H_3\text{-}i\text{-}Pr)\}(CCMe_3)\{OCMe(CF_3)_2\}_2 \quad (28)$$

To a stirred solution of Re{N(2,6-C_6H_3-*i*-Pr)}$_2$($CHCMe_3$)Cl [49] (2.0 g, 3.12 mmol) in diethyl ether (250 mL) is added via syringe HCl gas (210 mL, 0.4 mmol) at –78 °C. A fine orange precipitate of [N(2,6-C_6H_3-*i*-Pr)H_3][Re{NH(2,6-C_6H_3-*i*-Pr)}($CCMe_3$)Cl_4] forms after 5 min and the mixture is then warmed to 25 °C over 1h. The precipitate is collected by filtration and washed with diethyl ether (250 mL). 2.24 g (96%). IR (Nujol): ν(N—H of ArNH) = 3210 br; ν(N—H of $ArNH_3$) = 3065 v br cm^{-1}.

Re{N(2,6-C_6H_3-*i*-Pr)}$_2$($CHCMe_3$)Cl (10.62 g, 14.1 mmol) formed as above is partially dissolved in CH_2Cl_2. Over an hour, tetraethylammonium chloride (3.69 g, 21.2 mmol) is added as the mixture is stirred vigorously. The dark-red slightly cloudy solution is filtered and concentrated slowly until red crystals begin to form. Diethyl ether is added carefully until most of the product precipitate. The flask is cooled and left overnight at –40 °C to give red microcrystals of [NEt_4][Re{NH(2,6-C_6H_3-*i*-Pr)}($CCMe_3$)Cl_4] in two crops. 9.60 and 0.23 g (99%). IR (Nujol): (N–H) = 3190 cm^{-1}.

LiOC(CF_3)$_2$Me (1.71 g, 9.09 mmol) is added to [NEt4][Re{NH(2,6-C_6H_3-*i*-Pr)}($CCMe_3$)Cl_4] (2.00 g, 2.84 mmol) in 110 ml of CH_2Cl_2 at –30 °C. The mixture is allowed to warm to 25 °C over a period of 1 h to give a red solution containing suspended lithium chloride. The mixture is filtered through Celite and the solvent is removed from the filtrate *in vacuo*. The residue is extracted with pentane and the extract is reduced in volume to yield the red crystalline product, Re{N(2,6-C_6H_3-*i*-Pr)}($CCMe_3$){OCMe(CF_3)$_2$}$_2$, which can be recrystallized from cold pentane. 1.70 g (75%).

Properties: red crystals. $^{13}C\{^1H\}$NMR (benzene-d_6): δ(carbyne C_α) = 304.6 ppm. Re{N(2,6-C_6H_3-*i*-Pr)}($CCMe_3$){OCMe(CF_3)$_2$}$_2$ reacts with 3-hexyne to give a rhenacyclobutadiene derivative. 2,2-Dimethyl-3-hexyne is lost rapidly from the initially formed α-*tert*-butyl-substituted metallacycles.

$$(RO)_2Re(\equiv C\text{-}t\text{-}Bu)(=NAr) + EtC\equiv CEt \longrightarrow (RO)_2(ArN\equiv)Re[C(Et)C(Et)C(Et)] + t\text{-}BuC\equiv CEt \qquad (29)$$

RO– = Me–C(CF_3)$_2$–O–   Ar = 2,6-($CH(CH_3)_2$)$_2C_6H_3$

References

[1] J. Emsley, *The Elements*, Clarendon Press, Oxford (1989).

[2] C. R. Hammond, *CRC Handbook of Chemistry and Physics*, Ed. D. R. Lide (1995).

[3] R. B. King, *Organometallic Syntheses*, Volume 1, Academic Press, New York, pp. 89–93 (1965).

[4] (a) F. A. Cotton, G. Wilkinson, *Advanced Inorganic Chemistry, A Comprehensive Text*, 4th Ed., John Wiley, New York (1980).
(b) G. Wilkinson, F. G. A. Stone, E. W. Abel Ed., *Comprehensive Organometallic Chemistry*, Vol. 4, Sections 29.2 and 30.2, Pergamon Press, Oxford (1982).
(c) D. J. Cardin, S. A. Cotton, M. Green, J. A. Labinger, Ed., *Organometallic Compounds of*

the Lanthanides, Actinides and Early Transition Metals, Chapman and Hall, London (1985).
[5] G. L. Geoffroy, M. S. Wrighton, *Organometallic Photochemistry*, Academic Press, New York, p. 139 (1979).
[6] E. O. Fischer, R. Jira, *Z. Naturforsch., Teil B*, **9**, 618 (1954): T. S. Piper, F. A. Cotton, G. Wilkinson, *J. Inorg. Nucl. Chem.*, **1**, 165 (1955).
[7] R. B. King, *Organometallic Syntheses*, Volume 1, Academic Press, New York, pp. 111–114, (1965).
[8] R. B. King, *Organometallic Syntheses*, Volume 1, Academic Press, New York, pp. 163–164, (1965).
[9] H. Brunner, *Angew. Chem., Int. Ed. Engl.*, **8**, 382 (1969); H. Brunner, H. D. Schindler, *J. Organomet. Chem.*, **24**, C7 (1970).
[10] D. Sellmann, W. Weiss, *J. Organometal. Chem.*, **160**, 183 (1978).
[11] R. D. Closson, J. Kozikowski, T. H. Coffield, *J. Org. Chem.*, **22**, 598 (1957).
[12] K. Noack, F. Calderazzo, *J. Organomet. Chem.*, **10**, 101 (1967).
[13] R. A. Andersen, E. Carmona-Guzman, J. F. Gibson, G. Wilkinson, *J. Chem. Soc., Dalton Trans.*, 2204 (1976).
[14] K. Maruyama, T. Ito, A. Yamamoto, *Bull. Chem. Soc. Jpn.*, **52**, 849 (1979).
[15] A. Davison, J. A. McCleverty, G. Wilkinson, *J. Chem. Soc.*, 1133 (1963).
[16] M. A. Andrew, S. W. Kirtley, H. D. Kaesz, *Inorg. Synth.*, **17**, 66 (1977).
[17] J. R. Johnson, H. D. Kaesz, *Inorg. Synth.*, **18**, 60 (1978).
[18] H. D. Kaesz, R. Bau, D. Hendrickson, J. M. Smith, *J. Am. Chem. Soc.*, **89**, 2844 (1967).
[19] Yu. T. Struchkov, K. N. Anisimov, O. P. Osipova, N. E. Kolobova, A. N. Nesmeyanov, *Dokl. Akad. Nauk SSSR*, **172**, 107 (1967); *Dokl. Chem. (Engl. Transl.)*, **172**, 15 (1967).
[20] M. L. H. Green, G. Wilkinson, *J. Chem. Soc.*, 4314 (1958); C. P. Casey, M. A. Andrews, D. R. McAlister, J. E. Rinz, *J. Am. Chem. Soc.*, **102**, 1927 (1980).
[21] A. T. Patton, C. E. Strouse, C. B. Knobler, J. A. Gladysz, *J. Am. Chem. Soc.*, **105**, 5804 (1983).
[22] W. E. Buhro, S. Georgiou, J. M. Fernandez, A. T. Patton, C. E. Strouse, J. A. Gladysz, *Organometallics*, **5**, 956 (1986).
[23] C. F. Barrientos-Penna, A. B. Gilchrist, A. H. Klahn-Oliva, A. J. L. Hanlan, D. Sutton, *Organometallics*, **4**, 478 (1985).
[24] F. W. B. Einstein, A. H. Klahn-Oliva, D. Sutton, K. G. Tyers, *Organometallics*, **5**, 53 (1986).
[25] R. B. King, *Organometallic Syntheses*, Volume 1, Academic Press, New York, pp. 80–81, (1965).
[26] D. Baudry, M. Ephiritikhine, *J. Chem. Soc., Chem. Commun.*, 895 (1979).
[27] K. Mertis, G. Wilkinson, *J. Chem. Soc., Dalton Trans.*, 1488 (1976); I. R. Beattie, P. J. Jones, *Inorg. Chem.*, **18**, 2318 (1979).
[28] J. M. Huggins, D. R. Whitt, L. Lebioda, *J. Organomet. Chem.*, **312**, C15 (1986).
[29] W. A. Herrmann, R. Serrano, H. Bock, *Angew. Chem., Int. Ed. Engl.*, **23**, 383 (1984); W. A. Herrmann, R. Serrano, U. Küsthardt, M. L. Ziegler, E. Guggolz, T. Zahn, *Angew. Chem., Int. Ed. Engl.*, **23**, 515 (1984); W. A. Herrmann, R. Serrano, M. L. Ziegler, H. Pfisterer, B. Nuber, *Angew. Chem., Int. Ed. Engl.*, **24**, 50 (1985); H. J. R. de Boer, B. J. J. van de Heisteeg, M. Flöel, W. A. Herrmann, O. S. Akkerman, F. Bickelhaupt, *Angew. Chem., Int. Ed. Engl.*, **26**, 73 (1987); J. K. Felixberger, P. Kiprof, E. Herdtweck,W. A. Herrmann, R. Jazobi, P. Gütlich, *Angew. Chem., Int. Ed. Engl.*, **28**, 334 (1989); W. A. Herrmann, J. K. Felixberger, R. Anwander, E. Herdtweck, P. Kiprof, J. Riede, *Organometallics*, **9**, 1434 (1990).
[30] A. Fukuoka, S. Komiya, *Organic Chemistry of the Early Transition Metals*, Gakkai Shuppan Center, Tokyo, p. 64 (1993).
[31] S. C. Abrahams, A. P. Ginsberg, K. Knox, *Inorg. Chem.*, **3**, 558 (1964).
[32] T. F. Koetzle, R. K. McMullan, R. Bau, D. W. Hart, R. G. Teller, D. L. Tipton, R. D. Wilson, *Transition Metal Hydrides*, Ed. R. Bau, Chapter 5, American Chemical Society, New York (1978).
[33] R. Bau, W. E. Carroll, R. G. Teller, T. F. Koetzle, *I. Am. Chem. Soc.*, **99**, 3872 (1977).
[34] R. B. King, *Organometallic Syntheses*, Volume 1 Academic Press, New York, pp. 158–160, (1965); R. B. King, F. G. A. Stone, *Inorg. Synth.*, **7**, 198 (1963).

[35] W. Hieber, G. Wagner, *Z. Naturforsch.*, **12b**, 478 (1957); **13b**, 339 (1958).
[36] B. F. G. Johnson, R. D. Johnston, J. Lewis, B. H. Robinson, *Inorg. Synth.*, **12**, 43 (1970).
[37] H. D. Kaesz, *Inorg. Synth.*, **26**, 156 (1989); R. B. King, *Organometallic Syntheses*, Volume 1, Academic Press, New York, pp. 147–148, (1965).
[38] W. Hieber, G. Faulhaber, F. Theubert, *Z. Anorg. Allg. Chem.*, **314**, 125 (1962).
[39] W. Hieber, M. Hofler, J. Muschi, *Chem. Ber.*, **98**, 311 (1965).
[40] W. Beck, K. Raab, *Inorg. Synth.*, **26**, 107 (1989).
[41] E. O. Fischer and A. Maasböl, *Chem. Ber.*, **100**, 2445 (1967).
[42] T. T. Wenzel, R. G. Bergman, *J. Am. Chem. Soc.*, **108**, 4856 (1986).
[43] K. Mertis, G. Wilkinson, *J. Chem. Soc., Dalton Trans.*, 1488 (1976).
[44] I. R. Beattie, P. J. Jones, *Inorg. Chem.*, **18**, 2318 (1979).
[45] A. H. Klahn-Olivia, D. Sutton, *Organometallics*, **2**, 1313 (1984).
[46) E. H. K. Zeiher, D. G. DeWit, K. G. Caulton, *J. Am. Chem. Soc.*, **106**, 7006 (1984).
[47] G. W. Parshall, *Inorg. Synth.*, **17**, 110 (1977); N. P. Johnson, C. J. L. Lock, G. Wilkinson, *Inorg. Synth.*, **9**, 145 (1969).
[48] L. A. Weinstock, R. R. Schrock, W. M. Davis, *J. Am. Chem. Soc.*, **113**, 135 (1991).
[49] A. D. Horton, R. R. Schrock, *Polyhedron*, **7**, 1841 (1988).

10 Group 8 (Fe, Ru, Os) Metal Compounds

S. Komiya and M. Hurano, *Tokyo University of Agriculture and Technology*

10.1 Introduction

Organometallic complexes of Group 8 having carbon monoxide, cyclopentadienyl or tertiary phosphines as supporting ligands have been widely investigated. Because the starting materials for the preparation of Group 8 complexes are commercially available and fairly inexpensive, many alkyl, alkene, alkyne, alkylidene, and alkoxy complexes have been reported. In particular, relatively low valent complexes such as $Fe(CO)_5$, $FeCl_2$, $FeCp_2$ and $RuCl_3 \cdot 3H_2O$ are important for starting materials of organoiron and -ruthenium complexes. On the other hand, many organoosmium complexes are prepared from the complexes with higher oxidation states such as OsO_4 or Na_2OsCl_6.

The organoiron, -ruthenium and -osmium complexes generally have oxidation states of −2 to +4, −2 to +4 and −2 to +2, respectively [1]. Organoosmium chemistry has received less attention than that of iron and ruthenium, because of the high price of Os metal.

10.2 Iron Complexes

Organoiron complexes have played a central role in organometallic chemistry since $Fe(CO)_5$ was discovered in 1891 [2–3]. The Fischer–Tropsch reaction developed in 1925 using a heterogeneous iron complex encouraged interest in the chemistry of iron carbonyl and carbene complexes. On the other hand, the discovery and unexpected stability of ferrocene opened the modern era of organometallic chemistry [4]. Applications of organoiron complexes to organic synthesis have been developed since the 1960's, particularly the use of $Fe(CO)_5$, $Fe_2(CO)_9$, and $Na_2[Fe(CO)_4]$ in promoting isomerization of olefins, dehalogenations, carbonylations, and so on.

In the following sections, series of iron complexes are reviewed, divided into 4 groups according to their supporting ligands. The first group are homoleptic carbonyliron complexes. They are one of the most well investigated organoiron complexes for organic syntheses such as olefin isomerizations, olefin cycloadditions, and carbonylations. The simplest mononuclear carbonyliron complex is $Fe(CO)_5$. They are not only used for organic synthesis, but also used as starting materials of many carbonyliron

Synthesis of Organometallic Compounds: A Practical Guide. Edited by S. Komiya

derivatives. In this section, their chemical reactivities are described. The second group are iron complexes with a diene ligand. They frequently contain a carbonyl ligand as well. However, they should be described individually because they are extensively studied to date. The third group are iron complexes with cyclopentadienyl ligands. Ferrocene, $FeCp_2$, is one of the most important compounds in the rapid developement of organometallic chemistry in the last half century. The recent developments using asymmetric ferrocenyl phosphines are also noteworthy. A monocyclopentadienyliron fragment, $CpFe(CO)_2$, is interesting because it provides various kinds of organoiron complexes. The last group are iron complexes having phosphine ligands. Low valent iron complexes having phosphine ligands are of interest for activation of C–H bonds or small molecules. The chemistry of the other organoiron complexes are dealt with in reviews [5].

10.2.1 Iron Carbonyl Complexes

Iron carbonyls are among the most investigated compounds in both organic and organometallic chemistry. During studies of their chemical reactivities, several characteristic properties are revealed [2]. For example, carbonyliron complexes are strong reducing agents, and can abstract oxygen or halogen, and they are also capable of catalyzing carbonylation under a CO atmosphere. In carbonyliron complexes having an alkene, diene, or alkyne ligand, the unsaturated ligand is susceptible to addition of nucleophiles because of the strong electron withdrawing property of the $Fe(CO)_n$ fragment. On the other hand, it is shown that they react with some electrophiles such as H^+ or $MeCO^+$. By virtue of such various reactivities of carbonylirons, they are used for organic synthesis or preparation of organoiron complexes, in particular, $Fe(CO)_5$, $Fe_2(CO)_9$ and $Fe_3(CO)_{12}$ (Figure 10.1).

All three Group 8 metals form trinuclear clusters $M_3(CO)_{12}$. However, while all carbonyl ligands in ruthenium and osmium dodecacarbonyl complexes coordinate to the metal center as terminal carbonyls, there are two bridged carbonyl groups in iron dodecacarbonyl. This may be due to the smaller van der Waals radius of the iron atom. In this section, $Fe(CO)_5$, $Fe_2(CO)_9$ and $Na_2[Fe(CO)_4]$ are reviewed.

Pentacarbonyliron was prepared independently by Mond [2] and Berthelot [3] in 1891 by the direct reaction of finely ground iron with carbon monoxide. It is an air-stable liquid at room temperature which is used as a starting material for iron(0) chemistry (Scheme 10.1). Corresponding pentacarbonyls of ruthenium and osmium have found less application due to their facile decomposition to $M_3(CO)_{12}$. Since $Fe(CO)_5$ is commercially available and inexpensive, there is no need to synthesize it in the labora-

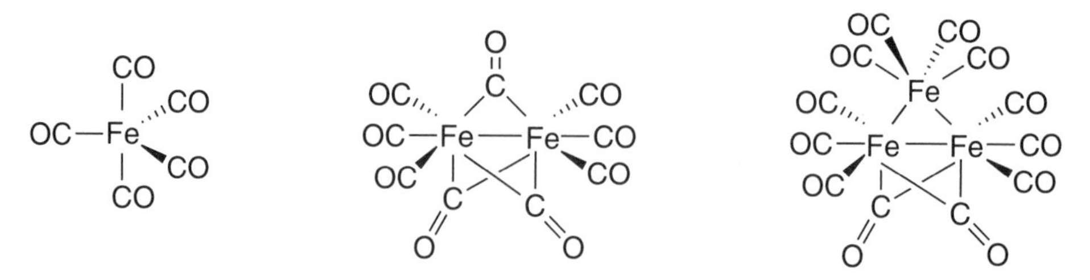

Figure 10.1 Structures of $Fe(CO)_5$, $Fe_2(CO)_9$ and $Fe_3(CO)_{12}$.

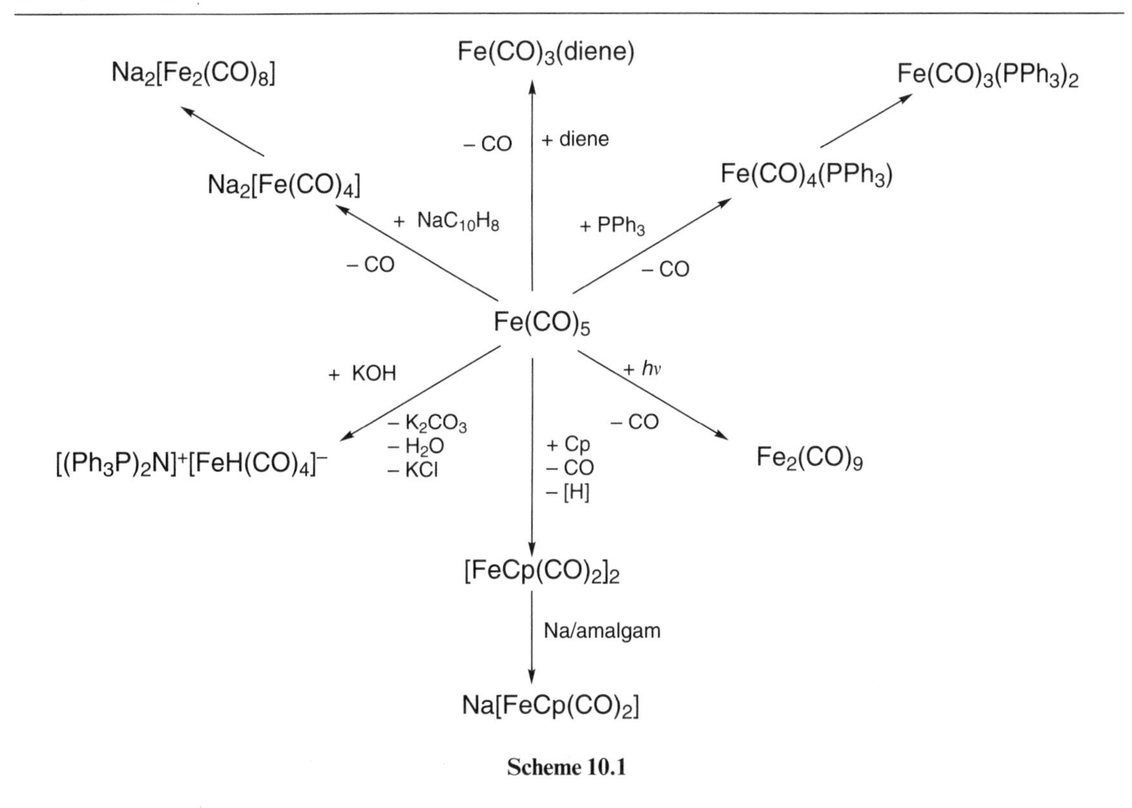

Scheme 10.1

tory except for preparing isotopically labeled compounds. Because $Fe(CO)_5$ is toxic (even though less toxic than $Ni(CO)_4$) with relatively high vapor pressure (21 mmHg, 20 °C), it should be handled with care under a well ventilated hood.

Dienes can displace two carbonyl ligands from $Fe(CO)_5$ giving $Fe(CO)_3$(diene) [diene = 1,3- cyclohexadiene [6], norbornadiene [7], cyclooctatetraene [8] or cyclobutadiene [9]]. Tertiary phosphines (L) give $Fe(CO)_4(L)$ and $Fe(CO)_3(L)_2$. On the other hand, if a carbonyl ligand is lost from $Fe(CO)_5$ in the absence of an appropriate 2e donor, another $Fe(CO)_5$ molecule can be trapped by the labile "$Fe(CO)_4$" species to form $Fe_2(CO)_9$ in high yield [10]. Interestingly, treatment of $Fe(CO)_5$ with dicyclopentadiene results in formation of the dinuclear complex $[FeCp(CO)_2]_2$ containing monovalent iron, which can be readily reduced to the Fe(0) anion complex $[Fe(CO)_2Cp]^-$ by sodium amalgam. Because the carbonyl ligand is a strong π-acid iron carbonyl complexes, its complexes can be isolated by the reduction of $Fe(CO)_5$ with base [11–12].

Enneacarbonyldiiron can be derived by the exposure of pentacarbonyliron to UV light and is also commercially available. It is an air-stable shiny yellow crystalline solid at room temperature. $Fe_2(CO)_9$ is less important as a precursor of iron complexes than as a reagent in organic synthesis. However, the following two reactions [13–14] are interesting because $Fe_2(CO)_9$ is used as a $Fe(CO)_4$ synthon (Scheme 10.2).

Pentacarbonyliron and enneacarbonyldiiron are also applied to many organic syntheses and show basically similar chemical reactivity. However, the former sometimes gives higher yields than the latter. Several typical reactions by pentacarbonyl iron and enneacarbonyldiiron are described.

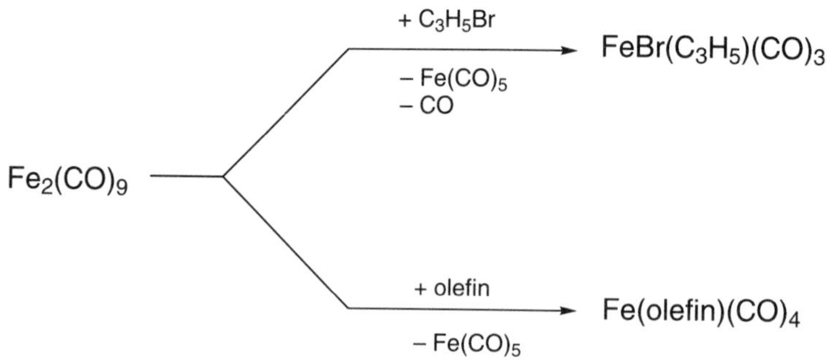

$$Fe_2(CO)_9 \xrightarrow[-\,Fe(CO)_5,\ -\,CO]{+\,C_3H_5Br} FeBr(C_3H_5)(CO)_3$$

$$Fe_2(CO)_9 \xrightarrow[-\,Fe(CO)_5]{+\,\text{olefin}} Fe(\text{olefin})(CO)_4$$

Scheme 10.2

10. 2.1.1. Isomerization of olefins

The reaction of pentacarbonyliron with mono–olefins is known to give olefin–$Fe(CO)_4$ complexes. These are probably intermediates in olefin isomerization, which generally occurs via a 1,3-hydrogen shift in a η^3-allylic intermediate (Scheme 10.3) [15].

Various kinds of olefin isomerizations are effectively promoted by $Fe(CO)_5$. However, because the isomerized product is usually released from the iron complex by oxidation, these isomerizations require a stoichiometric amount of pentacarbonyliron as described below (Scheme 10.4, eq (1)) [16–17].

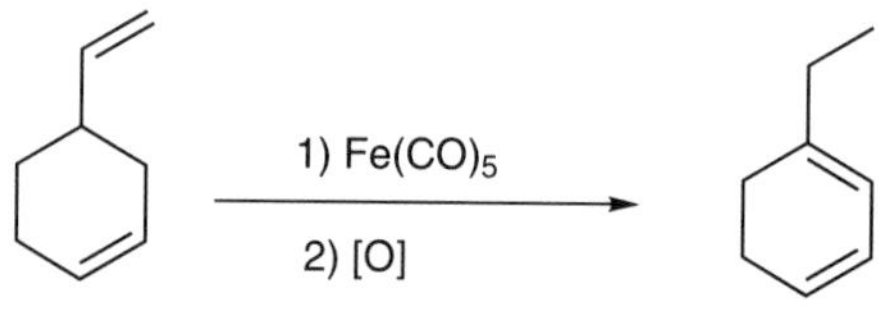

1) $Fe(CO)_5$

2) [O] (1)

On the other hand, catalytic isomerization of olefin by pentacarbonyliron can also occur (eq (2)) [18].

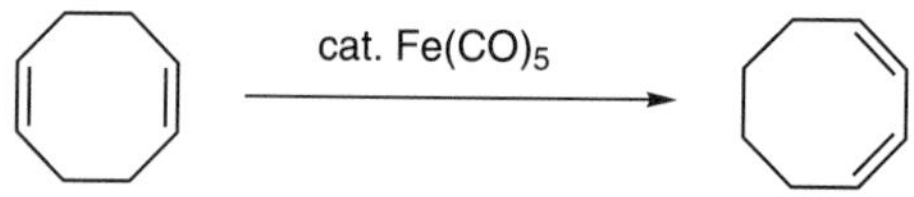

cat. $Fe(CO)_5$ (2)

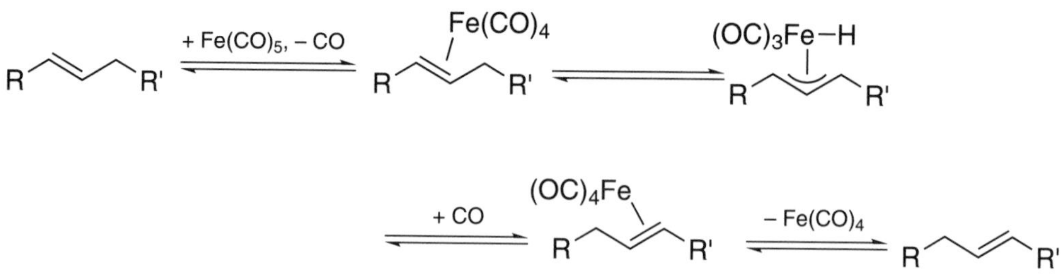

R R' + $Fe(CO)_5$, – CO ⇌ $Fe(CO)_4$ R R' ⇌ $(OC)_3Fe$–H R R'

+ CO ⇌ $(OC)_4Fe$ R R' – $Fe(CO)_4$ ⇌ R R'

Scheme 10.3

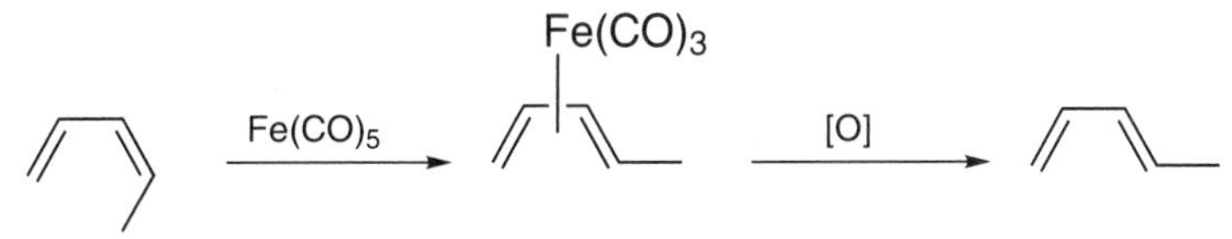

Scheme 10.4

The driving force for these reactions is considered to be the thermodynamic stability of the 1,3-diene-$Fe(CO)_3$ intermediate.

10.2.1.2 Carbonyl Insertion

Carbonyl insertion reactions are also carried out in the presence of pentacarbonyliron and are widely used for organic synthesis. Reaction of pentacarbonyliron with an organolithium followed by protonation gives an aldehyde derivative; if a suitable alkyl halide is used in place of acid, a ketone is obtained (Scheme 10.5) [19].
These reactions probably occur by the nucleophilic addition of the aryl anion to one of the CO groups of pentacarbonyliron followed by the addition of electrophiles. Pentacarbonyliron also causes carbonylation of electrophilic aryldiazonium salts [20].

Pentacarbonyliron is also capable of producing lactone from epoxides [21]. The reaction of dienemonoepoxide with $Fe(CO)_5$ gives a μ-π-allylcarbonyliron complex. Unsaturated lactone is obtained after treatment of the allyliron intermediate with cerium (Scheme 10.6).

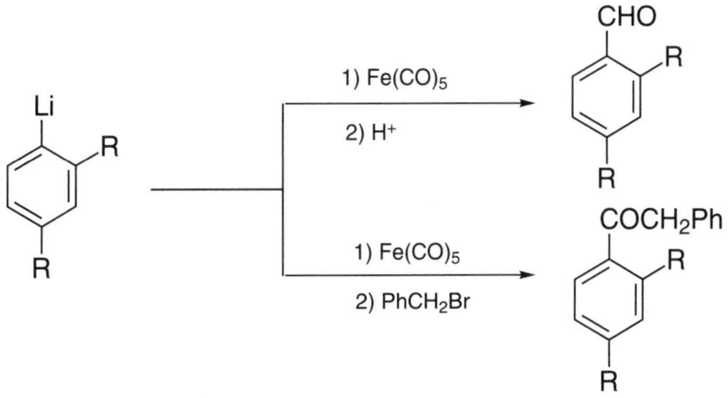

Scheme 10.5

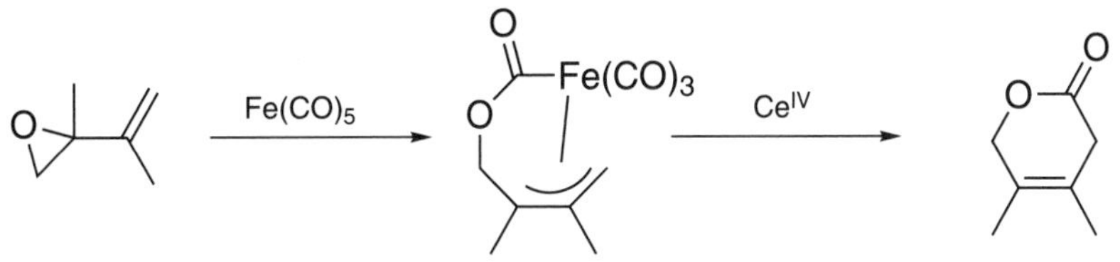

Scheme 10.6

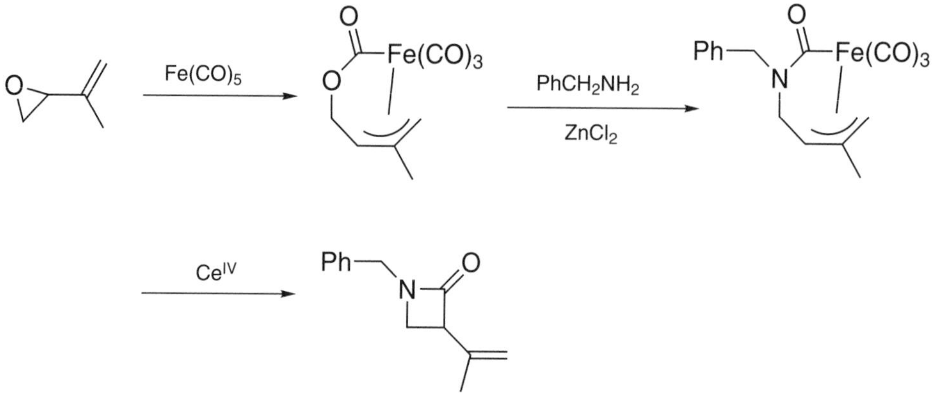

$$\text{(epoxide)} \xrightarrow{Fe(CO)_5} \text{(allyl-Fe(CO)}_3\text{ lactone)} \xrightarrow[ZnCl_2]{PhCH_2NH_2} \text{(allyl-Fe(CO)}_3\text{ lactam)} \xrightarrow{Ce^{IV}} \text{(}\beta\text{-lactam)}$$

Scheme 10.7

This reaction can also be applied to β-lactam synthesis as depicted below. Treatment of the μ-π-allylcarbonyliron intermediate with benzylamine followed by oxidation by cerium(IV) yields the β-lactam framework in high yield (Scheme 10.7) [22].

10.2.1.3 Dehalogenation

Treatment of an α-bromoketone with pentacarbonyliron results in coupling, loss of Br^- and formation of 1,4-diketones (eq (3)) [23].

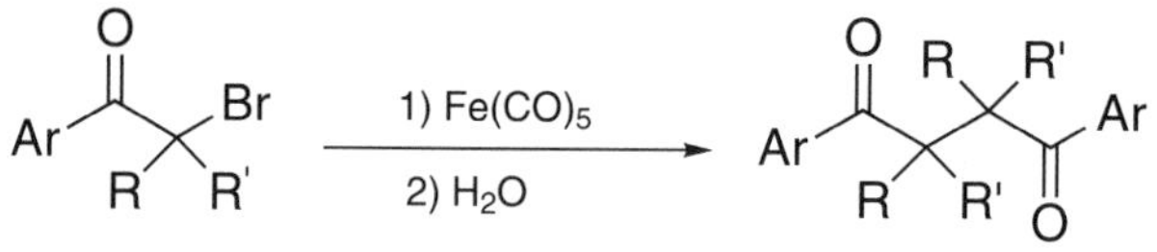

$$ArCOCRR'Br \xrightarrow[2)\ H_2O]{1)\ Fe(CO)_5} ArCOCRR'CRR'COAr \qquad (3)$$

10.2.1.4 Hydroquinone Synthesis

Reaction of acetylene and carbon monoxide with water in the presence of $Fe(CO)_5$ yields hydroquinone. Although this reaction occurs stoichiometrically at 50–80 °C, the reaction becomes catalytic and a yield of up to 70 % is obtained under high CO pressure (600–700 atm) (eq (4)) [24].

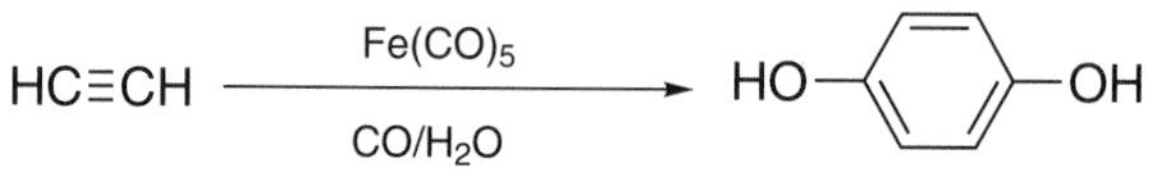

$$HC{\equiv}CH \xrightarrow[CO/H_2O]{Fe(CO)_5} HO{-}C_6H_4{-}OH \qquad (4)$$

Substituted acetylenes also react under similar conditions to form the corresponding alkylhydroquinones. Of particular interest is that a mixture of methyl acetylene and dimethyl acetylene yields trimethylhydroquinone, which is an important precursor of vitamin E (eq (5)) [25].

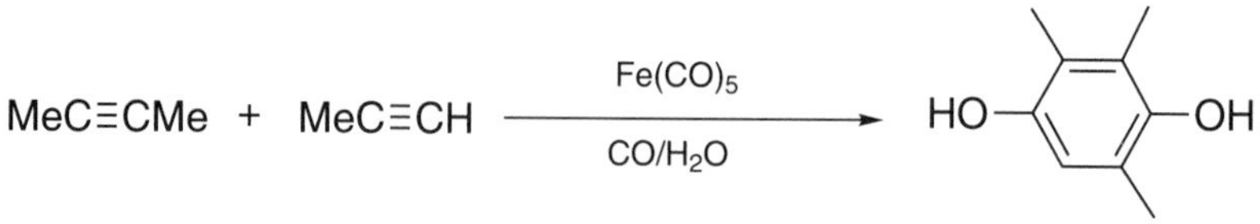

$$MeC{\equiv}CMe + MeC{\equiv}CH \xrightarrow[CO/H_2O]{Fe(CO)_5} HO{-}C_6HMe_3{-}OH \qquad (5)$$

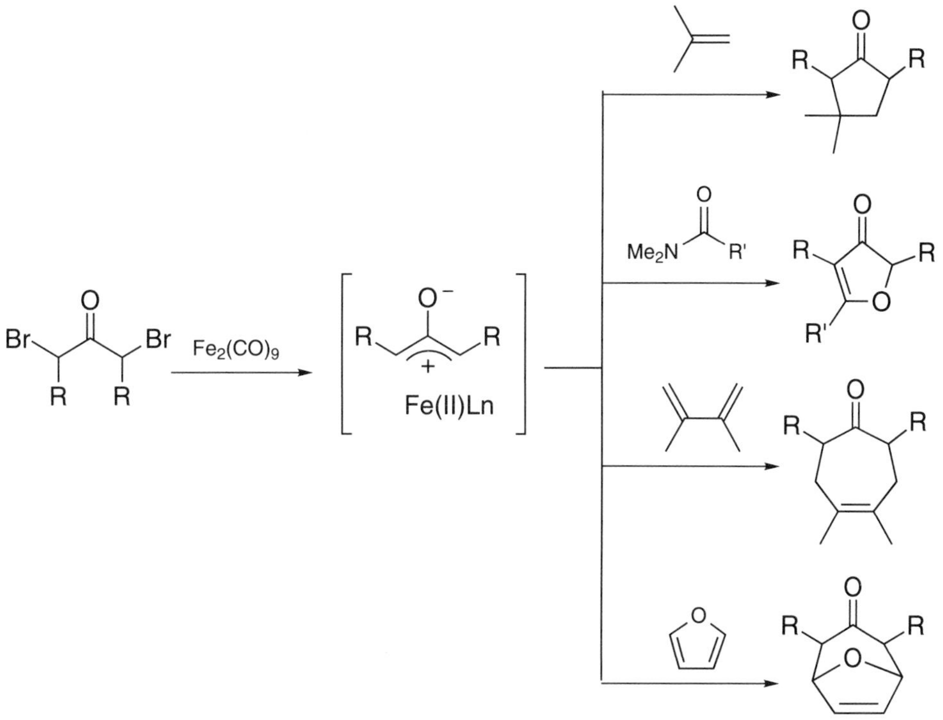

Br R O Br R $Fe_2(CO)_9$ [R O⁻ + R Fe(II)Ln] Me₂N O R' R O R R O R R' O R O R R O R O

Scheme 10.8

10.2.1.5 [3+2] and [3+4] Cycloadditions

Reaction of α,α'-dibromoketones with enneacarbonyldiiron gives oxoallyl cations which easily react with alkenes, enamines, and amides via [3+2] and with dienes via [3+4] cycloadditions, respectively, in high yield (Scheme 10.8) [26–29].
Although similar cycloadditions promoted by zinc–copper have been reported, they usually gave lower yields than those of the iron catalyst [30].

10.2.1.6 Desulfurization

Enneacarbonyldiiron acts as a strong reducing reagent. For example, desulfurization of episulfides has been carried out stereoselectively by $Fe_2(CO)_9$ (eq (6)) [31–32].

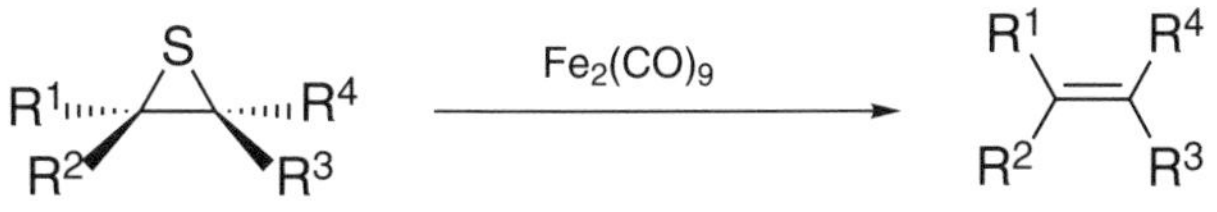

S R^1 R^4 R^2 R^3 $Fe_2(CO)_9$ R^1 R^4 R^2 R^3 (6)

10.2.1.7 Preparation of Organometallic Complexes by Dehalogenation

By using the reducing capability of enneacarbonyldiiron, several organoiron complexes have been prepared by dehalogenation. For example, treatments of enneacarbonyldiiron with dichloroisobutene [33–34] or 1,2-dichlorocyclobut-3-ene [35] generates trimethylenemethane or cyclobutadiene complexes of tricarbonyliron (eq (7) and (8)).

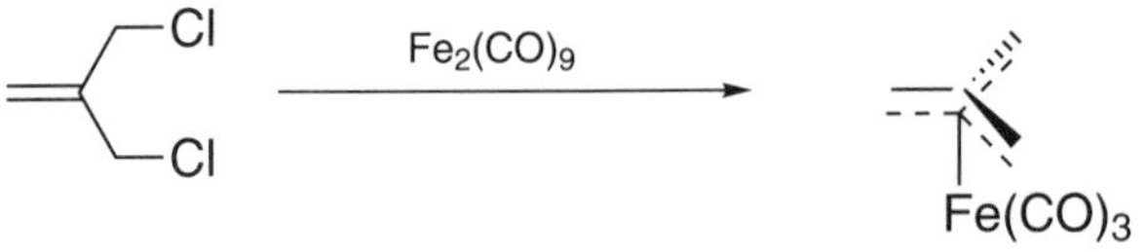

Cl, Cl; $Fe_2(CO)_9$ → $Fe(CO)_3$ (7)

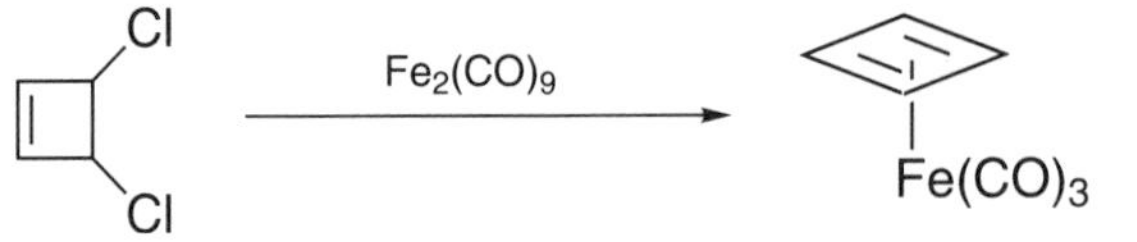

Cl, Cl; $Fe_2(CO)_9$ → $Fe(CO)_3$ (8)

These dihalo-compounds presumably coordinates to iron(0) and then the excess iron(0) species removes the halogen atoms from the coordinated dihalo-alkenes.

Disodium tetracarbonylferrate (Collman's reagent) is prepared by the reduction of $Fe(CO)_5$ with sodium naphthalene in THF or sodium benzophenone ketyl in dioxane. Though it can be isolated as a white precipitate, it is usually used in THF or dioxane solution without isolation because of its highly air-sensitive and pyrophoric character. It is reported that the solubility of $Na_2[Fe(CO)_4]$ is 7×10^{-3} M in THF and that it can be stored for moderate periods in an inert atmosphere in the dark [36]. X-ray crystallography of $Na_2[Fe(CO)_4]$ (Fig. 10.2) shows that the C—Fe—C bond angle opposite to the sodium cations is significantly distorted (129.7°) [37]. However, this distortion is reduced for the potassium analogue [38].

Oxidative addition reactions of $Na_2[Fe(CO)_4]$ have been the subject of many studies. It reacts with alkyl and acyl halides to yield the corresponding ketones, aldehydes, acids, amides and esters as shown below (Scheme 10.9) [39].

The reactions of $Na_2[Fe(CO)_4]$ with alkyl halides or alkyl *p*-toluenesulfonates proceed with inversion in stereochemistry via S_N2 mechanism (eq (9)) [39].

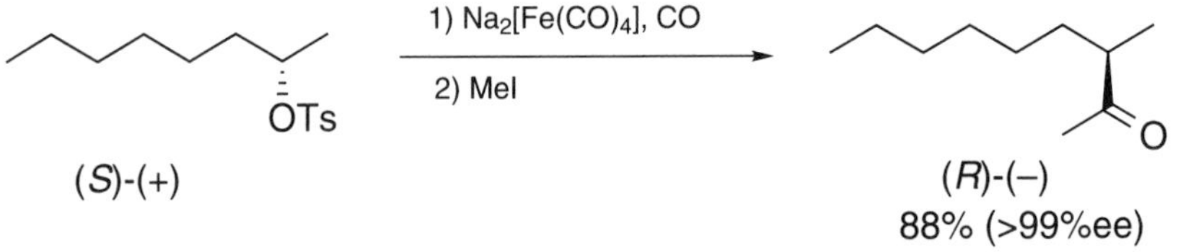

OTs (*S*)-(+); 1) $Na_2[Fe(CO)_4]$, CO; 2) MeI → O (*R*)-(−) 88% (>99%ee) (9)

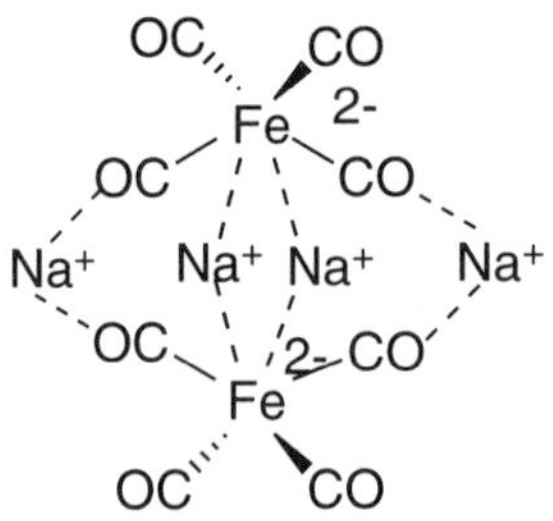

OC, CO, Fe, 2-, OC, CO, Na+, Na+, Na+, Na+, OC, 2-, CO, Fe, OC, CO

Figure 10.2 Structure of $Na_2[Fe(CO)_4]$.

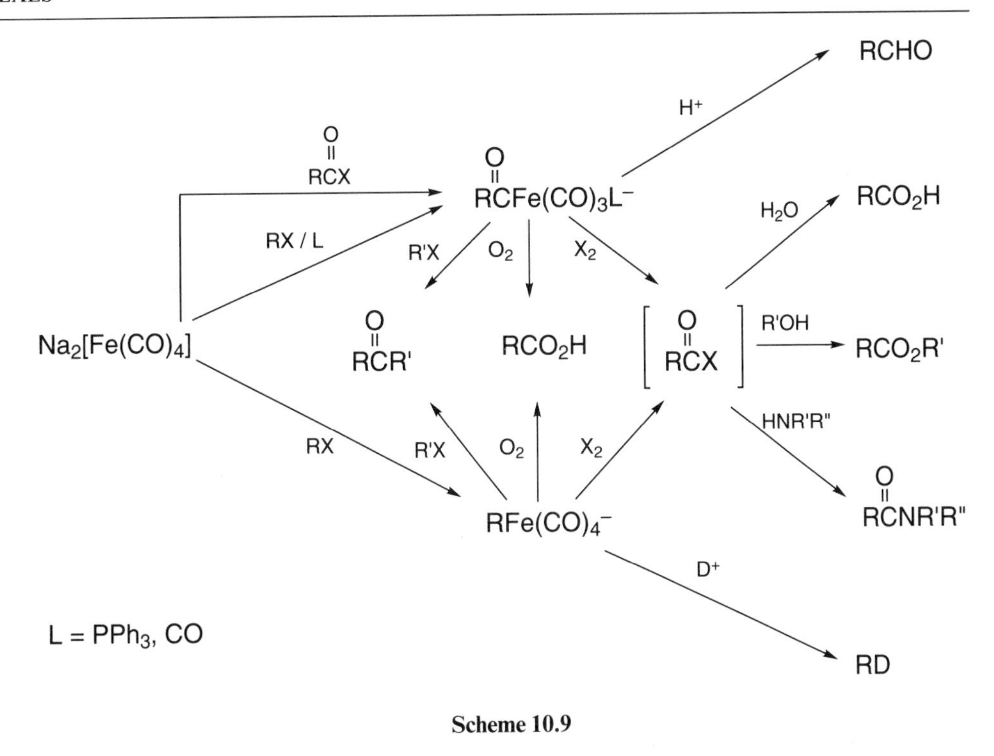

Scheme 10.9

$Na_2[Fe(CO)_4]$ is also used as a reductant, especially for the preparation of unsymmetric aldehydes or ketones. Reduction of maleic anhydride by $Na_2Fe(CO)_4$ gives maleinaldehydic acid (eq (10)) [40].

$$\xrightarrow{Na_2[Fe(CO)_4]} \qquad (10)$$

The unsymmetrical ketone $MeCOCH_2(CH_2)_4CO_2Me$ is obtained in reasonable yield from $BrCH_2(CH_2)_4CO_2Me$ and methyl iodide with $Na_2Fe(CO)_4$ (eq (11)) [41].

$$MeI + BrCH_2(CH_2)_4CO_2Me \xrightarrow{Na_2[Fe(CO)_4]} MeCO(CH_2)_5CO_2Me \qquad (11)$$

10.2.2 Diene Complexes of Iron

Diene complexes of iron are usually prepared from $Fe(CO)_5$ or $Fe_2(CO)_9$. The tricarbonyliron fragment acts as a protecting group of a coordinated diene for the hydrogenation, and Diels–Alder reactions. For example, (1,3-butadiene)$Fe(CO)_3$ given in equation (12) fails to react with hydrogen even in the presence of platinum catalyst, or to undergo cycloaddition with maleic anhydride [42]. On the other hand, it can act as an activating group of dienes toward nucleophilic addition, a reaction which does not proceed under usual conditions.

10.2.2.1 Reactions of Fe(CO)$_5$ with Dienes

Pentacarbonyliron releases its carbonyl ligands upon heating or exposure to UV light, and the resulting unsaturated species are known to react with various kinds of 1,3-dienes to form dieneiron tricarbonyl complexes. For example, (butadiene)tricarbonyliron is prepared by the direct reaction of 1,3-butadiene with $Fe(CO)_5$ at 130–140 °C in a closed system (eq (12)) [42].

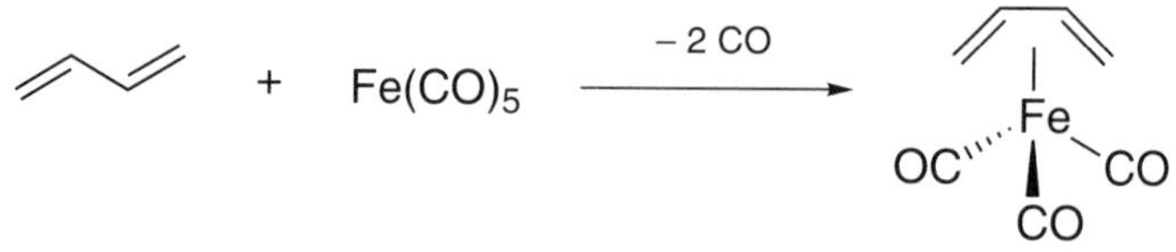

$$\text{butadiene} + Fe(CO)_5 \xrightarrow{-2\,CO} (\eta^4\text{-butadiene})Fe(CO)_3 \quad (12)$$

The diene ligands transform to the cisoid form because of the thermodynamic stability of the complex. Because the bond distances of $C_1—C_2$ and $C_2—C_3$ in the butadiene ligand are 1.45 and 1.46 Å, respectively, the double bond character between $C_2—C_3$ and the bond order alternation is still insignificant [43]. This may be due to the strong π-acidity of three carbonyl ligands which reduces π-back donation from iron to the butadiene ligand. (Cyclohexadiene)tricarbonyliron complexes are also noteworthy. Fe(1,3-cyclohexadiene)(CO)$_3$ was prepared by the reaction of 1,3-cyclohexadiene with $Fe(CO)_5$ [44] (eq (13)) and more stable Fe(1,4-cyclohexadiene)(CO)$_3$ was also prepared [45].

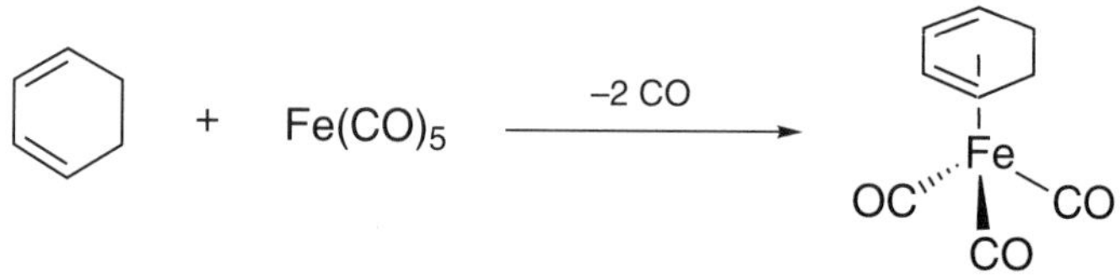

$$\text{1,3-cyclohexadiene} + Fe(CO)_5 \xrightarrow{-2\,CO} (\eta^4\text{-cyclohexadiene})Fe(CO)_3 \quad (13)$$

By use of cyclohexadiene complexes of iron, many stereo- or rigioselective reactions have been developed. For example, Fe(1,3-cyclohexadiene)(CO)$_3$ undergoes stereo-

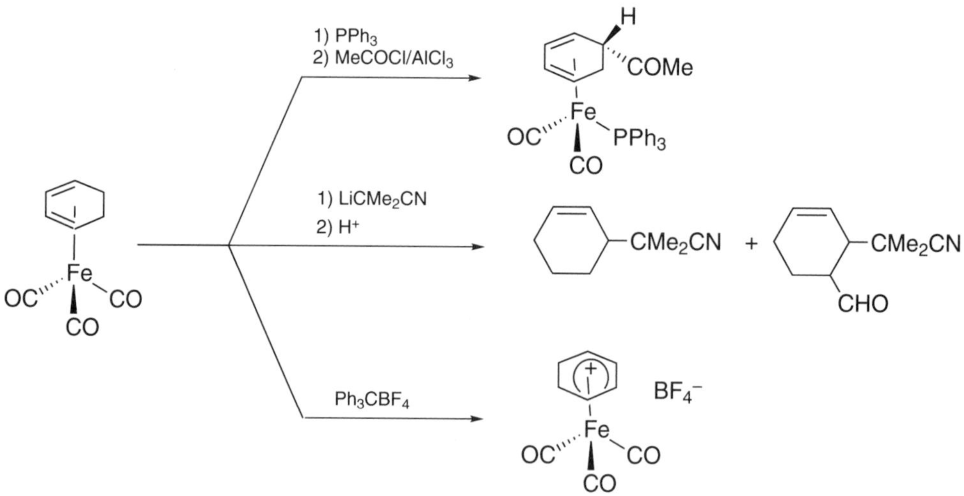

(η⁴-cyclohexadiene)Fe(CO)$_3$ → 1) PPh$_3$ 2) MeCOCl/AlCl$_3$ → (COMe-substituted cyclohexadiene, H)Fe(CO)$_2$(PPh$_3$)

(η⁴-cyclohexadiene)Fe(CO)$_3$ → 1) LiCMe$_2$CN 2) H$^+$ → cyclohexene–CMe$_2$CN + cyclohexene–CMe$_2$CN (with CHO)

(η⁴-cyclohexadiene)Fe(CO)$_3$ → Ph$_3$CBF$_4$ → [(η⁵-cyclohexadienyl)Fe(CO)$_3$]$^+$ BF_4^-

Scheme 10.10

selective Friedel–Crafts acylation to give an *endo*-acetyl complex, whose yield is much improved if one CO ligand was replaced by PPh_3 (Scheme 10) [46]. The reaction of $Fe(1,3\text{-cyclohexadiene})(CO)_3$ with 2-lithio-isobutylonitrile at –78 °C dominantly gives 3-(1-cyanoisopropyl)cyclohexene. If this reaction was carried out at room temperature, 3-(1-cyanoisopropyl)-4-formylcyclohexene was exclusively obtained. The tendency suggests that nucleophilic attack of the anion takes place at the C-2 position of the diene [47]. $Fe(1,3\text{-cyclohexadiene})(CO)_3$ also reacts with electrophiles such as triphenylmethyl tetrafluoroborate, giving a cationic cyclohexadienyl complex, which reacts with nucleophiles from the *exo*-face (Scheme 10.10).

10.2.2.2 Reactions of $Fe_2(CO)_9$ with Dienes

$Fe_2(CO)_9$ is sometimes used in place of $Fe(CO)_5$ for reactions with 1,3-dienes. It results under milder conditions because it readily generates a "$Fe(CO)_4$" species upon heating. It can also react with heterodienes under mild conditions as described below (eq (14)) [48].

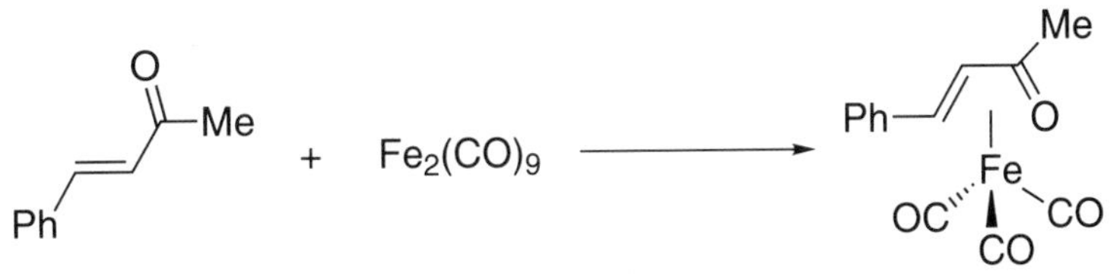

PhCH=CHC(O)Me + $Fe_2(CO)_9$ ⟶ (PhCH=CHC(O)Me)$Fe(CO)_3$ (14)

10.2.2.3 Reactions of $Fe_2(CO)_9$ with Methylenecyclopropane

Methylenecyclopropanes undergo C—C bond cleavage on reaction with $Fe_2(CO)_9$ to give conjugated diene complexes having electron withdrawing substituents (eq (15)) [49].

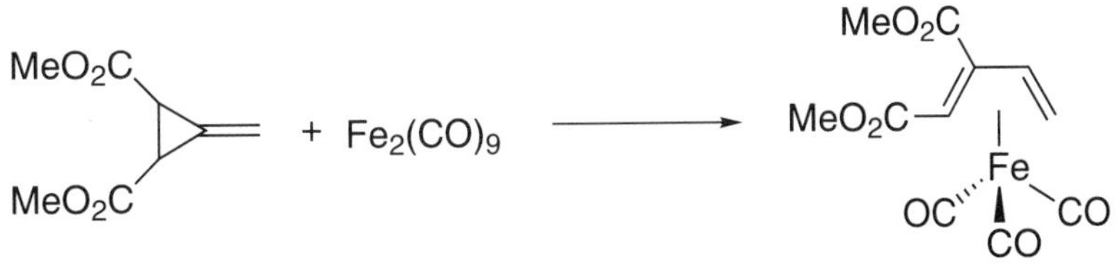

MeO_2C/MeO_2C-methylenecyclopropane + $Fe_2(CO)_9$ ⟶ $(MeO_2C)_2C{=}C(CH{=}CH_2)$–$Fe(CO)_3$ (15)

10.2.2.4 Cyclobutadiene Complexes of Iron

Cyclobutadiene complexes are prepared by the method described in equation (8); some members of the series can also be synthesized by dimerization of an acetylene on an iron tricarbonyl fragment (eq (16)) [50].

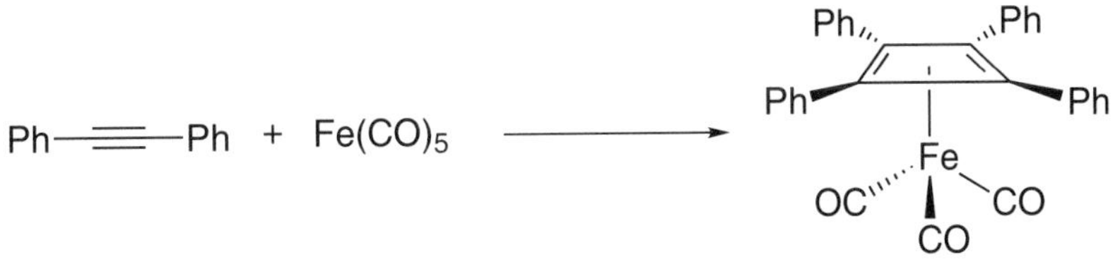

Ph—≡—Ph + $Fe(CO)_5$ ⟶ $(Ph_4C_4)Fe(CO)_3$ (16)

10.2.3 Iron Complexes with Cyclopentadienyl Ligands

Treatment of $Fe(CO)_5$ with dicyclopentadiene results in the formation of the dimeric iron(I) complex $[Fe(C_5H_5)(CO)_2]_2$, which is a convenient starting material for monocyclopentadienyl complexes. For examples, $[Fe(C_5H_5)(CO)_2]_2$ reacts with sodium amalgam to give the zero-valent ion salt $Na[Fe(C_5H_5)(CO)_2]$ via reductive Fe—Fe cleavage, while treatment with HCl leads to oxidative addition giving $FeCl(C_5H_5)(CO)_2$. There are many reports of stoichiometric reactions promoted by these complexes.

Bis(cyclopentadienyl)iron derivatives, ferrocenes, are remarkably stable against heat and air and undergo various kinds of chemical reactions. They are usually prepared by the reaction of $FeCl_2$ with an alkali metal salt of cyclopentadienyl or cyclopentadiene in the presence of base [51]. Although ferrocene derivatives basically have chemical reactivities similar to those of aromatic compounds, they have found only limited applications to organic synthesis so far. However, because chiral ferrocenylphosphines are capable of having both planar chirality and an asymmetry in the side chain in their rigid framework, they have been used recently in a number of asymmetric reactions. In this section, synthesis and developments of monocyclopentadienyl complexes and ferrocenylphosphines are described. The general chemistry of ferrocenes and half metallocene complexes is reviewed elsewhere [52–53].

10.2.3.1 Monocyclopentadienyliron Complexes

One of the most intriguing aspect of chemistry of monocyclopentadienyl complexes is a series of reactions starting from $Na[Fe(C_5H_5)(CO)_2]$. Cationic olefin complexes of iron are prepared by treatment of $Na[Fe(C_5H_5)(CO)_2]$ with chloroisobutene [54], epoxides [55], alkyl halides followed by β-hydrogen elimination [56], or alkyl halides followed by hydrogen abstraction [57]. $Fe(\eta^1\text{-allyl})(C_5H_5)(CO)_2$ is prepared by treatment of the cationic olefin complex with a tertiary amine [58]. The complexes obtained from reactions of the olefin complexes are η^1-carbon-bonded compounds, which have to be oxidized to release the organic products. The η^1-allyliron complex reacts with electrophilic olefins, enones, amines or isonitriles to give corresponding adducts (Scheme 10.11) [53,59].

10.2.3.2 Ferrocenylphosphines

Recently, a number of ferrocenylphosphines have been developed for asymmetric hydrogenation and C—C bond formation. A breakthrough in the chemistry of ferrocenylphosphines was the optical resolution of ferrocenylamines with tartaric acid [60]. Many optically active ferrocenylphosphines are derived as outlined in Scheme 10.12 [61].

It is worthwhile noting that Rh/BPPFAOH is capable of producing high asymmetric introduction in hydrogenation of aminoketones due to hydrogen bonding between the carbonyl group of the substrate and the OH group in the ligand. Furthermore, a series of ferrocenylphosphines having an aminoethyl group also resulted in high enantioselectivity because of interaction between the aminoethyl group and the substrate. For example, a gold(I) complex of a ferrocenylphosphine possessing a tertiary amine

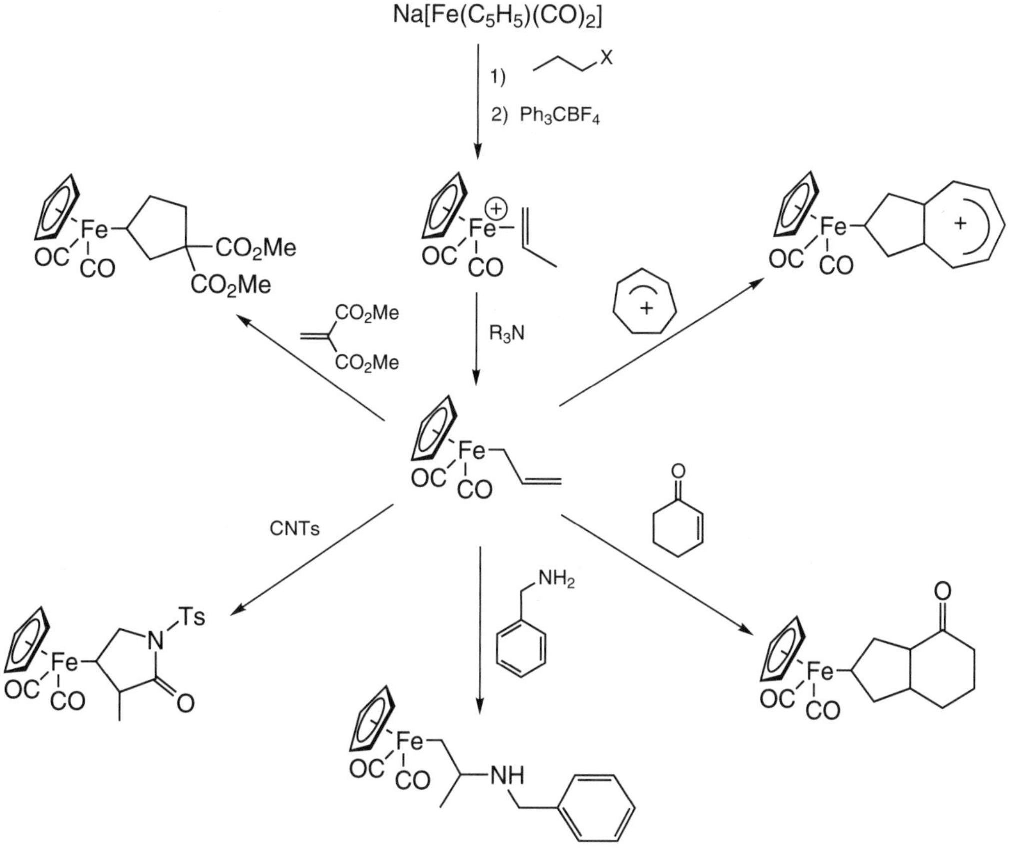

$Na[Fe(C_5H_5)(CO)_2]$

1) (propyl halide, X)
2) Ph_3CBF_4

R_3N

CNTs

Scheme 10.11

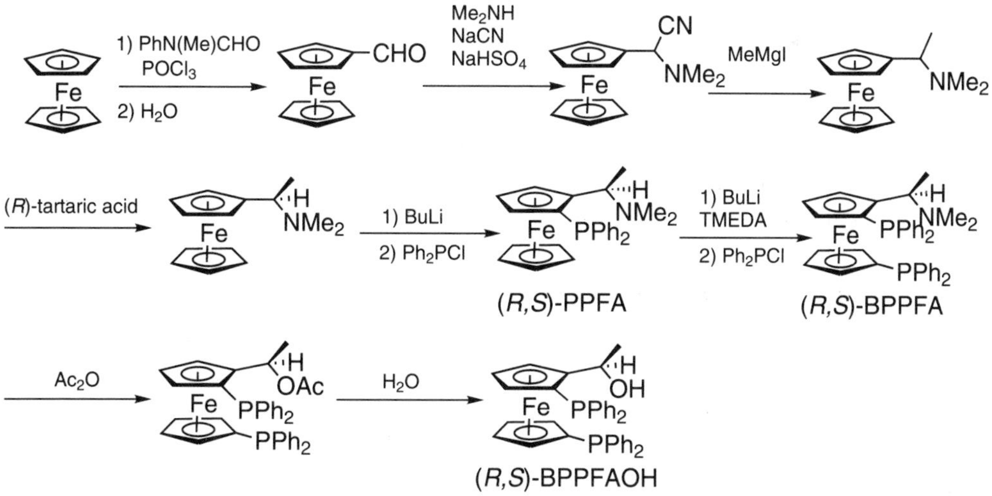

1) PhN(Me)CHO $POCl_3$ 2) H_2O

Me_2NH NaCN $NaHSO_4$

MeMgI

(*R*)-tartaric acid

1) BuLi 2) Ph_2PCl

(*R*,*S*)-PPFA

1) BuLi TMEDA 2) Ph_2PCl

(*R*,*S*)-BPPFA

Ac_2O

H_2O

(*R*,*S*)-BPPFAOH

Scheme 10.12

catalyzes the asymmetric aldol reaction of an isocyanoacetate with aldehydes to give oxazolines in high enantioselectivity (> 96 % ee) (eq (17)) [62].

$$RCHO + CNCH_2CO_2Me \xrightarrow[CH_2Cl_2\ 25\ ^\circ C]{[Au(c\text{-}HexNC)_2]BF_4/L} \text{trans-oxazoline (R, }CO_2Me) + \text{cis-oxazoline (R, }CO_2Me) \quad (17)$$

L = ferrocenyl ligand bearing PPh_2 on each Cp ring and a $CH(Me)N(Me)CH_2CH_2NR_2$ side chain

More recently, a *trans*-chelating asymmetric ferrocenyl phosphine (abbreviated as TRAP) has been developed [63]. The rhodium complex of TRAP catalyzes asymmetric Michael additions of α-cyanocarboxylates in high enantioselectivity (72–89 % ee) (eq (18)) [64]. Because these Michael reactions proceed via a *N*-bound enolato complex [65], the reaction center on the enolato ligand is far from the metal center. Thus, a *cis*-chelating phosphine cannot control the direction of electrophilic attack. The reaction with *cis*-chelating phosphines gives only a poor enantiomeric excess (BINAP, 17 % ee; DIOP, 12 % ee; CHIRAPHOS, 3 % ee).

$$RCOCH{=}CH_2 + NC\text{-}CH(Me)\text{-}CO_2R' \xrightarrow[(S,S)\text{-}(R,R)\text{-TRAP}]{RhH(CO)(PPh_3)_3} RCOCH_2CH_2C(Me)(CN)CO_2R' \quad (18)$$

(S,S)-(R,R)-TRAP = biferrocene bearing $CH(Me)PPh_2$ groups (Ph₂P, Me, H; H, PPh₂, Me; Fe, Fe)

These recent developments have reawakened interest in the fundamental chemistry of optically active ferrocenes [66].

10.2.4 Iron Complexes with Phosphine Ligands

Although a number of iron complexes having phosphine ligands have been prepared, of particular interest are a series of iron diphosphine complexes. They have been extensively studied because the iron(0) complexes generated from them are capable of activating C—H bonds or small molecules (Scheme 10.13). In this section, iron complexes with phosphines are briefly reviewed from the viewpoint of the chemistry of zerovalent iron complexes. Other recommended reading is given in ref. 5.

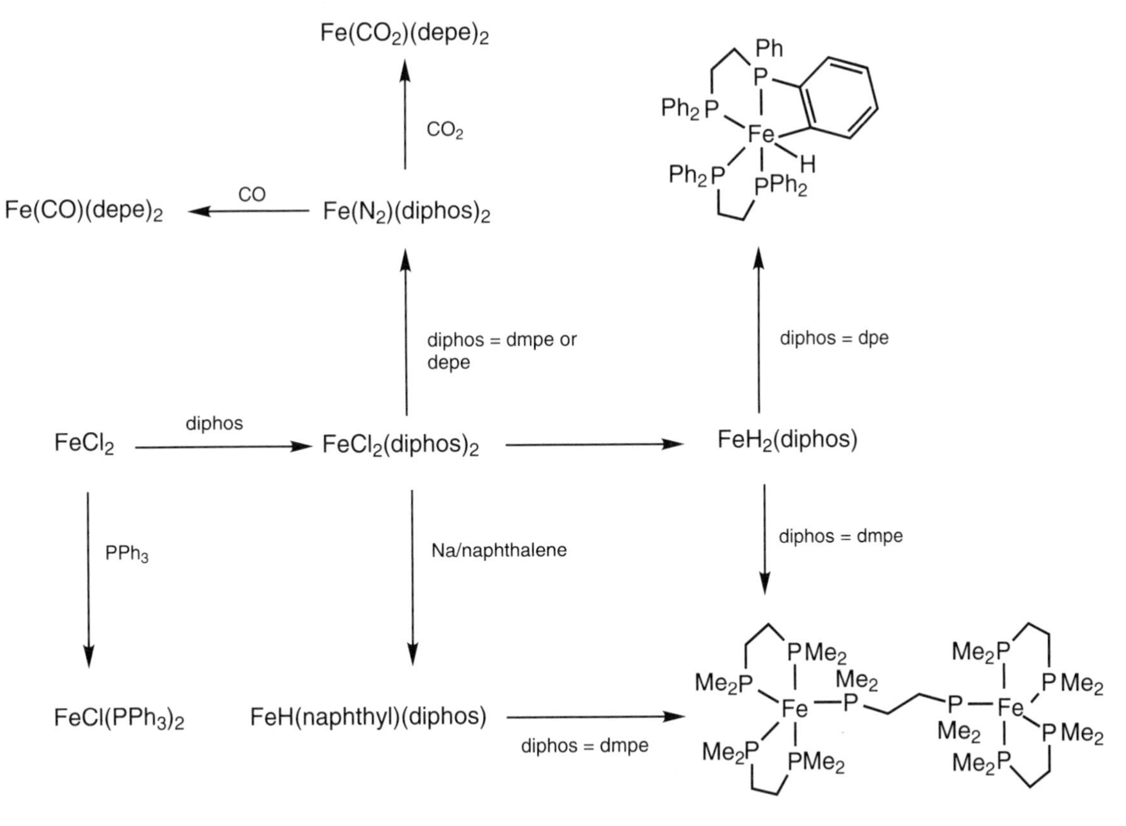

Scheme 10.13

Since Chatt and Davidson discovered C—H bond activation by reduction of $RuCl_2(dmpe)_2$ with a sodium/naphthalene system, the iron analogues have been widely investigated. Actually, reductions of $FeCl_2(diphos)_2$ or $FeH_2(diphos)_2$ have played the central role in this area and $FeCl_2(diphos)_2$ can be prepared by the reaction of purified $FeCl_2$ with diphosphine.

$FeH_2(diphos)_2$ is derived from the reduction of $FeCl_2(diphos)_2$. The hydrido ligands are known to be eliminated by exposure to UV light (313 nm) [67]. It is noteworthy that the exposure of $FeH_2(dmpe)_2$ to UV light gives a zero-valent iron dimer $Fe_2(dmpe)_5$ [68], while corresponding treatment of $FeH_2(dpe)_2$ leads to internal C—H activation [67,69]. Thus iron(0) species with phosphine ligands are good candidates for C—H bond activation. Indeed, C—H bond activation of arenes [70], ketones and nitriles [71], olefins [72], and even alkanes such as pentane [73] or methane [74] have been carried out by generation of Fe(0) species.

Zero-valent iron complexes can activate small molecules. For example, reduction of $FeCl_2(diphos)_2$ under nitrogen atmosphere resulted in formation of $Fe(N_2)(diphos)_2$ [diphos = dmpe or depe] (Scheme 10.13). Interestingly, protonolysis of $Fe(N_2)(dmpe)_2$ was reported to give ammonia under ambient conditions [75]. In addition, the coordinated dinitrogen in $Fe(N_2)(depe)_2$ [76–77] can be readily replaced with CO or CO_2.

10.3 Synthesis of Fe Compounds

(1) Pentacarbonyliron(0), $Fe(CO)_5$ [78]

(moderately air stable, toxic, hood)

$$Fe \xrightarrow{CO} Fe(CO)_5 \quad (19)$$

Because pure pentacarbonyliron(0) is commercially available at a reasonable price, it is not usually necessary to prepare it. However, this method is sometimes used for the preparation of ^{13}C or ^{18}O labeled $Fe(CO)_5$.

Iron powder (5.0 g) is suspended in dry heptane (30 mL) and is activated by shaking for 15 min in an atmosphere of H_2S. A slight gas evolution on the surface of the iron particles is observed during the activation. The suspension is then stored under nitrogen. After introduction of ^{13}CO or $C^{18}O$ at atmospheric pressure, the suspension is stirred at room temperature very vigorously for 20 h. The solution gradually turns red during the reaction. The solvent is then removed under reduced pressure. The title product is obtained by distillation at atmospheric pressure (bp 103 °C).

Properties: mp −20 °C. bp 103 °C. IR [$Fe(CO)_5$ in C_6H_{12}]: 2022, 2000 cm^{-1}. [$Fe(CO)_{5-n}(^{13}CO)_n$]: 2023, 2001, 1988, 1976, 1972, 1964, 1955 cm^{-1}; [$Fe(CO)_{5-n}(C^{18}O)_n$]: 2023, 2008, 2001, 1988, 1976, 1964, 1955 cm^{-1}.

Pentacarbonyliron is a straw-colored liquid with a musty odor, which is stable in air at room temperature. It should be handled in a hood because it is a toxic liquid with a relatively high vapor pressure even at room temperature (21 mmHg, 20 °C). It should be stored in the dark to prevent slow decomposition.

(2) Enneacarbonyldiiron(0), $Fe_2(CO)_9$ [79]

(moderately air sensitive, toxic, hood)

$$Fe(CO)_5 \longrightarrow Fe_2(CO)_9 \quad (20)$$

$Fe(CO)_5$ (146 g, 0.746 mmol) is dissolved in anhydrous glacial acetic acid (200 mL) in a 1 L three-necked pyrex flask with mechanical stirrer, stopper, and gas inlet under inert atmosphere. If the glacial acetic acid contains water, the reaction yields only a brown pyrophoric powder. The vessel is placed in a 5 L silver lined Dewar flask and is cooled continuously by running water. A bucket lined on the side and bottom with aluminum foil can also be used instead of a silver lined Dewar flask. A 125 W high pressure mercury lamp with a water-circulating glass jacket is placed as close as possible to the reaction vessel. The temperature of the reaction mixture should be kept between 20–25 °C during the irradiation. Moreover, vigorous stirring is required for this reaction because precipitation of diiron enneacarbonyl formed on the surface on the lamp side significantly reduces the product yield. On 24 h exposure of the reaction mixture to UV light with vigorous stirring, the title complex is precipitated. The solution layer is removed by cannula and the residuals are washed first with ethanol and then ether, repeatedly. The shiny yellow precipitate is dried *in vacuo*. 122 g (91%). The title complex is insoluble in almost all organic solvents and water and is slowly decomposed in THF and dichloromethane. Over a long period, it should be stored in the dark in an inert atmosphere, preferably carbon monoxide.

Properties: mp 100–120 °C (dec). IR(KBr) 2082, 2019, 1829 cm^{-1}.

(3) Disodium Tetracarbonylferrate(-II), $Na_2[Fe(CO)_4]$ [80]

(air sensitive, pyrophoric, hood)

$$Fe(CO)_5 \xrightarrow{NaC_{10}H_8} Na_2[Fe(CO)_4] \qquad (21)$$

A sodium dispersion in paraffin (1.73 g as Na, 75.0 mmol) is charged in a well-dried 1L three-necked round-bottomed flask with a magnetic stirrer bar equipped with a 200 mL dropping funnel and rubber serum stoppers under nitrogen atmosphere. The vessel is cooled to 0 °C and a THF solution (200 mL) of naphthalene (9.90 g, 77.0 mmol) is added by stainless cannula. Freshly distilled THF from sodium benzophenone ketyl should be used as solvent. The reaction mixture is stirred at 0 °C for 2 h. The resulting deep green solution is cooled at –78 °C. Freshly distilled iron pentacarbonyl (7.02 g, 36 mmol) in THF (100 mL) is added dropwise from the funnel over a 30 min period with vigorous stirring. It should be noted that low quality iron pentacarbonyl leads to poor yield and purity. The color changes from deep green to beige. At this point, the addition of iron pentacarbonyl is discontinued and the mixture is stirred for an additional hour before warming to room temperature. Then, dry pentane or hexane (200 mL) is added by cannula to the reaction mixture and stirs for 30 min. The white precipitate is isolated by a glass filter under nitrogen and washed with pentane (100 mL). It is finally dried *in vacuo*. Yield: 7.39 g (96 %).

Properties: mp 104–105 °C. IR(KBr) 1730 cm^{-1}.

(4) (η^3-Allyl)bromotricarbonyliron(II), $FeBr(\eta^3\text{-}C_3H_5)(CO)_3$ [13]

$$Fe_2(CO)_9 \xrightarrow{C_3H_5Br} FeBr(\eta^3\text{-}C_3H_5)(CO)_3 \qquad (22)$$

To a hexane suspension of $Fe_2(CO)_9$ (36.5 g, 0.1 mol) is added allyl bromide (12.0 g, 0.1 mol) under a nitrogen atmosphere. The reaction mixture is warmed at 40 °C for 1.5 h with vigorous stirring. The resulting dark red solution is filtered and the filtrate is concentrated under reduced pressure (20–30 °C/12 mmHg). Care should be taken because toxic pentacarbonyliron is present in the solution. Cooling the concentrated solution at 0 °C gives dark yellow prisms (*ca* 4 g). The solution layer is concentrated to give further crystals (*ca* 6 g). Recrystallization from petroleum ether (bp. 45–60 °C) gives dark yellow crystals. Yield: 9.9 g (38 %).

Chloro or iodo analogues are also prepared by the reaction of $Fe_2(CO)_9$ with corresponding allyl chloride and iodide.

Properties: $FeX(\eta^3\text{-}C_3H_5)(CO)_3$; (X = Cl): yield 25 %, yellow needles, mp 88–89 °C, IR(KBr): 2091, 2013 cm^{-1}. (X = Br): yield 38 %, yellow prisms, mp 86–87 °C (dec.), IR(KBr): 2095, 2020, 1991 cm^{-1}. (X = I): yield 67 %, yellow needles, mp 85–86 °C, IR(KBr): 2078, 2015 cm^{-1}.

(5) (η^4-1,3-Butadiene)tricarbonyliron(0), $Fe(\eta^4\text{-}1,3\text{-}C_4H_6)(CO)_3$ [42]

(under nitrogen, hood)

$$Fe(CO)_5 \xrightarrow{1,3\text{-}C_4H_6} Fe(\eta^4\text{-}C_4H_6)(CO)_3 \quad (23)$$

Though the preparation of the complex was first reported by Reihlen *et al.*, nowadays it was usually prepared by the following modified method [81].

$Fe(CO)_5$ (29 g, 0.15 mol) and butadiene (0.3 mol) is charged in a stainless autoclave (300 mL). The reaction mixture is heated at 135 °C for 24 h. After cooling the autoclave to room temperature, the generated carbon monoxide is discharged. This procedure is repeated 3 times and then unreacted $Fe(CO)_5$ is removed from the combined orange solution under reduced pressure at room temperature. Because this procedure generates toxic gaseous pentacarbonyliron, a cold trap (–78 °C) should be placed in the distillation system. After removal of $Fe(CO)_5$, distillation under reduced pressure (47–49 °C / 0.1 mmHg) gives a pure orange yellow oil: 4.6 g (16 %). Crystallization from ligroin or methanol at –78 °C affords pale yellow crystals.

This complex can also be prepared by the ligand exchange reaction of (η^4-4-phenyl-3-buten-2-one)(tricarbonyl)iron with 1,4-butadiene as described below [82].

$$PhCH{=}CHCOMe + Fe_2(CO)_9 \longrightarrow (\eta^4\text{-}PhCH{=}CHCOMe)Fe(CO)_3 \quad (24)$$

$$(\eta^4\text{-}PhCH{=}CHCOMe)Fe(CO)_3 + CH_2{=}CH{-}CH{=}CH_2 \longrightarrow Fe(\eta^4\text{-}1,3\text{-}C_4H_6)(CO)_3 \quad (25)$$

Properties: mp 19 °C. IR: 2051, 1978 cm^{-1}.

(6) Tricarbonyl(η^4-1,3-cyclohexadiene)iron, $Fe(\eta^4\text{-}C_6H_8)(CO)_3$ [81b]

(moderately air stable complex, but store under nitrogen)

$$Fe(CO)_5 \xrightarrow{1,4\text{-}C_6H_8} Fe(\eta^4\text{-}C_6H_8)(CO)_3 \quad (26)$$

Into a stainless autoclave (300 mL) is added $Fe(CO)_5$ (29.2 g, 0.149 mol) and 1,4-cyclohexadiene (12.9 g, 0.161 mol) under nitrogen. The reaction mixture is cooled to –78 °C, the pressure reduced to 1 mmHg and heated at 150 °C for 20 h. The autoclave is cooled to room temperature and the generated CO is discharged. The resulting yellow liquid is collected from the autoclave. This procedures are repeated for 3 times. The unreacted $Fe(CO)_5$ and 1,4-cyclohexadiene are removed from the combined yellow liquid under reduced pressure. Then, the title compound is distilled under reduced pressure (bp 50–66 °C/1 mmHg). Care should be taken because this procedure generates toxic $Fe(CO)_5$ vapor. A cold trap (–78 °C) should be therefore placed in the distillation system. Yield: 19.2 g (19 %).

Properties: mp 8–9 °C. IR: 2066, 1978 cm^{-1}.

(7) Bis(η^5-cyclopentadienyl)tetracarbonyldiiron(I), $[Fe(\eta^5\text{-}C_5H_5)(CO)_2]_2$ [83]

(fairly air stable complex, but prepare under nitrogen in hood)

$$Fe(CO)_5 \xrightarrow{C_5H_6} [Fe(C_5H_5)(CO)_2]_2 \quad (27)$$

Into a 2 L three-necked round-bottomed flask equipped with a three-way stopcock, a reflux condenser, and thermometer is added dicyclopentadiene (1 kg, >95 % purity) and iron pentacarbonyl (150 mL, 1.1 mol) under nitrogen. The mixture is heated at 140–150 °C with stirring until yellow vapors of iron pentacarbonyl are no longer observed (*ca* 8 h). Careful attention should be paid to keep the temperature exactly at 140 °C during the reaction because no product is obtained below 140 °C and extensive decomposition will occur over 150 °C. The deep red mixture is cooled to room temperature and then allowed to stand for several hours to give reddish purple crystals. The crystals are filtered and repeatedly washed with cold hexane or pentane. Yield: 76 g (38 %). These crystals are pure enough to use in further reactions. Analytically pure crystals are obtained by recrystallization from a mixture of pentane-dichlorom ethane or sublimation (110 °C / 0.1 mmHg). However, significant loss of the product occurs during purification.

(8) Alkyl(η^5-cyclopentadienyl)dicarbonyliron(II), $FeR(\eta^5\text{-}C_5H_5)(CO)_2$ [84]

(under nitrogen atmosphere)

$$Na[Fe(C_5H_5)(CO)_2] \xrightarrow{RX} FeR(C_5H_5)(CO)_2 \qquad (28)$$

Reaction of $[Fe(C_5H_5)(CO)_2]_2$ (see above) with sodium amalgam in THF for 2 h results in formation of orange $Na[Fe(C_5H_5)(CO)_2]$.

A two-necked round-bottomed flask equipped with a dropping funnel and magnetic stirrer bar is charged with THF solution (300 mL) of $Na[Fe(C_5H_5)(CO)_2]$ (20 g, 0.10 mol). The solution is cooled to 0 °C and is treated dropwise with a THF solution of methyl iodide (7.0 g, 0.12 mol). The mixture is stirred at room temperature for several hours and solvent is then removed under reduced pressure. The resulting solid is sublimed (60 °C / 0.1 mmHg) to give orange crystals of $Fe(CH_3)(\eta^5\text{-}C_5H_5)(CO)_2$. Yield: 14.3 g (70 %). Other alkyl iron complexes can be prepared similarly [84–91].

Properties: (R = Me): mp 78–82 °C. IR(CCl_4): 2010, 1955 cm^{-1}. ^1H NMR ($CDCl_3$): δ 0.17 (s, 3H), 4.78 (s, 5H).

(9) Chloro(η^5-cyclopentadienyl)dicarbonyliron(II), $FeCl(\eta^5\text{-}C_5H_5)(CO)_2$ [92]

(air stable)

$$[Fe(C_5H_5)(CO)_2]_2 \xrightarrow{HCl} FeCl(C_5H_5)(CO)_2 \qquad (29)$$

$[Fe(C_5H_5)(CO)_2]_2$ (15 g, 42.4 mmol), chloroform (180 mL), ethanol (120 mL) and finally 2M hydrochloric acid (120 mL) are added to an Erlenmeyer flask in that order. The reaction mixture is stirred for 5 min at room temperature. Concentrated hydrochloric acid (15 mL) is added to the mixture and then the solution is oxidized by gentle bubbling of air at room temperature. The air oxidation should be discontinued when the solution color changes from reddish purple to red (*ca* 1 h). Chloroform should be added to the reaction mixture if chloroform is significantly vaporized during the air oxidation. After air oxidation, the organic layer is separated and the water layer is extracted repeatedly with chloroform (5 × 20 mL). The combined organic layer is dried over anhydrous magnesium sulfate. Magnesium sulfate is filtered off and the filtrate is concentrated to dryness and dried in vacuo. The resulting dark red solid is dissolved in benzene (100 mL) and the insoluble black precipitate is removed by filtration. The filtrate is concentrated to *ca* 20 mL under reduced pressure and then hexane (150 mL) is added. The precipitate formed is filtered, washed with hexane and dried under vacuum. Yield: 14.4 g (80 %). The corresponding bromide and iodide are also obtained by an analogous procedure.

Properties: mp 85–86 °C. IR($CHCl_3$): 2057, 2012 cm^{-1}. ^1H NMR ($CDCl_3$): δ 5.05 (s, 5H).

(10) Bis(η^5-cyclopentadienyl)iron(II), (Ferrocene), $Fe(\eta^5\text{-}C_5H_5)_2$ [93]

(air stable complex, prepare under nitrogen atmosphere in hood)

$$FeCl_2 \xrightarrow{C_5H_6} Fe(C_5H_5)_2 \tag{30}$$

A mixture of cyclopentadiene (42 mL, 0.5 mol) and anhydrous iron dichloride (31.8 g, 0.25 mmol) in a 250 mL round-bottomed flask is cooled to 0 °C. To the reaction mixture is added diethylamine (100 mL, *ca* 1 mol) with vigorous stirring. The mixture is allowed to warm to room temperature and stirred for 6–8 h. The excess diethylamine is removed under reduced pressure and the resulting solid is extracted repeatedly with refluxing petroleum ether. The extract is purified by hot filtration and removal of the solvent gives orange solid. Recrystallization from pentane or cyclohexane, or sublimation (150 °C / 760 mmHg), gives pure ferrocene. Yield: 34–39 g (73–84 %).

Properties: mp 173–174 °C. ^1H NMR: δ 4.11 (in CS_2), δ 4.04 (in CCl_4), δ 4.01 (in C_6D_6).

(11) Dihydridobis{bis(1,2-diphenylphosphino)ethane}iron(II), $FeH_2(dpe)_2$ [94]

(under nitrogen atmosphere)

$$FeCl_2 \xrightarrow{dpe,\ NaBH_4} FeH_2(dpe)_2 \tag{31}$$

$$(DPE = Ph_2PC_2H_4PPh_2)$$

Highly reactive anhydrous iron dichloride should be used for this preparation. It is obtained by grinding $FeCl_2{\cdot}4H_2O$ under nitrogen atmosphere followed by drying under vacuum (110 °C / 0.1 mm Hg) for 3–4 h. Into a single-necked round-bottomed flask equipped with a reflux condenser and magnetic stirrer bar is added dpe [1,2-bis-(diphenylphosphino)ethane] (20 g, 0.05 mol), anhydrous iron dichloride (2.52 g, 0.02 mol) and dry THF (250 mL). The mixture is stirred at room temperature for 10 min and then solid $NaBH_4$ (4.0 g, 0.11 mol) is added. The purple-red reaction mixture is refluxed for 10 min and additional $NaBH_4$ (2.0 g, 0.055 mol) is added. Then, absolute ethanol (20 mL) is added dropwise. After gas evolution has caused, the chrome-yellow suspension is refluxed for a further 10 min. The insoluble materials are filtered off and the filtrate is concentrated to *ca* 50 mL. The resulting yellow powder is collected and dried *in vacuo*. The powder is dissolved in toluene (500 mL, its solution is filtered and the filtrate is concentrated to 200 mL. The resulting yellow crystals are collected and dried under reduced pressure at 50 °C for 8 h. Yield: 11.5 g (56%). The composition of the crystals is characterized as $FeH_2(dpe)_2 \cdot 2C_6H_5CH_3$.

Properties: mp 187 °C, IR(KBr): 1870, 1845 cm^{-1} (Fe—H).

(12) Chlorohydridobis{bis(1,2-diphenylphosphino)ethane}iron(II), $FeHCl\text{-}(dpe)_2$ [95]

(air stable complex but prepare under nitrogen atmosphere)

$$FeCl_2 \xrightarrow{\text{dpe, NaBH}_4} FeHCl(dpe)_2 \qquad (32)$$

Into a three-necked round-bottomed flask (250 mL) equipped with a dropping funnel, gas inlet and magnetic stirrer bar is charged $FeCl_2 \cdot 2H_2O$ (0.652 g, 4.1 mmol), 95 % ethanol (40 mL) and dpe [1,2-bis(diphenylphosphino)ethane] (3.2 g, 8.1 mmol) under a nitrogen atmosphere. The reaction mixture is warmed at 40 °C for 1 h. Then, an ethanol solution (20 mL) of $NaBH_4$ (160 mg, 4.2 mmol) is added dropwise over a 1 h period to the stirred solution, and then a reddish violet precipitates form. The reaction mixture is stirred at 40 °C for an additional hour. The precipitate is filtered on a glass frit. The precipitate is washed with deoxygenated water, then with warm ethanol (40–50 °C, 40 mL) and acetone (40 mL). The reddish violet compound is dried *in vacuo* to give the crude product. (2.8 g, 80 %). The product is dissolved in dry benzene (30 mL) and the extract is filtered to remove insoluble materials. The filtrate is concentrated under vacuum until crystals appears, then petroleum ether is added to the mixture. The reddish violet crystalline is filtered, washed with petroleum ether and dried in vacuum. Yield: 1.5 g (43 %).

Properties: mp 195 °C (dec.). IR(Nujol) 1955 cm^{-1} [ν(Fe—H)].

(13) Dichlorobis(triphenylphosphine)iron(II), $FeCl_2(PPh_3)_2$ [96]

(under nitrogen)

$$FeCl_2 \xrightarrow{PPh_3} FeCl_2(PPh_3)_2 \qquad (33)$$

Anhydrous $FeCl_2$ is obtained by grinding $FeCl_2 \cdot 4H_2O$ under nitrogen atmosphere and then heating under vacuum (110 °C / 0.1 mm Hg) for 3–4 h. Into a 100 mL Schlenk tube is charged anhydrous $FeCl_2$ (1.36 g, 1.07 mmol), triphenylphosphine (10.5 g, 4.00 mmol) and dry benzene (60 mL) under nitrogen atmosphere. The reaction mixture is refluxed for 4 h. After the reaction, the hot solution is filtered and the filtrate is allowed to stand for a day at room temperature. The colorless crystals (*ca* 0.9 g) are separated and washed with benzene. More crystals (*ca* 0.7 g) are obtained by concentration of the mother liquor to *ca* 30 mL. The combined crystals are dried in vacuum for 6 h (70 °C / 5 mmHg). Yield: 1.58 g (23 %).

Properties: $FeCl_2(PPh_3)_2$ has a tetrahedral geometry with high-spin configuration. μ_{eff} = 4.88 [97] (5.07 [98]) BM. Soluble in acetone, nitrobenzene, methanol, or dichloromethane. Insoluble in benzene, diethyl ether or chloroform.

10.4 Ruthenium Complexes

Recently, the chemistry of ruthenium complexes has been extensively explored. However, they have so far found less application in organic synthesis than palladium compounds, probably because their chemistry is more complicated [99]. Indeed, in contrast to palladium(II) compounds, ruthenium complexes generally have 5- or 6-coordinated geometry and their oxidation state can vary between –2 to 6. This complexity, however, leads to many interesting reactions and further developments in this field are expected.

A wide variety of organoruthenium complexes is known. They can be roughly divided into 4 groups according to their supporting ligands. The first group are carbonyl complexes which are generally derived from $Ru_3(CO)_{12}$. This complex is attractive because while it is an air stable orange complex and easy to handle, it is the precursor of an active catalyst for reduction of nitro groups, C—H bond activation or carbonylation. The second group are ruthenium complexes with tertiary phosphine ligands of which $RuCl_2L_4$, $RuHClL_4$,or RuH_2L_4 are noteworthy because they are useful for organic synthesis. They have been extensively explored and a number of catalytic reactions have been developed. Recent developments of ruthenium complexes having chiral ligands are also of interest. The third group are cyclopentadienyl complexes. Bis(cyclopentadienyl) complexes, ruthenocenes, undergo electrophilic substitutions, but half metallocene or multi-nuclear cyclopentadienyl complexes are more intriguing. Furthermore, both cyclopentadienyl and pentamethylcyclopentadienyl ligands effectively stabilize alkyl-ruthenium bonds, whereas in phosphine complexes the alkyl group tends to undergo β-hydrogen elimination.

The last group are ruthenium complexes having arenes or dienes. These π-stabilized complexes are not only used for low valent ruthenium starting materials via replacement of arene or diene ligands but are catalysts for olefin dimerization, hydrogenation of arenes, or C—C bond cleavage reaction.

For the preparation of these ruthenium complexes, $RuCl_3{\cdot}3H_2O$ and $Ru_3(CO)_{12}$ are frequently used as starting materials, though many ruthenium complexes are now commercially available. They are relatively inexpensive and stable against oxygen, and many ruthenium complexes are derived from them under ambient conditions.

In this section, we will review the nature and chemical reactivities of several kinds of ruthenium complexes under the following headings: ruthenium carbonyl complexes, dichlororuthenium complexes, chlorohydrido complexes, dihydridoruthenium complexes, ruthenium complexes with chiral ligands, ruthenium complexes with cyclopentadienyl ligands, and ruthenium arene/diene complexes.

10.4.1 Ruthenium Carbonyl Complexes

$Ru(CO)_5$ is less frequently used than $Fe(CO)_5$ for organic synthesis or as a starting material as a zero-valent ruthenium complex because of its ease of decomposition to $Ru_3(CO)_{12}$ [99]. Dodecacarbonyltriruthenium is very useful for these purposes. It has been shown to be an active catalyst for the hydrogenation of olefins [100], carbonylation of ethylene [101], hydroformylation of alkenes [102], water-gas shift reaction [103], and reduction of nitro groups [104], and recently, C—H bond activation [105] and coupling of diynes with CO [106].

Dodecacarbonylruthenium has been prepared by a number of methods. However, the following three methods are conveniently employed for laboratory scale preparations. It can be derived from the reaction of CO with $RuCl_3 \cdot 3H_2O$ in the presence [107] or absence [108] of zinc powder, or from $[Ru_3O(\mu\text{-}O_2CMe)_6(H_2O)_3](O_2CMe)$ [109]. It is obtained as orange crystals which are stable to light and oxygen. The structure of $Ru_3(CO)_{12}$ resembles that of $Fe_3(CO)_{12}$ in containing a triangle of metal atoms, but in contrast to $Fe_3(CO)_{12}$ it contains no bridging CO groups.

One noteworthy feature of $Ru_3(CO)_{12}$ is that it retains trinuclear framework during many reactions. Indeed, while treatment of $Fe_3(CO)_{12}$ with tertiary phosphines gives mononuclear phosphine complexes even under mild conditions, similar treatment of $Ru_3(CO)_{12}$ gives $Ru_3(CO)_9L_3$ (L = tertiary phosphine), and conversion of $Ru_3(CO)_9L_3$ into $Ru(CO)_4L$ requires severe conditions [L = PPh_3, 150 °C/ CO(80 atm)] [110]. This rigid triangular framework provides specific reaction sites, and several unique reactions have consequently been developed.

Typical chemical reactions of $Ru_3(CO)_{12}$ are outlined in Scheme 10.14. Reactions of olefins containing at least three carbon atoms give μ-H and μ-allyl trinuclear complexes [111]. Similarly, reaction of $Ru_3(CO)_{12}$ with terminal alkynes produces trinuclear

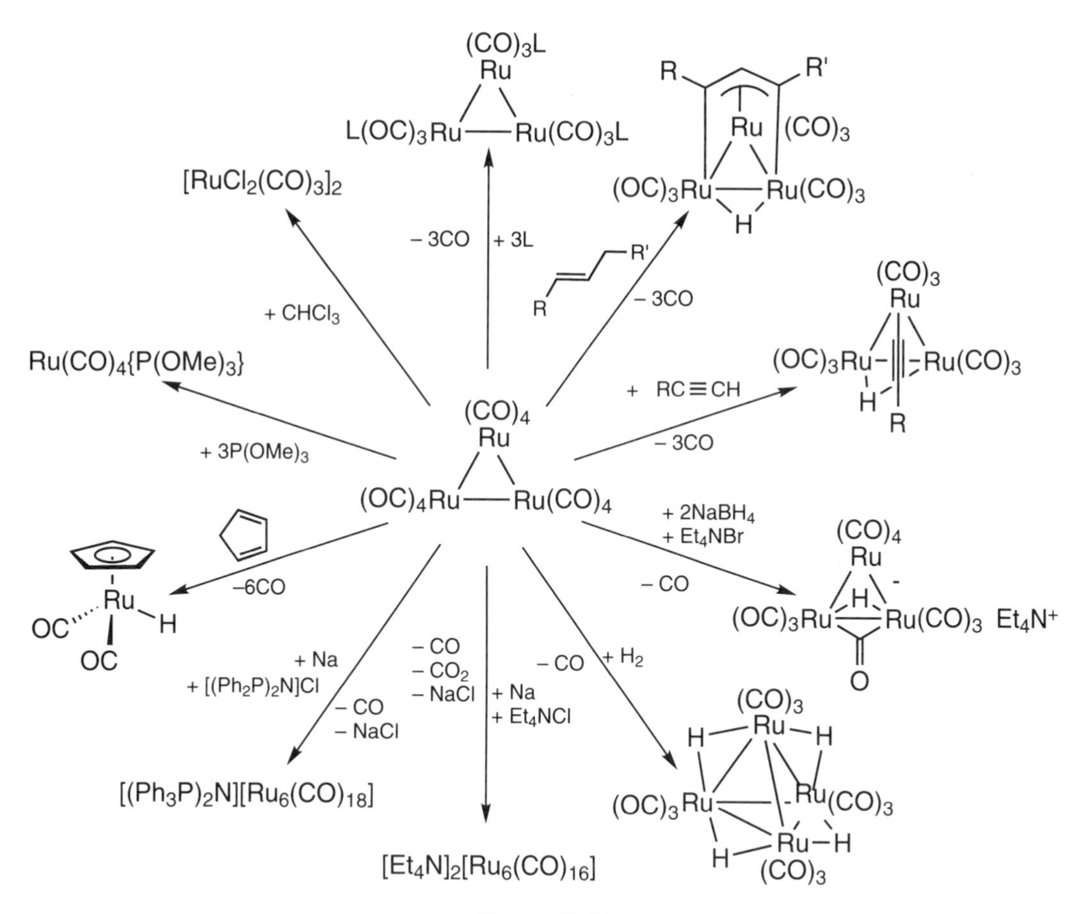

Scheme 10.14

μ-H and μ-acetylide complexes [112]. Treatment of $Ru_3(CO)_{12}$ with $NaBH_4$ followed by Et_4NBr [113] or with H_2[114] give μ-H complexes. Reductions of $Ru_3(CO)_{12}$ with sodium result in the formations of hexanuclear clusters [115–116]. Monomeric ruthenium complexes can also be produced by the treatment of $Ru_3(CO)_{12}$ with cyclopentadiene [117], phosphites [118] or chloroform [119]. Detailed reviews of the chemistry of $Ru_3(CO)_{12}$ are available [120–122].

$Ru_3(CO)_{12}$ is also used in several catalytic reactions as described below.

10.4.1.1 Reductions

It is used in a variety of reactions involving the conversion of CO to CO_2 under mild conditions. For example, nitro groups are readily reduced to amino groups by CO in the presence of $Ru_3(CO)_{12}$ (eq (34)) [104].

$$RNO_2 \xrightarrow[CO(1atm)]{Ru_3(CO)_{12}/\ NaOH} RNH_2 \tag{34}$$

Because $Ru_3(CO)_{12}$ is practically insoluble in water, the above reaction needs a phase transfer reagent. If this reaction is carried out in methanol under more severe conditions, reductive carbonylation takes place to give carbamate (eqs (35)) [123].

$$RNO_2 \xrightarrow[CO(80\ atm)]{Ru_3(CO)_{12},\ MeOH} RNHCO_2Me \tag{35}$$

10.4.1.2 C—H Bond Activation

Recently, a three-component coupling reaction of pyridine, olefin and CO by $Ru_3(CO)_{12}$ has been reported [105]. This is a rare example of catalytic functionalization of pyridine via C—H activation. The practical turnover frequency of the catalyst (300/h) is achieved under moderate conditions (eq (36)).

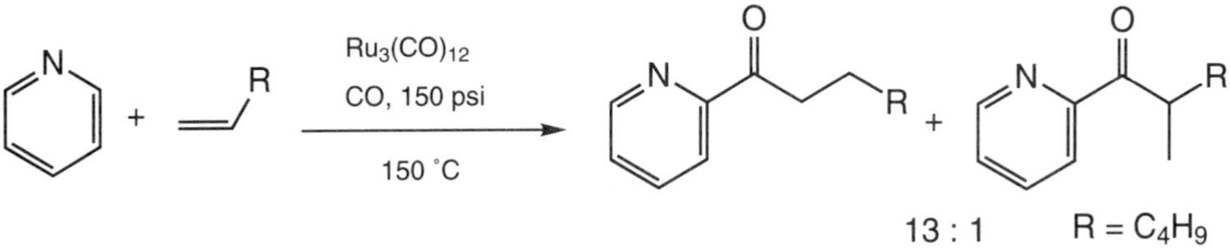

$$\text{pyridine} + \text{CH}_2\text{=CHR} \xrightarrow[150\ ^\circ C]{Ru_3(CO)_{12},\ CO,\ 150\ psi} \text{2-pyridyl-CO-CH}_2\text{CH}_2\text{R} + \text{2-pyridyl-CO-CH(CH}_3\text{)R} \tag{36}$$

13 : 1 R = C_4H_9

The mechanism of this interesting reaction is also discussed. Since kinetic studies suggest that the rate-determining step of the catalysis is C—H activation of pyridine, this reaction requires an excess of pyridine. An active key intermediate depicted below is isolable (Figure 10. 3). Although a mononuclear ruthenium complex cannot be ruled out as the active catalyst, the cluster framework remains intact during the course of the catalysis.

10.4.1.3 Reaction of Diynes with Carbon Monoxide

$Ru_3(CO)_{12}$ catalyzes an unprecedented C—C bond formation between a diyne and two molecules of CO (eq (37)) [106]. The carbonyl groups may be transformed *in situ* into carbyne ligands in this process.

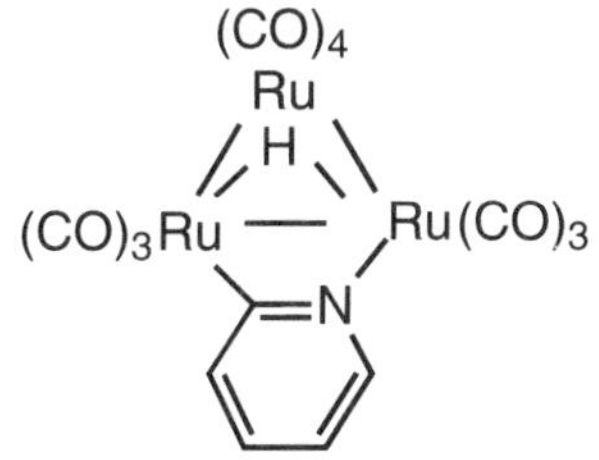

Figure 10.3 C—H Bond activation of pyridine.

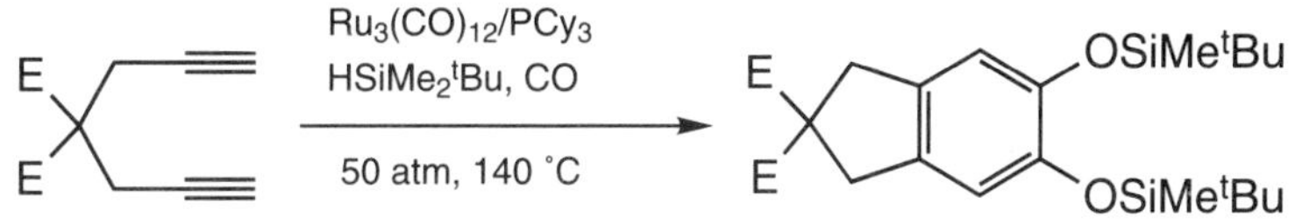

(37)

10.4.2 Dichlororuthenium Complexes

Halide ligands in dihaloruthenium(II) complexes can be replaced with alkyl, aryl, or hydrido ligands by treatment with an appropriate reagent. Furthermore, since La Placa and Ibers [124] discovered that the vacant site of $RuCl_2(PPh_3)_3$ is occupied by an agostic proton, and Chatt and Davidson demonstrated that C—H bond activation of naphthalene could be carried out by reduction of a dichlororuthenium complex [125], dihaloruthenium complexes have been attractive precursors of coordinatively unsaturated species. They have also been used as catalysts in various organic syntheses such as the *N*-alkylation of amines or oxidation reactions. In this section, the chemical reactivities of a widely used dihaloruthenium complex, $RuCl_2(PPh_3)_3$, and an example of a ruthenium(0) species isolated by the reduction of $RuCl_2(CO)_2(PMe^tBu_2)_2$ are described.

Most dichlororuthenium complexes are formed by the reduction of $RuCl_3{\cdot}3H_2O$ in the presence of the ligand. $RuCl_2(PPh_3)_3$ is obtained by treatment of $RuCl_3{\cdot}3H_2O$ with an excess of PPh_3 in methanol as air-stable shiny black crystals [126]. However, reaction of $RuCl_3{\cdot}3H_2O$ with PRR'_2 or PR_2R' (R = phenyl, R′ = alkyl) gives cationic dinuclear complexes $[Ru_2Cl_3(PR_nR'_{3-n})_6]Cl$ under similar conditions [126b]. The X-ray crystallography of $RuCl_2(PPh_3)_3$ showed that it has a distorted octahedral geometry with a vacant site which is occupied by an agostic proton of a phenyl group [124]. The PPh_3 ligand in $RuCl_2(PPh_3)_3$ is readily displaced by CO [126], norbornadiene [127], acetonitrile [128] or nitrosyl [129] and the Cl ligands are replaced by hydrides by treatment of the complex with H_2/NEt_3 or $NaBH_4$ (Scheme 10.15) [130–132].

10.4.2.1 N-Alkylation of Amines by Primary Alcohols

Divalent ruthenium complexes are efficient catalysts for *N*-alkylation of amines by a primary alcohol. $RuCl_2(PPh_3)_3$ or $RuCl_3{\cdot}3H_2O/P(OBu)_3$ effectively catalyze the *N*-alkylation of aromatic amines (eq (38)) [133–134]. On the other hand, *N*-alkylation of aliphatic amines with a primary alcohol is carried out in high yield by the use of $RuH_2(PPh_3)_4$ as catalyst [135].

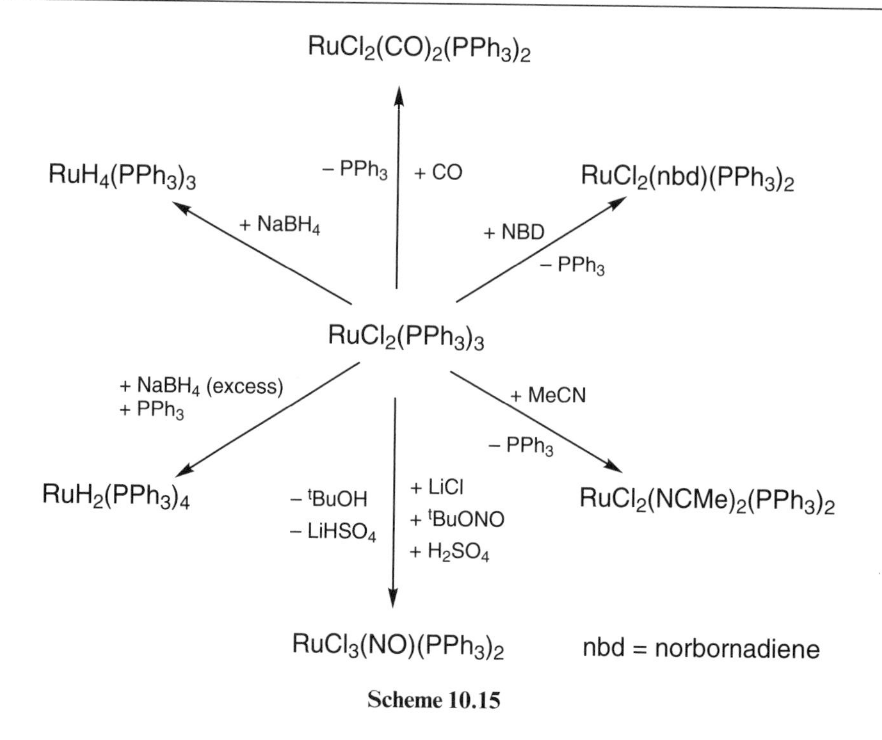

Scheme 10.15

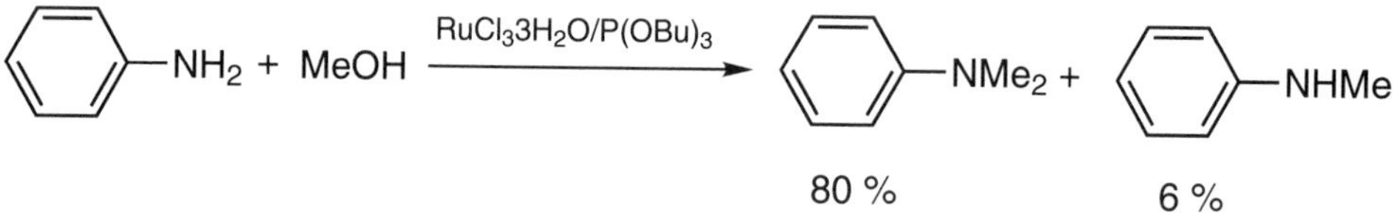

$$\text{PhNH}_2 + \text{MeOH} \xrightarrow{\text{RuCl}_3\text{3H}_2\text{O/P(OBu)}_3} \underset{80\,\%}{\text{PhNMe}_2} + \underset{6\,\%}{\text{PhNHMe}} \qquad (38)$$

These *N*-alkylations are potentially applicable to the preparation of nitrogen heterocycles. For examples, reactions of amines with 1,5-pentadiol [136], 1,4-pentadiol [137], or 2-butyne-1,4-diol [138] give *N*-alkyl piperidine, pyrrolidine, or pyrrole, respectively, in high yield (Scheme 10.16).

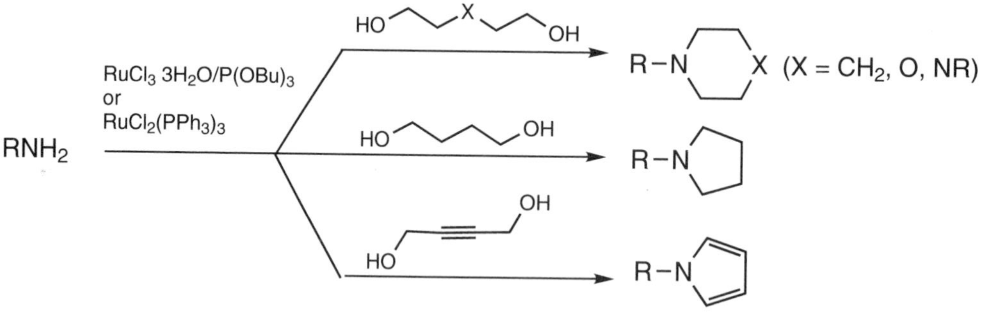

RNH_2 — $\text{RuCl}_3\ 3\text{H}_2\text{O/P(OBu)}_3$ or $\text{RuCl}_2(\text{PPh}_3)_3$ —

- $\text{HO}\sim\text{X}\sim\text{OH}$ → R–N(ring)X (X = CH_2, O, NR)
- HO~~~~OH → R–N (pyrrolidine)
- HO–CH$_2$–C≡C–CH$_2$–OH → R–N (pyrrole)

Scheme 10.16

10.4.2.2 Oxidation of Amines, Amides, and Diols

$RuCl_2(PPh_3)_3$ is also a catalyst for the oxidation of nitriles, amides and lactams under moderate conditions. In this respect, its activity resembles that of cytochrome P-450.

While reactions of secondary amines with ^tBuOOH give the corresponding imines (eq (39)), t-butyldioxylation at the α-position of *N*-alkyl groups has been carried out by the reaction of tertiary amines with ^tBuOOH in the presence of $RuCl_2(PPh_3)_3$ (eq (40)) [139].

$$RR'CHNHR'' \xrightarrow[RuCl_2(PPh_3)_3]{^tBuOOH} RR'C{=}NR'' \quad (39)$$

$$RR'N{-}CH_3 \xrightarrow[RuCl_2(PPh_3)_3]{^tBuOOH} RR'N{-}CH_2OO^tBu \quad (40)$$

Moreover, $RuCl_2(PPh_3)_3$ is an active catalyst for the oxidation of amides which are otherwise hard to oxidize except electrochemically.

Because $RuCl_2(PMe_3)_4$ is soluble in supercritical carbon dioxide, it catalyzes the formation of formate or formamide from carbon dioxide, hydrogen, and alcohol or amine under supercritical conditions with a maximum 370 000 $_{TON}$. These reactions may lead to a breakthrough in CO_2 activation [140].

Recently, an intriguing coordinatively unsaturated $16e^-$ ruthenium(0) complex has been reported [141]. Reduction of $RuCl_2(CO)_2(P^tBu_2Me)_2$ with magnesium affords an isolable 16e ruthenium(0) complex $Ru(CO)_2(P^tBu_2Me)_2$ (eq (41)). This complex is highly reactive toward hydrogen, acetylenes and phosphines to give coordinatively saturated complexes. The X-ray structure analysis showed that this complex has two phosphine ligands in trans configuration and that the carbonyl groups are significantly bent. An *ab initio* calculation for the model phosphine complex $Ru(CO)_2(PH_3)_2$ indicates that this bending lowers the total energy.

$$RuCl_2(CO)_2(P^tBu_2Me)_2 \xrightarrow[-\,MgCl_2]{+\,Mg} trans\text{-}Ru(CO)_2(P^tBu_2Me)_2 \quad (41)$$

Analogously a four coordinate styrene complex $Ru(\eta^2\text{-}CH_2{=}CHPh)_2(PPh_3)_2$ has been known, though one of the styrene ligands has an η^3 character [142].

10.4.3 (Carbonyl)(chloro)(hydrido)ruthenium Complexes

Carbonylchlorohydridoruthenium complexes are often formed by the reduction of ruthenium chloride with alcohol in the presence of tertiary phosphines. The carbonyl ligand is derived from the alcohol just as Vaska's complex, $IrCl(CO)(PPh_3)_2$, is formed from $(NH_4)_2IrCl_6$, PPh_3 and alcohols. The carbonyl ligand is always located *trans* to

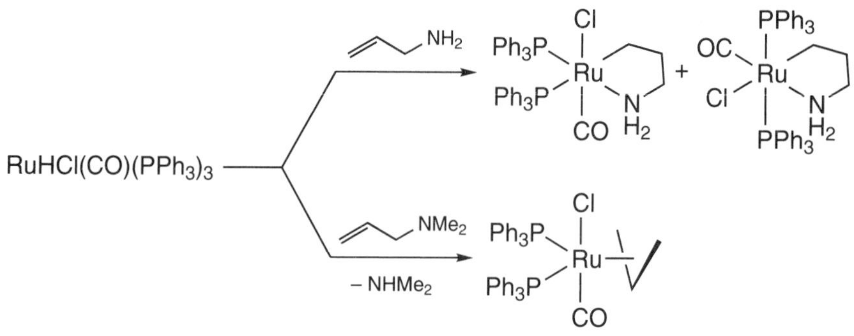

Scheme 10.17

chloride in these complexes. Because the hydrido ligand has a strong trans effect, a phosphine ligand trans to hydride is exclusively replaced by an incoming 2e donor [143]. In this section, two recent developments in the chemistry of these complexes are described.

One of the most interesting recent topics in ruthenium chemistry is undoubtedly C—H bond activation in which the generation of coordinatively unsaturated species may play an important role. These species are usually produced by thermal or photo-mediated reductive elimination of dihydrogen, alkanes, alkenes or arenes. Recently, dehydrochlorination from $RuHCl(CO)(P^tBu_2Me)_2$ is reported to give a π-allyl complex via C—H activation of propylene (eq (42)) [144].

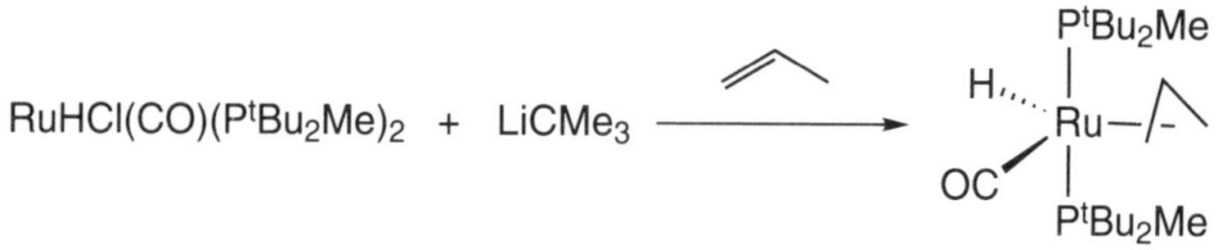

$$RuHCl(CO)(P^tBu_2Me)_2 + LiCMe_3 \longrightarrow \qquad (42)$$

Interesting chemical reactivity of $RuHCl(CO)(PPh_3)_3$ has recently been reported [145]. Treatment of $RuHCl(CO)(PPh_3)_3$ with *N,N*-dimethylallylamine resulted in formation of a π-allylruthenium complex via C—N cleavage, while reaction with allylamine gives a azaruthenacycle complex (Scheme 10.17).

10.4.4 Dihydridoruthenium Complexes

Dihydridoruthenum complexes are reported to be catalysts for either the direct [146] or transfer [147] hydrogenation of olefins and the detailed mechanism for the hydrogenation has been studied [148]. Ruthenium hydride complexes are also catalysts for organic reactions such as the coupling reaction of alkenes with terminal alkynes [149], the [2+2] cycloaddition of norbornene with alkynes [150], Tishchenko-type reactions [151], and the catalytic insertion of olefins into the *ortho* C—H bond of aromatic ketones [152]. This section concentrates on the chemistry of $RuH_2(PPh_3)_4$. The chemistry of other ruthenium dihydride complexes, including $RuH_2(PPh_3)_4$, is reviewed in detail by Geoffroy and Lehman [148d].

$RuH_2(PPh_3)_4$ is prepared by the reaction of $RuCl_2(PPh_3)_3$ with $NaBH_4$ in the presence of PPh_3 in refluxing methanol [130] or by the direct reaction of $RuCl_3{\cdot}3H_2O$ with

$NaBH_4$ and PPh_3 in refluxing ethanol [153]. It is formed as an off-yellow powder and should be kept under argon, not nitrogen, because a PPh_3 ligand is readily replaced by dinitrogen.

Though $RuH_2(PPh_3)_4$ is a coordinatively saturated complex, it easily releases a PPh_3 ligand in solution and it reacts with N_2 [154], CO_2 [155] or ethylene [156]. While hydrogenation of olefins is catalyzed by $RuH_2(PPh_3)_4$ [157–158], regioselective C—H activation of acrylate takes place at room temperature (Scheme 10.18) [159]. Additionally, treatment of $RuH_2(PPh_3)_4$ with crotyl esters gives π-allylruthenium complexes [160]. On the other hand, reaction of Ru-$H_2(PPh_3)_4$ with vinyl acetate results in C—O bond activation to yield RuH(OAc) $(PPh_3)_3$ and ethylene [161].

Reactions with active hydrogen compounds have also been extensively studied. For example, treatment of $RuH_2(PPh_3)_4$ with nitromethane gives a zwitterionic *O*-bound enolate complex, $RuH[O_2NCH_2](PPh_3)_3$ [156b]. Of particular interest are chemoselective aldol and Michael reactions catalyzed by ruthenium complexes [65a,162–163]. Though both ethyl cyanoacetate and acetylacetone have similar pK_a values ($pK_a = 9.0$), the former exclusively reacts with electrophiles in the presence of ruthenium complexes, even in a mixture of ethyl cyanoacetate and acetylacetone (eq (43)).

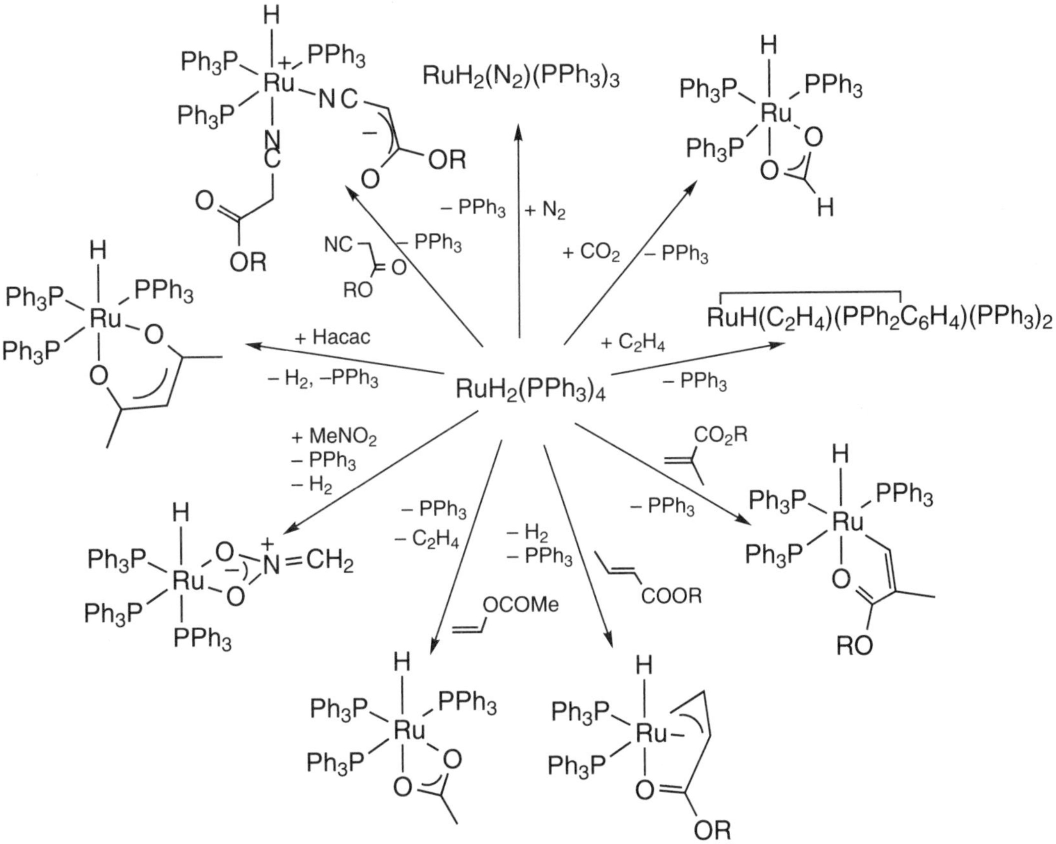

Scheme 10.18

$$\text{PhCHO} + \underset{pKa = 9}{\text{NCCH}_2\text{CO}_2\text{Et}} + \underset{pKa = 9}{\text{MeCOCH}_2\text{COMe}} \xrightarrow{\text{RuH}_2(\text{PPh}_3)_4} (\text{NC})(\text{EtO}_2\text{C})\text{C}{=}\text{CHPh} + \text{MeCOCH}_2\text{COMe} \quad (43)$$

While $RuH_2(PPh_3)_4$ dominantly reacts with acetylacetone in the mixture of acetylacetone and ethyl cyanoacetate, the resulting complex, $RuH[OCMeCHCOCMe](PPh_3)_3$ has little nucleophilicity. Indeed, it does not react even with MeI. On the other hand, a reaction of $RuH_2(PPh_3)_4$ with 2 equivalent amounts of ethyl cyanoacetate gives a zwitterionic *N*-bound enolate complex, $RuH(NCCHCO_2Et)(NCCH_2CO_2Et)(PPh_3)_3$ which is much more nucleophilic than $RuH[OCMeCHCOCMe](PPh_3)_3$. Thus, this chemoselectivity results from the kinetic factor. It was recently reported that a ruthenium polyhydride complex $RuH_4(PPh_3)_3$, also acts as a chemoselective catalyst for the analogous Knoevenagel reactions [164].

Ruthenium dihydride complexes catalyze coupling reactions of acetylenes with dienes [149]. The reaction of 1-octyne with 1,3-butadiene catalyzed by $RuH_2(PBu_3)_4$ affords 2- dodecen-5-yne (eq (44)). A similar coupling reaction is also catalyzed by $RuCl(C_5H_5)(C_8H_{12})$ [165].

$$C_6H_{13}C{\equiv}CH + CH_2{=}CHCH{=}CH_2 \xrightarrow{\text{RuH}_2(\text{PBu}_3)_4} C_6H_{13}C{\equiv}CCH_2CH{=}CHCH_3 \quad (44)$$

$RuH_2(PPh_3)_4$ reacts with aldehydes to give esters via Tishchenko-type dimerization. For example, benzaldehyde is converted to benzyl benzoate by $RuH_2(PPh_3)_4$ (eq (45)) [166–167]. This reaction involves C—H bond activation of the formyl proton followed by formation of a ruthenium acyl alkoxide complex $Ru(OCH_2Ph)(COPh)(PPh_3)_4$.

$$\text{PhCHO} \xrightarrow{\text{RuH}_2(\text{PPh}_3)_4} \text{PhCO}_2\text{CH}_2\text{Ph} \quad (45)$$

Treatment of $RuH_2(PMe_3)_4$ with alcohol gives a hydridoalkoxoruthenium complex $RuH(OR)(PMe_3)_4$. This reaction occurs by an associative mechanism via an intermediate $[RuH(H_2)(PMe_3)_4]OR$ [168].

An analogous ruthenium dihydride complex, $RuH_2(CO)(PPh_3)_3$, has recently attracted much attention because of its ability to catalyze olefin coupling reactions of aromatic ketones via C—H bond activation (eq (46)) [152].

$$2\text{-MeC}_6\text{H}_4\text{COMe} + \text{CH}_2{=}\text{CHSiMe}_3 \xrightarrow{\text{cat. RuH}_2(\text{CO})(\text{PPh}_3)_3} 2\text{-Me-6-(Me}_3\text{SiCH}_2\text{CH}_2)\text{C}_6\text{H}_3\text{COMe} \quad (46)$$

This reaction is carried out almost quantitatively under relatively mild conditions (refluxing toluene). Moreover, it can be applied to almost all aromatic ketones, including heterocyclic compounds such as acetylfuran or acetylthiophene. Although $RuH_2(PPh_3)_4$, $Ru(CO)_2(PPh_3)_3$, and $Ru(CO)_3(PPh_3)_2$ are also effective catalysts for this reaction, $RuH_2(CO)(PPh_3)_3$ is the most active. A plausible intermediate in this reaction is an orthometallated ruthenium complex; stoichiometric orthometallation by ruthenium complexes has been reported (Figure 10.4) [157, 169].

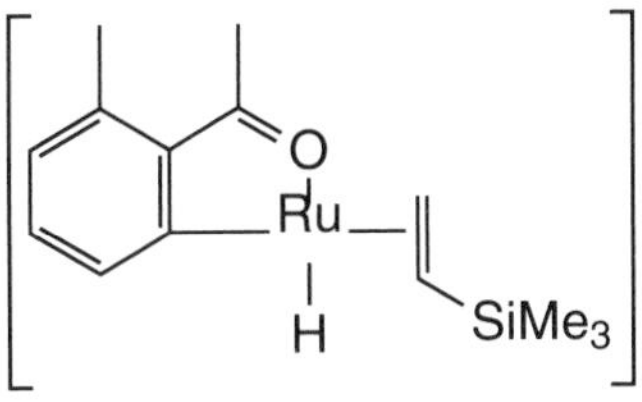

Figure 10.4 A possible intermediate in the olefin coupling reaction of aromatic ketone catalyzed by $RuH_2(CO)(PPh_3)_3$. Other ligands are omitted.

10.4.5 Ruthenium Complexes with Chiral Ligand

Ruthenium complexes have played an important role in asymmetric synthesis. Although rhodium and palladium complexes are also employed in asymmetric hydrogenation, they are usually less effective than ruthenium complexes for hydrogenation of prochiral olefins with functional groups such as amides, acids, or alcohols. Ruthenium complexes are also effective for hydrogenation of carbonyl group to give chiral alcohols. This behavior is of interest both academically and industrially. Indeed, chiral ruthenium complexes are used as catalysts in commercial processes such as the manufacture of Naproxen or of antibiotic precursors. In this section, recent developments in the chemistry of ruthenium complexes with the chiral ligands BINAP and PYBOX are described.

BINAP complexes of ruthenium are one of the most intriguing catalysts for asymmetric hydrogenation of olefins. Because both *R*- and *S*-forms of BINAP (Figure 10.5) are also commercially available, a series of ruthenium(binap) complexes can be prepared without difficulty [170–172]. For example, $Ru(OAc)_2$(binap) is obtained by the reaction of $[RuCl_2(cod)]_n$ with AcONa and BINAP in the presence of Et_3N [170].

Ruthenium(binap) complexes effectively catalyze asymmetric hydrogenation of α-amidocinnamic acids [172], allylic alcohols [173] and acrylic acids with almost quantitative enantiomeric excess [174]. For example, one of the largest-selling anti-inflammatory agents, Naproxen should be supplied as the enantiomerically pure *S*-isomer, because the *R*-isomer is expected to be toxic to the liver. Asymmetric hydrogenation of the precursor by $RuCl_2$[(*S*)- binap] produces *S*-Naproxen with 96–98 % ee (eq (47)) [175–176].

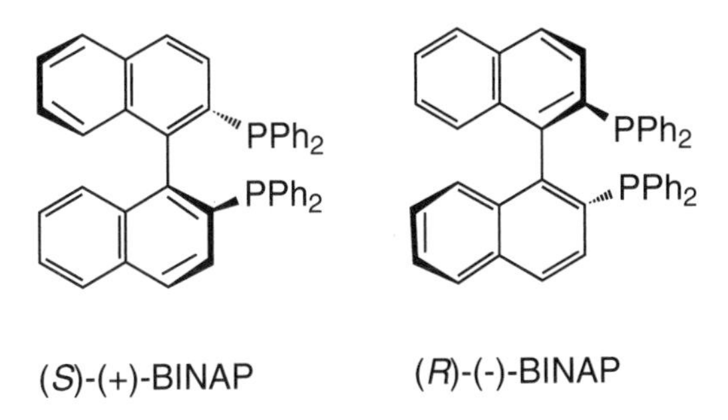

Figure 10.5 Atropisomers of the BINAP Ligand

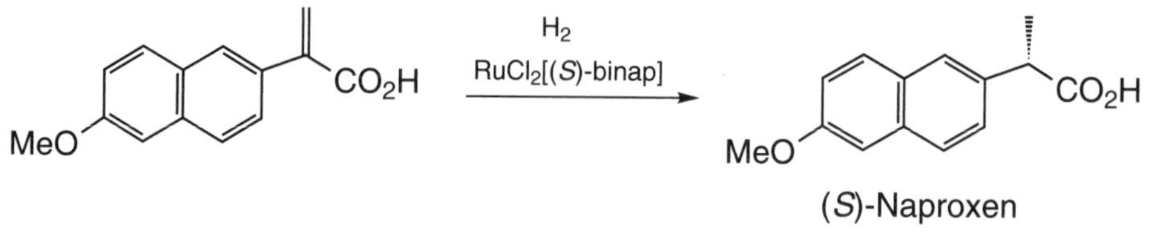

H_2, $RuCl_2[(S)\text{-binap}]$ (47)

(*S*)-Naproxen

Furthermore, these ruthenium(binap) complexes are remarkably effective catalysts for the hydrogenation of C=O bond of β-ketoesters giving optically active alcohols [175].

A ruthenium carbene complex in the presence of a chiral ligand is capable of catalyzing the formation of optically active cyclopropane derivatives from alkenes and diazo compounds in high enantiomeric excess [177]. A mixture of $[RuCl_2(p\text{-cymene})]_2$ in the presence of pybox-(*S,S*)-*ip* catalyzes the asymmetric cyclopropanation of styrene (eq (48)). The key intermediate is proposed to be a dichloro(pybox)ruthenium carbene complex.

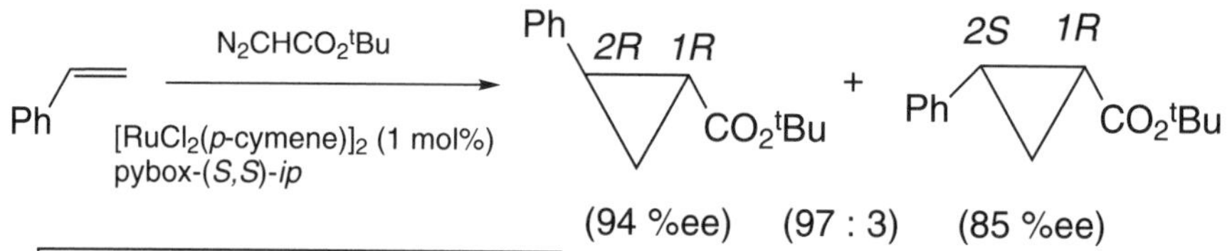

$N_2CHCO_2{}^tBu$, $[RuCl_2(p\text{-cymene})]_2$ (1 mol%), pybox-(*S,S*)-*ip* (48)

2*R* 1*R* (94 %ee) (97 : 3) 2*S* 1*R* (85 %ee)

possible intermediate

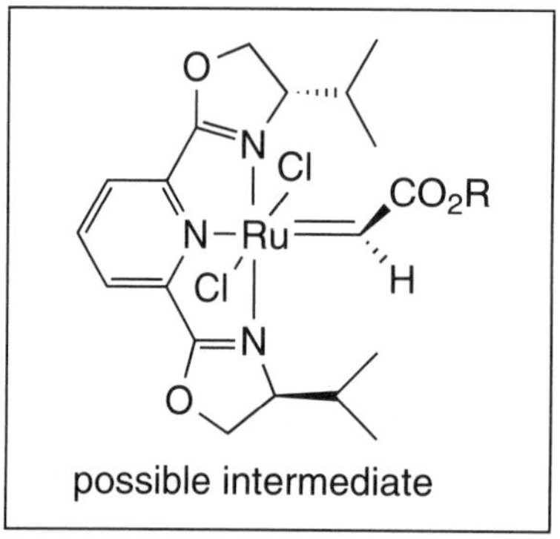

10.4.6 Ruthenium Complexes having Cyclopentadienyl Ligands

Ruthenium complexes having cyclopentadienyl ligands have also been extensively investigated. Because ruthenocene is relatively unreactive, much attention has been focused on mononuclear ruthenium complexes with one cyclopentadienyl ligand or ruthenium dimers having cyclopentadienyl ligands.

The dinuclear complex $[RuCl_2(C_5Me_5)]_2$ prepared by the reaction of $RuCl_3{\cdot}3H_2O$ with pentamethylcyclopentadiene in ethanol (Scheme 10.19) [178] is a versatile reagent. For example, it reacts with cyclooctadiene to give monomeric $RuCl(C_5Me_5)(C_8H_{12})$ [178] with chlorine in CH_2Cl_2 to give $[RuCl_3(C_5Me_5)]_2$ [179], and with CO to give $RuCl(CO)_2(C_5Me_5)$ [178], respectively. Recently, $RuCl(C_5H_5)(C_8H_{12})$ was reported to catalyze an Alder type coupling reaction of alkenes with alkynes under relatively mild conditions (DMF, 100 °C) [165, 180]. Treatment of $[RuCl_2(C_5Me_5)]_2$

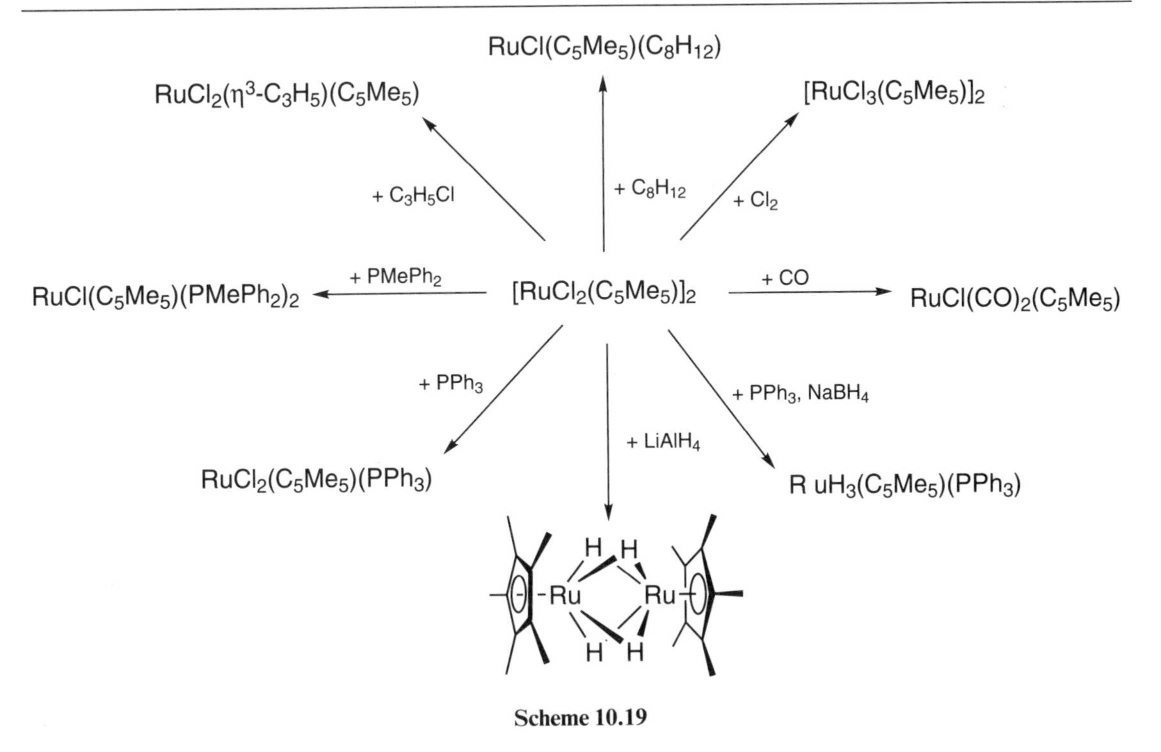

Scheme 10.19

with PPh_3 followed by reduction with $NaBH_4$ yields a mononuclear ruthenium trihydride complex $RuH_3(C_5Me_5)(PPh_3)$ in high yield [178, 181]. On the other hand, simple reduction of $[RuCl_2(C_5Me_5)]_2$ by $LiAlH_4$ followed by ethanolysis produced a tetrahydride ruthenium dimer [182]. Because the resulting hydridoruthenium complex, $[Ru(\mu\text{-}H)_2(C_5Me_5)]_2$ readily releases dihydrogen to give a coordinatively unsaturated complex, the chemical reactivity of the analogous ruthenium dimer has been extensively studied [183]. While reaction of $[RuCl_2(C_5Me_5)]_2$ with PPh_3 at room temperature gives (triphenylphosphine)ruthenium(III) dichloride complex [179], treatment with $PMePh_2$, DPE, DPPP, DPPB, DPPPh, or DPPH in hot ethanol give corresponding ruthenium(II) monochloride complexes [178]. Ruthenium allyl and diruthenium azaallyl complexes are also derived from $[RuCl_2(C_5Me_5)]_2$ in high yield [184].

Treatment of $Ru_2H_4(C_5Me_5)_2$ with ethylene results in the formation of a divinyl(ethylene)diruthenium complex under ambient conditions (Scheme 10.20) [183]. This is an

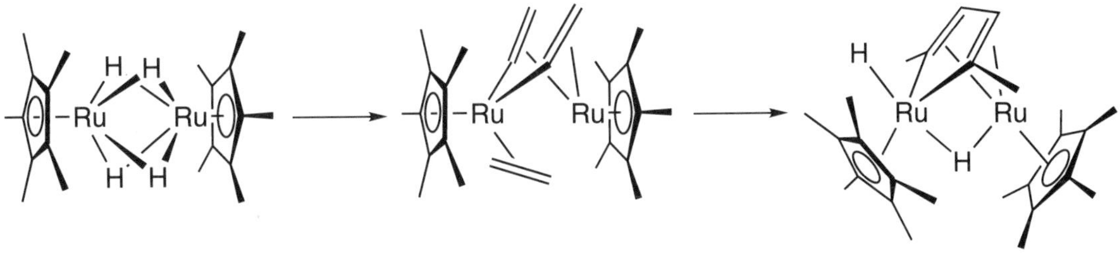

Scheme 10.20

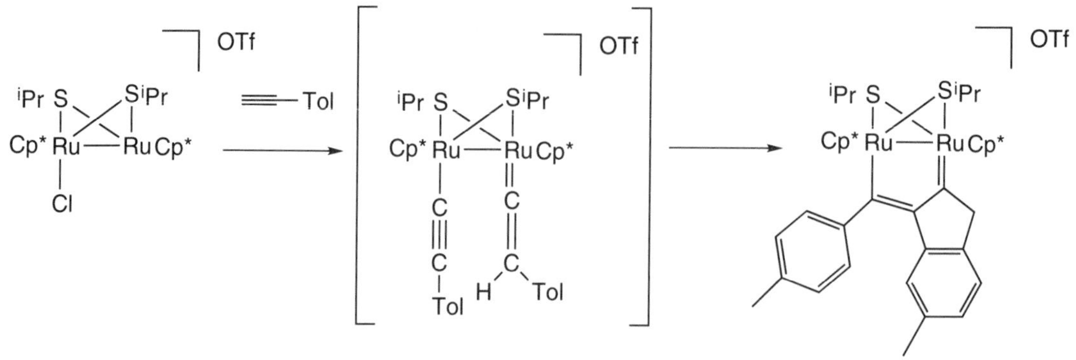

Scheme 10.21

interesting reaction because there are few examples of vinylic C—H bond activation with metal polyhydride complexes. Furthermore, a 2,5-dimethylruthenacyclopentadiene complex can be isolated upon warming the divinyl(ethylene)diruthenium complex. There is also an intriguing thiolato bridged cationic ruthenium(II)/(III) dimer which reacts with tolyl acetylene to give a metallacycle complex having an indane framework (Scheme 10.21) [185]. This unique reaction probably proceeds via an acetylide–vinylidene intermediate.

10.4.7 Ruthenium Complexes with Arene/Diene Ligands

Arenes and dienes are employed as supporting ligands of low valent ruthenium complexes. Ruthenium arene/diene complexes are starting materials for organometallic complexes of ruthenium. For example, reduction of $[RuCl_2(arene)]_2$ in the presence of cod and aqueous ethanolic Na_2CO_3 gives Ru(arene)(cod) complexes, which are also prepared by the reaction of Ru(cod)(cot) with arene under atmospheric hydrogen [186]. The coordinated arene or diene is cleanly replaced by phosphine ligands. Furthermore, these areneruthenium complexes are applied to catalytic reactions such as hydrogenation of arenes by $[Ru_2H_2Cl\,(arene)_2]Cl$ [187], and Ru(arene)(cod) gives tail-to-tail dimerization of acrolein [188], methyl acrylate [189] or acrylonitrile [190]. In this section, Ru(cod)(cot), $[RuCl_2(arene/diene)]_n$, and $Ru(benzyne)L_4$ are described. Other arene- or dieneruthenium complexes are reviewed in detail by Bennett, Bruce and Matheson [191].

Ru(cod)(cot) is prepared by the reduction of $RuCl_3{\cdot}3H_2O$ with zinc powder in the presence of 1,5-cyclooctadiene in methanol [192]. It is used in several catalytic reactions and as a convenient precursor to various zero- or multi-valent ruthenium complexes (Scheme 10.22). For example, it gives $Ru_3(CO)_{12}$ by the treatment with CO [193]. Reactions of Ru(cod)(cot) with tertiary phosphines or phosphites give Ru(cod)(cot)L [L = PMe_3 or $P(OMe)_3$] [194]. Reactions of Ru(cod)(cot) with hydrogen in the presence of tertiary phosphines give ruthenium polyhydride complexes [194]. However, when Ru(cod)(cot) reacts with hydrogen and cyclohexadiene or benzene in the absence of tertiary phosphines, an η^6-benzene complex is formed. No ligand displacement reactions take place in the absence of hydrogen [195]. Interestingly, Ru(cod)(cot) is reversibly protonated by HBF_4 to give $[RuH(\eta^5\text{-}C_8H_{11})_2]BF_4$ [196]. Treatment of

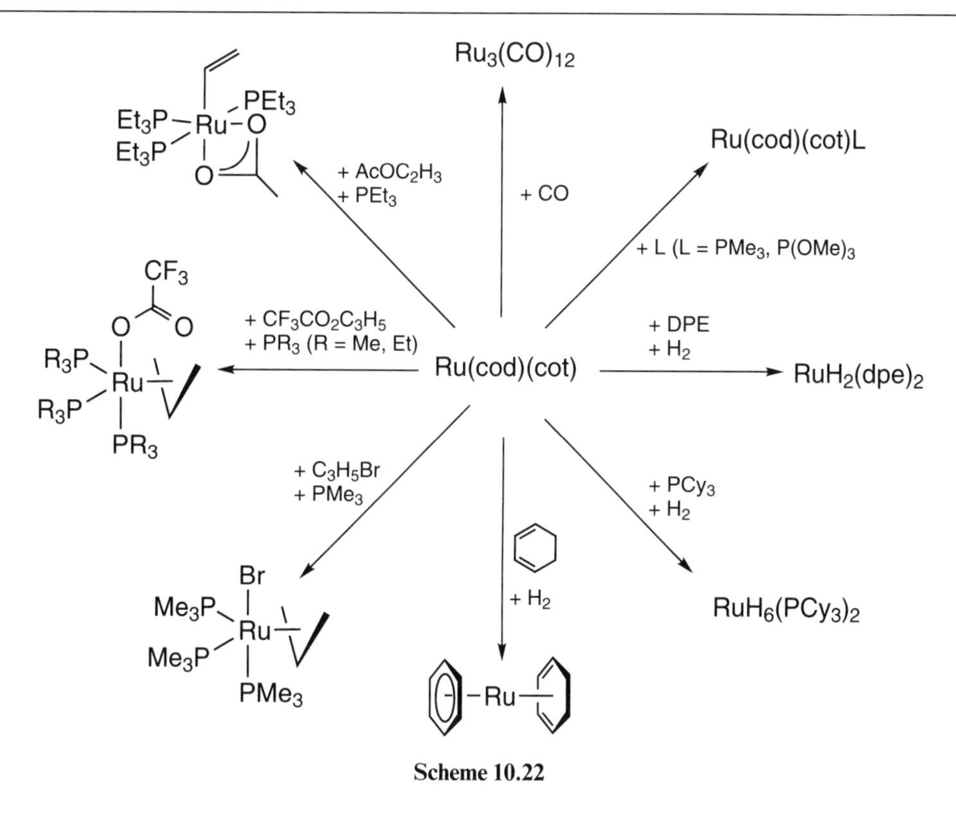

Scheme 10.22

Ru(cod)(cot) with allyl bromide in the presence of PMe_3 yields a π-allylruthenium complex [197]. Of particular interest is that reactions of Ru(cod)(cot) with allyl esters or vinyl esters result in the oxidative addition of the C—O bonds at room temperature giving allyl- and vinylruthenium(II) complexes [198-199].

Ru(cod)(cot) has been used as a catalyst for several unique processes as described below.

10.4.7.1 Dimerization of Norbornadienes

An unique dimerization of norbornadienes involving C—C bond activation by Ru(cod)(cot) has been reported recently (eq (49)) [200]. This reaction occurs under mild conditions (85 °C) to give a single species with high yield. This reaction is also applicable to substituted norbornadienes to give corresponding dimers.

2 [norbornadiene] —— Ru(cod)(cot) 80 °C, 10h; $CH_2{=}CHCONMe_2$ ——→ [dimer] (49)

10.4.7.2 Catalytic Allylation

Recently developed ruthenium-catalyzed allylations are often show quite different reactivities and selectivities from those of palladium-catalyzed allylations. For example, ruthenium complexes sometimes show ambiphilic reactivity (eqs (50) and (51)) [201].

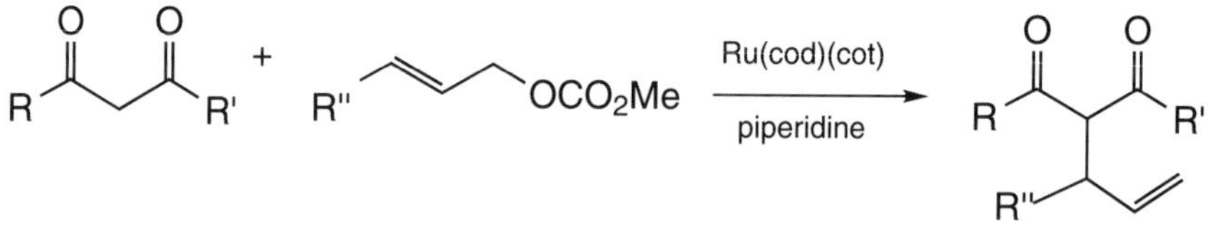

R–CO–CH$_2$–CO–R' + R''–CH=CH–CH$_2$–OCO$_2$Me —[Ru(cod)(cot) / piperidine]→ R–CO–CH(CH(R'')–CH=CH$_2$)–CO–R' (50)

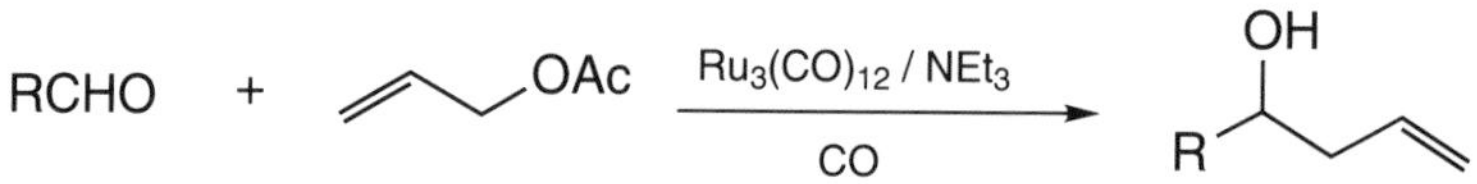

RCHO + CH$_2$=CH–CH$_2$–OAc —[$Ru_3(CO)_{12}$ / NEt_3 / CO]→ R–CH(OH)–CH$_2$–CH=CH$_2$ (51)

The (π-allyl)ruthenium complex $Ru(\pi\text{-}C_3H_5)(OAc)(CO)_3$, which is a model for a possible key intermediate of $Ru_3(CO)_{12}$ catalyzed allylations, shows ambiphilic reactivity [202].

Furthermore, allylation of ethyl acetoacetate with allyl carbonate in the presence of a ruthenium complex exclusively gives a branched isomer, whereas palladium complexes give predominantly the linear isomer (Scheme 10.23). However, the detailed mechanism of the regiocontrolling step is still unclear [203].

Reduction of $RuCl_3 \cdot 3H_2O$ with COD, or 1,3- or 1,4-cyclohexadiene in ethanol resulted in the formation of $[RuCl_2(cod)]_2$ and $[RuCl_2(C_6H_6)]_2$, respectively, in high yield (Scheme 10.24) [204]. Similarly, the corresponding *p*-cymene complex could be obtained from phellandrene. These complexes are particularly important for starting materials of divalent ruthenium catalysts. For example, $RuCl_2$(binap) complex can be cleanly derived by the reaction of $[RuCl_2(cod)]_n$ with BINAP.

In the presence of other π-acceptor ligands, these ruthenium complexes can be reduced by appropriate reagents to give the ruthenium(0) species [205]. Protonation of zero-valent Ru(arene)(diene) give cationic ruthenium(II) complexes possessing an agostic proton [206–207]. Deprotonation from cationic areneruthenium(II) complexes is of interest because double deprotonation from $[Ru(C_6Me_6)_2][PF_6]_2$ gives *endo*-coordinated tetramethyl-*o*-xylyleneruthenium(0) complex [208], while that from $[RuX(C_6Me_6)L_2][PF_6]$ (X = nitrate or trifluoroacetate, L = tertiary phosphine or phos-

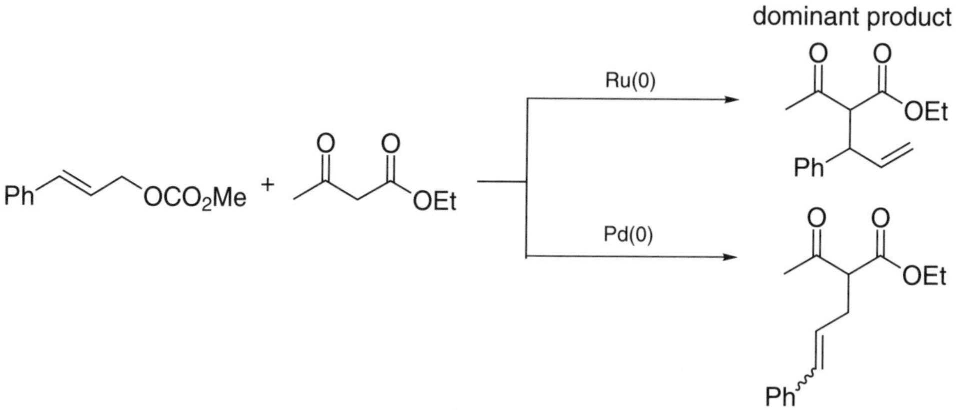

Ph–CH=CH–CH$_2$–OCO$_2$Me + CH$_3$–CO–CH$_2$–CO–OEt
—Ru(0)→ dominant product: CH$_3$–CO–CH(CH(Ph)–CH=CH$_2$)–CO–OEt
—Pd(0)→ CH$_3$–CO–CH(CH$_2$–CH=CH–Ph)–CO–OEt

Scheme 10.23

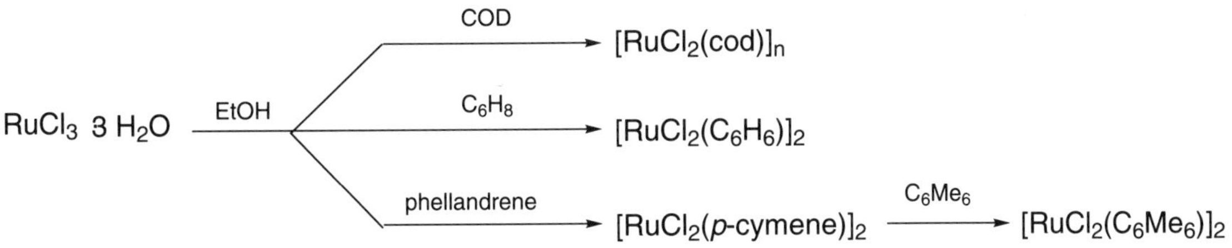

$$RuCl_3 \cdot 3H_2O \xrightarrow{EtOH} \begin{cases} \xrightarrow{COD} [RuCl_2(cod)]_n \\ \xrightarrow{C_6H_8} [RuCl_2(C_6H_6)]_2 \\ \xrightarrow{phellandrene} [RuCl_2(p\text{-cymene})]_2 \xrightarrow{C_6Me_6} [RuCl_2(C_6Me_6)]_2 \end{cases}$$

Scheme 10.24

phite) results in formation of *exo*-coordinated tetramethyl-*o*-xylyleneruthenium(0) complex [209].

$[Ru(nbd)Cl_2]_n$ or $[Ru(cod)Cl_2]_n$ are reduced by cyclooctatetraene dianion to give $Ru(nbd)(C_8H_8)$ and $Ru(cod)(C_8H_8)$, respectively, though the product yields are relatively low (eq (52)) [210].

$$[RuLCl_2]_x \xrightarrow[L = nbd,\ cod]{K_2[C_8H_8]} RuL(C_8H_8) \qquad (52)$$

A ruthenium benzyne complex is derived by the elimination of methane or benzene from *cis*-methylphenylruthenium or *cis*-diphenylruthenium complexes, respectively [211]. This complex is capable of activating C—H, O—H or N—H bonds under ambient conditions (Scheme 10.25).

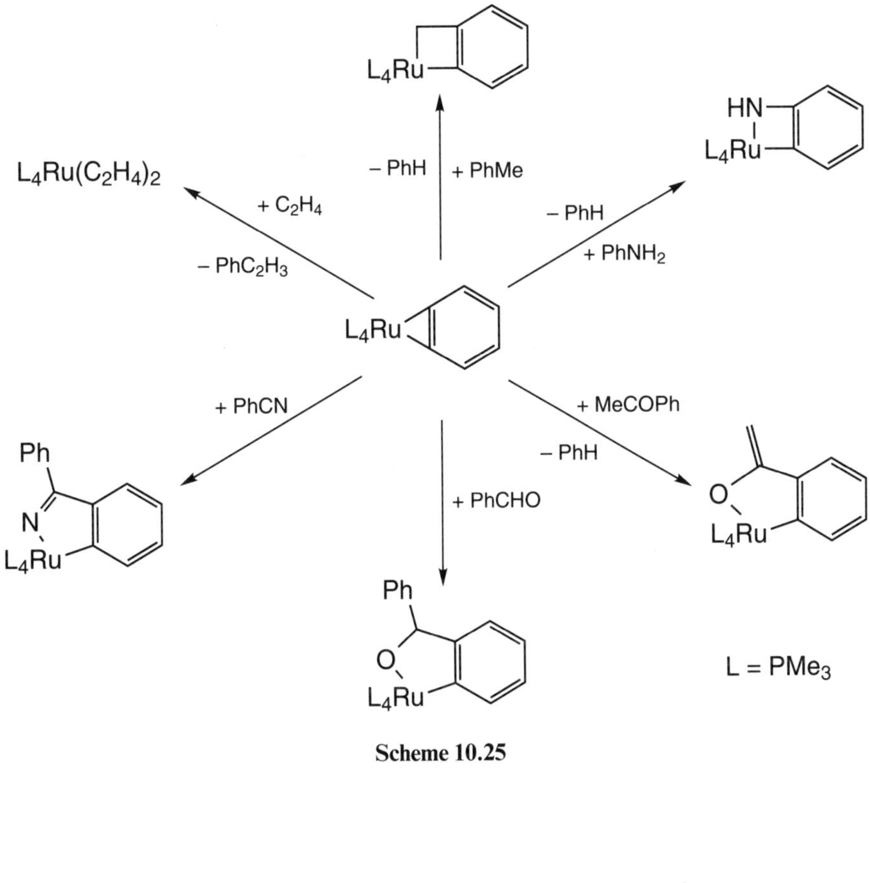

L4Ru(C2H4)2 ← + C2H4, – PhC2H3
– PhH, + PhMe
– PhH, + PhNH2
+ PhCN
+ PhCHO
+ MeCOPh, – PhH
L = PMe3

Scheme 10.25

10.5 Synthesis of Ru Compounds

(1) Dodecacarbonyltriruthenium(0), $Ru_3(CO)_{12}$ [119]

(air stable, hood)

$$RuCl_3 \cdot 3H_2O \xrightarrow{CO} Ru_3(CO)_{12} \quad (53)$$

Hydrated ruthenium trichloride (5.0 g, 19 mmol) is treated with carbon monoxide (50–65 atm) in methanol (use as received, 200 mL) at 125 °C for 8 h in an autoclave. The reaction temperature should be kept at 125 °C during the reaction because side reactions (reddish purple or dark blue precipitates) takes place at higher temperatures. It is therefore highly recommended to use an autoclave with small volume (200–300 mL) which allows facile thermal control. After cooling the reaction vessel and discharging carbon monoxide, the reddish orange precipitate is separated by decantation. The mother liqor can be used for several syntheses of $Ru_3(CO)_{12}$, if more $RuCl_3 \cdot 3H_2O$ is added to the solution. Recrystallization from hot hexane gives reddish orange crystals. Yield; 3.0 g (74 %).

Properties: IR(C_6H_{14}): ν(CO) = 2061 (vs), 2031 (s), 2011 (m) cm^{-1}. $^{13}C\{^1H\}$ NMR: δ 199.7(s).

(2) *trans*-Dichlorotris(triphenylphosphine)ruthenium(II), $RuCl_2(PPh_3)$ [126]

(air stable in solid state)

$$RuCl_3 \cdot 3H_2O \xrightarrow{PPh_3} RuCl_2(PPh_3)_3 \quad (54)$$

Hydrated ruthenium trichloride (1.0 g, 3.8 mmol) is dissolved in methanol (250 mL) and then the reddish brown solution is refluxed for 5 min under nitrogen. After cooling, an excess of triphenylphosphine (6.0 g, 23 mmol) is added. The reaction mixture is again refluxed for 3 h. The shiny black crystals are filtered, washed with ether several times and then dried under vacuum. Yield: 2.7 g (74 %).

$RuCl_2(PPh_3)_3$ is air sensitive in solution and moderately soluble in hot chloroform, acetone, benzene or ethyl acetate to give dark yellow solutions.

X-ray crystallography shows that the title complex has a distorted octahedral geometry with a vacant site which is occupied by an agostic proton of a phenyl group.

Properties: mp 132–134 °C.

(3) (Carbonyl)(chloro)(hydrido)tris(triphenylphosphine)ruthenium(II), $RuHCl(CO)(PPh_3)_3$ [212]

(air stable)

$$RuCl_3 \cdot 3H_2O \xrightarrow{PPh_3,\ HCHO} RuHCl(CO)(PPh_3)_3 \quad (55)$$

Although the title complex was originally prepared by the reaction of ruthenium trichloride with triphenylphosphine in basic alcohol solution, it can also be prepared conveniently in higher yield by the following procedure.

A 2-methoxyethanol solution (60 mL) of triphenylphosphine (1.6 g, 6.0 mmol) is stirred vigorously and refluxed. Hydrated ruthenium trichloride (0.26 g, 1.0 mmol) dissolved in a mixture of 2-methoxyethanol (20 mL) and aqueous formaldehyde (20 mL, 40 w/v %) is rapidly added. The reaction mixture is refluxed for 10 min and then is allowed to cool to room temperature. The resulting cream-white microcrystals are filtered off and washed with ethanol (2 × 10 mL), water (2 × 10 mL), ethanol (2 x 10 mL), and finally with hexane (10 mL) and then dried under vacuum. Yield; 0.89 g (93 %).

The title complex is insoluble in ethanol, ether, and hexane, and soluble in benzene and chloroform.

Properties: mp 209–210 °C (in air), mp 235–237 °C (under nitrogen). IR(Nujol): 2020(m), 1922(vs), 1903(sh) cm^{-1}. ^{1}H NMR(C_6D_6): δ -6.64 (dt, Ru—H).

(4) Dihydridotetrakis(triphenylphosphine)ruthenium(II), $RuH_2(PPh_3)_4$ [153]

(moderately air sensitive)

$$RuCl_3{\cdot}3H_2O \xrightarrow{PPh_3,\ NaBH_4} RuH_2(PPh_3)_4 \qquad (56)$$

Hydrated ruthenium trichloride (0.53 g, 2.0 mmol) in ethanol (20 mL) is added to a vigorously stirred hot ethanol solution (120 mL) of triphenylphosphine (3.14 g, 12.0 mmol) in a Schlenk tube with a reflux condenser and one inlet. Sodium borohydride (0.38 g, 10.0 mmol) in ethanol is warmed until it has just dissolved (sodium borohydride is decomposed rapidly in hot ethanol, therefore the solution should be used within 15 min). Portionwise addition of the sodium borohydride solution by a syringe into hot ethanol solution of ruthenium trichloride and triphenylphosphine gives a yellow precipitate, which is filtered, washed with ethanol (3 × 50 mL), water (3 × 50 mL) and finally ethanol (3 × 50 mL), and dried *in vacuo* to give a light yellow precipitate. Yield: 2.17 g (94 %). Care should be taken during the addition of sodium borohydride solution, since vigorous evolution of hydrogen with bubbling takes place.

The product obtained here is sufficiently pure to use for most reactions, but recrystallization from hot toluene gives pure, bright yellow microcrystals (20–50 %). The compound is very stable under argon and slowly decomposes in air in the solid state, but it decomposes rapidly in solution in air. In solution at room temperature, it dissociates one triphenylphosphine ligand to give $RuH_2(solvent)(PPh_3)_3$ [213]. To avoid formation of a dinitrogen complex, it is highly recommended to keep the complex under argon.

The compound can also be prepared by the reaction of dichloro(triphenylphosphine)ruthenium(II) (1.0 g, 0.87 mmol) with finely ground sodium borohydride (1.5 g, 40 mmol) in the presence of excess triphenylphosphine (6.0 g, 23 mmol) in a mixture of benzene (60 mL) and ethanol (100 mL). The preparation should be carried out at room temperature [214]. Yield: 85–90 %. Reduction of tris(2,4-pentanedionato)ruthenium(III) with triethylaluminum in the presence of triphenylphosphine also affords a low yield (24 %) of the product [213].

$$RuCl_2(PPh_3)_3 + PPh_3 + NaBH_4 \longrightarrow RuH_2(PPh_3)_4 \qquad (57)$$

$$Ru(acac)_3 + PPh_3 + AlEt_3 \longrightarrow RuH_2(PPh_3)_4 \qquad (58)$$

Properties: mp 220 °C (capillary tube). IR: ν(RuH) = 2080(m) cm^{-1}. ^{1}H NMR(C_6D_6): δ –10.1 (br, Ru—H)

(5) Bis(acetylacetonato)[(*S*)-2,2′-bis(diphenylphosphino)-1,1′-binaphthyl]-ruthenium(II), Ru[(*S*)-binap](OAc)$_2$ [170]

$$[RuCl_2(C_8H_{12})]_n \xrightarrow{\text{BINAP, AcONa, Et}_3\text{N}} Ru(OAc)_2(binap) \quad (59)$$

To a toluene solution of (*S*)-BINAP (1.37 g, 2.20 mmol) and $[RuCl_2(cod)]_n$ (0.56 g, 2.0 mmol per $RuCl_2$(cod)) in 250 mL Schlenk tube is added triethylamine (1.2 mL, 8.6 mmol) under argon. The brown solution is refluxed for 12 h with stirring and is then cooled to room temperature. The resultant reddish brown solution is evaporated to dryness under reduced pressure to give a brown solid, which is dissolved in dichloromethane (40 mL). The solution filtrates through a celite pad and the filtrate is evaporated to dryness. To the solid is added anhydrous sodium acetate (0.88 g, 11 mmol) and *tert*-butyl alcohol (107 mL). The mixture is refluxed for 12 h under argon. The solution is then evaporated to dryness under reduced pressure and the resulting solid is extracted with diethyl ether (3 × 20 mL). The combined extract is evaporated under reduced pressure to give brown solid. The solid is extracted with absolute ethanol (3 × 20 mL). Evaporation of the solvent gives a yellow brown solid (1.58 g). Recrystallization from a mixture of toluene (12 ml) and hexane (30 mL) yields fine yellow needles. 1.23 g (68 %).

This crystals are pure enough for use as Λ–(*S*)- Ru(binap)(OAc)$_2$ for asymmetric hydrogenations. The related complexes Ru(*p*-tolbinap)(OAc)$_2$, Ru(binap)[OCOCMe$_3$]$_2$ and their enantiomers can be also prepared by this method.

Properties: mp 185–187 °C (enantiomer). ^{1}H NMR ($CDCl_3$): δ 1.80 (s, 6H), 6.47–7.84 (m, 32H). $^{31}P\{^1H\}$ NMR ($CDCl_3$): δ 65.13 (s). IR (CH_2Cl_2): 1452, 1518 cm^{-1}.

(6) Dichloro(η^5-pentamethylcyclopentadienyl)ruthenium(III), [RuCl$_2$.{η^5-C$_5$Me$_5$}]$_2$ [178]

$$RuCl_3{\cdot}3H_2O \xrightarrow{C_5Me_5H} [RuCl_2(C_5Me_5)]_2 \quad (60)$$

To an ethanol solution (40 mL) of hydrated ruthenium trichloride (2.01 g, 7.7 mmol) is added pentamethylcyclopentadiene (2.70 g, 19.8 mmol) at room temperature under nitrogen. The reaction mixture is warmed at 50 °C for 5 h with vigorous stirring. The solution color gradually changes from dark green to dark brown. After the reaction, the solution is cooled to room temperature and set aside for an hour. The solution layer is removed to give dark brown microcrystals. The crystals are successively washed with abs. ethanol (4 × 10 mL) and diethyl ether (2 × 10 mL), and then dried under vacuum. Yield; 1.97 g (83 %).

Properties: mp 272 °C (dec.). ^{1}H NMR ($CDCl_3$): δ 4.90 (frequency at half band width = 17 Hz). μ_{eff} = 1.89 B.M. IR (KBr): 2983, 2906, 1478, 1375, 1023, 440 cm^{-1}.

(7) Trichloro(η^5-pentamethylcyclopentadienyl)ruthenium(IV), $[RuCl_3\text{-}\{\eta^5\text{-}C_5Me_5\}]_2$ [179]

(hood)

$$[RuCl_2(C_5Me_5)]_2 \xrightarrow{Cl_2} [RuCl_3(C_5Me_5)]_2 \qquad (61)$$

Into a dichloromethane solution (10 mL) of $[RuCl_2(C_5Me_5)]_2$ (0.132 g, 0.215 mmol) in a Schlenk tube (20 mL) is bubbled chlorine gas (1 atm) for 10 min at room temperature. Then, bubbling of chlorine is stopped. The reaction mixture is stirred for 2 h under a chlorine atmosphere, during which time a dark brown precipitate is formed. The solvent is removed by cannula and the resulting precipitate is successively washed with dichloromethane (5 × 5 mL) and diethyl ether (3 × 5 mL). The brownish black precipitate is dried under vacuum to give the title product. Yield: 0.134 g (91 %).

Properties: mp >280 °C. IR (KBr): 2901, 1464, 1423, 1366, 1078, 1008, 998, 333 cm^{-1}.

Related ruthenium tribromide or -triiodide complexes are also obtained by using bromine or iodine instead of chlorine, respectively.

$[RuBr_3(C_5Me_5)]_n$: mp > 280 °C. IR(KBr): 2988, 2950, 2905, 1476, 1369, 1076, 1003 cm^{-1}.
$[RuI_3(C_5Me_5)]_n$: mp > 181 °C. IR (KBr): 2942, 2877, 1472, 1447, 1421, 1372, 1073, 1006 cm^{-1}.

The poor solubility of these ruthenium trihalide complexes in most organic solvents suggests that they have an oligomeric structure bridged by halogen atoms.

Treatment of the tribromide with PPh_3 in CH_2Cl_2 gives a mononuclear ruthenium(IV) complex.

$$[RuBr_3(C_5Me_5)]_n + PPh_3 \longrightarrow RuBr_3(C_5Me_5)(PPh_3) \qquad (62)$$

Dark red microprisms, 55%. mp. 184 °C, IR(KBr): 3047, 2966, 2800, 1586, 1480, 1436, 1371, 1185, 1121, 1090, 1015, 742, 723, 695, 344 cm^{-1}. ^1H NMR($CDCl_3$):δ 1.57 (s, 15H), 7.3–7.8 (m, 15H), $^{31}P\{^1H\}$ NMR($CDCl_3$): δ 35.2 (s).

A reaction of $[RuBr_3(C_5Me_5)]_n$ with 5 equivalents of $NaBH_4$ in abs. ethanol in the presence of PPh_3 (1.3 equvalents) gave mononuclear ruthenium(IV) hydride complex.

$$[RuBr_3(C_5Me_5)]_n + PPh_3 \longrightarrow RuH_3(C_5Me_5)(PPh_3) \qquad (63)$$

Pale brown prisms, 18%. mp 136 °C, IR(KBr): 3052, 2972, 2896, 1974, 1960, 1479, 1434, 1375, 1091, 758, 744, 699, 538, 514 cm-^{1}. ^1H NMR(C_6D_6): δ –9.72 (3H, d, J = 20.5 Hz), 1.84 (15H, d, J = 1.3 Hz), 7.0–7.7 (15H, m). $^{31}P\{^1H\}$ NMR(C_6D_6): δ 79.3.

(8) Chloro(η^5-cyclopentadienyl)bis(triphenylphosphine)ruthenium(II), RuCl (η^5-C_5H_5)$(PPh_3)_2$ [215]

(air sensitive)

$$RuCl_3{\cdot}3H_2O \xrightarrow{C_5H_6,\ PPh_3} RuCl(C_5H_5)(PPh_3)_2 \qquad (64)$$

$RuCl_3 \cdot 3H_2O$ (0.52 g, 2 mmol) is solved in absolute ethanol (20 mL) under nitrogen and the resulting dark brown solution is filtrated. The filtrate is added into a refluxing ethanol (100 mL) solution of PPh_3 (2.10 g, 8.0 mmol), immediately. Then, an ethanol (20 mL) solution of freshly distilled cyclopentadiene (1–2 mL) is added into the reaction mixture. The reaction mixture is refluxed for 45–60 min. The solution color changes from dark brown to orange. The reaction mixture is cooled in a refrigerator for a night. The resulting orange crystals are collected by filtration and washed with cold ethanol (10 mL × 2), water (10 mL × 2), cold ethanol (10 mL × 2), and petroleum ether (10 mL× 2). The orange crystals are dried under vacuum to give the title product. Yield: 1.25 g (86 %).

Properties: mp 131–135 °C (dec.).

(9) Chloro(η^5-cyclopentadienyl)(η^4-1,5-cyclooctadiene)ruthenium(II), $RuCl(\eta^5\text{-}C_5H_5)(\eta^4\text{-}C_8H_{12})$ [216]

(under nitrogen)

$$RuH(C_5H_5)(C_8H_{12}) \xrightarrow{CCl_4} RuCl(C_5H_5)(C_8H_{12}) \quad (65)$$

$RuH(C_5H_5)(C_8H_{12})$ is prepared from $[RuH(C_8H_{12})(NH_2NMe_2)_3]PF_6$ by the published procedure [217]. To a pentane (150 mL) solution of $RuH(C_5H_5)(C_8H_{12})$ (5.51 g, 20.0 mmol) is added carbon tetrachloride (2 mL). The reaction mixture is stirred at room temperature for 5 min to give golden orange microcrystals. The microcrystals are collected by filtration and dried under vacuum. Yield: 5.1 g, (82 %).

Properties: mp 240–235 °C (dec.). ^1H NMR($CDCl_3$): δ 4.95 (5H); $^{13}C\{^1H\}$ NMR ($CDCl_3$): δ 27.6, 33.4 (CH_2), 78.6, 85.9 (CH), 87.5 (C_5H_5).

Related ruthenium bromide or -iodide complexes are also obtained by using CH_2Br_2 or CH_3I instead of CCl_4 in 75 and 60 % yield, respectively.

(10) Dichloro(η^4-cyclooctadiene)ruthenium(II), $[RuCl_2(\eta^4\text{-}C_8H_{12})]_n$ [218]

$$RuCl_3 \cdot 3H_2O \xrightarrow{C_8H_{12}} [RuCl_2(C_8H_{12})]_n \quad (66)$$

Into a Schlenk tube (50 mL) is charged $RuCl_3 \cdot 3H_2O$ (1.20 g, 4.60 mmol), 1,5-cyclooctadiene (1.60 g, 13.0 mmol), absolute ethanol (25 mL) and a magnetic stirrer bar. The reaction mixture is refluxed for 44 h under nitrogen atmosphere. The resulting brown precipitate is collected and then washed with absolute ethanol (6 x 20 mL). After drying under vacuum, $[RuCl_2(C_8H_{12})]_n$ can be isolated as a brown solid, quantitatively. This product is pure enough to be used for further reactions.

(11) Dichlorobis(η^6-*p*-cymene)diruthenium(II), $[RuCl_2(\eta^6\text{-}p\text{-}MeC_6H_4CHMe_2)]_2$ [219]

$$RuCl_3 \cdot 3H_2O \xrightarrow{p\text{-}MeC_6H_4CHMe_2} [RuCl_2(p\text{-}MeC_6H_4CHMe_2)]_2 \quad (67)$$

$RuCl_3 \cdot 3H_2O$ (0.91 g, 3.48 mmol) and *p*-phellandrene (4.93 g, 35.5 mmol) is dissolved in ethanol (70 mL). The reaction mixture is refluxed for 5 h. The resulting orange precipitate is collected by filtration and washed with ethanol (20 mL × 5) and dried in vacuo. Yield: 0.75 g (70 %).

Properties: ^1H NMR($CDCl_3$): δ 1.3 (d, 6H), 2.3 (s, 3H), 2.9 (q, 1H), 5.4 (d, 2H), 5.4 (d, 2H).

Related ruthenium iodide complex is also obtained by the following procedure. $RuCl_3 \cdot 3H_2O$ (0.48 g, 1.8 mmol) and *p*-phellandrene (1.35 g, 9.8 mmol) is dissolved in ethanol (40 mL). The reaction mixture is refluxed for 5 h. To the reaction mixture is added dropwise a water–ethanol (50 v/v %, 20 mL) solution of KI (1.54 g, 9.0 mmol). This solution is refluxed for 2 h. The resulting purple precipitate is collected by filtration, and washed with ethanol (5 mL × 3) and dried *in vacuo*. Yield: 0.80 g (89 %).

Properties: ^1H NMR($CDCl_3$): δ 1.3 (d, 6H), 2.3 (s, 3H), 3.0 (q, 1H), 5.5 (d, 2H), 5.5 (d, 2H).

(12) Tetrachlorobis(η^6-hexamethylbenzene)diruthenium(II), $[RuCl_2(\eta^6\text{-}C_6Me_6)]_2$ [220]

(under nitrogen)

$$[RuCl_2(p\text{-}MeC_6H_4CHMe_2)]_2 \xrightarrow{C_6Me_6} [RuCl_2(C_6Me_6)]_2 \qquad (68)$$

$[RuCl_2(p\text{-}MeC_6H_4CHMe_2)]_2$ (1.0 g, 1.6 mmol) is dissolved into fused hexamethylbenzene (10 g). After 2 h stirring, the reaction mixture is cooled to room temperature and excess hexamethylbenzene is washed out with hexane. The title compound is obtained as reddish brown microcrystals which are enough pure to be used further reactions. Yield: 0.87 g, (80 %). Recrystallization from chloroform followed by silicagel chromatography gives the analytically pure compound.

Properties: IR(Nujol) 299(s), 258(br) cm^{-1}. ^1H NMR($CDCl_3$): δ 2.03.

(13) Bis(η^5-cyclopentadienyl)ruthenium(II), (ruthenocene), $Ru(\eta^5\text{-}C_5H_5)_2$ [221]

(the procedure should be performed under inert gas, although the product is stable in air)

$$\text{(a) } [RuCl_2(C_8H_{12})]_n + \text{n-}Bu_3Sn(C_5H_5) \longrightarrow Ru(C_5H_5)_2 \qquad (69)$$

An ethanol (150 mL) suspension of $[RuCl_2(C_8H_{12})]_n$ (8.4 g, 30 mmol) and n-$Bu_3Sn(C_5H_5)$ (32.0 g, 90 mmol) is refluxed for 48 h under nitrogen atmosphere. After the reaction, the hot mixture is immediately filtered by a jacketed glass filter in air and the filtrate is cooled at 0 °C to give pale yellow microcrystals. The mother liquor is separated by catheter tube and the resulting crystals are washed with cooled ethanol (2 × 20 mL) and diethyl ether (2 × 20 mL). Recrystallization from either hot ethanol or acetone gives pure white crystals. Yield: 5.2 g (75 %).

$$\text{(b) } [RuCl_2(C_8H_{12})]_n + Tl(C_5H_5) \longrightarrow Ru(C_5H_5)_2 \qquad (70)$$

A dimethoxyethane suspension of $[RuCl_2(C_8H_{12})]_n$ (1.4 g, 5.0 mmol) and $Tl(C_5H_5)$ (2.7 g, 10 mmol) is refluxed for an hour under nitrogen. The hot reaction mixture is filtered rapidly in air, the filtrate is evaporated and the residue is dried *in vacuo*. The resulting solid is extracted with diethylether (4 × 20 mL) and the combined extracts are evaporated to yield crude product. Analytically pure product is obtained by recrystallization as described above. Yield: 0.9 g (80 %).

Properties: mp >250 °C (dec). ^1H NMR ($CDCl_3$) δ 4.55 (s); $^{13}C\{^1H\}$ NMR ($CDCl_3$) δ 70.1(s).

The pentamethylcyclopentadienyl analogue could be also prepared by the following procedure.

$$\text{(c) } [RuCl_2(\eta^4\text{-}C_8H_{12})]_n + n\text{-}Bu_3Sn(C_5Me_5) \longrightarrow Ru(C_5Me_5)_2 \quad (71)$$

(under nitrogen)

Into a Schlenk tube (300 mL) is charged $[RuCl_2(C_8H_{12})]_n$ (5.6 g, 20.0 mmol), n-$Bu_3Sn(C_5Me_5)$ (23.2 g, 44.0 mmol) and ethanol (150 mL) and a magnetic stirrer bar. The reaction mixture is refluxed for 1.5 h under nitrogen. The hot reaction mixture is filtrated rapidly, and the filtrate is cooled at 0 °C to give white crystals. The resulting crystals are collected by filtration, washed with cold ethanol (15 mL) and then diethyl ether (10 mL× 2). The pure title compound is obtained by the recrystallization from hot ethanol. Yield: 5.2 g (70 %).

Properties: mp >250°C (dec). ^1H NMR ($CDCl_3$): δ 1.63 (Me); $^{13}C\{^1H\}$ NMR ($CDCl_3$) δ 10.0(s), 82.9(s).

Reaction of $Ru(C_5Me_5)_2$ with HPF_6 or CF_3CO_2H in $CHCl_3$ gives hydride cations.

$$Ru(C_5Me_5)_2 + HX \longrightarrow [RuH(C_5Me_5)_2]^+X^- \quad (X = PF_6 \text{ or } CF_3CO_2) \quad (72)$$

^1H NMR($CDCl_3$): δ –8.30 (1H), 1.84 (15H).

(14) (η^4-1,5-cyclooctadiene)(η^6-1,3,5-cyclooctatriene)ruthenium(0), $Ru(\eta^4\text{-}C_8H_{12})(\eta^6\text{-}C_8H_{10})$ [192]

(moderately air-sensitive)

$$RuCl_3 \cdot 3H_2O \xrightarrow{1,5\text{-}C_8H_{12},\ Zn} Ru(C_8H_{12})(C_8H_{10}) \,. \quad (73)$$

An absolute methanol solution (18 mL) of 1,5-COD (45 mL) with finely ground zinc powder (17.03 g) is refluxed in two-necked flask (500 mL) equipped with a reflux condenser and a dropping funnel under nitrogen. At the same time, $RuCl_3 \cdot 3H_2O$ (1.98 g, 7.59 mmol) is dissolved in methanol (15 mL) in 20 mL Schlenk tube. The deep red (almost black) methanol solution of $RuCl_3 \cdot 3H_2O$ is transferred to the dropping funnel by cannula and then added dropwise. The reaction mixture is heated at 70 °C for 3 h with vigorous stirring. The color of the mixture changes from deep red to brownish yellow, and finally to reddish brown. Ultrasonic irradiation during the reaction increases the yield [222], while longer reaction time or large-scale reaction (*ca* > 5 g) tends to reduce the yield. After 3 h, the reaction mixture is set aside for 30 min to allow the zinc powder to settle. The reddish brown suspension is transferred to a Schlenk tube (500 mL) by a cannula (or a bridge filter with a G-3 glass frit). The gray residual is washed repeatedly with methanol (4 × 20 mL). The combined filtrate is evaporated under reduced pressure and the resultant reddish yellow powder is dried in vacuum. The crude products are extracted with dry hexane (3 × 30 mL) and the extract is concentrated to about one third. The red solution is purified on an alumina column (70–200 mesh, 10 mmφ × 300 mm) with hexane as eluent. The first yellow fraction is collected, evaporated to dryness and the resultant yellow powder is dried *in vacuo*. Recrystallization from pentane gives yellow needles. Yield: 1.54 g, (64%). By concentration of the filtrate, further crystals can be obtained.

Properties: mp 92–94 °C. ^{1}H NMR(C_6D_6): δ 0.90 (m, 2H), 1.64 (m, 2H), 2.22 (m, 8H), 2.92 (m, 4H), 3.79 (m, 2H), 4.78 (m, 2H), 5.22 (dd, 2H).

10.6 Osmium Complexes

Due to the cost of osmium, the chemistry of organoosmium complexes has received less attention than that of iron or ruthenium complexes, although, in general, it resembles the latter than the former. Thus, $Os_3(CO)_{12}$ is structurally similar to $Ru_3(CO)_{12}$ and OsO_4 is an oxidizing agent similar to, but less powerful than RuO_4. Organoosmium complexes are generally more stable to air and thermally than their ruthenium analogues. Osmium complexes have been widely studied for homogeneous oxidation reactions. For example, some olefins are effectively oxidized by stoichiometric amounts of OsO_4 to give 1,2-diols. However, because OsO_4 is highly toxic and is relatively expensive, use of a catalytic amount of OsO_4 with an oxidant such as H_2O_2, $NaClO_3$, ^{t}BuOOH, tertiary amine *N*-oxide, or $K_3Fe(CN)_6$ to give Os(VIII) from Os(VI) is generally preferred. Sharpless discovered that oxidation of olefins in the presence of a chiral amine ligand led to asymmetric dihydroxylation in high optical yield [223]. Because oxidation reactions are generally hard to control and the nature of the active species is usually unclear, these results gave fundamental information on stereospecific oxidation reactions.

Reviewed in this section is the chemistry of $Os_3(CO)_{12}$, $OsH_2Cl_2(P^iPr_3)_2$, Os(*p*-$MeC_6H_4CHMe_2$)(N^tBu), Na_2OsCl_6, OsO_4 and osmium carbene complexes. Reviews of other osmium complexes are available [224–225].

10.6.1 Osmium Carbonyl Complexes

A series of osmium carbonyl complexes have been prepared by the reaction of OsO_4 with CO or decomposition of $Os_3(CO)_{12}$ [226] and a mononuclear homoleptic osmium carbonyl complex, $Os(CO)_5$, is also known. It is a volatile, colorless liquid and is the most robust $M(CO)_5$ type complex of the iron triad against both oxidation and heat but it gradually loses CO to form $Os_3(CO)_{12}$. Multinuclear osmium carbonyl clusters such as $Os_3(CO)_{12}$, $Os_5(CO)_{16}$, $Os_5(CO)_{19}$, $Os_6(CO)_{18}$, $Os_7(CO)_{21}$, and $Os_8(CO)_{23}$ have also been reported [227]. In this section, several carbonyl complexes based on $Os_3(CO)_{12}$ are described.

Trinuclear $Os_3(CO)_{12}$ has the triangular framework common to the Group 8 trinuclear dodecacarbonyl complexes. However, the triangular framework of Os_3 is less readily broken down to mononuclear fragments than that of the Ru_3 framework.

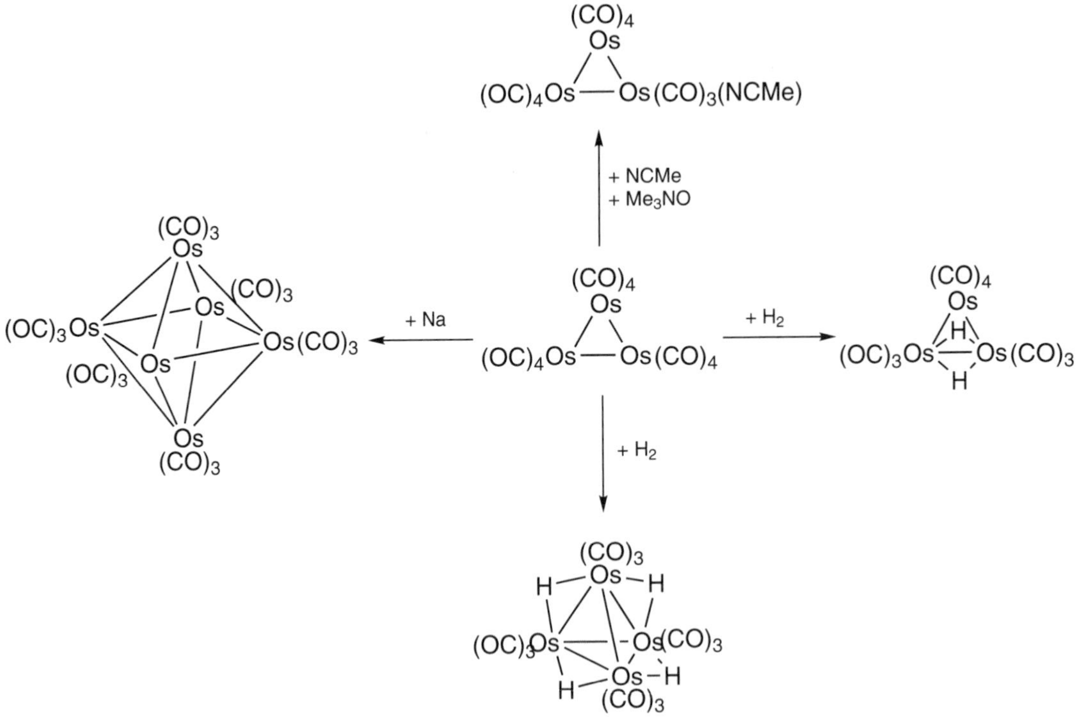

Scheme 10.26

Indeed, several chemical reactivities of $Os_3(CO)_{12}$ are depicted in Scheme 10.26 [228–234]. For example, $Os_3(CO)_{12}$ reacts with hydrogen at room temperature to give μ-H bridged $Os_3H_2(CO)_{10}$ in high yield [235–238]. Reaction of $Os_3(CO)_{12}$ with 25 atm of hydrogen at 140 °C gives tetranuclear μ-H bridged $Os_4H_4(CO)_{12}$ [239–240].

One of the most striking features of $Os_3(CO)_{12}$ is its ability to activate C—H bonds in alkyl or aryl groups. For example, the methyl group in $Os_3H(Me)(CO)_{10}$ gives a μ^3-CH group upon heating via α-elimination when the complex is heated [241]; β-[242], γ-[243], and δ-CH bond activations [244] have also been reported. Surprisingly, a tungsten-triosmium cluster having a μ^3-allyl ligand leads consecutive C—C bond scission to give a trialkylidyne complex under ambient conditions (eq (74)) [245].

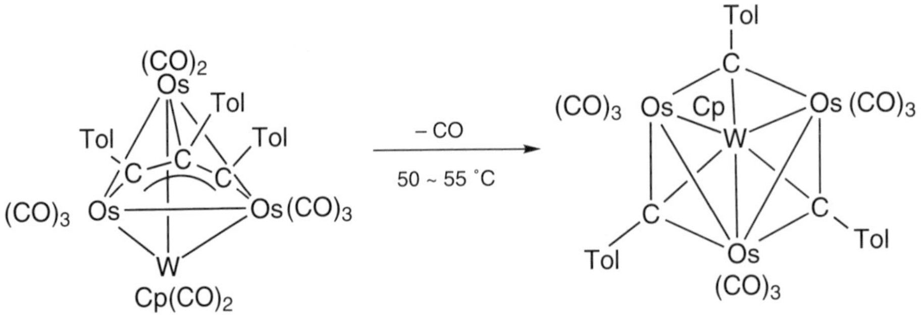

(74)

10.6.2 Hydridoomsmium(II) Complexes

Hydridochloro(carbonyl)osmium(II) complexes $OsHCl(CO)L_n$ (n = 2 or 3) are relatively well-investigated. Osmium hydride complexes are generally active in a number of insertion reactions of a hydrido ligand into unsaturated bonds. For examples, $OsHX(CO)_2(PPh_3)_2$ (X = Cl or Br) reacts with CS_2 [246], RNCO [247] or *p*-TolN=C=N*p*-Tol [248], to give corresponding hydrido insertion products, $OsCl(SCHS)(CO)(PPh_3)_2$, $OsBr(RNCHO)(CO)(PPh_3)_2$, and $OsCl(p\text{-TolNCHN}p\text{-Tol})(CO)_2(PPh_3)$, respectively. Recently, an interesting insertion reaction of alkynylalcohols into the Os—H bond has been reported. $OsHCl(CO)(P^iPr_3)_2$ reacted with 2-propyn-1-ol to give a vinylosmium complex while the reaction of the complex with 1-phenyl-2-propyn-1-ol formed vinylcarbene and heterocycle complexes (Scheme 10.27) [249].

Because little is known about d^4-complexes of coordination number six, the dihydridodichloroosmium complex $OsH_2Cl_2(P^iPr_3)_2$ is focused on the viewpoint of coordination chemistry. The shape of the complex is better to understand as a trigonal prism rather than by octahedral geometry on the basis of both experimental and computational results [250]. Interestingly, this complex reacts with terminal alkynes to give carbyne complexes (eq (75)) [251].

$$OsH_2Cl_2(P^iPr_3)_2 + \equiv\!\!-Ph \longrightarrow (^iPr_3P)_2Cl_2HOs\equiv\!\!-CH_2Ph \quad (75)$$

Cationic hydridoosmium complexes catalyze the regio- and stereoselective dimerization of terminal alkynes. $[OsH(N_2)(PP_3)]BPh_4$ [$PP_3 = P(CH_2CH_2PPh_2)_3$] is a catalyst precursor for the dimerization of 1-phenylacetylene to give (*Z*)-1,4-diphenylbut-3-en-1-yne [252].

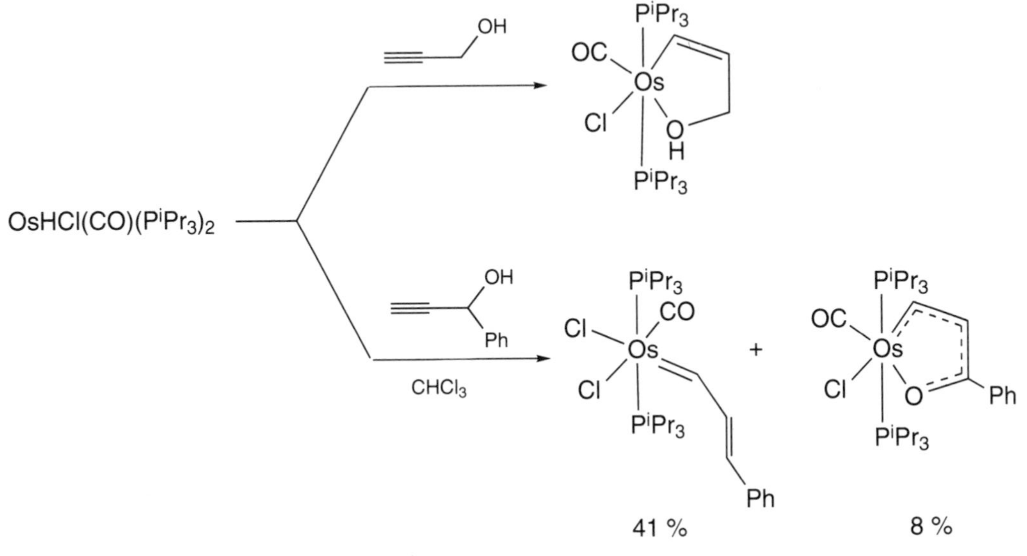

Scheme 10.27

10.6.3 Imidoosmium Complex

Treatment of $[(arene)OsCl_2]_2$ with lithium amide in THF results in the formation of imidoosmium complexes in high yield, which react with a with wide variety of active hydrogen compounds and polar compounds (Scheme 10.28) [253].

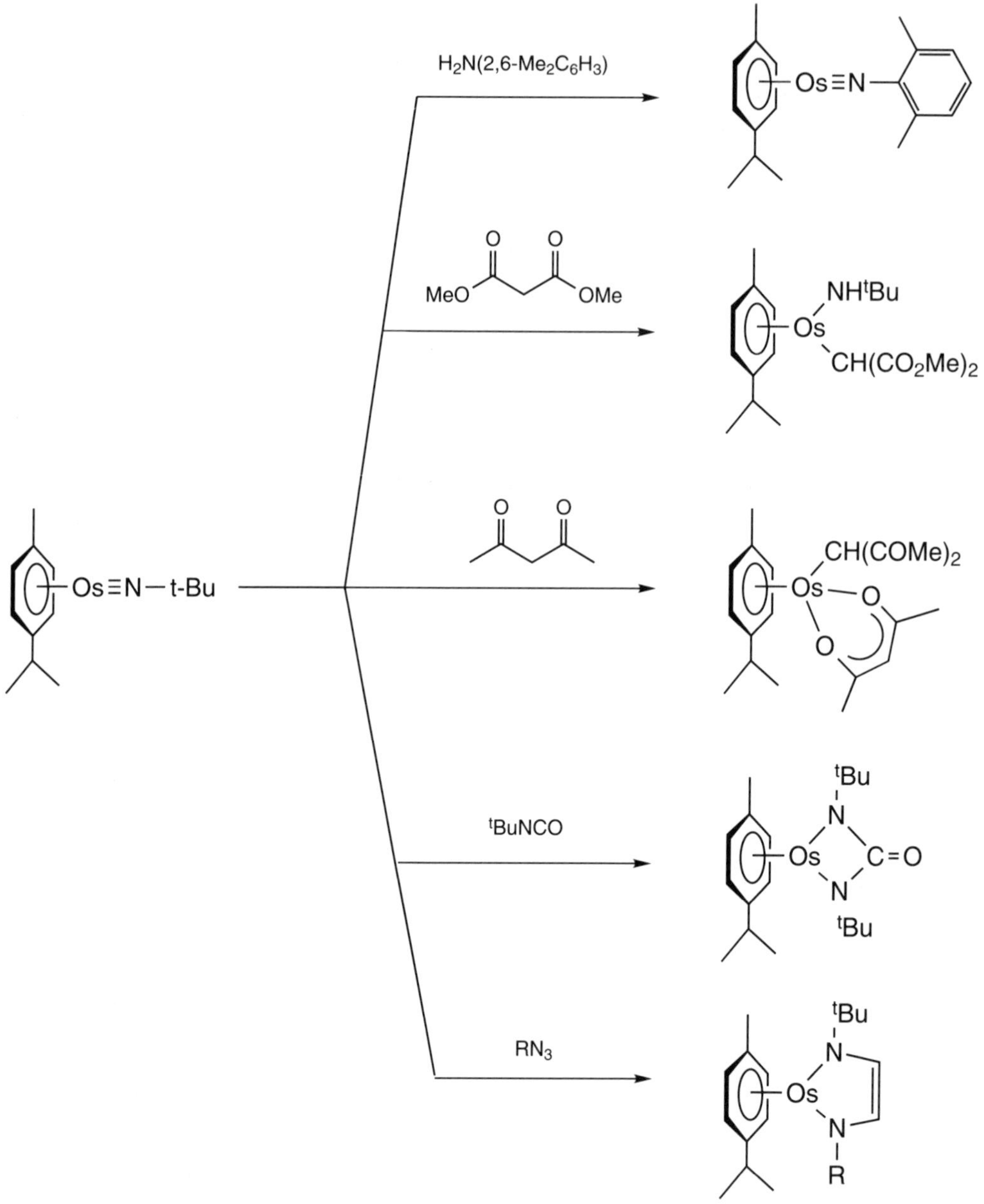

Scheme 10.28

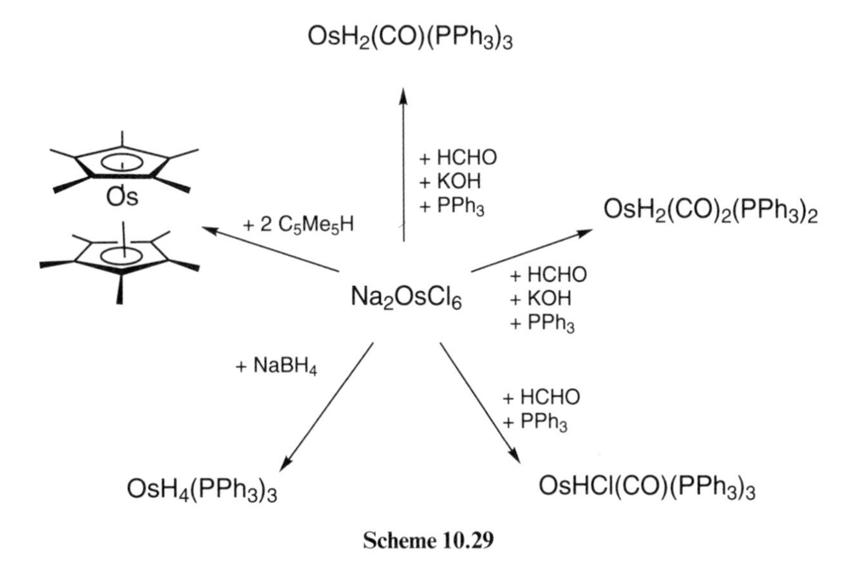

Scheme 10.29

10.6.4 Sodium hexachloroosmate

Na_2OsCl_6 is a commercially available but relatively expensive compound. This complex is sometimes used for preparation of mononuclear osmium complexes as outlined in Scheme 10.29 [254].

10.6.5 Osmium Oxide Complexes

Although OsO_4 is excluded from the definition of organometallic complexes, it should be briefly reviewed because it is a convenient reagent for the preparation of 1,2-diols from alkenes. OsO_4 is a colorless, volatile solid (mp 40 °C) which is very harmful especially to the eyes, hence reactions should be done in the hood. A number of systems consisting of catalytic amount of OsO_4 with an oxidant have been developed.

10.6.5.1 1,2-Diol Synthesis

In the presence of OsO_4 and $KClO_3$ or $NaClO_3$, mono- or disubstituted alkenes are effectively oxidized to the corresponding 1,2-diols. However, these catalytic systems are inactive for the oxidation of tri- or tetrasubstituted alkenes, and OsO_4-^tBuOOH is widely used for this purpose (eq (76)) [223].

OsO_4-^tBuOOH; 10 % Et_4NOH; ^tBuOH (76)

By combining OsO_4 and a natural alkaloid, asymmetric oxidation has been developed. Oxidation of *trans*-stilbene dissolved in aqueous acetone with morpholine-*N*-oxide, dihydroquinidine 4-chlorobenzoate as the ligand, and 0.4 mol % of OsO_4 as the catalyst quantitatively gives the *R,R*-diol in 90 % optical purity (eq (77)) [255–256]. The effects of the ligand [257], the oxidant [258], and the solvent [258] on the stereoselectivity have been studied in detail. Olefin oxidations by $OsO_4/K_3Fe(CN)_6$ system are significantly enhanced by the addition of a base such as DABCO (1,4-diazabicyclo[2.2.2]octane) or quinuclidine [258a].

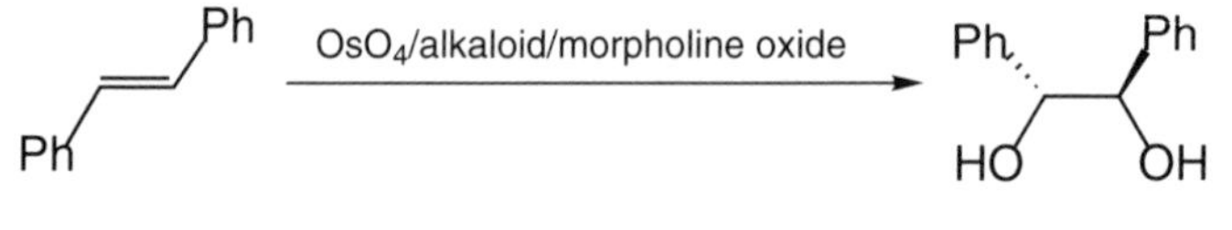

Ph–CH=CH–Ph —(OsO_4/alkaloid/morpholine oxide)→ PhCH(OH)–CH(OH)Ph (77)

10.6.5.2 *Hydroxyamination*

Similar treatment of cyclohexene with Chloramine-T (sodium *p*-toluenesulfonchloramide) in a polar solvent such as ^{t}BuOH in the presence of a catalytic amount of OsO_4 gives the optically active hydroxyaminated cyclohexane in good yield (80%) (eq (78)) [259].

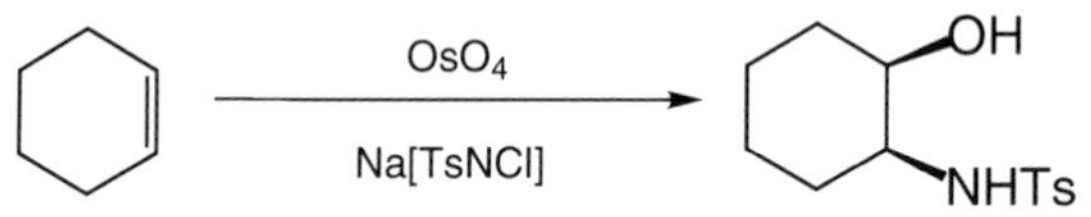

cyclohexene —(OsO_4 / Na[TsNCl])→ 2-(NHTs)cyclohexanol (OH, NHTs) (78)

However, the hydroxyamination of unsymmetrical olefins shows poor regioselectivity (eq (79)) [260].

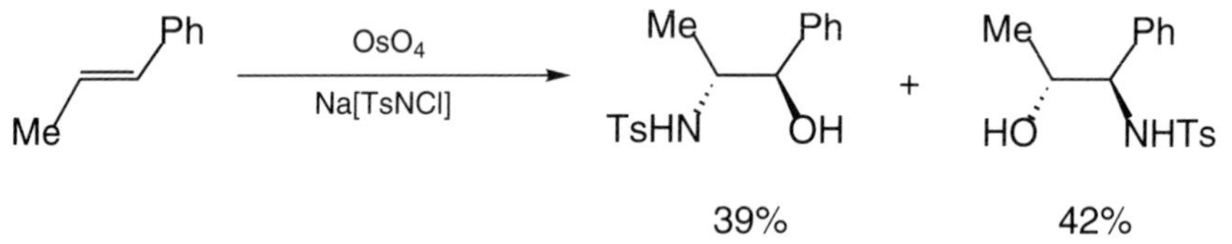

Me–CH=CH–Ph —(OsO_4 / Na[TsNCl])→ MeCH(NHTs)–CH(OH)Ph (39%) + MeCH(OH)–CH(NHTs)Ph (42%) (79)

10.6.5.3 *Olefin Metathesis*

The ring opening metathesis polymerization (ROMP) of norbornene by the OsO_4 complex has been reported recently [261]. The key intermediate is considered to be an oxaosmacyclobutane whose C—C and Os—O bonds can open to form an osmium alkylidene species (Scheme 10.30).

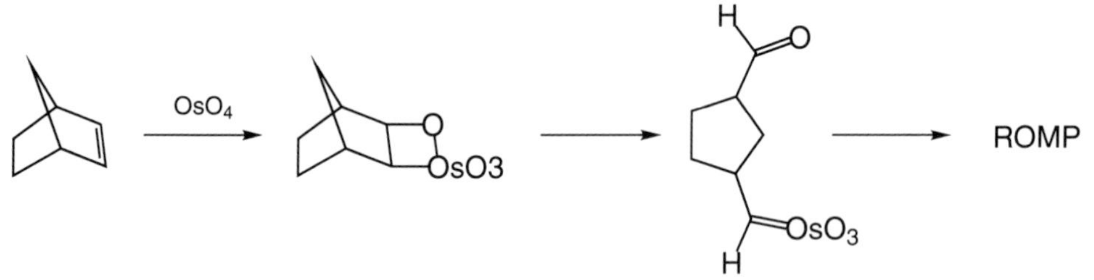

norbornene —(OsO_4)→ oxaosmacyclobutane (O–OsO3) → cyclopentane bearing CHO and CH=OsO_3 → ROMP

Scheme 10.30

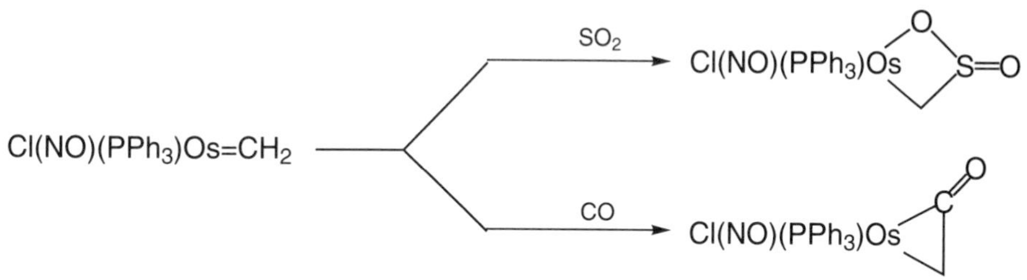

$$Cl(NO)(PPh_3)Os{=}CH_2 \xrightarrow{SO_2} Cl(NO)(PPh_3)Os(\text{O–S(=O)–CH}_2\text{ ring})$$

$$Cl(NO)(PPh_3)Os{=}CH_2 \xrightarrow{CO} Cl(NO)(PPh_3)Os(\text{C(=O)–CH}_2\text{ ring})$$

Scheme 10.31

10.6.6. Osmium Carbene Complexes

Although osmium carbene complexes have made little impact on organic synthesis, the osmium carbene complex $Cl(NO)(PPh_3)Os{=}CH_2$ has an interesting organometallic chemistry, because it reacts with both nucleophiles and electrophiles. For example, it can react with nucleophilic SO_2 and electrophilic CO to give the corresponding osmacycle complexes (Scheme 10.31). Because osmium carbene complexes sometimes show this ambiphilic property, they may be considered as being intermediate between Fischer- and Schrock-type carbene complexes [262].

10.7 Synthesis of Os Compounds

(1) Dodecarbonyltriosmium(0), $Os_3(CO)_{12}$ [229]

(hood)

$$OsO_4 \xrightarrow{CO} Os_3(CO)_{12} \tag{80}$$

In a 100 ml autoclave is charged OsO_4 (2.0 g, 7.8 mmol), methanol (30 ml) and carbon monoxide (75 atm). The reaction mixture is heated at 125 °C for 12 h with stirring. On cooling, yellow crystals are filtered off, washed with cold acetone, and dried *in vacuo* (20 °C, 0.1 mmHg). These orange crystals are usually sufficiently pure for further synthesis. Further purification is carried out by sublimation (130 °C / 0.01 mmHg) or recrystallization from hot benzene. 1.6 g (70 %).

Properties: IR(hexane): 2070 (s), 2036 (s), 2015 (m), 2003 (m) cm^{-1}.

(2) Dicarbonyldihydridotris(triphenylphosphine)osmium(II), $OsH_2(CO)$-$(PPh_3)_3$ [254]

$$Na_2OsCl_6 \xrightarrow{PPh_3} OsH_2(CO)(PPh_3)_3 \tag{81}$$

Into a single-necked round-bottomed flask (500 mL) equipped with a reflux condenser and a magnetic stirrer bar is charged triphenylphosphine (0.92 g, 3.5 mmol) and 2-methoxy-ethanol (100 mL). To the solution is rapidly added a 2-methoxyethanol solution (30 mL) of $Na_2OsCl_6{\cdot}6H_2O$ (1.18 g, 2.1 mmol), formaldehyde (55 mL, 40 w/v%) and 2-methoxy-ethanol solution (20 mL) of potassium hydroxide (2.0 g, 36 mmol) in this order with vigorous stirring. The reaction mixture is refluxed for 30 min during which time the mixture slowly lightens in color. The white precipitate is filtered off and washed successively with ethanol, water, ethanol, and n-hexane, and dried *in vacuo*. Yield: 0.93 g (57 %).

Properties: mp 243–244 °C (in air), 257–258 °C (in capillary). IR(Nujol) 2020 (s), 1990 (vs), 1926 (m), 1871 (s) cm^{-1}. ^{1}H NMR (C_6D_6): δ –7.08 (t, J = 22.75 Hz).

(3) Tetrahydridotris(triphenylphosphine)osmium(IV), $OsH_4(PPh_3)_3$ [263]

$$Na_2OsCl_6 \cdot 6H_2O \xrightarrow{PPh_3,\ NaBH_4} OsH_4(PPh_3)_3 \qquad (82)$$

In a single-necked round-bottomed flask equipped with a reflux condenser and a magnetic stirrer bar is placed triphenylphosphine (1.57 g, 6.0 mmol) and ethanol (80 mL). An ethanol solution (20 mL) of $Na_2OsCl_6 \cdot 6H_2O$ (0.57 g, 1.0 mmol) and an ethanol solution (20 mL) of $NaBH_4$ (0.2 g, 5.2 mmol) are rapidly added with vigorous stirring. The pink colored reaction mixture is refluxed for 15 min, during which period a white precipitate appears. The precipitate is filtered off, washed successively with ethanol, water, ethanol and hexane, and dried *in vacuo*. Yield: 0.95 g (95 %).

Properties: mp 172–175 °C (in air), 219–221 °C (in capillary). IR(Nujol) 2080 (w), 2020 (m), 1951 (m), 1890 (s) cm^{-1}. ^{1}H NMR (C_6D_6): δ –7.85 (q, J = 9 Hz).

References

[1] H. Suzuki, *Yuuki Kinzoku Kagoubutu*, Ed. A. Yamamoto, Tokyo Kagaku Dohjin, Tokyo, (1991).
[2] L. Mond and F. Quincke, *J. Chem. Soc.*, 59, 604 (1891).
[3] M. Berthelot, *C. R. Hebd. Seances Acad. Sci.*, **112**, 1343 (1891).
[4] (a) T. J. Kealy, P. L. Pauson, *Nature*, **168**, 1039 (1951).
(b) S. A. Miller, J. A. Tebboth, J. F. Tremaine, *J. Chem. Soc.*, 632 (1952).
(c) G. Wilkinson, M. Rosenblum, M. C. Whiting, R. B. Woodward, *J. Am. Chem. Soc.*, **74**, 2125 (1952).
(d) E. O. Fischer and W. Pfab, *Z. Naturforsch.*, **7B**, 377 (1952).
[5] (a) P. J. Harrington, *Transition Metal in Total Synthesis*, John Wiley (1990).
(b) D. F. Shriver, K. H. Whitmire, *Compr. Organomet. Chem.*, **4**, 243 (1982).
(c) M. D. Johnson, *Compr. Organoment. Chem.*, **4**, 331 (1982).
(d) A. J. Deeming, *Compr. Organomet. Chem.*, **4**, 377 (1982).
(e) W. P. Fehlhammer, H. Stolzenberg, *Compr. Organomet. Chem.*, **4**, 513 (1982).
(f) J. L. Davidson, *Compr. Organomet. Chem.*, **4**, 615 (1982).
(g) A. J. Pearson, *Iron Compounds in Organic Synthesis*, Academic, San Diego (1994).
(h) J. Silver, Ed., *Chemistry of Iron*, Blackie, London (1993).
(i) E. A. Koevner con Gustorf, I. Fischer, F. W. Grevels, Eds., *The Organic Chemistry of Iron*, Academic, New York (1978).
[6] R. B. King, *Organometallic Synthesis*, vol.1, Academic, San Diego, p. 128 (1965).
[7] R. Pettit, *J. Am. Chem. Soc.*, **81**, 1226 (1959).
[8] (a) A. Nakamura, N. Hagihara, *Bull. Chem. Soc. Jpn.*, **32**, 881 (1959).
(b) T. A. Manuel, F. G. A. Stone, *Proc. Chem. Soc.*, 90 (1959).
[9] (a) G. F. Emerson, L. Watts, R. Pettit, *J. Am. Chem. Soc.*, **87**, 131 (1965).

(b) R. Pettit and J. Henery, *Org. Synth.*, **50**, 21 (1970).
[10] (a) E. H. Braye, W. Hübel, *Inorg. Synth.*, **8**, 178 (1966).
(b) R. B. King, *Organometallic Syntheses*, vol.1, Academic, San Diego, p. 93 (1965).
[11] J. K. Ruff, W. J. Schlientz, *Inorg. Synth.*, **15**, 84 (1974).
[12] (a) M. P. Cook, Jr., *J. Am. Chem. Soc.*, **92**, 6080 (1970).
(b) H. Strong, P. J. Krusic, J. S. Filippo, *Inorg. Synth.*, **24**, 157 (1986).
[13] H. D. Murdoch, E. Weiss, *Helv. Chim. Acta*, **45**, 1927 (1962).
[14] E. Weiss, K. Stark, J. E. Lancaster, H. D. Murdoch, *Helv. Chim. Acta*, **46**, 288 (1963).
[15] M. Orchin, *Adv. Catal.*, **16**, 1 (1966).
[16] G. F. Emerson, J. E. Mahler, R. Kochhar, R. Pettit, *J. Org. Chem.*, **29**, 3620 (1964).
[17] H. Alper, J. T. Edward, *J. Organomet. Chem.*, **14**, 411 (1968).
[18] M. Cais, N. Maoz, *J. Organomet. Chem.*, **5**, 370 (1966).
[19] M. Rhyang, I. Rhee, S. Tsutsumi, *Bull. Chem. Soc. Jpn.*, **37**, 341 (1964).
[20] G. N. Schrauzer, *Chem. Ber.*, **94**, 1891 (1961).
[21] (a) R. Aumann, K. Frölich, H. Ring, *Angew. Chem., Int. Ed. Engl.*, **13**, 275 (1974).
(b) G. D. Annis, S. V. Ley, C. R. Self, R. Sivaramakrishnann, *J. Chem. Soc., Perkin Trans.*, **1**, 270 (1981).
(c) A. M. Horton, D. M. Hollinshead, S. V. Ley, *Tetrahedron Lett.*, **40**, 1737 (1984).
[22] G. D. Annis, E. M. Hebblethwaite, S. T. Hodgson, D. M. Hollinshead, and S. V. Ley, *J. Chem. Soc., Perkin Trans.*, **1**, 2851 (1983).
[23] H. Alper, E. C. H. Keung, *J. Org. Chem.*, **37**, 2566 (1972).
[24] (a) W. Reppe, H. Vetter, *Liebig's Ann. Chem.*, **582**, 133 (1953).
(b) W. Reppe, N. von Kutepow, A. Magin, *Angew. Chem. Int. Ed.*, **8**, 727 (1969).
[25] G. W. Parshall, S. D. Ittel, *Homogeneous Catalysis*, Wiley-Interscience, New York, pp. 206 (1992).
[26] R. Noyori, *Ann. N. Y. Acad. Sci.*, **295**, 225 (1977).
[27] R. Noyori, *Acc. Chem. Res.*, **12**, 61 (1979).
[28] R. Noyori, Y. Hayakawa, *Org. React.*, **29**, 163 (1983).
[29] R. Noyori, Y. Hayakawa, *Tetraheadron*, **41**, 5879 (1985).
[30] H. M. R. Hoffmann, K. E. Clemens, E. A. Schmidt, R. H. Smithers, *J. Am. Chem. Soc.*, **94**, 3201 (1972).
[31] R. B. King, *Inorg. Chem.*, **2**, 326 (1963).
[32] B. M. Trost and S. D. Ziman, *J. Org. Chem.*, **38**, 932 (1973).
[33] G. F. Emerson, K. Ehrlich, W. P. Giering, P. C. Lauterbur, *J. Am. Chem. Soc.*, **88**, 3172 (1966).
[34] K. Ehrlich, G. F. Emerson, *J. Am. Chem. Soc.*, **94**, 2464 (1972).
[35] G. F. Emerson, L. Watts, R. Pettit, *J. Am. Chem. Soc.*, **87**, 131 (1965).
[36] (a) J. P. Collman, R. G. Finke, J. N. Cawse, J. I. Brauman, *J. Am. Chem. Soc.*, **99**, 2515 (1977).
(b) J. P. Collman, R. G. Finke, P. L. Matlock, R. Wahren, R. G. Komoto, J. I. Brauman, *J. Am. Chem. Soc.*, **100**, 1119 (1978).
[37] H. B. Chin, R. Bau, *J. Am. Chem. Soc.*, **98**, 2434 (1976).
[38] R. G. Teller, R. G. Finke, J. P. Collman, H. B. Chin, R. Bau, *J. Am. Chem. Soc.*, **99**, 1104 (1977).
[39] J. P. Collman, R. S. Winter, D. R. Clark, *J. Am. Chem. Soc.*, **94**, 1788 (1972).
[40] Y. Watanabe, M. Yamashita, T. Mitsudo, M. Tanaka, Y. Takegami, *Tetrahedron Lett.*, 3535 (1973).
[41] (a) R. G. Finke and T. M. Sorrell, *Org. Synth.*, **VI**, 807 (1988).
(b) J. P. Collman, *Acc. Chem. Res.*, **8**, 342 (1975).
[42] H. Reihlen, A. Gruhl, G. von Hessling, O. Pfrengle, *Liebigs Ann. Chem.*, **482**, 161 (1930).
[43] O. S. Mills,G. Robinson, *Proc. Chem. Soc.*, 421 (1960).
[44] B. F. Hallam, P. L. Pauson, *J. Chem. Soc.*, 642 (1958).
[45] R. B. King, T. A. Manuel, F. G. A. Stone, *J. Inorg. Nucl. Chem.*, **16**, 23 (1961).
[46] A. J. Brich, S. Y. Hsu, A. J. Pearson, W. D. Raverty, *J. Organomet. Chem.*, **260**, C59 (1984).
[47] M. F. Semmelhack, J. W. Herndon, *Organometallics*, **2**, 363 (1983).
[48] K. Stark, J. E. Lancaster, H. D. Murdoch, E. Weiss, *Z. Naturforsch.*, **19**, 284 (1964).

[49] T. H. Whitesides, R. W. Slaven, *J. Organomet. Chem.*, **67**, 99 (1974).
[50] W. Hübel, E. H. Braye, *J. Inorg. Nucl. Chem.*, **10**, 250 (1959).
[51] M. Rosenblum, *Chemistry of the Iron Group Metallocenes*, Wiley, New York (1965).
[52] W. E. Watts, *Compr. Organometal. Chem.*, **8**, 1013 (1984).
[53] M. Rosenblum, *J. Organomet. Chem.*, **300**, 191 (1986).
[54] (a) A. Cutler, D. Ehntholt, W. P. Giering, P. Lennon, S. Raghu, A. Rosan, M. Rosenblum, J. Tancrede, D. Wells, *J. Am. Chem. Soc.*, **98**, 3495 (1976).
(b) W. P. Giering, M. Rosenblum, *J. Chem. Soc., Chem. Commun.*, 441 (1971).
(c) S. Samuel, S. R. Berryhill, M. Rosenblum, *J. Organomet. Chem.*, **166**, C9 (1979).
[55] (a) W. P. Giering, M. Rosenblum, J. Tanerede, *J. Am. Chem. Soc.*, **94**, 7170 (1972).
(b) M. Rosenblum, M. R. Sadi, M. Madhavarao, *Tetrahedron Lett.*, **46**, 4009 (1975).
[56] T. C. T. Chang, M. Rosenblum, N. Simms, *Org. Synth.*, **66**, 95 (1987).
[57] (a) M. L. H. Green, P. L. I. Nagy, *J. Organomet. Chem.*, **1**, 58 (1963).
(b) D. E. Laycock, J. Hartgerink, M. C. Baird, *J. Org. Chem.*, **45**, 291 (1980).
[58] M. Rosenblum, *Acc. Chem. Res.*, **7**, 125 (1974).
[59] (a) T. S. Abram, R. Baker, C. M. Exon, *Tetrahedron Lett.*, 4103 (1979).
(b) A. Rosan, M. Rosenblum, J. Tancrede, *J. Am. Chem. Soc.*, **95**, 3062 (1973).
(c) N. Genco, D. Marten, S. Raghu, M. Rosenblum, *J. Am. Chem. Soc.*, **98**, 848 (1976).
[60] D. Marquarding, H. Klusacek, G. Gokel, P. Hoffmann, I. Ugi, *J. Am. Chem. Soc.*, **92**, 5389 (1970).
[61] T. Hayashi, M. Kumada, *Acc. Chem. Res.*, **15**, 395 (1982).
[62] (a) Y. Ito, M. Sawamura, T. Hayashi, *J. Am. Chem. Soc.*, **108**, 6405 (1986).
(b) Y. Ito, M. Sawamura, T. Hayashi, *Tetrahedron Lett.*, **28**, 6215 (1987).
(c) Y. Ito, M. Sawamura, H. Hamashima, T. Emura, T. Hayashi, *Tetrahedron Lett.*, **35**, 4681 (1989).
(d) V. A. Soloshonok T. Hayashi, *Tetrahedron Lett.*, **35**, 2713 (1994).
[63] M. Sawamura, H. Hamashima, Y. Ito, *Tetrahedron : Asymmetry*, **2**, 593 (1991).
[64] Sawamura, H. Hamashima, Y. Ito, *J. Am. Chem. Soc.*, **114**, 8295 (1992).
[65] (a) Y. Mizuho, N. Kasuga, S. Komiya, *Chem. Lett.*, 2127 (1991).
(b) M. Hirano, Y. Ito, M. Hirai, A. Fukuoka, S. Komiya, *Chem. Lett.*, 2057 (1993).
[66] (a) A. Riant, O. Samuel, H. B. Kagan, *J. Am. Chem. Soc.*, **115**, 5835 (1993).
(b) D. Giullaneux, H. B. Kagan, *J. Org. Chem.*, **60**, 2502 (1995).
[67] P. Bergamini, S. Sostero, O. Traverso, *J. Organomet. Chem.*, **299**, C11 (1986).
[68] C. A. Tolman, S. D. Ittel, A. D. English, J. P. Jesson, *J. Am. Chem. Soc.*, **100**, 4080 (1978).
[69] H. Azizian, R. H. Morris, *Inorg. Chem.*, **22**, 6 (1983).
[70] C. A. Tolman, S. D. Ittel, A. D. English, J. P. Jesson, *J. Am. Chem. Soc.*, **101**, 1742 (1979).
[71] S. D. Ittel, C. A. Tolman, A. D. English, J. P. Jesson, *J. Am. Chem. Soc.*, **100**, 7577 (1978).
[72] M. V. Baker, L. D. Field, *J. Am. Chem. Soc.*, **108**, 7436 (1986).
[73] M. V. Baker, L. D. Field, *J. Am. Chem. Soc.*, **109**, 2825 (1987).
[74] M. K. Whittlesey, R. J. Mawby, R. Osman, R. N. Perutz, L. D. Field, M. P. Wilkinson, M. W. George, *J. Am. Chem. Soc.*, **115**, 8627 (1993).
[75] G. J. Leigh, M. Jimenez–Tenorio, *J. Am. Chem. Soc.*, **113**, 5862 (1991).
[76] S. Komiya, M. Akita, A. Yoza, N. Kasuga, A. Fukuoka, Y. Kai, *J. Chem. Soc., Chem. Commun.*, 787 (1993).
[77] C. Perthuisot, W. D. Jones, *New J. Chem.*, **18**, 621 (1994).
[78] (a) K. Noack, M. Ruch, *J. Organomet. Chem.*, **17**, 309 (1969).
(b) M. Poliakoff, *Chem. Soc. Rev.*, **7**, 527-540 (1978).
[79] E. H. Brayer, W. Hübel, *Inorg. Synth.*, **8**, 178 (1966).
[80] (a) V. M. Hieber, G. Braun, *Z. Naturforsh.*, **146**, 132 (1959).
(b) H. Strong, P. J. Krusic, J. S. Filippo, *Inorg. Synth.*, **24**, 157 (1986).
(c) M. P. Cook, Jr., *J. Am. Chem. Soc.*, **92**, 6080 (1970).
[81] (a) B. F. Hallam, P. L. Pauson, *J. Chem. Soc.*, 642 (1958).
(b) R. B. King, *Organometallic Synthesis*, vol. 1, Academic, New York, p. 128 (1965).
[82] (a) J. A. S. Howell, B. F. G. Johnson, P. L. Josty, J. Lewis, *J. Organomet. Chem.*, **39**, 329 (1972).
(b) G. Evans, B. F. G. Johnson, J. Lewis, *J. Organomet. Chem.*, **102**, 507 (1975).

[83] R. B. King, F. G. A. Stone, *Inorg. Synth.*, **7**, 110 (1963).
[84] T. S. Piper, G. Wilkinson, *J. Inorg. Nucl. Chem.*, **3**, 104 (1956).
[85] T. S. Piper, D. Lemal, G. Wilkinson, *Naturwissenshaften*, **43**, 129 (1956).
[86] R. B. King, D. M. Braitsch, *J. Organomet. Chem.*, **54**, 9 (1973).
[87] J. P. Bibler, A. Wojcicki, *J. Am. Chem. Soc.*, **88**, 4862 (1966).
[88] R. B. King, M. B. Bisnette, *J. Organomet. Chem.*, **2**, 15 (1964).
[89] R. B. King, *J. Am. Chem. Soc.*, **85**, 1918 (1963).
[90] R. B. King, W. M. Douglas, A. Efraty, *J. Organomet. Chem.*, **69**, 131 (1974).
[91] S. E. Jacobson, A. Wojcicki, *J. Am. Chem. Soc.*, **95**, 6962 (1973).
[92] E. O. Fischer, E. Moser, *Inorg. Synth.*, **12**, 36 (1971).
[93] G. Wilkinson, *Org. Synth.*, **IV**, 476 (1963).
[94] (a) M. Aresta, P. Giannoccaro, M. Rossi, A. Sacco, *Inorg. Chim. Acta*, **5**, 115 1971).
(b) W. G. Peet, D. H. Gerlach, *Inorg. Synth.*, **15**, 39 (1974).
[95] P. Giannoccaro, A. Sacco, *Inorg. Synth.*, **17**, 69 (1977).
[96] (a) G. Gooth, J. Chatt., *J. Chem. Soc.*, 2099 (1962).
(b) L. H. Pignolet, D. Forster, W. W. Horrocks, Jr., *Inorg. Chem.*, **7**, 828 (1968).
[97] L. Naldini, *Gazz. Chim. Ital.*, **90**, 922 (1968).
[98] Von K. Issleib, G. Doll, *Z. Anaorg. Chem.*, **305**, 1 (1969).
[99] (a) W. Manchot, W. J. Manchot, *Z. Anorg. Allg. Chem.*, **266**, 385 (1936).
(b) E. A. Seddon, K. R. Seddon, *The Chemistry of Ruthenium*, Elsevier, New York, 1984.
[100] J. Robertson, G. Webb, *Proc. R. Soc. London, Ser. A.*, **341**, 383 (1974).
[101] P. Pino, G. Braca, G. Sbrana, A. Cuccuru, *Chem. Ind. (London)*, 1732 (1968).
[102] G. Braca, G. Sbrana, P. Piacenti, P. Pino, *Chim. Ind. (Milan)*, **52**, 1091 (1970).
[103] R. M. Laine, R. G. Rinker, P. C. Ford, *J. Am. Chem. Soc.*, **99**, 252 (1977).
[104] H. Alper, S. Amaratunga, *Tetrahedron Lett.*, **21**, 2603 (1980).
[105] J. Moore, W. R. Pretzer, T. J. O'Connell, J. Harris, L. LaBounty, L. Chou, S. S. Grimmer, *J. Am. Chem. Soc.*, **114**, 5888 (1992).
[106] N. Chatani, S. Ikeda, K. Ohe, and S. Murai, *J. Am. Chem. Soc.*, **114**, 9710 (1992).
[107] A. Mantovani, S. Cenini, *Inorg. Synth.*, **16**, 47 (1976).
[108] A. Mantovani, S. Cenini, *Inorg. Synth.*, **16**, 51 (1976).
[109] B. R. James, G. L. Rempel, W. K. Teo, *Inorg. Synthe.*, **16**, 45 (1976).
[110] F. Piacenti, M. Bianchi, P. Frediani, E. Benedetti, *Inorg. Chem.*, **10**, 2759 (1971).
[111] (a) S. Aime, L. Milone, E. Sappa, *J. Chem. Soc., Dalton Trans.*, 838 (1976).
(b) S. Aime, L. Milone, D. Osella, M. Valle, E. W. Randall, *Inorg. Chim. Acta*, **20**, 217 (1976).
(c) M. Castiglioni, L. Milone, D. Osella, G. A. Vaglio, M. Valle, *Inorg. Chem.*, **15**, 394 (1976).
(d) G. Gervasio, D. Osella, M. Valle, *Inorg. Chem.*, **15**, 1221 (1976).
(e) M. Evans, M. Hursthouse, E. W. Randall, E. Rosenberg, L. Milone, M. Valle, *J. Chem. Soc., Chem. Commun.*, 545 (1972).
[112] S. Aime, G. Gervasio, L. Milone, E. Sappa, M. Franchini-Angela, *Inorg. Chim. Acta*, **26**, 223 (1978).
[113] G. Süss-Fink, *Inorg. Synth.*, **24**, 168 (1986).
[114] (a) S. A. Knox, J. W. Koepke, M. A. Andrews, H. D. Kaesz, *J. Am. Chem. Soc.*, **97**, 3942 (1975).
(b) M. I. Bruce, M. L. Williams, *Inorg. Synth.*, **26**, 262 (1989).
(c) R. D. Wilson, S. M. Wu, R. A. Love, R. Bau, *Inorg. Chem.*, **17**, 1271 (1978).
[115] (a) C.-M. T. Hayward, J. R. Shapley, *Inorg. Chem.*, **21**, 3816 (1982).
(b) G. B. Ansell, J. S. Bradley, *Acta Crystallogr.*, **36B**, 726 (1980).
(c) B. F. G. Johnson, J. Lewis, S. W. Sankey, K. Wong, M. McPartlin, W. J. H. Nelson, *J. Organomet. Chem.*, **191**, C3 (1980).
[116] R. F. Jackson, B. F. G. Johnson, J. Lewis, M. McPartlin, W. J. H. Nelson, *J. Chem. Soc., Chem. Commun.*, 735 (1979).
[117] A. P. Humphries, S. A. R. Knox, *J. Chem. Soc., Dalton Trans.*, 1710 (1975).
[118] R. E. Cobbledick, F. W. B. Einstein, R. K. Pomeroy, E. R. Spetch, *J. Organomet. Chem.*, **195**, 77 (1980).

[119] (a) A. Mantovani and S. Cenini, *Inorg.Synth.*, **16**, 51 (1976).
(b) P. Teulon, J. Roziere, *J. Organomet. Chem.*, **214**, 391 (1981).
[120] M. I. Bruce F. G. A. Stone, *Angew. Chem., Int. Ed. Engl.*, **7**, 427 (1968).
[121] M. Hidai, *Yuki Gosei Kagaku Kyokai Shi*, **27**, 1243 (1969).
[122] S. C. Tripathi, S. C. Srivastava, R. P. Manit, A. K. Shrimal, *Inorg. Chim. Acta*, **15**, 249 (1975).
[123] (a) H. Alper, K. E. Hashem, *J. Am. Chem. Soc.*, **103**, 6514 (1981).
(b) S. Cenini, C. Crotti, M. Pizzotti, F. Porta, *J. Org. Chem.*, **53**, 1243 (1988).
[124] S. J. LaPlaca, J. A. Ibers, *Inorg. Chem.*, **4**, 778 (1965).
[125] J. Chatt, J. M. Davidson, *J. Chem. Soc.*, 843 (1965).
[126] (a) T. A. Stephenson, G. Wilkinson *J. Inorg. Nucl. Chem.*, **28**, 945 (1966).
(b) J. Chatt, R. G. Hayter, *J. Chem. Soc.*, 896 (1961).
(c) P. S. Hallman, T. A. Stephenson, G. Wilkinson, *Inorg. Synth.*, **12**, 237 (1970).
[127] S. D. Robinson, G. Wilkinson, *J. Chem. Soc.*, A, 300 (1966).
[128] D. J. Cole-Hamilton, G. Wilkinson, *J. Chem. Soc., Dalton Trans.*, 1283 (1979).
[129] B. L. Haymore, J. A. Ibers, *Inorg. Chem.*, **14**, 3060 (1975).
[130] R. Young, G. Wilkinson, *Inorg. Synth.*, **17**, 75 (1977).
[131] R. A. Head, J. F. Nixon, *J. Chem. Soc., Dalton Trans.*, 885 (1978).
[132] W. H. Knoth, *J. Am. Chem. Soc.*, **94**, 104 (1972).
[133] Y. Watanabe, Y. Tsuji, M. Ohsugi, *Tetrahedron Lett.*, **23**, 229 (1982).
[134] K.-T. Huh, Y. Tsuji, M. Kobayashi, F. Okada, Y. Watanabe, *Chem. Lett.*, 452 (1988).
[135] S.-I. Murahashi, K. Kondo, T. Hakata, *Tetrahedron Lett.*, **23**, 229 (1982).
[136] Y. Tsuji, K.-T. Huh, Y. Ohsugi, F. Okada, Y. Watanabe, *J. Org. Chem.*, **50**, 1365 (1985).
[137] Y. Tsuji, Y. Yokoyama, K.-T. Huh, Y. Watanabe, *Bull. Chem. Soc. Jpn.*, **60**, 3456 (1987).
[138] Y. Tsuji, K.-T. Huh, Y. Watanabe, *J. Org. Chem.*, **52**, 1673 (1987).
[139] (a) S.-I. Murahashi, T. Naota, K. Yonemura, *J. Am. Chem. Soc.*, **110**, 8256 (1988).
(b) S.-I. Murahashi, T. Naota, H. Taki, *J. Chem. Soc., Chem. Commun.*, 613 (1988).
[140] (a) P. G. Jessop, T. Ikariya, R. Noyori, *Nature*, **368**, 231 (1994).
(b) P. G. Jessop, Y. Hsiao, T. Ikariya, R. Noyori, *J. Am. Chem. Soc.*, **116**, 8851 (1994).
(c) P. G. Jessop, T. Ikariya, R. Noyori, *Chem. Rev.*, **95**, 259 (1995).
(d) P. G. Jessop, T. Ikariya, R. Noyori, *Scienece*, **269**, 1065 (1995).
(e) P. G. Jessop, Y. Hsiao, T. Ikariya, R. Noyori, *J. Chem. Soc., Chem. Commun.*, 707 (1995).
(f) P. G. Jessop, Y. Hsiao, T. Ikariya, R. Noyori, *J. Am. Chem. Soc.*, **118**, 344 (1996).
[141] (a) M. Ogasawara, S. A. Macgregor, W. E. Streib, K. Folting, O. Eisenstein, K. G. Caulton, *J. Am. Chem. Soc.*, **117**, 8869 (1995).
(b) M. Ogasawara, K. Folting, W. E. Streib, K. G. Caulton, *Abstracts of Papers*, 210th National Meeting of the American Chemical Society, Chicago, Il; American Chemical Society: Washington, DC (1995); INOR 67.
(c) M. Ogasawara, K. G. Caulton, *Abstracts of Papers*, 69th Annual Meeting of the Chemical Society of Japan, Kyoto, 3C731 (1995).
[142] B. N. Chaudret, M. Carrondo, D. J. Cole-Hamilton, A. C. Skapski, G. J. Wilkinson, *J. Chem. Soc., Chem. Commun.*, **1978**, 463.
[143] P. G. Douglas, B. L. Shaw, *J. Chem. Soc., A*, 1556 (1970).
[144] R. H. Heyn, K. G. Caulton, *J. Am. Chem. Soc.*, **115**, 3354 (1993).
[145] K. Hiraki, T. Matsunaga, H. Kawano, *Organometallics*, **13**, 1878 (1994).
[146] R. E. Harmon, S. K. Gupta, D. J. Brown, *Chem. Rev.*, **73**, 21 (1973).
[147] (a) H. Imai, T. Nishiguchi, M. Kobayashi, K. Fukuzumi, *Bull. Chem. Soc. Jpn.*, **48**, 1585 (1975).
(b) H. Imai, T. Nishiguchi, K. Fukuzumi, *J. Org. Chem.*, **41**, 665 (1971).
(c) H. Imai, T. Nishiguchi, K. Fukuzumi, *J. Org. Chem.*, **41**, 2688 (1976).
[148] (a) R. W. Mitchell, A. Spencer, G. Wilkinson, *J. Chem. Soc., Dalton Trans.*, 846 (1973).
(b) R. A. Head, J. F. Nixon, *J. Chem. Soc., Dalton Trans.*, 913 (1978).
(c) B. R. James, *Inorg. Chim. Acta, Rev.*, **4**, 73 (1970).
(d) G. L. Geoffroy, J. R. Lehman, *Adv. Inorg. Chem. Radiochem.*, **20**, 189 (1977).

[149] T. Mitsudo, Y. Nakagawa, K. Watanabe, Y. Hori, H. Misawa, H. Watanabe, Y. Watanabe, *J. Org. Chem.*, **50**, 565 (1985).
[150] Y. Hayashi, S. Komiya, T. Yamamoto, A. Yamamoto, *Chem. Lett.*, 1363 (1984).
[151] H. Horino, T. Ito, A. Yamamoto, *Chem. Lett.*, 17 (1978).
[152] (a) S. Murai, F. Kakiuchi, S. Sekine, Y. Tanaka, A. Kamatani, M. Sonoda, N. Chatani, *Nature (London)*, **366**, 529 (1993).
(b) S. Murai, F. Kakiuchi, S. Sekine, Y. Tanaka, A. Kamatani, M. Sonoda, N. Chatani, *Pure Appl. Chem.*, **66**, 1527 (1994).
(c) F. Kakiuchi, Y. Tanaka, T. Sato, N. Chatani, S. Murai, *Chem. Lett.*, **679**, 681 (1995).
[153] A. Yamamoto, S. Kitazume, S. Ikeda, *J. Am. Chem. Soc.*, **90**, 1089 (1968).
[154] (a) W. H. Knoth, *J. Am. Chem. Soc.*, **94**, 104 (1972).
(b) W. H. Knoth, *Inorg. Synth.*, **15**, 31 (1974).
[155] S. Komiya, A. Yamamoto, *Bull. Chem. Soc. Jpn.*, **49**, 784 (1976).
[156] (a) S. Komiya, A. Yamamoto, S. Ikeda, *Bull. Chem. Soc. Jpn.*, **48**, 101 (1975).
(b) D. J. Cole-Hamilton, G. Wilkinson, *Nouveau J. Chem.*, **1**, 141 (1977).
[157] S. Komiya, A. Yamamoto, S. Ikeda, *J. Organomet. Chem.*, **42**, C65 (1972).
[158] S. Komiya, A. Yamamoto, *Bull. Chem. Soc. Jpn.*, **49**, 2553 (1976).
[159] (a) S. Komiya, A. Yamamoto, *Chem. Lett.*, 475 (1975).
(b) S. Komiya, T. Ito, M. Cowie, A. Yamamoto, J. A. Ibers., *J. Am. Chem. Soc.*, **98**, 3874 (1976).
[160] S. Komiya, Y. Aoki, Y. Mizuho, N. Oyasato, *J. Organomet. Chem.*, **463**, 179 (1993).
[161] (a) S. Komiya, A. Yamamoto, *J. Chem. Soc., Chem. Commun.*, 524 (1974).
(b) S. Komiya, A. Yamamoto, *J. Organomet. Chem.*, **87**, 333 (1975).
[162] T. Naota, H. Taki, M. Mizuno, S.-I. Murahashi, *J. Am. Chem. Soc.*, **111**, 5954 (1989).
[163] S.-I. Murahashi, T. Naota, H. Taki, M. Mizuno, H. Takaya, S. Komiya, Y. Ito, N. Oyasato, M. Hiraoka, M. Hirano, and A. Fukuoka, *J. Am. Chem. Soc.*, **117**, 12436 (1995).
[164] Y. Lin, X. Zhu, M. Xiang, *J. Organomet. Chem.*, **448**, 215 (1993).
[165] B. M. Trost, *Science*, **254**, 1471 (1991).
[166] H. Horino, T. Ito, A. Yamamoto, *Chem. Lett.*, 17 (1978).
[167] T. Ito, H. Horino, Y. Koshiro, A. Yamamoto, *Bull. Chem. Soc. Jpn.*, **55**, 504 (1982).
[168] K. Osakada, K. Ohshiro, A. Yamamoto, *Organometallics*, **10**, 404 (1991).
[169] M. F. McGuiggan, L. H. Pignolet, *Inorg. Chem.*, **21**, 2523 (1982).
[170] R. Noyori, M. Ohta, Y. Hisao, Y. Kitamura, T. Ohta, H. Takaya, *J. Am. Chem. Soc.*, **108**, 7117 (1986).
[171] R. Noyori, T. Ikedam T. Ohkuma, M. Widhelm, M. Kitamura, H. Takaya, S. Akutagawa, N. Sayo, T. Saito, T. Taketomi, H. Kumobayashi, *J. Am. Chem. Soc.*, **111**, 9134 (1989).
[172] T. Ikariya, Y. Ishii, H. Kawano, T. Arai, M. Saburi, S. Yoshikawa, S. Akutagawa, *J. Chem. Soc., Chem. Commun.*, 922 (1985).
[173] H. Takaya, T. Ohta, N. Sayo, H. Kumobayashi, S. Akutagawa, S. Inoue, I. Kasahara, R. Noyori, *J. Am. Chem. Soc.*, **109**, 1596, 4129 (1987).
[174] R. Noyori, H. Takaya, *Acc. Chem. Res.*, **23**, 345 (1990).
[175] K. Mashima, K. Kusano, T. Ohta, R. Noyori, H. Takaya, *J. Chem. Soc., Chem. Commun.*, 1208 (1989).
[176] T. Ohta, H. Takaya, M. Kitamura, K. Nagai, R. Noyori, *J. Org. Chem.*, **52**, 3174 (1987).
[177] H. Nishiyama, Y. Itoh, H. Matsumoto, S.-B. Park, K. Itoh, *J. Am. Chem. Soc.*, **116**, 2223 (1994).
[178] N. Oshima, H. Suzuki, Y. Moro-oka, *Chem. Lett.*, 1161 (1984).
[179] N. Oshima, H. Suzuki, Y. Moro-oka, H. Nagashima, K. Ito, *J. Organomet. Chem.*, **314**, C46 (1986).
[180] B. M. Trost, A. F. Indolese, T. J. J. Müller, B. Treptow, *J. Am.Chem. Soc.*, **117**, 615 (1995).
[181] H. Suzuki, D. H. Lee, N. Oshima, Y. Moro-oka, *Organometallics*, **6**, 1569 (1987).
[182] H. Suzuki, H. Omori, D. H. Lee, Y. Yoshida, Y. Moro-oka, **7**, 2243 (1988).
[183] H. Suzuki, H. Omori, D. H. Lee, Y. Yoshida, M. Fukushima, M. Tanaka, Y. Morooka, *Organometallics*, **13**, 1129 (1994).

[184] (a) H. Nagashima, K. Mukai, Y. Shiota, K. Yamaguchi, K. Ara, T. Fukuhori, H. Suzuki, M. Akita, Y. Moro-oka, K. Itoh, *Organometallics*, **9**, 799 (1990).
(b) R. Kuhlman, K. Streib, K. G. Caulton, *J. Am. Chem. Soc.*, **115**, 5813 (1993).
[185] M. Hidai, M. Mizobe, H. Matsuzaka, *J. Organomet. Chem.*, **473**, 1 (1994).
[186] G. Vitulli, P. Pertici, P. Salvadori, *J. Chem. Soc., Dalton. Trans.*, 2255 (1984).
[187] M. A. Bennett, T.-N. Huang, T. W. Turney, *J. Chem. Soc., Chem. Commun.*, 312 (1979).
[188] Y. Ohgomori, S. Ichikawa, N. Sumitani, *Organometallics*, **13**, 3758 (1994).
[189] P. Pertici, V. Ballantini, P. Salvadori, M. A. Bennett, *Organometallics*, **14**, 2565 (1995).
[190] A. Fukuoka, S. Furuta, T. Nagano, M. Hirano, S. Komiya, *Abstracts of Papers*, 71st Anual Meeting of the Chemical Society of Japan, Tokyo (1996); 4J406.
[191] M. A. Bennett, M. I. Bruce, T. W. Matheson, *Compr. Organomet. Chem.*, **4**, 691 (1982).
[192] P. Pertici, G. Vitulli, M. Paci, *J. Chem. Soc., Dalton. Trans.*, 1961 (1980).
[193] T. Kondo, M. Akazome, Y. Tsuji, Y. Watanabe, *J. Org. Chem.*, **55**, 1286 (1990).
[194] B. Chaudret, G. Commenges, R. Poilbanc., *J. Chem. Soc., Chem. Commun.*, 1388 (1982).
[195] P. Pertici, G. P. Simonelli, G. Vitulli, G. Deganello, P. L. Sandrini, A. Mantovani, *J. Chem. Soc., Chem. Commun.*, 132 (1977).
[196] F. Bouachir, B. Chaudret, F. Dahan, F. Agbossou, I. Tkatchenko, *Organometallics*, **10**, 455 (1991).
[197] Y. Maruyama, I. Shimizu, A. Yamamoto, *Chem. Lett.*, 1041 (1994).
[198] S. Komiya, T. Kabasawa, K. Yamashita, M. Hirano, A. Fukuoka, *J. Organomet. Chem.*, **471**, C6 (1994).
[199] S. Komiya, J. Suzuki, K. Miki, N. Kasai, *Chem. Lett.*, 1287 (1987).
[200] T. Mitsudo, S.-W. Zhang, Y. Watanabe, *J. Chem. Soc., Chem. Commun.*, 435 (1994).
[201] Y. Tsuji, T. Mukai, T. Kondo, Y. Watanabe, *J. Organomet. Chem.*, **369**, C51 (1989).
[202] T. Kondo, H. Ono, N. Satake, T. Mitsudo, Y. Watanabe, *Organometallics*, **14**, 1945 (1995).
[203] T. Mitsudo, N. Suzuki, T. Kondo, Y. Watanabe, *J. Org. Chem.*, **59**, 7759 (1994).
[204] M. A. Bennett, A. K. Smith, *J. Chem. Soc., Dalton Trans.*, 233 (1974).
[205] (a) J. Müller, E. O. Fischer, *J. Organometal. Chem.*, **5**, 275 (1965).
(b) J. Müller, C. G. Kreiter, B. Mertschenk, S. Schmitt, *Chem. Ber.*, **108**, 273 (1975).
[206] M. A. Bennett, I. J. McMahon, S. Pelling, *Organometallics*, **11**, 127 (1992).
[207] M. A. Bennett, X. -Q. Wang, *J. Organomet. Chem.*, **428**, C17 (1992).
[208] J. W. Hull, Jr., W. L. Gladfelter, *Organometallics*, **1**, 1716 (1982).
[209] M. A. Bennett, L. Y. Goh, I. J. McMahon, T. B. Mitchell, G. B. Robertson, T. W. Turney, W. A. Wickramasinghe, *Organometallics*, **11**, 3069 (1992).
[210] R. R. Schrock, J. Lewis, *J. Am. Chem. Soc.*, **95**, 4102 (1973).
[211] J. F. Hartwig, R. G. Bergman, R. A. Andersen, *J. Am. Chem. Soc.*, **113**, 3404 (1991).
[212] J. J. Levison, S. D. Robinson, *J. Chem. Soc.*, A, 2947 (1970).
[213] T. Ito, S. Kitazume, A. Yamamoto, S. Ikeda, *J. Am. Chem. Soc.*, **92**, 3011 (1970).
[214] R. O. Harris, N. K. Hota, L. Sadvoy, J. M. C. Yuen, *J. Organomet. Chem.*, **54**, 259 (1973).
[215] M. I. Bruce, N. J. Windsor, *Aust. J. Chem.*, **30**, 1601 (1977).
[216] M. O. Albers, J. D. Robinson, A. Shaver, Singleton, *Organometallics*, **5**, 2199 (1986).
[217] T, V, Ashworth, E. Singleton, J. J. Hough, *J. Chem. Soc., Dalton Trans.*, **1977**, 1809.
[218] M. A. Bennett, G. Wilkinson, *Chem. Ind.* (London), **1959**, 1516.
[219] M. A. Bennett, A. K. Smith, *J. Chem. Soc., Dalton Trans.*, **1974**, 233.
[220] M. A. Bennett, T. W. Matheson, G. B. Robertson, A. K. Smith, P. A. Tucker, *Inorg. Chem.*, **19**, 1014 (1980).
[221] (a) M. O. Albers, D. C. Liles, D. J. Robinson, A. Shaver, E. Singleton, M. B. Wiege, J. C. A. Boeyens, D. C. Lebendis, *Organometallics*, **5**, 2321 (1986).
(b) D. E. Bublitz, W. E. McEwen, J. Kleinberg, *Org. Synth.*, **41**, 96 (1961).
[222] K. Ito, H. Nagashima, T. Ohshima, N. Oshima, H. Nishiyama, *J. Organomet. Chem.*, **272**, 179 (1984).
[223] K. B. Sharpless K. Akashi, *J. Am. Chem. Soc.*, **98**, 1986 (1976).
[224] R. D. Adams, J. P. Selegue, *Compr. Organomet. Chem.*, **4**, 967 (1982).
[225] R. A. Johnson, K. B. Sharpless, *Catalytic Asymmetric Synthesis*, Ed. I. Ojima, VCH, New York, p. 227 (1993).

[226] S. C. Tripathi, S. C. Srivastava, R. P. Mani, A. K. Shrimal, *Inorg. Chim. Acta*, **15**, 249 (1975).
[227] (a) C. R. Eady, B. F. G. Johnson, J. Lewis, *J. Chem. Soc., Dalton Trans.*, 2606 (1975).
(b) D. H. Farrar, B. F. G. Johnson, J. Lewis, P. R. Raithby, M. J. Rosales, *J. Chem. Soc., Dalton Trans.*, 2051 (1982).
(c) R. J. Goudsmit, B. F. G. Johnson, J. Lewis, P. R. Raithby, K. H. Whitmire, *Chem. Commun.*, 640 (1982).
[228] S. C. Tripathi, S. C. Srivastava, R. P. Mani, A. K. Shrimal, *Inorg. Chim. Acta*, **15**, 249 (1975).
[229] B. F. G. Johnson, J. Lewis, *Inorg. Synth.*, **13**, 92 (1972).
[230] B. F. G. Johnson, J. Lewis, D. A. Pipard, *J. Chem. Soc., Dalton Trans.*, 407 (1981).
[231] J. N. Nicholls, M.D. Vargas, *Inorg. Synth.*, **26**, 289 (1989).
[232] P. A. Dawson, B. F. G. Johnson, J. Lewis, J. Puga, P. R. Raithby, M. J. Rosales, *J. Chem. Soc., Dalton Trans.*, 233 (1982).
[233] C. -M. T. Hayward, J. R. Shapley, *Inorg. Chem.*, **21**, 3816 (1982).
[234] M. McPartlin, C. R. Eady, B. F. G. Johnson, J. Lewis, *J. Chem. Soc., Chem. Commun.*, 883 (1976).
[235] S. A. R. Knox, J. W. Koepke, M. A. Andrews, H. D. Kaesz, *J. Am. Chem. Soc.*, **97**, 3942 (1975).
[236] E. Sappa, M. Valle, *Inorg. Synth.*, **26**, 365 (1989).
[237] M. R. Churchill, F. J. Hollander, J. P. Hutchinson, *Inorg. Chem.*, **16**, 2697 (1977).
[238] A. G. Orpen, A. V. Rivera, E. G. Brayan, D. Pippard, G. M. Sheldrick, K. D. S. Rouse, *J. Chem. Soc., Chem. Commun.*, 723 (1978).
[239] C. Zuccaro, *Inorg. Synth.*, **26**, 293 (1989).
[240] B. F. G. Johnson, J. Lewis, P. R. Raithby, C. Zuccaro, *Acta Cryst.*, **37B**, 1728 (1981).
[241] R. B. Calvert, J. R. Sharpley, *J. Am. Chem. Soc.*, **99**, 5225 (1977).
[242] A. J. Deeming, M. Underhill, *J. Chem. Soc., Dalton Trans.*, 1415 (1974).
[243] C. W. Bradford, R. S. Nyholm, *J. Chem. Soc., Dalton Trans.*, 529 (1973).
[244] A. J. Deeming, *et al.*, *J. Chem. Soc., Dalton Trans.*, 1201 (1978).
[245] J. T. Park, B. W. Woo, J.-H. Chung, S. C. Shim, J.-H. Lee, S.-S. Lim, and I.-H. Suh, *Organometallics*, **13**, 3384 (1994).
[246] G. W. Parshall, *Acc. Chem. Res.*, **3**, 139 (1970).
[247] A. Sahajpal, S. D. Robinson, *Inorg. Chem.*, **18**, 3572 (1979).
[248] (a) L. D. Brown, S. D. Robinson, A. Sahajpal, J. A. Ibers, *Inorg. Chem.*, **16**, 2728 (1977).
(b) A. D. Harris, S. D. Robinson, A. Sahajpal, M. B. Hursthouse, *J. Organomet. Chem.*, **174**, C11 (1979).
[249] M. A. Esteruelas, F. J. Lahoz, E. Onata, L. A. Oro, B. Zeier, *Organometallics*, **13**, 1662 (1994).
[250] D. G. Gusev, R. Kuhlman, J. R. Rambo, H. Berke, O. Eisenstein, K. G. Caulton, *J. Am. Chem. Soc.*, **117**, 281 (1995).
[251] J. Espuelas, M. A. Esteruelas, F. J. Lahoz, L. A. Oro, N. Ruiz, *J. Am. Chem. Soc.*, **115**, 4683 (1993).
[252] P. Barbaro, C. Bianchini, M. Peruzzini, A. Polo, F. Zanobini, P. Frediani, *Inorg. Chim. Acta*, **220**, 5 (1994).
[253] (a) R. I. Michelman, R. G. Bergman, R. A. Andersen, *Organometallics*, **12**, 2741 (1993).
(b) R. I. Michelman, G. E. Ball, R. G. Bergman, R. A. Andersen, *Organometallics*, **13**, 869 (1994).
[254) N. Ahmad, J. J. Levison, S. D. Robinson, M. F. Uttley, *Inorg. Synth.*, **15**, 45 (1974).
[255] B. H. Mckee, D. G. Gilheany, K. B. Sharpless, *Org. Syn.*, **70**, 47 (1990).
[256] G. W. Parshall, S. D. Ittel, Homogeneous Catalysis, Wiley-Interscience, New York, pp. 157 (1992).
[257] B. B. Lohray, T. H. Kalantar, B. M. Kim, C. Y. Park, T. Shibata, J. S. M. Wai, K. B. Sharpless, *Tetrahedron Lett.*, **30**, 2041 (1989).
[258] (a) M. Minato, K. Yamamoto, J. Tsuji, *J. Org. Chem.*, **55**, 766 (1990).
(b) H. L. Kwong, C. Sorato, V. Ogino, H. Chen, K. B. Sharpless, *Tetrahedron Lett.*, **31**, 2999 (1990).

[259] E. Herranz and K. B. Sharpless, *Org. Synth.*, **61**, 85 (1983).
[260] K. B. Sharpless, A. O. Chong, K. Ohshima, *J. Org. Chem.*, **41**, 177 (1976).
[261] J. G. Hamilton, O. N. D. Mackey, J. J. Rooney, D. G. Gilheany, *J. Chem. Soc., Chem. Commun.*, 1600 (1990).
[262) W. R. Roper, *Adv. Organomet. Chem.*, **25**, 121 (1986).
[263] (a) J. J. Levison, S. D. Robinson, *J. Chem. Soc., A*, 2947 (1970).
(b) N. Ahmad, J. J. Levison, S. D. Robinson, M. F. Uttley, *Inorg. Synth.*, **15**, 55 (1974).

11 Group 9 (Co, Rh, Ir) Metal Compounds

S. Komiya and A. Fukuoka, *Tokyo University of Agriculture and Technology*

11.1 Introduction

There are many kinds of organometallic compounds of the Group 9 metals: cobalt, rhodium, and iridium. The chemistry of monovalent and trivalent complexes of these elements has been extensively studied. Cobalt complexes of oxidation states 0 and –1 are also well known. Monovalent complexes of Group 9 metals adopt a d^8 configuration, and square planar 16-electron complexes are common for Rh(I) and Ir(I), which generally contain π-acid ligands such as CO, PR_3, and alkenes. Oxidative addition to the M(I) complexes gives M(III) complexes with octahedral geometry and a d^6 18-electron configuration.

Group 9 metal complexes can catalyze important homogeneous reactions such as hydrogenation and hydroformylation of alkenes. Some of them are employed in industry. The mechanisms of the catalytic reactions involve M(I) and M(III) intermediates having 16 and 18 electrons. In some cases, the catalytic intermediates have been isolated, and a mechanistic study of the catalytic reactions has greatly contributed to the progress of organometallic chemistry.

In this section, the applications of compounds of Group 9 metals and the related organometallic chemistry are reviewed [1].

11.2 Cobalt Complexes

The oxidation state of the metal in organocobalt compounds ranges from –1 to +4, the common being –1, 0, +1, and +3. Examples of Co(0) complexes are the homoleptic carbonyls: $Co_2(CO)_8$ and $Co_4(CO)_{12}$. Most Co(I) complexes contain π-acid ligands such as Cp, CO, RNC, and PR_3. Co(I) complexes are usually five-coordinate and trigonal bipyramidal, and in oxidative addition reactions to give Co(III) complexes one ligand is eliminated. Since the discovery of vitamin B_{12} coenzyme, much work has been devoted to organocobalt(III) complexes with a metal–carbon bond. Here the characteristic reactions of some cobalt complexes are presented.

Synthesis of Organometallic Compounds: A Practical Guide. Edited by S. Komiya

11.2.1 Carbonyls and Related Complexes

$Co_2(CO)_8$ is one of the important organometallic compounds, which is used as a homogeneous catalyst in hydroformylation of alkenes. In the solid state $Co_2(CO)_8$ is a dinuclear complex having a Co—Co single bond (2.52 Å) with two briding CO ligands [2]. However, in the solution state an equilibrium exists between bridged and non-bridged isomers (eq (1)).

(1)

$Co_2(CO)_8$ is thermally unstable, decomposing at above 50 °C, to give the tetranuclear cluster $Co_4(CO)_{12}$. $Co_2(CO)_8$ is a useful starting material; for example, it reacts with pyridine to give $[Co(py)_6][Co(CO)_4]_2$, which on treatment with sulfuric acid produces $HCo(CO)_4$ [3]. Reaction of $Co_2(CO)_8$ with PR_3 in non-polar solvents or at high temperature in polar solvents gives $Co_2(CO)_6(PR_3)_2$. On the other hand, the reaction at room temperature in polar solvents leads to the formation of $[Co(CO)_3(PR_3)_2][Co(CO)_4]$ [4].

$Co_2(CO)_8$ is a catalyst precursor in the hydroformylation (oxo reaction) of alkenes to give aldehydes [5]. The active catalytic species generated *in situ* is proposed to be $HCo(CO)_3$. In industry, production of butanal is performed by this homogeneous catalytic reaction. PBu_3 is added to increase a n/iso ratio of the aldehydes (eq (2)).

$$\text{CH}_2\text{=CHCH}_3 + CO + H_2 \xrightarrow{Co_2(CO)_8} \underset{n}{\text{CH}_3\text{CH}_2\text{CH}_2\text{CHO}} + \underset{iso}{(\text{CH}_3)_2\text{CHCHO}} \quad (2)$$

For the hydroformylation of internal alkenes, addition of $Ru_3(CO)_{12}$ as a cocatalyst increases the reaction rate (eq (3)) [6].

$$\text{cyclohexene} + CO + H_2 \xrightarrow{Co_2(CO)_8 + Ru_3(CO)_{12}} \text{cyclohexyl–CHO} \quad (3)$$

When $HSiR_3$ is used in place of H_2, a silyl enol ether is obtained in the presence of catalytic amount of $Co_2(CO)_8$ (eq (4)) [7].

$$\text{cyclohexene} + CO + HSiEt_2Me \xrightarrow{Co_2(CO)_8} \text{cyclohexylidene=CH–OSiEt}_2\text{Me} \quad (4)$$

Homologation of methanol to ethanol has attracted much attention in C_1 chemistry, since ethanol is a potential precursor to ethylene [8]. The reaction is catalyzed by $Co_2(CO)_8$ and the addition of $Ru_3(CO)_{12}$ is also effective in this reaction (eq (5)) [9].

$$CH_3OH + CO + 2H_2 \xrightarrow{Co_2(CO)_8 / I_2 / PR_3} CH_3CH_2OH + H_2O \quad (5)$$

Orthometallation and subsequent carbonylation of aryl azo compounds is catalyzed by $Co_2(CO)_8$ (eq (6)) [10].

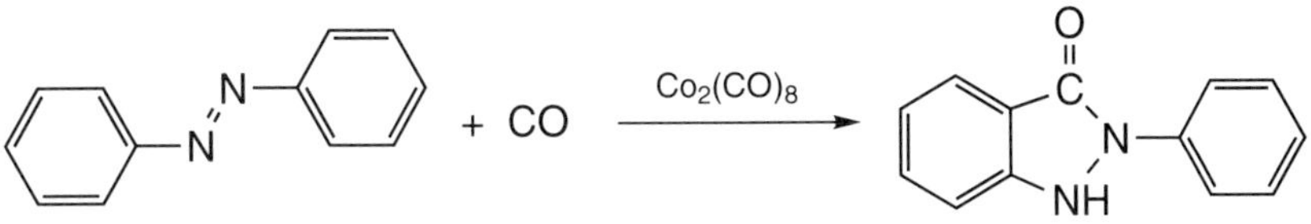

(6)

Reaction of $Co_2(CO)_8$ with a variety of alkynes forms compounds of the type $Co_2(CO)_6(RC_2R)$ (eq (7)).

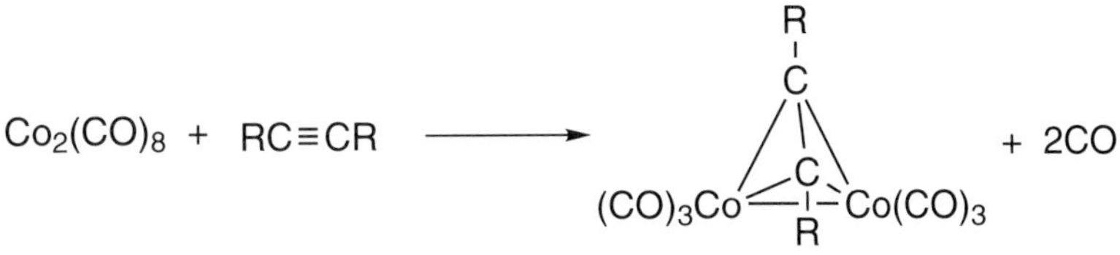

$$Co_2(CO)_8 + RC{\equiv}CR \longrightarrow (CO)_3Co(\mu\text{-}RC{\equiv}CR)Co(CO)_3 + 2CO \quad (7)$$

Accordingly, the $Co_2(CO)_6$ unit can be used as a protecting group for C—C triple bonds (eq (8)) [11].

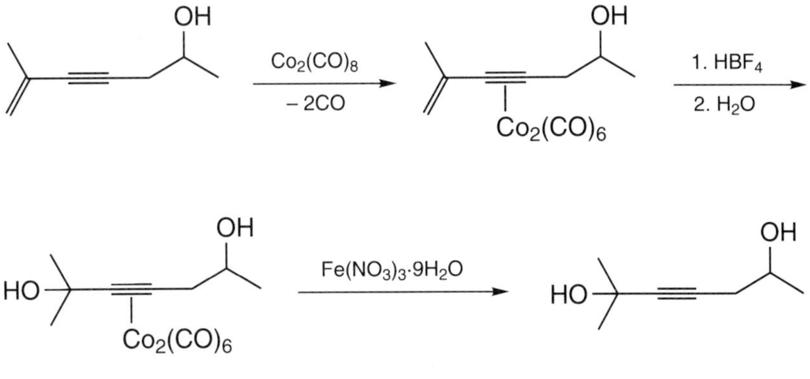

(8)

Hydrogen cyanide is added to butadiene to give 1,3-dicyanobutane in the presence of $Co_2(CO)_8$ catalyst (eq (9)) [12].

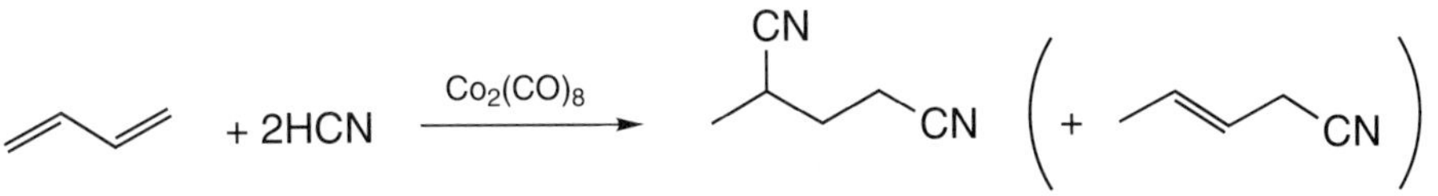

(9)

Reduction of $Co_2(CO)_8$ with sodium-amalgam in THF solution gives $Na[Co(CO)_4]$. $Na[Co(CO)_4]$ has high nucleophilicity, and the reaction with alkyl halides gives alkylcobalt complexes. It is generally difficult to isolate the alkylcobalt complexes, but in the presence of alkenes and alkynes, acylated compounds are obtained in good yields (eq (10)) [13].

$$\mathrm{RX + Na[Co(CO)_4] \xrightarrow{-NaX} [RCo(CO)_4] \xrightarrow{\text{butadiene}} RC(=O){-}Co(CO)_3(\eta^2\text{-butadiene}) \longrightarrow}$$

$$\mathrm{RC(=O)CH_2(\eta^3\text{-allyl})Co(CO)_3 \xrightarrow[-[NHEt_3][Co(CO)_4]]{NEt_3 / CO} RC(=O)CH{=}CHCH{=}CH_2} \quad (10)$$

11.2.2 Dinitrogen Complex

$CoH(N_2)(PPh_3)_3$ was the first dinitrogen–metal complex prepared directly from dinitrogen [14]. The dinitrogen ligand is easily displaced by other two electron donor ligands such as alkenes, NH_3, CO, CO_2, and RCN [15]. For example, reaction of $CoH(N_2)(PPh_3)_3$ with CO_2 liberates N_2 to form a formate complex $Co(O_2CH)(PPh_3)_3$ (eq (11)). Here CO_2 has inserted into the Co—H bond. Oxidative addition of H_2 results in the formation of a trihydride complex (eq (12)).

$$\mathrm{CoH(N_2)(PPh_3)_3 + CO_2 \longrightarrow Co(O_2CH)(PPh_3)_3 + N_2} \quad (11)$$

$$\mathrm{CoH(N_2)(PPh_3)_3 + H_2 \longrightarrow CoH_3(PPh_3)_3} \quad (12)$$

The following stoichiometric and catalytic reactions are based on the displacement of N_2 in $CoH(N_2)(PPh_3)_3$ by alkenes. For example, replacement of the N_2 ligand of $CoH(N_2)(PPh_3)_3$ to ethylene and the subsequent reaction of the ethylene complex with acyl chloride give ketone stoichiometrically (eq (13)) [16].

$$\mathrm{CoH(N_2)L_3 + CH_2{=}CH_2 \xrightarrow{-N_2} CoH(CH_2{=}CH_2)L_3 \rightleftharpoons Et{-}CoL_3}$$

$$\mathrm{\xrightarrow{RCOCl} RC(=O){-}Co(Et)(Cl){-}L_3 \longrightarrow RCOEt + CoClL_3} \qquad \mathrm{L = PPh_3} \quad (13)$$

Dimerization of ethylene to give 1- and 2-butenes is catalyzed by $CoH(N_2)(PPh_3)_3$ (Scheme 11.1) [17].

Decarbonylation and reduction of acid chlorides proceed with $CoH(N_2)(PPh_3)_3$ to give the corresponding hydrocarbons (eq (14)) [18].

$$\mathrm{RCOCl + CoH(N_2)(PPh_3)_3 \longrightarrow RH + CO + N_2 + CoCl_2(PPh_3)_3} \quad (14)$$

11.2.3 Cyclopentadienyl Complexes

The chemistry of monocyclopentadienyl complexes of cobalt has been extensively studied. Complexes of the type $CoCpL_2$ (L = PPh_3, CO, cod) are capable of catalyzing the cyclotrimerization of alkynes [19]. In this reaction, an η^2-acetylene complex is formed first and the reaction with the second alkyne gives a cobaltacyclopentadiene

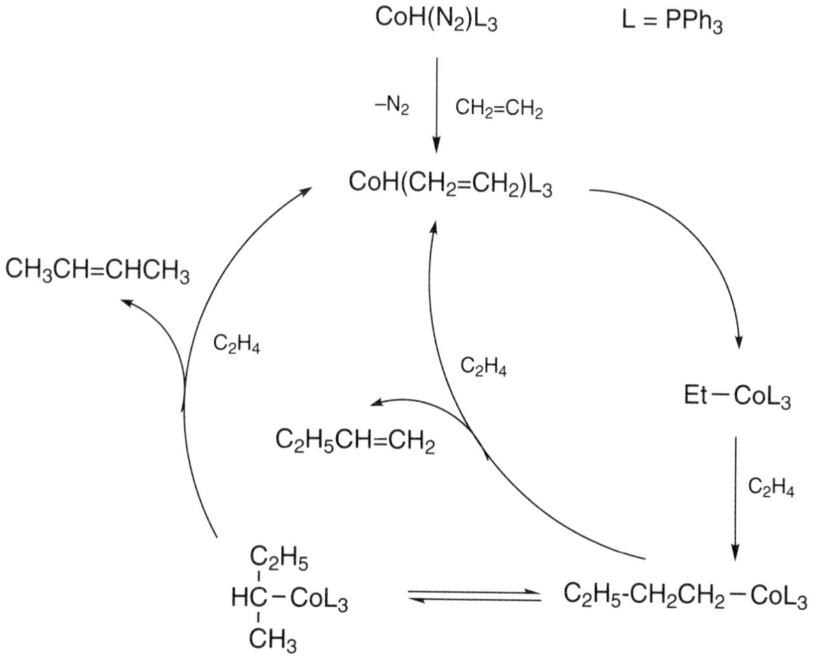

$CoH(N_2)L_3$ $L = PPh_3$

$-N_2$ $CH_2{=}CH_2$

$CoH(CH_2{=}CH_2)L_3$

$Et{-}CoL_3$

C_2H_4

$C_2H_5{-}CH_2CH_2{-}CoL_3$

$HC(C_2H_5)(CH_3){-}CoL_3$

C_2H_4

$C_2H_5CH{=}CH_2$

C_2H_4

$CH_3CH{=}CHCH_3$

Scheme 11.1

$R{-}C{\equiv}C{-}R$

$CpCoL$

$Cp = C_5H_5$, $L = PPh_3$

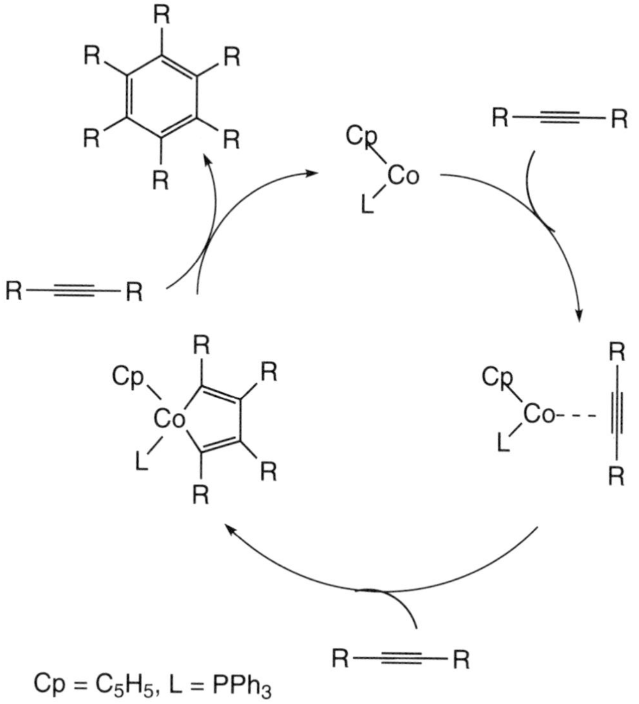

Scheme 11.2

complex (Scheme 11.2). This complex reacts further with the third alkyne to produce arene derivatives. In the case of $CoCp(PPh_3)_2$, the catalytic intermediate cobaltacyclopentadiene complex can be isolated.

In principle, use of unsymmetrical alkynes ($R^1C_2R^2$) could give four isomers of the metallacycle, but these are not observed (eq (15)). The alkyne carbon with a larger substituent tends to form a Co—C bond and the carbon with a smaller substituent to form a C—C bond. Thus, cyclization proceeds regioselectively [20]. However, sometimes a mixture of regioisomers is obtained, and column chromatography is needed to separate the isomers.

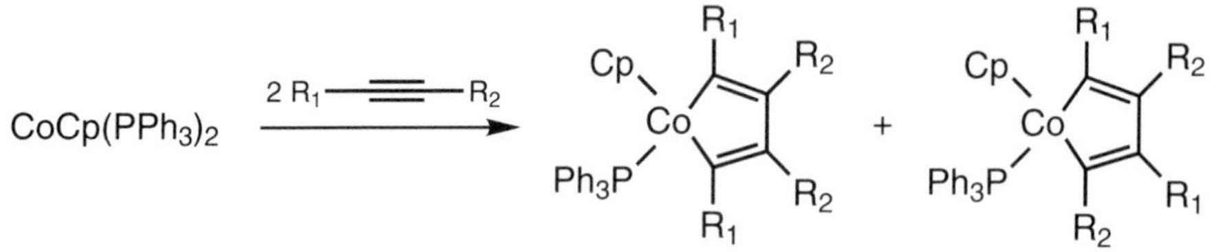

CoCp(PPh3)2 —(2 R1—≡—R2)→ Cp(Ph3P)Co[C(R1)=C(R2)–C(R2)=C(R1)] + Cp(Ph3P)Co[C(R1)=C(R2)–C(R1)=C(R2)] (15)

The cobaltacyclopentadiene complex reacts with a variety of compounds containing multiple bonds to give cyclic compounds (Scheme 11.3) [21]. In some cases, the organic products remain coordinated to Co; they can be removed by oxidation with Ce^{4+} or Fe^{3+}.

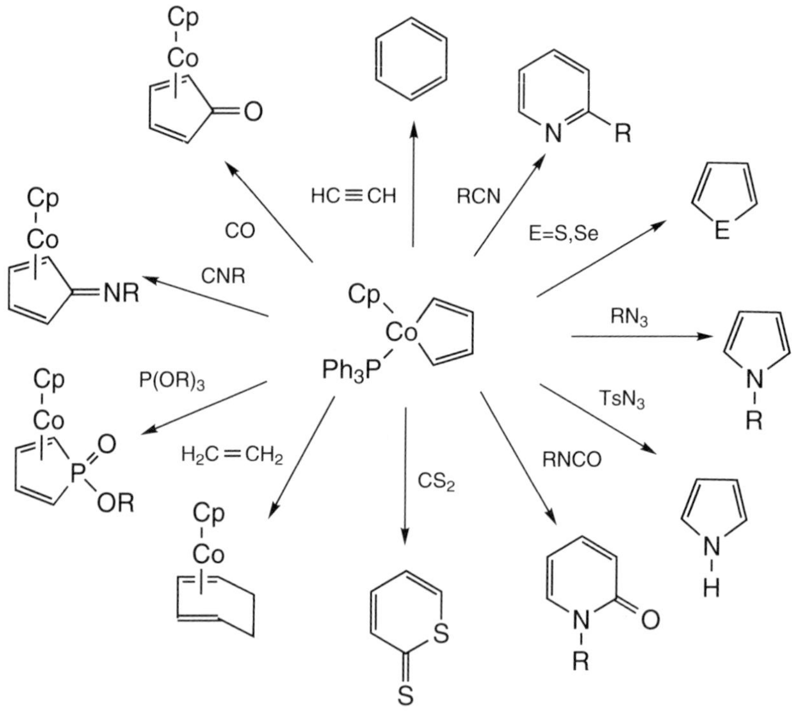

Cp Co CO =O HC≡CH RCN N R E=S,Se E CNR =NR Cp Co Ph3P RN3 N R P(OR)3 P O OR TsN3 N H H2C=CH2 CS2 S S RNCO N R O

Scheme 11.3

Substituents on the cyclopentadienyl group are omitted for clarity.

$CoCp(CO)_2$ also is a versatile synthetic precursor, which reacts with a wide variety of organic compounds. It serves as a source of the reactive CoCp unit, and this species catalyzes cyclization of alkynes as well. This reaction has been applied to the one step synthesis of steroid analogs [22].

SiMe3 / SiMe3 + [triyne] —CoCp(CO)2→ Me3Si / Me3Si [steroid analog] (16)

11.3 Synthesis of Co Complexes

(1) Octacarbonyldicobalt(0), $Co_2(CO)_8$ [23]

(hood)

$$2Co(OH)_2 + 8CO + 2H_2 \longrightarrow Co_2(CO)_8 + 4H_2O \quad (17)$$

$Co_2(CO)_8$ is commercially available. If some amount of $Co_2(CO)_8$ is in stock, it can be prepared easily on a large scale by using the compound itself as a catalyst. Cobalt(II) hydroxide (45 g, 484 mmol, red-pink) and 2-propanol (100 ml) are charged into a high pressure autoclave (inner volume 500 ml). Octacarbonyldicobalt(0) (10 g) is added to the mixture with CO bubbling, and the bubbling continues for a few minutes. Then the autoclave is closed, and CO (40 kg/cm^2) and H_2 (10 kg/cm^2) are added. The autoclave is heated to 55–60 °C with stirring. During the reaction, CO and H_2 are supplied periodically to keep the total pressure at ca. 50 kg/cm^2. After 6 h, the autoclave is cooled to room temperature and the pressure is released. The mixture is filtered under N_2 and the filtered sold is recrystallized from CO-saturated hexane to give red-orange crystals. Yield: 64 g (77 %). They are unstable in air, and decompose thermally to give black $Co_4(CO)_{12}$ even under N_2. Thus, they should be stored under CO below 0 °C.

Properties: mp 81–82 °C. IR(KBr): ν(CO) 2025(s, br), 1849(s), 1831(s) cm^{-1}.

(2) Sodium tetracarbonylcobaltate(-I), $Na[Co(CO)_4]$ [24a]

(under nitrogen atmosphere)

$$Co_2(CO)_8 + 2Na(Hg) \longrightarrow 2Na[Co(CO)_4] \quad (18)$$

Small portions of sodium (total 0.5 g) are added to mercury (50 g) to prepare 1 % sodium amalgam. After cooling to room temperature, anhydrous tetrahydrofuran (100 ml) is added. Octacarbonyldicobalt(0) (3.42 g, 10 mmol) is added and the mixture is vigorously stirred for 3 h. During this time, the solution turns from red-orange to gray-yellow. In some cases the solution is used without evaporation of THF, but it contains a small amount of $Hg[Co(CO)_4]_2$. The THF solution is also highly air-sensitive. Thus, filteration and evaporation of the THF solution is recommendable. The resulting white solid can be used for many purposes. $Na[Co(CO)_4]$ is also prepared from $Co_2(CO)_8$ and NaOH, where the use of Hg is avoided [24b].

Properties: IR(THF): ν(CO) 1889(s), 1886(vs), 1858(m) cm^{-1}.

(3) (η-Allyl)tricarbonylcobalt(I), $Co(CO)_3(\eta^3\text{-}C_3H_5)$ [25]

(under nitrogen atmosphere)

$$C_3H_5Br + Na[Co(CO)_4] \longrightarrow Co(CO)_3(C_3H_5) + NaBr + CO \qquad (19)$$

A solution of $Na[Co(CO)_4]$ (1.5 g) in diethyl ether (25 ml) is prepared according to the above method. Allyl bromide (0.5 ml, 5.8 mmol) is added to the solution at 0 °C, and the mixture is stirred overnight at 0 °C. The solution is evaporated to dryness at 0 °C, and the resulting solid is extracted with cold pentane (2 ml × 3). The extracted solution is concentrated to 2 ml, and the solution is cooled to –80 °C to give yellow crystals. The crystals are filtered at –80 °C and dried under vacuum. They are air-sensitive.

Properties: mp –33 to –32 °C. bp 39 °C / 15 mmHg. IR (CCl_4): 3058(w), 3003(w), 2924(vw), 2058(s), 2000(vs), 1488(w), 1471(w), 1387(w), 1224(w), 1199(w), 1188(w), 1126(w), 1017(w), 948(w), 930(w) cm^{-1}. ^1H NMR (C_6D_6): δ 2.38 (1H), 3.28 (2H), 5.25 (2H).

(4) Hydrido(dinitrogen)tris(triphenylphosphine)cobalt(I) [26]

(under nitrogen atmosphere)

$$Co(acac)_3 + Al(^iBu)_3 + 3PPh_3 + N_2 \longrightarrow CoH(N_2)(PPh_3)_3 \qquad (20)$$

Tris(acetylacetonato)cobalt(III) (3.56 g, 10 mmol) and triphenylphosphine (7.86 g, 30 mmol) are added to toluene (40 ml). The mixture is cooled to –30 to –40 °C, and tri-isobutylaluminum (4.6 ml) is added by using a hypodermic syringe. The mixture is stirred with N_2 bubbling. The temperature is held at –20 °C for 1 h, and then at 0–10 °C for 8–9 h. The reaction time is important to obtain a good yield. Petroleum ether (100 ml) is added to the red- brown solution, and the mixture is cooled to –20 to 0 °C to give orange crystals. The crystals are filtered, washed with petroleum ether, and dried under vacuum. Yield: 8.0 g (92 %). The crystals are purified by recrystallization from toluene to give golden orange crystals. They are air-sensitive, and soluble in tetrahydrofuran, benzene, toluene, ether, but insoluble in water.

Properties: mp 80 °C. IR (C_6H_6): ν(N_2) 2088 cm^{-1}.

(5) (η-Cyclopentadienyl)bis(triphenylphosphine)cobalt(I) [27]

(under nitrogen atmosphere)

$$\begin{aligned} CoCl_2\cdot 6H_2O + 3PPh_3 + NaBH_4 &\longrightarrow CoCl(PPh_3)_3 + NaCl + BH_3 + 6H_2O \\ CoCl(PPh_3)_3 + NaC_5H_5 &\longrightarrow Co(C_5H_5)(PPh_3)_2 + PPh_3 + NaCl \end{aligned} \qquad (21)$$

Cobalt(II) chloride hexahydrate (9.6 g) and triphenylphosphine (32.0 g) are added to ethanol (600 ml), and the mixture is vigorously stirred at 60–70 °C for 30 min to form blue $CoCl_2(PPh_3)_2$. The mixture is cooled to 30 °C, and sodium tetrahydridoborate (total 1.28 g, added in 10 small portions) is added over ca. 10 min. The reaction mixture turns dark-green, and finally minute brown crystals are formed. The crystals of chlorotris(triphenylphosphine)cobalt(I) are vacuum-filtered under air, washed with ethanol, water, ethanol, and petroleum ether, and dried under vacuum. Yield: 24.0 g (67 %).

To a suspension of $CoCl(PPh_3)_3$ (24 g, 27.2 mmol) in toluene (160 ml) is slowly added a THF solution of sodium cyclopentadienide (14 ml, 2 M) from a dropping funnel under N_2 (for preparation of NaCp, see $CoCp(C_2H_4)_2$). The mixture is stirred for 30 min, and a 10 % water solution of ammonium chloride (20 ml) is slowly added from a dropping funnel to decompose excess NaC_5H_5. The water layer is removed by means of a hypodermic syringe, and the resulting organic layer is transferred to another Schlenk flask and dried with anhydrous sodium sulfate for 1 h under N_2. After filtration, the solution is concentrated to 50 ml, and hexane (60 ml) is added. When small crystals appear, the flask is connected to a N_2 line with a pressure-releasing bubbler, and is moved into a water bath at 60 °C for 2 h to grow the crystals. Then hexane (30 ml) is added and the mixture is allowed to stand at room temperature overnight. The resulting large black crystals are washed with hexane (10 ml × 2), and dried under vacuum. They contain 1/2 molecule of hexane as crystal solvent. Yield: 9.9–13 g (53–69 %). The crystals should be stored under N_2 at low temperature, but big crystals can be handled in air for a short time.

Properties: mp 135–140 °C (Ar, dec). ^{1}H NMR(C_6D_6): δ 4.45 (s, Cp), 6.8–7.9 (m, Ph).

(6) (η-Cyclopentadienyl)bis(η²-ethylene)cobalt(I) [28]

(under nitrogen atmosphere) (hood)

$$\begin{aligned} CoCl_2 + 2NaC_5H_5 &\longrightarrow Co(C_5H_5)_2 + 2NaCl \text{ [28a]} \\ Co(C_5H_5)_2 + K + 2C_2H_4 &\longrightarrow Co(C_5H_5)(\eta^2\text{-}C_2H_4)_2 + KC_5H_5 \text{ [28b]} \end{aligned} \quad (22)$$

In a distillation apparatus, dicyclopentadinene is heated at ca. 180 °C in an oil bath and the distilled cyclopentadiene (bp 42–44 °C) is collected. The cyclopentadiene (98–123 ml, 1.2–1.5 mol) is slowly added to the mixture of Na (23 g, 1 mol) and THF (500 ml). To the NaC_5H_5 solution in THF is added anhydrous cobalt(II) chloride (65 g, 0.5 mmol) to give deep purple solution. The solution is refluxed for 2 h and then evaporated. The resulting solid is dried and dark purple crystals of cobaltocene is obtained by sublimation (60–200 °C / 10^{-2}–10^{-3} mmHg). Yield: ca. 85 g (90 %).

Small grains of potassium are prepared as follows. Potassium is added to hot benzene (ca. 70 °C) and the mixture is vigorously stirred until small grains of potassium are obtained. Stirring is stopped and the mixture is cooled to room temperature. The resulting solid is filtered, washed with pentane, and dried under vacuum.

Cobaltocene (31.3 g, 166 mmol) is added to diethyl ether (350 ml), and the solution is stirred and saturated with ethylene at −20 °C. The potassium grains (7.14 g, 183 mmol) are added to the solution, and the solution is stirred at −20 to 0 °C under ethylene until no more ethylene is absorbed (ca. 15 h). KC_5H_5 and excess K are filtered off at −30 °C, and washed with ether (3 × 50 ml). The filtered solution combined with the ether-wash is evaporated to dryness, and the resulting solid is dissolved in hexane. The solution is filtered, undissolved solid is washed with hexane, and the combined solutions are allowed to stand at –78 °C. The resulting brown crystals are filtered, and dried under vacuum. Yield: 25.3 g (85 %).

Properties: ^{1}H NMR ($C_6D_5CD_3$): δ 0.63 (=CH), 2.49(=CH), 4.23 (Cp). ^{13}C NMR (C_4H_8O): δ 37.7 (=CH_2), 85.6 (Cp).

(7) (η-Cyclopentadienyl)(2,3-dimethyl-1,4-diphenylcobaltacyclopentadiene) [29]

(under nitrogen atmosphere)

$$Co(C_5H_5)(PPh_3)_2 + 2PhC{\equiv}CMe \longrightarrow (C_5H_5)(PPh_3)\overline{Co\text{-}C(Ph){=}C(Me)\text{-}C(Me){=}C}(Ph) + PPh_3 \quad (23)$$

To a toluene solution (30 ml) of $Co(C_5H_5)(PPh_3)_2 \cdot \frac{1}{2}C_6H_{14}$ (1.82 g, 2.63 mmol) is slowly added a toluene solution (5 ml) of 1-phenyl-1-propyne (600 mg, 5.17 mmol) from a dropping funnel. The solution is stirred and the color turns from dark green to dark brown in 10–20 h. The dark brown solution is fairly stable to air. The reaction mixture is concentrated to ca. 5 ml under vacuum and separated by column chromatography (alumina, activity II–III, 3.5 × 23 cm). First, hexane is used as a moving phase to elute PPh_3, then toluene/hexane (1:3, 50 ml) is used to spread the brown layer, and finally the brown layer is eluted using toluene. The toluene solution is evaporated to dryness, and the resulting solid is recrystallized from dichloromethane/hexane to give dark brown crystals. The crystals are filtered and dried under vacuum. Yield: 660–845 mg (35–44 %). In the above procedures, benzene can be used in place of toluene.

Properties: mp 174–176 (dec). ^{1}H NMR ($CDCl_3$): δ 1.61 (CH_3), 4.64 (Cp).

11.3 Rhodium Complexes

Organometallic compounds of rhodium have the metal center in oxidation states ranging from +4 to –3, but the most common oxidation states are +1 and +3. The Rh(I) species have a d^8 electron configuration and both four coordinated square planar and five coordinated trigonal bipyramidal species exist. Oxidative addition reactions to Rh(I) form Rh(III) species with octahedral geometry. The oxidative addition is reversible in many cases, and this makes catalytic transformations of organic compounds possible. Presented here are important reactions of rhodium complexes in catalytic and stoichiometric transformations of organic compounds.

11.3.1 Carbonyls and Related Complexes

Representatives of homoleptic rhodium carbonyls are $Rh_4(CO)_{12}$ and $Rh_6(CO)_{16}$. $Rh_4(CO)_{12}$ decomposes at 130–140 °C to give $Rh_6(CO)_{16}$. Due to the low solubility of $Rh_6(CO)_{16}$ in organic solvents, $Rh_4(CO)_{12}$ is usually employed in reactions at low temperature. However, in reactions at high temperature and/or high pressure of CO, $Rh_4(CO)_{12}$ and $Rh_6(CO)_{16}$ give similar results.

There are many examples of carbonylation reactions using these two carbonyl complexes. $Rh_4(CO)_{12}$ is a good catalyst precursor for the hydroformylation of alkenes, and PPh_3 is frequently added to increase the n/iso ratio of the resulting aldehydes, as is true also for cobalt-catalyzed hydroformylation [30]. In this case, $RhH(CO)(PPh_3)_2$ is suggested to be an active species formed in the reaction system. Thus, $RhH(CO)(PPh_3)_3$ is a good catalyst precursor for hydroformylation and the reaction proceeds under much milder conditions: room temperature and 1 atmosphere (eq (24)) [31].

$$\text{R-CH=CH}_2 + CO + H_2 \xrightarrow[\text{or } RhH(CO)(PPh_3)_3]{Rh_4(CO)_{12}/PPh_3} \underset{n}{\text{R-CH}_2\text{CH}_2\text{CHO}} + \underset{iso}{\text{R-CH(CHO)CH}_3} \quad (24)$$

Nitrobenzene is reduced, carbonylated and coupled with benzene to give the corresponding amide in the presence of catalytic amount of $Rh_6(CO)_{16}$ (eq (25)) [32].

$$C_6H_5NO_2 + C_6H_6 + 3CO \xrightarrow{Rh_6(CO)_{16}} C_6H_5NHCOC_6H_5 + 2CO_2 \quad (25)$$

Unsymmetrical ketones are produced from alkynes, alkenes, CO, and H_2 by Rh carbonyl catalysts (eq (26)) [33].

$$RC{\equiv}CR' + H_2C{=}CH_2 + CO + H_2 \xrightarrow[\text{or } Rh_6(CO)_{16}]{Rh_4(CO)_{12}} \text{RCH=CR'-C(=O)-}C_2H_5 \quad (26)$$

Carbonylation of alkynes is catalyzed by $Rh_4(CO)_{12}$ to give unsaturated lactones, where two molecules of CO being incorporated (eq (27)) [34].

$$RC{\equiv}CR' + 2CO \xrightarrow{Rh_4(CO)_{12}} \text{lactone}(R_1, R_2) + \text{lactone}(R_2, R_1) \quad (27)$$

In C_1 chemistry, the Rh carbonyl complexes are homogeneous catalyst precursors for the conversion of synthesis gas to ethylene glycol under high pressure and high temperature (eq (28)) [35].

$$CO + H_2 \xrightarrow[(Ph_3P)_2N\text{ / amine solvent}]{Rh_4(CO)_{12}\text{ or }Rh_6(CO)_{16}} HOCH_2CH_2OH + CH_3OH \quad (28)$$

$[RhCl(CO)_2]_2$ is formed in the reaction of $RhCl_3$ and CO, and both $[RhCl(CO)_2]_2$ and $RhCl_3$ are used as catalyst precursors in the carbonylation of methanol to acetic acid, the so-called Monsanto process (eq (29)) [36]. In this reaction, methyl iodide is necessary as a promoter and $[RhI_2(CO)_2]^-$ is proposed as the active species.

$$CH_3OH + CO \xrightarrow{Rh\text{ / }CH_3I} CH_3COOH \quad (29)$$

Reaction of $[RhCl(CO)_2]_2$ with strained cycloalkanes induces oxidative addition of C—C bond to the Rh center, and subsequently CO insertion takes place. For example, reaction of $[RhCl(CO)_2]_2$ with cubane and further treatment with PPh_3 result in the reductive elimination of the corresponding ketone (eq (30)) [37].

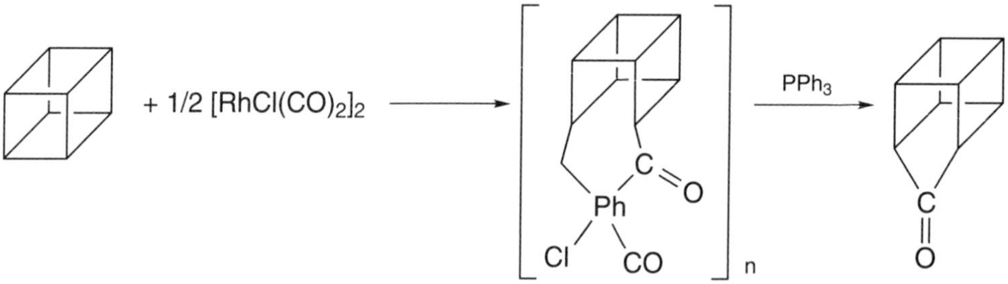

+ 1/2 $[RhCl(CO)_2]_2$ → [Ph, Cl, CO, C=O]$_n$ —PPh_3→ C=O (30)

The intermediacy of a metallacycle complex is supported by a study of the reaction with qaudricyclane, where a diacyl complex is isolated by the treatment with PPh_3 (eq (31)) [38].

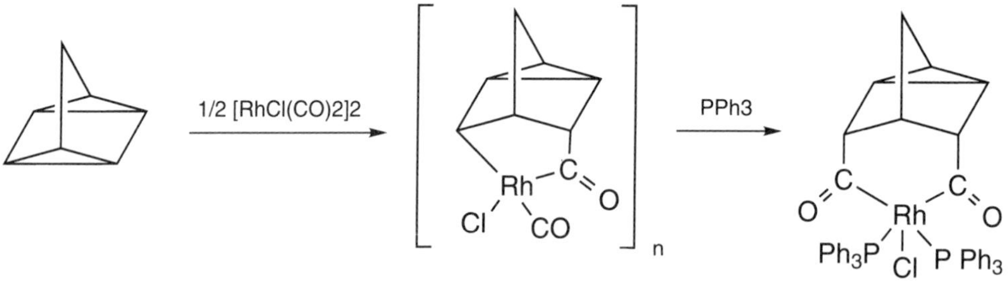

—1/2 [RhCl(CO)2]2→ [Rh, Cl, CO, C=O]$_n$ —PPh3→ O=C, Rh, C=O, Ph_3P, Cl, P Ph_3 (31)

11.3.2 Phosphine Complexes

Since the discovery of Wilkinson's complex $RhCl(PPh_3)_3$ in 1966 [39], many useful catalytic reactions have been reported which are catalyzed by this and related complexes. In particular, $RhCl(PPh_3)_3$ is a very useful catalyst precursor for hydrogenation of variety of alkenes. The chemical reactivity of $RhCl(PPh_3)_3$ is summarized in Scheme 11.4 [40].

Hydrogenations of C—C double and triple bonds are effectively catalyzed by $RhCl(PPh_3)_3$ under mild conditions (room temperature and 1 atm) (eq (32)) [39].

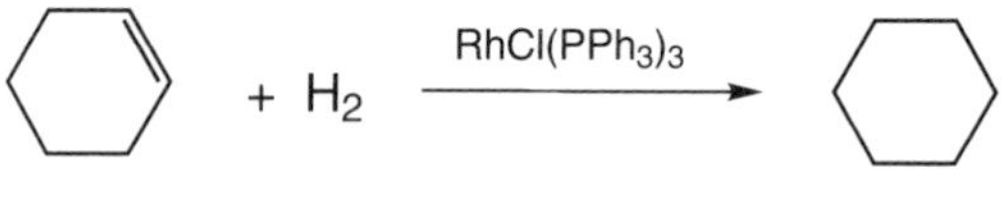

+ H_2 —$RhCl(PPh_3)_3$→ (32)

Selective hydrogenation of C—C double bonds is possible for a molecule having two C=C bonds (eq (33)) [41].

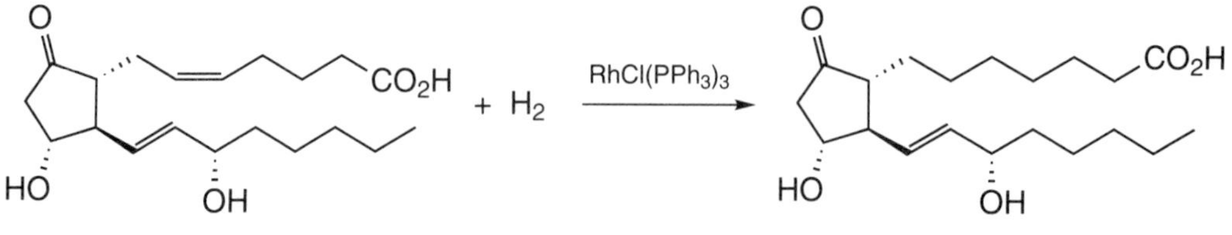

O, CO_2H, HO, OH + H_2 —$RhCl(PPh_3)_3$→ O, CO_2H, HO, OH (33)

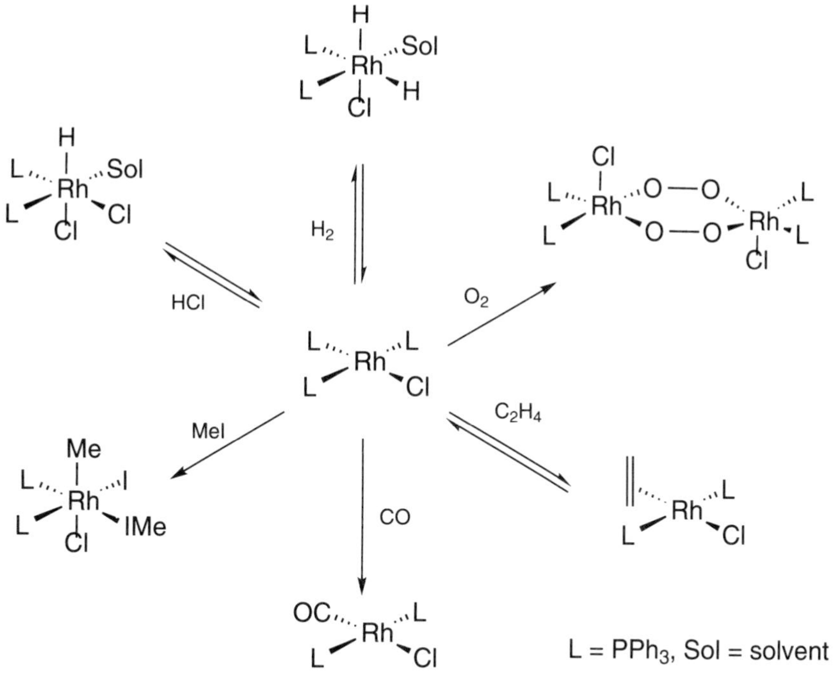

Scheme 11.4

This could be expressed more concisely as follws: $RhCl(PPh_3)_3$ also catalyzes transfer hydrogenations (eq (34)) [42], isomerization of C=C double bonds (eq (35)) [43] and hydrosilylation of α,β-unsaturated carbonyl compound (eq (36)) [44].

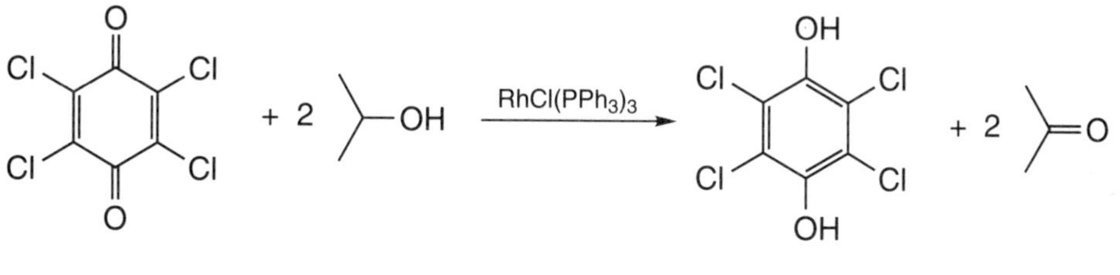

(34)

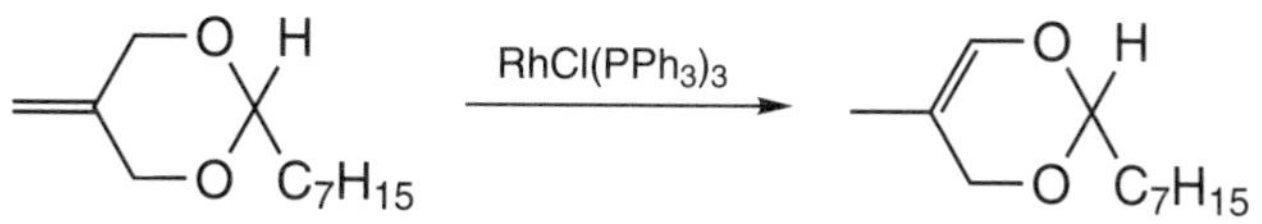

(35)

Reactions of $RhCl(PPh_3)_3$ with carbonyl compounds lead to stoichiometric decarbonylation to give the corresponding hydrocarbons and $RhCl(CO)(PPh_3)_2$ (eq (37)) [45]. Catalytic decarbonylation is possible at high temperature (eq 38) [45].

$RhCl(CO)(PPh_3)_2$ is also active for the decarbonylation reactions [46]. Cationic complexes like $[Rh(dppe)_2]^+Cl^-$ and $[Rh(dppp)_2]^+Cl^-$ (dppe = 1,2-bis(diphenylphosphi-

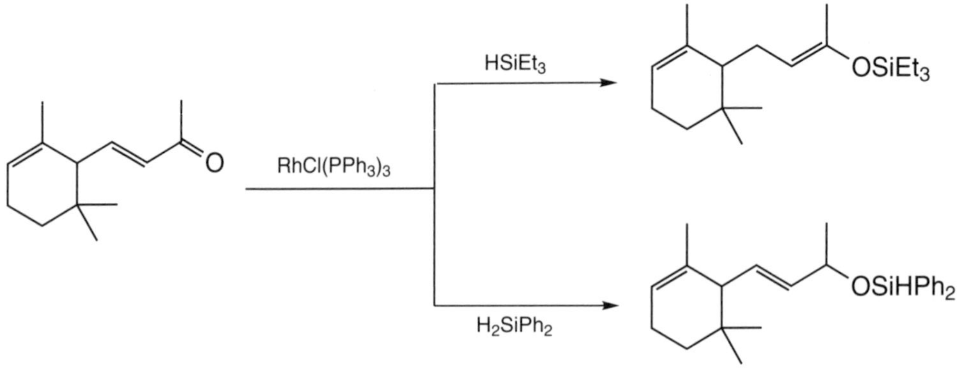

HSiEt3 OSiEt3 O RhCl(PPh3)3 (36) OSiHPh2 H2SiPh2

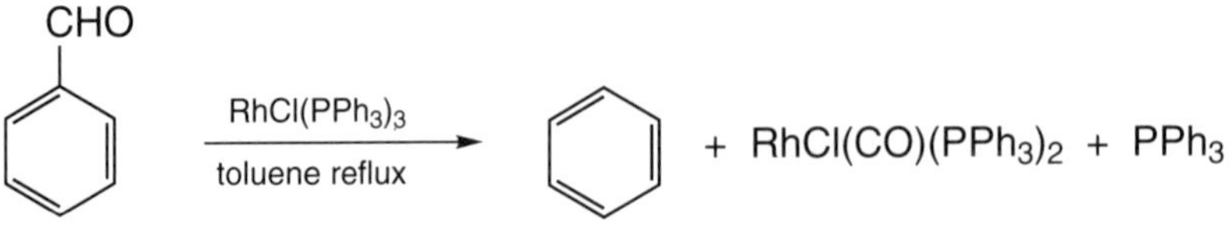

CHO RhCl(PPh3)3 toluene reflux + RhCl(CO)(PPh3)2 + PPh3 (37)

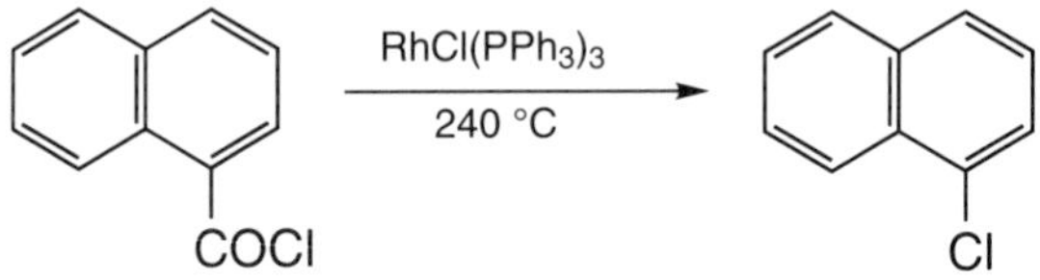

RhCl(PPh3)3 240 °C (38) COCl Cl

no)ethane; dppp = 1,3-bis(diphenylphosphino)propane) are more active catalysts to decarbonylate aldehyde under milder conditions [47].

$RhCl(PPh_3)_3$ decarbonylates cyclobutanones to give cyclopropanes, and in this reaction the activation of a C—C bond next to a carbonyl group is proposed to occur (eq (39)) [48].

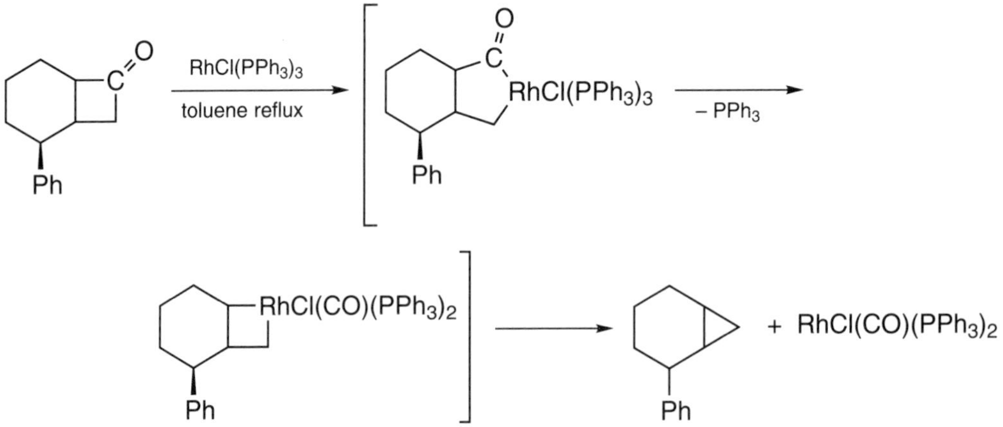

O C RhCl(PPh3)3 toluene reflux Ph RhCl(PPh3)3 – PPh3

RhCl(CO)(PPh3)2 Ph + RhCl(CO)(PPh3)2 (39)

Reactions of $RhCl(PPh_3)_3$ with organolithium or Grignard reagents gives $RhR(PPh_3)_3$ (R = Me, Ph), but analytically pure products have not been isolated [49a,b]. Similar reactions take place using $RhCl(CO)(PPh_3)_2$, where oxidative addition of alkyl halides or acyl halides and subsequent reductive elimination leads to C—C bond formation (eq (40)). In this case, the complex can be recovered and reused [49c].

$$RhCl(CO)(PPh_3)_2 + CH_2{=}CHCH_2MgBr \longrightarrow CH_2{=}CHCH_2Rh(CO)(PPh_3)_2 + MgClBr$$

$$\xrightarrow{C_{11}H_{23}COCl} CH_2{=}CHCH_2Rh(Cl)(CO)(PPh_3)_2(COC_{11}H_{23}) \longrightarrow CH_2{=}CHCH_2COC_{11}H_{23} + RhCl(CO)(PPh_3)_2 \quad (40)$$

Dimerization of terminal alkynes is promoted by $RhCl(PPh_3)_3$ (eq (41)) [50].

$$2\ CR{\equiv}CH \xrightarrow{RhCl(PPh_3)_3} RCH{=}CH{-}C{\equiv}CR \quad (41)$$

The activation of unreactive C—H bonds is a topic of current interest in organometallic chemistry [51]. Under irradiation, $RhCl(CO)(PMe_3)_2$ catalyzes the C—H activation and carbonylation of pentane to give hexanal (eq (42)) [52]. 1-Pentene is also formed by a Norrish type II reaction. The 14-electron Rh(I) species $RhCl(PMe_3)_2$ formed by irradiation is thought to be the active species.

$$\text{pentane} + CO \xrightarrow[h\nu]{RhCl(CO)(PMe_3)_2} \underset{2{,}730\%\,/\,Rh}{\text{hexanal (CHO)}} + \underset{9{,}620\%\,/\,Rh}{\text{1-pentene}} \quad (42)$$

Dehydrogenation of alkanes is catalyzed by $RhCl(PPh_3)_3$ under reflux (eq (43)). Effective removal of H_2 from the reaction system is essential for the catalytic reaction [53].

$$\text{cyclooctane} \xrightleftharpoons{RhCl(PPh_3)_3\,/\,\text{reflux}} \text{cyclooctene} + H_2 \quad (43)$$

Oxidative addition of an unreactive C—C bond occurs in the reaction of the ditertiary phosphine shown in eq (44) with $HRh(PPh_3)_4$, the reaction being preceded by the kinetically favored C—H activation [54]. The intermediates have been isolated and characterized. A similar reaction takes place using $RhCl(PEt_3)_3$.

$[RhCl(cod)]_2$ has a labile COD ligand, and reaction mixtures with tertiary phosphines at various PR_3/Rh ratios are used as homogeneous catalysts. For example, in a similar reaction to eq (39), hydrogenolysis of cyclobutanone is catalyzed by a $[RhCl(cod)]_2$-dppe system under H_2 pressure (eq (45)) [48].

Reactions of $[RhCl(cod)]_2$ with $AgClO_4$ and subsequent coordination of diphosphine give ionic complexes $[Rh(P)_2(cod)]^+ClO_4^-$. Complexes of this type having asymmetric diphosphine ligands, such as BINAP, BPPFOH, and DIOP (Fig. 11.1), are very effective catalysts for various asymmetric reactions.

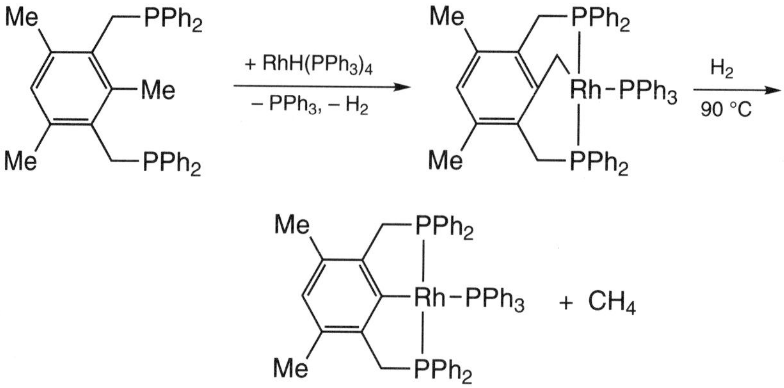

$$+ RhH(PPh_3)_4,\ - PPh_3,\ - H_2;\quad H_2,\ 90\ °C \quad \rightarrow Rh-PPh_3 + CH_4 \tag{44}$$

$$[RhCl(cod)]_2\text{-dppe},\ \text{toluene},\ 140\ °C;\ RhCl(PPh_3)_3;\ H_2;\ H_2 \rightarrow CH_2OH,\ Me,\ Ph \tag{45}$$

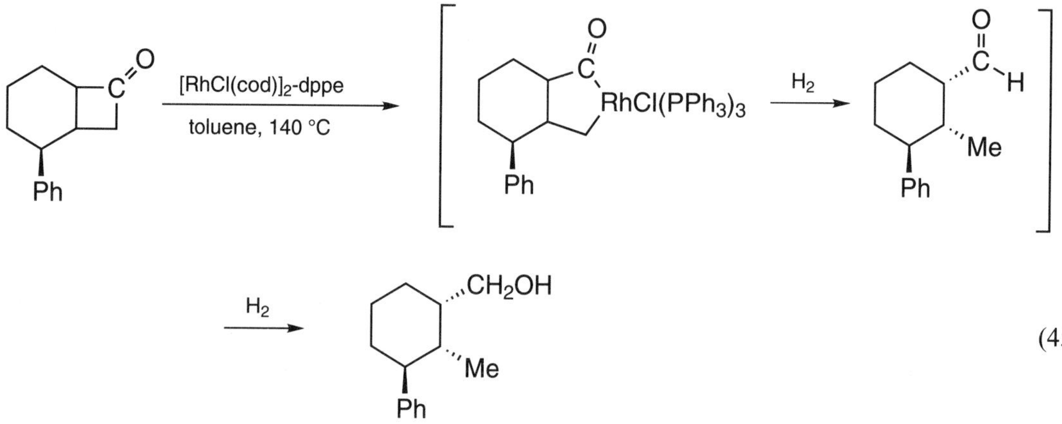

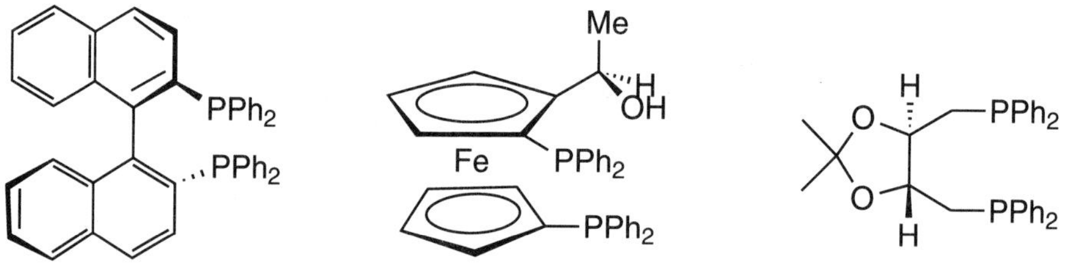

(*R*)-BINAP (*R*)-(*S*)-BPPFOH (*R,R*)-DIOP

Figure 11.1 Asymmetric diphosphines.

Catalytic asymmetric hydrogenation reactions are performed with high enantioselectivity (eq (46) and (47)) [55,56].

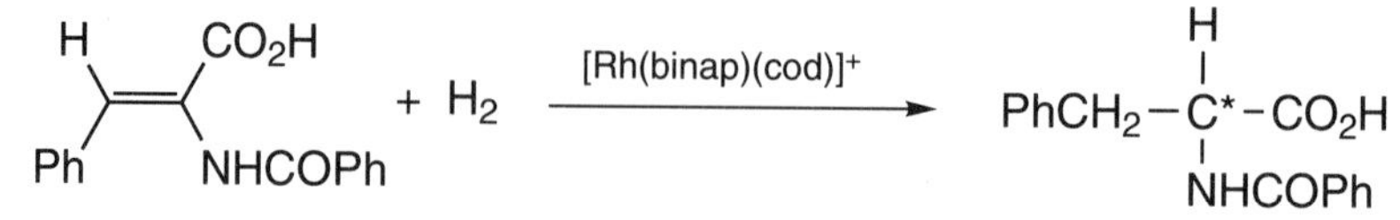

$$PhCH{=}C(CO_2H)(NHCOPh) + H_2 \xrightarrow{[Rh(binap)(cod)]^+} PhCH_2{-}C^*H(NHCOPh){-}CO_2H \tag{46}$$

$$\text{ArC(=O)CH}_2\text{NHMe} + H_2 \xrightarrow{\text{Rh}^+\text{-BPPFOH}} \text{Ar–C*H(OH)–CH}_2\text{NHMe} \quad (47)$$

Asymmetric isomerization of allylamines is performed with high enantioselectivity [57]. This process is applied to the industrial production of L-menthol from diethylgeranylamine by the Takasago Perfumery Company (eq (48)).

diethylgeranylamine (NEt$_2$) $\xrightarrow{[\text{Rh}((S)\text{-binap})(\text{cod})]^+}$ (NEt$_2$) $\xrightarrow{H^+}$ (CHO) $\xrightarrow{ZnBr_2}$ (OH) $\xrightarrow{H_2}$ (OH) L-menthol (48)

Asymmetric hydrosilylation of ketones gives silyl ethers and the resulting silyl ethers are readily hydrolyzed to give optically active alcohols (eq (49)) [58].

$$\text{MeC(=O)CO}_2{}^i\text{Pr} + \text{H}_2\text{SiPh(1-Np)} \xrightarrow{\text{Rh-diop}} \xrightarrow{\text{hydrolysis}} \text{Me–C*H(OH)–CO}_2{}^i\text{Pr} \quad (49)$$

11.3.3 Cyclopentadienyl Complexes

Rh complexes having a cyclopentadienyl ligand shows high reactivity in C—H bond activation. Irradiation of $\text{RhCp*(PMe}_3)(\text{H})_2$ ($\text{Cp*} = \text{C}_5\text{Me}_5$) in alkane solvent produced a coordinatively unsaturated $[\text{RhCp*(PMe}_3)]$ species, which undergoes activation of the sp^3 C—H bond of the alkane to give an alkyl hydride complex (eq (50)) [59,60].

$$\text{RhCp*(PMe}_3)(\text{H})_2 \xrightarrow{h\nu} [\text{RhCp*(PMe}_3)] \xrightarrow[-55\ \text{*C}]{\text{C}_3\text{H}_8} \text{RhCp*(PMe}_3)(\text{H})(\text{CH}_2\text{CH}_3\text{CH}_3) \quad (50)$$

$[RhCp^*(C_2H_4)(CH_2CH_2\text{-}\mu\text{-}H)]^+$, which is prepared by protonation of $RhCp^*(C_2H_4)_2$, is a highly active catalyst for tail-to-tail dimerization of methyl acrylate to dimethyl hexenedioates (eq (51)) [61].

$$2\ \text{CH}_2{=}\text{CHCO}_2\text{Me} \xrightarrow{[RhCp^*(C_2H_4)(CH_2CH_2\text{-}\mu\text{-}H)]^+} \text{MeO}_2\text{CCH}{=}\text{CHCH}_2\text{CH}_2\text{CO}_2\text{Me} + \text{MeO}_2\text{CCH}_2\text{CH}{=}\text{CHCH}_2\text{CO}_2\text{Me} \quad (51)$$

11.3.4 Synthesis of Rh Compounds

(1) Dodecacarbonyltetrarhodium(0), $Rh_4(CO)_{12}$ [62]

(hood)

$$RhCl_3{\cdot}3H_2O + 2Cu + 4CO + NaCl \longrightarrow Na[RhCl_2(CO)_2] + 2Cu(CO)Cl$$
$$4Na[RhCl_2(CO)_2] + 6CO + 2H_2O \longrightarrow Rh_4(CO)_{12} + 2CO_2 + 4NaCl + 4HCl \quad (52)$$

Copper powder is washed with a mixture of concentrated hydrochloric acid and acetone, and then rinsed with acetone. The powder is dried under vacuum. The activated copper powder (1.5 g, 24 mmol) and sodium chloride (0.6 g, 10 mmol) are added to water (200 ml), and the water is well degassed and saturated with CO. Under a stream of CO, the regorously stirred mixture is treated with a degassed solution of rhodium trichloride trihydrate (2.6 g, 9.9 mmol) in water (50 ml) from a dropping funnel (rate of dropping is about 1 ml/min). After the addition, the funnel is washed with degassed water (10 ml), and the mixture is stirred for 2 h. During this time, a small amount of $Rh_4(CO)_{12}$ separates as an orange powder. The funnel is charged with a solution of disodium citrate in degassed water (0.4 M, 50 ml), and the solution is added to the yellow mixture under a stream of CO over a period of 50 min. Then the mixture is stirred for 20 h, and the orange solid is filtered under CO. The solid is washed with water under CO and then dried under vacuum. The resulting solid is extracted with the minimum amount of dichlomethane (5 × 5 ml) under CO. The solution is quickly evaporated to dryness under vacuum on a water bath at room temperature and the resulting crystals are dried under vacuum for 2 h. Yield: 1.55–1.75 g (80–90 %). The compound is soluble in dichloromethane, chloroform, pentane, toluene, tetrahydrofuran, and acetone, and sparingly soluble in cyclohexane and methanol. It is slowly decomposed to dark and insoluble $Rh_6(CO)_{16}$ at room temperature under nitrogen, and should therefore be stored under CO below 0°C.

Properties: IR(heptane): 2101(w), 2075(s), 2069(s), 2059(w, sh), 2045(m), 2001(w), 1918(w), 1882(s), 1848(w) cm^{-1}.

(2) Chlorotris(triphenylphosphine)rhodium(I), $RhCl(PPh_3)_3$ [39,63]

(under nitrogen atmosphere)

$$RhCl_3{\cdot}3H_2O + 4PPh_3 \longrightarrow RhCl(PPh_3)_3 + 2H_2O + OPPh_3 + 2HCl \quad (53)$$

To a hot ethanol solution (350 ml) of triphenylphosphine (recrystallized from ethanol prior to use, 12 g, 46 mmol) is added a hot ethanol solution (70 ml) of rhodium trichloride trihydrate (2.0 g, 7.6 mmol). The solution is refluxed for 30 min, and red crystals are filtered from the hot solution. The crystals are washed with ether and dried under vacuum. Yield: 6.3 g (86 %). The compound is soluble in benzene, chloroform, and dichloromethane.

Properties: mp 157–158 °C.

(3) Carbonylhydridotris(triphenylphosphine)rhodium(I), $RhH(CO)(PPh_3)_3$ [64]

(under nitrogen atmosphere)

$$RhCl_3{\cdot}3H_2O + 3PPh_3 + HCHO + NaBH_4 \longrightarrow RhH(CO)(PPh_3)_3 \quad (54)$$

To a solution of triphenylphosphine (2.64 g, 10 mmol) in boiling and vigorously stirred ethanol (100 ml) is added a solution of rhodium trichloride trihydrate (0.26 g, 0.99 mmol) in ethanol (20 ml). After 15 s, aqueous formaldehyde (10 ml, 40 % w/v) and then a solution of potassium hydroxide (0.8 g) in hot ethanol (20 ml) are added quickly to the reaction mixture. The mixture is refluxed for 10 min, and then cooled to room temperature. The yellow crystals are filtered, washed with ethanol, water, ethanol, and hexane, then dried under vacuum. Yield: 0.85 g (94 %).

Properties: mp 120–122 °C (in air), 172–174 °C (under nitrogen). IR: ν(RhH) 2041(s), ν(CO) 1918(vs) cm^{-1}. ^1H NMR($CDCl_3$): δ-9.7 (br, RhH).

(4) Hydridotetrakis(triphenylphosphine)rhodium(I), $RhH(PPh_3)_4$ [65]

(under nitrogen atmosphere)

$$RhCl_3{\cdot}3H_2O + 3PPh_3 + KOH \longrightarrow RhH(PPh_3)_4 \quad (55)$$

To a refluxing solution of triphenylphosphine (2.62 g, 10 mmol) in ethanol (80 ml) are added quickly a solution of rhodium trichloride trihydrate (0.26 g, 0.99 mmol) in hot ethanol (20 ml) and then a solution of potassium hydroxide (0.4 g) in hot ethanol (20 ml). The mixture is further refluxed for 10 min and cooled to room temperature. The resulting yellow crystals are filtered; washed with ethanol, water, ethanol, and hexane; then dried under vacuum. Yield: 1.10 g (97 %). The crystals are soluble in benzene, chloroform, and dichloromethane, and the solutions are highly air-sensitive.

Properties: mp 145–147 °C (in air), 154–156 °C (under nitrogen). IR: ν(RhH) 2156(m) cm^{-1}.

(5) Di-μ-chlorotetrakis(η-ethylene)dirhodium(I), $Rh_2Cl_2(C_2H_4)_4$ [66]

$$2RhCl_3{\cdot}3H_2O + 6C_2H_4 \longrightarrow Rh_2Cl_2(C_2H_4)_4 + 4HCl + 2CH_3CHO + 4H_2O \quad (56)$$

Rhodium trichloride trihydrate (10 g, 38 mmol) is dissolved in hot water (15 ml), and the solution is added to ethanol (250 ml). The mixture is degassed with an aspirator and filled with 1 atm of ethylene. The reaction mixture is vigorously stirred at room temperature by bubbling ethylene (1 bubble/s) for 7 h. During this stirring, the product separates as red solid. Under nitrogen atmosphere, the solid is filtered and washed with methanol (50 ml), then dried under vacuum. Yield: 4.8–5.0 g (60–65 %). Some of the product reacts with HCl formed during the reaction and is dissolved in the filtered solution. To neutralize HCl, an aqueous solution (3 ml) of sodium hydroxide (1.5 g) is added to the filtered solution and washings. The mixture is bubbled with 1 atm of ethylene as before to recover a further 1.0–1.5 g of the product. The total yield is 6.0 g (75 %). The compound is slightly soluble in most organic solvents, and is thus impossible to purify by recrystallization. It can be handled in air for several minutes, but should be stored under nitrogen below 0 °C.

Properties: mp 115 °C (dec). IR(KBr): 3060, 2980, 1520, 1430, 1230, 1215, 999, 952, 930, 715 cm^{-1}.

(6) Di-μ-chlorobis(1,5-cyclooctadiene)dirhodium(I), $Rh_2Cl_2(C_8H_{12})_2$ [67]

(under nitrogen atmosphere)

$$2RhCl_3{\cdot}3H_2O + 2C_8H_{12} + 2CH_3CH_2OH \longrightarrow Rh_2Cl_2(C_8H_{12})_2 + 4HCl + 2CH_3CHO + 6H_2O \quad (57)$$

Rhodium trichloride trihydrate (1.0 g, 3.8 mmol) and 1,5-cyclooctadiene (2 ml, ca. 16 mmol) are added to ethanol (30 ml), and the mixture is stirred and refluxed for 3 h. After the mixture is cooled to room temperature, the orange solid is filtered, washed with ethanol, dried, and recrystallized from acetic acid. The crystals are dried under vacuum. Yield: 0.56 g (60 %). The crystals are stable under air. Soluble in dichloromethane, chloroform, acetic acid, and acetone, slightly soluble in ether, methanol, ethanol, and benzene, and insoluble in water.

Properties: IR(Nujol): 998, 964, 819 cm^{-1}. ^1H NMR($CDCl_3$): δ 4.3 (=CH), 1.7–2.6 (CH_2).

(7) [(1,5-cyclooctadiene){2,2′-bis(diphenylphosphino)-1,1′-binaphthyl}-rhodium(I)]perchlorate [57]

(under nitrogen atmosphere)

$$Rh_2Cl_2(C_8H_{12})_2 + 2AgClO_4 + 2BINAP \longrightarrow 2[Rh(binap)(C_8H_{12})]ClO_4 + 2AgCl \quad (58)$$

Silver perchlorate (0.104 g, 0.5 mmol) is added to an acetone (30 ml) solution of $Rh_2Cl_2(C_8H_{12})_2$ (0.123 g, 0.25 mmol), and the mixture is stirred at room temperature. The colorless precipitates are filtered and washed with acetone. To the filtered solution and washing is added solid (+ or –)-BINAP (0.311 g, 0.5 mmol), and the solution is stirred for 1 h at room temperature. After concentration of the solution to ca. 2 ml, ether (5 ml) is slowly added to give deep orange crystals. The crude product is again recrystallized from acetone/diethyl ether to give deep orange crystals. The crystals are filtered, washed with ether, and dried under vacuum. Yield: 75–80 %. Recrystallization from THF/acetone gives more well-formed crystals as THF solvated complex.

Properties of crystals from acetone/ether: 164 °C (in Ar, dec). ^1H NMR (CD_2Cl_2): δ

2.00–2.62 (m, 8H, CH_2), 4.58 (br, 2H, CH=), 4.84 (br, 2H, CH=), 6.42–8.22 (m, 32H, Ar).

(8) Di-μ-chlorotetracarbonyldirhodium(I), $Rh_2Cl_2(CO)_4$ [66]

(hood)

$$Rh_2Cl_2(C_2H_4)_4 + 4CO \longrightarrow Rh_2Cl_2(CO)_4 + 4C_2H_4 \quad (59)$$

This compound is obtained from the reaction of $RhCl_3{\cdot}3H_2O$ and CO at 100 °C [68], but in practice it does not always work well. The following method is recommended.

Di-μ-chlorotetrakis(ethylene)dirhodium(I) (1.0 g, 2.6 mmol) is suspended in diethyl ether (30 ml), and CO is bubbled (1 bubble/s) through the stirred mixture for 1 h. The mixture is then filtered, and the filtered solution is concentrated to 15 ml with a rotary evaporator. The solution is cooled to 0 °C to give red crystals, and the crystals are filtered and washed quickly with cold ether. Yield: 0.50 g (50 %). The crystals are soluble in most organic solvents. The compound is stable in air as solid, but should be stored under CO below 0 °C.

Properties: mp 126–127 °C. IR(hexane): 2105(m), 2089(s), 2080(vw), 2035(s), 2003(w) cm^{-1}.

11.4 Iridium Complexes

The chemistry of Ir complexes resembles that of Rh, and Ir(I) and Ir(III) complexes are common species. However, in contrast to Rh (I), Ir(I) often undergoes oxidative addition irreversibly. Thus, the corresponding Ir(III) complexes are more stable and can be isolated in many cases. Much work has been devoted to the oxidative addition to $IrX(CO)(PPh_3)_2$ complexes. Another characteristic of Ir is the extensive formation of polyhydrides.

Due to the stability of Ir complexes, catalytic activity of Ir complexes are usually lower than the corresponding Rh ones. Thus, the reported examples of catalytic reactions by Ir complexes are fewer than those of Co and Rh.

11.4.1 Carbonyls

Like Rh carbonyls, stable homoleptic Ir carbonyls are $Ir_4(CO)_{12}$ and $Ir_6(CO)_{16}$. $Ir_4(CO)_{12}$ catalyzes the water-gas shift reaction, which can be used to hydroformylate alkenes (eq (60) and (61)) [69].

$$CO + H_2O \rightleftharpoons CO_2 + H_2 \quad (60)$$

$$R\text{-}CH{=}CH_2 + 2\,CO + H_2O \xrightarrow{Ir_4(CO)_{12}} R\text{-}CH_2CH_2\text{-}CHO + CO_2 \quad (61)$$

Like Rh/I systems, the $Ir_4(CO)_{12}/I_2$ system catalyzes the carbonylation of methanol to acetic acid [70]. In homogeneous hydrogenation of CO (Fischer–Tropsch reaction), $Ir_4(CO)_{12}$ shows a relatively high catalytic activity compared with other transition metal carbonyls (eq (62)) [71].

$$CO + H_2 \xrightarrow[140\ °C,\ 2\ atm]{Ir_4(CO)_{12}} CH_4 + C_2H_6 + \cdots \quad (62)$$

11.4.2 Phosphine Complexes

The most important Ir(I) complexes are *trans*-$IrCl(CO)(PPh_3)_2$ (Vaska's complex) and its phosphine derivatives, since they provide clear examples of oxidative addition reactions. Scheme 11.5 summarizes the addition reactions to $IrCl(CO)(PPh_3)_2$ [72]. Alkyl halides, acyl halides, H_2, and $SnCl_4$ oxidatively add to the Ir center, and reactions with O_2, alkyne, CO, and SO_2 give their coordinated complexes without complete cleavage of the appropriate bond.

$IrCl(CO)(PPh_3)_2$ catalyzes the hydrogenation of alkene [73] and the transfer hydrogenation reaction of diene (eq (63)) [74].

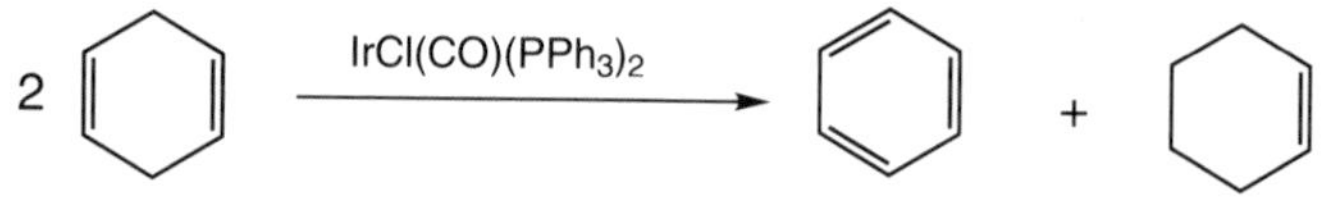

2 cyclohexa-1,4-diene $\xrightarrow{IrCl(CO)(PPh_3)_2}$ benzene + cyclohexene (63)

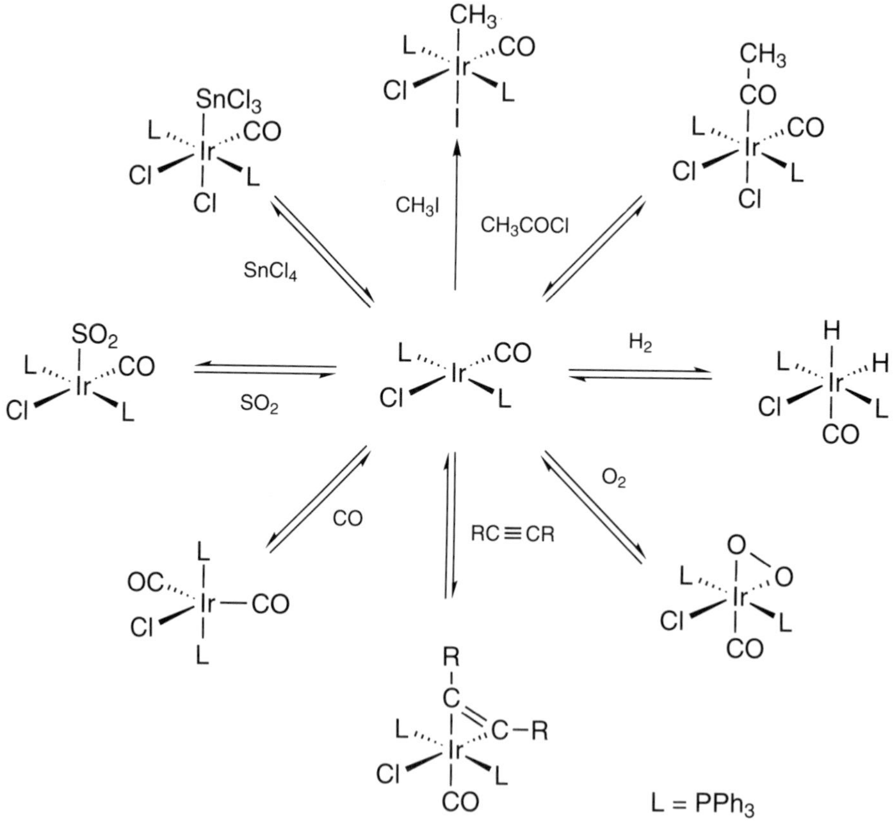

Scheme 11.5

Oxidation of alkenes and ketones with oxygen is catalyzed by $IrCl(CO)(PPh_3)_2$ (eqs (64) and (65)) [75].

$$\text{cyclohexene} + O_2 \xrightarrow{IrCl(CO)(PPh_3)_2} \text{2-cyclohexen-1-one} + \text{2-cyclohexen-1-ol} \quad (64)$$

$$\text{2-cyclohexen-1-one} + O_2 \xrightarrow{IrCl(CO)(PPh_3)_2} HO_2C(CH_2)_4CO_2H \quad (65)$$

The Ir analog of Wilkinson's complex, $IrCl(PPh_3)_3$, is relatively inactive in alkene hydrogenation, since the reaction of $IrCl(PPh_3)_3$ with H_2 gives stable $IrH_2Cl(PPh_3)_3$, which does not readily lose PPh_3. However, $[Ir(cod)(\mu\text{–}Cl)]_2$ is active when two mole of PR_3 are added per Ir. Another useful catalyst is $[Ir(cod)(PCy_3)(py)]^+PF_6^-$, which catalyzes hydrogenation of hindered alkenes with a higher rate than that of $RhCl(PPh_3)_3$ (eq (66)) [76].

$$Me_2C{=}CMe_2 + H_2 \xrightarrow[CH_2Cl_2,\ 0\ °C]{[Ir(cod)(PCy_3)(py)]^+PF_6^-} Me_2CH{-}CHMe_2 \quad (66)$$

An Ir(III) complex having tridentate phosphine ligand, $[Ir(triphos)(H)_2(C_2H_4)]BPh_4$, is capable of catalyzing cyclotrimerization of alkynes (eq (67)) [77]. The Ir(triphos) systems show a variety of reactions such as activation of C—H and C—S bonds [78].

$$3\ C_2H_2 \xrightarrow{[Ir(triphos)(H)_2(C_2H_4)]BPh_4} C_6H_6 \qquad \text{triphos} = CH_3C(CH_2PPh_2)_3 \quad (67)$$

11.4.3 Polyhydrides

Polyhydrides are complexes with a H:M ratio exceeding 3 [79]. Complexes of the type $IrH_5(PR_3)_2$ are polyhydrides of Ir(V), but their protonation products are not hexahydrides of Ir(VII) but Ir(III) complexes having two dihydride and two neutral dihydrogen ligands (eq (68)).

$$IrH_5(PCy_3)_2 \xrightarrow{H^+} [IrH_2(\eta^2\text{-}H_2)_2(PCy_3)_2]^+ \quad (68)$$

A typical polyhydride is $IrH_5(P^iPr_3)_2$. It reacts reversively with ethylene, releasing H_2 to give a cyclometallated complex (eq (69)).

$$IrH_5(P^iPr_3)_2 + 2\,C_2H_4 \rightleftharpoons Ir(C_2H_4)_2\{P(CMe_2)^iPr_2\}(P^iPr_3) + 3\,H_2 \qquad (69)$$

As a consequence of this type of reactions, $IrH_5(P^iPr_3)_2$ catalyzes hydrogenation–dehydrogenati on reaction of linear alkenes (eq (70)) [80].

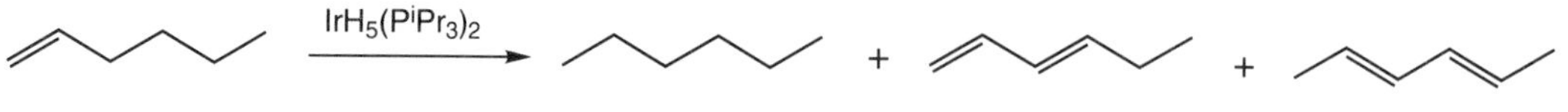

$$\text{1-hexene} \xrightarrow{IrH_5(P^iPr_3)_2} \text{hexane} + \text{hexadiene} + \text{hexadiene} \qquad (70)$$

$IrH_5(P^iPr_3)_2$ can activate the C—H bonds of benzene and catalyze the H—D exchange reaction of benzene and hydrogen (eq (71)) [81].

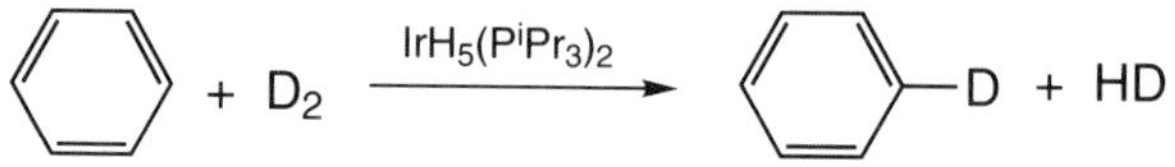

$$C_6H_6 + D_2 \xrightarrow{IrH_5(P^iPr_3)_2} C_6H_5D + HD \qquad (71)$$

11.4.4 Cyclopentadienyl Complexes

The first well-characterized example of oxidative addition of the unreactive C—H bond of an alkane was based on Ir complexes containing a Cp* ligand (eq (72) and (73) [82,83,51]. Photoirradiation is needed to create active IrCp*L species. However, a cationic complex $[IrCp^*(PMe_3)(Me)(ClCH_2Cl)]^+[B(Ar_f)_4]^-$ can thermally activate methane and alkane at 10 °C (eq (74)) [84].

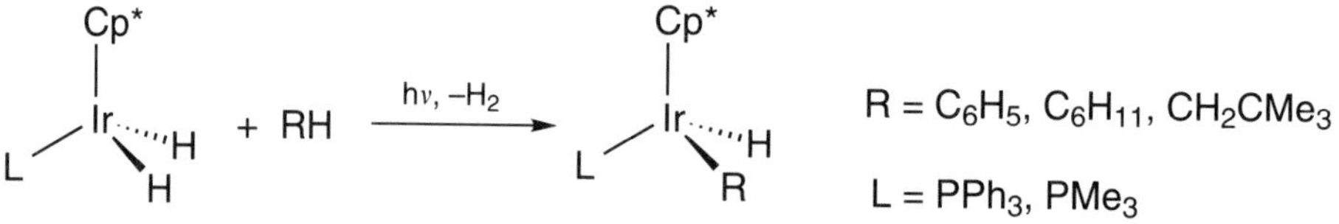

$$Cp^*IrL(H)_2 + RH \xrightarrow{h\nu,\,-H_2} Cp^*IrL(H)(R) \qquad (72)$$

R = C_6H_5, C_6H_{11}, CH_2CMe_3

L = PPh_3, PMe_3

$$Cp^*Ir(CO)_2 \xrightarrow{h\nu} [Cp^*Ir(CO)] \xrightarrow{RH} Cp^*(CO)Ir(R)(H) \qquad (73)$$

R = C_6H_{11}, CH_2CMe_3

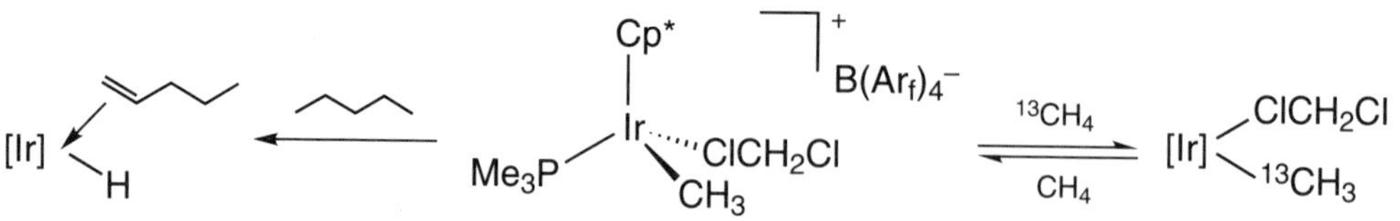

$$[Ir](\text{alkenyl})(H) \xleftarrow{\text{pentane}} [Cp^*Ir(PMe_3)(CH_3)(ClCH_2Cl)]^+ B(Ar_f)_4^- \underset{CH_4}{\overset{^{13}CH_4}{\rightleftharpoons}} [Ir](ClCH_2Cl)(^{13}CH_3) \qquad (74)$$

[Ir] = $Cp^*(PMe_3)Ir$, $B(Ar_f)_4 = B(3,5\text{-}C_6H_3(CF_3)_2)_4$

In the presence of a hydrogen acceptor, it is possible to generate an unsaturated iridium species thermally that can activate C—H bonds (eq (74)) [85].

$$[IrH_2S_2L_2]BF_4 + C_5H_6 + 3\ \text{CH}_2{=}\text{CHBu}^t \longrightarrow [(\eta^5\text{-}C_5H_5)IrHL_2]BF_4 + 3\ \text{CH}_3\text{CH}_2\text{Bu}^t \quad (75)$$

S = acetone, L = PPh_3

11.4.4 Synthesis of Ir Compounds

(1) Dodecacarbonyltetrairidium(0), $Ir_4(CO)_{12}$ [86]

(hood)

$$Na_3IrCl_6 + CO \longrightarrow Ir_4(CO)_{12} \quad (76)$$

After sodium hexachloroiridate(III) (5.0 g, 8.9 mmol) is dissolved in ethanol (100 ml), the solution is quickly filtered. To the filtered solution is added a solution of sodium iodide (1.4 g, 9.4 mmol) in ethanol. Fast reduction from Ir(IV) to Ir(III) takes place and solid sodium hexachloroiridate Na_3IrCl_6 separates almost quantitatively. The solid is filtered, washed with ethanol, and dried at 100 °C. The obtained Na_3IrCl_6 (ca. 4.8 g), NaI (7.9 g, 53 mmol), and water (5 ml) are added to methanol (100 ml). The mixture is well degassed and saturated with CO, and then refluxed with vigorous stirring under a CO stream (2 bubble/s). In 4 h, a red-brown solution is obtained, and then the solution is cooled to room temperature under CO (2 bubble/s). Finely ground potassium carbonate (2.9 g, 17.7 mmol) is added to the solution, and the mixture is vigorously stirred for 40 h with CO stream (2 bubble/s) to give a yellow solid. The solid is filtered under nitrogen, washed with water (4 × 2 ml), ethanol (2 × 10 ml), and hexane, and dried under vacuum. Yield: 1.8–1.9 g (74–78 %). The compound is slightly soluble in most organic solvents. It is inert to air and water.

Properties: IR(CH_2Cl_2): 2067, 2027 cm^{-1}. IR(KBr): 2084, 2053, 2020 cm^{-1}.

(2) Di-μ-chlorobis(1,5-cyclooctadiene)diiridium(I) [87]

(under nitrogen atmosphere)

$$2IrCl_3 + 2C_8H_{12} + 2CH_3CH_2OH \longrightarrow Ir_2Cl_2(C_8H_{12})_2 + 4HCl + 2CH_3CHO \quad (77)$$

Iridium trichloride (2.0 g, 5.7 mmol) is added to a mixture of 95 % ethanol (34 ml), water (17 ml), and 1,5-cyclooctadiene (6 ml). Under nitrogen atmosphere the solution is stirred and refluxed for 24 h, during which time the product separates as brick-red solid. After the mixture is cooled to room temperature, the solid is filtered and washed with ice-cold methanol. The orange-red solid is dried under vacuum. Yield: 1.5 g (72 %). The compound is air-stable. Soluble in chloroform and benzene, sparingly soluble in acetone, and insoluble in ether.

Properties: mp > 200 °C (dec). ^{1}H NMR (C_6D_6): δ 1.20 (CH_2, 8H), 1.96 (CH_2, 8H), 4.28 (=CH, 8H).

(3) Chlorocarbonylbis(triphenylphosphine)iridium(I) [88]

(under nitrogen atmosphere)

$$IrCl_3{\cdot}3H_2O + PPh_3 + HCONMe_2 \longrightarrow IrCl(CO)(PPh_3)_2 \quad (78)$$

Iridium trichloride trihydrate (3.52 g, 10 mmol) and triphenylphosphine (13.1 g, 50 mmol) is added to *N,N*-dimethylformamide (150 ml). The mixture is stirred and refluxed for 12 h. The red-brown solution is filtered while hot, and immediately warm methanol (300 ml) is added to the filtered solution. After the mixture is cooled in an ice bath, the yellow crystals are filtered and washed with ice-cold methanol. The crystals are dried under vacuum. Yield: 6.8–7.0 g (87–90 %). The crystals are air-stable, but take up oxygen in solution. Soluble in benzene and chloroform but insoluble in alcohols.

Properties: IR(nujol): ν(CO) 1961 cm^{-1}.

(4) Pentahydridobis(triisopropylphosphine)iridium(V) [80]

(under nitrogen atmosphere)

$$H_2IrCl_6{\cdot}6H_2O + 3P^iPr_3 \longrightarrow [PH^iPr_3][IrCl_4(P^iPr_3)_2]$$
$$[PH^iPr_3][IrCl_4(P^iPr_3)_2] + LiAlH_4 \longrightarrow IrH_5(P^iPr_3)_2 \quad (79)$$

Hydrogen hexachloroiridate(IV) hexahydrate (17.4 g, 34 mmol) is added to a mixture of ethanol (160 ml) and concentrated hydrochloric acid (20 ml). The mixture is refluxed for 2 h until the color turns from brown to green. After cooled to room temperature, an ethanol solution (70 ml) of triisopropylphosphine (18 ml) is slowly added to the mixture with vigorous stirring. The mixture is refluxed for 4 h and allowed to stand overnight at 0 °C. The resulting purple crystals of $[PH^iPr_3][IrCl_4(P^iPr_3)_2]$ are filtered, washed with cold ethanol, and dried under vacuum. Yield: 16 g (58 %).

To a stirred solution of $[PH^iPr_3][IrCl_4(P^iPr_3)_2]$ (3.15 g, 3.9 mmol) in THF (110 ml) is added lithium tetrahydridoaluminate in small portions until the violet solution turns to a gray suspension. The suspension is stirred for 30 min, and the residual $LiAlH_4$ is hydrolyzed with water/THF (5 ml/15 ml). The mixture is evaporated to dryness, the resulting solid is extracted with petroleum ether (100 ml), and the extracted solution is concentrated to small volume. Addition of methanol gives white crystals, which are filtered, washed with cold methanol, and dried under vacuum. Yield: 1.8 g (90 %).

Properties: IR: ν(Ir—H) 1950(s), δ(Ir—H) 875(m) cm^{-1}. ^1H NMR (C_6D_6): δ –10.8 (t, 5H, J_{PH} = 11 Hz, IrH), 1.13 (d, 18H, J_{HH} = 7 Hz, CH_3), 1.27 (d, 18 H, J_{HH} = 7 Hz, CH_3), 1.78 (m, br, 6H, CH).

References

[1] *Comprehensive Organometallic Chemistry*, vol. 5, Ed G. Wilkinson, F. G. A. Stone, E. W. Abel, Pergamon Press, Oxford (1982); *Comprehensive Organometallic Chemistry II*, vol. 8, Pergamon Press, Oxford (1995).
[2] G. G. Sumner, H. P. Klug, L. E. Alexander, *Acta Crystallogr.*, **17**, 732 (1964).
[3] H. W. Sternberg, I. Wender, M. Orchin, *Inorg. Synth.*, **5**, 192 (1957).
[4] T. J. Baillie, B. L. Booth, C. A. McAuliffe, *J. Organomet. Chem.*, **54**, 275 (1973).
[5] R. L. Pruett, *Adv. Organomet. Chem.*, **17**, 1 (1979); J. A. Davies, *The Chemistry of the Metal–Carbon Bond*, vol. 3, Ed. F. R. Hartley and S. Patai, Wiley, New York, p. 361 (1985).
[6] M. Hidai, A. Fukuoka, Y. Koyasu, Y. Uchida, *J. Chem. Soc., Chem. Commun.*, 516 (1984); *J. Mol. Catal.*, **35**, 29 (1986).

[7] Y. Seki, A. Hidaka, S. Murai, N. Sonoda, *Angew. Chem., Int. Ed. Engl.*, **16**, 174 (1977); Y. Seki, A. Hidaka, S. Makino, S. Murai, N. Sonoda, *J. Organomet. Chem.*, **140**, 361 (1977); Y. Seki, S. Murai, A. Hidaka, N. Sonoda, *Angew. Chem., Int. Ed. Engl.*, **16**, 881 (1977).
[8] P. Pino and G. Braca, *Organic Syntheses via Metal Carbonyls*, Ed. I. Wender and P. Pino, vol. 2, p. 419, Wiley, New York (1977).
[9] M. Hidai, M. Orisaku, M. Ue, Y. Koyasu, T. Kodama, Y. Uchida, *Organometallics*, **2**, 292 (1983).
[10] S. Murahashi, S. Horiie, *J. Am. Chem. Soc.*, **78**, 4816 (1956).
[11] K. M. Nicholas, M. O. Nestle, D. Seyferth, *Transition Metal Organometallics*, vol. II, p. 1, Academic Press, New York (1978).
[12] P. Arthur, Jr., D. C. England, B. C. Pratt, G. M. Whitman, *J. Am. Chem. Soc.*, **76**, 5364 (1954).
[13] R.F. Heck, *Adv. Organomet. Chem.*, **4**, 243 (1966).
[14] A. Yamamoto, S. Kitazume, L. S. Pu, S. Ikeda, *Chem. Commun.*, 79 (1967); A. Misono, Y. Uchida, T. Saito, *Bull. Chem. Soc. Jpn.*, **40**, 700 (1967); A. Sacco and M. Rossi, *Chem. Commun.*, 316 (1967).
[15] A. Misono, Y. Uchida, M. Hidai, T. Kuse, *J. Chem. Soc., Chem. Commun.*, 981 (1968); L.S. Pu, A. Yamamoto, S. Ikeda, *J. Am. Chem. Soc.*, **90**, 3896 (1968); I.S. Kolomnikov, T.S. Lobeeva, M.E. Vol'pin, *Izv. Akad. Nauk. SSR, Ser. Khim.*, **19**, 2650 (1970); A. Yamamoto, S. Kitazume, L.S. Pu, S. Ikeda, *J. Am. Chem. Soc.*, **93**, 371 (1971).
[16] J. Schwartz, J. B. Cannon, *J. Am. Chem. Soc.*, **96**, 4721 (1974).
[17] L. S. Pu, A. Yamamoto, S. Ikeda, *J. Am. Chem. Soc.*, **90**, 7170 (1968).
[18] S. Tyrlik, H. Stepowska, *Tetrahedron Lett.*, 3593 (1969).
[19] H. Yamazaki, N. Hagihara, *J. Organomet. Chem.*, **7**, P22 (1967); **21**, 431 (1970).
[20] Y. Wakatsuki, O. Nomura, K. Kitaura, K. Morokuma, H. Yamazaki, *J. Am. Chem. Soc.*, **105**, 1907 (1983).
[21] K. Yasufuku, A. Hamada, K. Aoki, H. Yamazaki, *J. Am. Chem. Soc.*, **102**, 4363 (1980) and references cited therein.
[22] K. P. C. Vollhardt, *Acc. Chem. Res.*, **10**, 1 (1977); R. L. Funk, K. P. C. Vollhardt, *J. Am. Chem. Soc.*, **101**, 215 (1979).
[23] *Japanese Patent*, S55-22418.
[24] (a) J. K. Ruff and W. J. Schlientz, *Inorg. Synth.*, **15**, 87 (1974).
(b) W. F. Edgell, J. Lyford, *Inorg. Chem.*, **9**, 1932 (1970).
[25] R. F. Heck, D. S. Breslow, *J. Am. Chem. Soc.*, **83**, 1097 (1961); W.R. McClellan, H. H. Hoehn, H. N. Cripps, E. L. Muetterties, B. W. Howk, *J. Am. Chem. Soc.*, **83**, 1601 (1961).
[26] A. Misono, *Inorg. Synth.*, **12**, 12 (1970); A. Yamamoto, S. Kitazume, L. S. Pu, S. Ikeda, *J. Am. Chem. Soc.*, **93**, 371 (1971).
[27] Y. Wakatsuki, H. Yamazaki, *Inorg. Synth.*, **26**, 190 (1989); **26**, 191 (1989).
[28] (a) G. Wilkinson, F.A. Cotton, J. M. Birmingham, *J. Inorg. Nucl. Chem.*, **2**, 96 (1956).
(b) K. Jonas, E. Deffense, D. Habermann, *Angew. Chem. Int. Ed. Engl.*, **22**, 716 (1983); *Angew. Chem. Suppl.*, 1005 (1983).
[29] Y. Wakatsuki, H. Yamazaki, *Inorg. Synth.*, **26**, 195 (1989).
[30] P. Chini, S. Martinengo, G. Garlaschelli, *J. Chem. Soc., Chem. Commun.*, 709 (1972).
[31] C. K. Brown, G. Wilkinson, *J. Chem. Soc. A*, 2753 (1970).
[32] T. Mise, P. Hong, H. Yamazaki, *Chem. Lett.*, 439 (1980).
[33] T. Mise, P. Hong, H. Yamazaki, *Chem. Lett.*, 401 (1982).
[34] K. Doyama, T. Joh, K. Onitsuka, T. Shiohara, S. Takahashi, *J. Chem. Soc., Chem. Commun.*, 649 (1987).
[35] J. L. Vidal, E. G. Walker, *Inorg. Chem.*, **19**, 896 (1980); B. D. Dombek, *Adv. Catal.*, **22**, 325 (1983).
[36] D. Forster, *Adv. Organomet. Chem.*, **17**, 255 (1979).
[37] L. Cassar, P. E. Eaton, J. Halpern, *J. Am. Chem. Soc.*, **92**, 3515 (1970).
[38] L. Cassar, J. Halpern, *Chem. Commun.*, 1082 (1970).
[39] J. A. Osborn, F. H. Jardine, J. F. Young, G. Wilkinson, *J. Chem. Soc. A*, 1711 (1966).
[40] Y. Wakatsuki, *Yuuki Kinzoku Kagoubutsu*, Ed. A. Yamamoto, Tokyo Kagaku Dojin, Tokyo, p. 206 (1991).

[41] F. H. Lincoln, W. P. Schneider, J. E. Pike, *J. Org. Chem.*, **38**, 951 (1973).
[42] T. Nishiguchi, A. Kurooka, K. Fukuzumi, *J. Org. Chem.*, **39**, 2403 (1974).
[43] E. J. Corey, J. W. Suggs, *Tetrahedron Lett.*, 3775 (1975).
[44] I. Ojima, T. Kogure, Y. Nagai, *Tetrahedron Lett.*, 5035 (1972).
[45] K. Ohno, J. Tsuji, *J. Am. Chem. Soc.*, **90**, 99 (1968).
[46] J. Tsuji, *Organic Syntheses via Metal Carbonyls*, vol. 2, Ed. I. Wender and P. Pino, Wiley, New York (1977).
[47] D. H. Doughty, L. H. Pignolet, *J. Am. Chem. Soc.*, **100**, 7083 (1978).
[48] M. Murakami, H. Amii, Y. Ito, *Nature*, **370**, 540 (1994).
[49] (a) W. Keim, *J. Organomet. Chem.*, **8**, P25 (1967).
(b) W. Kein, *J. Organomet. Chem.*, **14**, 179 (1968).
(c) L. S. Hegedus, P. M. Kendall, S. M. Lo, J. R. Sheats, *J. Am. Chem. Soc.*, **97**, 5448 (1975).
[50] H. Singer, G. Wilkinson, *J. Chem. Soc. A*, 849 (1968); J. Ohshita, K. Furumori, A. Matsuguchi, M. Ishikawa, *J. Org. Chem.*, **55**, 3277 (1990).
[51] A. E. Shilov, *The Activation of Saturated Hydrocarbons by Transition Metal Complexes*, D. Reidel, Dordrecht (1984); R. H. Crabtree, *Chem. Rev.*, **85**, 245 (1985); W. D. Jones, *Activation and Functionalization of Alkanes*, Ed. C. L. Hill, Wiley, New York (1989); E. P. Wasserman, C. B. Moore, R. G. Bergman, *Science*, **255**, 315 (1992).
[52] T. Sakakura, T. Sodeyama, K. Sasaki, K. Wada, M. Tanaka, *J. Am. Chem. Soc.*, **112**, 7221 (1990).
[53] T. Fujii, Y. Saito, *J. Chem. Soc., Chem. Commun.*, 757 (1990).
[54] M. Gozin, A. Weisman, Y. Ben-David, D. Milstein, *Nature*, **364**, 699 (1993); M. Gozin, A. Weisman, M. Aizenberg, S.-Y. Liou, A. Weisman, Y. Ben-David, D. Milstein, *Nature*, **370**, 42 (1994); S. -Y. Liou, D. Milstein, *J. Am. Chem. Soc.*, **117**, 9774 (1995).
[55] A. Miyashita, H. Takaya, K. Toriumi, T. Ito, T. Souchi, R. Noyori, *J. Am. Chem. Soc.*, **102**, 7932 (1980).
[56] T. Hayashi, A. Katsumura, M. Konishi, K. Kumada, *Tetrahedron Lett.*, 425 (1979).
[57] K. Tani, T. Yamagata, S. Akutagawa, H. Kumobayashi, T. Takemori, H. Takaya, A. Miyashita, R. Noyori, S. Otsuka, *J. Am. Chem. Soc.*, **106**, 5208 (1984).
[58] I. Ojima, T. Kogure, M. Kumagai, *J. Org. Chem.*, **42**, 1671 (1977).
[59] W. D. Jones, F. J. Feher, *Organometallics*, **2**, 562 (1983); *J. Am. Chem. Soc.*, **106** 1650 (1984).
[60] R. A. Periana, R. G. Bergman, *Organometallics*, **3**, 508 (1984).
[61] M. Brookhart, S. Sabo-Etienne, *J. Am. Chem. Soc.*, **113**, 2777 (1991); M. Brookhart, E. Hauptman, *J. Am. Chem. Soc.*, **114**, 4437 (1992); E. Hauptman, S. Sabo- Etienne, P. S. White, M. Brookhart, J. M. Garner, P. J. Fagan, J. C. Calabrese, *J. Am. Chem. Soc.*, **116**, 8038 (1994).
[62] S. Martinengo, G. Giordano, P.Chini, *Inorg. Synth.*, **20**, 209 (1908).
[63] J. A. Osborn, G. Wilkinson, *Inorg. Synth.*, **10**, 67 (1967).
[64] N. Ahmad, J. J. Levison, S. D. Robinson, M. F. Uttley, *Inorg. Synth.*, **15**, 59 (1974).
[65] N. Ahmad, J. J. Levison, S. D. Robinson, M. F. Uttley, *Inorg. Synth.*, **15**, 58 (1974).
[66] R. Cramer, *Inorg. Synth.*, **15**, 14 (1974).
[67] J. Chatt, L. M. Venanzi, *J. Chem. Soc.*, 4735 (1957).
[68] J. A. McCleverty, G. Wilkinson, *Inorg. Synth.*, **8**, 211 (1966).
[69] H. Kang, C. H. Mauldin, T. Cole, W. Slegeir, K. Cann, R. Pettit, *J. Am. Chem. Soc.*, **99**, 8323 (1977).
[70] F. E. Paulik, J. F. Roth, *J. Chem. Soc., Chem. Commun.*, 1578 (1986).
[71] M. G. Thomas, B. F. Beier, E. L. Muetterties, *J. Am. Chem. Soc.*, **98**, 1296 (1976); G. C. Demitras, E. L. Muetterties, *J. Am. Chem. Soc.*, **99**, 2796 (1977).
[72] A. Yamamoto, *Organotransition Metal Chemistry—Fundamental Concepts and Applications*, Wiley, New York, p. 223 (1986).
[73] W. Strohmeier, H. Steigerwald, M. Lukacs, *J. Organomet. Chem.*, **144**, 135 (1978).
[74] J. E. Lyons, *J. Catal.*, **30**, 490 (1973).
[75] W. Strohmeier, E. Eder, *J. Organomet. Chem.*, **94**, C14 (1975).
[76] R. H. Crabtree, *Acc. Chem. Res.*, **12**, 331 (1979).
[77] C. Bianchini, K. G. Caulton, T. J. Johnson, A. Meli, M. Peruzzini, F. Vizza, *Organometallics*, **14**, 933 (1995).

[78] C. Bianchini, P. Barbaro, A. Meli, M. Peruzzini, A. Vacca, F. Vizza, *Organometallics*, **12**, 2505 (1993); C. Bianchini, A. Meli, M. Peruzzi, F. Vizza, P. Frediani, V. Herrera, R. A. Sanchez-Delgado, *J. Am. Chem. Soc.*, **115**, 2731 (1993).
[79] G. G. Hlatky, R. H. Crabtree, *Coord. Chem. Rev.*, **65**, 1, 1985.
[80] M. G. Clerici, S. DiGioacchino, F. Maspero, E. Perrotti, A. Zanobi, *J. Organomet. Chem.*, **84**, 379 (1975).
[81] U. Klabunde, G. W. Parshall, *J. Am. Chem. Soc.*, **94**, 9081 (1972).
[82] A. H. Janowicz, R. G. Bergman, *J. Am. Chem. Soc.*, **104**, 352 (1982); **105**, 3929 (1983).
[83] J. K. Hoyano, W. A. G. Graham, *J. Am. Chem. Soc.*, **104**, 3723 (1982).
[84] B. A. Arndtsen, R. G. Bergman, *Science*, **270**, 1970 (1995).
[85] R. H. Crabtree, M. F. Mellea, J. M. Mihelcic, J. M. Quirk, *J. Am. Chem. Soc.*, **104**, 107 (1982).
[86] L. Malatesta, G. Caglio, M. Angoletta, *Inorg. Synth.*, **13**, 95 (1972).
[87] J. L. Herde, J. C. Lambert, C. V. Senoff, *Inorg. Synth.*, **15**, 18 (1974).
[88] K. Vrieze, J. P. Collman, C. T. Sears, Jr., M. Kubota, *Inorg. Synth.*, **11**, 101 (1968).

12 Group 10 (Ni, Pd, Pt) Metal Compounds

F. Ozawa, *Osaka City University*

12.1 Introduction

Organometallic complexes in this group possess two stable oxidation states, the +2 state and the zerovalent state. For platinum the +4 state is also common. Organoplatinum and palladium complexes are relatively stable toward oxidation, while organonickel complexes readily decompose in air, particularly in solution.

It has been well documented that organonickel and -palladium complexes are powerful tools for organic transformations in catalytic as well as stoichiometric systems, particularly for carbon–carbon bond formation [1,2]. In practice, such complexes are prepared *in situ* from stable metal species and organic reagents and are employed in organic synthesis without isolation. However, it is possible to isolate reactive organometallic species by introducing appropriate supporting ligands such as tertiary phosphines and aromatic nitrogen compounds. Studies on the structures and reactivity of the isolated organometallic species provide crucial informations which clarify the mechanisms of organic transformations and assist in the design of new synthetic organic reactions.

12.2 Nickel Complexes

12.2.1 Nickel(0) Olefin and Phosphine Complexes

The most common synthetic route to zerovalent organonickel complexes is the reduction of nickel(II) salts in the presence of appropriate ligands. Bis(2,4-pentanedionato)nickel(II) ($Ni(acac)_2$) is a typical starting material. Treatment of $Ni(acac)_2$ with triethylaluminum produces diethylnickel(II) species, which may be isolated in the presence of stabilizing ligands such as 2,2′-bipyridine (bipy) and 1,2-bis(diphenylphosphino)ethane (dppe) [3,4]. However, diethylnickel complexes thus formed are generally very unstable and readily decompose to give Ni(0) complexes (eq (1)) [5–7].

The reduction of nickel(II) dichloride with trialkyl phosphite is another synthetic route to Ni(0) complexes (eq (2)) [8]. This reaction involves the oxidation of trialkyl phosphite to trialkyl phosphate.

Synthesis of Organometallic Compounds: A Practical Guide. Edited by S. Komiya

$$\mathrm{Ni(acac)_2 + AlEt_3 + nL} \xrightarrow{[L = bipy]} \mathrm{(bipy)NiEt_2};\quad \xrightarrow{[L = PPh_3]} \mathrm{Ni(PPh_3)_4};\quad \xrightarrow{[L = cod]} \mathrm{Ni(cod)_2};\quad \xrightarrow{[L = PPh_3,\ C_2H_4]} \mathrm{(Ph_3P)_2Ni(C_2H_4)};\quad \xrightarrow{[L = dppe]} \mathrm{(dppe)NiEt_2} \quad (1)$$

$$\mathrm{NiCl_2 + 5P(OEt)_3 + 2Et_2NH \longrightarrow Ni\{P(OEt)_3\}_4 + (EtO)_3PO + 2Et_2NH_2Cl} \quad (2)$$

Bis(1,5-cyclooctadiene)nickel(0) ($Ni(cod)_2$) prepared in eq (1) is a versatile precursor to Ni(0) complexes. The cod ligand is readily replaced by group 15 donors and/or unsaturated organic molecules such as olefins, acetylenes, and isocyanides to give a wide variety of Ni(0) complexes (eq (3)) [9,10].

Nickel(0) complexes have been utilized as stoichiometric reagents for homocoupling of organic halides. Aryl, alkenyl, allyl, and benzyl halides can be employed in this reaction. Oxidative addition of organic halide to a Ni(0) species forms an organonickel(II) halide (**1**), which further reacts with another molecule of organic halide to give the coupling product and a nickel(II) dihalide (eq (4)). This reaction was originally developed with $Ni(CO)_4$ (eq (5)) [11], but nowadays is carried out with the less toxic compounds $Ni(cod)_2$ and $Ni(PPh_3)_4$ (eq (6) and (7)) [12,13].

In the presence of an excess of zinc, the homocoupling reaction proceeds catalytically with respect to nickel (eq (8)) [14]. A typical example is given in eq (9) [15].

Zerovalent nickel complexes promote the intramolecular coupling of organic halides with enolates and enones (eq (10) and (11)) [16,17]. These reactions are initiated by oxidative addition of organic halides to Ni(0) species to give monoorganonickel(II) complexes.

$$Ni(cod)_2 \xrightarrow{+\,bipy} Ni(cod)(bipy)$$
$$Ni(cod)_2 \xrightarrow{+\,4RNC} Ni(RNC)_4 \quad [R = cyclo\text{-}C_6H_{11},\ tert\text{-}Bu]$$
$$Ni(cod)_2 \xrightarrow{+\,2L} [NiL_2] \quad [L = bipy, phen, dmpe, dppe]$$
$$Ni(cod)_2 \xrightarrow{+\,4PR_3} Ni(PR_3)_4 \quad [R = Me, Et, Bu, OMe, OEt, OPh]$$
$$Ni(cod)_2 \xrightarrow{+\,olefin,\ 2PR_3} [Ni(olefin)(PR_3)_2] \qquad (3)$$
$$Ni(cod)_2 \xrightarrow{+\,3nbd} Ni(nbd)_3$$

$$RX + Ni(0)L_n \longrightarrow \underset{\mathbf{1}}{L_nNi(R)(X)} \xrightarrow{R'X} R{-}R' + NiX_2L_n \qquad (4)$$

$$\xrightarrow[70\text{–}75\%]{Ni(CO)_4} \qquad (5)$$

$$\xrightarrow[DMF,\ 52\%]{Ni(PPh_3)_4} \qquad (6)$$

$$n\ Br{-}C_5H_3N{-}Br \xrightarrow[PPh_3]{Ni(cod)_2} {-}(C_5H_3N)_n{-} \qquad (7)$$

$$L_nNi(0) \underset{Zn \;\; ZnX_2}{\overset{R\text{–}X \;\; R\text{–}R}{\rightleftarrows}} L_nNiX_2 \qquad (8)$$

$$2\ \text{MeO}_2\text{C–C}_6\text{H}_4\text{–Br} \xrightarrow[\text{Zn/KI/NMP}]{\text{NiCl}_2(\text{PEt}_3)_2\ (4\ \text{mol\%})} \text{MeO}_2\text{C–C}_6\text{H}_4\text{–C}_6\text{H}_4\text{–CO}_2\text{Me}\ \ 96\% \qquad (9)$$

$$\xrightarrow{\text{Ni(PPh}_3)_4} \qquad (10)$$

$$\xrightarrow[39\%]{\text{Ni(PPh}_3)_4} \qquad (11)$$

Catalytic co-oligomerization of unsaturated hydrocarbons is another important application of nickel(0) complexes [1a,b]. The reaction of 1,3-butadiene has been extensively studied by Wilke and his co-workers and it has been established that the structure of intermediate nickel species and the reaction course may be effectively controlled by the proper choice of ligand added to the catalytic system (eq (12)). The catalytically active nickel species can be prepared either by reduction of a nickel(II) salt (e.g., $Ni(acac)_2$) with alkylaluminum in the presence of butadiene and a ligand, or by ligand displacement of a nickel(0) complex such as $Ni(cod)_2$ with butadiene and a ligand. A nickel(0) complex containing two butadiene molecules and one ligand (L) (complex **2** in eq (12)) initially undergoes intramolecular rearrangement to give either a bis(π-allyl)nickel (**3**) or a (π-allyl)(σ-allyl)nickel intermediate (**4**). A complex of type **3** has been isolated using $P(OC_6H_4$-*o*-$Ph)_3$ and a type **4** complex using tricyclohexylphosphine as the supporting ligand. The π-allyl and σ-allyl forms are interconvertible depending upon the reaction conditions and the reaction products are formed from the intermediate nickel(II) species via reductive elimination and β-hydrogen elimination processes.

Intramolecular co-oligomerization of dienes and/or alkynes has been applied to the construction of bicyclic carbon skeletons (eq (13) and (14)) [18,19].

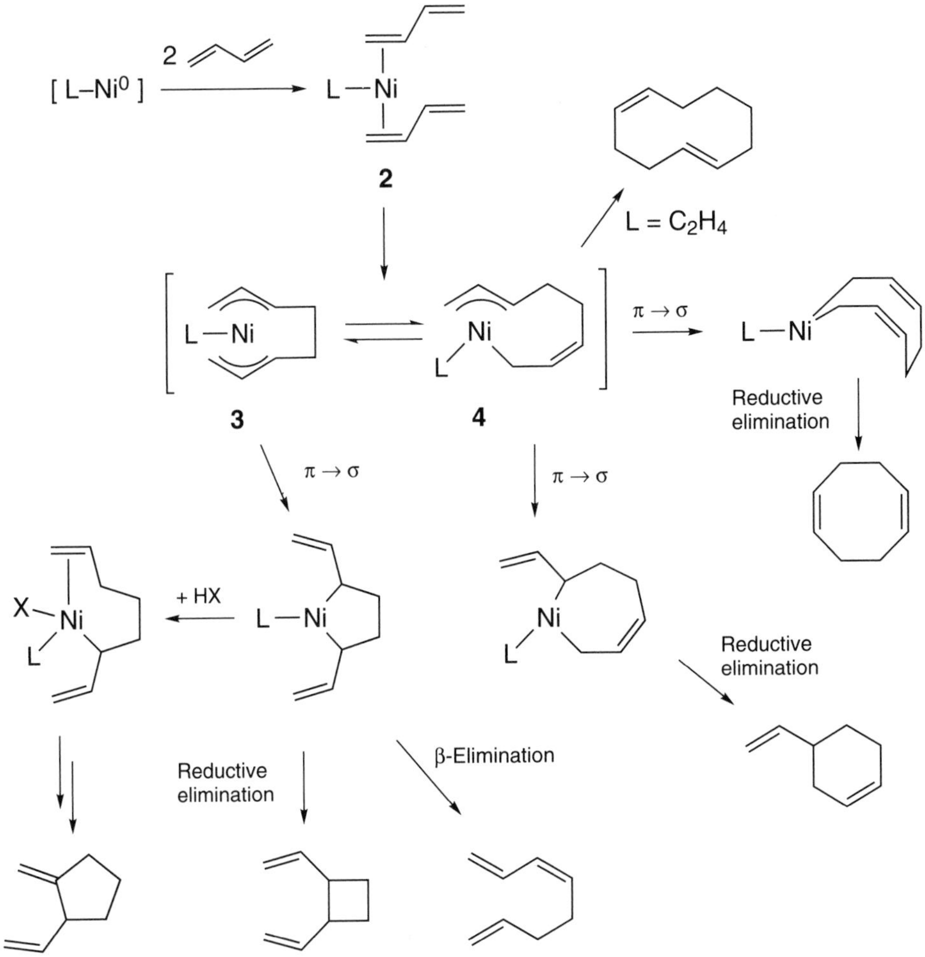

2
[L–Ni⁰]
L—Ni
2
L = C2H4
L—Ni
Ni
L
3
4
π → σ
L—Ni
Reductive elimination
π → σ
π → σ
X
Ni
L
+ HX
L—Ni
Ni
L
Reductive elimination
Reductive elimination
β-Elimination

(12)

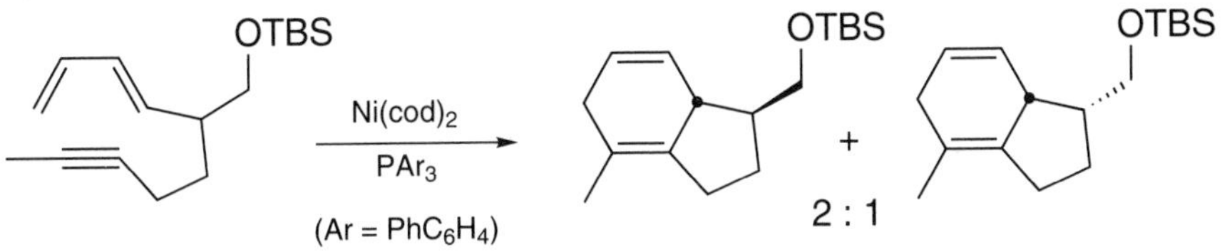

OTBS
Ni(cod)2
PAr3
(Ar = PhC6H4)
OTBS
+
2 : 1
OTBS

(13)

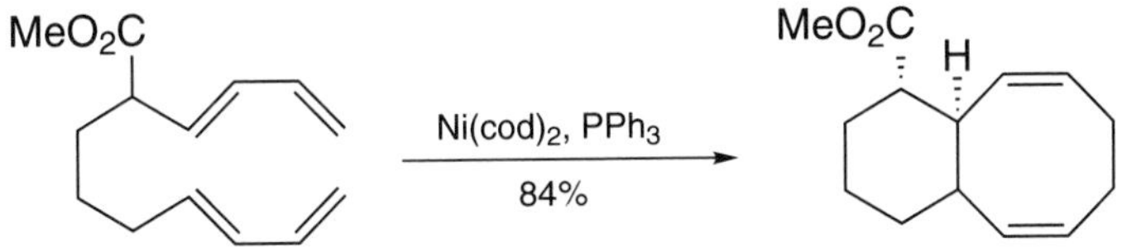

MeO2C
Ni(cod)2, PPh3
84%
MeO2C
H

(14)

12.2.2 Dialkylnickel(II) Complexes

Dialkylnickel(II) complexes in the type NiR_2L_2 have been prepared by the treatment of $Ni(acac)_2$ with $AlR_2(OEt)$ and a supporting ligand or by alkylation of bis(tertiary phosphine)nickel(II) dihalides with alkyllithium reagents (Table 12.1). The complexes bearing monophosphine ligands have a trans configuration and those with bidentate ligands have a cis geometry.

Alkylation of dibromobis(triethylphosphine)nickel(II) has been conducted stepwise using arylmagnesium bromide and methyllithium, leading to unsymmetrical *trans*-aryl(methyl)nickel complex **5** [24], which further reacts with 1,2-bis(dimethylphosphino)ethane (dmpe) to give *cis*-aryl(methyl)nickel complex **6** (eq (15)) [25].

$$trans\text{-}NiBr_2(PEt_3)_2 \xrightarrow{+\ ArMgBr} trans\text{-}NiAr(Br)(PEt_3)_2 \xrightarrow{+\ MeLi} trans\text{-}NiAr(Me)(PEt_3)_2\ (\mathbf{5}) \xrightarrow[-\ 2PEt_3]{+\ dmpe} cis\text{-}NiAr(Me)(Me_2PCH_2CH_2PMe_2)\ (\mathbf{6}) \quad (15)$$

(Ar = Ph, MeC_6H_4-, $MeOC_6H_4$-, etc.)

Reductive elimination of **5** and **6** has been studied in detail in connection with the nickel-catalyzed cross-coupling reaction of aryl halides (ArX) and alkylmagnesium halides (RMgX) (eq (18)). The reaction of *trans*-**5** is induced by addition of aryl halides (ArX) to the system, where the reductive elimination process involving para-

Table 12.1 Examples of NiR_2L_2 Complexes

R	L	method	ref
Me	PMe_3	a, b	20a, 21
$2\text{-}MeC_6H_4$	PMe_3	a	22
$Bu^tC{\equiv}C$	PMe_3	a	20b
Me	PEt_3	b	4
Me	dppe	b	4
Et	dppe	b	4
Me	bipy	b	23
Et	bipy	b	3

Method a: $NiCl_2L_2 + 2RLi$. Method b: $Ni(acac)_2 + AlR_2(OEt) + 2L$.
dppe = 1,2-bis(diphenylphosphino)ethane, bipy = 2,2'-bipyridine.

magnetic Ni(I) and Ni(III) species is operative (eq (16)) [24]. On the other hand, the reaction of *cis*-**6** is effectively accelerated by addition of tertiary phosphine ligands (L). In this case, the reductive elimination mechanism via a five-coordinate intermediate has been proposed (eq (17)) [25].

$$\textbf{5}\ trans\text{-}Ni(Ar)(Me)(PEt_3)_2 \xrightarrow[-\ ArNiX(PEt_3)_2]{+\ ArX} Ar\text{–}Me \qquad (16)$$

$$\left[\ \textbf{5} \longrightarrow [Ar_2Ni^{III}Me] \longrightarrow Ar\text{–}Me + [ArNi^{I}];\quad [ArNi^{I}] \xrightarrow{+\ ArX} [Ar_2Ni^{III}X];\quad [Ar_2Ni^{III}X] + \textbf{5} \longrightarrow ArNi^{II}X(PEt_3)_2 + [Ar_2Ni^{III}Me]\ \right]$$

$$\textbf{6}\ cis\text{-}Ni(Ar)(Me)(Me_2PCH_2CH_2PMe_2) \underset{-\ L}{\overset{+\ L}{\rightleftharpoons}} \left[Ni(Ar)(Me)(L)(Me_2PCH_2CH_2PMe_2) \right] \longrightarrow Ar\text{–}Me + (Me_2PCH_2CH_2PMe_2)Ni\text{–}L \qquad (17)$$

$$Ar\text{–}X + RMgX \xrightarrow{NiX_2(PR'_3)_2} Ar\text{–}R + MgX_2 \qquad (18)$$

In the catalytic cross-coupling reactions, nickel(II) dihalides bearing tertiary phosphine ligands are generally employed as the catalyst precursors. The dppp-coordinated complex $NiCl_2$(dppp) is among the most efficient catalysts, particularly for the reactions using alkylmagnesium reagents bearing β-hydrogens (eq (19) and (20)) [26,27].

Similar cross-coupling reactions can be conducted with palladium catalysts (see Section 13.4.2.2). While the palladium systems are more widely applicable, the nickel

$$o\text{-}C_6H_4Cl_2 + 2BuMgBr \xrightarrow[79\text{–}83\%]{NiCl_2(dppp)} o\text{-}C_6H_4Bu_2 \qquad (19)$$

Cl N Cl + $BrMgCH_2CH(CH_2)_8MgBr$ Me — $NiCl_2(dppp)$, 20% → N (20)

systems are still useful because of their high reactivity toward aryl chlorides and vinyl ethers, which are poorly reactive in the palladium systems (eq (21) and (22)) [28,29].

O + MeMgBr — $NiCl_2(PPh_3)_2$ → HO 96% (21)

S S $SiMe_3$ + MeMgBr — $NiCl_2(PPh_3)_2$ → $SiMe_3$ (22)

The Grignard cross-coupling reactions have been applied to asymmetric syntheses using chiral phosphine ligands. For the asymmetric alkylation using alkylmagnesium halides, chiral aminophosphine ligands **7** and **8** lead to high enantioselectivity (up to 94% ee) (eq (23)) [30].

MgCl Me + Br — $NiBr_2$–L* → Me (23)

up to 94% ee

L* = Me H Me_2N Ph_2P Fe **7** or *t*-Bu H Me_2N PPh_2 **8**

12.2.3 (π-Allyl)nickel(II) Complexes

(π-Allyl)nickel(II) complexes are synthesized by the oxidative addition of allylic halides or acetates to Ni(0) complexes, typically $Ni(cod)_2$ (eq (24) and (25)) [31,32]. The complex formed by the reaction of $Ni(cod)_2$ with allyl acetate is very unstable in solution and readily undergoes disproportionation to give bis(π-allyl)nickel(II) and nick-

$$Ni(cod)_2 + CH_2{=}CHCH_2Br \xrightarrow{\text{benzene}} [(\pi\text{-allyl})NiBr]_2 \quad (24)$$

$$Ni(cod)_2 + CH_2{=}CHCH_2OAc \xrightarrow[Et_2O]{+L} (\pi\text{-allyl})Ni(OAc)L \quad (25)$$

$L = PPh_3, PEtPh_2, PCy_3$

el(II) diacetate [32]. In the presence of tertiary phosphine ligands, on the other hand, the π-allyl acetate complexes can be successfully isolated as yellow crystalline solids (eq (25)). Bis(π-allyl)nickel(II) has been prepared by the treatment of nickel(II) dichloride with allylmagnesium chloride [33].

The complexes derived from allyl halides have been utilized as stoichiometric reagents for allylation of organic halides and ketones (eq (26) and 27)) [34–37].

(26) [scheme: MeO, MeO, N, Br, OMe, MeO + MeO–(π-allyl)NiBr dimer (2) → MeO, MeO, N, OMe, OMe, MeO; H^+/H_2O → MeO, MeO, N, O, OMe, MeO; 73%]

(27) [scheme: cyclohexanone (O) + EtO_2C–(π-allyl)NiBr dimer (2) → spiro α-methylene lactone (O, O); 80%]

12.2.4 Nickelacycle Complexes

Nickelacarbocyclic compounds can be synthesized by the addition of alkenes or alkynes to η^2-alkene or η^2-alkyne complexes of nickel (eq (28)) [38], the oxidative addition of a α,ω-dibromoalkane to a nickel(0) complex (eq (29)) [39], and the reaction of nickel(II) dihalides with organodilithium or organodimagnesium compounds (eq (30)) [40]. Treatment of a nickelacycle with carbon monoxide affords a cyclic ketone (eq (28)). The reaction with 1,1-dibromoalkane forms a polycyclic hydrocarbon (eq (29)).

$$(Ph_3P)_2Ni(CH_2{=}CH_2) + \text{1,7-octadiene} \xrightarrow{-CH_2=CH_2} (Ph_3P)_2Ni(\text{bicyclic nickelacycle}) \xrightarrow{+CO} \text{bicyclic ketone} \quad (28)$$

$$Br(CH_2)_4Br \xrightarrow[4bipy]{2Ni(cod)_2} \text{(CH}_2)_4\text{Ni(bipy)} \xrightarrow{\text{7,7-dibromobicyclo[4.1.0]heptane}} \text{spiro product (69\%)} \quad (29)$$

$$NiCl_2(PPh_3)_2 + Li(CH_2)_4Li \longrightarrow (CH_2)_4Ni(PPh_3)_2 \quad (30)$$

The thermolysis of nickelacyclopentane shows interesting behavior (eq (31)) [40]. The thermolysis product varies significantly depending upon the number of phosphine ligands bound to nickel.

$$(CH_2)_4NiL \underset{-L}{\overset{+L}{\rightleftharpoons}} (CH_2)_4NiL_2 \underset{-L}{\overset{+L}{\rightleftharpoons}} (CH_2)_4NiL_3 \quad (31)$$

$$(CH_2)_4NiL \rightarrow \text{1-butene}; \quad (CH_2)_4NiL_2 \rightarrow \text{cyclobutane}; \quad (CH_2)_4NiL_3 \rightarrow 2\ CH_2{=}CH_2$$

$Ni(cod)_2$ reacts with α,β-unsaturated amides in the presence of PCy_3 to give an azanickelacyclic compound (eq (32)) [41]. A similar complex has been synthesized by the co-cyclization of propylene and phenyl isocyanate with a nickel(0) phosphine complex (eq (33)) [42].

$$Ni(cod)_2 + PCy_3 + CH_2{=}C(CH_3)CONH_2 \longrightarrow Cy_3PNi[\text{metallacycle}] \quad (32)$$

$$Ni(cod)_2 + 2PEt_3 + PhN{=}C{=}O + CH_2{=}CHCH_3 \longrightarrow (Et_3P)_2Ni[\text{metallacycle}] \quad (33)$$

The nickel(0)-promoted co-cyclization of unsaturated organic molecules has been extended to useful catalytic reactions, where nickelacycle formation similar to eq (28) and (33) has been postulated (eq (34)–(36)) [43–45].

$$\text{enyne} + ArNC \xrightarrow[81\%]{Ni(cod)_2,\ PBu_3} \text{bicyclic imine (=NAr)} \quad (34)$$

$$\text{diyne} + PrCHO \xrightarrow[90\%]{Ni(cod)_2,\ PCy_3} \text{pyran product} \quad (35)$$

$$\text{diyne} + CO_2 \xrightarrow[88\%]{Ni(cod)_2,\ PCy_3} \text{pyrone product} \quad (36)$$

12.3 Synthesis of Ni Compounds

12.3.1 Starting Nickel Complexes

(1) Bis(2,4-pentanedionato)nickel(II), $Ni(acac)_2$

While the title complex can be obtained from commercial sources, it generally has water of crystallization. The anhydrous complex, which is required for the syntheses of nickel complexes described below, can be obtained by azeotropic distillation of the water with toluene using a Dean–Stark apparatus. The dark-green slurry is filtered while still hot through Celite under nitrogen, and the filtrate is evaporated to dryness on a rotary evaporator. The green solid residue is crushed in a mortar and dried at 80 °C under vacuum (< 0.1 mmHg) for 16 h to give the anhydrous complex.

Table 12.2. Preparation of NiX_2L_2 Complexes

X	L	mp (°C)	ref	X	L	mp (°C)	ref
Cl	PMe_3	199–200	47	Br	PPh_3	222–225	46,54
Cl	PEt_3	112–113	48	I	PPh_3	218–220	54
Cl	PPr_3	92–93	48	Cl	dmpe		55
Cl	$P(i\text{-}Pr)_3$	185–187	49,50	Cl	dppe		55
Cl	PBu_3	48–49	48	Br	dppe		55,56
Cl	PCy_3	227	50	I	dppe		56
Cl	PMe_2Ph	152–153	51,52	Cl	dppp		56
Cl	$PMePh_2$	148–150	53	Br	dppb		57
Cl	PPh_3	247–250	46	Cl	dppf		58

dmpe = $Me_2P(CH_2)_2PMe_2$; dppe = $Ph_2P(CH_2)_2PPh_2$; dppp = $Ph_2P(CH_2)_3PPh_2$; dppb = $Ph_2P(CH_2)_4PPh_2$; dppf = 1,1′-bis(diphenylphosphino)ferrocene.

(2) Dibromobis(triphenylphosphine)nickel(II), $NiBr_2(PPh_3)_2$ [46]

$$NiBr_2{\cdot}3H_2O + 2PPh_3 \longrightarrow NiBr_2(PPh_3)_2 + 3H_2O \qquad (37)$$

To a hot solution of $NiBr_2{\cdot}3H_2O$ (5.45 g, 20 mmol) in BuOH (100 mL) is added a hot solution of PPh_3 (10.4 g, 40 mmol) in BuOH (100 mL). Cooling the mixture results in the precipitation of dark green crystals of the title compound, which is collected by filtration, washed with BuOH and then with Et_2O, and dried under vacuum (72%). Further purification may be performed by recrystallization from hot BuOH.

Properties: mp 222–225 °C.

A series of $NiX_2(PR_3)_2$ complexes can be prepared by mixing an alcoholic solution of nickel(II) dihalide with a solution of tertiary phosphine ligand in alcohol or benzene. Typical examples are listed in Table 12.2.

12.3.2 Nickel(0) Complexes

(3) Bis(η^4-1,5-cyclooctadiene)nickel(0), $Ni(cod)_2$ [5]

$$Ni(acac)_2 + 2cod + 2AlEt_3 \longrightarrow Ni(cod)_2 + 2AlEt_2(acac) + C_2H_4 + C_2H_6 \qquad (38)$$

Peroxide impurities are removed from 1,5-cyclooctadiene by filtration through grade 1 neutral alumina; the filtration is repeated until the alumina is no longer colored yellow. Further purification is not necessary, but the liquid should be used immediately. Butadiene is dried by passing it through a U trap containing Type 3A molecular sieves, condensing the gas at –78 °C and storing it in a small stainless-steel cylinder.

A 1 L, four-necked flask is equipped with a gas inlet tube, a mechanical stirrer, and a dry ice condenser topped with a three-way stopcock which is connected to a source of a dry nitrogen gas and a mineral-oil bubbler. $Ni(acac)_2$ (103 g, 0.40 mol) is placed in the flask, and the system is replaced with nitrogen. Toluene (250 mL) and 1,5-cyclooctadiene (216 g, 2.0 mol) are added, and the mixture is stirred and cooled to –10 °C. The condenser is filled with a mixture of dry ice and acetone, and approximately 18 g of anhydrous butadiene is introduced through the gas inlet tube and dissolved to the cold mixture. The gas inlet tube is replaced with a nitrogen-flushed 250 mL dropping funnel, and a solution of triethylaluminum (103 g, 0.9 mol) in 100 mL of toluene is placed in the dropping funnel. The triethylaluminum is added dropwise to the stirred mixture with maintaining the temperature at –10 to 0 °C. During the addition, the green slurry becomes yellow-brown, and a yellow crystalline solid is formed. The mixture is stirred at 0 °C for 1 h and then allowed to warm to room temperature and is stirred overnight. The yellow slurry is cooled to –30 °C, and the solid is collected by filtration, washed several times with cold Et_2O, and dried under vacuum. The crude product is dissolved in toluene (500–700 mL) and 1,5-cyclooctadiene (*ca.* 10 mL) at 50–60 °C, filtered though a glass-fiber extraction thimble, and cooled to –20 °C to give yellow crystals of the title compound (*ca.* 60%).

Properties: ^1H NMR (C_6D_6) δ 1.38 (s, 8H, CH_2), 3.64 (s, 4H, CH)

The solid compound is decomposed after several minutes in the air.

(4) (η^2-Ethylene)bis(tricyclohexylphosphine)nickel(0), $Ni(C_2H_4)(PCy_3)_2$ [6]

$$\begin{aligned}&Ni(acac)_2 + C_2H_4 + 2PCy_3 + AlEt_2(OEt)\\&\longrightarrow Ni(C_2H_4)(PCy_3)_2 + Al(OEt)(acac)_2 + C_2H_4 + C_2H_6\end{aligned} \quad (39)$$

Anhydrous $Ni(acac)_2$ (4.45 g, 17.3 mmol) and tricyclohexylphosphine (9.7 g, 35 mmol) are added to a three-necked flask equipped with a gas inlet tube, a 100 mL dropping funnel, and a three-way stopcock which is connected to a source of argon and a mineral-oil bubbler. The system is replaced with argon. Anhydrous benzene (*ca.* 15 mL), the amount of which is sufficient to dissolve the reactants is added, followed by 70 mL of anhydrous Et_2O. The solution is stirred magnetically and a slow stream of ethylene is bubbled through the solution. A solution of diethylaluminum ethoxide (6.5 mL) in Et_2O (30 mL) is placed in the dropping funnel and added dropwise over the period of 30 min. The color of the solution changes from green to yellow-brown. The mixture is stirred overnight, and the resulting bright yellow precipitate of the title compound is collected by filtration, washed with Et_2O (2 × 25 mL), and dried under vacuum (7.1 g, 63%). Nickel(0) ethylene complexes bearing triethylphosphine, triphenylphosphine, and triarylphosphine ligands are prepared by similar procedures [59,60].

(5) Tetrakis(triphenylphosphine)nickel(0), $Ni(PPh_3)_4$ [7]

$$Ni(acac)_2 + 4PPh_3 + 2AlEt_3 \longrightarrow Ni(PPh_3)_4 + 2AlEt_2(acac) + C_2H_4 + C_2H_6 \quad (40)$$

Anhydrous $Ni(acac)_2$ (21.3 g, 83 mmol) and triphenylphosphine (125 g, 480 mmol) are added to a 2 L four-necked flask equipped with a mechanical stirrer, a 250 mL dropping funnel, a thermometer, and a three-way stopcock which is connected to a source of dry nitrogen gas and a mineral-oil bubbler. The atmosphere is replaced with nitrogen gas, and Et_2O (800 mL) is added. The mixture is stirred and cooled to 0 °C. A solution of triethylaluminum (28 g, 250 mmol) in Et_2O (100 mL) is place in the dropping funnel and added dropwise to the mixture over the period of 1–2 h while maintaining the temperature of mixture below +5 °C. (Note: Care must be taken in preparation of the Et_2O solution of triethylaluminum since an exothermic reaction occurs.) During the course of the reaction, the color of the mixture changes from green to brick-red, and a reddish brown solid forms. After the completion of addition, the mixture is stirred for 30 min at 5–10 °C and then for 1–2 h at 25 °C. The resulting precipitate is collected by filtration, washed several times with Et_2O, and dried under vacuum. The product is sufficiently pure for the use a catalyst and synthetic reagent.

Further purification is carried out by extraction with benzene (400–500 mL) containing triphenylphosphine (40 g) at 60 °C. To the dark red extract is added heptane (200 mL), and the solution is concentrated to *ca.* 300 mL. An additional 200 mL of heptane is added, and the resulting precipitate is collected by filtration and washed with heptane and Et_2O. This purification is repeated and the reddish brown crystalline solid is dried under vacuum for *ca.* 16 h (50 g, 55%).

The above reaction may be carried out using diethylaluminum ethoxide in place of triethylaluminum. The title compound can also be prepared by the reaction of $Ni(cod)_2$ with triphenylphosphine [9] or by the reduction of $NiCl_2$ with Zn powder in the presence of triphenylphosphine [61].

12.3.3 Dialkylnickel(II) Complexes

(6) *cis*-Diethyl(2,2′-bipyridine)nickel(II), $NiEt_2(bipy)$ [3]

$$Ni(acac)_2 + bipy + AlEt_2(OEt) \longrightarrow cis\text{-}NiEt_2(bipy) + Al(OEt)(acac)_2 \quad (41)$$

Anhydrous $Ni(acac)_2$ (10.0 g, 39 mmol) and 2,2′-bipyridine (10.0 g, 64 mmol) are placed in a 200 mL Schlenk tube equipped with a rubber septum, and the system is filled with nitrogen gas. Et_2O (100 mL) is added, and the mixture is stirred magnetically at 0 °C. Diethylaluminum ethoxide (30 mL) is added by syringe through the septum over the period of 10 min and then the system is allowed to warm to room temperature. After 8 h, the resulting dark green crystalline solid is collected by filtration, washed with Et_2O, and dried under vacuum. The crude product is recrystallized from hot acetone to give dark green needles of the title compound (55–65%).

Properties: ^1H NMR (DMF) δ 1.15 (t, 6H, CH_3), 0.82 (q, 4H, CH_2).

The solid compound is decomposed after several minutes in the air.

(7) *trans*-Dimethylbis(trimethylphosphine)nickel(II), $NiMe_2(PMe_3)_2$ [20a]

$$NiCl_2(PMe_3)_2 + 2MeLi \longrightarrow trans\text{-}NiMe_2(PMe_3)_2 + 2LiCl \quad (42)$$

Dichlorobis(trimethylphosphine)nickel(II) (1.0 g, 3.5 mmol) is placed in a 50 mL Schlenk tube and the flask is then filled with nitrogen. THF (20 mL) is added, and the mixture is stirred magnetically at –70 °C. A solution of MeLi in Et_2O (6 M, 12 mL) is added by syringe, and the system is allowed to warm to 0 °C. The solvent is removed by pumping at the same temperature, and the residue is extracted with cold pentane. The extract is gradually concentrated under reduced pressure to give yellow needles of the title compound (82%). Further purification may be performed by recrystallization from Et_2O.

(8) *trans*-Bromo(phenyl)bis(triethylphosphine)nickel(II), $NiPhBr(PEt_3)_2$ [24]

$$NiBr_2(PEt_3)_2 + PhMgBr \longrightarrow \textit{trans}\text{-}NiPhBr(PEt_3)_2 + MgBr_2 \quad (43)$$

A solution of PhMgBr in THF (1.0 M, 20 mL, 20 mmol) is placed in a 200 mL Schlenk tube and diluted with 50 mL of THF. The solution is cooled to 5 °C, and a solution of $NiBr_2(PEt_3)_2$ (6.0 g, 13 mmol) in benzene (40 mL) is added dropwise with stirring. The system is stirred for 30 min at the same temperature, and an aqueous solution of HBr (10%, 10 mL) is added dropwise. The organic layer is separated, dried over $MgSO_4$, and concentrated to dryness by pumping. The resulting solid is recrystallized from hot acetone to give reddish brown crystals of the title compound (3.8 g, 63%).

The complex is converted into *trans*-$NiMe(Ph)(PEt_3)_2$ on treatment with MeLi [24]. A series of monoorganonickel(II) halides of the type *trans*-$NiR(X)L_2$ (L = PPh_3, PEt_2Ph, PEt_3, PPr_3, PMe_2Ph; X = Cl, Br, I; R = *o*-MeC_6H_4, 2,6-$Me_2C_6H_3$, 2,4,6-$Me_3C_6H_2$) are prepared by similar procedures [62]. *trans*-$NiPh(Cl)(PPh_3)_2$ can be synthesized by the oxidative addition of PhCl to $Ni(PPh_3)_4$ [63]. A variety of monoalkylnickel(II) complexes are obtained by the treatment of dialkylnickels with an equimolar amount of HCl, phenol, and carboxylic acids [64].

12.3.4 π-Allylnickel(II) Complexes

(9) Di(μ-bromo)bis[η^3-2-(ethoxycarbonyl)allyl]dinickel(II), $[NiBr\{\eta^3\text{-}2\text{-}(EtO_2C)C_3H_4\}]_2$ [31]

$$2Ni(cod)_2 + 2CH_2{=}CH(CO_2Et)CH_2Br \longrightarrow EtO_2C\text{-}(\eta^3\text{-allyl})\text{-}Ni(\mu\text{-}Br)_2Ni\text{-}(\eta^3\text{-allyl})\text{-}CO_2Et + 2cod \quad (44)$$

$Ni(cod)_2$ (8.88 g, 32.4 mmol) is placed in a 100 mL Schlenk tube under a nitrogen atmosphere and dissolved in benzene (50 mL) at room temperature. Ethyl 2-(bromomethyl)acrylate (4.40 mL, 6.26 g, 32.4 mmol) is added dropwise to the stirred solution over the period of 30 min. During the addition, the color of the solution quickly turns from yellow to deep red. After the completion of addition, the mixture is stirred for an additional 30 min and filtered. The resulting red solution is gradually concentrated at room temperature under reduced pressure (ca. 25 mmHg) until a small amount of crystalline solid appears. Pentane (25 mL) is carefully added and the mixture is allowed to stand at 0 °C overnight to form red crystals of the title compound, which are filtered, washed two times with pentane, and dried under vacuum at room temperature (6.17 g, 70%). A variety of π-allylnickel halides are similarly prepared [31].

(10) Bis(η^3-allyl)nickel(II), $Ni(\eta^3\text{-}C_3H_5)_2$ [33]

$$NiCl_2 + 2C_3H_5MgCl \longrightarrow Ni(\eta^3\text{-}C_3H_5)_2 + 2MgCl_2 \quad (45)$$

Anhydrous nickel dichloride (4 g, 31 mmol) is added to a 500 mL Schlenk tube equipped with a dropping funnel, and the system is filled with argon. Et_2O (150 mL) is added and the resulting heterogeneous mixture is stirred and cooled to –30 °C. A solution of allylmagnesium chloride in Et_2O (0.5 M, 150 mL, 75 mmol) is added dropwise over a period of 2 h, and the mixture is stirred at –20 to –30 °C for 3 h, and then at –78 °C overnight. The solution is filtered at –60 °C, and the filtrate is concentrated to dryness. Owing to the volatility of the π-allylnickel, which co-distils with ether very readily, the solvent is best removed at –78 °C and a pressure of 10–2 mmHg. The resulting yellow-orange solid is extracted with pentane at –40 °C, and the extract is concentrated at –78 °C to give yellow crystals of the title compound (80%).

Properties: mp *ca* 0 °C; IR 1520 and 1493 cm^{-1}.

Since the compound is very air- and temperature-sensitive, it should be stored under argon or nitrogen at low temperature.

12.3.5 Nickelacycles

(11) Tetramethylenebis(triphenylphosphine)nickel(II), $Ni(CH_2)_4(PPh_3)_2$ [40]

$$NiCl_2(PPh_3)_2 + Li(CH_2)_4Li \longrightarrow (CH_2)_4Ni(PPh_3)_2 + 2LiCl \qquad (46)$$

Dichlorobis(triphenylphosphine)nickel(II) (5.0 g, 7.5 mmol) is placed in a 200 mL Schlenk tube equipped with a dropping funnel, and the system is filled with a nitrogen gas. Et_2O (100 mL) is added, and the mixture is stirred magnetically at –50 °C. A solution of 1,4-dilithiobutane in Et_2O (13 mmol, 63 mL) is added dropwise from the dropping funnel. The system is maintained at –50 °C during the addition and then slowly warmed to 0 °C. A bright yellow solid appears at –10 to 0 °C and is collected by filtration at –10 °C. The yellow solid is dissolved in toluene and filtered at –15 °C to remove LiCl and starting material. The toluene extract is concentrated under reduced pressure, diluted with hexane, and allowed to stand at –70 °C to give yellow crystals of the title compound (36%).

Properties: ^1H NMR δ 1.67 (m, 4H, CH_2), 1.87 (m, 4H, CH_2), 6.5 (m, 30H).

(12) (η^2-C^3,*N*-2-methylpropanamidato)(tricyclohexylphosphine)nickel(II), $Ni\{CH_2CH(Me)CONH\}(PCy_3)$ [41]

$$Ni(cod)_2 + PCy_3 + CH_2{=}CH(Me)CONH_2 \longrightarrow Ni\{CH_2CH(Me)C(O)NH\}PCy_3 + 2cod \qquad (47)$$

$Ni(cod)_2$ (1.7 g, 6.2 mmol), tricyclohexylphosphine (1.7 g, 6.2 mmol), and methacrylamide (1.1 g, 13 mmol) are placed in a 50 mL Schlenk tube and dissolved in THF (50 mL) at room temperature. The solution is stirred for 6 h to give a greenish yellow solid. After the mixture is allowed to stand at –30 °C for 2 h, the resulting solid is collected by filtration, washed several times with Et_2O, and dissolved in toluene at 60 °C. The solution is filtered and cooled to room temperature to provide yellow crystals of the title compound (1.8 g, 68%).

Properties: IR 3400 and 1572 cm^{-1}.

12.4 Palladium Complexes

12.4.1 Palladium(0) Complex

12.4.1.1 Palladium(0) Phosphine Complexes

The most widely used methods of preparation of palladium(0) phosphine complexes are the reduction of palladium(II) salts in the presence of tertiary phosphine ligands (eq (48)–(50)) [65–67]. Palladium(II) dichloride and bis(2,4-pentanedionato)palladium(II) ($Pd(acac)_2$) are the typical starting materials. A variety of reducing agents including hydrazine [65], sodium borohydride [68], sodium alkoxide [69], and triethylalminum [66] were employed.

$$PdCl_2 + 4PPh_3 + N_2H_4 \xrightarrow{DMF} Pd(PPh_3)_4 \tag{48}$$

$$Pd(acac)_2 + 4PPh_3 + AlEt_3 \xrightarrow{Et_2O} Pd(PPh_3)_4 \tag{49}$$

$$K_2[PdCl_4] + 5P(OEt)_3 + 2Et_2NH \longrightarrow Pd\{P(OEt)_3\}_4 + (EtO)_3PO + 2Et_2NH_2Cl \tag{50}$$

Another convenient method to synthesize palladium(0) phosphine complexes involves the reduction of CpPd(π-allyl) [70]. The starting π-allyl complex can be readily prepared and stored as a relatively stable crystalline solid. In the presence of tertiary phosphine ligands in solution, this complex undergoes reductive elimination to give palladium(0) phosphine complexes. The reaction proceeds cleanly under mild conditions without formation of inorganic by-products, leading to a facile synthesis of kinetically unstable species. This method was used to prepare coordinatively unsaturated bis(tertiary phosphine)palladium(0) complexes (eq (51)) [71].

$$CpPd(\pi\text{-allyl}) + 2L \longrightarrow PdL_2 \qquad L = PCy_3,\ PPh(t\text{-Bu})_2,\ P(t\text{-Bu})_3 \tag{51}$$

12.4.1.2 Palladium(0) Olefin Complexes

The synthetic reactions of palladium(0) olefin complexes bearing tertiary phosphine ligands may be classified into four types (eq (52)–(55)) [72–75]. The first type of reaction involves the reduction of $Pd(acac)_2$ with $AlEt_2(OEt)$ in the presence of olefins and phosphine ligands and it has been applied to the synthesis of palladium(0) ethylene complexes bearing PPh_3, PCy_3 and dppe ligands (eq (52)) [72].

$$Pd(acac)_2 + 2L + AlEt_2(OEt) + C_2H_4 \xrightarrow{Et_2O} L_2Pd{-}(C_2H_4) + C_2H_6 \quad (52)$$

$L_2 = (PPh_3)_2$, $(PCy_3)_2$, $\{P(OC_6H_4Me)_3\}_2$, $Ph_2PCH_2CH_2PPh_2$

The reaction in eq (52) proceeds via a diethylpalladium(II) intermediate, which undergoes β-hydrogen elimination to give Pd(ethylene)L_2 and ethane. This reaction sequence can be conducted stepwise using more basic and compact tertiary phosphine ligands. Thus the diethyl complexes have been synthesized with PMe_3, PEt_3, PMe_2Ph, PEt_2Ph, $PMePh_2$, and $PEtPh_2$ ligands [76]. The diethylpalladium(II) complexes decompose in solution at room temperature to provide palladium(0) ethylene complexes [76a]. In the presence of added olefins, the ethylene complex successively undergoes ligand displacement with olefins to provide a variety of Pd(olefin)L_2 complexes (eq (53)) [73].

$$trans\text{-}PdEt_2L_2 + \text{olefin} \longrightarrow Pd(\text{olefin})L_2 + C_2H_4 + C_2H_6 \quad (53)$$

olefin = $CH_2{=}CHPh$, $CH_2{=}CHCOOMe$, $CH_2{=}C(Me)COOMe$, $CH_2{=}C(Me)CN$

The third type of reaction involves ligand displacement of tetrakis(tertiary phosphine)-palladium(0) complexes with olefins (eq (54)) [74]. Since the coordination of olefin competes with the phosphine coordination in the reaction system, this route is only applicable to the olefins bearing highly electron-withdrawing substituents.

$$PdL_4 + \text{olefin} \longrightarrow Pd(\text{olefin})L_2 + 2L \quad (54)$$

olefin = MeOOC–CH=CH–COOMe, maleic anhydride, p-benzoquinone (O=C₆H₄=O)

The fourth type of reaction proceeds via the reductive elimination of a bis(allyl)-palladium(II) species (eq (55)) [75a]. Palladium(0) olefin complexes bearing bidentate phosphine ligands have been synthesized by this method. A similar reaction can be performed with CpPd(π-allyl) [75b].

$$(\eta^3\text{-allyl})_2Pd \xrightarrow{+\ R_2P(CH_2)_2PR_2} (R_2P(CH_2)_2PR_2)Pd(\eta^1\text{-CH}_2CH{=}CH_2)_2 \xrightarrow{+\ CH_2{=}CHR} (R_2P(CH_2)_2PR_2)Pd(CH_2{=}CHR) \quad (55)$$

As described in the previous section, $Ni(cod)_2$ is used as a versatile precursor to organonickel complexes (see Section 12.2). On the other hand, the corresponding palladium complex $Pd(cod)_2$ is too unstable to be handled [77]. Therefore, the more stable palladium(0) olefin complex $Pd_2(dba)_3$ (dba = dibenzylideneacetone) is generally employed instead (eq (56)) [78]. The dba complex can be isolated as air-stable purple crystals and stored at room temperature without decomposition. This complex was applied to the synthesis of various palladium(0) olefin complexes [79] and mono-organopalladium(II) complexes.

$$PdCl_2 + PhCH{=}CHCOCH{=}CHPh \xrightarrow{MeCOONa} Pd_2(PhCH{=}CHCOCH{=}CHPh)_3 \quad (56)$$

12.4.2 Palladium(II) Alkyl, Aryl, and Alkenyl Complexes

12.4.2.1 Synthesis via Oxidative Addition Reactions

Oxidative addition of organic halide to a palladium(0) complex is among the most frequently employed methods to prepare monoorganopalladium(II) complexes (eq (57)) [80]. Besides alkyl, benzyl, aryl and alkenyl halides, pseudo halides having aryl and alkenyl groups (e.g., triflates) also exhibit high reactivity toward the oxidative addition [81]. Detailed mechanistic studies on the reactions of aryl halides have revealed that the reactivity significantly decreases in the order iodides > triflates > bromides >> chlorides [81,82]. Although aryl chlorides are generally inactive toward the oxidative addition, their reactions proceed satisfactorily with palladium(0) complexes bearing a basic diphosphine ligand [82d,e].

$$PdL_n + R{-}X \longrightarrow RPd(X)L_2 + (n\text{-}2)L$$
$$R = \text{alkyl, aryl, alkenyl, benzyl; } X = Cl, Br, I, OTf \quad (57)$$

Monoorganopalladium(II) complexes are also synthesized by the transmetallation of dihalobis(tertiary phosphine)palladium(II) complexes with Grignard reagents (eq (58)) [83] or by the treatment of diorganopalladium complexes with inorganic and organic acids (eq (59)) [84]. The former reaction is useful to prepare alkylpalladium(II) halides and the latter to synthesize organopalladium complexes bearing anionic ligands other than halides.

$$PdBr_2(PEt_3)_2 + MeMgBr \longrightarrow trans\text{-}PdMe(Br)(PEt_3)_2 + MgBr_2 \quad (58)$$

$$PdMe_2(PMe_3)_2 + HOPh \longrightarrow trans\text{-}PdMe(OPh)(PMe_3)_2 + CH_4 \quad (59)$$

A variety of palladium-catalyzed organic reactions involve the oxidative addition process. A typical example is seen in the catalytic arylation and alkenylation of olefins (eq (60)) [85]. Aryl- and alkenylpalladium(II) complexes (**9**) formed by oxidative addition undergo olefin insertion into the palladium–carbon bond to give an alkylpalladium species (**10**), which provides arylated and alkenylated olefins via β-hydrogen elimination. The hydridopalladium species **11** thus generated is reduced to a Pd(0) species upon its interaction with a base and carries the sequence of reactions catalytically. This catalytic reaction was first discovered by Mizoroki [86] and Heck [87] independently and has often been called the Heck reaction. Typical examples are shown in eq (61) and (62) [88,89]. The catalytically active palladium(0) species are most frequently generated *in situ* from $Pd(OAc)_2$ and tertiary phosphine ligands [90].

$$[Pd(0)L_2] \xrightarrow{+RX} L_2Pd(R)(X)\ \mathbf{9} \xrightarrow{+\ R'CH=CH_2} L_2Pd(X)CH(R')CH_2R\ \mathbf{10} \longrightarrow L_2Pd(H)(X)\ \mathbf{11} + R'CH=CHR;\quad \mathbf{11} \xrightarrow[-\ base\cdot HX]{+\ base} [Pd(0)L_2] \tag{60}$$

$$C_6H_5I + CH_2=C(CH_3)CH_2OH \xrightarrow[MeCN]{Pd(OAc)_2,\ Et_3N} C_6H_5CH_2CH(CH_3)CHO\ \ (82\%) \tag{61}$$

$$\text{2-methylcyclohexenyl-OTf} + CH_2=CHCO_2Me \xrightarrow[Et_3N]{PdCl_2(PPh_3)_2} \text{2-methylcyclohexenyl-CH=CHCO}_2\text{Me}\ \ (84\%) \tag{62}$$

Intramolecular Heck reactions are useful for synthesizing cyclic compounds. When the reaction is performed with polyolefins, successive insertion of olefins into a Pd—C bond takes place, leading to polycyclic compounds (eq (63)) [91].

The Heck reaction has been applied to a catalytic asymmetric synthesis using a chiral diphosphine ligand (eq (64) and (65)) [92,93]. Generation of a cationic organopalladium intermediate is of particular importance to gain high enantioselectivity. Such a species can be produced by the oxidative addition of organic triflates or by the anionic exchange between organopalladium halides and silver salts [85c].

Under a CO atmosphere, monoorganopalladium(II) complexes undergo CO insertion into the Pd—C bond to give acylpalladium complexes (eq (66)) [94]. A kinetic study showed that the CO-insertion into *trans*-$PdAr(X)L_2$ type complexes proceeds

Cl — $Pd(PPh_3)_4$, NEt_3 → [PdCl] → [PdCl] — $-Pd^0$, $-HCl$ → 57% (63)

OTBDMS, OTf — $PdCl_2[(R)\text{-BINAP}]$, $i\text{-}Pr_2NEt$ → OTBDMS, H, 92% ee (64)

PPh_2, PPh_2, (*R*)-BINAP

O, NMe, O, O, I — $Pd_2(dba)_3$, (*R*)-BINAP, Ag_3PO_4 → O, NMe, O, O, 71% ee (65)

via a four- or five-coordinate intermediate, $PdAr(CO)(X)L$ or $PdAr(CO)(X)L_2$; the reaction course depends upon the electronic property of the aryl ligand. Acylpalladium complexes have been prepared also by the oxidative addition of acyl halides to palladium(0) phosphine complexes [80].

$$\text{Ph-Pd(PPh}_3)_2\text{-I} + \underset{(1\ \text{atm})}{\text{CO}} \xrightarrow{\text{room temp.}} \text{PhCO-Pd(PPh}_3)_2\text{-I} \quad \mathbf{12} \tag{66}$$

The benzoylpalladium iodide (**12**) prepared by eq (66) has been proven to be involved in the palladium-catalyzed single and double carbonylation of phenyl iodide, giving benzoic and benzoylformic acid derivatives. For example, complex **12** reacts with diethylamine under CO pressure to give $PhCOCONEt_2$ as the double carbonylation product exclusively, whereas the reaction of **12** with ethanol in the presence of a base affords PhCOOEt as the single carbonylation product. Detailed mechanistic studies indicated the double carbonylation process illustrated in eq (67) [95]. Coordination of CO to **12** followed by nucleophilic attack of amine on the carbonyl ligand in **13** forms a bis-acyl intermediate (**14**), which reductively eliminates the double carbonylation product. The presumed intermediates **13** and **14** have been isolated using basic tertiary phosphine ligands [96]. On the other hand, single carbonylation (ester formation) was found to involve a benzoyl(alkoxo)palladium intermediate **15** (eq. (68)) [95a].

$$PhCO{-}PdL_2{-}I \ (\mathbf{12}) \xrightarrow{+CO} [PhCO{-}PdL_2{-}CO]^+ \ (\mathbf{13}) \xrightarrow[-\,Et_2NH\cdot HI]{+\,2Et_2NH} PhCO{-}PdL_2{-}CONEt_2 \ (\mathbf{14}) \longrightarrow PhCOCONEt_2 + Pd(0)L_2 \quad (67)$$

$$PhCO{-}PdL_2{-}I \ (\mathbf{12}) \xrightarrow[+\,base]{+\,EtOH} PhCO{-}PdL(OEt){-}L \ (\mathbf{15}) \longrightarrow PhCOOEt + Pd(0)L_2 \quad (68)$$

The palladium-catalyzed carbonylation of organic halides is a convenient means of synthesizing carbonyl compounds. For example, the single carbonylation of aryl iodides has been applied to synthesize lactones and lactams (eq (69) and (70)) [97,98].

$$\text{2-iodobenzyl alcohol} \xrightarrow[CO,\ K_2CO_3]{PdCl_2(PPh_3)_2} \text{phthalide (100\%)} \quad (69)$$

$$\text{(R = CH}_2\text{OMe)} \xrightarrow[\text{Bu}_3\text{N, CO}]{\text{Pd(OAc)}_2\text{, PPh}_3} \text{90\%} \quad (70)$$

The double carbonylation product of *o*-haloacetanilide is readily converted into isatin and quinoline derivatives in high yields (eq (71)) [99].

$$o\text{-IC}_6\text{H}_4\text{NHAc} + \text{CO} + \text{Et}_2\text{NH} \xrightarrow{\text{PdCl}_2\text{(PMePh}_2)_2} o\text{-(Et}_2\text{NCOCO)C}_6\text{H}_4\text{NHAc} \quad (71)$$

$$\xrightarrow{\text{1) NaOH, 2) Me}_2\text{CO}} \text{2-methylquinoline, 75\%}; \quad \xrightarrow{\text{H}^+} \text{isatin, 93\%}$$

The successive insertion of CO and olefin into a Pd—C bond leads to an alternating copolymerization of CO and olefins [100]. When the reaction is carried out with an isolated, cationic acetylpalladium catalyst, living polymerization giving a singly dispersed polymer proceeds (eq (72)) [101].

$$\text{L}_2\overset{+}{\text{Pd}}(\text{COMe})(\text{CO}) \xrightarrow{+\text{CO}, \text{ CH}_2\text{=CHR}} \text{L}_2\overset{+}{\text{Pd}}(\text{CO}\{\text{CH(R)CH}_2\text{CO}\}_n\text{COMe})(\text{CO}) \quad (72)$$

12.4.2.2 Diorganopalladium(II) Complexes

The two geometrical isomers of dialkylpalladium(II) complexes, *trans*- and *cis*-PdR_2L_2, have been synthesized independently by the proper choice of transmetallation reagents [76,83a,102]. Treatment of palladium(II) acetylacetonate with dialkylaluminum ethoxide in the presence of tertiary phosphine ligands affords trans isomers (eq (73)), whereas the reaction of palladium(II) dihalides bearing appropriate supporting ligands with an excess of alkyllithium followed by hydrolysis provides cis isomers (eq (74)). The selective formation of the cis isomers in eq (74) involves a trialkylpalladate intermediate, which gives the cis isomer on hydrolysis [103].

The behavior of the *trans*- and *cis*-dialkylpalladium(II) complexes having β-hydrogens on thermolysis offer a striking contrast [76b]. Thus the trans isomers decompose selectively via a β-hydrogen elimination process, whereas the cis isomers undergo reductive elimination leading to the coupling of two alkyl groups. Dialkylpalladium(II) complexes bearing bidentate nitrogen ligands such as 2,2′-bipyridine and *N*,*N*,*N*′,*N*′-tetramethylethylenediamine undergo oxidative addition of alkyl halides to give trialkylpalladium(IV) complexes [102d].

$$\mathrm{Pd(acac)_2 + AlR_2(OEt) + 2L \longrightarrow } \ trans\text{-}\mathrm{PdR_2L_2} \quad (73)$$

R = Me, Et, Pr, Bu; L = PMe_3, PEt_3, PMe_2Ph, PEt_2Ph, $PMePh_2$, $PEtPh_2$

$$\mathrm{PdCl_2L_2 + xs.\ RLi \xrightarrow{-L} [PdR_3L]^-\,Li^+ \xrightarrow[-RH]{+H^+,\ L} PdR_2L_2} \quad (74)$$

R = Me; L = PMe_3, PEt_3, PMe_2Ph, PEt_2Ph, $PMePh_2$, $PEtPh_2$, PPh_3,
R = Me; L = dppe, bipy, tmeda, cod
R = Et; L = PMe_2Ph, PEt_2Ph

The syntheses of trans and cis isomers of unsymmetrical diorganopalladium(II) complexes have been reported [104]. For example, the reaction of arylpalladium(II) halides, which have been prepared either by the oxidative addition of aryl halides to palladium(0) phosphine complexes or by the treatment of diarylpalladium(II) with HCl, with 1 equivalent of methyllithium provides *trans*-PdAr(Me)L_2 complexes (eq (75)). On the other hand, treatment of arylpalladium(II) halides with excess methyllithium followed by methanolysis of the resulting dimethyl(aryl)palladate intermediates leads to *cis*-PdAr(Me)L_2 complexes (eq (76)).

Reductive elimination of the unsymmetrical complexes is assumed to be the product-forming step in palladium-catalyzed cross-coupling reactions of organic halides

$$trans\text{-}\mathrm{ArPdXL_2 + MeLi \longrightarrow } \ trans\text{-}\mathrm{ArPd(Me)L_2} \quad (75)$$

$$trans\text{-}\mathrm{ArPdXL_2 + xs.\ MeLi \xrightarrow{-L} [ArPdMe_2L]^-\,Li^+ \xrightarrow[-CH_4]{+H^+,\ +L}} \ cis\text{-}\mathrm{ArPd(Me)L_2} \quad (76)$$

and organometallic reagents [104,105]. Unlike the nickel(II) complexes (see Section 12.2.2), the reductive elimination of palladium analogs has been found to proceed without participation of organic halides or added phosphine ligands [104]. Therefore, the entire catalytic cycle of palladium-catalyzed cross-coupling reaction may be illustrated by eq (77).

[Pd(0)L2] —(+ RX)→ L2Pd(R)(X); L2Pd(R)(X) —(+ R'M, − MX)→ L2Pd(R)(R'); L2Pd(R)(R') → [Pd(0)L2] + R–R' (77)

The palladium-catalyzed cross-coupling reaction is superior to the nickel-catalyzed one because a wide variety of transmetallation reagents including Mg [106,107], Sn [108], Zn [109], B [110–112], Al [113], Zr [114], and Cu [115] can be employed (eq (78)–(84)). The reactions using aluminum, boron, and zirconium reagents in eq (80)–(83) are particularly useful because these transmetallation reagents are readily accessible by hydrometallation of olefins and acetylenes [111–114]. Triphenylphosphine catalysts are satisfactory for the arylation and alkenylation reactions. On the other hand, for the alkylation reactions, dppf complexes prevent undesirable side reactions involving β-hydrogen elimination, and provide the cross-coupling products in high yields (eq (78) and (80)) [107].

PhBr + sec-BuMgCl —($PdCl_2(dppf)$), 80%→ sec-BuPh (78)

4-*t*-Bu-cyclohexenyl-OTf + $CH_2{=}CHSnMe_3$ —($Pd(PPh_3)_4$, LiCl)→ 4-*t*-Bu-1-vinylcyclohexene, 78–79% (79)

Palladium-catalyzed cross-coupling of acyl halides with organometallic reagents of zinc and tin is a convenient synthetic route to ketones and aldehydes. Typical examples are shown in eq (85) and (86) [116,117].

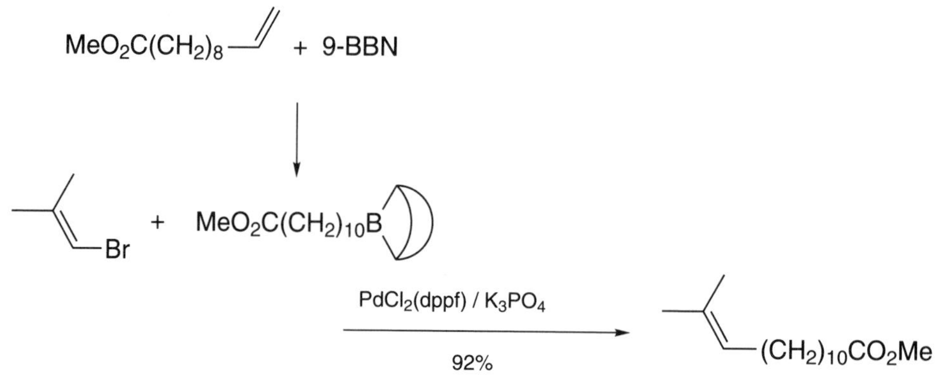

$MeO_2C(CH_2)_8$ + 9-BBN

+ $MeO_2C(CH_2)_{10}B$

Br

$PdCl_2(dppf)$ / K_3PO_4

92%

$(CH_2)_{10}CO_2Me$ (80)

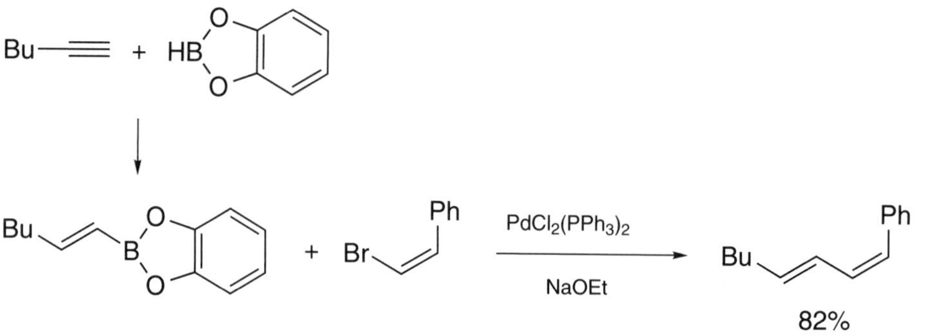

Bu + HB O O

Bu B O O + Br Ph

$PdCl_2(PPh_3)_2$

NaOEt

Bu Ph

82% (81)

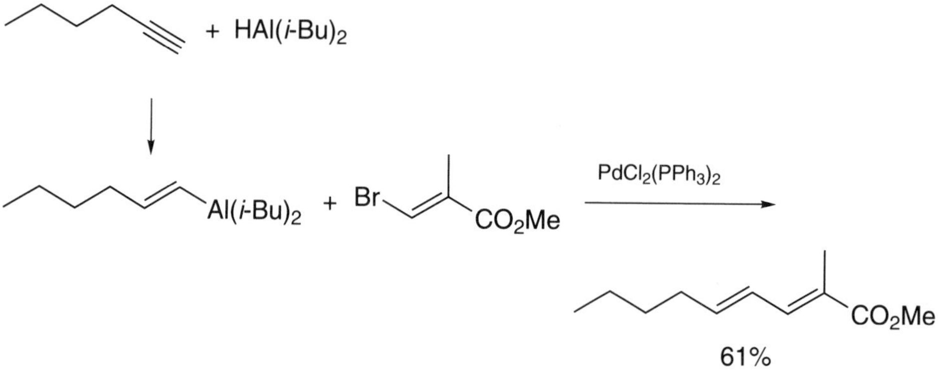

+ $HAl(i\text{-}Bu)_2$

$Al(i\text{-}Bu)_2$ + Br CO_2Me

$PdCl_2(PPh_3)_2$

CO_2Me (82)

61%

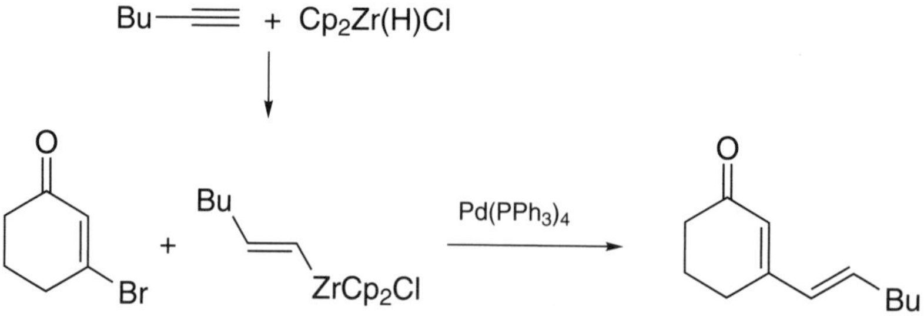

Bu + $Cp_2Zr(H)Cl$

O

Br + Bu $ZrCp_2Cl$

$Pd(PPh_3)_4$

O

Bu (83)

$Me_3Si{-}C{\equiv}C{-}H$ + $Br{-}C_6H_4{-}NO_2$ $\xrightarrow[\text{CuI, Et}_2\text{NH}]{\text{PdCl}_2(\text{PPh}_3)_2}$ $Me_3Si{-}C{\equiv}C{-}C_6H_4{-}NO_2$ 92% (84)

$CH_2{=}C(CH_3)COCl$ + $IZn(CH_2)_3CO_2Et$ $\xrightarrow{\text{Pd(PPh}_3)_4}$ $CH_2{=}C(CH_3)CO(CH_2)_3CO_2Et$ 87–88% (85)

$(CH_3)_2CHCOCl$ + $Bu_3Sn{-}C{\equiv}C{-}Ph$ $\xrightarrow{\text{PdCl}_2(\text{PPh}_3)_2}$ $(CH_3)_2CHCOC{\equiv}CPh$ 94% (86)

12.4.2.3 Synthesis via Orthopalladation Reactions

Aromatic compounds react with palladium(II) salts such as $Pd(OAc)_2$ and Na_2PdCl_4 via an electrophilic aromatic substitution process to give arylpalladium complexes. This type of reaction is most commonly observed with aromatic rings bearing a substituent that makes a five- or six-membered chelate ring with palladium in the metallation products (eq (87)) [118]. In this case, the electrophilic substitution occurs only at the ortho position to the chelating substituent.

$C_6H_5{-}A{-}B$: $\xrightarrow[\text{(X = OAc, Cl)}]{\text{PdX}_2}$ $[C_6H_4(A{-}B{\rightarrow})Pd(\mu{-}X)]_2$ (87)

(A–B = CH_2NMe_2, CH=NR, $CH_2N{=}CR_2$, NHN=CHR, CH=NOH, $CH{=}NNR_2$, 2-pyridyl, 2-oxazoyl)

The palladation products exhibit reactivity similar to that of the arylpalladium complexes formed by oxidative addition of aryl halides to Pd(0) species, although the reactions are stoichiometric with respect to palladium. Representative examples include vinylation via an olefin insertion process (eq (88)) [119], double and single carbonylation (eq (89) and (90)) [120,121], and alkylation via a transmetallation process (eq (91)) [122].

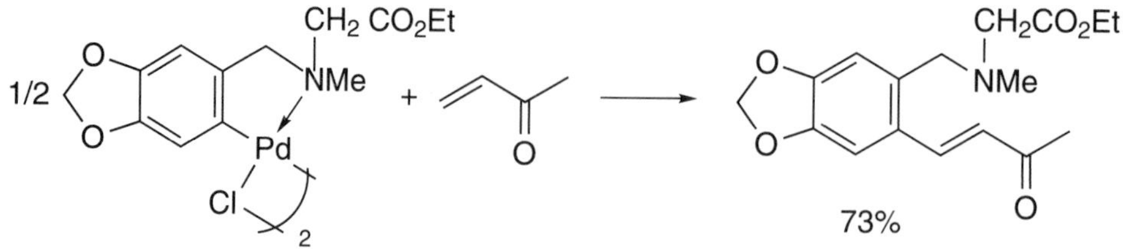

(88)

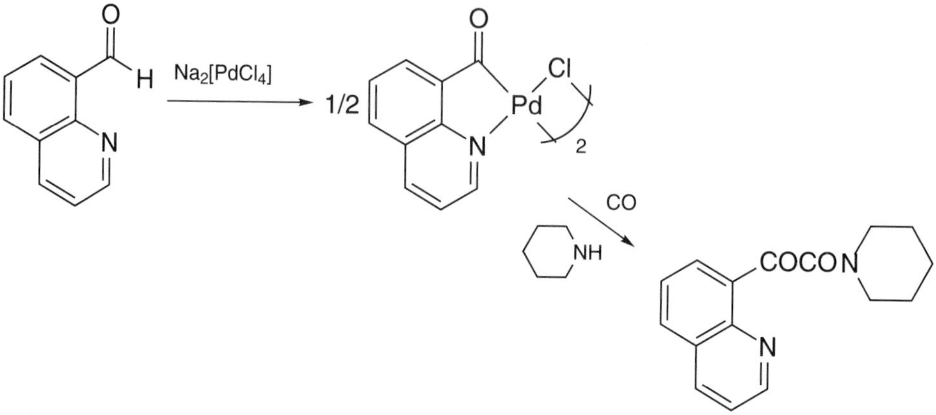

(89)

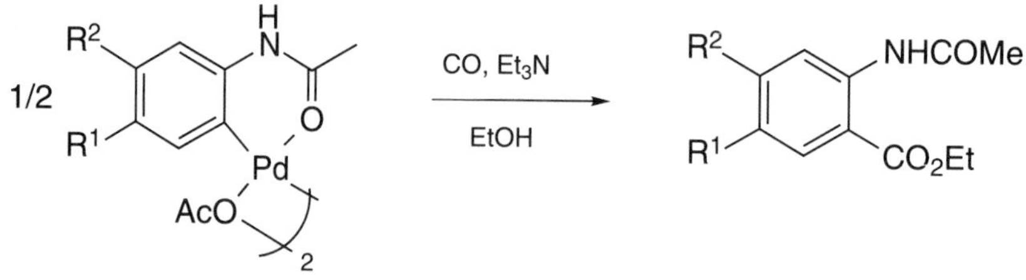

(90)

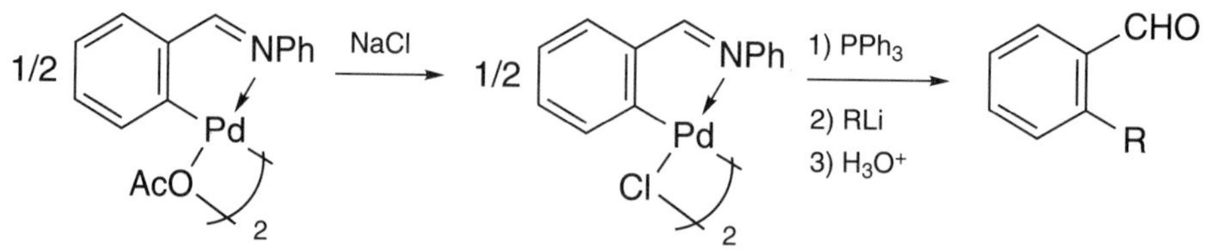

(91)

12.4.2.4 Synthesis from Palladium(II)-Olefin Complexes

Since palladium(II) is electrophilic, olefins are activated toward nucleophiles by coordination to palladium(II) species. The attack of nucleophiles occurs at the more substituted vinylic carbon from the anti side of palladium to give alkylpalladium complexes (eq (92)) [123].

The process involving an oxygen nucleophile is termed oxypalladation. A representative example is seen in eq (93), where nucleophilic attack of hydroxide ion on the

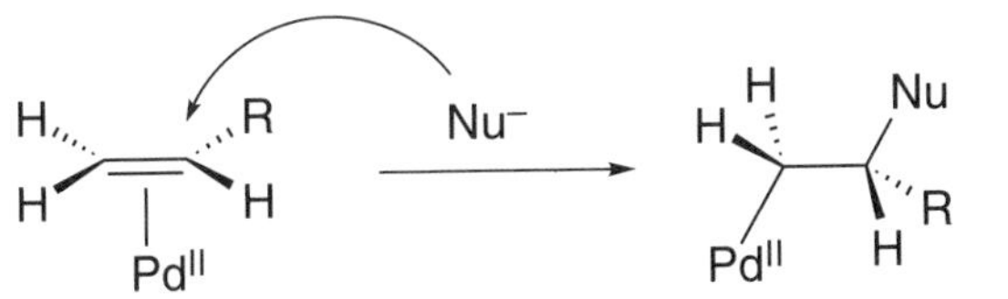

(92)

cyclooctadiene ligand give an alkylpalladium(II) complex [124]. The starting palladium(II) olefin complex can be prepared by the treatment of palladium(II) compounds such as $[PdCl_4]^{2-}$ and $PdCl_2(PhCN)_2$ with olefin [125a–c].

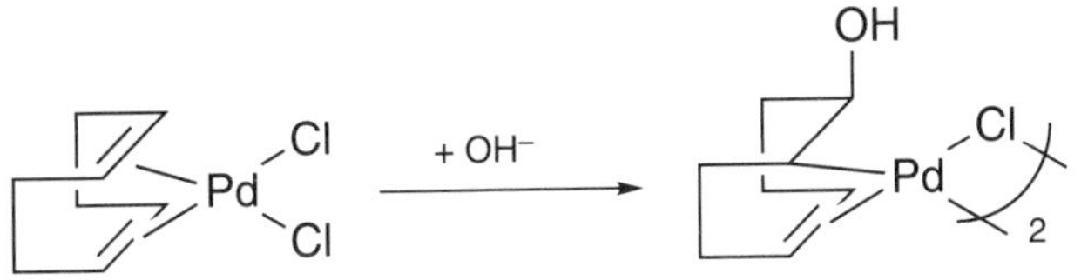

(93)

The oxypalladation process is assumed to occur in the Wacker type oxidation of olefins (eq (94) and (95)) [126–128].

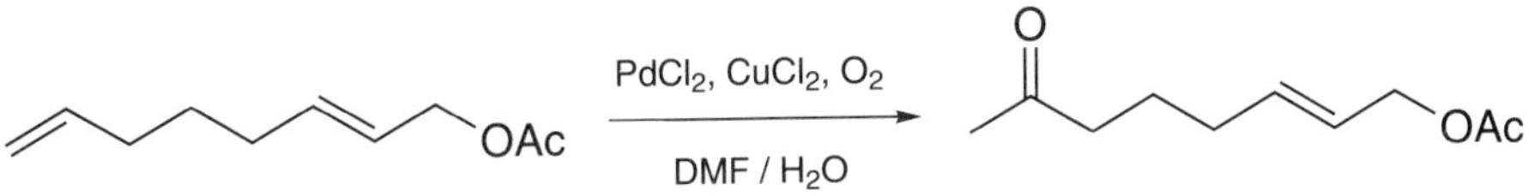

(94)

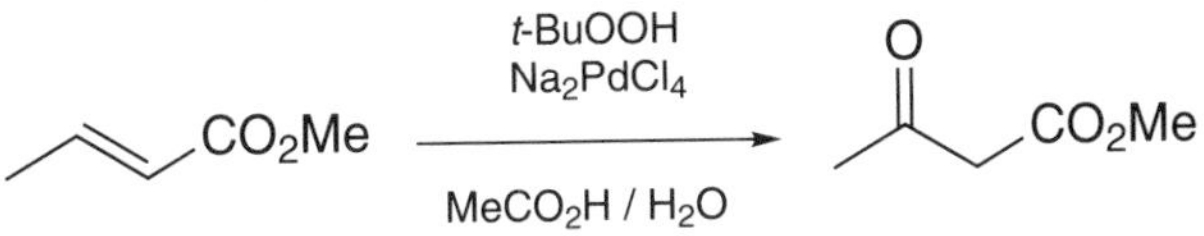

(95)

When an aliphatic amine is used as nucleophile, aminopalladation of olefins takes place (eq (96)) [129]. The resulting β-aminoalkylpalladium complex undergoes CO insertion into the Pd—C bond to give a β-aminoacyl complex, which has been isolated as a stable solid.

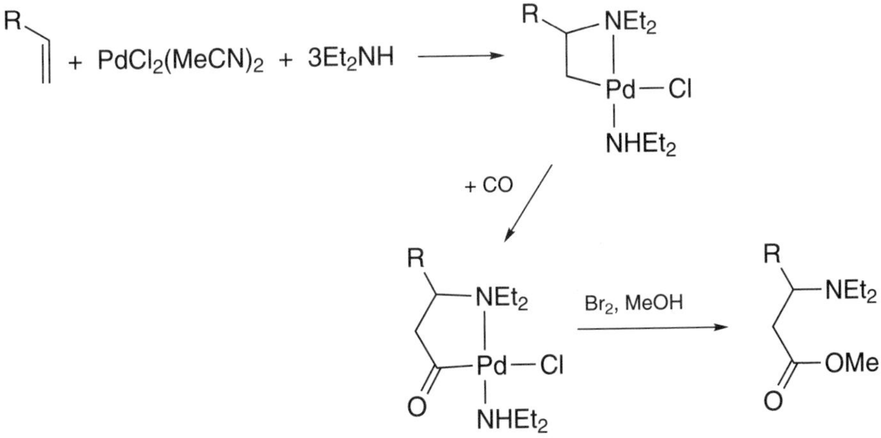

(96)

Stabilized carbon nucleophiles such as malonates attack the olefins bound to palladium(II) similarly to O- and N-nucleophiles (eq (97) and (98)) [130,131]. The alkylpalladium species thus formed undergoes β-hydrogen elimination, hydrogenolysis, and carbonylation (eq (98)).

$$CpPd^{+}(PPh_3)(\eta^2\text{-}CHD{=}CHD) \xrightarrow{^{-}CH(CO_2Me)_2} CpPd(PPh_3)\text{-}CHD\text{-}CHD\text{-}CH(CO_2Me)_2 \quad (97)$$

$$CH_2{=}CHR + Na^{+}[CH(CO_2Et)_2]^{-} + PdCl_2(MeCN)_2 \xrightarrow{2Et_3N} (EtO_2C)_2CH\text{-}CHR\text{-}CH_2\text{-}PdCl(NEt_3)_2$$

$$\xrightarrow{\beta\text{-H elimination}} (EtO_2C)_2CH\text{-}CR{=}CH_2; \quad \xrightarrow{H_2} (EtO_2C)_2CH\text{-}CHR\text{-}CH_3; \quad \xrightarrow{CO,\ MeOH} (EtO_2C)_2CH\text{-}CHR\text{-}CH_2\text{-}CO_2Me \quad (98)$$

12.4.3 π-Allylpalladium(II) Complexes

12.4.3.1. Synthesis via Oxidative Addition Reactions

(π-Allyl)palladium(II) complexes are synthesized by oxidative addition of allylic compounds to palladium(0) complexes. The starting palladium(0) species are often generated *in situ* from palladium(II) salts and reducing agents (eq (99)) [132]. Allylic chlorides react with coordinatively saturated palladium(0) phosphine complexes (e.g., $Pd(PPh_3)_4$) to give (π-allyl)palladium(II) chlorides in good yields (eq (100)) [133]. On the other hand, the oxidative addition of allylic carboxylates toward a $Pd(0)L_4$ type complex is a reversible process and often provides no or only a small amount of π-allyl complexes. In this case, coordinatively unsaturated $Pd(PCy_3)_2$ affords π-allyl complexes in good yields (eq (101)) [134]. The complex Pd(styrene)L_2 generated from *trans*-$PdEt_2L_2$ and styrene also serves as a useful starting material. The syntheses of (π-allyl)palladium(II) formates and carbonates have been achieved by this method (eq (102)) [135]. The oxidative addition of allylic substrates generally proceeds with inversion of the stereochemistry at the allylic carbon (eq (103)) [136a], although retention of the stereochemistry has also been observed as a result of pre-coordination of the Pd(0) center to the leaving group [137].

$$Na_2PdCl_4 \xrightarrow{+\,CO,\ H_2O} \text{“ Pd(0) ”} \xrightarrow{+\ \text{allyl-Cl}} [(\eta^3\text{-allyl})Pd(\mu\text{-Cl})]_2 \quad (99)$$

$$Pd(PPh_3)_4 + \text{allyl-Cl} \longrightarrow (\eta^3\text{-allyl})PdCl(PPh_3) \quad (100)$$

$$Pd(PCy_3)_2 + \text{allyl-OAc} \longrightarrow (\eta^3\text{-allyl})Pd(OAc)(PCy_3) \quad (101)$$

$$L_2Pd(\text{CH}_2{=}\text{CHPh}) + \text{allyl-OCOY} \longrightarrow [(\eta^3\text{-allyl})PdL_2]^+\ {}^-OCOY \quad (102)$$

Y = H, OR

$$\text{MeCH(OAc)CH{=}CHPh} \xrightarrow[\text{2) } NaBF_4]{\text{1) } PdCl_2(dppe),\ PPh_3;\ i\text{-}Bu_2AlH} [(\eta^3\text{-MeCHCHCHPh})Pd(Ph_2PCH_2CH_2PPh_2)]^+\ BF_4^- \quad (103)$$

(π-Allyl)palladium(II) complexes react with a variety of nucleophiles and two types of process have been identified [136]. The first involves the external attack of a nucleophile on the π-allyl ligand from the opposite side of the palladium center (eq (104)). Thus the reaction gives rise to the inversion of stereochemistry at the allylic carbon. This type of process is observed with stabilized carbon nucleophiles as well as nitrogen and oxygen nucleophiles [136a,c]. On the other hand, when nonstabilized carbon nucleophiles (e.g., main group organometallics) and metal hydrides are employed, the π-allyl complex undergoes transmetallation with the nucleophile to form a diorganopalladium intermediate (**16**), which reductively eliminates the allylation product (eq (105)) [136b,138]. In this case, since the nucleophile is introduced to the π-allyl ligand from the palladium center, retention of stereochemistry at the allylic carbon is observed.

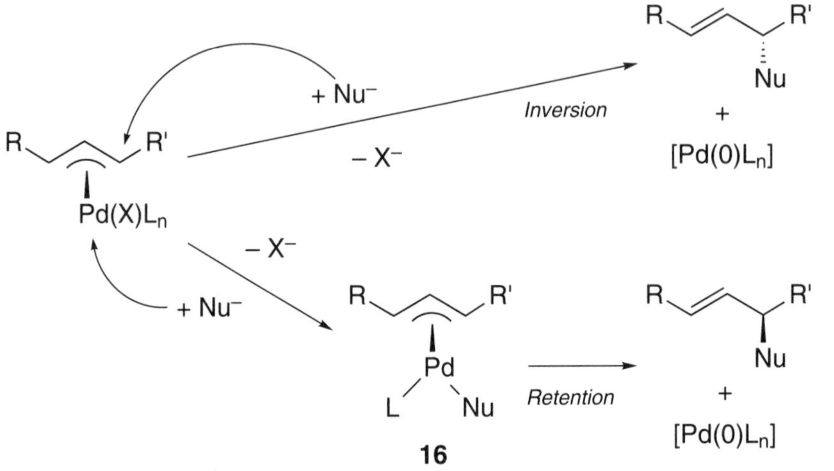

(104)

(105)

The (π-allyl)palladium(II) complexes formed by oxidative addition have been extensively applied to catalytic organic synthesis. Representative examples are shown in eq (106)–(111) [139–142]. A variety of allylic substrates have been employed as the source of the allyl ligand, including allylic acetates, chlorides, epoxides, and carbonates.

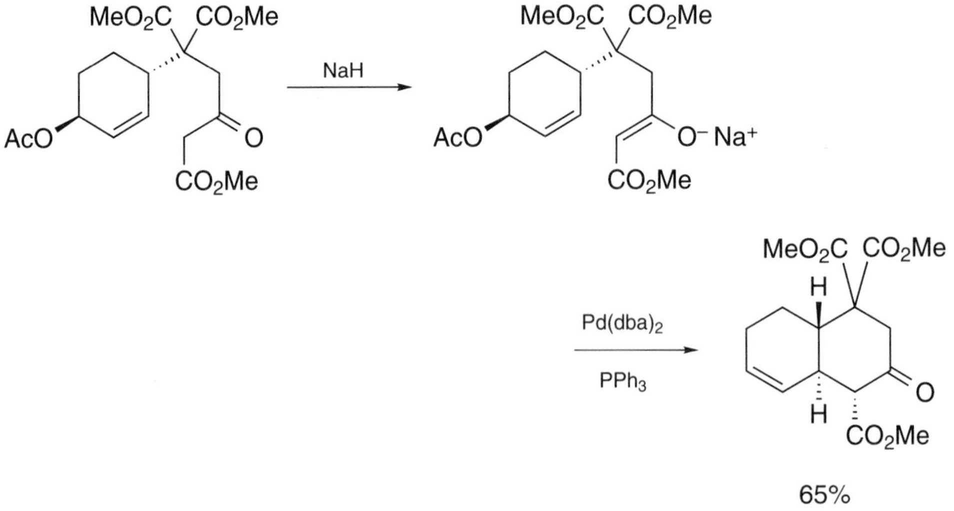

(106)

Allyl carbonates are particularly useful allylic substrates. Thus, in addition to their high reactivity in oxidative addition, allyl carbonates have the advantage that the allylation reaction can be conducted under neutral conditions without an added base. The (π-allyl)palladium intermediate (**17**) derived from an allylic carbonate has a carbonate anion that serves as a masked base to generate carbon nucleophiles (eq (112)) [143].

OAc
AcO + H_2N N H $\xrightarrow[>90\%]{Pd(PPh_3)_4}$ OAc HN N H (107)

$+ MeCO_2H \xrightarrow[72\text{–}76\%]{Pd(PPh_3)_4}$ O_2CMe OH (108)
O

OAc + Bu_3Sn O OBn

$\xrightarrow[LiCl]{Pd(dba)_2}$ OBn O (109)
81%

Cl + Me_2Al $\xrightarrow[83\%]{Pd(PPh_3)_4}$ (110)

O $CO_2Et + HCO_2H \xrightarrow[PBu_3,\ Et_3N]{Pd(OAc)_2}$ CO_2Et (111)
OH 91%

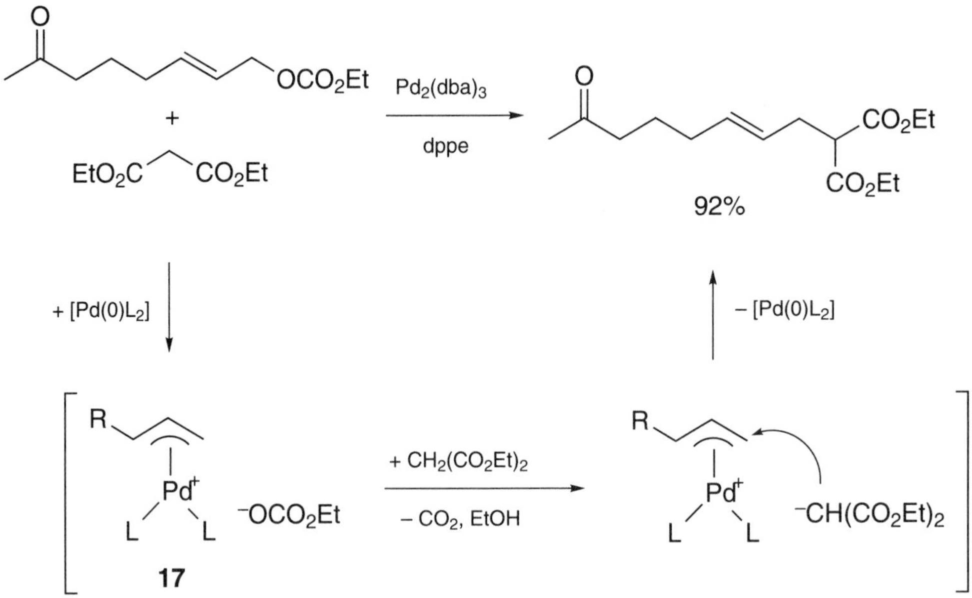

(112)

In the absence of a nucleophile, π-allylpalladium complexes bearing alkyl substituent(s) at the allylic carbon(s) undergo β-hydrogen elimination to give dienes (eq (113)) [144].

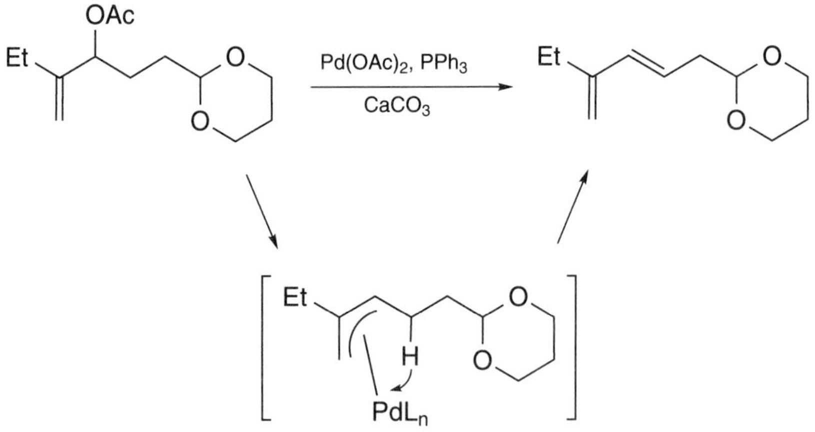

(113)

(π-Allyl)palladium species undergo intramolecular insertion of olefins to give cyclic compounds (eq (114)) [145]. This reaction has been termed a "metalo-ene" reaction, although the exact mechanism differs from that of the classical organic ene reaction.

The oxidative addition of an allylic acetate having an allylsilane structure (**18**) to a Pd(0) complex provides a (trimethylenemethane)palladium species (**19**), which undergoes [3+2] cycloaddition to a variety of electron-deficient olefins (eqs (115) and (116)) [146].

MeO_2C CO_2Me AcO — $Pd(PPh_3)_4$, PPh_3 / AcOH → MeO_2C CO_2Me (114)

+ [Pd(0)] — Insertion — − [HPd(OAc)]

MeO_2C CO_2Me Pd OAc → MeO_2C CO_2Me H Pd–OAc

$[Pd(0)L_n]$ — Me_3Si OAc **18** → ^{-}OAc Me_3Si $\overset{+}{Pd}L_n$ — $- Me_3SiOAc$ → $\overset{+}{Pd}L_n$ **19** — Z — Z (115)

Me_3Si OAc + O O MeO_2C — $Pd(PPh_3)_4$ → O O CO_2Me (116)

87%

12.4.3.2 Synthesis via Palladation Reactions

Palladium(II) salts such as $PdCl_2$ and $Pd(O_2CCF_3)_2$ cause the activation of allylic C—H bond of olefins, leading to π-allylpalladium complexes (eq (117) and (118)) [147]. The more electron-rich olefins tend to show the higher reactivity. When the olefin has more than two allylic positions, the regioselectivity obeys the Markovnikov rule.

$$\text{methylenecyclohexane} + PdCl_2 \xrightarrow[\text{HOAc, Ac}_2\text{O, NaCl}]{\text{CuCl}_2\text{, NaOAc}} \pi\text{-allyl-}PdCl/2 \quad (117)$$

$$\text{ketone} + Pd(O_2CCF_3)_2 \xrightarrow[\text{2) Bu}_4\text{NCl}]{\text{1) MeCOMe, r.t.}} \pi\text{-allyl-}PdCl/2 \quad (118)$$

Conjugated dienes coordinated to palladium(II) undergo nucleophilic addition of oxygen and nitrogen nucleophiles from the opposite side of the palladium center to give π-allylpalladium complexes. Subsequent reaction of the π-allyl complexes with another molecule of nucleophile results in the 1,4-addition of two kinds of nucleophiles to dienes. The direction of the second nucleophilic attack is dependent on the nature of the nucleophiles as well as the reaction conditions. Representative examples

(119): π-allyl(OMe)Pd(OAc) complex, with benzoquinone (O=C₆H₄=O) → AcO/OMe cyclohexene; formed from **20** by + AgOAc, − AgCl

(120): 1,3-cyclohexadiene $\xrightarrow[\text{MeOH}]{Na_2PdCl_4}$ π-allyl(OMe)PdCl/2 (**20**) $\xrightarrow{\text{LiOAc, LiCl, benzoquinone}}$ AcO/OMe cyclohexene

(121): **20** $\xrightarrow{Me_2NH,\ PPh_3}$ Me_2N/OMe cyclohexene

are shown in eq (119)–(121) [148]. The reaction of π-allyl complex **20** with silver acetate forms the *trans*-1,4-adduct (eq (119)). In this case, anionic displacement of the chloride in **20** with acetate gives a π-allylpalladium acetate, which reductively eliminates the trans adduct. On the other hand, treatment of **20** with lithium acetate in the presence of lithium chloride affords *cis*-1,4-adduct, because the presence of chloride ion in the system prevents the prior coordination of acetate to palladium, leading to external attack of acetate on the π-allyl ligand (eq (120)). Nitrogen nucleophiles also react with the π-allyl ligand from the anti side of palladium to give *cis*-1,4-adducts (eq (121)).

12.5 Synthesis of Palladium Compounds

12.5.1 Starting Palladium Complexes

(1) Bis(2,4-pentanedionato)palladium(II), $Pd(acac)_2$ [149]

$$Na_2PdCl_4 + 2CH_2(COMe)_2 + 2NaOH \longrightarrow Pd(acac)_2 + 4NaCl + 2H_2O \quad (122)$$

To an aqueous solution of Na_2PdCl_4 (0.5 M, 10 mL), which is prepared *in situ* from $PdCl_2$ and NaCl (2.5 equiv.) in water, are added 2,4-pentanedione (2.5 mL) and an aqueous solution of NaOH (5 M, 4 mL). The mixture is magnetically stirred for several hours at room temperature to give a yellow solid, which is filtered off with suction, washed successively with water, methanol, and then with Et_2O. The solid product thus obtained is dissolved in chloroform and filtered, and the filtrate is evaporated to dryness (1.37 g, 90%). Further purification may be performed by recrystallization from hot benzene, yielding orange needles of the title compound.

(2) Dichlorobis(acetonitrile)palladium(II), $PdCl_2(MeCN)_2$

$$PdCl_2 + 2MeCN \longrightarrow PdCl_2(MeCN)_2 \quad (123)$$

Anhydrous palladium(II) dichloride (2 g) is suspended in acetonitrile (50 mL) and the mixture is stirred until the brown palladium dichloride disappears. Since palladium dichloride sometimes coagulates, it should be broken up with a spatula. Filtration followed by drying on the filter gives a quantitative yield of the title compound, which is pure enough to use as a catalyst and a starting material for organometallic synthesis. A related complex $PdCl_2(PhCN)_2$ is prepared by the treatment of palladium dichloride with benzonitrile [125d].

The compounds $PdCl_2(MeCN)_2$ and $PdCl_2(PhCN)_2$ are convenient starting materials for the synthesis of dichlorobis(tertiary phosphine)palladium(II) complexes. In a typical procedure, $PdCl_2(MeCN)_2$ is suspended in benzene (for arylphosphines) or ether (for alkylphosphines) under nitrogen. Two equivalents of tertiary phosphine is added and the mixture is stirred at room temperature for several hours. A yellow or orange precipitate of the desired compound is formed. If necessary, hexane or ether is added to complete the precipitation. The product is collected by filtration, washed with ether, and dried under vacuum. $PdCl_2(PR_3)_2$ type complexes are also prepared by treatment of $[PdCl_4]^{2-}$ with phosphine [150].

12.5.2 Palladium(0) Complexes

(3) Tetrakis(triphenylphosphine)palladium(0), $Pd(PPh_3)_4$ [65]

$$2PdCl_2 + 8PPh_3 + 5N_2H_4{\cdot}H_2O \longrightarrow 2Pd(PPh_3)_4 + 4N_2H_4{\cdot}HCl + N_2 + 5H_2O \quad (124)$$

Palladium(II) dichloride (4.43 g, 25 mmol) and triphenylphosphine (32.8 g, 125 mmol) are placed in a Schlenk tube containing a magnetic stirring bar and the system is filled with nitrogen. Dimethylsulfoxide (DMSO) (300 mL) which has been degassed by bubbling a dry nitrogen gas at least 30 min, is added, and the mixture is heated by means of an oil bath at 140 °C with stirring until complete solution occurs. The bath is then taken away, and the solution is rapidly stirred. Hydrazine hydrate (5 g, 100 mmol) is then rapidly added to the vigorously stirred solution over approximately 1 min by means of a pipette. A vigorous reaction takes place with evolution of nitrogen. The dark solution is then allowed to cool without stirring, resulting in gradual precipitation of yellow plates of the title compound. After the mixture has reached room temperature, the yellow crystals are collected by filtration, washed successively with EtOH and then with Et_2O, and dried under vacuum (90–94%).

Properties: mp 116 °C (under N_2 with decomposition).

Since the complex is gradually decomposed in the air, it should be stored under an inert atmosphere at low temperature.

(4) Bis(tricyclohexylphosphine)palladium(0), $Pd(PCy_3)_2$ [71]

$$(\eta^5\text{-}C_5H_5)(\eta^3\text{-}C_3H_5)Pd + 2PCy_3 \longrightarrow Pd(PCy_3)_2 + C_8H_{10} \quad (125)$$

In a 50 mL Schlenk tube containing a magnetic stirring bar are placed (η^3-allyl)(η^5-cyclopentadienyl)palladium (0.34 g, 1.60 mmol) and a toluene solution of tricyclohexylphosphine (0.99 g, 3.54 mmol). The dark red mixture is stirred with heating at 75–80 °C for 3 h. The brown solution is concentrated to dryness under reduced pressure. The brown crystalline solid is washed with two 10 mL portions of MeOH to remove a slight excess of the phosphine. The solid is dissolved in hot toluene (5 mL), and then MeOH (5 mL) is added to give crystals. After standing the mixture in a freezer (–35 °C) overnight, the crystals are isolated by removing the mother liquid with a syringe, washed with MeOH (two 5 mL portions), and dried under vacuum (0.84 g, 79%). The product is pure enough to prepare organometallic complexes. Further purification can be performed by recrystallization from a toluene (5 mL)–MeOH (5 mL) mixture.

Properties: mp 185–189 °C (under N_2 in a sealed capillary tube).

Several bis(tertiary phosphine)palladium(0) complexes can be synthesized by similar procedures [71].

(5) (η^2-Methyl methacrylate)bis(methyldiphenylphosphine)palladium(0), $Pd\{CH_2{=}CH(Me)COOMe\}(PMePh_2)_2$ [73]

$$\textit{trans}\text{-}PdEt_2(PMePh_2)_2 + CH_2{=}CH(Me)COOMe \longrightarrow Pd\{CH_2{=}CH(Me)COOMe\}(PMePh_2)_2 + C_2H_4 + C_2H_6 \qquad (126)$$

The complex *trans*-$PdEt_2(PMePh_2)_2$ (0.21 g, 0.37 mmol) is placed in a Schlenk tube and the system is filled with nitrogen. Toluene (5 mL) and methyl methacrylate (80 mL, 0.76 mmol) are added at –20 °C, and the heterogeneous mixture is stirred at room temperature for 5 h to give a pale brown solution. The solution is concentrated to *ca.* 1 mL and cooled to –78 °C. Hexane (5 mL) is added to the stirred solution to precipitate a pale brown solid, which is collected by filtration, washed with a small amount of hexane, and dried under vacuum. The crude product is dissolved in a minimum amount of hot acetone containing methyl methacrylate (1–2 equiv.). Cooling the solution to –20 °C results in the formation of fine crystals of the title compound (0.18 g, 76%).

Properties: mp 115–119 °C.

Bis(tertiary phosphine)palladium(0) complexes having styrene, methyl acrylate, methacrylonitrile ligands are prepared by similar procedures [73]. The palladium(0)-olefin complexes are stable under an inert atmosphere in the solid as well as in a solution containing an excess amount of olefin, while they rapidly decompose in solution in the absence of free olefin even under a nitrogen or argon atmosphere. Ethylene complexes bearing PPh_3, PCy_3, $P(OC_6H_4Me\text{-}o)_3$, and PBu_3 ligands are synthesized by the reduction of palladium(II) salts in the presence of ethylene [72]. Palladium(0) complexes bearing highly electron-deficient olefins such as maleic anhydride and dimethyl maleate can be prepared by the ligand displacement of $Pd(PPh_3)_4$ or $Pd_2(dba)_3$ with the olefins [74].

(6) Tris(dibenzylideneacetone)dipalladium(0), $Pd_2(dba)_3{\cdot}CHCl_3$ [78]

$$2PdCl_2 + 3PhCH{=}CHCOCH{=}CHPh + MeCO_2Na \longrightarrow Pd_2(dba)_3 \qquad (127)$$

Palladium(II) dichloride (1.05 g, 5.92 mmol) is added to a hot (*ca.* 50 °C) methanol solution containing dibenzylideneacetone (dba) (4.60 g, 19.6 mmol) and sodium acetate (3.90 g, 47.5 mmol). The mixture is stirred for 4 h at 40 °C to give a reddish purple precipitate and allowed to cool to room temperature to complete the precipitation. The precipitate is collected by filtration, washed successively with water and acetone, and dried under vacuum (3.39 g). This product having the composition of $Pd(dba)_2$ is dissolved in hot chloroform and filtered to give a deep violet solution. Diethyl ether is slowly added to the solution and the mixture is allowed to stand at room temperature to give deep red needles of the title compound (80%).

Properties: mp 122–124 °C.

The compound is stable in the air.

12.5.3 Monoorganopalladium(II) Complexes

(7) *trans*-Bromo(methyl)bis(triethylphosphine)palladium(II), *trans*-$PdMe(Br)(PEt_3)_2$ [83a]

$$PdBr_2(PEt_3)_2 + MeMgBr \longrightarrow trans\text{-}PdMe(Br)(PEt_3)_2 + MgBr_2 \quad (128)$$

Dibromobis(triethylphosphine)palladium (10.0 g, 20 mmol) is suspended in Et_2O at −65 °C, and an Et_2O solution of MeMgBr (40 mL, 45 mmol) is added dropwise over a period of 30 min. The mixture is stirred at −65 °C for 15 min and then at room temperature for 20 min. The mixture is hydrolyzed with water (100 mL) at 0 °C, and the Et_2O phase is separated, dried over $MgSO_4$, and concentrated to dryness. The crude product is recrystallized from hexane to give crystals of the title compound (6.7 g, 78%).

Properties: mp 73–74 °C.

(8) *trans*-Bromo(phenyl)bis(triphenylphosphine)palladium(II), *trans*-$PdPh(Br)(PPh_3)_2$ [80c]

$$Pd(PPh_3)_4 + PhBr \longrightarrow trans\text{-}PdPh(Br)(PEt_3)_2 + 2PPh_3 \quad (129)$$

A mixture of tetrakis(triphenylphosphine)palladium(0) (1 g) and bromobenzene (1 g) in degassed benzene (15 mL) is heated overnight at 80 °C under nitrogen. After cooling to room temperature, the benzene is removed *in vacuo*, and the resulting solid is triturated with ether and the ether discarded. The product is further purified by recrystallization from methylene chloride–hexane. The yield of colorless air-stable crystals (mp 216–220 °C) is 95% of theory.

(9) Di(μ-chloro)bis[2-(dimethylaminomethyl)phenyl-C^1,*N*]dipalladium(II) [151]

$$Li_2PdCl_4 + C_6H_4CH_2NMe_2 + NEt_3 \longrightarrow 1/2\ [\text{cyclopalladated } C_6H_4CH_2NMe_2\text{–Pd–Cl}]_2 + NEt_3HCl + LiCl \quad (130)$$

Lithium tetrachloropalladate(II) can be prepared by the treatment of $PdCl_2$ and two equivalents of LiCl in boiling water. After dissolution of $PdCl_2$, the water is removed under vacuum to afford Li_2PdCl_4 as a brown solid. The Li_2PdCl_4 thus obtained (13.12g, 50 mmol) is dissolved in methanol (300 mL) at room temperature and *N*,*N*-dimethylaminomethylbenzene (7.42 g, 55 mmol) is added to the well-stirred solution. After *ca.* 5 min, a cream precipitate begins to form. Then Et_3N (0.05 g, 50 mmol) dissolved in methanol (50 mL) is added dropwise over the period of 1 h. The mixture is stirred for 5 h, after which time a bright yellow precipitate is obtained, along with a pale yellow supernatant solution. The precipitate is collected by filtration, washed with methanol (3 × 50 mL) and Et_2O (2 × 50 mL), and dried under vacuum. 12.95 g (94% yield) of the title compound is thus obtained, which is sufficiently pure for most purposes.

Properties: IR (KBr) 745 and 736 cm^{-1}; ^{1}H NMR ($CDCl_3$) δ 3.92 (CH_2), 2.86 and 2.84 ($N(CH_3)_2$).

The compound is air stable and is slightly soluble in $CHCl_3$, acetone, or THF.

12.5.4 Dialkylpalladium(II) Complexes

(10) *trans*-Diethylbis(diphenylmethylphosphine)palladium(II), *trans*-$PdEt_2(PMePh_2)_2$ [76a,b]

$$Pd(acac)_2 + 2PMePh_2 + Al_2Et_3(OEt)_3 \longrightarrow trans\text{-}PdEt_2(PMePh_2)_2 \quad (131)$$

The ethylaluminum compound $Al_2Et_3(OEt)_3$ is prepared by slow addition of absolute ethanol (1.5 equiv.) to well stirred triethylaluminum (neat) at –20 to –10 °C. Care must be taken in this preparation because a vigorous reaction takes place with evolution of ethane.

Bis(2,4-pentanedionato)palladium (1.5 g, 5 mmol) is suspended in Et_2O (30 mL) at –70 °C. $PMePh_2$ (2.1 mL, 11 mmol) is added, and then $Al_2Et_3(OEt)_3$ (6.0 mL) added dropwise by means of a syringe. The reaction mixture is allowed to warm to 0 °C with stirring. As the reaction proceeds, orange crystals of $Pd(acac)_2$ dissolve at about –20 °C to give a red homogeneous solution, then a white solid of the title compound precipitates from the solution. After the temperature has reached 0 °C, the mixture is stirred for an additional 4 h at 0 °C, and the resulting precipitate is collected by filtration, washed successively with Et_2O, and dried under vacuum (40–50%). The product is relatively stable in the solid, but it rapidly decomposes in solution via β-hydrogen elimination [76c].

(11) *cis*-Dimethylbis(triethylphosphine)palladium(II), *cis*-$PdMe_2(PEt_3)_2$ [83a]

$$PdBr_2(PEt_3)_2 + 2MeLi \longrightarrow cis\text{-}PdMe_2(PEt_3)_2 + 2LiBr \quad (132)$$

Dibromobis(triethylphosphine)palladium (10.0 g, 20 mmol) is suspended in an Et_2O solution (200 mL) containing a small amount of free PEt_3 (*ca.* 0.1 equiv.) and the mixture is cooled to –60 °C. An Et_2O solution of MeLi (38 mL, 45 mmol), which has been prepared from MeBr and Li in Et_2O, is added dropwise over the period of 15 min. The resulting mixture is stirred at room temperature for 1 h, and then hydrolyzed with water (50 mL) at 0 °C. The Et_2O phase is separated and the aqueous phase is extracted with Et_2O. The combined Et_2O solution is dried over $MgSO_4$ and concentrated to dryness. The crude product is recrystallized from hexane to give colorless crystals of the title compound (7.4 g, 90%).

Properties: mp 47–49 °C.

12.5.5 π-Allylpalladium(II) Complexes

(12) Di(μ-chloro)bis(η^3-allyl)dipalladium(II), $[PdCl(\eta^3\text{-}C_3H_5)]_2$ [132]

$$2Na_2PdCl_4 + 2CH_2{=}CHCH_2Cl + 2CO + 2H_2O \longrightarrow (\eta^3\text{-}C_3H_5)_2Pd_2Cl_2 + 4NaCl + 2CO_2 + 4HCl \quad (133)$$

The preparation should be performed in a well ventilated hood. A 200 mL, two-necked, round-bottom flask equipped with a magnetic stirring bar, a gas inlet tube, and a condenser topped with a bubbler is charged with an aqueous solution of palladium(II) dichloride (4.44 g, 25 mmol) and sodium chloride (2.95 g, 50 mmol) in 10 mL of H_2O, followed by methanol (60 mL) and allyl chloride (6.0 g, 67 mmol). Carbon monoxide is passed slowly (2–2.5 L/h) with stirring through the reddish-brown solution by way of a gas-inlet tube for 1 h. The bright yellow suspension thus obtained is poured into water (300 mL) and extracted with chloroform (2 × 100 mL). The extract is washed with water (2 × 150 mL portions) and dried over calcium chloride. Evaporation under reduced pressure (20 mmHg) gives yellow crystals (4.3 g, 93%), which is pure enough to use as a catalyst and a starting material of organometallic synthesis. The analytically pure sample can be obtained by recrystallization from a mixture of dichloromethane/hexane.

Properties: mp 155–156 °C (with decomposition).

A variety of π- allylpalladium complexes may be prepared in similar procedures [132].

(13) (η^3-Allyl)(η^5-cyclopentadienyl)palladium(II), [Pd(η^3-C_3H_5)(η^5-C_5H_5)] [70]

$$(\eta^3\text{-}C_3H_5)_2Pd_2Cl_2 + 2NaC_5H_5 \longrightarrow 2Pd(\eta^3\text{-}C_3H_5)(\eta^5\text{-}C_5H_5) + 2NaCl \quad (134)$$

A THF solution of sodium cyclopentadienide is prepared by addition of freshly distilled cyclopentadiene [152] to a sodium suspension in THF. The concentration of the resulting pale pink solution can be determined by titration with acid.

In a 300 mL, three-necked flask equipped with a three-way stopcock, a pressure-equalizing dropping funnel, and a Teflon-coated magnetic stirring bar, is placed bis(η^3-allyl)di-(μ-chloro)dipalladium (9.9 g, 27 mmol). The flask is evacuated and filled with nitrogen three times. THF (100 mL) and benzene (100 mL) are added through the three-way stopcock under nitrogen by means of syringe, to give clear yellow solution. The flask is then cooled with an ice–sodium chloride mixture to –20 °C. A THF solution of sodium cyclopentadienide (54 mmol in 28 mL of THF) is transferred by syringe to a nitrogen-flushed dropping funnel and is then added dropwise to the cooled solution with stirring at –20 °C. The solution changes slowly from yellow to dark red. After 1 h, the ice bath is removed and the temperature of the reaction mixture is allowed to reach room temperature with stirring. The stirring is continued for an additional 30 min. The solvents are removed by distillation *in vacuo* (30–60 mmHg) to give a dark red solid. If the pressure is lower than 30 mmHg, a considerable amount of the palladium complex sublimes at 25 °C. The solid residue is extracted with hexane (80 mL) and the extract is filtered through a dried filter paper under a nitrogen atmosphere. The red filtrate is evaporated *in vacuo* (30–60 mmHg), affording red needles of the title compound (9.2 g, 80%). This compound can be further purified by sublimation at 40 °C under 30 mmHg.

12.6 Platinum Complexes

Organoplatinum complexes have a strong resemblance to the palladium analogs in their synthetic methodology and reaction patterns, but because of their high stability, platinum compounds have been much less widely used in catalytic organic synthesis. The relative inertness, however, often allows the chemistry of palladium catalysis to be modeled using platinum complexes. In this section the methods of preparation of organoplatinum complexes are briefly described. Further details of the chemistry have been appeared in review articles [153].

12.6.1 Platinum(0) Complexes

Tables 12.3 and 12.4 list the examples of platinum(0) complexes together with their syntheses. The complexes are generally prepared by the reduction of K_2PtCl_4 with alcoholic potassium hydroxide, sodium amalgam, or hydrazine in the presence of appropriate ligands.

Table 12.3 Preparation of $[Pt(0)L_n]$ Complexes

L	n	synthetic method	ref
PPh_3	4	$K_2PtCl_4 + L + KOH + EtOH \rightarrow$	154
	4	$PtCl_2L_2 + L + N_2H_4 \rightarrow$	155
	3	$PtL_4\ (\Delta) \rightarrow$	154
PEt_3	4	$K_2PtCl_4 + L + KOH + EtOH \rightarrow$	156
	3	$PtL_4\ (\Delta) \rightarrow$	157
$P(i\text{-}Pr)_3$	3	$K_2PtCl_4 + L + Na(Hg) \rightarrow$	157
PCy_3	2	$K_2PtCl_4 + L + Na(Hg) \rightarrow$	71
$P(t\text{-}Bu)_3$	2	$K_2PtCl_4 + L + Na(Hg) \rightarrow$	71
$P(OPh)_3$	4	$Pt(PPh_3)_4 + L \rightarrow$	158
$P(OEt)_3$	4	$K_2PtCl_4 + L + KOH \rightarrow$	8
cod	2	$PtCl_2(cod) + Li_2cot \rightarrow$	159
dba	2	$K_2PtCl_4 + L + + EtOH + MeCO_2Na \rightarrow$	160
$CH_2{=}CH_2$	3	$Pt(cod)_2 + L \rightarrow$	161

cod = 1,5-cyclooctadiene, dba = dibenzylideneacetone.

Table 12.4 Preparation of Platinum(0) Alkene and Alkyne Complexes $[Pt(0)(un)L_2]$

un	L	synthetic method	ref
$CH_2{=}CH_2$	PEt_3	$PtCl_2L_2$ + un + Na(naphthalene) $\rightarrow$	162
	PPh_3	$PtCl_2L_2$ + un + Na(naphthalene) $\rightarrow$	162
	PPh_3	$Pt(O_2)L_2$ + un + $NaBH_4 \rightarrow$	163
	1/2dppe	$PtCl_2L_2$ + un + Na(naphthalene) $\rightarrow$	162
PhC≡CPh	PPh_3	$PtCl_2L_2$ + un + $N_2H_4 \rightarrow$	164
PhCH=CHPh	PPh_3	$PtCl_2L_2$ + un + $N_2H_4 \rightarrow$	165
dba	PPh_3	$Pt(dba)_2 + L \rightarrow$	160

dppe = 1,2-bis(diphenylphosphino)ethane, dba = dibenzylideneacetone.

12.6.2 Platinum(II) Olefin Complexes

While Zeise's salt, $K^+[PtCl_3(C_2H_4)]^-$, the oldest compound in organometallic chemistry, was originally prepared by the reaction of ethanol with a mixture of $PtCl_2$ and $PtCl_4$ in the presence of KCl [166], nowadays a more convenient synthetic route starting from K_2PtCl_4 and ethylene (eq (135)) is commonly used [167]. The reaction is catalyzed by $SnCl_2$. Zeise's salt is converted to the dimeric complex $[PtCl_2(C_2H_4)]_2$ according to the process shown in eq (136) [168]. This complex induces C—C bond cleavage of cyclopropanes to give platinacyclobutanes (eq (137) and (138)) [169,170].

$$K_2PtCl_4 + C_2H_4 \xrightarrow{SnCl_2} [PtCl_3(C_2H_4)]^- K^+ \quad (135)$$

$$[PtCl_3(C_2H_4)]^- K^+ \xrightarrow{+\text{ pyridine}} PtCl_2(py)(C_2H_4) \xrightarrow{+H^+} [PtCl(\mu\text{-}Cl)(C_2H_4)]_2 \quad (136)$$

$$[PtCl_2(C_2H_4)]_2 + \text{R-cyclopropane} \xrightarrow{+\text{ pyridine}} PtCl_2(py)_2(\text{CH(R)CH}_2\text{CH}_2) + PtCl_2(py)_2(\text{CH}_2\text{CH(R)CH}_2) \quad (137)$$

$$[PtCl_2(C_2H_4)]_2 + \text{bicyclobutane} \xrightarrow{+\text{ pyridine}} PtCl_2(py)_2(\text{bicyclobutane}) \quad (138)$$

The complex $PtCl_2(cod)$, a useful starting material for the synthesis of organoplatinum complexes, can be synthesized by heating $[PtCl_4]^{2-}$ with 1,5-cyclooctadiene (eq (139)) [125a,171]. The platinum(II)-cyclooctadiene complex undergoes addition of various nucleophiles, including alkoxides, amines, and carboxylates, to give the corresponding σ-alkylplatinum(II) complexes (eq (139)) [124b].

$$[PtCl_4]^{2-} + \text{cod} \longrightarrow PtCl_2(\text{cod}) \xrightarrow{+\text{ Nu}} [Pt(\text{C}_8\text{H}_{12}\text{Nu})(\mu\text{-Cl})]_2 \quad (139)$$

12.6.3 Platinum(II) Alkyl Complexes

A number of diorganoplatinum(II) complexes of the type PtR_2L_2 have been synthesized by transmetallation of dihaloplatinum(II) complexes with organometallic reagents (Table 12.5). The cod ligand in *cis*-$PtR_2(cod)$ [172] and the SMe_2 ligand in *cis*-$PtMe_2(SMe_2)_2$ [173] are readily replaced by a variety of phosphine ligands, leading to dialkylbis(tertiary phosphine)-platinum(II) complexes.

Table 12.5 Examples of Diorganoplatinum(II) Complexes [PtR_2L_2]

R	L	geometry	ref
Me	1/2 cod	cis	172
Me	SMe_2	cis	173
Me	PMe_3	cis	174
Me	PMe_2Ph	cis	175
Me	$PMePh_2$	cis	176
Me	PPh_3	cis	177
Et	PPh_3	cis	178
Et	1/2 dppe	cis	178
Bu	PPh_3	cis	179
Et, Pr	PPh_3	cis	180
1/2 $(CH_2)_4$	PPh_3	cis	179
Ph	PEt_3	cis	174
Ph	PEt_3	trans	174
C≡CH	PEt_3	cis	181
C≡CH	PEt_3	trans	181

cod = 1,5-cyclooctadiene, dppe = 1,2-bis(diphenylphosphino)ethane.

Dialkylplatinum(II) complexes bearing β-hydrogens undergo exclusive β-hydrogen elimination on thermolysis [179,180,182]. On the other hand, the di-neopentyl complex, which has no β–hydrogens, undergoes γ-C—H bond activation to provide a platinacyclobutane (eq (140)) [183].

(140)

Alkyl(hydrido)platinum(II) complexes decompose via reductive elimination of alkane [184,185]. When the complex has a bidentate phosphine ligand bearing bulky cyclohexyl groups, the platinum(0) species generated by reductive elimination causes C—H bond activation of a variety of hydrocarbons (eq (141)) [184].

Monoorganoplatinum(II) complexes can be synthesized either by oxidative addition of organic halides to platinum(0) complexes [153], by protonolysis of diorganoplatinum(II) complexes (eq (142)) [174], or by monoalkylation of dihalobis(tertiary phosphine)platinum(II) complexes with 1 equivalent of Grignard reagents (eq (143)) [174,175]. When monodentate phosphines are employed as supporting ligands, the products generally have a trans configuration, whereas the treatment of *cis*-PtR_2L_2 with hydrogen chloride often provides cis complexes. Hydridoplatinum(II) complexes are prepared by the reaction of dichloroplatinum phosphine complexes with hydrazine or $NaBH_4$ [186,187].

$$Cy_2P(CH_2)_2PCy_2Pt(H)(CH_2CMe_3) \xrightarrow{-\,CMe_4} [Cy_2P(CH_2)_2PCy_2Pt]$$

$$[Cy_2P(CH_2)_2PCy_2Pt] \xrightarrow{CH_4} Cy_2P(CH_2)_2PCy_2Pt(H)Me \qquad (141)$$

$$[Cy_2P(CH_2)_2PCy_2Pt] \xrightarrow{C_5H_{10}} Cy_2P(CH_2)_2PCy_2Pt(H)(C_5H_9)$$

$$[Cy_2P(CH_2)_2PCy_2Pt] \xrightarrow{SiMe_4} Cy_2P(CH_2)_2PCy_2Pt(H)(CH_2SiMe_3)$$

$$[Cy_2P(CH_2)_2PCy_2Pt] \xrightarrow{C_6H_6} Cy_2P(CH_2)_2PCy_2Pt(H)Ph$$

$$Me_2Pt(PMe_2Ph)_2 + HCl \xrightarrow[-\,CH_4]{Et_2O} ClPt(Me)(PMe_2Ph)_2 \qquad (142)$$

$$Cl_2Pt(PEt_3)_2 + MeMgI \xrightarrow[-\,MgClI]{benzene} IPt(Me)(PEt_3)_2 \qquad (143)$$

12.7 Synthesis of Platinum Compounds

(1) Dichloro(1,5-cyclooctadiene)platinum(II), $PtCl_2(cod)$ [125a]

$$H_2PtCl_6{\cdot}(H_2O)_x + C_8H_{12} \longrightarrow PtCl_2(C_8H_{12}) \qquad (144)$$

Hydrated chloroplatinic acid (5.0 g, 8.41 mmol) is dissolved in 15 mL of glacial acetic acid, and the solution is heated to 75 °C. Then 1,5-cyclooctadiene (6 mL, 5.32 g,

49.3 mmol) is added to the warm solution and the mixture swirled gently, cooled to room temperature, and diluted with 50 mL of water. The black suspension is allowed to stand at room temperature for 1 h, and the crude product is collected on a Büchner funnel by suction, washed with 50 mL of water, and then with 100 mL of Et_2O. The crude product is suspended in dichloromethane (400 mL), and the mixture is heated to the boiling point and kept at this temperature for 5 min. The solution is cooled, mixed with 5 g of chromatographic-grade silica gel, and allowed to settle. The supernatant should be colorless; if not, additional silica gel (*ca.* 1 g) is added until the solution is clear. The mixture is filtered and the residue is washed with dichloromethane (2 × 50 mL). The combined CH_2Cl_2 solution (*ca.* 500 mL) is concentrated to ca. 75 mL, and the hot solution is poured into 200 mL of petroleum ether (bp 60–70 °C), giving a finely divided white solid (2.55g, 80%). Further purification may be carried out by recrystallization from hot dichloromethane.

Properties: mp 220–278 °C (with decomposition); ^1H NMR ($CDCl_3$) δ 5.62 (=CH, $J_{Pt\text{-}H}$ = 65 Hz) and 2.71 (CH_2); IR 1334, 1179, 1009, 871, 834 and 782 cm^{-1}.

The title compound can be prepared from K_2PtCl_4 and cod in *n*-propanol (Note: cannot use MeOH or EtOH [171].) The related complexes $PtBr_2$(cod) and Pt I_2(cod) can be prepared by similar procedures.

(2) Bis(1,5-cyclooctadiene)platinum(0), Pt(cod)$_2$ [159]

$$2Li + C_8H_8 \longrightarrow Li_2C_8H_8 \qquad (145)$$

$$Li_2C_8H_8 + PtCl_2(C_8H_{12}) + 3C_7H_{10} \longrightarrow Pt(C_7H_{10})_3 + 2LiCl + C_8H_8 + C_8H_{12} \qquad (146)$$

$$Pt(C_7H_{10})_3 + 2C_8H_{12} \longrightarrow Pt(C_8H_{12})_2 + 3C_7H_{10} \qquad (147)$$

Lithium foil (1.0 g, 144 mmol) is suspended under nitrogen in dry Et_2O (200 mL) with magnetic stirring at 0 °C. A 5.0 g sample (48 mmol) of 1,3,5,7-cyclooctatetraene is added and mixture is stirred for 16 h. The small quantity of white precipitate is allowed to settle, and an aliquot of the orange solution is removed with a syringe, and the molarity is checked by hydrolysis and titration against standard acid. A saturated solution of (1,3,5,7-cyclooctatetraene)dilithium is approximatly 0.24 mol L^{-1}.

A 100 mL three-necked round-bottomed flask, fitted with a pressure-equalized dropping funnel, a nitrogen inlet, and a magnetic stirring bar, is placed in a cold bath at –30 °C. Finely powdered $PtCl_2$(cod) (13.1 g, 35 mmol) and 2-norbornene (25 g, 266 mmol) are added through the third neck. These are slurried with diethyl ether (40 mL) at –30 °C. Freshly prepared solution of (1,3,5,7-cyclooctatetraene)dilithium (140 mL of a 0.24 mol dm^{-3} solution) is transferred with a syringe to the dropping funnel, and then added over a 1 h period to the stirred slurry while the temperature is maintained at approximately –30 °C. The tan reaction mixture is then allowed to warm to room temperature and the volatile material removed *in vacuo*. The tan residue is dried at 0.05 mmHg for 1 h to remove any traces of cyclooctatetraene. The flask is filled with nitrogen and the solid scraped from the walls before extraction with hexane (200 mL and then 2 × 50 mL). A few crystals of 2-norbornene are added to each extraction. The combined extracts is filtered through an alumina pad (5 mL Brockman activity II) under a nitrogen atmosphere. A positive pressure may be applied to assist the filtration.

The colorless or pale yellow filtrate obtained is evaporated *in vacuo*. A mass of fine needles precipitate as the volatiles are carefully removed. The product, which should be almost colorless, is dried at room temperature under vacuum (0.05 mmHg) for 10 min (11.0–12.5g, 65–74%). The compound may be recrystallized as long fine colorless needles by slowly cooling a filtered saturated solution in hexane to –20 °C.

Tris(2-norbornene)platinum (1.2 g, 2.5 mmol) is dissolved in petroleum ether (bp 100–120 °C) (25 mL) in a Schlenk tube connected to a N_2 line and 1,5-cyclooctadiene (3 mL) is added. If necessary this solution should be filtered though a fine glass frit using standard Schlenk techniques into another tube. A small magnetic stirring bar is added to the tube, which is then connected to a vacuum line and placed in a water bath at 30 °C on a magnetic stirring hotplate. The volatile compounds are evaporated slowly (1 h) at reduced pressure until approximately 2 mL of solution remain. During the evaporation, off-white crystals of $Pt(cod)_2$ are deposited. The mixture is cooled to 20 °C, nitrogen is readmitted to the tube, and the supernatant liquid is decanted with a syringe. The crystals are washed with hexane (2 × 1 mL) from a syringe and dried under vacuum (0.05 mmHg, 15 min) (0.8g, 78%).

Properties: ^1H NMR (C_6D_6) δ 4.20 (m, 8H, CH, J_{PtH} = 55 Hz) and 2.19 (m, 16H, CH_2); $^{13}C\{^1H\}$ NMR (C_6D_6) δ 73.3 (C=C, J_{PtC} = 143 Hz) and 33.2 (CH_2, J_{PtC} = 15 Hz).

(3) Dimethyl(1,5-cyclooctadiene)platinum(II), $PtMe_2(cod)$ [172]

$$PtI_2(C_8H_{12}) + 2MeLi \longrightarrow PtMe_2(C_8H_{12}) + 2LiI \qquad (148)$$

The complex $PtI_2(cod)$ (11.8 g, 21.2 mmol) is dissolved in Et_2O (100 mL) and the solution is cooled to 0 °C. An Et_2O solution of MeLi (30 mL, 1.95 M, 58.5 mmol) is added and the mixture is stirred for 2 h. The mixture is hydrolyzed with an aqueous solution of ammonium chloride (saturated) at 0 °C. The Et_2O phase is separated, dried over $MgSO_4$ containing a small amount of charcoal and concentrated to dryness, yielding white crystals of the title compound (6.05 g, 87%).

Properties: mp 94–95 °C.

(4) Potassium Trichloro(ethylene)platinate(II), $K[PtCl_3(C_2H_4)]$ [159]

$$K_2PtCl_4 + C_2H_4 + H_2O \xrightarrow{SnCl_2} K[PtCl_3](C_2H_4)\cdot H_2O + KCl \qquad (149)$$

Aqueous hydrochloric acid (5 M, 45 mL) is placed in a 150 mL, three-necked flask equipped with a gas inlet tube, and potassium tertrachloroplatinate(II) (4.5 g, 1.08 mmol) is added. The flask is sealed with rubber septums and the system is deoxygenated immediately by passage of nitrogen or ethylene for 30 min through the solution. Hydrated tin(II) chloride ($SnCl_2\cdot 2H_2O$; 40 mg, 0.2 mmol) is placed in a Schlenk tube, which is sealed with a rubber septum and deoxygenated by flushing nitrogen. With a syringe, deoxygenated distilled water is added to the tin(II) chloride, and the resulting suspension is transferred into the flask containing the chloroplatinate solution. A stream of ethylene is bubbled slowly through the resulting reaction mixture, which is shaken periodically. During the course of reaction (2–4 h), the initially red-brown suspension turns yellow, and most of the solid dissolved as reaction proceeds. The reaction mixture is warmed to 40–45 °C, filtered through a sintered glass filter (paper should not be used), and cooled in an ice bath, yielding yellow needles of Zeise's salt. They are collected by filtration, washed with a small amount of cold water, and air-dried at room temperature (3.6 g, 86%). Pumping *in vacuo* for 16 h removes the water of hydration, giving $K[PtCl_3(C_2H_4)]$.

(5) Di(μ-chloro)dichlorobis(ethylene)diplatinum(II), $[PtCl_2(C_2H_4)]_2$ [168]

$$K[PtCl_3(C_2H_4)] + C_5H_5N \longrightarrow \textit{trans}\text{-}PtCl_2(C_2H_4)(C_5H_5N) + KCl \quad (150)$$

$$\textit{trans}\text{-}PtCl_2(C_2H_4)(C_5H_5N) + 2R\text{–}SO_3H \longrightarrow [PtCl_2(C_2H_4)]_2 + 2R\text{–}SO_3\text{–}C_5H_5NH^+ \quad (151)$$

Potassium trichloro(ethylene)platinate (2.00 g, 5.43 mmol) is dissolved in water (50 mL) and pyridine (0.44 mL, 5.46 mmol) is added dropwise by means of a syringe with stirring. A yellow precipitate forms immediately and the mixture is allowed to stir for 30 min. The precipitate is collected by filtration, washed with water (10 mL x 3), and dried *in vacuo* overnight (1.82 g, 90%).

Properties: mp 125–127 °C.

The complex *trans*-$PtCl_2(C_2H_4)(C_5H_5N)$ thus prepared is fairly stable to air and moisture.

The pyridine complex (400 mg, 4.07 mmol) is dissolved in Et_2O (25 mL) and added to 5.0 g of Dowex 50 W-X8 ion exchange resin in a 50 mL flask equipped with a magnetic stirrer. The mixture is vigorously stirred for 1 h and filtered through a sintered glass filter. The residue is washed with Et_2O (2 × 5 mL), and the combined Et_2O solution is again treated with the ion exchange resin (5 g). This procedure is repeated once again. The yellow filtrate is dried overnight over sodium sulfate at 0 °C. The drying agent is removed by filtration and washed three times with 5 mL of benzene. The combined filtrates are evaporated to dryness under reduced pressure to give an orange solid. The crude product is dissolved in hot benzene with a small amount of charcoal and filtered. On cooling the filtrate, orange crystals of the title compound separate and are collected (130 mg, 40%).

Properties: mp 180–185 (decomposition).

References

[1] (a) P. W. Jolly, G. Wilke, *The Organic Chemistry of Nickel*; Academic Press, New York (1974, 1975), Vols. I and II.
(b) P. W. Jolly, *Comprehensive Organometallic Chemistry*, Ed. by G. Wilkinson, F. G. A. Stone, E. W. Abel, Pergamon Press, Oxford (1982), Vol. 8, p. 613; C. P. Kubiak, A. K. Smith, M. J. Chetcuti, *Comprehensive Organometallic Chemistry II*, Pergamon Press, Oxford (1995), Vol. 9, Ed. E. W. Abel, F. G. A. Stone, G. Wilkinson and R. J. Puddephatt.
(c) K. Tamao, M. Kumada, *The Chemistry of the Metal–Carbon Bond*, Ed. F. R. Hartley, Jola Wiley, New York (1987), p. 819.

[2] (a) L. S. Hegedus, *Organometallics in Synthesis. A Manual*, Ed. M. Schlosser, Jola Wiley, Chichester (1994), p. 383.

(b) Heck R. F. *Palladium Reagents in Organic Synthesis*, Academic Press, London (1985).
(c) J. Tsuji, *Organic Synthesis with Palladium Compounds*, Springer-Verlag, Heidelberg (1980).

[3] T. Saito, Y. Uchida, A. Misono, A. Yamamoto, K. Morifuji, S. Ikeda, *J. Am. Chem. Soc.*, **88**, 5198 (1966).
[4] T. Yamamoto, M. Takamatsu, A. Yamamoto, *Bull. Chem. Soc. Jpn.*, **55**, 325 (1982).
[5] R. A. Schunn, S. D. Ittel, M. A. Cushing, *Inorg. Synth.*, **28**, 94 (1990).
[6] P. W. Jolly, K. Jonas, *Inorg. Synth.*, **15**, 29 (1974).
[7] R. A. Schunn, *Inorg. Synth.*, **13**, 124 (1972).
[8] M. Meier, F. Basolo, *Inorg. Synth.*, **28**, 104 (1990); R. S. Vinal, L. T. Reynolds, *Inorg. Chem.*, **3**, 1062 (1964).
[9] S. D. Ittel, *Inorg. Synth.*, **28**, 102 (1990).
[10] P. W. Jolly, *Comprehensive Organometallic Chemistry*; ed. by G. Wilkinson, F. G. A. Stone, and E. W. Abel, Pergamon Press, Oxford (1982), Vol. 6, p. 101.
[11] E. J. Corey, H. A. Kirst, *J. Am. Chem. Soc.*, **94**, 667 (1972).
[12] M. F. Semmelhack, P. Helquist, L. D. Jones, L. Keller, L. Mendelson, L. S. Ryono, J. G. Smith, and R. D. Stauffer, *J. Am. Chem. Soc.*, **103**, 6460 (1981).
[13] T. Yamamoto, T. Ito, K. Kubota, *Chem. Lett.*, 153 (1988).
[14] M. Zembayashi, K. Tamao, M. Kumada, *Tetrahedron Lett.*, 409 (1977).
[15] K. Takagi, N. Hayama, K. Sasaki, *Bull. Chem. Soc. Jpn.*, **57**, 1887 (1984).
[16] M. F. Semmelhack, B. P. Chong, R. D. Chong, R. D. Stauffer, T. D. Rogerson, A. Chong, L. D. Jones, *J. Am. Chem. Soc.*, **97**, 2507 (1975).
[17] M. Mori, Y. Ban, *Heterocycles*, **9**, 391 (1978).
[18] P. A. Wender, T. E. Jenkins, *J. Am. Chem. Soc.*, **111**, 6432 (1989).
[19] P. A. Wender, N. C. Ihle, *J. Am. Chem. Soc.*, **108**, 4678 (1986).
[20] (a) H. F. Klein, H. H. Karsch, *Chem. Ber.*, **105**, 2628 (1972).
(b) H. F. Klein, *Chem. Ber.*, **345**, 383 (1988).
[21] Y.-J. Kim, K. Osakada, A. Takenaka, A. Yamamoto, *J. Am. Chem. Soc.*, **112**, 1096 (1990).
[22] M. Wada, *J. Organomet. Chem.*, **259**, 245 (1983); *J. Chem. Soc., Dalton Trans.*, 1443 (1982).
[23] T. Yamamoto, A. Yamamoto, S. Ikeda, *J. Am. Chem. Soc.*, **93**, 3350 (1971).
[24] D. G. Morrel, J. K. Kochi, *J. Am. Chem. Soc.*, **97**, 7262 (1975); G. Smith, J. K. Kochi, *J. Organomet. Chem.*, **198**, 199 (1980).
[25] S. Komiya, Y. Abe, A. Yamamoto, T. Yamamoto, *Organometallics*, **2**, 1466 (1983); K. Tatsumi, A. Nakamura, S. Komiya, T. Yamamoto, A. Yamamoto, *J. Am. Chem. Soc.*, **106**, 8181 (1984).
[26] K. Tamao, K. Sumitani, M. Kumada, *Org. Synth.*, *Coll. Vol. VI*, 407 (1988).
[27] K. Tamao, S, Kodama, T. Nakatsuji, Y. Kiso, M. Kumada, *J. Am. Chem. Soc.*, **97**, 4405 (1975).
[28] S. Wadman, R. Whitby, C. Yeates, P. Kocienski, K. Cooper, *J. Chem. Soc., Chem. Commun.*, 241 (1987).
[29] Z.-J. Ni, P.-F. Yang, D. K. P. Ng, Y.-L. Tzeng, T.-Y. Luh, *J. Am. Chem. Soc.*, **112**, 9356 (1990).
[30] T. Hayashi, M. Konishi, M. Fukushima, K. Kanehira, T. Hioki, M. Kumada, *J. Org. Chem.*, **48**, 2195 (1983); T. Hayashi, M. Konishi, M. Fukushima, T. Mise, M. Kagotani, M. Tajika, M. Kumada, *J. Am. Chem. Soc.*, **104**, 180 (1982).
[31] M. F. Semmelhack, *Org. React.*, **19**, 115 (1972).
[32] T. Yamamoto, J. Ishizu, A. Yamamoto, *J. Am. Chem. Soc.*, **103**, 6863 (1981).
[33] S. O'Brien, M. Fishwick, B. McDermott, M. G. H. Wallbridge, G. A. Write, *Inorg. Synth.*, **13**, 79 (1972).
[34] M. F. Semmelhack, P. M. Helquist, *Org. Synth.*, **VI**, 722 (1988).
[35] L. S. Hegedus, R. K. Stiverson, *J. Am. Chem. Soc.*, **96**, 3250 (1974).
[36] L. S. Hegedus, S. Varaprath, *Organometallics*, **1**, 259 (1982).
[37] L. S. Hegedus, S. D. Wagner, E. L. Waterman, K. Sjoberg, *J. Org. Chem.*, **40**, 593 (1975).
[38] R. H. Grubbs, A. Miyashita, *J. Organomet. Chem.*, **161**, 371 (1978).
[39] S. Takahashi, Y. Suzuki, N. Hagihara, *Chem. Lett.*, 1363 (1974); S. Takahashi, Y. Suzuki, K. Sonogashira, N. Hagihara, *J. Chem. Soc., Chem. Commun.*, 839 (1976).

[40] R. H. Grubbs, A. Miyashita, M. Liu, P. Burk, *J. Am. Chem. Soc.*, **100**, 2418 (1978).
[41] T. Yamamoto, *Inorg. Synth.*, **26**, 204 (1989); T. Yamamoto, K. Sano, K. Osakada, S. Komiya, A. Yamamoto, Y. Kushi, T. Toda, *Organometallics*, **9**, 2396 (1990), references cited therein.
[42] H. Hoberg, K. Sümmermann, E. Hernandez, C. Ruppin, D. Guhl, *J. Organomet. Chem.*, **344**, C35 (1988).
[43] K. Tamao, Y. Ito, *J. Am. Chem. Soc.*, **110**, 1286 (1988).
[44] T. Tsuda, T. Kiyoi, T. Miyano, T. Saegusa, *J. Am. Chem. Soc.*, **110**, 8570 (1988).
[45] T. Tsuda, S. Morikawa, N. Hasegawa, T. Saegusa, *J. Org. Chem.*, **55**, 2978 (1990).
[46] L. M. Venanzi, *J. Chem. Soc.*, 719 (1958).
[47] M. A. A. Beg, H. C. Clark, *Can. J. Chem.*, **39**, 595 (1961).
[48] K. A. Jensen, *Z. Anorg. Allg. Chem.*, **229**, 265 (1936).
[49] G. Giacometti , A. Turco, *J. Inorg. Nucl. Chem.*, **15**, 242 (1960).
[50] T. Saito, H. Munakata, H. Imoto, *Inorg. Synth.*, **17**, 84 (1977).
[51] R. C. Cass, G. E. Coates, R. G. Hayter, *J. Chem. Soc.*, 4007 (1955).
[52] E. C. Alyea, D. W. Meek, *J. Am. Chem. Soc.*, **91**, 5761 (1969).
[53] R. G. Hayter, F. S. Humiec, *Inorg. Chem.*, **4**, 1701 (1965).
[54] K. Yamamoto, *Bull. Chem. Soc. Jpn.*, **29**, 501 (1954).
[55] G. Booth, J. Chatt, *J. Chem. Soc.*, 3239 (1965).
[56] G. R. van Hecke, W. D. Horrocks, Jr., *Inorg. Chem.*, **5**, 1968 (1966).
[57] L. Sacconi, J. Gelsomini, *Inorg. Chem.*, **7**, 291 (1968).
[58] T. Hayashi, M. Konishi, K. Yokota, M. Kumada, *Chem. Lett.*, 767 (1980).
[59] E. O. Greaves, C. J. L. Lock, P. M. Maitlis, *Can. J. Chem.*, **46**, 3879 (1968).
[60] G. Wilke, G. Nerrman, *Angew. Chem.*, **74**, 693 (1962).
[61] C. A. Tolman, W. C. Seidel, D. H. Gerlach, *J. Am. Chem. Soc.*, **94**, 2669 (1972).
[62] J. Chatt, B. L. Show, *J. Chem. Soc.*, 1718 (1960).
[63] M. Hidai, T. Kashiwagi, T. Ikeuchi, Y. Uchida, *J. Organomet. Chem.*, **30**, 279 (1971).
[64] T. Yamamoto, T. Kohara, A. Yamamoto, *Bull. Chem. Soc. Jpn.*, **54**, 2010 (1981); S, Komiya, Y. Akai, K. Tanaka, T. Yamamoto, A. Yamamoto, *Organometallics*, **4**, 1130 (1985).
[65] D. R. Coulson, *Inorg. Synth.*, **28**, 107 (1990).
[66] E. O. Greaves, C. J. L. Lock, P. M. Maitlis, *Can. J. Chem.*, **46**, 3879 (1968).
[67] M. Meier, F. Basolo, *Inorg. Synth.*, **28**, 104 (1990).
[68] J. Chatt, F. A. Hart, H. R. Watson, *J. Chem. Soc.*, 2537 (1962).
[69] R. Roffia, F. Conti, G. Gregorio, *Chim. Ind. (Milan)*, **54**, 317 (1962).
[70] Y. Tatsuno, T. Yoshida, S. Otsuka, *Inorg. Synth.*, **28**, 343 (1990).
[71] T. Yoshida, S. Otsuka, *Inorg. Synth.*, **28**, 113 (1990).
[72] A. Visser, R. van der Linde, R. O. de Jongh, *Inorg. Synth.*, **16**, 127 (1976); R. van der Linde, R. O. de Jongh, *J. Chem. Soc., Chem. Commun.*, 563 (1971); R. S. Paonessa, A. L. Prignano, W. C. Trogler, *Organometallics*, **4**, 647 (1985).
[73] F. Ozawa, T. Ito, Y. Nakamura, A. Yamamoto, *J. Organomet. Chem.*, **169**, 375 (1979).
[74] S. Takahashi, N. Hagihara, *J. Chem. Soc. Jpn.*, **88**, 1306 (1967); H. Minematsu, Y. Nonaka, S. Takahashi, N. Hagihara, *J. Organomet. Chem.*, **59**, 395 (1973); K. Ito, S. Hasegawa, Y. Takahashi, Y. Ishii, *J. Organomet. Chem.*, **73**, 401 (1974).
[75] (a) J. Kraus, W. Bonrath, K. R. Pörschke, *Organometallics*, **11**, 1158 (1992).
(b) N. Carr, B. J. Dunne, L. Mole, A. G. Orpen, J. L. Spencer, *J. Chem. Soc., Dalton Trans.*, 863 (1991).
[76] (a) T. Ito, H. Tsuchiya, A. Yamamoto, *Bull. Chem. Soc. Jpn.*, **50**, 1319 (1977).
(b) F. Ozawa, T. Ito, Y. Nakamura, A. Yamamoto, *Bull. Chem. Soc. Jpn.*, **54**, 1868 (1981).
(c) F. Ozawa, T. Ito, A. Yamamoto, *J. Am. Chem. Soc.*, **102**, 6457 (1980).
(d) F. Ozawa, T. Son, S. Ebina, K. Osakada, A. Yamamoto, *Organometallics*, **11**, 171 (1992).
[77] M. Green, J. A. K. Howard, J. L. Spencer, F. G. A. Stone, *J. Chem. Soc., Dalton Trans.*, 271 (1977).
[78] T. Ukai, H. Kawazura, Y. Ishii, *J. Organomet. Chem.*, **65**, 253 (1974).
[79] K. Ito, F. Ueda, K. Hirai, Y. Ishii, *Chem. Lett.*, 877 (1977).

[80] (a) A. J. Canty, in *Comprehensive Organometallic Chemistry II*, ed. E. W. Abel, F. G. A. Stone, G. Wilkinson, R. J. Puddephatt, Pergamon Press, Oxford (1995), Vol. 9, p. 233.
(b) J. F. Fauvarque, F. Pflüger, M. Troupel, *J. Organomet. Chem.*, **208**, 419 (1981).
(c) P. Fitton, E. A. Rick, *J. Organomet. Chem.*, **28**, 287 (1971).
(d) P. F. Fitton, M. P. Johnson, J. E. McKeon, *J. Chem. Soc., Chem. Commun.*, 6 (1968).
(e) J. K. Stille, K. S. Y. Lau, *Acc. Chem. Res.*, **10**, 434 (1977).
[81] A. Jutand, A. Mosleh, *Organometallics*, **14**, 1810 (1995), and references cited therein.
[82] (a) C. Amatore, A. Jutand, F. Khalil, M. A. M'Barki, L. Mottier, *Organometallics*, **12**, 3168 (1993).
(b) C. Amatore, M. Azzabi, A. Jutand, *J. Am. Chem. Soc.*, **113**, 8375 (1991), **113**, 1670 (1991).
(c) C. Amatore, F. Pflüger, *Organometallics*, **9**, 2276 (1990).
(d) M. Portnoy, D. Milstein, *Organometallics*, **12**, 1665 (1993).
(e) V. V. Grushin, H. Alper, *Chem. Rev.*, **94**, 1047 (1994).
[83] (a) G. Calvin, G. E. Coates, *J. Chem. Soc.*, 2008 (1960).
(b) Y. Yamamoto, H. Yamazaki, *Inorg. Chem.*, **13**, 438 (1974).
[84] W. de Graaf, J. Boersma, W. J. J. Smeets, A. L. Spek, G. van Koten, *Organometallics*, **8**, 2907 (1989); D. R. Schaad, C. R. Landis, *Organometallics*, **11**, 2024 (1992); F. Ozawa, T. Sugimoto, Y. Yuasa, M. Santra, T. Yamamoto, A. Yamamoto, *Organometallics*, **3**, 683 (1984); S. Komiya, Y. Akai, T. Yamamoto, A. Yamamoto, *Organometallics*, **4**, 1130 (1985); Y.-J. Kim, K. Osakada, T. Yamamoto, A. Yamamoto, *Organometallics*, **7**, 2182 (1988); Y.-J. Kim, K. Osakada, A. Takenaka, A. Yamamoto, *J. Am. Chem. Soc.*, **112**, 1096 (1990).
[85] (a) R. F. Heck, *Org. React.*, **27**, 345 (1982).
(b) G. D. Daves, Jr., A. Hallberg, *Chem. Rev.*, **89**, 1433 (1989).
(c) W. Cabri, I. Candiani, *Acc. Chem. Res.*, **28**, 2 (1995).
[86] T. Mizoroki, K. Mori, A. Ozaki, *Bull. Chem. Soc. Jpn.*, **44**, 581 (1971); K. Mori, T. Mizoroki, A. Ozaki, *Bull. Chem. Soc. Jpn.*, **46**, 1505 (1973).
[87] R. F. Heck, J. P. Nolley, Jr., *J. Org. Chem.*, **37**, 2320 (1972); H. A. Dieck, R. F. Heck, *J. Am. Chem. Soc.*, **96**, 1133 (1974).
[88] S. A. Buntin, R. F. Heck, *Org. Synth.*, **61**, 82 (1983).
[89] W. J. Scott, M. R. Pena, K. Swärd, S. J. Stoessel, J. K. Stille, *J. Org. Chem.*, **50**, 2302 (1985).
[90] C. Amatore, E. Carré, A. Jutand, M. A. M'Barki, *Organometallics*, **14**, 1818 (1995); C. Amatore, A. Jutand, M. A. M'Barki, *Organometallics*, **11**, 3099 (1993); F. Ozawa, A. Kubo, T. Hayashi, *Chem. Lett.*, 2177 (1992); V. V. Grushin, H. Alper, *Organometallics*, **12**, 1890 (1993).
[91] G. Z. Wu, F. Lamaty, E. Negishi, *J. Org. Chem.*, **54**, 2507 (1984).
[92] Y. Sato, S. Watanabe, M. Shibasaki, *Tetrahedron Lett.*, **33**, 2589 (1992); Y. Sato, M. Sodeoka, M. Shibasaki, *Chem. Lett.*, 1953 (1990).
[93] A. Ashimori, L. E. Overman, *J. Org. Chem.*, **57**, 4571 (1992).
[94] P. E. Garrou, R. F. Heck, *J. Am. Chem. Soc.*, **98**, 4115 (1976); G. K. Anderson, R. J. Cross, *Acc. Chem. Res.*, **17**, 67 (1984), and references cited therein.
[95] (a) F. Ozawa, N. Kawasaki, H. Okamoto, T. Yamamoto, A. Yamamoto, *Organometallics*, **6**, 1640 (1987).
(b) F. Ozawa, H. Soyama, H. Yanagihara, I. Aoyama, H. Takino, K. Izawa, T. Yamamoto, A. Yamamoto, *J. Am. Chem. Soc.*, **107**, 3235 (1985).
(c) J. Chen, A. Sen, *J. Am. Chem. Soc.*, **106**, 1506 (1984).
(d) A. Sen, J. Chen, E. M. Vetter, R. R. Whittle, *J. Am. Chem. Soc.*, **109**, 148 (1987).
(e) W. R. Moser, A. W. Wang, N. K. Kildahl, *J. Am. Chem. Soc.*, **110**, 2816 (1988).
[96] L. Huang, F. Ozawa, A. Yamamoto, *Organometallics*, **9**, 2603, 2612 (1990).
[97] A. Cowell, J. K. Stille, *J. Am. Chem. Soc.*, **102**, 4193 (1980); M. Mori, K. Chiba, N. Inotsume, Y. Ban, *Heterocycles*, **12**, 921 (1979).
[98] M. Mori, Y. Uozumi, M. Kimura, Y. Ban, *Tetrahedron*, **42**, 3793 (1986).
[99] F. Ozawa, H. Yanagihara, A. Yamamoto, *J. Org. Chem.*, **51**, 415 (1986).
[100] A. Sen, *Acc. Chem. Res.*, **26**, 303 (1993); B. A. Markies, D. Kruis, M. H. P. Rietveld, K. A. N. Verkerk. J. Boersma, H. Kooijman, M. T. Lakin, A. L. Spek, G. van Koten, *J. Am. Chem. Soc.*, **117**, 5263 (1995), and references cited therein.

[101] M. Brookhart, F. C. Rix, J. M. DeSimone, *J. Am. Chem. Soc.*, **114**, 5894 (1992); M. Brookhart, M. I. Wagner, *J. Am. Chem. Soc.*, **116**, 3641 (1994); K. Nozaki, N. Sato, H. Takaya, *J. Am. Chem. Soc.*, **117**, 9911 (1995).
[102] (a) R. Tooze, K. W. Chiu, G. Wilkinson, *Polyhedron*, **3**, 1025 (1984).
(b) W. de Graaf, J. Boersma, W. J. J. Smeets, A. L. Spek, G. van Koten, *Organometallics*, **8**, 2907 (1989).
(c) A. Gillie, J. K. Stille, *J. Am. Chem. Soc.*, **102**, 4933 (1980).
(d) A. J. Canty, *Acc. Chem. Res.*, **23**, 155 (1992).
[103] H. Nakazawa, F. Ozawa, A. Yamamoto, *Organometallics*, **2**, 241 (1983).
[104] (a) M. K. Loar, J. K. Stille, *J. Am. Chem. Soc.*, **103**, 4174 (1981).
(b) J. M. Brown, N. A. Cooley, *Organometallics*, **9**, 353 (1990).
(c) F. Ozawa, K. Kurihara, M. Fujimori, T. Hidaka, T. Toyoshima, A. Yamamoto, *Organometallics*, **8**, 180 (1989).
[105] J. M. Brown, N. A. Cooley, *Chem. Rev.*, **88**, 1031 (1988).
[106] T. Hayashi, M. Konishi, H. Ito, M. Kumada, *J. Am. Chem. Soc.*, **104**, 4962 (1982); T. Hayashi, M. Konishi, M. Kumada, *J. Org. Chem.*, **51**, 3772 (1986).
[107] T. Hayashi, M. Konishi, Y. Kobori, M. Kumada, T. Higuchi, K. Hirotsu, *J. Am. Chem. Soc.*, **106**, 158 (1984).
[108] W. J. Scott, G. T. Crisp, J. K. Stille, *Org. Synth.*, **68**, 116 (1989).
[109] E. Negishi, T. Takahashi, A. O. King, *Org. Synth.*, **66**, 67 (1987).
[110] W. J. Thompson, J. Gaudion, *J. Org. Chem.*, **49**, 5237 (1984).
[111] N. Miyaura, T. Ishiyama, H. Sasaki, M. Ishikawa, M. Satoh, A. Suzuki, *J. Am. Chem. Soc.*, **111**, 314 (1989).
[112] N. Miyaura, A. Suzuki, *Org. Synth.*, **68**, 130 (1989).
[113] S. Baba, E. Nigishi, *J. Am. Chem. Soc.*, **98**, 6729 (1976).
[114] N. Okukado, D. E. van Horn, W. L. Klima, E. Negishi, *Tetrahedron Lett.*, 1027 (1978); E. Negishi, D. E. van Horn, *J. Am. Chem. Soc.*, **99**, 3168 (1977).
[115] S. Takahashi, Y. Kuroyama, K. Sonogashira, N. Hagihara, *Synthesis*, 627 (1980).
[116] Y. Tamaru, H. Ochiai, T. Nakamura, Z. Yoshida, *Org. Synth.*, **67**, 98 (1988).
[117] M. W. Logue, K. Teng, *J. Org. Chem.*, **47**, 2549 (1982).
[118] I. Omae, *Coord. Chem. Rev.*, **32**, 235 (1980); A. D. Ryabov, *Synthesis*, 233 (1985).
[119] N. Barr, S. F. Dyke, S. N. Quessy, *J. Organomet. Chem.*, **253**, 391 (1983).
[120] F. Ozawa, I. Yamagami, M. Nakano, F. Fujisawa, A. Yamamoto, *Chem. Lett.*, 125 (1989).
[121] H. Horino, N. Inoue, **46**, 4416 (1981).
[122] S.-I. Murahashi, Y. Tamba, M. Yamamura, N. Yoshimura, *J. Org. Chem.*, **21**, 4009 (1978).
[123] L. S, Hegedus, *Tetrahedron*, **40**, 2415 (1984); O. Eisenstein, R. Hoffmann, *J. Am. Chem. Soc.*, **103**, 4308 (1981).
[124] (a) J. K. Stille, D. E. James, *J. Organomet. Chem.*, **108**, 401 (1976).
(b) L. A. P. Kane-Maguire, E. D. Honig, D. A. Sweigart, *Chem. Rev.*, **84**, 525 (1984).
[125] (a) D. Drew, J. R. Doyle, *Inorg. Synth.*, **13**, 47 (1972); **28**, 346 (1990).
(b) J. Chatt, L. M. Vallarino, L. M. Venanzi, *J. Chem. Soc.*, 3413 (1957).
(c) M. S. Kharasch, R. C. Seyler, F. R. Mayo, *J. Am. Chem. Soc.*, **60**, 822 (1938).
(d) J. R. Doyle, P. E. Slade, H. B. Jonassen, *Inorg. Synth.*, **6**, 218 (1960).
[126] T. Takahashi, S. Hashiguchi, K. Kasuga, J. Tsuji, *J. Am. Chem. Soc.*, **100**, 7424 (1978).
[127] J. Tsuji, H. Nagashima, K. Hori, *Chem. Lett.*, 257 (1980); J. Tsuji, *Synthesis*, 369 (1984).
[128] An alternative catalytic process via insertion of olefin into a XPd–OOH intermediate has been proposed: H. Mimoun, *Angew. Chem., Int. Ed. Engl.*, **21**, 734 (1982); T. Hosokawa, S.-I. Murahashi, *Acc. Chem. Res.*, **23**, 49 (1990).
[129] L. S. Hegedus, B. Åkermark, K. Zetterberg, L. F. Olsson, *J. Am. Chem. Soc.*, **106**, 7122 (1984); L. S. Hegedus, O. P. Anderson, K. Zetterberg, G. Allen, K. Siirala-Hansen, D. J. Olsen, A. B. Packard, *Inorg. Chem.*, **16**, 1887 (1977).
[130] H. Kurosawa, T. Majima, N. Asada, *J. Am. Chem. Soc.*, **102**, 6996 (1980).
[131] L. S. Hegedus, R. E. Williams, M. A. McGuire, T. Hayashi, *J. Am. Chem. Soc.*, **102**, 4973 (1980); L. S. Hegedus, W. H. Darlington, *J. Am. Chem. Soc.*, **102**, 4980 (1980).
[132] Y. Tatsuno, T. Yoshida, S. Otsuka, *Inorg. Synth.*, **28**, 343 (1990); W. T. Dent, R. Long, A. J.

Wilkinson, *J. Chem. Soc.*, 1585 (1964); B. Åkermark, B. Krakenberger, S. Hansson, A. Vitagliano, *Organometallics*, **6**, 620 (1987).
[133] J. Powell, B. L. Show, *J. Chem. Soc.* (*A*), 774 (1968).
[134] T. Yamamoto, O. Saito, A. Yamamoto, *J. Am. Chem. Soc.*, **103**, 5600 (1981); T. Yamamoto, M. Akimoto, O. Saito, A. Yamamoto, *Organometallics*, **5**, 1559 (1986).
[135] M. Oshima, I. Shimizu, A. Yamamoto, F. Ozawa, *Organometallics*, **10**, 1221 (1991); F. Ozawa, T.-I. Son, S. Ebina, K. Osakada, A. Yamamoto, *Organometallics*, **11**, 171 (1992).
[136] (a) T. Hayashi, T. Hagihara, M. Konishi, M. Kumada, *J. Am. Chem. Soc.*, **105**, 7767 (1983).
(b) T. Hayashi, M. Konishi, M. Kumada, *J. Chem. Soc., Chem. Commun.*, 107 (1984).
(c) B. M. Trost, *Acc. Chem. Res.*, **13**, 385 (1980).
(d) C. G. Frost, J. H. Howarth, J. M. J. Williams, *Tetrahedron Asymmetry*, **3**, 1089 (1992).
(e) G. Consiglio, R. M. Waymouth, *Chem. Rev.*, **89**, 257 (1989).
[137] I. Stary, P. Kocovsky, *J. Am. Chem. Soc.*, **111**, 4981 (1989); H. Kurosawa, H. Kajimura, S. Ogoshi, H. Yoneda, K. Miki, N. Kasai, S. Murai, I. Ikeda, *J. Am. Chem. Soc.*, **114**, 8417 (1992).
[138] For detailed mechanism of the reductive elimination of intermediate **16**, see: H. Kurosawa, M. Emoto, H. Ohnishi, K. Miki, N. Kasai, K. Tatsumi, A. Nakamura, *J. Am. Chem. Soc.*, **109**, 6333 (1987); H. Kurosawa, H. Ohnishi, M. Emoto, Y. Kawasaki, S. Murai, *J. Am. Chem. Soc.*, **110**, 6272 (1988).
[139] J.-E. Bäckvall, J.-O. Vågberg, K. L. Granberg, *Tetrahedron Lett.*, **30**, 617 (1989).
[140] B. M. Trost, E. Keinan, *J. Am. Chem. Soc.*, **100**, 7779 (1978); *J. Org. Chem.*, **44**, 3451 (1979).
[141] D. R. Deardroff, D. C. Myles, *Org. Synth.*, **67**, 114 (1988).
[142] L. Del Valle, J. K. Stille, L. S. Hegedus, *J. Org. Chem.*, **55**, 3019 (1990); E. Negishi, H. Matsushita, *Org. Synth.*, **62**, 31 (1984); M. Oshima, H. Yamazaki, I. Shimizu, M. Nisar, J. Tsuji, *J. Am. Chem. Soc.*, **111**, 6280 (1989).
[143] J. Tsuji, I. Shimizu, I. Minami, Y. Ohashi, T. Sugiura, K. Takahashi, *J. Org. Chem.*, **50**, 1523 (1985).
[144] F. M. Hauser, R. Tommasi, P. Hewawasam, Y. S. Rho, *J. Org. Chem.*, **53**, 4886 (1988).
[145] W. Oppolzer, *Angew. Chem., Int. Ed. Engl.*, **28**, 38 (1989); *Pure Appl. Chem.*, **62**, 1941 (1990).
[146] B. M. Trost, S. M. Mignani, *Tetrahedron Lett.*, **27**, 4137 (1986); B. M. Trost, J. Lynch, P. Renant, D. H. Steinman, *J. Am. Chem. Soc.*, **108**, 284 (1986); B. M. Trost, *Angew. Chem., Int. Ed. Engl.*, **25**, 1 (1986); *Pure Appl. Chem.*, **60**, 1615 (1988).
[147] B. M. Trost, P. E. Strege, L. Weber, T. J. Fullerton, T. J. Dietsche, *J. Am. Chem. Soc.*, **100**, 3407 (1978); B. M. Trost, P. J. Metzer, *J. Am. Chem. Soc.*, **102**, 3572 (1980); B. M. Trost, *Acc. Chem. Res.*, **13**, 385 (1980).
[148] J.-E. Bäckvall, R. E. Nordberg, K. Zatterberg, B. Åkermark, *Organometallics*, **2**, 1625 (1983); J.-E. Bäckvall, R. E. Nordberg, D. Wilhelm, *J. Am. Chem. Soc.*, **107**, 6892 (1985). See also: J.-E. Bäckvall, *Advances in Metal–Organic Chemistry*, Ed. L. S. Liebeskind, JAI Press, Greenwich (1989), p. 135; J. E. Nyström, T. Rein, J.-E. Bäckvall, *Org. Synth.*, **67**, 105 (1988); J.-E. Bäckvall, J. O. Vågberg, *Org. Synth.*, **69**, 38 (1990).
[149] S. Okeya, S. Ooi, K. Matsumoto, Y. Nakamura, S. Kawagura, *Bull. Chem. Soc. Jpn.*, **54**, 1085 (1981).
[150] J. M. Jenkins, B. L. Shaw, *J. Chem. Soc.* (*A*), 770 (1966); F. G. Mann, D. Purdie, *J. Chem. Soc.*, 1549 (1935).
[151] A. C. Cope, E. C. Friedrich, *J. Am. Chem. Soc.*, **90**, 909 (1968).
[152] R. B. Moffett, *Org. Synth. Coll. Vol. IV*, 238 (1963); R. B. King, F. G. A. Stone, *Inorg. Synth.*, **7**, 99 (1963).
[153] (a) E. W. Abel, F. G. A. Stone, G. Wilkinson, R. J. Puddephatt, eds., *Comprehensive Organometallic Chemistry II*; Pergamon Press, Oxford (1995), Vol. 9, Chapters 7–9.
(b) P. F. R. Hartley, *Comprehensive Organometallic Chemistry*, Ed. G. Wilkinson, F. G. A. Stone, E. W. Abel, Pergamon Press, Oxford (1982), Vol. 8, p. 471.
(c) G. K. Anderson, *Chemistry of the Platinum Group Metals. Recent Developments*, Ed. F. R. Hartley, Elsevier, Amsterdam (1991).

[154] R. Ugo, F. Cariati, G. La Monica, *Inorg. Synth.*, **28**, 123 (1990).
[155] L. Malatesta, C. Cariello, *J. Chem. Soc.*, 2323 (1958).
[156] T. Yoshida, T. Matsuda, S. Otsuka, *Inorg. Synth.*, **28**, 122 (1990).
[157] T. Yoshida, T. Matsuda, S. Otsuka, *Inorg. Synth.*, **28**, 119 (1990).
[158] J. J. Levison, S. D. Robinson, *Inorg. Synth.*, **13**, 105 (1972).
[159] L. E. Crascall, J. L. Spencer, *Inorg. Synth.*, **28**, 126 (1990).
[160] W. J. Cherwinski, B. F. G. Johnson, J. Lewis, *J. Chem. Soc., Dalton Trans.*, 1405 (1974).
[161] L. E. Crascall, J. L. Spencer, *Inorg. Synth.*, **28**, 129 (1990).
[162] R. A. Head, *Inorg. Synth.*, **28**, 132 (1990).
[163] C. D. Cook, G. S. Jauhal, *J. Am. Chem. Soc.*, **90**, 1464 (1968).
[164] D. M. Blake, D. M. Roundhill, *Inorg. Chem.*, **18**, 122 (1979).
[165] J. Chatt, B. L. Shaw, A. A. Williams, *J. Chem. Soc.*, 3269 (1962).
[166] W. C. Zeise, *Pogg. Ann.*, **9**, 632 (1827); W. C. Zeise, *Mag. Pharm.*, **35**, 105 (1830).
[167] P. B. Chock, J. Halpern, F. E. Paulik, *Inorg. Synth.*, **28**, 349 (1990).
[168] P. J. Buss, B. Greene, M. Orchin, *Inorg. Synth.*, **20**, 181 (1980).
[169] F. J. McQuillin, K. G. Powell, *J. Chem. Soc., Dalton Trans.*, 2123 (1972); R. J. Al-Essa, R. J. Puddephatt, M. A. Quyser, C. F. H. Tipper, *J. Am. Chem. Soc.*, **101**, 364 (1979); C. P. Casey, D. M. Scheck, A. J. Shusterman, *J. Chem. Soc.*, **101**, 4233 (1979); R. J. Al-Essa, S. S. M. Ling, R. J. Puddephatt, *Organometallics*, **6**, 951 (1987), and references cited therein.
[170] A. Miyashita, M. Takahashi, H. Takaya, *J. Am. Chem. Soc.*, **103**, 6257 (1981); A. Miyashita, Y. Watanabe, H. Takaya, *Tetrahedron Lett.*, 2959 (1983).
[171] J. Chatt, L. M. Vallarino, L. M. Venanzi, *J. Chem. Soc.*, 2496 (1957); H. C. Clark, L. E. Manzer, *J. Organomet. Chem.*, **59**, 411 (1973).
[172] H. C. Clark, L. E. Manzer, *J. Organomet. Chem.*, **59**, 411 (1973).
[173] J. D. Scott, R. J. Puddephatt, *Organometallics*, **2**, 1643 (1983).
[174] J. Chatt, B. L. Shaw, *J. Chem. Soc.*, 4020 (1959).
[175] J. D. Ruddick, B. L. Shaw, *J. Chem. Soc. (A)*, 2801 (1969).
[176] M. A. Bennett, H. Chee, G. B. Robertson, *Inorg. Chem.*, **18**, 1061 (1979).
[177] J. Chatt, B. L. Shaw, *J. Chem. Soc.*, 705 (1959).
[178] D. A. Slack, M. C. Baird, *Inorg. Chim. Acta*, **24**, 277 (1977).
[179] G. M. Whitesides, J. F. Gassch, E. R. Stedronsky, *J. Am. Chem. Soc.*, **94**, 5258 (1972); G. B. Young, G. M. Whitesides, *J. Am. Chem. Soc.*, **100**, 5808 (1978); J. X. McDermott, J. F. White, G. M. Whitesides, *J. Am. Chem. Soc.*, **98**, 6521 (1976).
[180] S. Komiya, Y. Morimoto, A. Yamamoto, T. Yamamoto, *Organometallics*, **1**, 1528 (1982).
[181] K. Sonogashira, Y. Fujikura, T. Yatake, N. Toyoshima, S. Takahashi, N. Hagihara, *J. Organomet. Chem.*, **145**, 101 (1978).
[182] G. M. Whitesides, *Pure Appl. Chem.*, **53**, 287 (1981).
[183] P. Foley, R. DiCosimo, G. M. Whitesides, *J. Am. Chem. Soc.*, **102**, 6713 (1980).
[184] M. Hackett, J. A. Ibers, G. M. Whitesides, **110**, 1436, 1449 (1988).
[185] L. Abis, A. Sen, J. Halpern, *J. Am. Chem. Soc.*, **100**, 2915 (1979).
[186] G. W. Parshall, *Inorg. Synth.*, **12**, 28 (1970).
[187] H. C. Clark, H. Kurosawa, *J. Organomet. Chem.*, **36**, 399 (1972).

13 Groups 11 and 12 (Cu, Au, Zn) Metal Compounds

F. Ozawa, *Osaka City University*

13.1 Organocopper Complexes

Organocopper complexes are among the most extensively used organometallic reagents in organic synthesis [1]. These complexes are generally prepared *in situ* from copper(I) salts and organometallic compounds of Li, Mg and Zn, and are used in organic synthesis without prior isolation. Scheme 13.1 summarizes the synthetic routes to organocopper complexes using organolithium compounds. It is seen that an extremely wide variety of complexes have been prepared. The reactivity may be inge-

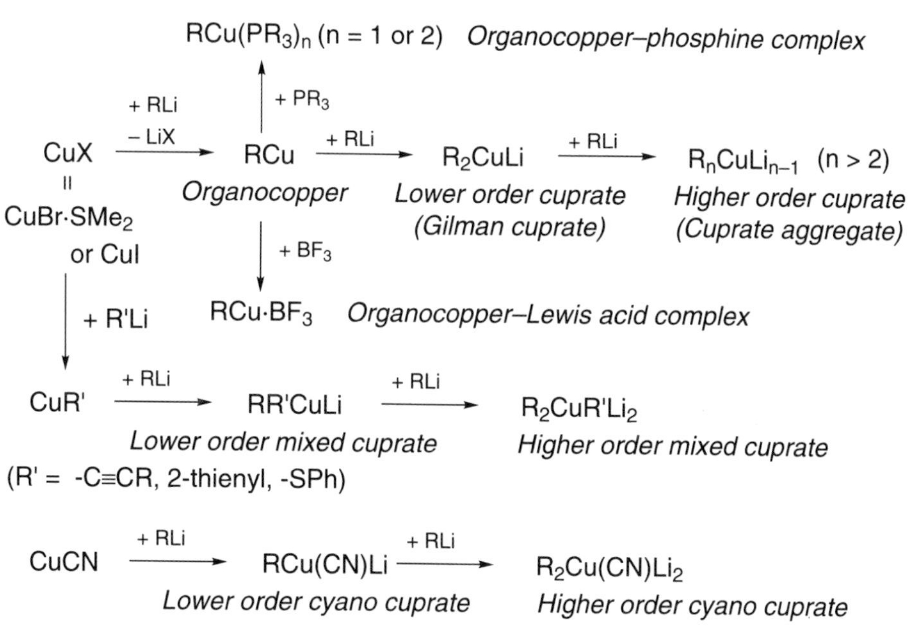

$RCu(PR_3)_n$ (n = 1 or 2) *Organocopper–phosphine complex*

CuX (= $CuBr{\cdot}SMe_2$ or CuI) $\xrightarrow[-\,LiX]{+\,RLi}$ RCu (*Organocopper*) $\xrightarrow{+\,RLi}$ R_2CuLi (*Lower order cuprate (Gilman cuprate)*) $\xrightarrow{+\,RLi}$ R_nCuLi_{n-1} (n > 2) (*Higher order cuprate (Cuprate aggregate)*)

$RCu \xrightarrow{+\,PR_3} RCu(PR_3)_n$

$RCu \xrightarrow{+\,BF_3} RCu{\cdot}BF_3$ *Organocopper–Lewis acid complex*

$CuX \xrightarrow{+\,R'Li} CuR'$

$CuR' \xrightarrow{+\,RLi} RR'CuLi$ (*Lower order mixed cuprate*) $\xrightarrow{+\,RLi} R_2CuR'Li_2$ (*Higher order mixed cuprate*)

(R' = -C≡CR, 2-thienyl, -SPh)

$CuCN \xrightarrow{+\,RLi} RCu(CN)Li$ (*Lower order cyano cuprate*) $\xrightarrow{+\,RLi} R_2Cu(CN)Li_2$ (*Higher order cyano cuprate*)

Scheme 13.1

Synthesis of Organometallic Compounds: A Practical Guide. Edited by S. Komiya

niously tuned by variation of the CuX/RM ratio, the choice of starting copper salts and organometallic reagents, and the additional components such as Lewis acids and bases. Despite their extensive applications in organic synthesis, the exact structures of most organocopper complexes are still not well understood; some of the complexes are apparently not a single species, but an equilibrium mixture of several components. Since detailed information on the preparation and usage of organocopper complexes has appeared in a recent review article [1a], this section briefly presents the synthetic procedures for some isolable, well characterized complexes.

13.2 Synthesis of Cu Compounds

(1) Methyltris(triphenylphosphine)copper(I), $CuMe(PPh_3)_3 \cdot THF$ [2a]

$$Cu(acac)_2 + 3PPh_3 + 2AlEt_2(O^iPr) \longrightarrow CuMe(PPh_3)_3 \quad (1)$$

Bis(2,4-pentanedionato)copper(II) obtained from a commercial source should be dried under vacuum before use. The methyl aluminum reagent $AlMe_2(O^iPr)$ can be prepared by slow addition of *i*-PrOH to a hexane solution of $AlMe_3$ at –78 °C and purified by vacuum distillation (bp 34–36 °C/0.05 mmHg).

Diethyl ether (20 mL) is added to a Schlenk tube containing $Cu(acac)_2$ (0.500 g, 1.91 mmol) and PPh_3 (1.75 g, 6.68 mmol). This dissolves the PPh_3 and forms a light blue suspension of $Cu(acac)_2$. The flask is cooled to –78 °C, and $AlMe_2(O^iPr)$ (0.888 g, 7.64 mmol) is added. The system is allowed to warm slowly to –25 °C at which point the color of the suspension begins to lighten to yield a yellow solid and a yellow solution. The solution is cooled to –100 °C and filtered as cold as possible to yield a yellow solid. The yellow solid is dissolved in a minimum amount of THF, carefully layered with an equal volume of pentane, and held at –15 °C for several days, to yield yellow needles. The solid can be handled briefly at 25 °C without significant decomposition.

Properties: ^1H NMR (C_6D_6) δ 7.42 (br s, *o*-Ph), 7.01 (br s, m, *p*-Ph), 3.57 (m, THF), 1.39 (m, THF) and 0.44 (br s, Cu—Me); ^{31}P NMR (C_6D_6) δ 4.3 (br s); IR 2770 cm^{-1} (CuMe, ν(C—H)). The compound has a tetrahedral structure as confirmed by X-ray crystallography [2a].

(2) [Tetrakis(trimethylphosphine)copper(I)] Dimethylcuprate(I), $[Cu(PMe_3)_4]$ $[CuMe_2]$ [3]

$$Cu_2(OAc)_4 + 4PMe_3 + Me_2Mg \longrightarrow [Cu(PMe_3)_4][CuMe_2] + 2Mg(OAc)_2 \quad (2)$$

To a suspension of anhydrous tetrakis(acetato)dicopper(II) (0.93 g, 2.56 mmol) in diethyl ether (300 mL) at 0 °C are added a diethyl ether solution of dimethylmagnesium (0.45 M, 11.3 mL) and trimethylphosphine (2.10 mL, 21.3 mmol). The slurry slowly turns colorless upon stirring for 4 h. After being warmed to room temperature, the solvent is removed by pumping. The residue is washed with pentane (30 mL) and extracted with toluene (30 mL). The filtered extract is cooled to –20 °C, resulting in the formation of colorless prisms (0.64 g, 55%).

Properties: mp 89 °C (decomposition); ^{1}H NMR (C_6D_6) δ 0.93 (br, PMe_3), 0.15 (s,CuMe). The X-ray analysis showed that the cationic and anionic parts have tetrahedral and linear geometries, respectively [3].

(3) (η^5-Pentamethylcyclopentadienyl)(triphenylphosphine)copper(I), $Cp^*Cu(PPh_3)_3$ [4]

$$Cu_4Cl_4(PPh_3)_4 + 4Li(C_5Me_5) \longrightarrow 4Cu(\eta^5\text{-}C_5Me_5)(PPh_3) + 4LiCl \quad (3)$$

The starting copper(I) complex $Cu_4Cl_4(PPh_3)_4$ is prepared by mixing CuCl and 1.5 equivalents of triphenylphosphine in ethanol under reflux (70%) [5].

Lithium pentamethylcyclopentadienyl (0.86 g, 6.0 mmol) and $Cu_4Cl_4(PPh_3)_4$ (2.19 g, 1.5 mmol) are added to a 250 mL round-bottom flask with a side arm. The flask is cooled to –10 °C and *ca.* 100 mL of THF is added. The reaction mixture is stirred magnetically for 6 h while warming slowly to 25 °C. The volatile materials are removed under vacuum and the resulting residue extracted with diethyl ether. The combined extracts are filtered through decolorizing charcoal and Celite. The filtrate is concentrated under vacuum and then cooled overnight at –78 °C, yielding off-white crystals of the title compound (1.21 g, 43%).

Properties: mp 145 °C (decomposition); ^{1}H NMR (C_6D_6) δ 2.31 (s, C_5Me_5), 6.88–7.20 (m, Ph).

(4) Mesitylcopper(I), $Cu_5(Mesityl)_5$ [6]

$$5CuCl + 5MesitylMgBr \longrightarrow Cu_5(Mesityl)_5 + 5MgBrCl \quad (4)$$

A THF solution (178 mL) of mesitylmagnesium bromide (0.895 M, 0.159 mol) is added to a cooled (–20 °C) suspension of CuCl (15.8 g, 0.159 mol) in 280 mL of THF. The gray-brown reaction mixture is stirred for 20 h and warmed to room temperature. *p*-Dioxane (95 mL, 1.08 mol) is added dropwise, and stirring is continued for 12 h. The magnesium halides are filtered off, and the green filtrate is brought to dryness, the residue being dried under high vacuum for 2 h. It is re-dissolved in warm (50 °C) toluene (200 mL), the remaining magnesium halides are filtered off, and the solution is carefully concentrated to 80 mL and allowed to stand in the refrigerator. Yellow crystals of the title compound (15.32 g) can be isolated. From the mother liquor another crop of 3.74 g of the compound separates after concentration. The crystals are washed with a small volume of cooled toluene and dried under vacuum for a short time. Total yield: 19.06 g (65%). This product has a pentameric structure in the solid state ($Cu_5(Mesityl)_5$), while in solution it is in an equilibrium with the dimeric form ($Cu_5(Mesityl)_5$).

Properties: ^{1}H NMR (toluene-d_8) δ 6.54 (s, 2H, C_6H_2), 2.88 (s, 6H, *o*-Me), 1.88 (s, 3H, *p*-Me) for [$Cu_5(Mesityl)_5$]; δ 6.62 (s, 2H, C_6H_2), 2.99 (s, 6H, *o*-Me), 1.99 (s, 3H, *p*-Me) for [$Cu_2(Mesityl)_2$].

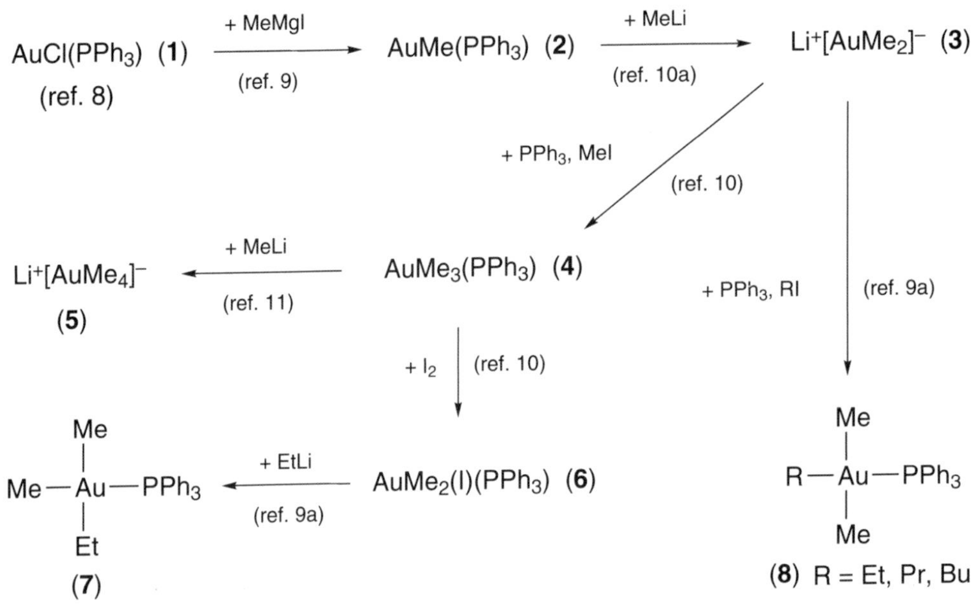

$AuCl(PPh_3)$ (**1**) (ref. 8) — + MeMgI (ref. 9) → $AuMe(PPh_3)$ (**2**) — + MeLi (ref. 10a) → $Li^+[AuMe_2]^-$ (**3**)

(**3**) — + PPh_3, MeI (ref. 10) → $AuMe_3(PPh_3)$ (**4**)

(**4**) — + MeLi (ref. 11) → $Li^+[AuMe_4]^-$ (**5**)

(**4**) — + I_2 (ref. 10) → $AuMe_2(I)(PPh_3)$ (**6**)

(**6**) — + EtLi (ref. 9a) → Me—Au(Me)(Et)—PPh_3 (**7**)

(**3**) — + PPh_3, RI (ref. 9a) → R—Au(Me)(Me)—PPh_3 (**8**) R = Et, Pr, Bu

Scheme 13.2

13.3 Organogold Complexes

Organogold complexes commonly exist in +1 and +3 oxidation states [7]. Gold(I) usually forms linear two-coordinate species, although three- and four-coordinate complexes having trigonal planar and tetrahedral structures, respectively, are also known. On the other hand, organogold(III) complexes having the d^8-electron configuration strongly prefer the coordination number four with square planar stereochemistry. Most organogold compounds so far reported are σ–bonded complexes. The π complexes having alkene or alkyne ligands are generally very unstable and readily decompose at room temperature.

Scheme 13.2 shows typical synthetic routes to alkylgold complexes. Treatment of gold(I) chloride **1** with methylmagnesium iodide forms methylgold(I) complex **2**. A variety of monoorganogold(I) complexes of the type $AuR(PPh_3)$ (R = Et, Pr, Bu) have been similarly prepared [9]. The methyl complex **2** releases the PPh_3 ligand on reaction with methyllithium to form dimethylaurate **3**, which can be isolated as a relatively stable solid after treatment with a tridentate nitrogen ligand, $Me_2NCH_2CH_2N(Me)CH_2CH_2NMe_2$ [10a]. The anionic dimethyl species **3** undergoes oxidative addition of alkyl iodides in the presence of PPh_3 to afford trialkylgold(III) bearing PPh_3 (**4** and **8**). In this case, when alkyl iodides other than MeI are employed, the *trans* isomer **8** is formed selectively [9a]. The *cis* isomer **7**, on the other hand, can be obtained by the reaction of *cis*-$AuMe_2(I)(PPh_3)$ (**6**) with alkyllithium.

The trialkylgold(III) complexes (**4**, **7**, and **8**) were employed in the earliest mechanistic study on reductive elimination of σ-bonded organometallic complexes [12]. A detailed kinetic investigation, together with a molecular orbital calculation, indicated the following mechanism involving T- and Y-shaped three-coordinate species

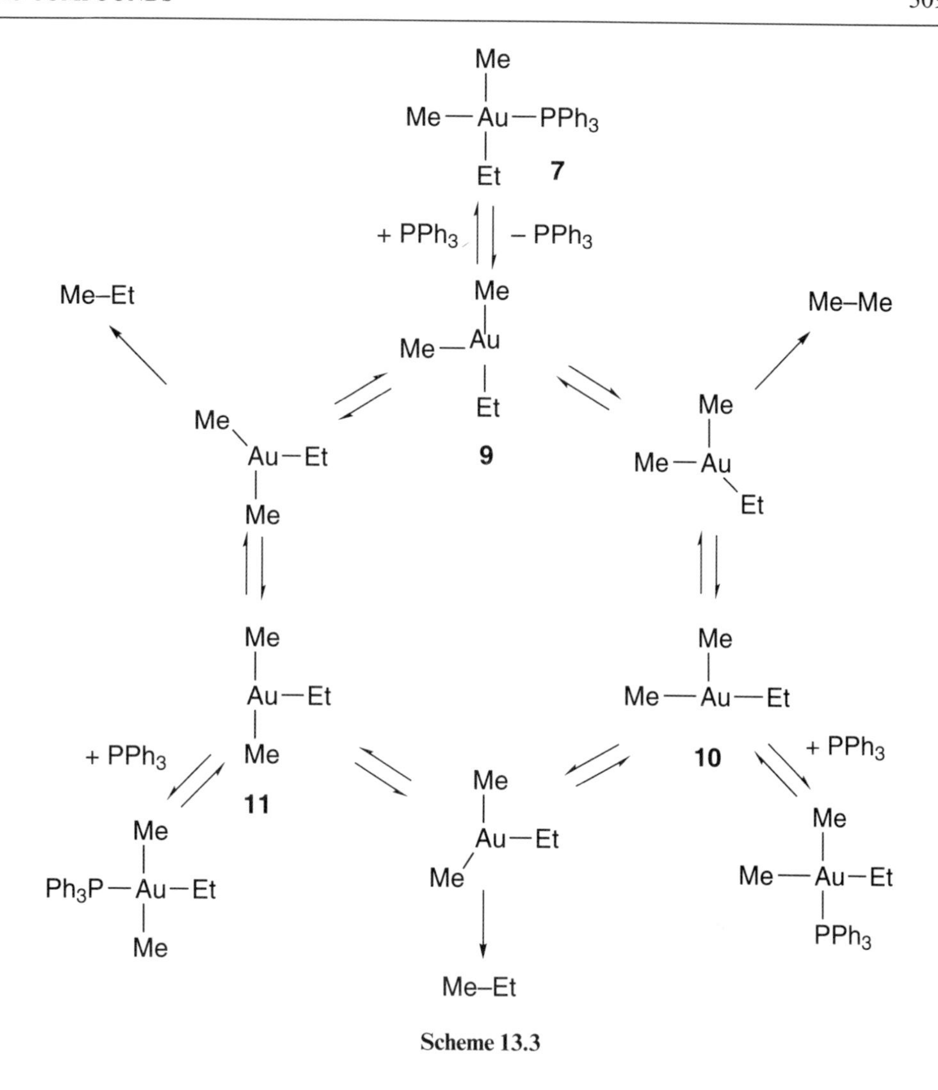

Scheme 13.3

(Scheme 13.3). Dissociation of the PPh_3 ligand from **7** forms a T-shaped intermediate **9**, which undergoes relatively rapid interconversion with other T-shaped intermediates **10** and **11** via transient Y-shaped species. The reductive elimination proceeds from the Y-shaped species. This study provided a basis for subsequent studies on the thermolysis of a variety of organotransition metal complexes.

13.4 Synthesis of Au Compounds

(1) Methyl(triphenylphosphine)gold(I), $AuMe(PPh_3)$ [9]

$$AuCl(PPh_3) + MeMgI \longrightarrow AuMe(PPh_3) + MgICl \qquad (5)$$

A solution of MeMgI prepared from Mg (1.0 g, 43 mmol) and MeI (2.5 mL, 40 mmol) in Et_2O (15 mL) is added dropwise to a suspension of $AuCl(PPh_3)$ (3.00 g, 6.06 mmol) in Et_2O (25 mL) at –5 °C during 15 min. The mixture is stirred for 1 h under reflux and then added to 150 mL of ice-cold 0.5% H_2SO_4. Et_2O (20 mL) is added and the mixture is filtered. The remaining white solid is washed with benzene (100 mL × 3). The ether layer and the benzene washings are combined, washed with water, dried over Na_2SO_4, and concentrated to about 80 mL. Pentane (200 mL) is then added and the mixture is chilled at –20 °C, yielding colorless fine crystals (2.23 g, 78%).

Properties: mp 167–168 °C; ^{1}H NMR ($CDCl_3$) δ 1.10 (d, $J = 8.0$ Hz).

(2) *trans*-Ethyldimethyl(triphenylphosphine)gold(III), *trans*-$AuEtMe_2(PPh_3)$ [9a]

$$AuMe(PPh_3) + MeLi \longrightarrow Li[AuMe_2] + PPh_3 \quad (6)$$

$$Li[AuMe_2] + EtI + PPh_3 \longrightarrow \textit{trans}\text{-}AuEtMe_2(PPh_3) + LiI \quad (7)$$

A solution of dimethylaurate(I) prepared from $AuMe(PPh_3)$ (474 mg, 1.00 mmol) and MeLi (1.00 mmol) in Et_2O (1.86 mL) is treated with ethyl iodide (1.25 mmol) at room temperature for 20 h. The solvent is removed under vacuum and the residue is extracted with 100 mL of pentane. The pentane extract is concentrated to 20 mL and chilled to –20 °C to provide colorless crystals of the title compound (56%).

Properties: ^{1}H NMR (dioxane) δ 0.29 (d, $J = 7.0$ Hz, Me), 0.9–2.4 (m, Et), 7.4–8.1 (m, Ph).

(3) Chloro(carbonyl)gold(I), Au(CO)Cl [6]

$$AuCl_3 + CO + SOCl_2 \longrightarrow Au(CO)Cl \quad (8)$$

A thionyl chloride suspension (155 mL) of anhydrous $AuCl_3$ (15.58 g, 51.3 mmol) is placed under a dry CO atmosphere and stirred magnetically. After 1 h the gold(III) chloride is dissolved and the red solution is filtered. CO uptake continues for 18 h while very slow agitation is maintained and Au(CO)Cl crystallizes. Hexane (200 mL) is then added dropwise (under CO atmosphere); Au(CO)Cl is collected and washed with four portions of CO-saturated hexane and dried under a stream of CO. Yield: 11.03 g (8.5%). Au(CO)Cl is best stored under CO atmosphere and should not be dried under vacuum as CO loss is facile and promotes decomposition of material.

(4) Mesitylgold(I), $Au_5(Mesityl)_5$ [6]

$$5Au(CO)Cl + 5MesitylMgBr \longrightarrow Au_5(Mesityl)_5 + 5MgBrCl + 5CO \quad (9)$$

A THF solution (17.1 mL) of mesitylmagnesium bromide (1.15 M, 19.7 mmol) is added dropwise under exclusion of light to a cold (–40 °C) THF suspension (110 mL) of Au(CO)Cl, which is prepared at –50 °C to avoid decomposition of Au(CO)Cl. As the reaction proceeds, CO evolves gently from the yellow-grayish suspension which is allowed to slowly reach room temperature after all the mesitylmagnesium bromide have been added. *p*-Dioxane (15 mL, 170 mmol) is added, and stirring is continued for 2 h. Decantation (3 h) and filtration give a yellow-orange solution. It is carefully concentrated to 75 mL on the water bath (40–45 °C). When the solution is cooled in the refrigerator, yellow crystals of the title compound begin to separate. The product (2.01 g) is isolated, and the mother liquor yields another 0.51 g of product. Total yield: 2.52 g (40.5%). All the operations must be carried out under exclusion of light. The compound can be recrystallized from toluene, THF, and benzene.

Properties: ^{1}H NMR (toluene-d_8) δ 6.57 (m, 2H, C_6H_2), 2.83 (s, 6H, *o*-Me), 1.85 (s, 3H, *p*-Me).

13.5 Organozinc Complexes

Organozinc chemistry originated in the synthesis of diethylzinc by Frankland in 1849, which was the first organometallic compound having a metal–carbon σ-bond [13]. Despite the early discovery, the synthetic utility of organozinc compounds has been limited until very recently because of their air-sensitivity as pure liquids and their low reactivity as compared with the other organometallic compounds such as Grignard reagents; only the Reformatsky reagent has been routinely used by organic chemists for a long time [14].

Recently, the utility of organozinc compounds has been significantly expanded by combining them with transition metal reagents in transformations such as nickel- or palladium-catalyzed cross-coupling reactions, copper-mediated substitution and addition reactions, and catalytic asymmetric addition to aldehydes [15]. Except for commercially available diethyl- and dimethylzincs, organozinc compounds are generally employed in organic synthesis without isolation. The most remarkable aspect of organozinc reagents is their high functional group tolerance, resulting from their low nucleophilicity. An extremely wide variety of functional groups can be introduced to the organic moiety of zinc reagents, including esters, nitriles, halides, ketones, phosphates, thioethers, sulfoxides, sulfones and amides.

Monoorganozinc halides (RZnX) can be synthesized by oxidative addition of organic halides to zinc metal. The oxidative addition rate is strongly affected by the reaction conditions (solvent, concentration) [16] and by activation of the zinc [15,17]. Zinc powder or zinc foil, which is activated by treatment with 1,2-dibromoethane and then with trimethylsilyl chloride, will oxidatively add alkyl iodides [18]. The reaction of alkyl bromides, on the other hand, requires more active zinc, which may be prepared by the reduction of zinc chloride with either lithium naphthalenide [19] or lithium metal under ultrasonic irradiation [20, 21].

Other convenient methods to prepare mono-oganozinc halides include zinc–halogen exchange between Et_2Zn [22] and alkyl bromides or iodides (eq (10)) [23], catalyzed by $PdCl_2(dppf)$, $Ni(acac)_2$, and $MnBr_2/CuCl$. When organic halides bearing a terminal vinyl group are employed, intramolecular carbozincation of the vinyl group takes place (eq (11)) [23a,c].

$$\text{n-C}_8\text{H}_{17}\text{I} + Et_2Zn \xrightarrow[\text{THF, 25 °C, 2 h}]{PdCl_2(dppf)\ (1.5\ mol\%)} \text{n-C}_8\text{H}_{17}\text{ZnI} \quad (10)$$

$$\text{(Ph-substituted iodoalkene)} + Et_2Zn \xrightarrow[\text{THF, 25 °C, 2 h}]{\text{PdCl}_2\text{(dppf) (1.5 mol\%)}} \text{(2-phenylcyclopentyl)methyl-ZnI} \quad (11)$$

Dialkylzincs can be prepared by transmetallation of zinc salts with alkylating reagents of the main group elements. Transmetallation of Me_2Zn[23] with alkylborons is a particularly useful route to dialkylzincs (eq (12)) [24]. Since alkylborons are readily accessible by well documented hydroboration reactions of alkenes, a wide variety of dialkylzincs can be synthesized by this route. Dialkylzincs can be also obtained by the reaction of Et_2Zn [21] with olefins in the presence of a catalytic amount of $Ni(acac)_2$ (eq (13)) [25].

$$R^2C(R^1)=CH_2 \xrightarrow{+ HBX_2} R^2CH(R^1)CH_2BX_2 \xrightarrow[- ZnX_2,\ Me_3B]{+ Me_2Zn} (R^2CH(R^1)CH_2)_2Zn \quad (12)$$

$$2\ RCH=CH_2 + Et_2Zn \xrightarrow[\text{neat, 50 °C}]{\text{Ni(acac)}_2 + \text{cod (1–2 mol\%)}} (RCH_2CH_2)_2Zn \quad (13)$$

13.6 Synthesis of Zn Compounds

(1) Iodo(butyl)zinc(II), Zn(Bu)I [17]

$$Zn + BuI \longrightarrow Zn(Bu)I \quad (14)$$

A suspension of 1.7 g (26 mmol) of zinc (99.99% purity) in 2 mL of THF containing 190 mg (1.0 mmol) of 1,2-dibromoethane is heated to 65 °C for a minute, cooled to 25 °C, and treated with 0.1 mL (0.8 mmol) of chlorotrimethylsilane. After 15 min at 25 °C, a solution of butyl iodide (25 mmol) in 10 mL of THF is slowly added at 30 °C. After the end of addition, the reaction mixture is stirred for 12 h at 35–40 °C. Usually, less than 100 mg of zinc remains, indicating a yield of 90%.

(2) Bis{2-(1-methylcyclohexen-4-yl)propyl}zinc(II) [24]

$$2\ \text{(1-methylcyclohexen-4-yl)CH(Me)CH}_2\text{BEt}_2 + Me_2Zn \longrightarrow \text{(1-methylcyclohexen-4-yl)CH(Me)CH}_2)_2Zn + 2MeBEt_2 \quad (15)$$

A 25 mL Schlenk flask equipped with a rubber septum is charged under argon with borane (522 mg, 2.53 mmol) in hexane (5 mL) and cooled to 0 °C. Me_2Zn (1.43 g, 1.48 mL, 15 mmol) is added at once via syringe and the solution is stirred at 0 °C for 10 min. The solvent, the excess Me_2Zn and the formed trialkylborane are pumped off carefully at 0 °C. The resulting solid is left at room temperature under vacuum for 1 h to give the title compound in over 94% yield.

The treatment of hexylboronic pinacol ester with Et_2Zn (5 equiv) at 70 °C for 17 h without solvent gives dihexylzinc, which is isolated as a pure compound after distilling off the excess Et_2Zn (0.1 mmHg, 50 °C).

References

[1] (a) B. H. Lipshutz, *Organometallics in Synthesis. A Manual*, Ed. M. Schlosser, Wiley, Chichester (1994), p. 283.
(b) G. H. Posner, *An Introduction to Synthesis Using Organocopper Reagents*, Wiley, New York (1980).
(c) B. H. Lipshutz, S. Sengupta, *Org. React.*, **41**, 135 (1992).
(d) G. H. Posner, *Org. React.*, **19**, 1 (1972) **22**, 253 (1975).
(e) B. H. Lipshutz, *Synthesis*, 325 (1987).
(f) Y. Yamamoto, *Angew. Chem., Int. Ed. Engl.*, **25**, 947 (1986).
(g) R. J. K. Taylor, *Synthesis*, 364 (1985).
(h) R. J. K. Taylor, *Synthesis*, 364 (1985).
(i) B. H. Lipshutz, R. S. Wilhelm, J. A. Kozlowski, *Tetrahedron*, **40**, 5005 (1984).
(j) E. Erdik, *Tetrahedron*, **40**, 641 (1984).
(k) J. F. Normant and A. Alexakis, *Synthesis*, 841 (1981).
(l) W. Carruthers, *Comprehensive Organometallic Chemistry*, Ed. G. Wilkinson, F. G. A. Stone, E. W. Abel, Pergamon Press, Oxford (1982), Vol. 7, p. 685.
(m) G. van Koten, S. L. James, J. T. B. H. Jastrzenski, *Comprehensive Organometallics Chemistry II*, Ed. E. W. Abel, F. G. A. Stone, G. Wilkinson, J. L. Wardell, Pergamon Press, Oxford (1995), Vol. 3, p. 57.
[2] (a) P. S. Coan, K. Folting, J. C. Huffman, K. G. Caulton, *Organometallics*, **8**, 2724 (1989).
(b) A. Miyashita A. Yamamoto, *Bull. Chem. Soc. Jpn.*, **50**, 1102 (1977).
[3] D. F. Dempsey, G. S. Girolami, *Organometallics*, **7**, 1208 (1988).
[4] D. W. Macomber, M. D. Rausch, *J. Am. Chem. Soc.*, **105**, 5325 (1983).
[5] F. H. Jardine, L Rule, A. G. Vohra, *J. Chem. Soc. (A)*, 237 (1970).
[6] E. M. Meyer, S. Gambarotta, C. Floriani, A. Chiesi-Villa, C. Guastin, *Organometallics*, **8**, 1067 (1989).
[7] R. J. Puddepatt, *Comprehensive Organometallic Chemistry*, Ed. G. Wilkinson, F. G. A. Stone, E. W. Abel, Pergamon Press, Oxford (1982), Vol. 2, p. 765; A. Grohmann and H. Schmidbaur, *Comprehensive Organometallics Chemistry II*, Ed. E. W. Abel, F. G. A. Stone, G. Wilkinson, J. L. Wardell, Eds. Pergamon Press, Oxford Vol. 3, p. 1. (1995).
[8] C. A. McAuliffe, R. V. Parish, P. D. Randall, *J. Chem. Soc., Dalton, Trans.*, 1730 (1979).
[9] (a) A. Tamaki, S. A. Magennis, J. K. Kochi, *J. Am. Chem. Soc.*, **96**, 6140 (1984).
(b) A. Tamaki, J. K. Kochi, *J. Organomet. Chem.*, **61**, 441 (1973).
[10] (a) A. Tamaki, J. K. Kochi, *J. Chem. Soc., Dalton Trans.*, 2620 (1973).
(b) F. H. Brain, C. S. Gibson, *J. Chem. Soc.*, 762 (1936).
[11] G. W. Rice, R. S. Tobis, *Inorg. Chem.*, **14**, 2402 (1975); **15**, 489 (1976).
[12] S. Komiya, T. A. Albright, R. Hoffmann, J. K. Kochi, *J. Am. Chem. Soc.*, **99**, 8440 (1977).
[13] E. Frankland, *Liebigs Ann. Chem.*, **71**, 171 (1949).
[14] P. L. Shriner, *Org. React.*, **1**, 1 (1942); M. W. Rathke, *Org. React.*, **22**, 423 (1975).
[15] P. Knochel, R. D. Singer, *Chem. Rev.*, **93**, 2117 (1993).
[16] T. N. Majid, P. Knochel, *Tetrahedron Lett.*, **31**, 4413 (1990); C. Jubert, P. Knochel, *J. Org. Chem.*, **57**, 5425 (1992).
[17] E. Erdik, *Tetrahedron*, **43**, 2203 (1987).

[18] M. C. P. Yeh, H. G. Chen, P. Knochel, *Org. Synth.*, **70**, 195 (1991); P. Knochel, M. C. P. Yeh, S. C. Berk, J. Talbert, *J. Org. Chem.*, **53**, 2390 (1988).
[19] L. Zhu, R. M. Wehmeyer, R. D. Rieke, *J. Org. Chem.*, **56**, 1445 (1991).
[20] C. Petrier, J. C. de Souza Barbora, C. Dupuy, J.-L. Luche, *J. Org. Chem.*, **50**, 5761 (1985).
[21] For other methods to prepare activated zinc metal, see: S. C. Berk, M. C. P. Yeh, N. Jeong, P. Knochel, *Organometallics*, **9**, 3053 (1990); H. Stadtmüller, B. Greve, K. Lennick, A. Chair, P. Knochel, *Synthesis*, 69 (1995); A. Fürstner, R. Singer, P. Knochel, *Tetrahedron Lett.*, **35**, 1047 (1994).
[22] The compound is available from commercial sources. For detailed procedure of preparation, see: C. R. Noller, *Org. Synth., Col. Vol. II,* 184 (1943).
[23] (a) H. Stadtmüller, R. Lentz, C. E. Tucker, T. Stüdemann, W. Dörner, P. Knochel, *J. Am. Chem. Soc.*, **115**, 7027 (1993).
(b) I. Klement, P. Knochel, K. Chau, G. Cahiez, *Tetrahedron Lett.*, **35**, 1177 (1994).
(c) A. Vaupel, P. Knochel, *Tetrahedron Lett.*, **35**, 8349 (1994).
[24] The compound is available from commercial sources. For detailed procedure of preparation, see: A. L. Galyer, G. Wilkinson, *Inorg. Synth.*, **19**, 253 (1979); D. Paker, *J. Chem. Soc., Perkin Trans.*, **2**, 83 (1983).
[25] F. Langer, J. Waas, P. Knochel, *Tetrahedron Lett.*, **34**, 5261 (1993).
[26] S. Vettel, A. Vaupel, P. Knochel, *Tetrahedron Lett.*, **36**, 1023 (1995); I. Klement, H. Lütjens, P. Knochel, *Tetrahedron Lett.*, **36**, 3161 (1995).

14 Group 1 (Li, Na, K) Metal Compounds

K. Maruoka, *Hokkaido University*

14.1 Introduction

Organometallics containing Group 1 metals are characterized by the presence of polar metal–carbon bonds [1, 2]. The high polarity of the metal–carbon bonds of the organoalkali metals is primarily responsible for both their high nucleophilicity, i.e., the high reactivity toward carbonyl and related polar functional groups, as well as their high basicity.

Many organolithium compounds are soluble in hydrocarbons; exceptions are methyllithium and phenyllithium which are associated in these solvents. Butyllithium is mostly hexameric and *tert*-butyllithium is tetrameric in cyclohexane. A Lewis basic solvent can interact with an organolithium oligomer, thereby decreasing the degree of association. Thus, methyllithium, which is tetrameric in the solid phase, becomes a solvated tetramer in ether, and BuLi, hexameric in hydrocarbons, becomes tetrameric in ether. In the more basic THF, BuLi has a degree of association between dimeric and tetrameric at –108 °C, and phenyllithium is between monomeric and dimeric [3].

14.2 Deprotonation with Organolithium Compounds

The α-protons that are less sterically hindered are most rapidly removed by a bulky base. Thus, addition of an unsymmetrical ketone to an excess of lithium diisopropylamide (LDA) gives the enolate anion on the less substituted side as the result of kinetic control. 2-Methylcyclohexanone has been specifically benzylated in the β-position in this manner [4].

A specific enolate, generated under aprotic conditions, could be annulated [5, 6]. The silicon atom stabilizes the anion resulting initially from the conjugate addition, thus

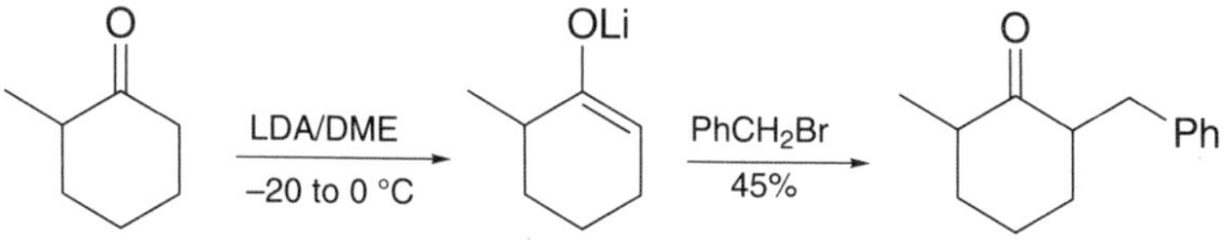

O OLi O
LDA/DME PhCH$_2$Br Ph
–20 to 0 °C 45% (1)

Synthesis of Organometallic Compounds: A Practical Guide. Edited by S. Komiya

preventing it from competing with the remainder of the original enolate anion for the α,β-unsaturated substrates. It is cleaved under the basic conditions of the subsequent aldol cyclization.

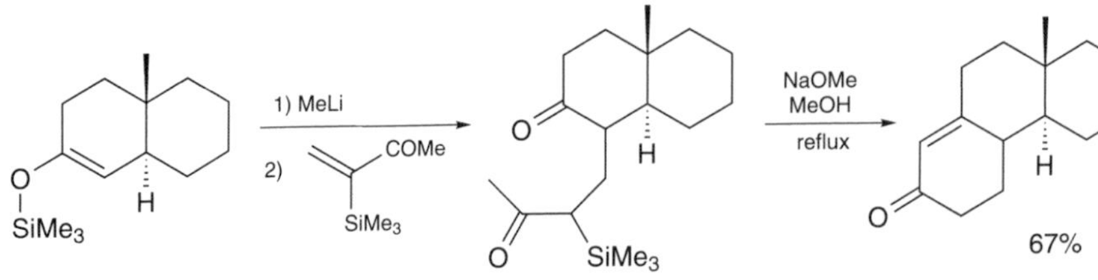

1) MeLi
2) COMe SiMe3
NaOMe MeOH reflux
67%
(2)

When α,β-unsaturated esters are treated with a 1:1 complex of LDA and HMPA, a γ-proton is removed giving the conjugated enolate anion. This anion is rapidly alkylated exclusively at the α-position giving the unconjugated ester [7, 8].

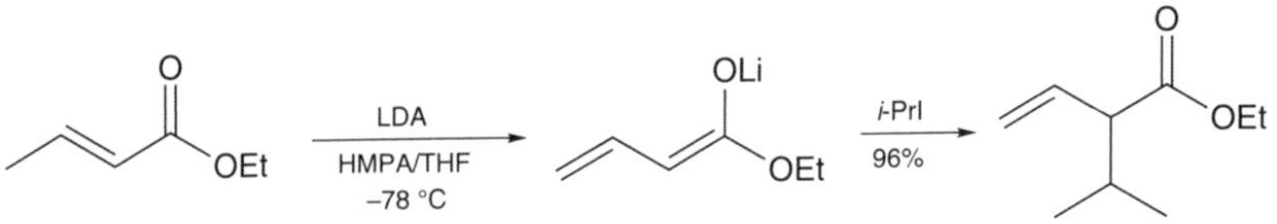

LDA HMPA/THF −78 °C
i-PrI 96%
(3)

Since nitrogen is less electronegative than oxygen, imines are less reactive toward carbanions than ketones or aldehydes are. Therefore imine enolate anions can be prepared without self- condensation, yet they will rapidly add to carbonyl compounds [9]. This amounts to a directed cross aldol condensation and can even be used to add aldehyde enolate equivalents to ketones.

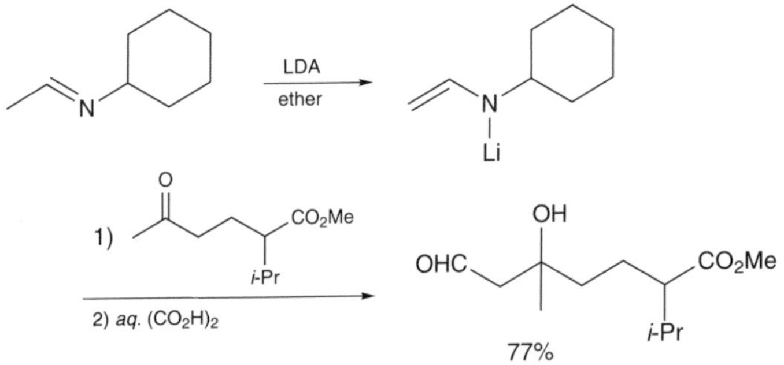

LDA ether
1) CO_2Me i-Pr
2) aq. $(CO_2H)_2$
OHC OH CO_2Me i-Pr
77%
(4)

14.3 Addition of Organolithiums to C—C Multiple Bonds

The highly nucleophilic character of delocalized organometallics towards carbon–carbon double bonds is the origin of the anionic polymerization of butadiene, isoprene, and styrene [10–12]. This is the major industrial application of organolithium compounds. Organolithium compounds are generally recommended to initiate the polymerization. Since each addition step to diene or styrene restores an allylic or benzylic type structure, the polymer chain grows rapidly until the monomer has been consumed.

$$PhCH{=}CH_2 \xrightarrow[\text{slow}]{\text{BuLi}} Li{-}CH(Ph){-}CH_2{-}Bu \xrightarrow[\text{fast}]{n\ PhCH{=}CH_2} Li{-}CH(Ph){-}CH_2{-}(CH(Ph){-}CH_2)_n{-}Bu \quad (5)$$

The combination of organolithiums with chiral coordination ligands such as sparteine allows the asymmetric polymerization to furnish optically active polymers with very high enantioselectivity, especially when trityl methacrylate is utilized as a monomer [13, 14].

$$CH_2{=}C(Me){-}C({=}O){-}O{-}CPh_3 \xrightarrow[\text{toluene, } -78\ ^\circ C]{\text{sparteine·Li–Bu}} {-}CH_2{-}C(Me)(C({=}O){-}O{-}CPh_3){-} \text{(helical polymer)} \quad (6)$$

Such nucleophilic additions of organolithiums to olefins are of limited use in laboratories. Nevertheless, there are certain categories of alkenes and alkynes that undergo ready addition [15]. A secondary organolithium reagent is easily transformed to the more stable primary reagent via the rapid equilibration [16].

$$CH_2{=}CH{-}CH_2OH \xrightarrow[\text{TMEDA}]{\text{BuLi}} LiCH_2{-}CH(Bu){-}CH_2OLi \xrightarrow{H_2O} CH_3{-}CH(Bu){-}CH_2OH \ \ (64 - 74\%) \quad (7)$$

$$CH_2{=}CH{-}CH_2{-}CH_2{-}CH(Li){-}Me \rightleftharpoons [\text{2-methylcyclobutylmethyllithium}] \rightleftharpoons LiCH_2{-}CH_2{-}CH_2{-}CH(Me){-}CH{=}CH_2 \quad (8)$$

Addition to coordinated arenes is a reliable method for achieving overall aromatic nucleophilic substitution with formal displacement of hydride [17]. This method illustrates the use of nucleophilic addition to an arenetricarbonyl-chromium for the synthesis of aromatic compounds with unusual substitution patterns.

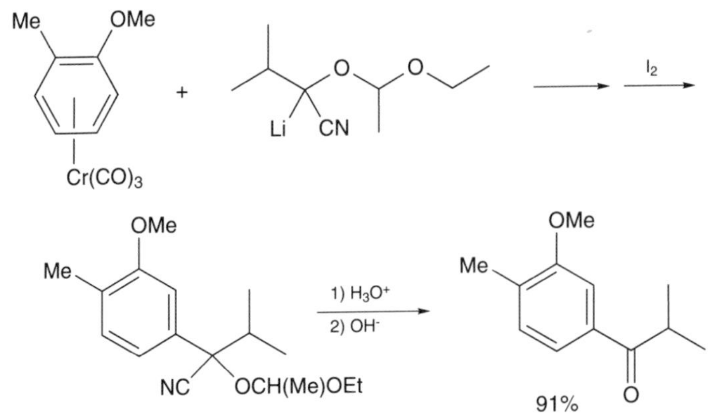

(9)

14.4 Addition of Organolithiums to C—N Multiple Bonds

The addition of organolithium compounds to simple imines is less satisfactory as a general synthetic method, particularly when α-hydrogens are present, as already described in the α-deprotonation of imines with organolithium reagents.

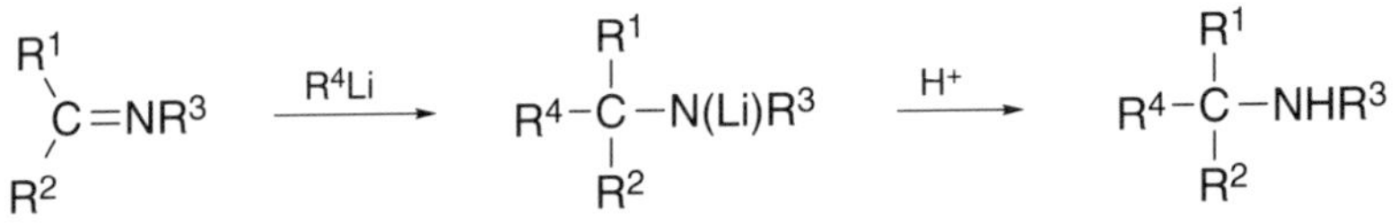

$$R^1R^2C{=}NR^3 \xrightarrow{R^4Li} R^4{-}C(R^1)(R^2){-}N(Li)R^3 \xrightarrow{H^+} R^4{-}C(R^1)(R^2){-}NHR^3 \quad (10)$$

The organolithium addition to pyridines or other nitrogen aromatic heterocycles is a well-established general synthetic method. It is usually used to achieve overall substitution on the ring, via lithium hydride elimination or oxidation of the dihydro-intermediate [18].

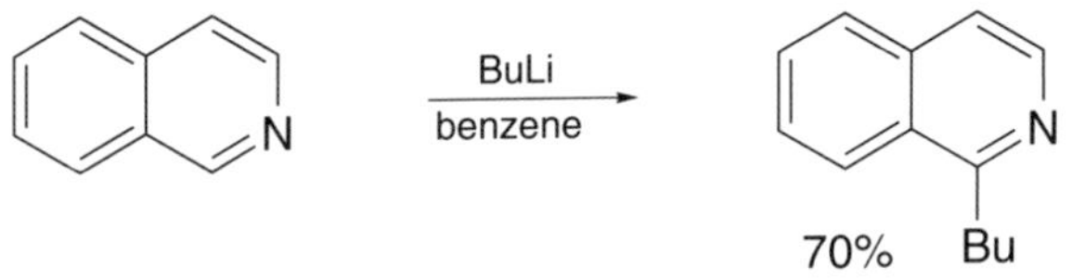

(11)

The addition of organolithiums to nitriles also causes various side reactions including α-deprotonation. Thus, aromatic nitriles are susceptible toward straightforward addition giving the N-lithioketimine predominantly, while acetonitrile and particularly phenylacetonitrile undergo extensive deprotonation. Acidic workup of the resulting N-lithioketimine gives a ketone [19].

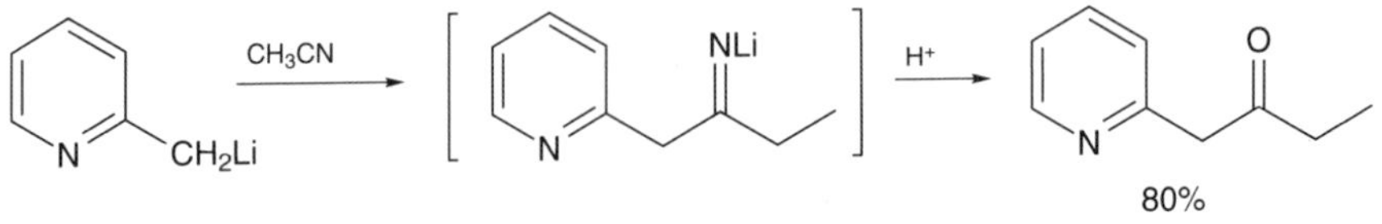

(12)

14.5 Addition of Organolithiums to Carbonyl Groups

In general, the experimental procedures for nucleophilic addition of organolithiums to aldehydes or ketones are straightforward [20]. Reduction via β-hydride transfer and via ketyl formation is usually less troublesome in the case of organolithiums. On the other hand, α-deprotonation i.e., enolization of carbonyl compounds resulting from the high basicity of many organolithium compounds can be a real problem.

$$\text{2-thienyl-Li} + C_5H_{11}\text{-CHO} \longrightarrow \text{2-thienyl-CH(OH)-}C_5H_{11}\ (63\%) \qquad (13)$$

Recently, the conversion of an organolithium compound to an organotitanium or an organocerium reagent *in situ* has been utilized to avoid deprotonation since such reagents are less basic than organolithiums, yet retain sufficient nucleophilicity [21].

$$\text{2-tetralone} \xrightarrow[100\%]{MeLi/TiCl_4} \text{2-methyl-2-tetralol (Me, OH)} \qquad (14)$$

Organolithium compounds normally add to α,β-unsaturated carbonyl compounds in a 1,2-manner [22]. In certain circumstances, however, conjugate addition of organolithiums takes precedence over 1,2-addition. This process has been demonstrated in the addition of 2-lithio-2-phenyl-1,3-dithiane to 2-cyclohexenone [23]. When the addition was carried out in THF at –78 °C and the mixture was warmed to 25 °C, the conjugate addition product was formed in high yield. When a hexane-THF solvent mixture is used and quenched at –78 °C, the allylic alcohol was 95% of the product. 2-Lithio-1,3-dithiane gives only 1,2-addition to cyclohexenone; thus without the stabilization of the phenyl group, reversal of the initial attack is apparently energetically unfavorable.

Phosphorus ylids are prepared by treatment of phosphonium salts with bases such as phenyllithium, butyllithium, dimsyl anion, or potassium *t*-butoxide. They are not isolated but used directly after preparation because they are sensitive to oxygen and moisture. The requisite phosphonium salts are available in great variety from the reaction of triphenylphosphine and primary or secondary alkyl halides, usually bromides [24].

The addition of an α-silylorganolithium reagent to a carbonyl compound and subsequent elimination of lithium trialkylsilyloxide gives an alkene [25]. Alternatively, hydrolysis of the intermediate adduct, a β-hydroxysilyl compound, converts it into an alkene [26]. The overall processes are called "Peterson Olefination"; they are highly stereoselective and complementary to the Wittig reaction. The use of organolithium compounds in the Horner–Wadsworth–Emmons synthesis may be regarded as analogous.

$$CH_2{=}CH{-}CHO + (PhS)_2CHLi \longrightarrow CH_2{=}CH{-}CH(OH){-}CH(SPh)_2$$

PH, HO (2-phenyl-1,3-dithian-2-yl)cyclohexenol, 95% $\xrightarrow[-78\ °C]{\text{hexane/THF}}$ cyclohexenone + 2-Ph-2-Li-1,3-dithiane $\xrightarrow[-78\text{ to }25\ °C]{\text{THF}}$ 3-(2-Ph-1,3-dithian-2-yl)cyclohexanone, 93%

$$Ph_3P + CH_3Br \xrightarrow[25\ °C]{\text{benzene}} Ph_3P^+CH_3Br^- \xrightarrow[\text{DMSO}]{CH_3SOCH_2Na}$$

$$Ph_3P^+\text{-}CH_2^- \xrightarrow[60\text{ to }78\%]{\text{cyclohexanone}} \text{methylenecyclohexane} + Ph_3PO \quad (15)$$

$$Ph_2P(S)CH_2SiMe_3 \xrightarrow[\text{THF}]{\text{BuLi}} Ph_2P(S)CH(Li)(SiMe_3) \xrightarrow{Ph_2CO} Ph_2P(S)CH{=}CPh_2 \quad 80\%$$

(H)(Pr)C=C(Pr)(H) $\xleftarrow[\text{THF}]{\text{KH}}$ *syn* — Me$_3$Si, H, Pr / Pr, OH, H — $\xrightarrow{BF_3OEt_2}$ *anti* (H)(Pr)C=C(H)(Pr)

14.6 Addition of Organolithiums to N,N-Disubstituted Amides

The addition of organolithiums to N,N-dimethylformamide (DMF) works well giving aldehydes in synthetically useful yields [27]. N-Methylformanilide, Comins' reagent, and N-formylpiperidine are possible alternatives.

$$\text{1-Cl-2-}OCH_2OMe\text{-benzene} \xrightarrow[\text{TMEDA}]{\text{BuLi}} \text{(Cl, }OCH_2OMe\text{, Li)} \xrightarrow{\text{DMF}} \text{(Cl, }OCH_2OMe\text{, CHO)}\ 85\% \quad (16)$$

α,β-Unsaturated amides are more prone to undergo conjugate addition than are α,β-unsaturated esters and ketones [28, 29].

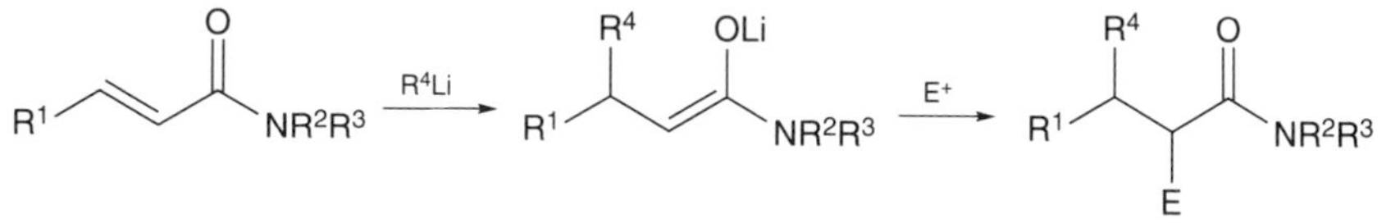

(17)

14.7 Addition of Organolithiums to Acids

The reaction of an organolithium compound with a carboxylic acid or a carboxylate salt gives an insoluble dilithium salt which is stable in the absence of protic solvents and resists any addition that might lead to tertiary alcohol [30, 31]. This salt must be quenched by pouring it into aqueous acid with vigorous stirring to avoid local contact of the rapidly formed ketone with any remaining lithium reagent, which would lead to tertiary alcohol.

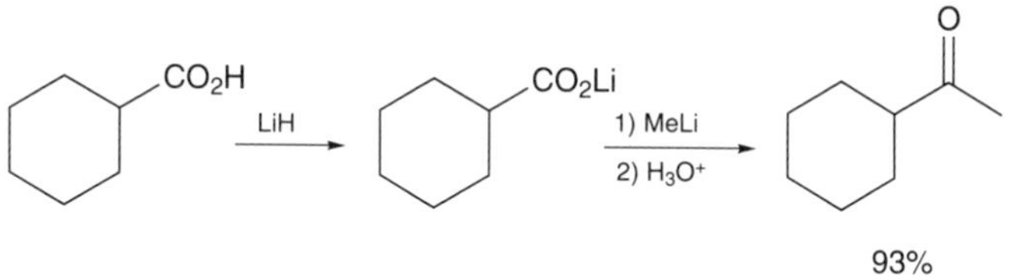

(18)

Addition of organolithiums to carbon dioxide affording lithium carboxylates, thence carboxylic acids, is a very familiar reaction [32].

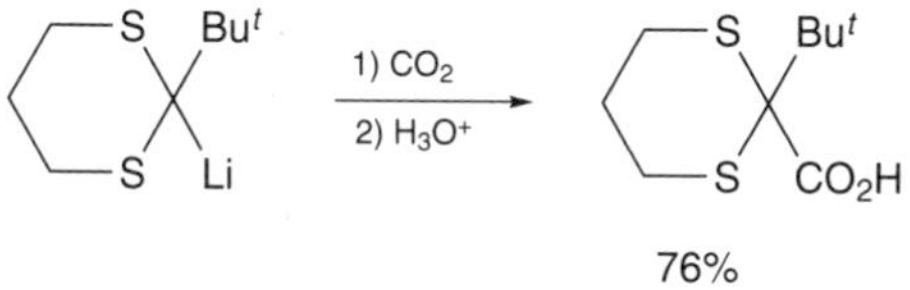

(19)

To achieve the better yields of a carboxylic acid, it is necessary to avoid the simultaneous presence of the organolithium compound and the carboxylate salt particularly at other than low temperatures.

14.8 Substitution at Carbon by Organolithium and Organosodium Compounds

The substitution reactions of organolithiums with alkyl halides are susceptible to several side-reactions involving elimination and metal–halogen exchange, and their mechanism and stereochemistry are often complicated. Satisfactory results are usually obtained by combination of organolithium compounds containing a delocalyzed carbanion with primary alkyl bromides and iodides. [33]. Iodides are more susceptible to metal-halogen exchange, though methyl iodide usually works well.

The apparently higher electronegativity of sp-hybridized carbon gives relatively greater acidity to 1-alkynes so that bases such as butyllithium may be used to generate the carbanion, which can be used for smooth alkylation [34].

$$\text{3-methylpyridine} \xrightarrow{\text{LDA}} [\text{3-}(CH_2Li)\text{pyridine}] \xrightarrow[77\%]{\text{Pr-Br}} \text{3-butylpyridine} \quad (20)$$

$$THPO(CH_2)_5C{\equiv}CH \xrightarrow[\text{THF, 0 °C}]{\text{BuLi}} [\,THPO(CH_2)_5C{\equiv}CLi\,] \xrightarrow[\text{HMPA, 25 °C, 30 min}]{C_7H_{15}\text{-X}} \underset{92\%}{THPO(CH_2)_5C{\equiv}C(CH_2)_6CH_3} \quad (21)$$

Allyllithium reagents may be prepared directly from the hydrocarbon with the strong base combination, butyllithium and TMEDA [35]. These complexed reagents give high yields too, when used in two-fold excess with primary halides.

$$\text{isobutylene} \xrightarrow[\text{ether/hexane, −78 to 25 °C}]{\text{BuLi/TMEDA}} [\text{2-methylallyllithium·TMEDA}]\ (\textit{precipitate}) \xrightarrow[\text{−78 to 25 °C; 96\%}]{C_6H_{13}\text{-Br}} \text{2-methyl-1-nonene} \quad (22)$$

The more resonance-stabilized cyclopentadienide anion is readily alkylated with alkyl halides, tosylates, or epoxides. The sodium reagent often gives best results [36]. The initially formed 5-alkylcyclopentadiene rapidly isomerizes at room temperature under the reaction conditions to the 1-allyl isomer in good yield and isomeric purity.

$$C_5H_5Na + \text{TsO-alkyl} \xrightarrow[\text{0 to 25 °C, 4 h}]{\text{THF}} \text{1-alkylcyclopentadiene}\ (85\%) \quad (23)$$

14.9 Nucleophilic Cleavage of Epoxides with Organolithiums

The nucleophilic reaction of organolithium compounds with ethylene oxide is a useful and general method for two-carbon homologation [37].

$$(Et)(Bu)C{=}C(I)(H) \xrightarrow{\text{BuLi}} [(Et)(Bu)C{=}C(Li)(H)] \xrightarrow{\text{ethylene oxide}} \underset{88\%}{(Et)(Bu)C{=}C(CH_2CH_2OH)(H)} \quad (24)$$

The corresponding reaction with substituted epoxides is more susceptible to side-reactions, including deprotonation. Recently, activation of such epoxides by complexation with BF_3OEt_2 enabled the smooth ring opening of epoxides [38].

BuLi/BF_3OEt_2 97% Bu OH (25)

14.10 Reaction of Organolithiums with Polyhaloalkanes

A halogen atom, especially bromine, gives sufficient stabilization to an attached carbanion to allow preparation by metal halogen exchange from 1,1-dihalo compounds [39, 40]. These α-halocarbanions decompose rapidly via a carbene intermediate, since the halide is a good leaving group. The preparatively useful reactions involve *in situ* reactions of the transient carbanions in the presence of aldehydes or ketones [41].

BuLi/CH_3CHBr_2 THF −78 to 25 °C OLi Br 95% (26)

The α,α-dihalocarbanions can be prepared by removal of a proton from a dihalo compound using lithium dialkylamide or butyllithium-TMEDA bases [42].

+ CH_2Cl_2 + LiN( )$_2$ 0 °C HO $CHCl_2$ 89% (27)

14.11 Oxidation of Organolithium Compounds

The reaction of organolithium compounds with oxygen can be controlled to give, at low temperatures, a hydroperoxide or at higher temperatures an alcohol or phenol [43].

Esters, lactones, ketones, and nitriles with *secondary* α-carbons can all be α-hydroxylated using the peroxide MoO_5/Pyridine/HMPA [44–46]. This method does not involve hydroperoxide intermediates so that elimination to the ketone does not occur.

$$PhCH_2CO_2H \xrightarrow{BuLi} [\,Ph\bar{C}HC\bar{O}_2\,]\,2Li^+ \xrightarrow[20\ °C]{O_2} Ph{-}CH(OH){-}CO_2H \quad (86\%) \tag{28}$$

$$[\,Ph\bar{C}HC\bar{O}_2\,]\,2Li^+ \xrightarrow{O_2,\ -70\ °C} Ph{-}CH(OOH){-}CO_2H \quad (82\%)$$

$$\text{THPO-steroid-17-one} \xrightarrow[2)\ MoO_5Py,\ HMPA]{1)\ LDA,\ -78\ °C} \text{THPO-steroid-16-OH-17-one} \quad (75\%) \tag{29}$$

Oxidation of alkenyllithiums with bis(trimethylsilyl)peroxide gives trimethylsilyl enol ethers, which may be easily hydrolyzed to the corresponding ketones [47].

$$\text{1-cyclooctenyllithium} \xrightarrow{Me_3SiOOSiMe_3} \text{1-}(OSiMe_3)\text{cyclooctene}\ (52\%) \xrightarrow{H^+} \text{cyclooctanone}\ (59\%)$$

14.12 Synthesis of Group 1 Metal Compounds

(1) Butyllithium, LiBu [48]

$$Br{-}Bu + Li \longrightarrow Li{-}Bu \tag{30}$$

In a 500 mL three-necked flask equipped with a stirrer, a low-temperature thermometer, and a dropping funnel is placed 200 mL of anhydrous ether. After the apparatus has been swept with dry, oxygen-free nitrogen, 8.6 g (1.25 gram atoms) of lithium wire (or any other convenient form of lithium metal) is cut into small pieces which are allowed to fall directly into the reaction flask in a stream of nitrogen. With the stirrer started, about 30 drops of a solution of 68.5 g (0.50 mol) of n-butyl bromide in 100 ml of anhydrous ether is added from the dropping funnel. The reaction mixture is then cooled to –10 °C by immersing the flask in a dry ice–acetone bath kept at about –30 to – 40 °C. The solution becomes slightly cloudy and bright spots appear on the lithium when the reaction has started. The remainder of the n-butyl bromide solution is then added at an even rate over a period of 30 min while the internal temperature is maintained at –10 °C. After addition is complete, the reaction mixture is allowed to warm up to 0 to 10 °C with stirring during 1 to 2 h. The reaction mixture is then filtered under an atmosphere of nitrogen by decantation through a narrow tube loosely plugged with glass wool into a graduated dropping funnel previously flushed with nitrogen.

The yield is 80–90%, determined as follows. A 5 or 10 mL aliquot of the solution is withdrawn by means of a pipet connected to a rubber suction bulb, and hydrolyzed by adding to distilled water (10 mL). This is titrated with standard acid to determine the total alkali, using phenolphthalein as indicator. A second 5 or 10 mL aliquot is withdrawn and run into a solution of anhydrous ether (10 mL) containing benzyl chloride (1 mL). The mixture is allowed to stand for 1 min after the addition and is then hydrolyzed with water (10 mL) and titrated with standard acid. Care must be taken not to overstep the end point since the aqueous layer becomes decolorized before the ether layer. To overcome this the mixture should be shaken vigorously near the end point. The second titration determines the alkali present in the form of compounds other than n-butyllithium. The difference between the two titration values represents the concentration of n-butyllithium.

(2) Vinyllithium, $LiCH{=}CH_2$ [49]

$$ClCH{=}CH_2 + Li \longrightarrow LiCH{=}CH_2 \quad (31)$$

Vinyl chloride from a cylinder is condensed into a calibrated trap. Six grams (0.1 mol) of the chloride is entrained in a steam of argon and bubbled into a stirred flask containing lithium–2% sodium dispersion (1.5 g) in THF (250 mL) under an argon atmosphere. A reaction begins after about one-fourth of the vinyl chloride has been added, and the remainder of the addition is carried out at 0–10 °C. After addition the mixture is stirred for 2 h at 0 °C and then allowed to warm to room temperature with stirring. Filtration in an argon atmosphere gives a clear colorless solution of vinyllithium in THF. The yield of vinyllithium is 60–65% as estimated by titration with base and by reaction with vanadium pentoxide followed by titration with standard permanganate solution. A derivative from acetone yields approximately 20% of dimethylvinylcarbinol, bp 97–99 °C, n_D^{25} 1.4178.

Vinyllithium can be also prepared by the reaction of teravinyltin with phenyllithium in ether.[50]

$$Sn(CH{=}CH_2)_4 + 4\,PhLi \longrightarrow 4\,LiCH{=}CH_2 + 4\,SnPh_4 \quad (32)$$

(3) Trichloromethyllithium, $LiCCl_3$ [51]

$$HCCl_3 + BuLi \longrightarrow LiCCl_3$$

Chloroform (12.46 g, 104 mmol) is dissolved in THF (130 mL) and cooled to –105 °C. Butyllithium in hexane (73 mL, 104 mmol), is then added very slowly over a 45 min period. The reaction mixture darkens rapidly (after first few drops) and changes from purple finally to a greenish slurry. After 1.25 h from start of the addition, the reaction is poured onto stirred dry ice (amber-green color is not discharged). After evaporation of the dry ice and solvent on a steam bath, a semicrystalline brown residue remains. Water (150 mL) is added and most of the residue dissolved. After extraction with ether, the aqueous phase is acidified and extracted with ether. The ether phases are combined, dried, and vacuum stripped to yield 6.7 g of a red oil, which is vacuum distilled.

The liquid gives a few drops of forerun on distillation and then yields a clear liquid distillate (4 g) boiling at 66–69 °C (0.6 mm), which partially crystallizes on standing. About half is pure white crystals and the other half clear liquid (which crystallizes slightly below room temperature). The infrared spectra of both are very similar to a standard spectrum of trichloroacetic acid. The solid is deliquescent and is identified as trichloroacetic acid. Anal. Calcd. for $C_2HO_2Cl_3$: Cl, 65.1. Found: Cl, 63.9. The liquid fraction contains 55.3 % Cl, which is close to theoretical for dichloroacetic acid, but its infrared spectrum shows it to be predominantly trichloroacetic acid, contaminated with lesser amounts of dichloroacetic acid (yield *ca.* 25%).

The compound can be also prepared as follows [52]:

$$CCl_4 + BuLi \longrightarrow LiCCl_3 \qquad (33)$$

Carbon tetrachloride (5.6 g, 36.5 mmol) is dissolved in THF (100 mL) and cooled to –105 °C. Butyllithium in hexane (26 mL, 37 mmol), is then added over a 30 min period. There occurs a mild exotherm and a very gradual change from colorless to pale pink. After about half the butyllithium is added, a white precipitate becomes visible. After 1 h from the start, the reaction is a white slurry (very clean looking). After 1.5 h, the reaction is poured onto stirred powdered Dry Ice and worked up as before. All of the acid distilled at 69.5 °C (0.4 mm) and crystallized in the condenser. 4.5 g of trichloroacetic acid is isolated (76% yield). The infrared spectum is identical with that of an authentic sample.

(4) Methallyllithium, $LiCH_2C(Me)CH_2$ [35]

BuLi/TMEDA, ether/hexane, –78 to 25 °C → [Me₂N–Li–NMe₂ complex] *precipitate* → C_6H_{13}-Br, –78 to 25 °C → 96% (34)

By adding isobutylene (100 mL) to a stirred solution of the BuLi/TMEDA reagent (from BuLi (33 mol) in hexane and TMEDA (33 mol) in ether (20 mL)) at –78 °C, then stirring overnight under nitrogen atmosphere at room temperature, there results a white crystalline precipitate as methallyllithium/TMEDA complex, and a pale yellow supernatant. Subsequent addition of 1-bromohexane (16.5 mol) dissolved in ether (*ca.* 0.8 M) at –78 °C, warming the resultant mixture to room temperature for 3 h prior to adding saturated ammonium chloride solution, leads after washing the pentane extract with diluted HCl solution and drying over Na_2SO_4 to acceptable yields of methallated coupling product.

(5) 2-Lithio-1,3-dithiane, $LiCH(SCH_2CH_2CH_2S)$ [32]

$$\text{1,3-dithiane} + BuLi \xrightarrow{THF} \text{2-lithio-1,3-dithiane} \qquad (35)$$

A round-bottomed flask with ST neck and side arm is equipped with a magnetic spin bar, a three-way stopcock, and a serum cap. Solid dithianes are weighed into the flask prior to subsequent flushing with nitrogen or argon. The reaction vessel is kept under positive inert gas pressure until workup. Solvents, liquid reagents, and solutions of reagents are introduced, and samples are withdrawn through the serum cap by hypodermic syringes. To avoid loss of pressure, pierced caps are sealed with parafilm tape.

The amount of freshly distilled THF necessary to obtain a 0.1–0.5 M solution of dithiane is added. A 5% excess of BuLi in hexane (1.5–2.5 M) is added at a rate of 3–5 mL/min to the solution stirred at –40 °C. After 1.5–2.5 h at –25 to –15 °C, most dithianes are metalated quantitatively as determined by deuteration of an aliquot of the solution containing 50–100 mg of dithiane. This is done by injecting the withdrawn solution into 1–3 mL of D_2O in a small separatory funnel and extracting with ether, methylene chloride, or pentane; the orange layer is dried for a few minutes with K_2CO_3 and concentrated evaporatively. Integration of the dithiane C2-proton NMR signal vs. any other well-defined and separated peak of the particular dithiane thus provides the extent of deuteration with an accuracy of ±5% within 15 min.

The anion solutions of dithiane are clear and colorless, and can be stored for a few hours at room temperature without decomposition. After two weeks at –25 °C, the anion solution showed no decomposition.

Instead of butyllithium, *tert*-butyllithium can be used to metalate dithiane at lower temperatures or within shorter periods of time.

(6) Lithium Diisopropylamide, $LiN(i\text{-}Pr)_2$ [52]
Lithium 6-methylcyclohexenolate, $LiO(C{=}CHCH_2CH_2CH_2CH(Me))$ [52]

$$HN(i\text{-}Pr)_2 + MeLi \xrightarrow{DME} LiN(i\text{-}Pr)_2 \qquad (36)$$

$$\text{2-methylcyclohexanone} \xrightarrow[DME]{LiN(i\text{-}Pr)_2} \text{lithium 6-methylcyclohexenolate} \xrightarrow[DME]{PhCH_2Br} \text{2-benzyl-6-methylcyclohexanone (58-61\%)} \qquad (37)$$

A 1 L three-necked flask is equipped with a nitrogen-inlet tube fitted with a stopcock, a glass joint fitted with a rubber septum, a 125 mL pressure-equalizing dropping funnel, a thermometer, and a glass-covered magnetic stirring bar. After the apparatus has been dried in an oven, 2,2′-bipyridine (45 mg) is added to the flask and the apparatus is thoroughly flushed with anhydrous, oxygen-free nitrogen. A static nitrogen atmosphere is maintained in the reaction vessel throughout subsequent operations involving organometallic reagents. An ethereal solution containing methyllithium (0.2 mol) is added to the reaction flask with a hypodermic syringe. After the ether is removed under reduced pressure, the reaction vessel is refilled with nitrogen and DME (400 mL) is added to the vessel with a hypodermic syringe or a stainless steel cannula. The resulting purple solution of methyllithium and the methyllithium–bipyridyl charge-transfer complex is cooled to –50 °C with a dry ice–methanol bath before diisopropylamine (29.2 mL, 0.208 mol) is added with a hypodermic syringe, dropwise and with stirring. During this addition, which requires 2–3 min, the temperature of the reaction solution should not be allowed to rise above –20 °C. The resulting reddish-purple solution of lithium diisopropylamide is stirred at –20 °C for 2–3 min before 2-methylcyclohexanone (21.3 g, 0.19 mol) in DME (50 mL) is added dropwise and with stirring. During this addition the temperature of the reaction solution should not be allowed to rise above 0 °C. After the addition of the ketone, the solution of the lithium enolate must still retain a pale reddish-purple color indicating the presence of a slight excess of lithium diisopropylamide. The enolate solution is stirred and warmed to 30 °C with a water bath before benzyl bromide (68.4 g, 0.4 mol) is added, rapidly and with vigorous stirring, from a hypodermic syringe. The temperature of the reaction mixture rises to about 50 °C within 2 min and then begins to fall. After a total reaction period of 6 min, the reaction mixture is poured into saturated sodium hydrogen carbonate (500 mL) and extracted with three 150 ml portions of pentane. The combined organic extracts are washed successively with two 100 mL portions of 5N HCl and 100 mL of saturated sodium hydrogen carbonate, dried over anhydrous magnesium sulfate, and concentrated with a rotary evaporator. The residual yellow liquid is fractionally distilled under reduced pressure, separating 31–32 g of forerun fractions, bp 67–92 °C (20 mm) and 40–91 °C (0.3 mm), and 21.3–23.3 g of crude 2-benzyl-6-methylcyclohexanone as a colorless liquid, bp 91–97 ° (0.3 mm.), n_D(25 °C) 1.5282–1.5360. The crude products contain 2- benzyl-6-methylcyclohexanone (86–90%) and crude 2-benzyl-2-methylcyclohexanone (10–14%) accompanied by in some cases by small amounts of *trans*-stilbene.

$$\mathrm{BuLi + HN}(i\text{-Pr})_2 \xrightarrow[\mathrm{THF}]{} \mathrm{LiN}(i\text{-Pr})_2 \qquad (38)$$

Lithium diisopropylamide can be conveniently prepared by the reaction of diisopropylamine with BuLi in THF solvent at 0 °C for 15 min.

(7) Lithium Bis(trimethylsilylamide), $LiN(SiMe_3)_2$ [53]

$$\mathrm{BuLi + HN(SiMe_3)_2 \longrightarrow LiN(SiMe_3)_2} \qquad (39)$$

A dry 500 mL three-necked flask, fitted with a pressure-equalizing dropping funnel and a stopcock in each side-neck, is equipped for magnetic stirring and maintained under a static nitrogen pressure by attaching a nitrogen source to one stopcock and a mercury bubbler to the other. In the flask is placed BuLi (0.25 mol) in hexane (153 mL), and stirring is started. The flask is immensed in an ice–water bath, and hexamethyldisilazane (42.2 g, 0.263 mol) is added dropwise over a period of 10 min. The ice bath was removed and the solution is stirred for 15 min longer. The hexane is removed under reduced pressure by replacing the mercury bubbler with heavy rubber tubing connected to a dry-ice condenser and an oil pump. During this step, the flask is immensed in a water bath at 40–50 °C, and stirring is continued as long as possible. After complete evaporation of the hexane, white crystals of lithium bis(trimethylsilyl)amide appear. The flask is again subjected to a static pressure of nitrogen, and THF (225 mL) is added to dissolve the crystals.

(8) Cyclopentadienyl Sodium, $Na(C_5H_5)$ [36]

$$C_5H_6 + Na \longrightarrow C_5H_5Na \quad (40)$$

$$C_5H_5Na + \text{TsO-R} \xrightarrow[\text{0 to 25 °C, 4 h}]{\text{THF}} C_5H_5\text{-R} \quad 85\% \quad (41)$$

Sodium spheres (2.88 g, 0.125 mol) are refluxed in dry xylene (35 mL) under nitrogen until sodium sand has formed. After cooling to room temperature the xylene is decanted, and the sand is washed with dry THF (2 × 25 mL) and suspended in THF (75 mL). Freshly distilled cyclopentadiene (10 mol) is added to the stirred mixture in four portions, resulting in a deep red-purple solution in which all the sodium has been consumed (3 h). The solution is cooled in an ice bath, and the tosylate (31.3 g, 0.11 mol) in dry THF (100 mL) is added dropwise over 0.5 h. After the addition is complete, the reaction mixture is allowed to warm to room temperature and stirring continues for a further 4 h. The light-brown mixture is transferred to a separatory funnel, diluted with brine (100 mL), and extracted with ether (4 × 700 mL), the combined extracts are washed with brine and dried, and the solvent is removed to afford an oil which is purified by column chromatography on silica gel (10:1) elution with hexane to give the hydrocarbon (16.5 g, 85%) as a clear oil.

Properties: IR 1660, 1601 cm-1; ^1H NMR 5.8–6.5 (3H, m, cyclopentadiene), 5.02 (1H, br t, $J = 6$ Hz, CH=CMe_2), 2.80 (2H, d of m, $J = 5$, 1.3, 1 Hz, C=CCH^2C=C), 1.09 (3H, d, $J = 7$ Hz, CH_3CH), 1.55, 1.63 (6H, s, Me_2C=C), 1.09 (3H, d, $J = 7$ Hz, CH_3CH) ppm; mass spectrum M* 176.

(9) Triphenylmethylpotassium, KCPh3 [54]

$$Ph_3CH + K \longrightarrow KCPh_3 \quad (42)$$

A 500 mL three-necked round-bottomed reaction flask having ground-glass joints is equipped with a mercury-sealed stirrer, a dry-ice condenser (having a soda-lime tube), and an inlet tube of wide diameter (12 mm) leading from a second 500 mL round-bottomed flask containing commercial anhydrous ammonia (about 300 mL). The ammonia (150 mL) is distilled from sodium into the reaction flask and the inlet tube replaced by a ground glass stopper. To the stirred ammonia is added in small pieces (7.8 g, 0.2 g-atom) of clean potassium which is converted to potassium amide by means of a piece of rusty iron gauze. The ground-glass stopper is replaced by a dropping funnel containing triphenylmethane (48.8 g, 0.2 mol) dissolved in ether (250 mL). This solution is added to the stirred 0.2 mol of potassium amide as rapidly as the ammonia is efficiently condensed. The mixture acquires an orange color. After 1 h, the dry-ice in the condenser is replaced by crushed ice. The reaction mixture is allowed to come to room temperature and finally refluxed on a steam-bath for 2 h to expel the ammonia, sufficient ether being added gradually so that the volume of the mixture remains approximately 300 mL. The potassium triphenylmethide reagent is obtained as a blood-red suspension in ether.

The conversion of the potassium to potassium triphenylmethide is assumed to be quantitative, producing 0.2 mol of the reagent. Carbonation of the reagent gives a 91% yield (based on either triphenylmethane or potassium) of triphenylacetic acid, mp 263–265 °.

References

[1] W. N. Setzer, P. von. R. Schleyer, *Adv. Organomet. Chem.*, **24**, 353 (1985).
[2] P. V. R. Schleyer, *Pure Appl. Chem.*, **56**, 151 (1984).
[3] W. Bauer, D. Seebach, *Helv. Chim. Acta*, **67**, 1972 (1984).
[4] H. O. House, L. J. Czuba, M. Gall, H. D. Olmstead, *J. Org. Chem.*, **34**, 2324 (1969).
[5] G. Stork, J. Singh, *J. Am. Chem. Soc.*, **96**, 6181 (1974).
[6] R. K. Boeckman, Jr., *J. Am. Chem. Soc.*, **96**, 6179 (1974).
[7] A. B. Smith, III, S. J. Branca, B. H. Toder, *Tetrahedron. Lett.*, 4225 (1975).
[8] J. L. Herrmann, G. R. Kieczykkwski, R. H. Sohlessinger, *Tetrahedron. Lett.*, 2433 (1973).
[9] W. G. Dauben G. H. Beasley, M. D. Broadhurst, B. Muller, D. J. Peppard, P. Peshelle, C. Suter, *J. Am. Chem. Soc.*, **97**, 4973 (1975).
[10] A. V. Tobolsky, C. E. Rogers, *J. Polymer Sci.*, **40**, 73 (1959).
[11] S. Bywater, D. J. Warsfold, *Canad. J. Chem.*, **42**, 2884 (1964).
[12] W. K. R. Barnikol, G.V. Schulz, *Makromolek. Chem.*, **86**, 298 (1965).
[13] Y. Okamoto, K. Suzuki, K. Ohta, K. Hatada, H. Yuki, *J. Am. Chem. Soc.*, **101**, 4769 (1979).
[14] Y. Okamoto, K. Suzuki, H. Yuki, *J. Polym. Sci. Poly. Chem. Ed.*, **18**, 3043 (1980).
[15] J. K. Crandall, A. J. Rajas, *Org. Syhth.*, **55**, 1 (1976).
[16] E. A. Hill, H. G. Richey, Jr, T. C. Rees, *J. Org. Chem.*, **28**, 2161 (1963).
[17] M. F. Semmelhack, *New Applications of Organometallic Reagents in Organic Synthesis*, ed., D. Seyferth, Elsevier, Amsterdam (1978), p. 361.
[18] K. Ziegler, H. Zeiser, *Justus Liebigs Ann. Chem.*, **485**, 174 (1931).
[19] J. Büchi, F. Kracher, G. Schmidt, *Helv. Chim. Acta*, **45**, 729 (1962).
[20] G. Van Zyl, R. J. Langenbcrg, H. M. Tan, R. N. Schut, *J. Am. Chem. Soc.*, **78**, 1955, (1956).
[21] M. T. Reetz, *Organotitanium Reagents in Organic Synthesis*, Springer-Verlag, Berlin, 1986.
[22] T. Cohen, R. J. Ruffner, D. W. Shull, E. R. Fogel, J. R. Falok, *Org. Synth.*, **59**, 202 (1980).
[23] M. C. Roux, L. Wartski, J. Seyden-Penne, *Tetrahedron*, **37**, 1927 (1981).
[24] G. Wittig, U. Schoelkopf, *Org. Synth., Coll.*, **5**, 751 (1973).
[25] D. J. Peterson, *J. Org. Chem.*, **33**, 780 (1968).
[26] P. F. Hudrlik, D. Peterson, R. J. Rona, *J. Org. Chem.*, **40**, 2263 (1975).
[27] H. Christensen, *Synth. Commun.*, **5**, 65 (1975).
[28] J. E. Baldwin, W. A. Dupont, *Tetrahedron Lett.*, **21**, 1881 (1980).
[29] G. P. Mpango, K. K. Mahalanabis, Z. Mahdavi-Damghani, V. Snieckus, *Tetrahedron Lett.*, **21**, 4823 (1980).
[30] M. J. Jorgenson, *Org. React.*, **18**, 1 (1970).
[31] T. M. Bare, H. O. House, *Org. Synth.*, **49**, 81 (1969).
[32] D. Seebach, E. J. Corey, *J. Org. Chem.*, **40**, 231 (1975).
[33] E. M. Kaiser, J. D. Pettie, *Synthesis*, 705 (1975).
[34] M. Schwarz, R. M. Waters, *Synthesis*, 567 (1972).
[35] S. Akiyama, J. Hooz, *Tetrahedron Lett.*, 4115 (1973)

[36] E. J. Breithalle, A. G. Fallis, *J. Org. Chem.*, **43**, 1964 (1978).
[37] G. Cahiez, D. Bernard, J. F. Normant, *Synthesis*, 245 (1976).
[38] M. J. Eis, J. E. Wrobel, B. Ganem, *J. Am. Chem. Soc.*, **106**, 3693 (1984).
[39] G. Köbrich, *Angew. Chem. Int. Ed.*, **6**, 41 (1967).
[40] G. Köbrich, *Angew. Chem. Int. Ed.*, **11**, 473 (1972).
[41] G. Cainelli, N. Tangari, A. V. Ronchi, *Tetrahedron*, **28**, 3009 (1972).
[42] H. Taguchi, H. Yamamoto, H. Nozaki, *J. Am. Chem. Soc.*, **96**, 3010 (1974).
[43] W. Adam, O. Cueto, *J. Org. Chem.*, **42**, 38 (1977).
[44] E. Vedejs, *J. Am. Chem. Soc.*, **96**, 5944 (1974).
[45] E. Vedejs, J. E. Telshow, *J. Org. Chem.*, **41**, 740 (1976).
[46] E. Vedejs, D. A. Engler, J. E. Telschow, *J. Org. Chem.*, **43**, 188 (1978).
[47] H. Neumann, D. Seebach, *Chem. Ber.*, **111**, 2785 (1978).
[48] R. G. Jones, H. Gilman, *Org. React.*, **6**, 352 (1951).
[49] R. West, W. H. Glaze, *J. Org. Chem.*, **26**, 2096 (1961).
[50] D. Seyferth, M. A. Weiner, *J. Am. Chem. Soc.*, **87**, 4147 (1965).
[51] D. F. Hoeg, D. I. Lusk, A. L. Crumbliss, *J. Am. Chem. Soc.*, **83**, 3583 (1961).
[52] M. Gall, H. O. House, *Org. Synth., Coll.*, **6**, 121 (1988).
[53] M. W. Rathke, *Org. Synth., Coll.* **6**, 598 (1988).
[54] R. Levine, E. Baumgarten, C. R. Hauser, *J. Am. Chem. Soc.*, **66**, 1230 (1944).

15 Group 2 (Mg) Metal Compounds

N. Miyaura, *Hokkaido University*

15.1 Introduction

Organomagnesium compounds are one of the most readily available, easily prepared, and easily handled of the metal reagents, which were discovered by Barbier and Grignard in 1900. The most commonly prepared organomagnesium reagents are the solvated Mg(R)X, which are usually prepared by the classical technique of direct reaction of organic halides and magnesium metal. The reagents are directly used as intermediates for the syntheses of organic and organometallic compounds.

There are a number of pertinent reviews and books [1] for the preparation of organomagnesium compounds and their reactions. Thus, this chapter summarizes the representative methods for the preparations of organomagnesium (Grignard) reagents.

15.2 Synthesis of Organomagnesium Compounds

15.2.1 Preparation from Organic Halides and Magnesium Metal

Many simple alkyl, aryl, and 1-alkenylmagnesium compounds are readily prepared by the slow addition of organic halides to a stirred suspension of a slight excess of magnesium turnings in an ether solvent (eq. (1)).

$$\text{R-X} + \text{Mg} \longrightarrow \text{R-Mg-X} \qquad (1)$$

The formation of the Grignard reagent is generally slow to start; however, when it is established in ethereal solvents, it is exothermic. Thus, care must be taken during the preliminary stage to avoid the addition of too much halide before it has been observed that the reaction is well started.The halide is then added at such a rate as to maintain a gentle reflux of the ether. The induction period at the beginning is due to the presence of moisture and a thin oxide film coated on the surface of magnesium metal. Moisture can be removed by heating the flask containing magnesium turnings with an electric heat gun, while maintaining a slow stream of dry nitrogen through the flask. Activation of the magnesium surface can be carried out by the addition of a small amount of iodine or 1,2-dibromoethane. The activated magnesium slurries, prepared

Synthesis of Organometallic Compounds: A Practical Guide. Edited by S. Komiya

by reduction of magnesium halides with potassium metal [2] or sodium naphthalene radical anion [3] (Rieke's magnesium), are useful for alkyl and aryl halides that do not react with magnesium turnings or for the synthesis of unstable Grignard reagents at low temperature [4]. Diethyl ether and THF are the most popular solvents. The use of THF is essential to achieve high yields for vinyl halides and other inert organic halides such as chlorobenzene and bromomesitylene; however, the more polar nature of THF may accelerate the formation of the Wurtz coupling products for alkyl halides. Other ether solvents such as di-*n*-propyl ether, di-*n*-butyl ether, diisopropyl ether, and DME are also suitable for the reactions at higher temperatures.

The formation of Grignard reagent is initiated by a rate-determining electron transfer from the magnesium metal to the σ* anti-bonding orbital of the carbon–halogen bond [5]. Most of the organic radicals thus formed remain adsorbed on the magnesium surface to form the Grignard reagent. The surface radical can also undergo dimerization and disproportionation and some can escape the surface and react in the solution.

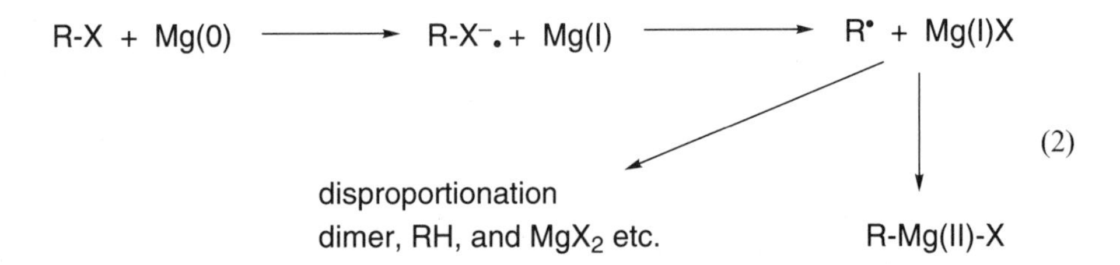

(2)

The order of reactivity of organic halides is I > Br > Cl > F. Although organic iodides are the most reactive, they produce more Wurtz coupling products through the radical oxidative addition process. Thus, it is usually advisable to use the chlorides or bromides, except in the case of aromatic iodides and methyl iodide. 1-Alkenyl Grignard compounds are obtained by the direct reaction of magnesium and vinylic halides in THF [6], but the reaction is not completely stereospecific and the retention of the stereochemistry of haloalkenes varies from 60–90% [7]. The difficulty associated

Table 15.1 Synthesis of Grignard Reagents from Organic Halides and Magnesium Metal

Halide	Solvent	Grignard Reagent	Yield/%[a]
$(CH_3)_2CHX$	ether	$(CH_3)_2CHMgX$[b]	X=Cl (94), X=Br (82), X=I (57)
$(CH_3)_3CCl$	ether	$(CH_3)_3CMgCl$[c]	
C_6H_5Cl	THF	C_6H_5MgCl[d]	95
C_6H_5Br	ether	C_6H_5MgBr[e]	—
2,4,6-$(CH_3)_3C_6H_2Br$	THF	2,4,6-$(CH_3)_3C_6H_2MgBr$[f]	84
$(CH_3)_2C{=}CHBr$	THF	$(CH_3)_2C{=}CHMgBr$[g]	90
$CH_3CH{=}CHCH_2Br$	ether	$CH_3CH{=}CHCH_2MgBr$[h]	90
$CH_2{=}CHCH_2CH_2Cl$	ether	$CH_2{=}CHCH_2CH_2MgCl$[i]	>70

[a]Yields are estimated by titration or by reactions with carbonyl compounds. [b]Houben, J.; Boedler, J.; Fischer, W. *Chem. Ber.*, **1936**, *69*, 1766. [c]Puntambeken, S.V.; Zoellner, E.A. *Org. Synth.* **1941**, *I*, 524. [d]Gilman, H.; Zolner, E.A.; Dickey, J.B, *J. Am. Chem. Soc.* **1929**, *51*, 1576. [e]Allen, F.H.; Converse, S. *Org. Synth.* **1941**, *I*, 226, 550. [f]Bowen, D.M. *Org. Synth.*, **1955,** *III*, 533. [g]H. Normant, J. Ficini, *Bull. Soc. Chim. Fr.*, **1956**, 1441.. [h]Benkeser, R. A. *Synthesis*, **1971**, 347. [i]R. I. Trust, R. E. Ireland, Org. *Synth.*, **1988**, VI, 606.

with the Wurtz reaction in the syntheses of allyl- [8] and benzylmagnesium chlorides or bromide [9] can be limited by the very slow addition of the halides to a large excess of magnesium in an ether solvent at a low temperature. The use of fluorides is generally not recommended; however, Rieke's magnesium is active enough to convert alkyl and aryl fluorides to the corresponding magnesium fluorides in high yields [2].

Representative organomagnesium reagents and the reaction conditions for their preparation are summarized in Table 15.1.

15.2.2 Metallation of Acidic C—H Bonds

The replacement of hydrogen in an organic compound by magnesium is convenient for the preparation of Grignard reagents from sufficiently acidic hydrocarbons. The simplest metallations are those of relatively strong hydrocarbon acids ($pKa < 25$) such as 1-alkynes ($pKa = 25$) [10], cyclopentadiene ($pKa = 16$) [11], and indene ($pKa = 20$) [12], which may be regarded as straightforward acid–base reactions. The reaction is accelerated by increasing the nucleophilicity of the organic group on the magnesium atom. Thus, the reaction is slow in hydrocarbon solvents, but is faster in ethereal solvents and faster still in the presence of up to two moles of HMPA. Cyclopentadiene, indene, fluorene, and 9,10-dihydroanthracene are readily metallated with isopropylmagnesium chloride in HMPA below 40 °C [13].

$$\text{RC}{\equiv}\text{CH} + \text{Et–Mg–Br} \longrightarrow \text{RC}{\equiv}\text{C–Mg–Br} + \text{EtH} \quad (3)$$

$$\text{2,4-dimethylpyrrole (N–H)} \xrightarrow{C_2H_5MgBr} \text{2,4-dimethyl-1-(MgBr)pyrrole} \rightleftarrows \text{3,5-dimethyl-2-(MgBr)pyrrole (N–H)} \quad (4)$$

$$\text{CHX}_3 + {}^i\text{Pr–Mg–Cl} \longrightarrow \text{CX}_3\text{–Mg–Cl} + {}^i\text{PrH} \quad (5)$$

$$X = Cl\ (-78\ °C),\ X = Br\ (-95\ °C)$$

Heterocyclic aromatic compounds such as pyrrole are readily metallated with Grignard reagents. The resulting compounds have N—Mg bonds and are, therefore, not organometallic compounds, but on reaction with electrophiles give 2-substituted pyrroles [14] (eq (4)). The reaction of chloroform or bromoform with ^{i}PrMgCl at –78 °C in THF-HMPA (4:1) is mild and convenient method for the generation of an unstable carbenoid in the solution [15] (eq (5)).

15.2.3 Transmetallation

When the preparation of a halide-free diorganomagnesium compound is desired, the method of choice is the reaction between magnesium and an diorganomercury compound [16] (eq (6)). The reaction of dimethylmercury or divinylmercury with magnesium metal proceeds at *ca* 60 °C, but diphenylmercury requires heating at 150 °C for

40 h without a solvent. Extraction of the reaction mixture with ether or THF produces a pure solution of diorganomagnesium. The reaction is most suited for critical physical studies, but the reaction has limited use because of the need to handle extremely poisonous organomercury compounds.

$$R_2Hg + Mg \longrightarrow R\text{-}Mg\text{-}R \tag{6}$$

$$LiR + MgX_2 \longrightarrow R\text{-}Mg\text{-}X \tag{7}$$

Alternatively, the reaction of an organolithium compound with one equivalent of magnesium halide, is a useful route to Grignard reagents that are difficult to prepare directly (eq (7)). The reaction occurs with retention of configuration for optically active organolithiums [17]. The preparation of diorganomagnesium in benzene or hydrocarbon solvents is carried out by the addition of two equivalents of organolithium to a solution of activated $MgCl_2$ [18].

15.2.4 Carbometallation and Hydrometallation

The transition metal-catalyzed reactions of Grignard reagents provide a valuable technique for the addition of Grignard reagents (R=Me, Ph) and magnesium hydrides to alkenes and alkynes. Nickel(II) chloride catalyzes the addition of MeMgBr or PhMgBr to alkenes [19]. A combination of $Ni(acac)_2$ and Me_3Al is effective for the *cis*-addition of methylmagnesium bromide to alkynes [20] (eq (8)). On the other hand, the intramolecular addition of the carbon–magnesium bond to alkenes and alkynes proceeds without catalyst. The formation of cyclic compounds is favoured for five- and seven-membered rings, but not for three- and four-membered rings [21]. The reaction may occur through the radical cyclization process during the preparation of the Grignard reagents.

$$R\text{-}C{\equiv}C\text{-}SiMe_3 + Me\text{-}Mg\text{-}Br \xrightarrow{Ni(acac)_2\ /\ Me_3Al} (R)(Me)C{=}C(SiMe_3)(MgBr) \tag{8}$$

(9) [cyclic alkynyl Grignard with Ph and MgBr → cyclohexylidene bearing Ph and MgBr]

The titanium-catalyzed addition of magnesium hydrides to alkanes was first reported by Ashby [22], but this reaction can be more conveniently carried out using alkylmagnesium halides (^{i}Bu, ^{n}Pr), because β-hydride elimination from organotransition metal complexes to produce transition metal hydrides is generally very rapid ($RCH_2CH_2MgX + M{-}X \rightarrow RCH_2CH_2\text{-}M \rightarrow H{-}M + RCH{=}CH_2$). Thus, dimerization and subsequent cyclization of butadiene is achieved with ^{n}PrMgBr in the presence of $NiCl_2(PPh_3)_2$ [23] (eq (10)). The catalytic hydrometallation of alkynes [24] and 1,3-alkadienes [25] with ^{n}PrMgBr proceeds with Cp_2TiCl_2 catalyst (eq (11) and (12)). The reaction may also involve a Ti—H intermediate resulting from β-hydride elimination.

$$2\ \text{CH}_2{=}\text{CH-CH}{=}\text{CH}_2 \xrightarrow[\text{NiCl}_2(\text{PPh}_3)_2]{^n\text{PrMgBr}} \text{BrMg-CH}_2\text{-CH}{=}\text{CH-CH}_2\text{-CH}_2\text{-CH}_2\text{-CH}{=}\text{CH}_2 \longrightarrow \text{(cyclopentane bearing vinyl and CH}_2\text{MgBr, H, H)} \quad (10)$$

$$\text{R-C}{\equiv}\text{C-SiMe}_3 + {}^i\text{Bu-Mg-Cl} \xrightarrow{\text{Cp}_2\text{TiCl}_2} \text{R(H)C}{=}\text{C(SiMe}_3\text{)(MgBr)} \quad (11)$$

$$\text{CH}_2{=}\text{C(CH}_3\text{)-CH}{=}\text{CH}_2 + {}^n\text{Pr-Mg-Br} \xrightarrow{\text{Cp}_2\text{TiCl}_2} \text{BrMg-CH}_2\text{-C(CH}_3\text{)}{=}\text{CH-CH}_3 \quad (12)$$

The addition of a magnesium hydride to propargylic alcohols proceeds with a Cp_2TiCl_2 catalyst, in which the oxomagnesium species may provide anchimeric assistance for the addition [26] (see Section 15.3(10)).

15.2.5 Other Methods for Preparation

A simple fractional crystallization of magnesium halide allows the separation of the diorganomagnesium compounds because the Schlenk equilibrium produces a quantity of MgR_2 in the Grignard solution [27] (eq (13)). The magnesium halide forms an insoluble MgX_2·(1,4-dioxane)$_2$ complex. Thus, the preparation of a diorganomagnesium solution is simply achieved by the addition of 5–10% excess of a 1:1 molar ratio of 1,4-dioxane to an ethereal Grignard solution [28, 29].

$$2\ \text{R-Mg-X} + 2\ \text{(1,4-dioxane)} \longrightarrow \text{MgX}_2(\text{1,4-dioxane})_2 + \text{R-Mg-R} \quad (13)$$

The reaction of diethylmagnesium with sodium triethylhydroborate in ether at –78 °C gives a solution of Mg(H)Et, which disproportionates to $MgEt_2$ and MgH_2 above –20 °C [29] (eq (14)). The reaction of MgR_2 (R = Et, Ph) with lithium aluminum hydride produces insoluble MgH_2 accompanied by Li[$AlH_{4-n}R_n$] (n =1–4) [30] (eq (15)).

$$\text{MgEt}_2 + \text{Na[BHEt}_3] \xrightarrow{-78\ ^\circ\text{C}} \text{H-Mg-Et} \xrightarrow{>-20\ ^\circ\text{C}} \text{MgH}_2 + \text{MgEt}_2 \quad (14)$$

$$2\ \text{MgR}_2 + \text{LiAlH}_4 \xrightarrow[\text{R = Me, Ph}]{\text{THF}} 2\ \text{MgH}_2 + \text{Li[AlR}_4] \quad (15)$$

15.3 Synthesis of Organomagnesium Compounds

15.3.1 General Remarks

For the preparation of Grignard reagents, all reagents, solvents, and apparatus must be dry and moisture and oxygen should be excluded. If a clear stock solution of Grignard reagent, the reaction mixture should be filtered through a dry Celite pad on a sintered-glass filter. Alternatively, the reaction mixture can be allowed to stand overnight so that the unreacted magnesium residues settle, and the supernatant solution is then transferred to a bottle charged with nitrogen by means of a cannula and nitrogen pressure.

The Grignard solution is standardized by the following procedure [31]. 2 mL of Grignard solution is added to 5 mL of water that contains four drops of aqueous methyl orange solution as an indicator. The ether is removed by evaporation and an excess of 1 N hydrochloric acid is added until the color of the solution changes from yellow to red (X mL). The excess acid is titrated with a 0.5 N sodium hydroxide until the solution turns to yellow (Y mL). The molar concentration of Grignard reagent = $(X\text{-}0.5Y)/2$.

(1) Butylmagnesium bromide, $Mg(C_4H_9)Br$ [32]:

General procedure from organic halides and magnesium metal is given below:

$$\text{R-X} + \text{Mg} \xrightarrow{\text{ether or THF}} \text{R-Mg-X} \tag{16}$$

RX = C_4H_9Br, 3-chlorobromobenzene, $CH_2{=}CHCH_2Cl$, $CH_2{=}CHBr$, $HC{\equiv}CCH_2Br$, 1-chloronorbornane

A flask is assembled with a magnetic stirring bar or a mechanical stirrer, a dropping funnel, and a reflux condenser to which an oil bubbler and a nitrogen inlet are connected. The flask is charged with magnesium turnings, and then the flask is heated with an electric heat gun or a Bunsen burner while maintaining a slow stream of nitrogen through the flask to remove traces of moisture. Use the above general procedure for all preparations of Grignard reagents from magnesium metal and organic halides. Three times repetition of evacuation and filling nitrogen procedure is also used for this purpose.

The flask is charged with magnesium turnings (36.5 g, 1.5 mol) and ether (500 mL). The dropping funnel is charged with a solution of butyl bromide (206 g, 1.5 mol) in ether (250 mL). Add 10–15 mL of the bromide solution to the flask. The reaction generally begins within a few minutes (if necessary, add a crystal of iodine or small amount of 1,2-dibromoethane). As soon as refluxing is vigorous, immerse the flask in an ice–water bath. Add dropwise the butyl bromide solution at a rate necessary to maintain a gentle reflux (30–40 min). After all solution has been added, stir the mixture at room temperature until almost all of unreacted magnesium residue disappears.

(2) 3-Chlorophenylmagnesium bromide, $Mg(3\text{-}ClC_6H_4)Br$ [33]:

The flask is charged with magnesium turnings (29.1 g, 1.2 mol), a crystal of iodine, and ether (50 mL). A solution of 3-chlorobromobenzene (229 g, 1.2 mol) in ether (850 mL) is added dropwise with stirring at a rate which maintains rapid refluxing. The reaction begins after 20–50 mL of the ether solution is added. If the reaction does not start spontaneously, heat the mixture to reflux before more than 50 mL of the ether solution is added. After all of the ether solution has been added, stir and heat to reflux with an electric heating mantle for 1 h.

(3) Allylmagnesium bromide, $Mg(CH_2CH{=}CH_2)Br$ [8]:

Magnesium turnings (153 g, 6.28 mol), ether (360 mL), and a few crystals of iodine are placed in the flask. A solution of allyl bromide (351 g, 2.9 mol in 2.6 L of ether) is added in small portions until the reaction begins, and then at such a rate as to maintain a gentle reflux of the ether (*ca* 3 h). After the addition has been completed, the mixture is refluxed for 1 h to give an allylmagnesium bromide solution.

(4) Vinylmagnesium bromide, $Mg(CH{=}CH_2)Br$ [34]:

A dry-ice–acetone reflux condenser is used. The flask is charged with magnesium turnings (29.2 g, 1.2 mol) and enough THF to cover magnesium turnings. About 5 mL of vinyl bromide (redistil prior to use and collect it in a receiver cooled with a dry ice–acetone bath) is added. After the reaction has started, THF (350 mL) is added to the flask. In cases where the reaction does not begin within a few minutes, methyl iodide (0.5 mL) can be added to initiate attack on the magnesium. The rest of the vinyl bromide (140 g, 1.3 mol) is dissolved in THF (120 mL), and then added at such a rate that a moderate reflux is maintained. After the addition has been completed, the solution is refluxed for 30 min to give a vinylmagnesium bromide solution.

(5) Allenylmagnesium bromide, $Mg(CH{=}C{=}CH_2)Br$ [35]:

The flask is charged with mercury(II) chloride (0.5 g, 0.002 mol), magnesium (crush the magnesium turning with a mortar and pestle, 29.2 g, 1.2 mol), and ether (160 mL). Propargyl bromide (7.6 g, 5.0 mL, 0.064 mol) is added to the flask. The ether will begin to reflux within 1 min. If the reaction does not begin, the flask is heated by an electric heat gun. After the reaction has been initiated, the flask is cooled to 5 °C in an ice–salt bath. A solution of propargyl bromide (135 g, 1.13 mol) in ether (560 mL) is added under vigorous stirring. The addition rate is adjusted so as to maintain the internal temperature between 5–10 °C. The cooling bath is removed and then the dark green mixture is stirred for 45 min at room temperature to give a solution of allenylmagnesium bromide. Some unreacted magnesium turnings will remain in the flask, however, do not heat the mixture to achieve further conversion, since the allenylmagnesium bromide will isomerize to 1-propynylmagnesium bromide.

(6) 1-Norbornanemagnesium chloride [2] using Rieke's activated magnesium (reduction of MgX_2 with K)

The flask is charged with potassium[1] (1.5 g, 0.038 mol), anhydrous magnesium chloride[2] (2.01 g, 0.0211 mol), and anhydrous potassium iodide[3] (3.55 g, 0.0214 mol), and THF (50 mL) under argon. The mixture is heated to reflux for 3 h under vigorous stirring to give active magnesium as a black powder.[4] 1-Chloronorbornane

[1] Use freshly cut potassium which is free from oxide coating. Cut it into two or three pieces under hexane and place the wet metal in a tared flask that has been purged with argon. Evacuate to remove the solvent, fill again with argon, and to determine the exact amount of potassium. Potassium metal free from solvent ignites in air!

[2] It may be stored in a desiccator over calcium sulfate. If required, dry it overnight in an oven at 120 °C.

[3] It was finely ground with a motor and pestle, dried over night in an oven at 120 °C, and stored in a desiccator.

[4] For an alternative and, presumably, more convenient laboratory procedure for the preparation of active magnesium, see Section 15.3(8).

(1.25 g, 0.009 58 mol) is added to the flask at room temperature. The mixture is refluxed for 6 h to give a solution of 1-norbornanemagnesium chloride solution.

(7) Ethynylmagnesium chloride, Mg(C≡CH)Cl [10]

$$\text{H-C}{\equiv}\text{C-H} + {}^{n}\text{C}_4\text{H}_9\text{MgCl} \xrightarrow{\text{THF}} \text{H-C}{\equiv}\text{C-MgCl} + \text{C}_4\text{H}_{10} \qquad (17)$$

The flask is charged with magnesium turnings (39.6 g, 1.65 mol) and THF (150 mL), and then the mixture is heated to reflux temperature. The dropping funnel is filled with 1-chlorobutane (152.5 g, 1.65 mol), and then a portion (*ca* 15 mL) is added to the boiling THF mixture. The source of heat is removed. After the reaction has commenced, a further volume of THF (400 mL) is added to the reaction mixture. The remainder of the chlorobutane is added dropwise at a rate sufficient to maintain the reaction under reflux. Finally, the mixture is heated and stirred until all magnesium has been consumed. The hot solution (*ca* 60 °C) of butylmagnesium chloride is transferred into a 1-L dropping funnel using a cannula.

A 2L flask is assembled with a mechanical stirrer, a thermometer, a dropping funnel, and two swan neck adapters to which a gas inlet and an outlet are connected. The flask is filled with THF (500 mL), and then it is saturated with acetylene[1] by bubbling the gas for 0.5–1 h. The flask is cooled to –5 °C in a dry ice–acetone bath. The dropping funnel is charged with the above Grignard solution using a cannula and is then added to the stirred mixture at a rate sufficient to maintain the internal temperature below 20 °C[2] (*ca* 1 h). The rapid flow of acetylene is kept during the addition, and further 0.5 h after the addition has been completed. The solution of ethynylmagnesium chloride thus obtained is homogeneous at 30 °C.

(8) Mg(CHPhCH=CHCHPh) from Rieke's active magnesium (Reduction of MgX_2 with Li/naphthalene) [3]

$$\text{PhCH}{=}\text{CHCH}{=}\text{CHPh} + \text{“Mg”} \xrightarrow{\text{THF}} \text{Ph–}\overbrace{\text{CH–CH=CH–CH}}^{}\text{–Ph (ring with Mg)} \qquad (18)$$

All glassware is dried in a 120 °C electric oven before use, and a hot two- necked 50 mL flask is assembled with a magnetic stirring bar (coated with glass), a septum inlet, and a T-piece to which an argon inlet and an oil bubbler are connected. The apparatusis cooled while maintaining a flow of dry argon through the flask. Naphthalene (2.82 g, 22 mmol) and lithium[3] (0.14 g, 20 mmol) are added to the flask from the neck capped with the septum, while maintaining a slow stream of argon through the flask. THF

---

[1] Acetylene is introduced at the rate of about 20 L/h, and is purified by passing the gas through two cold traps (–78 °C) and a tower of 10 mesh alumina to remove acetone.

[2] At the higher temperature, ethynylmagnesium halides can rapidly disproportionate to acetylene and bis(chloromagnesium) acetylene precipitating white solids.

[3] Use a freshly cut lithium under hexane which is free from oxide coating.

(30 mL) is added to the flask using a syringe by puncturing the septum. The mixture is stirred at room temperature for 2 h to give a dark green solution of lithium naphthalide.

The same apparatus as above is assembled, except that a 200mL flask is used. The flask is charged with anhydrous magnesium chloride[1] (0.91 g, 9.6 mmol), and then flush throughly with argon. THF (20 mL) is added to the flask using a syringe. The above lithium naphthalide solution is transferred to the flask by means of a cannula and nitrogen pressure. At the beginning the green color of naphthalide disappears quickly at the beginning, and a dark gray to black suspension is finally obtained. The mixture is stirred for 30 min, and then let it stand for more than 3 h so that fine magnesium powder settles out. Tetrahydrofuran is removed with a syringe to give a gray to black powder of active magnesium[2].

THF (40 mL) and then a solution of (*E,E*)-1,3-diphenylbutadiene (1.65 g, 8.0 mmol) in THF (20 mL) are added to the above flask by syringe. The mixture is stirred for 2 h at room temperature. The solution turns purple at the beginning, and finally red. The Grignard solution thus obtained is used for the next reaction.

The title compound is also prepared using Mg turnings activated with iodine [3c,d]. Charge the flask with magnesium turnings (6.1 g, 250 mmol), a crystal of iodine, and THF (100 mL). A solution of (*E,E*)-1,3-diphenylbutadiene (5 g, 25 mmol) in THF (20 mL) is added to the flask and then the mixture is stirred for 5 days at 15 °C using a magnetic stirring bar to give a deep red solution of the title compound. The crystallization from THF gives Mg(CHPhCH=CHCHPh).3THF as a crystalline solid in 80 % yield. The coordination geometry of the magnesium atom is trigonal bipyramidal [3d].

(9) Di-*t*-butylmagnesium, $Mg({}^tC_4H_9)_2$ [29]

$$2\,{}^tC_4H_9MgCl + 2\ \text{dioxane} \xrightarrow{\text{ether}} ({}^tC_4H_9)_2Mg + MgCl_2(\text{dioxane})_2 \qquad (19)$$

A solution of *t*-butylmagnesium chloride in ether is prepared, in the usual way (Section 3.2). A dry centrifuge tube assembled with a rubber septum and a magnetic stirring bar, and flush with nitrogen. The tube is charged with the *t*-butylmagnesium chloride solution (1 M, 30 mL) by syringe through the septum. Dioxane (3.7 g, 0.042 mol) is added to the vigorously stirred solution. White solid will precipitate immediately. Three similar centrifuge tubes are also prepared, and all four tubes are spinned at 3000 rpm for 30 min. The supernatant solution is transferred to a nitrogen-filled flask using a cannula. Analysis of the solution shows it to be 0.36 M in *t*-butyl, 0.177 M in magnesium, and 0.007 M in chloride. The presence of a little dioxane is also detected by its NMR spectrum.

---

[1]Commercially available anhydrous magnesium chloride is directly used. If required, dry it in an oven at 120 °C overnight.

[2]This operation can be ignored when the presence of naphthalene does not cause any troubles for the next reaction.

(10) $Mg(Cl)\{C(C_5H_{11})=CHCH_2O\}MgCl$ [26] by Catalytic Hydrometallation

$$C_5H_{11}C{\equiv}CCH_2OH + 2\ {}^iC_4H_9MgCl \xrightarrow[\text{ether}]{Cp_2TiCl_2} (ClMg)(C_5H_{11})C{=}CH(CH_2OMgCl) \tag{20}$$

A 500 mL three-necked flask is assembled with a magnetic stirring bar, a septum inlet, a glass stopper, a T-piece connected one end to a supply of argon and the other to an oil bubbler. The flask is charged with a solution of isobutylmagnesium chloride in ether (320 mL, 0.75 M, 0.24 mol)[1] via cannula through the septum inlet. The flask is cooled to 0 °C in an ice–water bath. Titanocene dichloride (1.3 g, 5.2 mmol) is added at once from the neck stoppered with a glass cap, while maintaining a slow stream of argon through the flask. Then, the mixture is stirred 10 min at 0 °C to give a purple solution[2]. The dropping funnel is charged with 2-octyn-1-ol (13.2 g, 0.105 mol) and ether (30 mL), and the octynol is add dropwise to the flask over 20 min at 0 °C. The mixture stirred for 4 h at room temperature to complete the reaction[3].

References

[1] M. S. Kharasch, O. Reinmuth, *Grignard Reactions of Nonmetallic Substances*, Prentice Hall, New York (1954); S. T. Ioffe, A.Nesmeyanov, *The Organic Compounds of Magnesium, Berylium, Calcium, Strontium and Barium*, North-Holland, Amsterdam (1967); W. E. Lindsell, *Comprehensive Organometallic Chemistry*, Vol. 1, Ed. G. Wilkinson, F. G. A. Stone, Pergamon Press, Oxford (1982).

[2] R. D. Rieke, S. E. Bales, *J. Am. Chem. Soc.*, **96**, 1775 (1974); R. D. Rieke, *Acc. Chem. Res.*, **10**, 301 (1977). R. D. Rieke, S. E. Bales, P. M. Hudnall, G. S. Poindexter, *Org. Synth.*, **59**, 85 (1980).

[3] (a) R. D. Rieke, P. T.-J. Li, T. P. Burns, S. T. Uhm, *J. Org. Chem.*, **46**, 4324 (1981).
(b) R. D. Rieke, H. Xiong, *J. Org. Chem.*, **56**, 3109 (1991).
(c) H. Yasuda, Y. Kajihara, K. Mashima, K. Nagasuna, K. Lee, A. Nakamura, *Organometallics*, **1**, 388 (1982).
(d) Y. Kai, N. Kanehisa, K. Miki, N. Kasai, K. Mashima, H. Yasuda, A. Nakamura, *Chem. Lett.*, 1277 (1982).

[4] T. P. Burns, R. D. Rieke, *J. Org. Chem.*, **48**, 4141 (1983).

[5] H. M. Walborsky, M. Topolski, *J. Am. Chem. Soc.*, **114**, 3455 (1992).

[6] H. Normant, *C. R. Hebd. Seances Acad. Sci.*, **239**, 1510 (1954); H. Sakurai, K. Tominaga, M. Kumada, *Bull. Chem. Soc. Jpn.*, **39**, 1279 (1966).

[7] B. Méchin, N. Naulet, *J. Organomet. Chem.*, **39**, 229 (1972); T. Yoshino,Y. Manabe, Y. Kikuchi, *J. Am. Chem. Soc.*, **86**, 4670 (1964); D. S. Matteson, J. D. Liedtke, *J. Am. Chem. Soc.*, **87**, 1526 (1965).

[8] S. O'Brien, M. Fishwick, B. McDermott, M. G. H. Wallbridge, G. A.Wright, *Inorg. Synth.*, **13**, 73 (1971); J. C. H. Hwa, H. Sims, *Org. Synth.*, *V*, 452, 608 (1973).

[9] R. B. Moffett, *Org. Synth. IV*, 605 (1963).

[10] L. Sakattebol, E. R. H. Jones, M. C. Whiting, *Org. Synth.*, IV, 793 (1963); A. B. Holmes, C. N. Sporikou, *Org. Synth.*, **65**, 61 (1987); R. A. Earl, L. B. Townsend, *Org. Synth.*, **VII**, 334, 456 (1990).

---

[1] A solution of isobutylmagnesium chloride is obtained from isobutyl chloride and magnesium turnings in ether. Determine its concentration by titration before use.

[2] The active catalyst formed in the solution is highly sensitive to oxygen and moisture. Thus, contact with air should be rigorously excluded.

[3] The reaction with hexanal (0.137 mol) gives a 72% yield of (*E*)-3-pentyl-2- nonene-1,4-diol.

[11] V. Grignard, Courtot, *Compt. Rend.*, **152**, 272 (1911).
[12] Datta, I. T. Mitter, *J. Am. Chem. Soc.*, **41**, 287 (1919) and Ref. 11.
[13] H. Normant, *Bull. Soc. Chim. Fr.* (1968) 791; H. Normant, *Angew. Chem. Int. Ed. Engl.*, **6**, 1046 (1967);C. Chevrot, K. Kham, J. Périchon, J. F. Fauvarque, *J. Organomet. Chem.*, **161**, 139 (1978).
[14] H. Fischer, *Org. Synth.* **II**, 198 (1943).
[15] J. Villiéras, *Organomet. Chem. Res., (A)*, **7**, 81 (1971).
[16] E. C. Ashby, R. C. Arnott, *J. Organomet. Chem.*, **14**, 1 (1968).
[17] H. M. Walborsky, A. E. Young, *J. Am. Chem. Soc.*, **86**, 3288 (1964); C. Agami, M. Chauvin, J. Levisalles, *Bull. Soc. Chim. Fr.*, 2713 (1970).
[18] C. W. Kameienski, J. F. Eastham, *J. Org. Chem.*, **34**, 1116 (1969); C. W. Kameienski, J. F. Eastham, *J. Organomet. Chem.*, **8**, 542 (1967); W. H. Glaze, C. R. McDaniel, *J. Organomet. Chem.*, **51**, 23 (1973).
[19] L. Garády, L. Markó, *J. Organomet. Chem.*, **28**, 159 (1971); J.-G. Duboudin, B. Jousseaume, *J. Organomet. Chem.*, **44**, C1 (1972).
[20] B. B. Snider, M. Karras, R. S. E. Conn, *J. Am. Chem. Soc.*, **100**, 4624 (1978); B. B. Snider, R. S. E. Conn, M. Karras, *Tetrahedron Lett.*, 1679 (1979).
[21] A review for cyclization: E. A. Hill, *Adv. Organomet. Chem.*, **16**, 131 (1977).
[22] E. C. Ashby, T. Smith, *J. Chem. Soc. Chem. Commun.*, 30 (1978).
[23] H. Felkin, G. Swierczewski, *Tetrahedron*, **31**, 2735 (1975).
[24] Y. Kobayashi, H. Uchiyama, H. Kanbara, F. Sato, *J. Am. Chem. Soc.*, **107**, 5541 (1985). For a review: F. Sato, *J. Organomet. Chem.*, **285**, 53 (1985).
[25] F. Sato, H. Watanabe, Y. Tanaka, T. Yamaji, M. Sato, *Tetrahedron Lett.*, **24**, 1041 (1983).
[26] F. Sato, H. Ishikawa, H. Watanabe, T. Miyake, M. Sato, *J. Chem. Soc. Chem. Commun.*, (1981) 718; F. Sato, Y. Kobayashi, *Org. Synth.*, **69**, 106 (1990).
[27] E. C. Ashby, W. E. Becker, *J. Am. Chem. Soc.*, **85**, 118 (1963).
[28] W. Schlenk, *Chem. Ber.*, **62**, 920 (1929).
[29] G. E. Coates, J. A. Heslop. *J. Chem. Soc., (A)*, 514 (1968).
[30] E. C. Ashby, R. G. Beach, *Inorg. Chem.*, **16**, 1441 (1977); E. C. Ashby, A. B. Goel. *Inorg. Chem.*, **16**, 1441 (1977).
[31] R. Lespieau, M. Bourguel, *Org. Synth.*, **I**, 187 (1941).
[32] G. H. Coleman, D. Craig, *Org. Synth.*, **II**, 179 (1943).
[33] C. G. Overberger, J. H. Saunders, R. E. Allen, *Org. Synth.*, **III**, 200 (1955).
[34] D. Seyferth, *Org. Synth.*, IV, 258, 260 (1963).
[35] H. Hopf, I. Böhm, J. Kleinschroth, *Org. Synth.*, **VII**, 485 (1990).

16 Group 13 (B, Al) Metal Compounds

N. Miyaura and K. Maruoka, *Hokkaido University*

16.1 Boron Compounds

16.1.1 Introduction

Organoboron compounds, although known for a hundred years, attracted little attention until the pioneering work of two Nobel laureate chemists; the discovery of hydroboration by Herbert C. Brown in 1956 at Purdue University, and the impressive predictions about structure and bonding in the boron hydride compounds by W. N. Lipscomb. The ready formation of the carbon–boron bonds by addition of borane reagents to alkenes and alkynes has given new and important organic preparative procedures through the agency of simple boron hydride reactants. The theory of structure and bonding in B—H compounds is fundamental to any discussion of boranes, carboranes, and related metal hydrides.

Now, organoboron compounds have a very rich organic chemistry and are among the most readily available and easily handled of metal reagents. Their real synthetic utility lies in the very wide range of organic transformations specific to boron compounds, and in the specificity and functional group tolerance of most of these transformations.

Although several important classes of boron compounds will not be discussed here because of space limitation, this chapter attempts to discuss the representative transformations which are likely to be useful for researchers with little experience in organoboron chemistry. No attempt is made to cover in detail physical properties; however, information on boron–carbon bond lengths, nuclear magnetic resonance (^{11}B and ^{1}H), vibrational spectroscopy, electronic transitions, mass spectrometry, and thermodynamic properties of organoboron compounds has been summarized in earlier texts [1].

16.2 Synthesis of Organoborane Compounds

16.2.1 Transmetallation

Transmetallation refers to the transfer of an organic group from a metal to boron by the reaction of an organometallic reagent with an appropriate boron compound. The

Synthesis of Organometallic Compounds: A Practical Guide. Edited by S. Komiya

common procedure is to mix both reagents, usually well diluted in hydrocarbon or ether solvents under nitrogen, with cooling in a Grignard-type apparatus. The work-up procedure is either by direct distillation of the filtrate solution or hydrolysis to the air-stable boronic acids. On the laboratory scale, Grignard reagents or lithium reagents are most useful from the standpoint of availability or easy preparation. Other organometallic derivatives of Al, Zn, Sn, Hg have been also used, depending on availability. The boron substrate may be chosen from a wide range of compounds which contain at least one halogen atom or alkoxy group, e.g., $BF_3{\cdot}O(CH_2CH_3)_2$, BCl_3, and $B(OR)_3$ (R=Me, ^{i}Pr, ^{n}Bu). However, care must be taken to avoid the use of ether solvents with compounds containing the B—X (X=Cl, Br, I) bonds, since haloboranes often react rapidly with ethers.

The combination of an organometallic reagent and the boron substrate to be adopted depends very much upon the final compound required. A very general procedure for the monoalkylation or the monoarylation of boron substrates is by interaction of a stoichiometric amount of Grignard or lithium reagent and a boronic ester [2] (eq (1)–(5)). Alkyl-, aryl- [3], vinyl- [4], and ethynylboronic acids [5] and their esters can be synthesized by the Grignard method. The first stereocontrolled synthesis of 1-alkenylboronates involve the reaction of a (*Z*)- or (*E*)-2-buten-2-ylmagnesium bromides with trimethyl borate [6] (eq (4)). Metallation of (*E*)- or (*Z*)-2-butene with BuLi/^{t}BuOK followed by trapping of the intermediate with haloboranes or trialkyl borates is an efficient and stereoselective method for the synthesis of (*E*)- or (*Z*)-crotylboronate in larger quantities [7] (eq (5)). Alkylation of B—Br with allyl- or propargyl-stannanes, followed by evaporation of the tributyltin bromide *in vacuo* is a convenient *in situ* preparation of allyl- and propargylboronates [8] (eq (6)).

$$ArMgBr + B(OMe)_3 \longrightarrow \xrightarrow{H_2O} ArB(OH)_2 \qquad (1)$$

$$CH_2{=}CHMgBr + B(OBu)_3 \longrightarrow \xrightarrow{HCl} CH_2{=}CHB(OBu)_2 \qquad (2)$$

$$HC{\equiv}CMgBr + B(OBu)_3 \longrightarrow \longrightarrow HC{\equiv}C\text{-}B(OBu)_2 \qquad (3)$$

$$(CH_3)(H)C{=}C(Br)(CH_3) \xrightarrow{Mg} \xrightarrow[2.\ H_3O^+]{1.\ B(OMe)_3} (CH_3)(H)C{=}C(B(OH)_2)(CH_3) \qquad (4)$$

$$MeCH{=}CHMe \xrightarrow[{^t}BuOK]{{^n}BuLi} [Me\text{-allyl-}Li] \xrightarrow{ClB(NMe_2)_2} MeCH{=}CHCH_2\text{-}B(NMe_2)_2 \qquad (5)$$

$$\text{1,3-bis}(SO_2Ar)\text{-4,5-diphenyl-1,3,2-diazaborolidine-}B\text{-}Br \xrightarrow{HC{\equiv}CCH_2SnBu_3} \text{1,3-bis}(SO_2Ar)\text{-4,5-diphenyl-1,3,2-diazaborolidine-}B\text{-}CH_2C{\equiv}CH \qquad (6)$$

The transmetallation of $(R'O)_3B$ with R—M (M=Li, MgX) at low temperature (usually at –78 °C) proceeds by initial formation of a relatively unstable teracoordinated complex $[RB(OR')_3]M$, which is in equilibrium with $RB(OR')_2$ and R'OM. If the monoalkyl(trialkoxy)borate can be cleanly formed, and if equilibrium favors this complex, the boronic ester will be formed selectively. Otherwise, successive steps will give rise to the di-, tri-, or tetraalkylborates (eq (7)). Triisopropyl borate is shown to be the best of the available alkyl borates to prevent such side reactions thus allowing the syntheses of a number of alkyl, aryl, 1-alkenyl [9], and 1-alkynylboronates [10] in high yields, often over 90% (eq (8)).

$$\begin{aligned} B(OR')_3 + R\text{-}M &\longrightarrow [R\text{-}B(OR')_3]M \longrightarrow R\text{-}B(OR')_2 + MOMe \\ &\xrightarrow{R\text{-}M} [R_2B(OR')_2]Li \longrightarrow R_2BOR' + MOMe \\ &\longrightarrow R_3B \xrightarrow{R\text{-}M} [R_4B]\,M \end{aligned} \quad (7)$$

$$RLi + B(O^iPr)_3 \longrightarrow [R\text{-}B(O^iPr)_3]Li \xrightarrow{HCl} R\text{-}B(O^iPr)_2 \quad (8)$$

R = alkyl, aryl, allyl, 1-alkenyl, 1-alkynyl

The best methods for attracting two organic groups to boron are the reaction of boron trihalides with the relatively poorly nucleophilic organometallic reagents, as expected, of Al, Hg, and Sn, and the reaction between magnesium or lithium reagents and trialkyl borates [2]. It is possible to obtain intermediate mono- and diorganoboron products and is anticipated these products are favored over the triorganoboron product when low ratios of organometallic reagent to boron substrate are used [11] (eq (9)).

$$2\ PhMgBr\ or\ Li \xrightarrow[2.\ H_3O^+]{1.\ B(OBu)_3} Ph_2BOH \xrightarrow{HOCH_2CH_2NH_2} Ph_2B(OCH_2CH_2NH_2)\ \text{(cyclic)} \quad (9)$$

The synthesis of triorganoboron compounds by means of zinc alkyls was first achieved by Frankland. A similar procedure using Grignard reagents and boron trifluoride etherate has been used for a number of trialkyl- [12] and arylboron compounds [13] (eq (10) and (11)).

$$3\ RMgX + BF_3 \longrightarrow R_3B + 3\ MgXF \quad (10)$$

$$Me_3Al + B(OMe)_3 \longrightarrow Me_3B + [Al(OMe)_3]_n \quad (11)$$

Although there is no report of a systematic examination of operational conditions, the use of exactly stoichiometric amount of Grignard reagent to boron trifluoride diethyl etherate at low temperature is essential to achieve high yields. However, the synthesis often results in low yields due to the formation of gummy nonvolatile residues (R_4B^- anions). The triorganoboron product is highly electrophilic, thus such *ate*-complex formation occurs easily in both aryl and alkyl series when organolithium or magnesium reagents are used. In such cases, the use of the less nucleophilic organometallic reagents of Hg, Sn, Al may have advantages. For example, triallylboranes are prepared in high yields by the reaction of allylaluminum sesquibromide and boron trifluoride etherate or trialkyl borates [14] (eq (12)).

$$3\ CH_2{=}CHCH_2Br + Al \longrightarrow (C_3H_5)_3Al_2Br_3 \xrightarrow{BF_3} (CH_2{=}CHCH_2)_3B \quad (12)$$

16.2.2 Hydroboration

Under mild experimental conditions, compounds containing B—H bonds add to alkenes or alkynes to form organoboron compounds. For the preparation of trialkylboranes and boronic acids the hydroboration procedure is the method of choice when the starting alkenes or alkynes are readily available [15]. As a way of preparing simple organoboron compounds, the hydroboration reaction is not as generally useful as the above transmetallation method. However, the reaction is important because of its relevance to organic synthesis, though the boron intermediates are usually not isolated for this purpose. A variety of borane reagents are now available for selective hydroboration (Scheme 16.1).

For the synthesis of trialkylboranes, hydroboration is carried out with a borane solution in THF or with the borane–methyl sulfide complex (BMS) in THF, diethyl ether, or dichloromethane. Diborane reacts rapidly and quantitatively with alkenes to produce a solution of trialkylborane [16] (eq (13)). The addition of dialkylboranes, such as 9-borabicyclo[3.3.1]nonane (9-BBN, **3**), disiamylborane (**1**), or dicyclohexylborane (**2**), to alkenes or alkynes gives mixed alkylboron compounds. The high regio-, stereo-, or chemoselectivity in the additions of these borane reagents unsaturated C—C bonds have been extensively used in organic syntheses.

The addition of the B—H bond across a carbon–carbon multiple bond proceeds in an *anti*-Markovnikov manner. Many functional groups tolerate hydroboration, so it is

$$3\ RCH{=}CH_2 + H_3B{\cdot}THF \text{ or } H_3B{\cdot}SMe_2 \longrightarrow (RCH_2CH_2)_3B \quad (13)$$

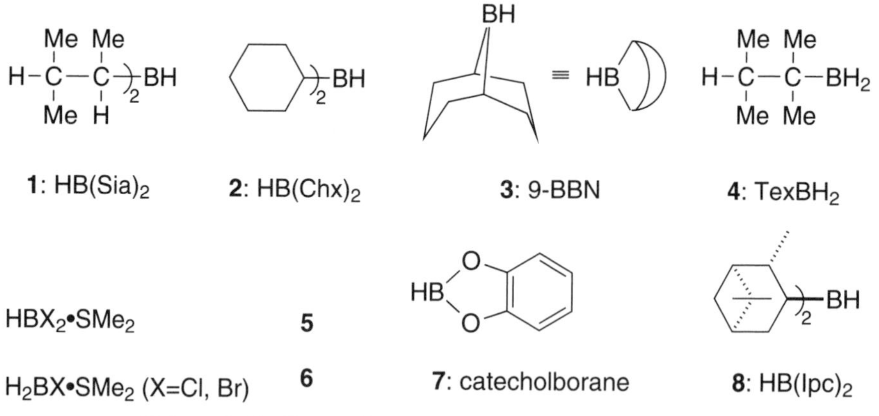

1: $HB(Sia)_2$ **2**: $HB(Chx)_2$ **3**: 9-BBN **4**: $TexBH_2$

$HBX_2{\bullet}SMe_2$ **5**

$H_2BX{\bullet}SMe_2$ (X=Cl, Br) **6**

7: catecholborane **8**: $HB(Ipc)_2$

Scheme 16.1 Borane Reagents for Selective Hydroboration

possible to synthesize reactive intermediates containing those functional groups and to directly utilize those intermediates for organic synthesis. The details of the scope, stoichiometry, regioselectivity and stereoselectivity in the hydroboration of alkenes and alkynes have been extensively reviewed [15, 16].

Hydroboration is especially valuable for the synthesis of stereodefined 1- alkenylboronic acids. A general method is the hydroboration of terminal alkynes with catecholborane **7** [16, 17] (eq (14)). The reaction is generally carried out at 70 °C without solvent, but it is very slow in THF solvent. More recent results demonstrate that the hydroboration of alkenes or alkynes with catecholborane is strongly accelerated in the presence of palladium [18], rhodium [19], or nickel catalysts [20], thus allowing the reaction to proceed below room temperature.

$$\mathrm{RC{\equiv}CH} + \mathrm{HBX_2}\ (\mathbf{1}, \mathbf{2}, \mathbf{3}, \mathbf{5}, \text{or } \mathbf{7}) \longrightarrow \mathrm{R{-}CH{=}CH{-}BX_2} \tag{14}$$

$$\mathrm{RC{\equiv}CH} \xrightarrow{\mathbf{8}} \mathrm{R{-}CH{=}CH{-}B(Ipc)_2} \xrightarrow{\mathrm{CH_3CHO}} \mathrm{R{-}CH{=}CH{-}B(OEt)_2} \tag{15}$$

$$\mathrm{R^1 = CH_2Cl\ (73\%);\ CO_2Et\ (70\%);\ CH(OSiMe_3)CH_3\ (74\%)}$$

The hydroboration of alkynes with dihaloboranes **5** followed by hydrolysis with water gives boronic acids in high yields [21]. Diisopinocampheylborane has been used as a reagent for asymmetric hydroboration, and additionally it has attractive features as a hydroboration reagent for alkynes, e.g., high regioselectivity resulting from its bulkiness, and ease of dealkylation to boronic esters with acetaldehyde under neutral conditions [22] (eq (15)). These methods have been complementarily used for the various 1-alkenylboronic acid synthesis depending on the availability of borane reagents and their limitation to functionalized alkynes.

Terminal and internal (*Z*)-1-alkenylboronates are prepared from (*Z*)-(1-halo-1-alkenyl)boronates [23], which can be readily obtained by hydroboration of 1-halo-1-alkynes (Scheme 16.2). The internal S_N2-like displacement of the halogen with hydrides [24] or organolithiums [25] takes place with complete inversion of configuration at the sp^2 carbon. On the other hand, the palladium-catalyzed alkylation of the C—X bond with organozinc reagents provides (*E*)-1-alkenylboronates [26] which are not available by conventional hydroboration of internal alkynes.

16.2.3 Haloboration

Terminal 2-bromo-1-alkenylboronates are readily obtained by bromoboration of terminal alkynes with tribromoborane. The subsequent displacement of the bromine atom with organozinc reagents proceeds with strict retention of configuration in the presence of a palladium catalyst [27] (eq (16)).

Haloboranes add to terminal alkynes in a *cis anti*-Markovnikov manner; however, the bromoboration of acetylene itself exceptionally provides a *trans* adduct [28] (eq (17)). A sequence of haloboration with boron tribromide and then cross-coupling with organozincs is useful for a formal carboboration of alkynes with various organic groups. Addition of boron triboromide to allene is convenient for the synthesis of (2-bromo-2-propenyl)borate [29] (eq (18)).

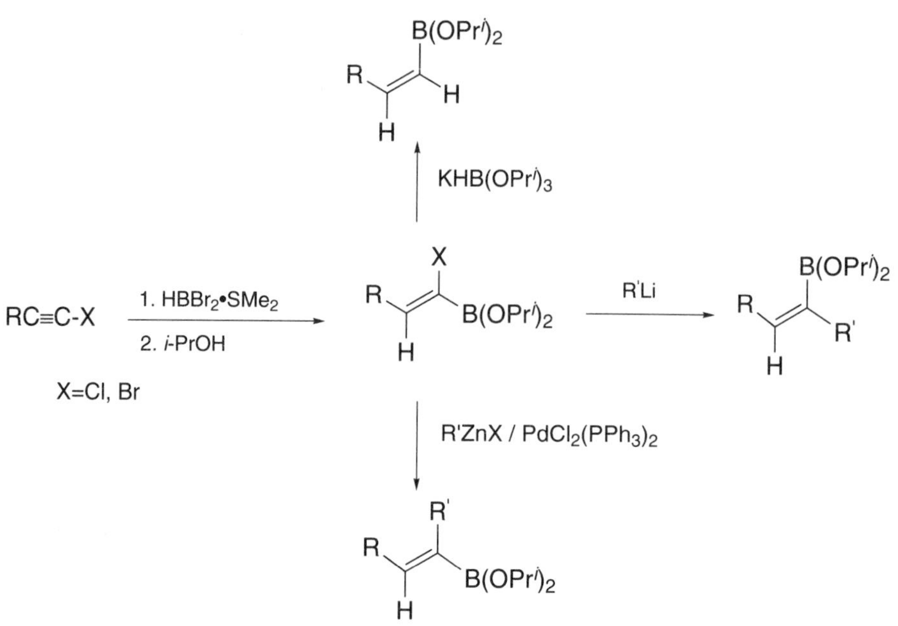

$$RC{\equiv}C\text{-}X \xrightarrow[\text{2. } i\text{-PrOH}]{\text{1. } HBBr_2{\bullet}SMe_2} (Z)\text{-}RCH{=}C(X)B(OPr^i)_2 \qquad X = Cl, Br$$

$$(Z)\text{-}RCH{=}C(X)B(OPr^i)_2 \xrightarrow{KHB(OPr^i)_3} RCH{=}CHB(OPr^i)_2$$

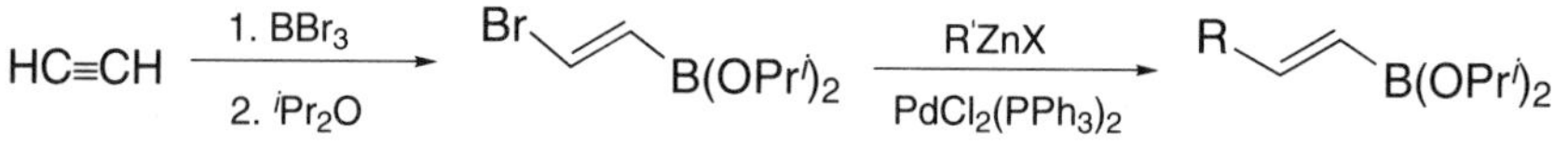

$$(Z)\text{-}RCH{=}C(X)B(OPr^i)_2 \xrightarrow{R'Li} RCH{=}C(R')B(OPr^i)_2$$

$$(Z)\text{-}RCH{=}C(X)B(OPr^i)_2 \xrightarrow{R'ZnX\ /\ PdCl_2(PPh_3)_2} RCH{=}C(R')B(OPr^i)_2$$

Scheme 16.2 (*E*)- and (*Z*)-1-Alkenylborates

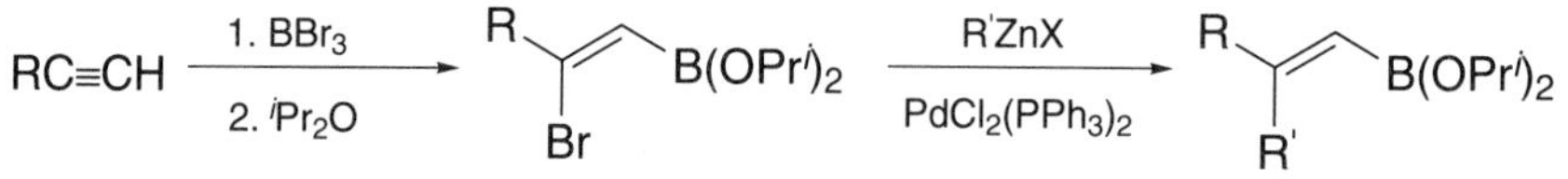

$$RC{\equiv}CH \xrightarrow[\text{2. } {}^iPr_2O]{\text{1. } BBr_3} RC(Br){=}CHB(OPr^i)_2 \xrightarrow[PdCl_2(PPh_3)_2]{R'ZnX} RC(R'){=}CHB(OPr^i)_2 \qquad (16)$$

$$HC{\equiv}CH \xrightarrow[\text{2. } {}^iPr_2O]{\text{1. } BBr_3} BrCH{=}CHB(OPr^i)_2 \xrightarrow[PdCl_2(PPh_3)_2]{R'ZnX} RCH{=}CHB(OPr^i)_2 \qquad (17)$$

$$H_2C{=}C{=}CH_2 + BBr_3 \longrightarrow \xrightarrow{PhOMe} H_2C{=}C(Br){-}CH_2B(OPh)_2 \qquad (18)$$

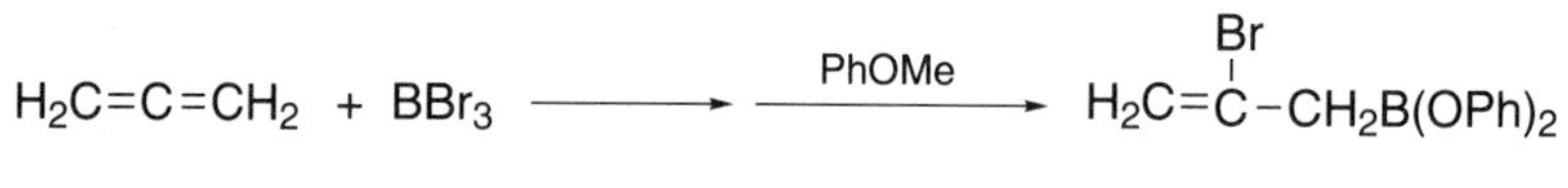

Iodo- or bromoboration of terminal alkynes has been also carried out with 9-X-9-BBN (X=I, Br) to provide (*Z*)-(2-halo-1-alkenyl)boron derivatives, but adducts are too susceptible to dehaloboration for their isolation. Thus, the adducts obtained *in situ* have been directly used for the next reaction in most cases [27].

16.2.4 From α-Haloboronic Esters

The α–haloboronic esters are readily available by addition of hydrogen halides to ethenylboronates, a radical halogenation of alkylboronates, or hydroboration of 1-halo-1-alkenes [30]. By far the easiest laboratory preparation of chloromethyl- [31] or dichloromethylboronates [32] is alkylation of triisopropyl borate with the corresponding lithium reagents generated *in situ* at low temperature (eq (19) and (20)). For those preparations, the use of relatively hindered triisopropyl borate is essential to achieve high yields.

$$ClCH_2I \xrightarrow{BuLi} ClCH_2Li \xrightarrow{B(O^iPr)_3} \xrightarrow{HCl} ClCH_2B(O^iPr)_2$$

$$\xrightarrow{RLi} Cl{-}CH_2{-}\overset{R}{\overset{|}{B^-}}(O^iPr)_2 \longrightarrow RCH_2B(O^iPr)_2 \qquad (19)$$

$$CH_2Cl_2 + B(O^iPr)_3 \xrightarrow[2.\ HCl]{1.\ BuLi} Cl_2CHB(O^iPr)_2 \xrightarrow{RLi} R{-}\underset{Cl}{\underset{|}{C}}H{-}B(O^iPr)_2 \qquad (20)$$

R = alkyl, aryl, 1-alkenyl, 1-alkynyl, R_3Sn, RO, R_2N, RS

The α-haloalkylboronates are of considerable interest as reagents for preparing various α-substituted alkyboronates. Addition of organolithium, Grignard reagent, oxygen or nitrogen nucleophiles at low temperature forms the borate complex and a migrating group then displaces the halide by S_N2 substitution with complete inversion of configuration at the α-carbon. The procedure is quite general and proceeds under very mild conditions, thus allowing the syntheses of a variety of allyl, propargyl, benzylboronates having alkyl, amino, trimethylstanyl, alkoxy, and alkylthio groups at the α-carbon [30] (eq (19) and (20)).

$$CH_2{=}CHCH_2Cl + LDA + B(O^iPr)_3 \longrightarrow \xrightarrow{HO(CH_2)_3OH} CH_2{=}CH{-}\underset{Cl}{\underset{|}{C}}HB\langle O(CH_2)_3O\rangle \qquad (21)$$

In situ generation of $CH_2{=}CHCH(Li)Cl$ and its trapping with triisopropyl borate provides new access to 1-chloroallylboronate [33] which is a valuable intermediate for the synthesis of variously substituted allylboronates via the reaction with organolithiums (eq (21)).

Alternatively, a similar borate intermediate can be obtained by the reaction of organoborate with chloromethyl- or (dichloromethyl)lithiums (eq (22) and (23)). When a chiral diol ester reacts with (dichloromethyl)lithium, an optically active α–chloroalkylboronate is produced with an excellent enantiomeric excess [34] (eq (23)). The C–Cl bond is then displaced readily with alkyl, aryl, or 1-alkenyl nucleophiles with inversion of configuration. The procedure has been extensively used for the preparation of chiral boronates, which have been used in organic syntheses [30].

$$RB(OR')_2 + ClCH_2I \xrightarrow{RLi} ClCH_2{-}\overset{R}{\overset{|}{B^-}}(OR')_2 \longrightarrow RCH_2B(OR')_2 \qquad (22)$$

$$R{-}B(\text{pinanediol}) \xrightarrow{LiCHCl_2} Cl(R)CH{-}B(\text{pinanediol}) \xrightarrow{R'MgX} R(R')CH{-}B(\text{pinanediol}) \qquad (23)$$

The (dialkoxyboryl)methylzinc reagent [35], the most stable boron stabilized carboanion, was prepared by Knochel from the pinacol ester of iodomethylboronic acid and activated zinc metal. The reagent couples with 1-alkenyl or aryl halides in the presence of a palladium catalyst to afford benzylic [36] and allylic boronates [37] in high yields (eq (24)). The reagent is readily converted to the corresponding copper complex, which is an excellent reagent for Michael addition to α,β-unsaturated carbonyl compounds [35] (eq (25)) or for coupling with allyl and propargyl halides [38].

(pinacol)B-CH$_2$I —Zn→ (pinacol)B-CH$_2$Zn-I —PhX, Pd catalyst→ (pinacol)BCH$_2$Ph

(pinacol)B-CH$_2$Zn-I —RCH=CHX, Pd catalyst→ (pinacol)B-CH$_2$CH=CHR (24)

(pinacol)B-CH$_2$Zn-I —1. CuCN / LiCl; 2. CH$_2$=CHCO$_2$Me; 3. H$_2$O→ (pinacol)B-CH$_2$CH$_2$CH$_2$CO$_2$Me (25)

16.2.5 Miscellaneous Methods

For the preparation of 1-alkenylboronates from ketones or aldehydes, the boron–Wittig reaction is the method of choice when the the starting carbonyl compounds are readily available. An efficient route to (*E*)-1-alkenylboronates [39] from carbonyl compounds is achieved by the reaction with lithio(boryl)methanes. The (*E*)/(*Z*) isomeric ratio is reported to be 20/1 (eq (26)). Lithio(triboryl)methane with aldehydes or ketones yields diborylalkenes [40] (eq (27)).

HC[B(OCH$_2$CH$_2$O)]$_3$ —BuLi→ LiHC[B(OCH$_2$CH$_2$O)]$_2$ —RCHO→ R-CH=CH-B(OCH$_2$CH$_2$O) (26)

C[B(OCH$_2$CH$_2$O)]$_4$ —BuLi→ LiC[B(OCH$_2$CH$_2$O)]$_3$ —cyclohexanone→ cyclohexylidene=C[B(OCH$_2$CH$_2$O)]$_2$ (27)

Lithium 1-alkynylboronates are attacked by many electrophiles at the position β to the boron atom. The subsequent rearrangement gives a variety of functionalized 1-alkenylboronanes (eq (28)). The stereochemistry can be either *E* or *Z*, or a mixture of the two [15]. The reaction with chloro(diethyl)borane or chloro(trimethyl)silane and -stanane stereoselectively produces *cis*-bimetallic compounds [41].

$$R_3B^{-}\text{-}C{\equiv}CR' + EX \longrightarrow R_2B(R)C{=}C(R')E \quad (28)$$

$EX = ClBEt_2$, $ClSnMe_3$, $NCCH_2I$, EtO_2CCH_2Br, oxirane etc

Allylboranes add to alkynes and electron-rich alkenes to give *cis* addition products through a six-center transition state and allylic rearrangement [42]. Allylboration of alkynes can be followed by cyclization though intramolecular allylboration of the terminal double bond. The cyclization proceeds slowly at room temperature and rapidly at 40–100 °C (eq (29)).

$$RC{\equiv}CH + (C_3H_5)_3B \xrightarrow{rt} \xrightarrow{\Delta} \quad (29)$$

9-(1-Alkenyl)- or 9-(1-alkynyl)-9-BBN are exceptionally reactive dienophiles with high regio- and endoselectivity for the Diels–Alder reaction [43]. *Ab intio* calculation predicts a unique [4 atom + 3 atom] transition state. Although the reaction of vinylboronic esters is rather slower than that of dialkyl or dichloro derivatives, the procedure is quite useful for the synthesis of cyclic allyl- [44] or cycloalkenylboronates [45] (eq (30) and (31)).

The addition of diboron compounds to alkynes is an excellent method for the preparation of *cis*-bisboryl alkenes. Tetrachlorodiboron adds to alkenes and alkynes without catalyst, although the addition of teraalkoxydiboron to alkynes is efficiently catalyzed by platinum(0) complexes [46] (eq (32)).

$$B(OR)_2\text{-diene} + \text{maleic anhydride} \longrightarrow (RO)_2B\text{-cycloadduct} \quad (30)$$

$$(RO)_2B\text{-diene} + \text{N-Ph maleimide} \xrightarrow{50\ °C} (RO)_2B\text{-cycloadduct (N-Ph)} \quad (31)$$

$$R\text{-}C{\equiv}C\text{-}R' + (RO)_2B\text{-}B(OR)_2 \xrightarrow[\text{DMF, 80 °C}]{\text{Pt-catalyst}} (RO)_2B(R)C{=}C(R')B(OR)_2 \quad (32)$$

16.3 Synthesis of Boron Compounds

16.3.1 General Remarks

All procedures should be carried out in a well-ventilated hood using a safety shield. The borane reagents are extremely sensitive to moisture and air. Thus, dry all glassware in a 120 °C electric oven before use, assemble quickly the hot apparatus, and cool the apparatus while maintaining a flow of dry nitrogen or argon through the flask. The hot syringes and needles should be cooled to room temperature in a desicator.

Except for allylic and propargylic derivatives, organoboronic acids and their esters are thermally stable and relatively inert to water and oxygen. They are isolated, stored without special precautions, and not particularly toxic. However, trialkylboranes, particularly the low molecular weight compounds, are highly reactive with oxygen and spontaneously ignite in air. The hazards associated with ignition can be limited by using the reagents diluted with solvents.

(1) *p*-Tolylboronic Acid, $CH_3C_6H_4B(OH)_2$ [47]

$$CH_3-C_6H_4-Br \xrightarrow[\text{2. } B(OCH_3)_3 \quad \text{3. } H_2O]{\text{1. Mg}} CH_3-C_6H_4-B(OH)_2 \qquad (33)$$

The Grignard reagent is prepared in a 500 ml round-bottomed flask fitted with a pressure-equalizing addition funnel, a mechanical stirrer, and a reflux condenser to which a nitrogen inlet tube and an oil bubbler are attached. Magnesium turnings (11 g, 0.45 mol) are added to the flask, and then the apparatus is dried by heating with an electronic heat gun under the stream of dry nitrogen. 200 ml of ether and 1 mL of 4-boromotoluene are added to the flask at room temperature. If there is no immediate change, the flask is heated with an electronic heat gun or a small amount of 1,2-dibromoethane (*ca* 0.1–0.2 ml) is added to initiate the reaction. When this is done properly, the liquid will become slightly cloudy and ebullition occurs at the surface of the turnings. A solution of 4-bromotoluene (75.2 g, 0.44 mol) in ether (100 ml) is added in such a rate necessary to maintain a gentle reflux. After the addition has been completed, the mixture is refluxed for an additional 1 h.

The above Grignard solution is cooled to lower than –70 °C with a dry ice–acetone bath keeping the flask under a slightly positive nitrogen pressure and vigorous stirring. A white solid of the Grignard reagent will precipitate. Under vigorous stirring and cooling, a solution of trimethyl borate (51.8 ml, 0.44 mol) in ether (100 ml) is slowly added over 1 h. Allow the temperature to rise to ambient temperature without removing the cooling bath, and then the mixture is stirred for 2 h at room temperature. The flask is cooled to 0 °C with an ice bath, and then the mixture is quenched with an aqueous sulfuric acid (12 ml of H_2SO_4 in 240 ml of water) by carefully controlling the exothermic reaction. The mixture is poured into a separatory funnel with an aid of small amount of ether. The lower aqueous phase is removed, and the aqueous phase is extracted with 100 ml of ether. The combined extracts are wash with saturated brine, and is finally concentrated under reduced pressure using a rotary evaporator to give a white solid of the boronic acid.

The boronic acid in 1200 ml of water in a 2 L beaker is heated and stirred for 15–20 min to azeotropically remove a trace of bitolyl and other organic materials. The hot solution is filtered with suction to remove any insoluble matter, and then the beaker and filter cake is rinsed with two 30 mL portions of hot water. The filtrate is slowly cooled to room temperature and then to 0 °C for a few hours. The precipitated white crystals are collected in a Buchner funnel with suction, and are then allowed to stand overnight in air or in a desiccator to obtain p-tolylboronic acid. ~50 g, 80–90% yield, mp 240 °C[1].

A variety of alkyl- and arylboronic acids can be obtained in large quantity by the above general procedure. The reaction of organolithiums with triisopropyl borates at –78 °C is an alternative and convenient method to achieve high yields of boronic acids or their esters [9, 10] (eq (8)).

p-Tolylboronic acid is a useful precursor for the syntheses of a variety of functionalized arylboronic acids. Oxidation of methyl group with potassium permanganate provides p-carboxyphenylboronic acid [48]. Bromination with bromine under irradiation of sun light gives *p*-bromomethylphenylboronic acid [49]. 3-Amino-4-methylphenylboronic acid is prepared by nitration with fuming nitric acid, followed by reduction of the nitro group with hydrogen in the presence of platinum oxide [50].

(2) (R,R)-Diisopropyl Tartrate Crotylborate, $CH_3CH{=}CHCH_2B\{OCH(CO_2{}^iPr)CH(CO_2{}^iPr)O\}$ [7]

CH3 / CH3 — 1. BuLi-KO^{t}Bu, 2. B(O^{i}Pr)$_3$, 3. tartarate → CH3 ... B(O...CO$_2$^{i}Pr)(O...CO$_2$^{i}Pr) (34)

A 1L three-necked flask fitted with a magnetic stirring bar is assembled with a septum inlet, a –100 °C thermometer, and an outlet to which a nitrogen inlet and an oil bubbler are connected. Charge the flask with KO^{t}Bu (48.0 g, 425 mmol) and THF (350 mL). The flask is cooled with a dry ice–acetone bath. By means of a cannula and nitrogen pressure, *trans*-2-butene (42 mL, 450 mmol), which is collected in a rubber stoppered 25 mL graduated cylinder immersed in a dry ice–acetone bath, is added. A solution of n-BuLi in hexane (2.5 M, 170 mL, 425 mmoL) is added dropwise at a rate such that the internal temperature does not rise above –65 °C (*ca*. 2 h). The reaction mixture is warmed until the temperature reaches –50 °C and is stirred for exactly 15 min at –50 °C, and is then recooled to –78 °C.

Triisopropyl borate (98.2 mL, 425 mmol) is slowly added to the flask via cannula, by controlling the internal temperature lower than –65 °C (*ca* 2 h). After stirring for 10 min at –78 °C, the reaction mixture is rapidly poured into a 2 L separatory funnel containing 1N HCl (800 mL) saturated with NaCl. The aqueous layer is adjusted to pH 1 by addition of additional 1N HCl. A solution of (R,R)- diisopropyl tartrate (100 g, 425 mmol) in ether (150 mL) is added. The aqueous layer is removed, and the aqueous

[1]Since the boronic acids are readily dehydrated to the boroxine $(RBO)_3$, they are generally a mixture of boronic acid and boroxine, depending on the conditions used for the drying. Thus, NMR spectra exhibits two signals corresponding to the both species. Different melting points have been reported depending on the ratio of the two species.

phase is extracted four times with ether (200 mL × 4). The combined extracts are dried with $MgSO_4$ for at least 2 h. After filtration with suction, the filtrate is concentrated by a rotary evaporator to a colorless thick liquid, and is finally pumped to a constant weight (125 g) at 0.50–1.0 mmHg. The product is contaminated with tartrate butylborate (12%); however the isomeric purity of (*E*)-crotylborate is 99% by GC analysis (a fused silica capillary column, SE-54, 50 m × 0.25 mm). Although the compound can be purified by a shot-path distillation (95 °C/0.01 mmHg), the isomeric purity decreases to 96%.

Properties: ^1H NMR (300 MHz, $CDCl_3$) δ, 5.53–5.63 (m, 1 H), 5.37–5.48 (m, 1 H), 4.83–5.00 (m, 1 H), 4.89 (s, 2 H), 1.86 (broad d, 2 H, *J*=6.4 Hz), 0.90 (d, 12 H, *J*=6.3 Hz).

By a similar method, the (*Z*)-crotylborate is synthesized from *cis*-2-butene in 70–75% yield with a 98% isomeric purity. The tartrate esters of allylboronic acids are an excellent reagent for asymmetric allylboration of carbonyl compounds. Allyl(diisopinocampheyl)borane [51] and the allylic boron derivatives of ester and amide, such as camphordiol [52], pinanediol [53], 1,2-diphenyl-1,2-ethylenediamine [54], have also been successfully used for asymmetric allylboration of carbonyls.

(3) Allenylboronic Acid, $CH_2{=}C{=}CHB(OH)_2$ [55]

$$HC{\equiv}CCH_2Br \xrightarrow[\substack{2.\ B(OCH_3)_3 \\ 3.\ H_2O}]{1.\ Mg} CH_2{=}C{=}CHB(OH)_2 \qquad (35)$$

A four-necked 500 mL flask fitted with a mechanical stirrer is assembled with two dropping funnels, a thermometer, a side arm connected to an oil bubbler and a nitrogen inlet. Add ether (225 mL) to the flask. Then, one dropping funnel is charged with a solution of trimethyl borate (17.03 mL, 150 mmol) in ether (30 mL), and the other funnel with a solution of allenylmagnesium bromide in ether (see Chapter 15, section 15.3(5), 160 mL, *ca.* 150 mmol) via cannula. The flask is cooled in a dry ice–acetone bath. Under vigorous stirring, both reactants are added to the flask alternately in small portions, first $\frac{1}{12}$ of methyl borate and then $\frac{1}{12}$ of Grignard reagent. The rate of addition should be as rapid as is possible but the internal temperature should not be above –70 °C (*ca.* 30 min). The mixture is stirred for 30 min, and then the flask is immersed in an ice–water bath. Water (10.5 mL) is added during 1 min, and then the mixture is neutralized by addition of dilute sulfuric acid (75 mL, prepared from concentrated sulfuric acid 5.56 mL + water 100 mL) during 10 min. The reaction mixture is transferred to a separatory funnel. The ether layer is separated and the aqueous layer is extracted with ether (2 × 20 mL). The combined extracts are dried over $MgSO_4$ and the filtrate is concentrated on a rotary evaporator. The evaporator is charged with argon after approximately one-tenth of ether has been removed. A small amount of ether is added if some crystals of boronic acid have appeared. Hexane (70 mL) is added to the flask to precipitate the boronic acid under an argon atmosphere. The solvent is removed by a syringe or a cannula. The boronic acid is dissolved in ether (60 mL) and then the solution is transferred to another dry flask via cannula using argon pressure. The solution is concentrated until some crystals of allenylboronic acid precipitate. A minimum amount of ether is added to dissolve the solid, and then hexane (70 mL) is added to precipitate the boronic acid again. The solvent is removed by a cannula. The same operation is repeated once again to provide pure allenylboronic acid.

Properties: 40-60%, mp 150 °C (dec). ^{1}H NMR ($CDCl_3$—Me_2SO-d^6) δ, 4.48 (d, 1 H, J = 5.8 Hz), 4.51 (d, 1 H, J = 7.2 Hz), 4.83 (dd, 1 H, J = 5.8 and 7.2 Hz), 6.52 (br s, 2 H).

The compound, when suspended in hexane (10 mL), can be stored at –20 °C for several weeks without isomerization to propargylboronic acid. Allenylborates add to carbonyl compounds to provide homo-propargylic alcohols with high enantioselectivity.

(4) Triallylborane, $B(CH_2CH{=}CH_2)_3$ [13]

$$3\ CH_2{=}CHCH_2Br \xrightarrow[2.\ F_3B\cdot OEt_2]{1.\ Al} (CH_2{=}CHCH_2)_3B \qquad (36)$$

A dry three-necked 300 mL flask fitted with a mechanical stirrer is assembled with a vacuum seal, a dropping funnel, and a reflux condenser to which a nitrogen inlet and an oil bubbler are attached. The flask is charged with aluminum granules (30–60 mesh, 15 g, 0.556 mol) and mercuric chloride (0.1 g), and then the apparatus is thoroughly flushed with dry nitrogen. Ether (75 mL) and allyl bromide (freshly distilled, 10 g, 0.0827 mol) are added succesively under stirring. The start of reaction should be confirmed by the obsevation of ebullition at the surface of the metal and the formation of a slightly cloudy solution. If there is no noticeable change, the addition of small piece of iodine may help to initiate the reaction. The rest of allyl bromide (81 g, 0.67 mol) is slowly added at a rate sufficient to maintain a gentle reflux. The mixture is stirred for 3 h under reflux to dissolve the aluminum almost completely. Boron trifluoride etherate (purified by distillation from calcium hydride, 28.4 g, 0.2 mol) is added dropwise at such a rate that the temperature remains at 55–60 °C. The mixture is heated to reflux for 3 h and then cooled to room temperature. While keeping a stream of nitrogen in both pieces of apparatus, the condenser is quickly displaced with a Claisen head distillation apparatus to which nitrogen inlet and a vacuum line are attached. The ether is evaporated *in vacuo* (~20 mmHg) at room temperature and then the flask is heated with an oil bath to distill triallylborane over the range of 45–74°C/20–25 mmHg. Redistillation gives pure triallylborane (21 g, 78%). bp 59–61 °C / 21 mmHg.

Tricrotylborane, in 75% yield, and tri(2-methylallyl)borane have been synthesized by the above procedure. Tributyl borate can be also used for these preparations in place of boron trifluoride etherate.

The ^{1}H NMR spectrum of triallylborane at low temperature (at –65 °C) is similar to the spectrum of organic allylic compounds. It exhibits a doublet at 2.05 ppm (CH_2)

[1] Catecholborane (1,3,2-benzodioxaborole) purchased from Aldrich is purified by distillation under nitrogen. bp 58 °C/52 mmHg. It is a liquid at room temperature and a solid in a cold room. The neat material is 9.0 M in catecholborane. It is recommended to store the purified material in refrigerator as a solid, in a bottle purged with nitrogen and capped with a rubber septum. Wash the used syringe with acetone without delay because the plunger becomes very sticky on standing.

[2] Hydroboration of terminal alkynes is complete within 1 h, but internal alkynes require approximately 4 h at 70 °C. Although the reaction is generally carried out without solvent, a similar result can be obtained in benzene, but it is very slow in THF solution.

[3] The catechol ester of 1-octenylboronic acid can be isolated by distillation. bp 112–120 °C/0.2 mmHg.

and signals of terminal CH_2 protons in a region of 4.70 ppm, and a multiplet of methine proton at 5.65 ppm. However, the spectrum broadens at higher temperature due to an allylic rearrangement. At 135 °C, two CH_2 groups become equivalent which exhibits a AH_4 type spectrum containing a quintet at 5.00 ppm and a doublet at 3.2 ppm.

(5) (E)-1-Octenylboronic Acid, $C_6H_{13}CH{=}CHB(OH)_2$ [17]

$$CH_3(CH_2)_5C{\equiv}CH \xrightarrow[2.\ H_2O]{1.\ \text{catecholborane}} CH_3(CH_2)_5CH{=}CHB(OH)_2 \qquad (37)$$

A two-necked 200 mL flask is assembled with a magnetic stirring bar, a septum inlet, and a reflux condenser to which a nitrogen inlet and an oil bubbler are attached. The flask is immersed in the water bath (*ca* 20 °C). 1-Octyne (5.9 mL, 40 mmol) is added to the flask by puncturing the septum with a syringe. Catecholborane (4.9 mL, 44 mmol)[1] is added to the flask by syringe, over a 1 min period with stirring. The mixture is stirred for 30 min at room temperature, and then the flask temperature is raised slowly to 70 °C with an oil bath. Temperature should not be raised quickly because hydroboration is often very exothermic. The temperature is kept at 70 °C for 2 h to complete the reaction[2,3]. The flask is cooled with an ice–water bath. 1–2 mL of water is added very carefully to hydrolyze the excess borane. After the evolution of hydrogen gas has ceased, the nitrogen source is disconnected. An additional 40 mL of water is added and then the milky suspension is heated at 70–80 °C for 1 h under vigorous stirring. The flask is cooled slowly to room temperature and then left to stand at 0 °C for 2 h. The precipitate is collected by filtration through a filter paper with suction and washed three times with 12 mL of ice-cooled water to remove catechol. The suction is continued for additional 1 h. The solid in a 50 mL-beaker is dried by standing in air over night or in a desiccator. The 1-octenylboronic acid thus obtained is light gray due to the contamination by a trace of catechol. 5.63 g (90%). The product is pure enough to use for most of the reaction, but recrystallization from hot water gives pure white crystals.

The 1-alkenylboronic acids are also obtained by hydroboration of alkynes with diboroborane–methyl sulfide complex followed by hydrolysis [21] Dihydroboration of terminal alkyne with two equivalent of 9-BBN followed by dehydroboration with benzaldehyde stereoselectively provides 9-[(E)-1-alkenyl]-9- BBN in high yields [56].

(6) 2-[(R)-1-Phenylethyl]-1,3,2-benzodioxaborole, $C_6H_5(Me)(H)CB(OC_6H_4O)$ [57]

$$PhCH{=}CH_2 + HB(O_2C_6H_4) \xrightarrow{Rh(COD)(BINAP)^+} Ph{-}\overset{*}{C}H(CH_3){-}B(O_2C_6H_4) \qquad (38)$$

A one-necked 20 mL flask is assembled with a magnetic stirring bar, a septum inlet, and an outlet connected to an oil bubbler and a nitrogen inlet. The flask is charged with [Rh(cyclooctadiene){(R)-2,2′-bis(diphenylphosphino)-1,1′-binaphthyl}]BF_4

(0.1 mmol) and is flushed with nitrogen. Styrene (5 mmol) and dimethoxyethane (2 mL) are added by syringes. The flask is cooled with a dry ice–acetone bath. Catecholborane (5.5 mmol) is added by syringe, and then the mixture is stirred for 2 h at –78 °C. The disappearance of styrene is analyzed by GC. The cooling bath is removed and the temperature is allowed to raise to room temperature. The flask is attached to the Kugelrohr distillation apparatus, and then the product is distilled under reduced pressure to give a colorless and moderately air stable liquid. 98%, bp 120°C/0.15 mmHg, $[\alpha]_D = -55.59°$ ($c = 1.7$, benzene). The optical purity of 1-phenylethanol, obtained by alkaline hydrogen peroxide oxidation, is 94% ee.

The procedure represents one of the most convenient methods for the synthesis of optically active alkylboronates. However, the procedure can be of limited use for the hydroboration of styrene derivatives. The optically active organoborates are also obtained by hydroboration of internal alkenes with diisopinocampheylborane, followed by dealkylation of two diisopinocampheyl groups with acetaldehyde [58].

(7) Diisopropyl (Z)-1-Hexenylboronate, (Z)-$C_4H_9CH{=}CHB(O^iPr)_2$ [23, 24]

$$C_4H_9C{\equiv}CBr \xrightarrow[2.\ ^iPrOH]{1.\ HBBr_2\cdot SMe_2} (Z)\text{-}C_4H_9(H)C{=}C(Br)B(O^iPr)_2 \xrightarrow{KIPBH} (Z)\text{-}C_4H_9(H)C{=}C(H)B(O^iPr)_2 \quad (39)$$

A two-necked 50 mL flask is assembled with a magnetic stirring bar, a septum inlet and a reflux condenser to which a nitrogen inlet and oil bubbler are connected. The flask is charged with borane–methyl sulfide (7.2 mL, 72 mmol) and methyl sulfide (10.6 mL, 144 mmol) by means of syringes. The flask is cooled with a –20–30 °C bath. Boron tribromide (freshly purified by distillation, 13.6 mL, 144 mmol) is added dropwise using syringe. A very dry syringe should be used and the plunger should not be pull back: this tends to cause the plunger to stick. Care must be also taken during the addition because the reaction of boron tribromide with methyl sulfide forming a white solid adduct is highly exothermic. The mixture is stirred at 40 °C for 12 h to provide a light brown liquid. The borane thus obtained is sufficiently pure to use for hydroboration, but distillation under reduced pressure gives a pure colorless liquid. 93%, bp 98 °C/0.1 mmHg and mp 30–35 °C. Make a 2 M solution in dichloromethane[1].

The same apparatus as above is assembled, except that a 100 mL flask is used. The flask is charged with 1-bromo-1-hexyne (4.03 g, 25 mmol) and dichloromethane (10 mL). The flask is immersed in a water bath (*ca.* 25 °C). The above dibromoborane solution (25 mmol) is slowly added from a syringe, and then the mixture is stirred for 8 h. The flask is cooled with an ice–water bath. Under vigorous stirring, pentane (25 mL) is added and then 2-propanol (7.7 mL, 100 mmol). The stirring is stopped after 15 min to separate the mixture into two phases. The upper layer is transferred to the distillation flask by means of a cannula and nitrogen pressure. The lower layer is washed with 10 mL of pentane, and again the pentane layer is transferred to the distillation flask. This operation is repeated one more time to extract the product. The combined extracts are distilled under reduced pressure to give diisopropyl (1-bromo-1-hexenyl)boronate. 87%, bp 64–65 °C/0.2 mmHg.

The same apparatus is assembled as above. The flask is charged with diisopropyl

[1] The reagent is also available from Aldrich.

(1-bromo-1-hexenyl)boronate (8.35 g, 25 mmol) and ether (25 mL). Cool the flask with an ice–water bath. A solution of potassium triisopropoxyborohydride in ether (1 M, 25 mmol) is slowly added[1], and the mixture is stirred for 15 min at 0 °C, and then the flask is allowed to come to room temperature over 30 min. Potassium bromide will precipitate during the addition. By using a cannula, the mixture is filtered through Celite pad on a sintered-glass filter, and the flask and the filter cake is rinsed with small amount of ether. The filtrate is distilled under reduced pressure to give diisopropyl (*Z*)-1-hexenylboronate. 90%, bp 70–71 °C/6 mmHg.

An alternative procedure for the synthesis of (1-halo-1-alkenyl)boronates (X=Br, I) is hydroboration of haloalkynes with diisopinocanphylborane (eq (14)).

(8) Diisopropyl [(*E*)-2-Methyl-1-hexenyl]borate, $C_4H_9(CH_3)C{=}CHB(O^iPr)_2$ [59]

$$C_4H_9C{\equiv}CH \xrightarrow[2.\ ^iPr_2O]{1.\ BBr_3} (Z)\text{-}C_4H_9(Br)C{=}CHB(O^iPr)_2 \xrightarrow[PdCl_2(PPh_3)_2]{CH_3ZnCl} (E)\text{-}C_4H_9(CH_3)C{=}CHB(O^iPr)_2 \quad (40)$$

A two-necked 200 mL flask is assembled with a magnetic stirring bar, a septum inlet, and a reflux condenser to which a nitrogen inlet and an oil bubbler are connected. The flask is charged with boron tribromide (freshly distilled, 2.84 mL, 30 mmol) and dichloromethane (50 mL) by syringes. The flask is immersed in a dry ice–acetone bath. 1-Hexyne (3.5 mL, 30 mmol) is slowly added and the mixture is stirred for 30 min at –78 °C. The flask temperature is raised slowly without removing the cooling bath, and the mixture is stirred for 2 h at room temperature. The reaction flask is recooled to –78 °C, and then diisopropyl ether (9.9 mL, 70 mmol)[1] (exothermic) is very slowly added. After being stirred for 30 min at –78 °C, the mixture is allowed to warm to room temperature, and stirred overnight. The septum is displaced with a glass stopper and the reflux condenser with a Claisen head semi-micro short-path distillation apparatus. Distillation under reduced pressure gives diisopropyl (*Z*)-2-bromo-1-hexenylboronate (73%, bp 68 °C/0.07 mmHg) as a colorless oil (The bath temperature should be controlled lower than 150 °C to minimize thermal decomposition.)

The same apparatus is assembled with the above synthesis. The flask is charged with powdered zinc chloride (dried at 100 °C *in vacuo*, 0.3 g, 22 mmol). Add 40 ml of THF and stir for 1 h. The flask is cooled to 0 °C and then a solution of MeMgI in ether (0.44 M, 21.4 mmol) is added. After being stirred for 30 min, $PdCl_2(PPh_3)_2$ (0.22 g, 0.28 mmol) and diisopropyl (*Z*)-2-bromo-1-hexenylborate (3.11 g, 10.7 mmol) is added to the flask. Heat the mixture to reflux for 2 h. By means of a cannula and nitrogen pressure, the mixture is filtered through a Celite pad on a sintered-glass filter to remove any inorganic precipitates, and the flask and the filter cake is rinsed with three 30 mL of ether. the filtrate is concentrated and the residu is distilled *in vacuo* to give diisopropyl (*E*)-2-methyl-1-hexenylboronate as a colorless liquid. 65%, bp 60 °C/0.07 mmHg (Keep the bath temperature at lower than 150 °C).

Properties: ^1H NMR ($CDCl_3$) δ, 0.90 (t, 3 H, *J*=6.0 Hz), 1.16 (d, 12 H, *J*=6.2 Hz), 1.40 (m, 4 H), 1.85 (s, 3 H), 2.07 (t, 3 H, *J*=6.3 Hz), 4.41 (q, 2 H, *J*=6.2 Hz), and 5.18 (s, 1 H).

[1] The B–Br bond can be conveniently converted to the corresponding isopropyl esters with diisopropyl ether. However, the reaction with the corresponding methyl sulfide complex is very slow.

(9) 2-Iodomethyl-4,4,5,5-tetramethyl-1,3,2-dioxaboronane, $ICH_2B(OCMe_2CMe_2O)$ [31, 38]

$$ClCH_2I \xrightarrow[\text{2. AcOH; 3. pinacol}]{\text{1. B(O}^i\text{Pr)}_3\text{, BuLi}} ClCH_2\text{-B(OCMe}_2\text{CMe}_2\text{O)} \xrightarrow[\text{acetone}]{\text{NaI}} ICH_2\text{-B(OCMe}_2\text{CMe}_2\text{O)} \quad (41)$$

The 500 mL two necked flask is assembled with a magnetic stirring bar, a septum inlet and a outlet connected to an oil bubbler and a nitrogen inlet. Chloroiodomethane (31.7 g, 0.18 mol), triisopropyl borate (40.4 mL, 0.175 mol), and THF (200 mL) are added to the flask by puncturing the septum with a syringe. The flask is immersed in the dry ice–acetone bath. Butyllithium (1.55 M solution in hexane, 0.165 mol) is slowly added. The mixture is stirred for 30 min at –78 °C, and then allowed to warm to room temperature over a 1–2 h period without displacing the cooling bath. The mixture is stirred for additional 3 h at room temperature. The flask is cooled to –78 °C. Acetic acid (10 mL, 0.175 mol) and a solution of pinacol (20.7 g, 0.175 mol) in THF (20 mL) are succesively added. The mixture stirred for 30 min at –78 °C and for 3 h at room temperature by displacing the cooling bath. The reaction mixture is washed with brine and dried over $MgSO_4$ overnight. The solution is filtered to remove the drying agent, concentrated on a rotary evaporator, and finally distilled under reduced pressure using a semi-micro short-path distillation apparatus to give a mixture of chloromethyl- and iodomethylboronic acid pinacol esters (29.6 g). bp 73–85 °C/10 mmHg.

The above boronic ester (29.6 g) is added to the mixture of NaI (dried *in vacuo*, 38 g) and acetone (a middle fraction is collected by distillation, 280 mL) under nitrogen. The mixture is refluxed for 4 h. After cooling to room temperature, the solvent removed *in vacuo* using a rotary evaporator. The product is extracted with ether (100 mL X 2), washed the extract with brine, and dried over $MgSO_4$. The concentrated residue is distilled *in vacuo* to give an air-stable slightly yellow oil in a 82% total yield (38.5 g). bp 93–95 °C/10 mmHg (a colorless oil can be obtained by distillation from small amount of copper powder).

Properties: ^{1}H NMR (60 MHz, $CDCl_3$) δ, 1.30 (s, 12 H), 2.17 (s, 2 H); ^{13}C NMR δ, 24.35, 84.02.

(10) 3,3,4,4-Tetramethyl-2-(2-formylphenylmethyl)-1,3,2-dioxa- boronane, $2\text{-OHCC}_6H_4B(OCMe_2CMe_2O)$ [36]

$$ICH_2\text{-B(OCMe}_2\text{CMe}_2\text{O)} \xrightarrow{\text{Zn}} IZnCH_2\text{-B(OCMe}_2\text{CMe}_2\text{O)} \xrightarrow[\text{PdCl}_2\text{(PPh}_3\text{)}_2]{2\text{-IC}_6\text{H}_4\text{CHO}} 2\text{-OHCC}_6\text{H}_4\text{-CH}_2\text{B(OCMe}_2\text{CMe}_2\text{O)} \quad (42)$$

A two-necked 100 mL flask is assembled with a magnetic stirring bar, a septum inlet, and a reflux condenser. The flask is charged with zinc powder (50 mesh, 5.67 g, 86.7 mmol) and then the apparatus and zinc are dried by heating the flask with an electric heat gun while maintaining a stream of nitrogen. THF (15 mL) and 1,2-dibromoethane (0.3 mL) are added to the flask, and the mixture is stirred for 30 min to activate the zinc surface. The flask is cooled in an ice–water bath. A solution of

2-iodomethyl-4,4,5,5-teramethyl-1,3,2-dioxaboronane (7.5 g, 28 mmol) in THF (20 mL) are added dropwise via cannula or syringe. The bath temperature is allowed to warm slowly to room temperature, and the mixture is stirred for an additional 2 h. By means of a cannula and nitrogen pressure, the reaction mixture is filtered through a Celite pad on a fritted glass funnel, and the flask and filter cake is rinsed with a small amount of THF to give a solution of the zinc compound. The zinc ion concentration is titrated by a EDTA standard solution. 0.5–0.6 M, the yield generally exceed 90%[1].

The same flask is assembled as above. The flask is charged with $PdCl_2(PPh_3)_2$ (180 mg, 0.23 mmol) and flushed with nitrogen. THF (30 mL), 2- iodobenzaldehyde (1.35 g, 5.8 mmol), and finally the above zinc reagent (0.6 M, 8.2 mmol) are added to the flask by syringes. The mixture is heated and stirred for 3 h at 50 °C. The flask is cooled with an ice–water bath. Benzene (35 mL) and then water (20 mL) are added. The mixture is stirred for 30 min at room temperature. The mixture is transferred into a separatory funnel. The lower aqueous phase is discarded and the organic layer is washed with brine (20 mL × 2), and finally dried over $MgSO_4$ over night. After filtering off drying agent, the filtrate is concentrate on a rotary evaporator, and then the residue is dilstilled by Kugelrohr to give a colorless thick liquid which slowly crystallizes at room temperature. 65%, 90–100 °C/0.3 mmHg (The compound is thermally unstable, thus distillation at temperature lower than 100 °C is recommended)[2].

Properties: ^1H NMR (90 MHz, $CDCl_3$) δ, 1.20 (s, 12 H), 2.55 (s, 2 H), 7.05–7.35 (m, 3 H), 7.55–7.70 (broad d, 1 H), 10.12 (s, 1 H).

References

[1] T. Onak, *Organoborane Chemistry*, Academic Press, London (1975);G. Wilkinson, F. G. A. Stone, E. W. Abel, *Comprehensive Organometallic Chemistry*, Vol. 1, Pergamon Press, Oxford (1982).

[2] A. N. Nesmeyanov, R. A. Sokolik, *Methods of Elemento-Organic Chemistry*, Vol. 1, North-Holland, Amsterdam (1967); R. Köster, *Houben-Wey Methoden der Organischen Chemie*, Georg Thieme, Verlag Stuttgart (1984); D. S. Matteson, *The Chemistry of the Metal Carbon Bond*, Ed. by Hartley, S. Patai, Vol. 4, p 307, Wiley, New York, (1987).

[3] Alkylboronic acids: P. A. McCusker, E. C. Ashby, H. S. Makowski, *J. Am.Chem. Soc.*, **79**, 5179 (1957); H. R. Snyder, J. A. Kuck, J. R. Johnson, *J. Am. Chem. Soc.*, **60**, 105 (1938). Arylboronic acids: F. R. Bean, J. R. Johnson, *J. Am. Chem. Soc.*, **54**, 4415 (1932); M. F. Howthorne, *J. Am. Chem. Soc.*, **80**, 4291 (1958); R. T. Hawkins, W. J. Lennarz, H. R. Snyder, *J. Am. Chem. Soc.*, **82**, 3053 (1960); H. A. Staab, B. Meissner, *Liebigs Ann. Chem.*, **753**, 80 (1971) and ref. 2.

[4] D. S. Matteson, *J. Am. Chem. Soc.*, **82**, 4228 (1960).

[5] D. S. Matteson, K. Peacock, *J. Am. Chem. Soc.*, **82**, 5759 (1960);W. G. Woods, P. L. Strong, *J. Organomet. Chem.*, **7**, 371 (1967).

[6] D. S. Matteson, J. D. Liedtke, *J. Am. Chem. Soc.*, **87**, 1526 (1965).

[7] W. R. Roush, K. Ando, D. B. Powers, A. D. Palkowitz, R. L. Halterman, *J. Am. Chem. Soc.*, **112**, 6339 (1990).

[1] The solution can be stored in refrigerator (–15 °C) for several weeks in a bottle purged with nitrogen and capped with a glass stopcock.

[2] The compound is a good precursor to *ortho*-quinodimethane, formed by a 1,5-migration of boron to the carbonyl oxygen atom under heating at 110–180 °C or irradiation of the benzene solution with a high pressure Hg light.

[8] E. J. Corey *et al J. Am. Chem. Soc.*, **112**, 878 (1990); E. J. Corey, C-M. Yu, S. S. Kim. *J. Am. Chem. Soc.*, **111**, 5495 (1989).
[9] H. C. Brown, T. E. Cole, *Organometallics*, **2**, 1316 (1983).
[10] H. C. Brown, N. G. Bhat, M. Srebnik, *Tetrahedron Lett.*, **29**, 2631 (1988).
[11] P. Jacob, III, *J. Organomet. Chem.* **150**, 101 (1978); T. P. Povlock, W. T. Lippincott, *J. Am. Chem. Soc.*, **80**, 5409 (1958).
[12] H. C. Brown, *J. Am. Chem. Soc.*, **67**, 374 (1945); R. Köster, *Ann.* **618**, 31 (1958); H. C. Brown, *J. Am. Chem. Soc.*, **67**, 374 (1945); G. F. Hennion, P. A. McCusker, E. C. Ashby, A. J. Rutkowski, *J. Am. Chem. Soc.*, **79**, 5190 (1957).
[13] G. Wittig, G. Keicher, A. Rückert, P. Raff, *Ann.* **563**, 110 (1949);. H. C. Brown, V. H. Dodson, *J. Am. Chem. Soc.*, **79**, 2303 (1957).
[14] L. I. Zakharkin, V. I. Stanko, *Izv. Akad. Nauk SSSR, Otd. Khim. Nauk*, (1960); 1896. B. M. Mikhailov, V. F. Pozdnev, *Izv. Akad. Nauk SSSR, Ser Kim.* (1964) 1477; V. S. Bogdanov, Yu. N. Bubnov, M. N. Bochkarea, B. M. Mikhailov, *Dokl. Akad. Nauk SSSR*, **201**, 605 (1971).
[15] H. C. Brown, *Hydroboration*, W. A. Benjamin, Inc., New York (1962); G. M. L. Cragg, *Organoboranes in Organic Synthesis*, Marcel Dekker, New York (1973); A. Pelter, K. Smith, H. C. Brown, *Borane Reagents*, Academic Press, London (1988).
[16] H. C. Brown, *Organic Syntheses via Boranes*, John Wiley, New York (1975); K. Smith, *Organometallics in Synthesis*, Ed. M. Schlosser, John Wiley, New York (1994).
[17] H. C. Brown, S. K. Gupta, *J. Am. Chem. Soc.*, **94**, 4370 (1972); N. Miyaura, A. Suzuki, *Org. Synth.*, **68**, 130 (1989).
[18] Unpublished results. The catalytic hydroboration of 1-octyne with **7** in benzene in the presence of $Pd(PPh_3)_4$ (2 mol%) proceeds at room temperature to give a 4:96 mixture of 1-octenylborate favoring the terminal boron addition product.
[19] D. Mannig, H. Nöth, *Angew. Chem. Int. Ed. Engl.*, **24**, 878 (1985). For areview, see: K. Burgess, M. J. Ohlmeyer, *Chem. Rev.*, **91**, 1179 (1991).
[20] I. D. Gridnev, N. Miyaura, A. Suzuki, *Organometallics,* **12**, 589 (1993).
[21] H. C. Brown, J. B. Campbel, *J. Org. Chem.*, **45**, 389 (1980); H. C. Brown, N. G. Bhat, V. Somayaji, *Organometallics*, **2**, 1311 (1983).
[22] A. Kamabuchi, T. Moriya, N. Miyaura, A. Suzuki, *Synth. Commun.*, **23**, 2851 (1993).
[23] H. C. Brown, T. Imai, *Organometallics*, **3**, 1329 (1984), and refs 22 and 24.
[24] H. C. Brown, V. Somayaji, *Syntheis*, 919 (1984).
[25] H. C. Brown, T. Imai, N. G. Bhat, *J. Org. Chem.*, **51**, 5277 (1986).
[26] T. Moriya, N. Miyaura, A. Suzuki, *Chem. Lett.*, 1429 (1993).
[27] General reviews, see: A. Suzuki, *Pure Appl. Chem.*, **59**, 629 (1986); A. Suzuki, *Haloboration and Its Application to Organic Synthesis* in *Reviews on Heteroatom Chemistry*, Ed. S. Oae, Myu, Tokyo (1988). S. Hara, *Journal of Synthetic Organic Chemistry*, **48**, 1125 (1990).
[28] S. Hyuga, Y. Chiba, S. Hara, A. Suzuki, *Chem. Lett.*, (1987) 1767; S. Hyuga, N. Yamashina, S. Hara, A. Suzuki, *Chem. Lett.* (1988), 809.
[29] S. Hara, A. Suzuki, *Tetrahedron Lett.*, **32**, 6749 (1991).
[30] Reviews for α–haloboronic esters: D. S. Matteson, *Chem. Rev.*, **89**, 1535 (1989); D. S. Matteson, *Tetrahedron*, **45**, 1859 (1989).
[31] K. M. Sadhu, D. S. Matteson, *Organometallics*, **4**, 1687 (1985) and ref. 38.
[32] M. W. Rathke, E. Chao, G. Wu. *J. Organomet. Chem.*, **122**, 145 (1976).
[33] H. C. Brown, M. V. Rangaishenvi, *Tetrahedron Lett.*, **49**, 7113 (1990) ; **49**, 7115 (1990).
[35] P. Knochel, *J. Am. Chem. Soc.*, **112**, 7431 (1990).
[36] G. Kanai, N. Miyaura, A. Suzuki, *Chem. Lett.*, 845 (1993).
[37] T. Watanabe, M. Sakai, N. Miyaura, A. Suzuki, *J. Chem. Soc. Chem. Commun.*, 467 (1994).
[38] I. Gridnev, G. Kanai, N. Miyaura, A. Suzuki, *J. Organomet. Chem.*, **481**, C4 (1994).
[39] D. S. Matteson, P. K. Jesthi, *J. Organomet. Chem.*, **110**, 25 (1976); D. S. Matteson, R. J. Moody, *J. Org. Chem.*, **45**, 1091 (1980); D. S. Matteson, R. J. Moody, *Organometallics*, **1**, 20 (1982).
[40] D. S. Matteson, P. B. Tripathy, *J. Organomet. Chem.*, **69**, 53 (1974); D. S.Matteson, L. Hagelee, *J. Organomet. Chem.*, **93**, 21 (1975).

[41] P. Binger, R. Köster, *Tetrahedron Lett.*, 1902 (1965); P. Binger, R. Köster, *Synthesis*, 309 (1973); P. Binger, R. Köster, *J. Organomet. Chem.*, **73**, 205 (1974); J. Hooz, R. Mortimer, *Tetrahedron Lett.*, 805 (1976).
[42] Mikhailov, B. M., Bubnov, Yu. N., Korobeinikova, S. A., Frolov, S. *I. Izv. Akad. Nauk SSSR, Ser. Khim.*, (1968) 1923; Frolov, S. I., Bubnov, Yu. N., Mikhailov, B. M. *Izv. Akad. Nauk SSSR, Ser. Khim.*, 1996 (1969); Yu. N. Bubnov, S. I. Frolov, V. G. Kiselev, V. S. Bogdanov, B. M. Mikhailov, *Organometal. Chem. Synth.*, **1**, 37 (1970).
[43] D. A. Singleton, J. P. Martinez, *J. Am. Chem. Soc.*, **112**, 7423 (1990); D. A. Singleton, *J. Am. Chem. Soc.*, **114**, 6563 (1992); N. Noiret, A.Youssofi, B. Carboni, M. Vaultier, *J. Chem. Soc. Chem. Commun.*, 1105 (1992).
[44] M. Valtier, F. Truchet, B. Carboni, R. Hoffmann, I. Denne, *Tetrahedron Lett.*, **28**, 4169 (1987).
[45] A. Kamabuchi, N. Miyaura, A. Suzuki, *Tetrahedron Lett.*, **34**, 4827 (1993); D. A. Singleton, S-W. Leung, *J. Org. Chem.*, **57**, 4796 (1992).
[46] T. Ishiyama, N. Matsuda, N. Miyaura, A. Suzuki, *J. Am. Chem. Soc.*, **115** 11018 (1993).
[47] M. F. Howthorne, *J. Am. Chem., Soc.*, **80**, 4291 (1958).
[48] A. Michaelis, *Ann.*, **315**, 33 (1901).
[49] H. R. Snyder, A. J. Reedy, W. J. Lennarz, *J. Am. Chem. Soc.*, **80**, 835 (1958).
[50] F. R. Bean, J. R. Johnson, *J. Am. Chem. Soc.*, **54**, 4415 (1932).
[51] H. C. Brown, K. S. Bhat, *J. Am. Chem. Soc.*, **108**, 293 (1986); H. C. Brown, K. S. Bhat, R. S. Randad, *J. Org. Chem.*, **52**, 320 (1987).
[52] V. R. W. Hoffman, H. J. Heiß, W. Landner, S. Tabche, Ber., **115**, 2357 (1982).
[53] D. S. Matteson, D. Majumdar, *J. Am. Chem. Soc.*, **102**, 7588 (1980).
[54] E. J. Corey, C-M. Yu, S. S. Kim. *J. Am. Chem. Soc.*, **111**, 5495 (1989).
[55] N. Ikeda, I. Arai, H. Yamamoto, *J. Am. Chem. Soc.*, **108**, 483 (1986).
[56] J. C. Colberg, A. Rane, J. Vaquer, J. A. Soderquist, *J. Am. Chem. Soc.*, **115**, 6065 (1993).
[57] T. Hayashi, Y. Matsumoto, Y. Itoh, *J. Am. Chem. Soc.*, **111**, 3426 (1989).
[58] H. C. Brown, P. K. Jadhav, M. C. Desai, *J. Am. Chem. Soc.*, **104**, 4303 (1982); M. Srebnik, P. V. Ramachandran, *Aldrichimica. Acta.*, **20**, 9 (1987).
[59] S. Hara, S. Hyuga, M. Aoyama, M. Satoh, A. Suzuki, *Tetrahedron Lett.*, **31**, 247 (1990).

16.4 Aluminum Complexes

16.4.1 Introduction

Organoaluminum compounds, little known until the 1950s, have become widely accepted and increasingly important in the field of industry and in the laboratory, particularly after be brilliant work of K. Ziegler and colleagues, who discovered the direct synthesis of trialkylaluminums and their application to olefin polymerization [1,2]. A comprehensive review of organoaluminum chemistry appeared in 1972, covering a variety of literature up to late 1971 [3]. Other important review articles have been published since then [4–13]. The chemistry of organoaluminum compounds can be understood in terms of the dynamic nature of the C—Al bond and the high Lewis acidity of the monomeric species. This is directly related to the pronounced tendency of the aluminum atom to build up an octet of electrons. The C—Al bonds exhibit a unique set of properties: (i) their ability to alkylate certain metals and to reduce transition metal salts, (ii) their tendency to form bridged complexes containing other metals and organometallics, and (iii) their facile reactions with olefins under certain conditions. Organoaluminum compounds also possess a strong affinity for various heteroatoms in organic molecules, particularly oxygen. They generate 1:1 coordination complexes even with neutral bases such as ethers. Utilization of this property (heterogenophilicity including oxygenophilicity) in organic synthesis allows facile reactions with hetero atoms particularly oxygen- and carbonyl-containing compounds. However, in sharp contrast to classical Lewis acids such as $BF_3{\cdot}OEt_2$, $AlCl_3$, $SnCl_4$, and $TiCl_4$, they are endowed with a latent nucleophilic character which can emerge prominently on coordination with a heteroatom-containing functional group. The aluminum atom serves primarily as the coordination site for the substrate, while the nucleophilic center attached to the aluminum atom can be activated by the formation of the coordination complex, facilitating nucleophilic attack on the substrate.

This process is illustrated by the successive Beckmann rearrangement–alkylation sequence using trialkylaluminums [14]. Occasionally the nucleophilic center may behave as a proton scavenger. These characteristic features are of great interest to synthetic organic chemists.

Although organoaluminums have found numerous applications in organic synthesis, there have been virtually no synthetically useful reactions of organogalliums and organoindiums. In contrast, organothalliums have emerged as uniquely useful organometallics within the last decade. While thallium compounds share many features with the lighter members of the Group 13 elements, their chemistry is dominated by certain features unique to the heavy elements.

H H N OTs — Pr_3Al → [H H N + $Pr_3\bar{A}l$-OTs] ↓ H H N ← DIBAH — H H H H N

Pumiliotoxin C   60% overall yield   (43)

16.4.2 Reactions with Hydrocarbons

There is a far-reaching analogy between the addition of dialkylaluminum hydrides to olefins (hydroalumination) and Brown's hydroboration reaction, and in particular the third stage of hydroboration. Hydroboration is more facile than hydroalumination, and proceeds by *cis* addition to the double bond. *Cis* addition of R_2Al—H to the double bond may reasonably be presumed but is not definitely established, since there is no test of *cis* addition for the terminal olefins which are most commonly subjected to hydroalumination.

In view of the weak affinity of the ionic Al—H bond for olefins [15–21], the hydroalumination of olefins has been studied extensively in the presence of transition metal catalysts. Among these, only Ti and Zr catalysts have proved to be effective for obtaining hydroalumination products and their subsequent functionalization with various

$$R_2Al{-}H + \;>C{=}C< \longrightarrow R_2Al{-}\overset{|}{\underset{|}{C}}{-}\overset{|}{\underset{|}{C}}{-}H \quad (44)$$

$$R_2B{-}H + \;>C{=}C< \longrightarrow R_2B{-}\overset{|}{\underset{|}{C}}{-}\overset{|}{\underset{|}{C}}{-}H \quad (45)$$

electrophiles has been studied [22, 23]. Recently, several other transition metals have been utilized. UCl_3 or UCl_4 is an effective catalyst for the preparation of organoaluminates from terminal olefins and $LiAlH_4$ [24]. The active species in this reaction is thought to be $U(AlH_4)_3$.

$$x\ RCH{=}CH_2 + LiAlH_4 \xrightarrow{UCl_3} (RCH_2CH_2)_xAlH_{4-x}Li \xrightarrow{H_2O} RCH_2CH_3 \quad (46)$$

$$[\,R = C_nH_{2n+1}\ (n = 0-6),\ x = 0-4\,]$$

Nickel(0)—olefin complexes such as tris(ethylene)nickel(0) and tris(bicycloheptene)nickel(0) are highly active homogeneous catalysts for the transalkylation of trialkylaluminum with terminal olefins [25]. After completion of the reaction, the "naked" nickel(0) can be removed in the form of $Ni(CO)_4$ by reaction with carbon monoxide.

$$\text{(1,5-hexadiene)} + i\text{-}Bu_3Al \xrightarrow[0\ °C]{Ni(0)} i\text{-}BuAl(CH_2\text{—cyclopentyl})_2 + 2\ C_4H_8 \quad (47)$$

Extensive studies have shown that only terminal olefins can be successfully utilized and hydroalumination of internal olefins proceeds reluctantly even in the presence of transition metal catalysts [16, 22]. In this context, the possibility of organoborane-catalyzed hydroalumination has been explored, because hydroboration is far superior to other hydrometalation reactions. This expectation has been realized by combining organoboranes in catalytic quantities and Cl_2AlH as the hydrometalation agent [26]. Cl_2AlH (or its synthetic equivalent) can be generated *in situ* from $LiAlH_4$ and $AlCl_3$ in ether or R_2AlH (R=Et or *i*-Bu) and $AlCl_3$ in 1,2-dichloroethane. The organoborane-catalyzed hydroalumination using the catalytic $PhB(OH)_2/Cl_2AlH$ or catalytic Et_3B-/Cl_2AlH systems in ether is applicable to the regio- and chemoselective functionalization of monosubstituted olefins. Internal olefins as well as terminal olefins are readily hydrometalated with catalytic Et_3B/Cl_2AlH in 1,2-dichloroethane. Apparently, the coordination of an ethereal oxygen to a Lewis acidic aluminum center in ether solvent significantly lowers the reactivity of the catalytic system.

$$\text{(4-vinylcyclohexene)} \xrightarrow[Cl_2AlH,\ ether]{cat\ R\text{-}B<} \text{(cyclohexenyl)}CH_2CH_2AlCl_2 \xrightarrow[60\text{-}80\%]{E^+} \text{(cyclohexenyl)}CH_2CH_2E \quad (48)$$

$$E = H, Br, I, OH$$

$$\xrightarrow[\text{Et}_2\text{AlH-AlCl}_3,\ \text{ClCH}_2\text{CH}_2\text{Cl}]{cat\ \text{Et}_3\text{B}} \text{AlCl}_2 \xrightarrow[80\%]{\text{O}_2} \text{OH} \qquad (49)$$

A new carbon–carbon bond formation can be realized by the selective coupling between the intermediary alkylaluminum dichloride, generated with an Et_3B/Cl_2AlH system, and certain organic electrophiles [26, 27].

$$\text{C}_{10}\text{H}_{21}\text{CH=CH}_2 \xrightarrow[\text{Cl}_2\text{AlH}]{cat\ \text{Et}_3\text{B}} \text{C}_{12}\text{H}_{25}\text{AlCl}_2 \xrightarrow{\text{CH}_3\text{COCl}} \text{C}_{12}\text{H}_{25}\text{COCH}_3\ \ 82\% \qquad (50)$$

$$\text{C}_{12}\text{H}_{25}\text{AlCl}_2 \xrightarrow{\text{MeOCH(Cl)COOMe}} \text{C}_{12}\text{H}_{25}\text{CH(OMe)COOMe}\ \ 71\%$$

Quite recently, a novel inter–intra cascade carboalumination reaction of dienes and trienes with dialkylaluminum chlorides (R_2AlCl) has been developed in the presence of a catalytic amount of $Ti(OPr^i)_4$ [28]. In the presence of pre-existing chiral centers in the starting dienes, the formation of cyclopentane derivatives is highly *trans*-selective with respect to the aluminomethyl and R groups, while that of cyclohexane derivative is highly *cis*-selective. The stereochemical outcome can be rationalized in terms of chair cyclohexane-like transition states. Since the alumino group in the products can be readily functionalized, the reaction provides a novel and synthetically useful route to stereo-defined cyclopentane and cyclohexane derivatives.

$$\xrightarrow[cat.\ \text{Ti(OPr}^i)_4]{\text{R}_2\text{AlCl}} \left[\ \text{R, H, L}_m\text{M, H}\ \right] \longrightarrow \text{R, H, AlL}_m\text{, H} \qquad (51)$$

$$\xrightarrow[cat.\ \text{Ti(OPr}^i)_4]{\text{R}_2\text{AlCl}} \left[\ \text{R, L}_m\text{M, H, H}\ \right] \longrightarrow \text{R, H, AlL}_m\text{, H} \qquad (52)$$

The zirconocene-catalyzed carboalumination reaction of alkynes provides a new route to stereo- and regio-defined trisubstituted olefins [29]. Phenylacetylene and several other alkynes on treatment with Cp_2ZrCl_2/Me_3Al undergo *cis*-addition to yield the corresponding alkenylaluminums almost exclusively. Such alkenylaluminums have already proven to be versatile intermediates for the preparation of a wide variety of trisubstituted olefins. Although the Cp_2ZrCl_2—catalyzed reaction of alkynes with tri-akylaluminum possessing β-hydrogens is complicated by competitive hydrometalation [29] and diminished regioselectivity (70–80%), the hydrometalation can be avoided by using dialkylaluminum chloride in place of trialkylaluminum. Based on the mechanistic studies, the Zr-catalyzed carboalumination reaction probably involves direct Al—C bond addition assisted by Zr.

$$PhC{\equiv}CH \xrightarrow[2)\ D_2O]{1)\ n\text{-}BuLi} PhC{\equiv}CD \xrightarrow[Me_3Al]{Cp_2ZrCl_2} (Ph)(Me)C{=}C(D)(AlMe_2) \xrightarrow{H_2O} (Ph)(Me)C{=}C(D)(H)\ \ 98\%\ (>98\%\ Z) \quad (53)$$

$$PhC{\equiv}CH \xrightarrow{Cp_2ZrCl_2,\ Me_3Al} (Ph)(Me)C{=}C(H)(AlMe_2) \xrightarrow{D_2O} (Ph)(Me)C{=}C(H)(D)\ \ (96\%\ E)$$

The bimetallic species, $Bu_2Mg/2Et_3Al$ was found to be effective for the carboalumination of silylacetylene [30]. The compound, $Bu_2Mg/2Et_3Al$, has the bridged structure **A** which is essential for enhancing the reactivity of the C-Al bond. A similar bimetallic bridged species **B** is also involved in the Cp_2ZrCl_2-catalyzed carboalumination of terminal acetylenes with Me_3Al [31]. The new reaction proceeds in a regiospecific but nonstereoselective manner. However, substrates carrying unsaturated groups in conjugation with the triple bond exhibit high or exclusive *trans* selectivity.

$$Et_2Al(\mu\text{-}Bu)(\mu\text{-}Et)Mg(\mu\text{-}Et)(\mu\text{-}Bu)AlEt_2 \quad \textbf{A}$$

$$Me_2Al(\mu\text{-}Cl)(\mu\text{-}Me)ZrCp_2(\mu\text{-}Me)(\mu\text{-}Cl)AlMe_2 \quad \textbf{B}$$

$$C_{10}H_{21}C{\equiv}CSiMe_3 \xrightarrow[2)\ H_3O^+]{1)\ Bu_2Mg\text{-}2Et_3Al} (C_{10}H_{21})(Et)C{=}C(SiMe_3)(H) + (C_{10}H_{21})(Bu)C{=}C(SiMe_3)(H) \quad 80\%\ (5:1) \quad (54)$$

$$C_6H_{13}CH{=}CHC{\equiv}CSiMe_3 \xrightarrow[2)\ I_2;\ 3)\ Bu_4NF]{1)\ Hex_2Mg\text{-}Et_3Al} C_6H_{13}CH{=}CHC(Et){=}C(H)(I) \quad 85\% \quad (55)$$

16.4.3 Aldehydes and Ketones

Oxophilic organoaluminum compounds are capable of reacting with various aldehydes and ketones as alkylating and reducing agents. A good summary of this area is given by Mole and Jeffery [3].

Pronounced solvent as well as temperature effects were observed in the course of trialkylaluminum-induced cyclization of unsaturated aldehydes [32]. Thus, unimolecular decomposition of the 1:1 complex of citronellal/Me_3Al at –78 °C to room temperature yielded the acyclic compound **1** in hexane, whereas isopregol (**2**) was produced exclusively in 1,2-dichloroethane. Furthermore, the cyclization–methylation product **3** was formed with high selectivity using excess Me_3Al in CH_2Cl_2 at low temperature. The 1:1

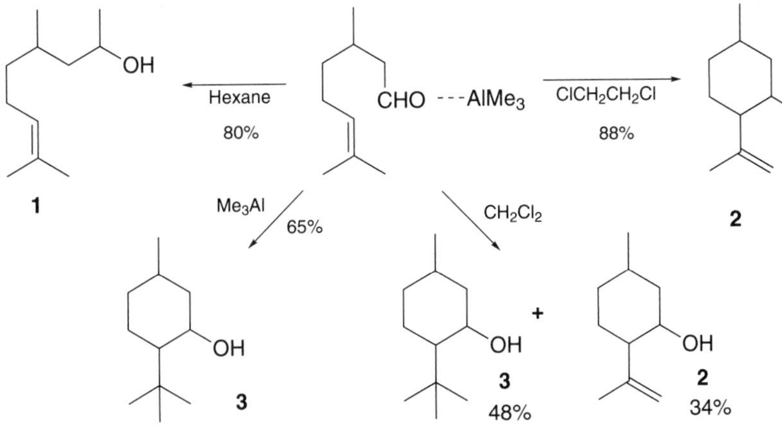

(56)

complex of other trialkylaluminum–citronellal decomposed upon warming to room temperature to furnish a reduction product, citronellol, as a major product.

The studies of $MeAlCl_2$-induced cyclization of unsaturated ketones indicate the advantage of alkylaluminum chloride over $AlCl_3$ in Lewis acid catalyzed reactions, since these reagents are capable of acting as proton scavengers as well as Lewis acids [33]. The reaction is interpreted as a $MeAlCl_2$-promoted cyclization of the γ,δ-unsaturated ketone followed by the sequential hydride and methyl shift as illustrated below.

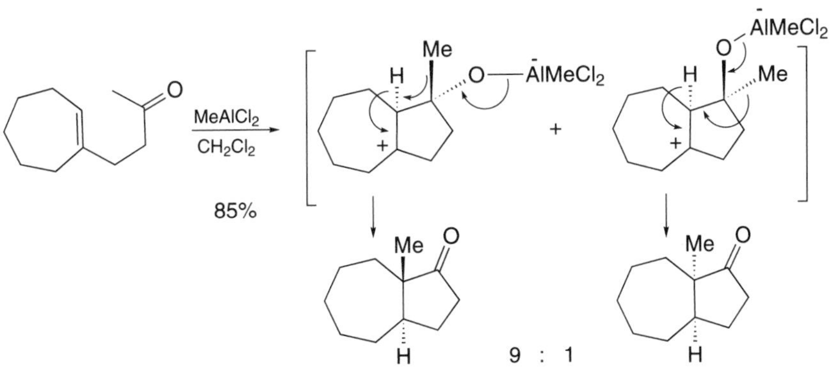

(57)

The aldol condensation of aldehydes and silyl enol ethers has been realized in the presence of various organoaluminum catalysts [34]. Me_2AlCl (0.05–0.5 equiv) is most effective and other organoaluminum reagents such as Me_3Al, $EtAlCl_2$, and $MeAlCl_2$ gave lower yields of aldol products.

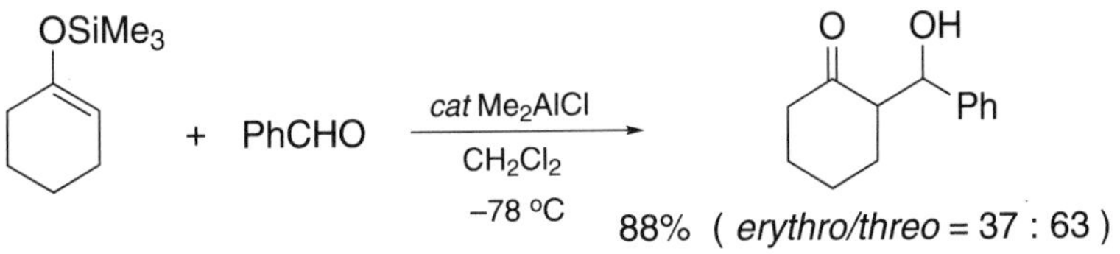

(58)

The aluminum enolate can be generated regiospecifically by treatment of α-halo carbonyl compounds with reagents of the type $Bu_3SnAlEt_2$, $Bu_3PbAlEt_2$, or $Ph_3PbAlEt_2$. Reaction of the resulting adduct with aldehydes or ketones gave β-hydroxy carbonyl compounds under mild conditions [35]. This aldol reaction is accelerated by adding a catalytic amount of $Pd(PPh_3)_4$.

2-bromo-2-methylcyclohexanone —($Et_2AlSnBu_3$, THF, 0 °C, 30 min)→ aluminum enolate ($OAlEt_2$, Me) —(PhCHO, 55%)→ 2-methyl-2-(hydroxy(phenyl)methyl)cyclohexanone (59)

The standard Wittig reagents can function as strong bases and remove the acidic α-protons of carbonyl compounds. With easily enolizable ketones, proton abstraction becomes the dominant reaction. However, Tebbe's reagent, $Cp_2TiCH_2{\cdot}AlMe_2Cl$,which cleanly converts esters and lactones into their corresponding enol ethers [36], was found to be highly effective for the methylenation of enolizable ketones [37]. The titanium methylidene fragment, $Cp_2Ti{=}CH_2$ is an active species which reacts chemoselectively with ketones over esters.

β-tetralone —($Cp_2Ti(\mu\text{-}CH_2)(\mu\text{-}Cl)AlMe_2$, THF)→ 2-methylene-tetralin, 84% (60)

Organoaluminum compounds are endowed with high oxophilic character, and hence are capable of forming long-lived monomeric 1:1 complexes with carbonyl substrates. For example, the reaction of benzophenone with Me_3Al in a 1:1 molar ratio gives a yellow, long-lived monomeric 1:1 species which decomposed unimolecularly to dimethylaluminum 1,1-diphenylethoxide during some minutes at 80 °C or many hours at 25 °C [38]. This unique property may be utilized for stereoselective activation of the carbonyl group. Among various organoaluminum derivatives examined, exceptionally bulky, oxophilic organoaluminum reagents such as methylaluminum bis(2,6-di-*tert*-butyl-4-alkylphenoxide) (MAD and MAT), have shown excellent diastereofacial selectivity in carbonyl alkylation [39,40]. Thus, treatment of 4-*tert*-butylcyclohexanone with MAD or MAT in toluene produced a 1:1 complex which on subsequent treatment with methyllithium or Grignard reagents in ether at –78°C afforded the equatorial alcohol almost exclusively. Methyllithium or Grignard reagents undergo preferential equatorial attack yielding axial alcohols as the major product. MAD and MAT have played a crucial role in the stereoselective synthesis of hitherto inaccessible equatorial alcohols from cyclohexanones.

This approach has been quite useful in the stereoselective alkylation of steroidal ketones. Reaction of 3-cholestanone with MeLi gave predominantly 3β–methyl-

$$Ph_2C{=}O \xrightarrow[25^{\circ}C]{Me_3Al} [\,Ph_2C{=}O\cdots AlMe_3\,] \xrightarrow{80^{\circ}C} Ph_2C(Me){-}OAlMe_2 \quad (61)$$

yellow complex

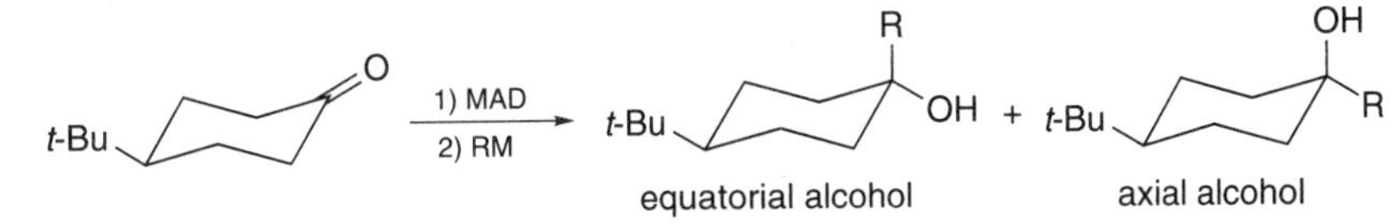

t-Bu (4-tert-butylcyclohexanone) → 1) MAD, 2) RM → equatorial alcohol + axial alcohol (62)

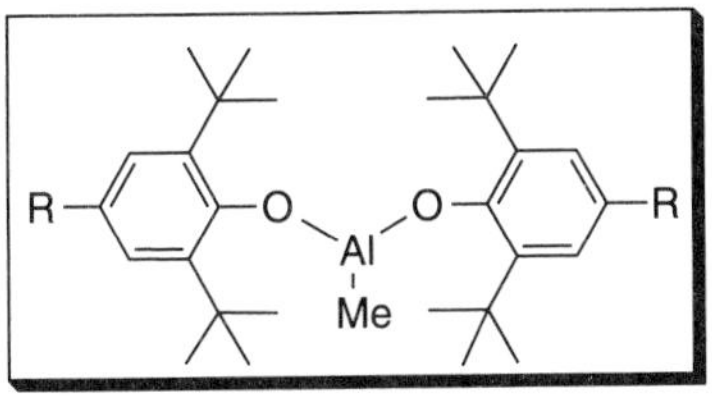

MAD : R = Me
MAT : R = t-Bu

RM =		
MeLi	84%	(99 : 1)
EtMgBr	91%	(100 : 0)
BuMgBr	67%	(100 : 0)
AllylMgBr	90%	(91 : 9)

cholestan-3α-ol (axial alcohol),whereas amphiphilic alkylation of the ketone with MAD/MeLi or MAT/MeLi afforded 3α- methylcholestan-3β-ol (equatorial alcohol)exclusively. In addition, unprecedented *anti*-Cram selectivity was achievable in the MAD- or MAT-mediated alkylation of α-chiral aldehydes.

Based on the concept of the diastereoselective activation of carbonyl groups with MAD or MAT, the bulky, chiral organoaluminum reagent, (*R*)-BINAL or (*S*)-BINAL has been devised for enantioselective activation of carbonyl groups.

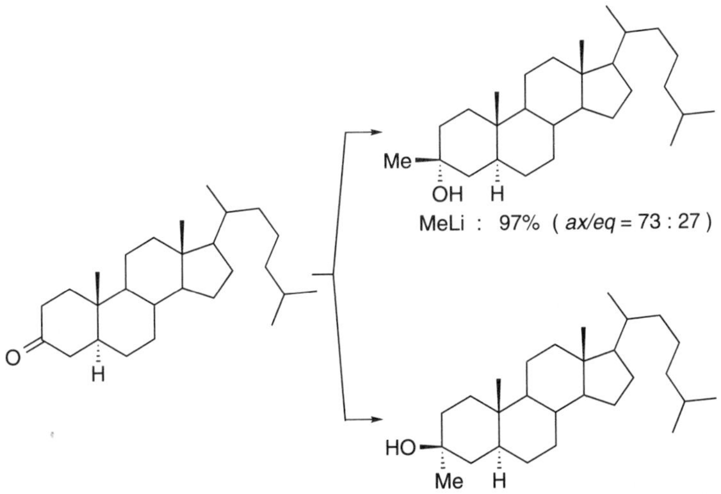

(63)

MeLi : 97% (*ax/eq* = 73 : 27)

MAD/MeLi : 98% (*ax/eq* = 2 : 98)
MAT/MeLi : 99% (*ax/eq* = 1 : 99)

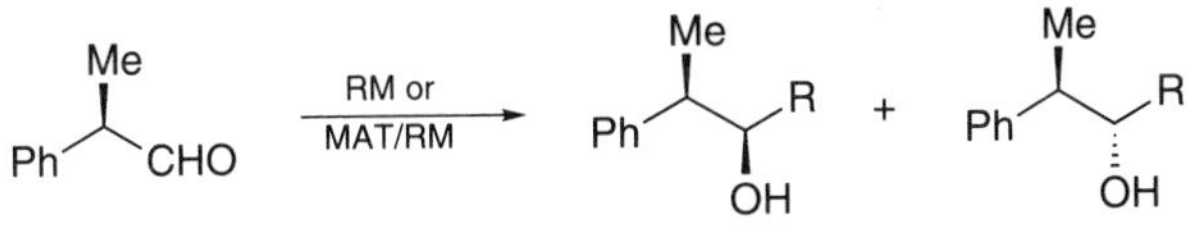

Ph(Me)CHO → RM or MAT/RM → Cram + *anti*-Cram (64)

MeMgI	64% (72 : 28)
MAT/MeMgI	96% (7 : 93)
EtMgBr	78% (84 : 16)
MAT/EtMgBr	90% (13 : 87)

(*R*)-BINAL (*S*)-BINAL

The modified organoaluminum reagent, BINAL, can be used as a chiral Lewis acid catalyst in the asymmetric hetero-Diels–Alder reaction [41]. Reaction of various aldehydes with activated dienes under the influence of catalytic BINAL (5–10 mol%) at –20 °C gave, after exposure of the resulting hetero-Diels–Alder adducts to trifluroacetic acid, predominantly *cis*-dihydropyrone in high yield with excellent enantioselectivity.

1) (*R*)-BINAL (5-10 mol%) 2) CF_3COOH (65)

Ar = Ph	77% (95% ee)	7%
Ar = 3,5-Xylyl	90% (97% ee)	3%

Reactive aldehydes such as formaldehyde and α-chloroaldehydes can be successfully generated by treatment of readily available trioxane and α-chloroaldehyde trimers, respectively, with methylaluminum bis(2,6-diphenylphenoxide) (MAPH), and stabilized as their 1:1 coordination complexes with MAPH. The resulting $CH_2{=}O{\cdot}MAPH$ complex reacts with a variety of olefins to furnish ene-reaction products with excellent regio- and stereoselectivities [42]. In addition, this complex as well as α-chloroaldehyde-MAPH complexes can be utilized as a stable source of gaseous formaldehyde and reactive α-chloroaldehydes, respectively, for the nucleophilic addition of various carbanions (organometallics, enolates, etc.) [43]. Formation of reactive aldehyde–MAPH complexes is firmly confirmed by ^1H NMR spectroscopy. A space-filling model of

trioxane (MAPH) $CH_2{=}O\cdots Al$ OCuBuLi Bu OH 63 % MAPH (66)

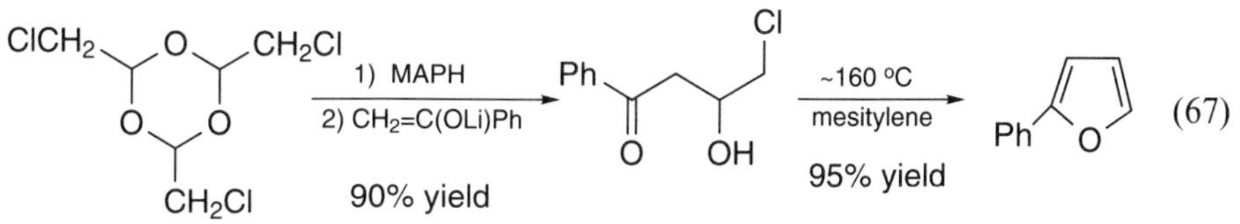

(67)

aldehyde–MAPH complexes implies that formaldehyde and α-chloroaldehydes coordinated with MAPH may be electronically stabilized by two parallel phenyl groups of aluminum ligands.

This unusual stabilization effect of MAPH has been successfully utilized for discrimination between two structurally different aldehyde carbonyls, thereby allowing the chemoselective functionalization of sterically more hindered aldehydes [44].

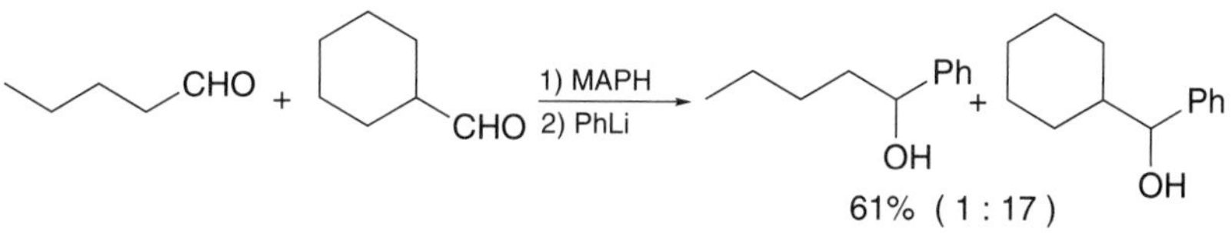

(68)

In contrast to MAPH, aluminum tris(2,6-diphenylphenoxide) (ATPH) has been introduced for effective blocking of aldehyde carbonyls, thereby allowing the hitherto difficult conjugate addition of reactive nucleophiles to α,β-unsaturated aldehydes [45]. In particular, Michael additions of lithium alkynides and thermally unstable lithium carbenoids, which are very difficult to achieve in organocopper chemistry, are realized with α,β-unsaturated aldehyde/ATPH complexes. A similar regiochemical reversal was also observed in the addition of enolates and silyl ketene acetals to α,β-unsaturated aldehydes in the presence of ATPH. Based on the X-ray structure of the DMF–ATPH complex, a plausible cinnamaldehyde–ATPH complex structure is suggested which shows that ATPH blocks the cinnamaldehyde carbonyl effectively, thereby serving as a tight aldehyde pocket for the selective 1,4-addition of carbon nucleophiles to α,β-unsaturated aldehydes. This approach is also applicable to the conjugate addition to α,β-unsaturated ketones [46].

Regioselective olefin formation from ketones has been exploited. Conversion of ketones to the corresponding enol diphenylphosphates and subsequent coupling of

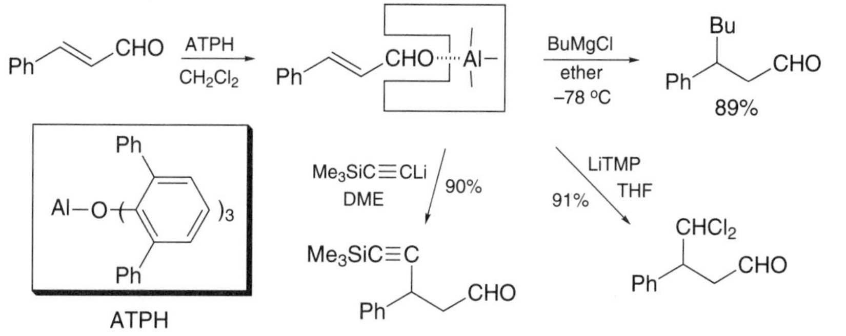

(69)

enol phosphates by stereospecific C—O bond fission with alkylaluminum in the presence of catalytic $Pd(PPh_3)_4$ yields alkenes [47]. In this coupling, alkenyl and alkynyl groups are introduced selectively in preference to alkyl substituents. The reaction does not affect a co-existing vinyl sulfide group. This feature enabled 1,2- and 1,3-carbonyl transposition, with or without alkylation, via phenylthio-substituted enol phosphates.

1) LDA; 2) $ClPO(OPh)_2$ → $OPO(OPh)_2$ (t-Bu) → cat $Pd(PPh_3)_4$, $Et_2AlC{\equiv}CPh$, 70% → $C{\equiv}CPh$ (t-Bu) (70)

1) LDA/PhSSPh; 2) NaH/$ClPO(OPh)_2$, 64% → $OPO(OPh)_2$, SPh, Ph, Me; 71%, cat $Pd(PPh_3)_4$, Me_3Al → Me, SPh, Ph, Me; $TiCl_4$, H_2O/AcOH, 95% → Me, O, Ph, Me (71)

The tandem aldol condensation–radical cyclization sequence for the elaboration of functionalized bicyclo[3.3.0]octane systems has been developed [48]. Conjugate addition of $Me_2AlSePh$ to dimethylcyclopentenone followed by trapping of the resultant enolate with aldehyde afforded the *trans,erythro* aldol predominantly which then underwent radical cyclization with Bu_3SnH and catalytic AIBN yielding the bicyclic ketol stereospecifically. This approach represents a highly convergent method for the annulation of carbocycles leading to the polyquinane sesquiterpenes.

1) $Me_2AlSePh$; 2) $CH_2{=}CHC(Me)_2CHO$, 63% → (O, H, OH, Se, Ph); cat AIBN, Bu_3SnH, 80% → (O, H, OH, H, Me) (72)

Diisobutylaluminum hydride (DIBAH) is undoubtedly one of the most common reducing agents in organic synthesis and recent interest in the synthetic utility of DIBAH has been directed toward diastereoselective reduction of carbonyl substrates. High 1,3-*syn* diastereoselectivity has been achieved in the chelation-controlled reduction of β-hydroxy ketones with DIBAH in THF [49]. The choice of solvents strongly affects the selectivity. Use of CH_2Cl_2 or toluene in place of THF did not show any diastereoselectivity.

OH O Ph — DIBAH, THF, -78 °C, 87% (93 : 7) → OH OH Ph (*syn*) + OH OH Ph (*anti*) (73)

The reducing reactivity of DIBAH can be dramatically modified by adding a transition-metal compound as exemplified by the methylcopper(I)-catalyzed highly efficient and selective conjugate reduction of α,β-unsaturated carbonyl compounds with DIBAH [50]. HMPA is an indispensable component for the present conjugate reduction, and it functions as a ligand rather than a cosolvent. The newly developed MeCu/DIBAH/HMPA system exhibits high chemoselectivity and even in the presence of saturated carbonyl groups, the selective conjugate reduction of α,β-unsaturated carbonyl compounds took place efficiently.

The asymmetric reduction of prochiral α- and β-hydroxy ketones has been realized with a regent generated from $SnCl_2$, a chiral diamine, and DIBAH [51]. Use of MEM ethers as a substrate and (*S*)-1-ethyl-2-(piperidinomethyl)pyrrolidine as a chiral ligand is the most effective to achieve high enantioselection.

cat MeCu, DIBAH, HMPA, THF, –50 °C, 2 h → 84% + 6% (74)

Ph–CO–CH₂–OMEM — $SnCl_2$/DIBAH, (N-Et pyrrolidinylmethyl)piperidine, –100 °C, 20 min → Ph–CH(OH)–CH₂–OMEM, 74%, 93% ee (75)

16.4.4 Acid Derivatives

Asymmetric Diels–Alder reactions are little affected by solvent changes. However, Lewis acids exert a strong catalytic effect and induce higher optical yields. As a consequence of faster rates, increased stereoselectivity, and enhanced regioselectivity, Lewis acid-catalyzed [4+2] cycloadditions offer many attractive synthetic advantages. Homogeneous alkylaluminum chlorides have now been accepted as most reliable reagents. Chiral α,β-unsaturated N-acyl oxazolidones exhibit high diastereofacial selection in Diels–Alder reactions, particularly those conducted in the presence of Et_2AlCl [52]. Reaction of the chiral acrylate and crotonate imides with cyclopentadiene furnished *endo* -adducts almost exclusively with diastereoselection of about 95%. The exceptional reactivity of these dienophile–Lewis acid complexes allowed the use of less reactive acyclic dienes with high diastereoselectivity (>95% de). The chiral auxil-

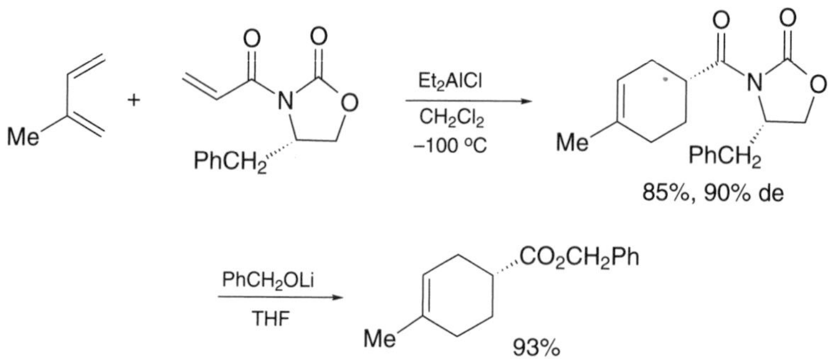

(76)

(77)

iary is cleaved by transesterification with lithium benzyloxide to the corresponding benzyl ester in 85–95% yield.

(*E,E*)-Trienecarboximides derived from chiral oxazolidones undergo Me_2AlCl-catalyzed intramolecular Diels–Alder reactions yielding bicyclic compounds with high *endo*- and diastereoselectivity (*endo*/*exo* =~100 : 1) [53]. The stereochemistry is controlled by the stereogenic at C_4 of the chiral auxiliary.

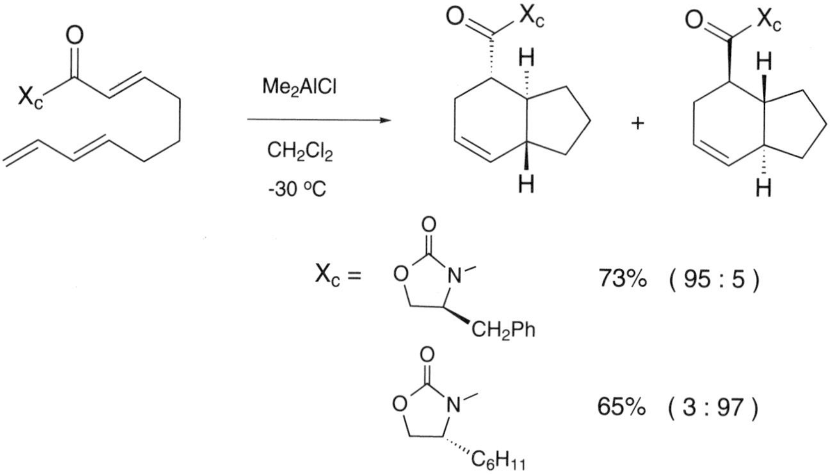

(78)

The distinct advantage of homogeneous alkylaluminum chloride over $AlCl_3$ was clearly demonstrated by the organoaluminum-catalyzed asymmetric Diels–Alder reaction of (–)-dimenthyl fumarate with various cyclic as well as acyclic dienes with remarkably high diastereofacial selectivity [54]. Here a single reaction species may be responsible for the cycloaddition, since a straight line of the observed enantioselectivity, ln (*S,S*)/(*R,R*) against the reciprocal of the temperature, $1/T$ (ln K) was obtained a temperatures ranging from 25 to –40 °C.

$EtAlCl_2$-catalyzed Diels–Alder reactions between alkyl-substituted 1,3-butadienes and (η^1-acrylol) (η^5-cyclopentadienyl) diacarbonyliron(II) complexes have been found, where the observed regio- and stereochemistry were consistent with that generally observed in Diels–Alder reactions [55]. Conventional Lewis acids such as $BF_3{\cdot}OEt_2$ and $TiCl_4$ were not effective in promoting the desired cycloaddition.

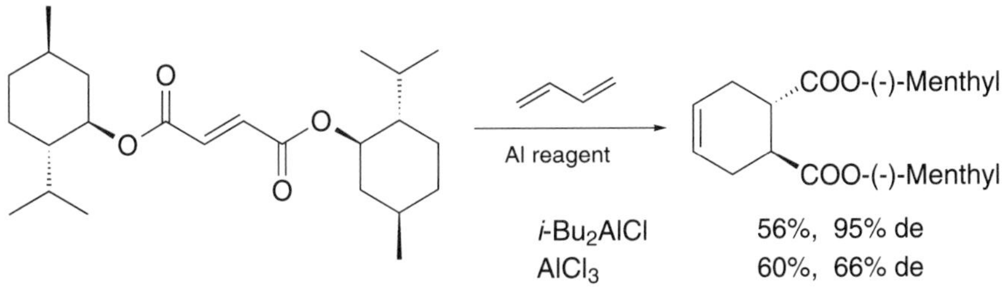

(79)

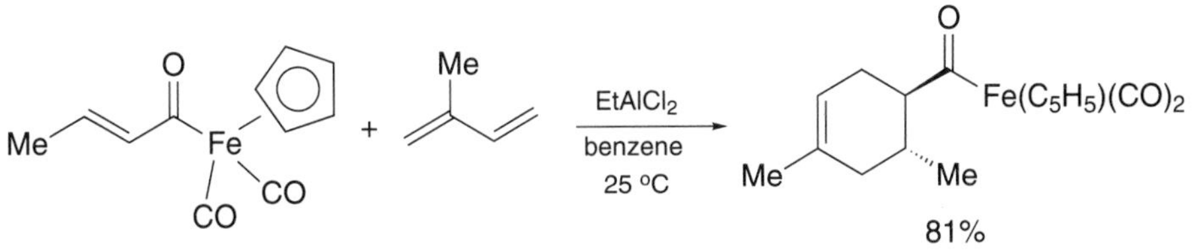

Al reagent

i-Bu_2AlCl 56%, 95% de
$AlCl_3$ 60%, 66% de

(80)

$EtAlCl_2$, benzene, 25 °C

81%

An interesting metal effect was observed in the aldol condensations of the enolate derived from the iron acetyl complex (η^5-C_5H_5)Fe(CO)(PPh_3)(COMe) with aldehydes [56,57]. Although the lithium enolate did not show any selectivity, the corresponding aluminum enolate by transmetalation with Et_2AlCl exhibited exceptionally high diastereoselectivity (>99% de). The resultant β-hydroxy acyl complexes are transformed to β-hydroxy acids on decomplexation with Br_2.

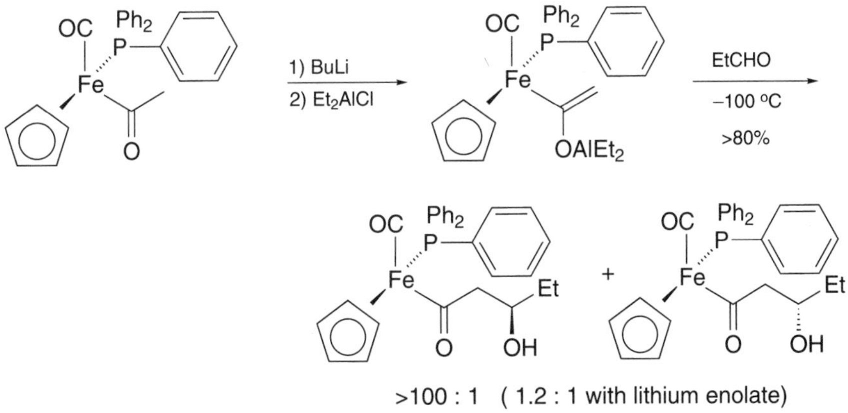

(81)

1) BuLi 2) Et_2AlCl

EtCHO, −100 °C, >80%

>100 : 1 (1.2 : 1 with lithium enolate)

16.4.5 Alcohol Derivatives

Organoaluminum-promoted asymmetric pinacol-type rearrangements have been studied extensively; chiral β-mesyloxy tertiary alcohols, when treated with excess Et_2AlCl or Et_3Al, undergo a stereospecific pinacol-type rearrangement to furnish optically pure α–alkyl, α-alkenyl, or α-aryl ketones [58–62]. The reaction is particularly useful for preparation of optically pure α–methyl-β,γ-unsaturated ketones by migration of an alkenyl group, which occurs with retention of the olefin geometry. The resultant ketones can be reduced by lithium tri-*sec*-butylborohydride with high *threo*-selectivity.

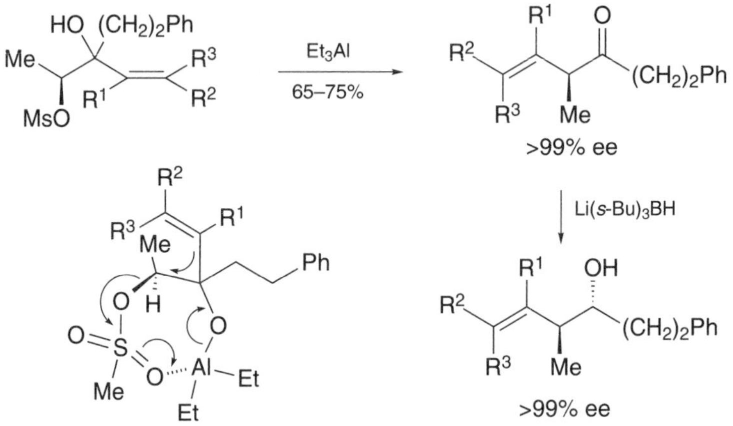

(82)

Reductive pinacol-type rearrangement of chiral α-mesyloxy ketones was executed by *in situ* reduction with DIBAH followed by addition of Et_3Al or Et_2AlCl. The resulting aldehyde is reduced as formed to an optically pure 2-aryl- or 2-alkenylpropanol [63].

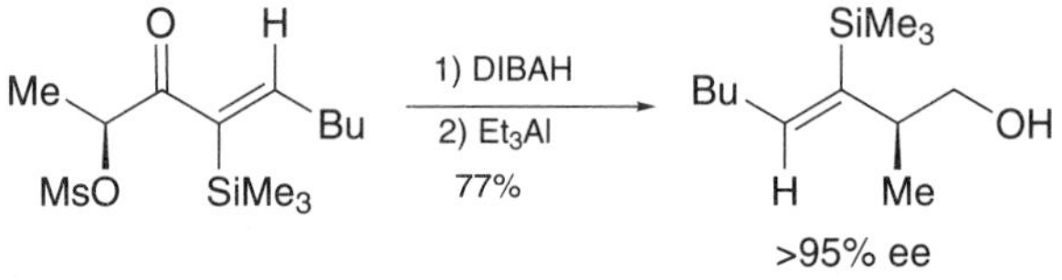

(83)

Cationic rearrangement of 3-(trimethylsilyl)methylcyclohexyl mesylates was effected with excess Me_2AlOTf [64]. In this reaction, the silyl group exhibited a remarkable directing effect to induce successive rearrangement of a hydride and an alkyl group in order to generate a stable β-silyl cationic species which finally afforded the corresponding olefins. As observed in the usual cationic 1,2-rearrangement process, successive migrations of two *anti*-periplanar substituents on axial positions may be greatly favored in these silicon-directing rearrangements.

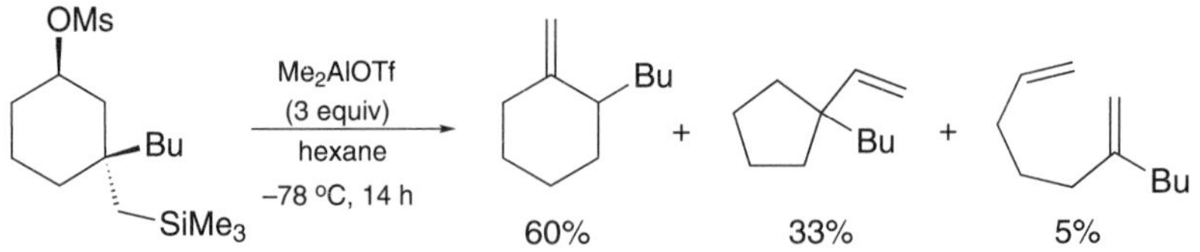

(84)

A mild and efficient preparation of functionalized allylic stannanes is performed from allylic acetates with $Et_2AlSnBu_3$ in THF in the presence of catalytic $Pd(PPh_3)_4$ [65]. This functional group interconversion represents a net conversion of the electronic nature of the allyl acetate from electrophile to nucleophile. The stannylating agent, which is readily available by reaction of Bu_3SnLi with Et_2AlCl, exhibits a high degree of regioselectivity for the less-substituted carbon of the allyl system. The reaction also proceeds with a remarkably high chemoselectivity. Enone, ketone and ester functionalities remain totally intact even in the presence of excess stannyl–aluminum reagent.

COOMe, OAc; cat Pd(PPh3)4, Et2AlSnBu3, THF → COOMe, SnBu3 + COOMe, SnBu3; 73% (7 : 3) (85)

16.4.6 Ethers, Epoxides, and Acetals

Aliphatic Claisen rearrangements have been accomplished under very mild conditions in the presence of organoaluminum reagents [66]. Treatment of simple ally vinyl ether substrates with trialkylaluminums resulted in the [3,3] sigmatropic rearrangement and subsequent alkylation on the aldehyde carbonyl group. The rearrangement-reduction product was obtained exclusively with *i*-Bu_3Al or DIBAH. The aluminum thiolate, Et_2AlSPh or a combination of Et_2AlCl and PPh_3 was effective for the rearrangement providing the normal Claisen products, γ,δ-unsaturated aldehydes, however, without any stereoselectivity. Accordingly, a new approach for the stereocontrolled Claisen rearrangement of allyl vinyl ethers has been developed based on a molecular recognition technique using the Lewis acid receptor, ATPH, giving Claisen products with high *E*/*Z* ratio on treatment of 1-alkyl-2-propenyl vinyl ether with ATPH in CH_2Cl_2 [67]. Use of sterically more hindered aluminum tris(2-α-naphthyl-6-phenylphenoxide) (ATNP) exhibited better selectivity (*E*/*Z* ratio = 47:1). The present chemistry is further extended to the asymmetric Claisen rearrangement of allyl vinyl ethers with the designing of chiral ATPH analogues, aluminum tris((*R*)-1-α-naphthyl-3-phenylnaphthoxide) ((*R*)-ATBN) or aluminum tris((*R*)-3-(*p*-fluorophenyl)-1-α-naphthylnaphthoxide) ((*R*)-ATBN-F) with high enantioselectivity [67].

The regioselectivity of epoxide-opening reactions using alkynylaluminum reagents has been studied for prostaglandin synthesis [68]. With two different cyclopentane oxide derivatives, the simple substitution of an aluminum ate complex for the usual trialkylaluminum can sometimes be useful in achieving desired regioselectivity.

A selective ring-opening of 2,3-epoxy alcohols has been attained with organoaluminum reagents [69,70]. More recently, a highly regio- and stereoselective addition of

$Et_2AlC{\equiv}CPh$, $ClCH_2CH_2Cl$, 25 °C, 15 min → C≡CPh, OH; 88% (86)

Bu, O; Al reagent → Bu, CHO + Bu, CHO (87)

Al reagent	Yield	
Et_2AlSPh	84%	(*E*/*Z* = 39 : 61)
$Et_2AlCl + PPh_3$	81%	(*E*/*Z* = 43 : 57)
ATPH	87%	(*E*/*Z* = 94 : 6)
ATNP	90%	(*E*/*Z* = 98 : 2)

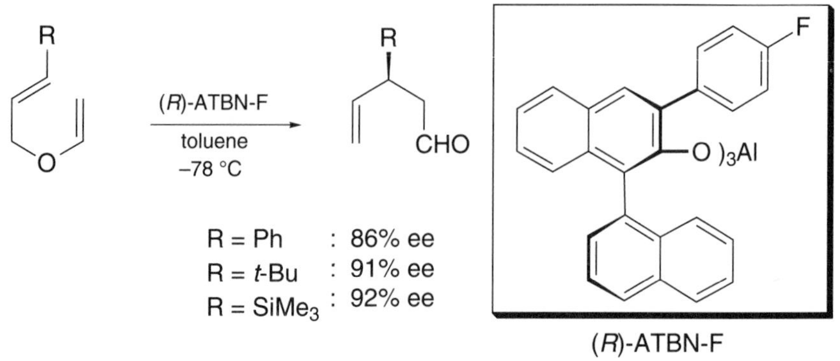

(R)-ATBN-F, toluene, −78 °C (88)

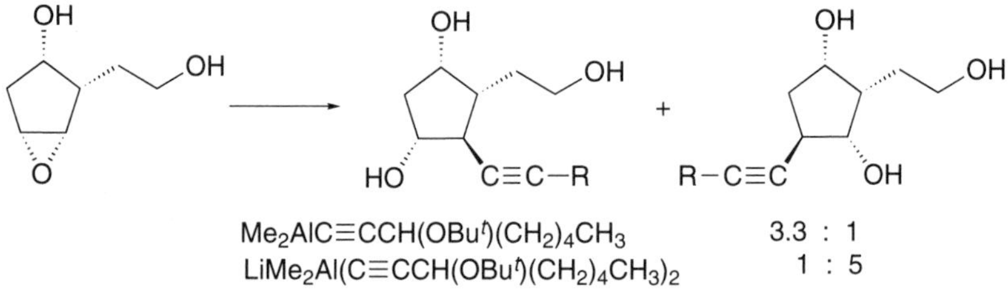

R = Ph : 86% ee
R = t-Bu : 91% ee
R = $SiMe_3$: 92% ee

(R)-ATBN-F

(89)

$Me_2AlC{\equiv}CCH(OBu^t)(CH_2)_4CH_3$ 3.3 : 1
$LiMe_2Al(C{\equiv}CCH(OBu^t)(CH_2)_4CH_3)_2$ 1 : 5

the azido group to 2,3-epoxy alcohols has been reported with Me_3SiN_3—Et_2AlF system. This method is superior to the conventional one using azide anion which strongly reflects the steric effect of all of the epoxide substituents [71].

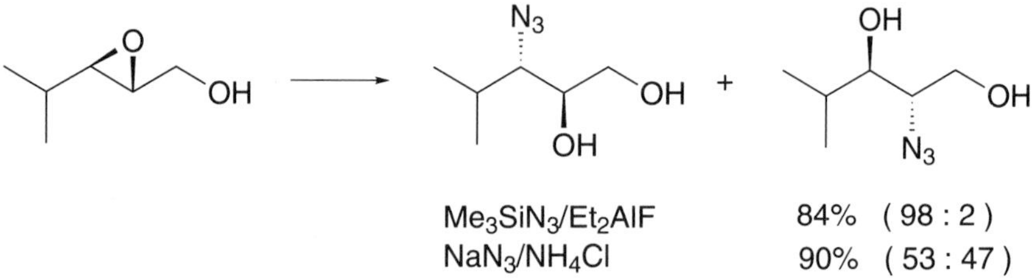

(90)

Me_3SiN_3/Et_2AlF 84% (98 : 2)
NaN_3/NH_4Cl 90% (53 : 47)

The reaction of chiral acetals with organoaluminum reagents has been thoroughly investigated [72,73]. Noteworthy is the unprecedented regio- and stereochemical control observed in the addition of trialkylaluminums to chiral α,β-unsaturated acetals derived from optically pure tartaric acid diamide [74–76]. The course of the reaction appeared to be strongly influenced by the nature of substrates, solvents, and temperature. These findings provide easy access to optically active α-substituted aldehydes, β-substituted aldehydes, α-substituted carboxylic acids, or allylic alcohols. Since both (*R*,*R*)- and (*S*, *S*)-tartaric acid diamide are readily obtainable in optically pure form, this method allows the synthesis of both enantiomers of substituted aldehydes, carboxylic acids, and allylic alcohols from α,β-unsaturated aldehydes in a predictable manner.

This asymmetric reaction possesses vast potential in natural product synthesis. This is illustrated by the short synthesis of the side-chain alcohol which is present in the biologically important vitamins E and K [74].

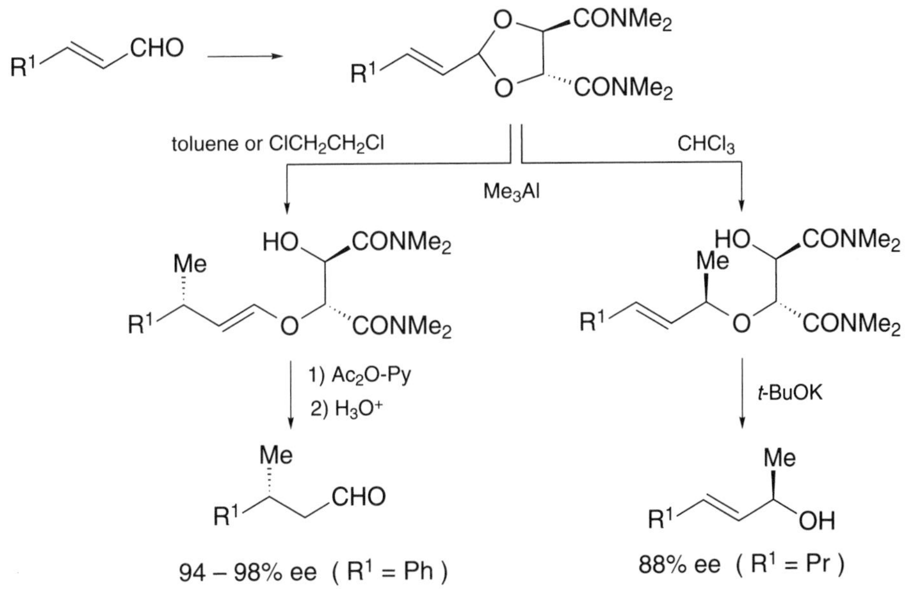

(91)

toluene or $ClCH_2CH_2Cl$ — Me_3Al — $CHCl_3$

1) Ac_2O-Py
2) H_3O^+

t-BuOK

94 – 98% ee (R^1 = Ph)

88% ee (R^1 = Pr)

16.4.7 Halohydrocarbons

A highly convenient and versatile cyclopropanation method has been recently developed which involves treatment of olefins with various organoaluminum compounds and alkylidene iodide under mild conditions [77]. Cyclopropane formation using Et_3Al–methylene iodide in cyclohexene was previously found to proceed in quite disappointing yields [78]. However, making a survey of a range of aluminum reagents as well as manipulating experimental conditions, the intermediary dialkyl (iodomethyl)aluminum species is revealed to be responsible for the cyclopropanation of olefins and it easily undergoes decomposition in the absence of olefins or by excess trialkylaluminum. Hence, the use of equimolar amounts of trialkylaluminum and methylene iodide in the presence of olefins is essential for the achievement of reproducible results in the cyclopropanation process. In addition, since dialkylaluminum halide is also employable as a cyclopropanation agent, the use of a half equivalent of trialkylaluminum is not detrimental.

The organoaluminum-mediated cyclopropanation exhibited unique selectivity which is not observable in the Simmons–Smith reaction and its modifications [79]. Treatment of geraniol with *i*-Bu_3Al/methylene iodide in CH_2Cl_2 at room temperature

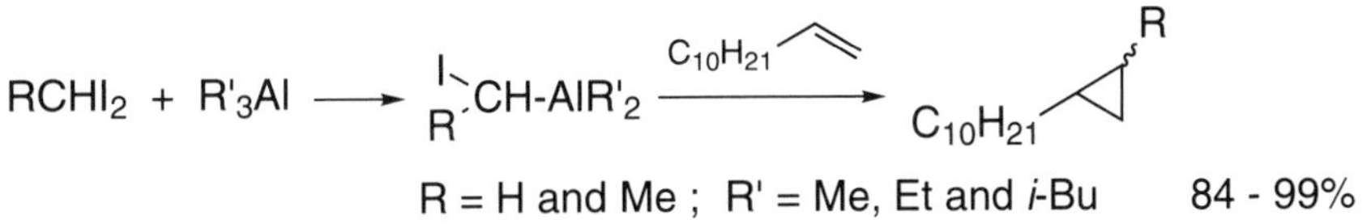

$$RCHI_2 + R'_3Al \longrightarrow \text{I(R)CH-AlR}'_2 \xrightarrow{C_{10}H_{21}CH=CH_2} C_{10}H_{21}\text{-cyclopropane-R} \qquad (92)$$

R = H and Me ; R' = Me, Et and *i*-Bu 84 - 99%

$$C_{10}H_{21}CH=CH_2 \xrightarrow[\text{CH}_2\text{Cl}_2,\ 25\ ^\circ\text{C},\ 10-13\ \text{h}]{Et_2AlX/CH_2I_2} C_{10}H_{21}\text{-cyclopropane} \qquad (93)$$

85 - 94%

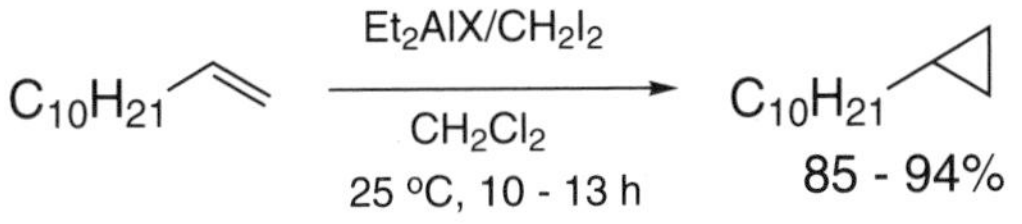

for 5 h gave rise to cyclopropanation products in 75% combined yields in a ratio of 76:1:4. Consequently, the methylene transfer by the aluminum method takes place almost exclusively at the C(6)—C(7) olefinic site far from the hydroxy group of geraniol and the C(2)—C(3) olefinic bond was left intact. In sharp contrast the zinc method showed opposite regioselectivity via hydroxyl-assisted cycopropanation.

geraniol

i-Bu_3Al/CH_2I_2	76	:	1	:	4	(94)
Et_2Zn/CH_2I_2	2	:	74	:	3	

16.4.8 Industrial Applications

Alkylaluminum compounds are used widely in Ziegler–Natta polymerization of olefins and dienes. It has been estimated that approximately half of the worldwide production of high density polyethylene (HDPE) is made with Ziegler catalysts [80]. Most of the worldwide production of low density polyethylene (LDPE) is currently made by free radical high-pressure polymerization of ethylene, but significant inroads into LDPE markets have been made in recent years by linear low density polyethylene (LLDPE) which can also be produced with Ziegler–Natta catalysts. Polypropylene is manufactured solely by Ziegler–Natta catalysts [81]. Excluding the captive use of aluminum alkyls for α-olefin and α-alcohol manufacture, Ziegler–Natta polymerization is by far the largest volume application of aluminum alkyls in the commercial market.

In the broadest sense, Ziegler–Natta catalysts may be regarded as combinations of Group IV–VIII transition metal compounds with Group I–III organometallics which can effect polymerization of olefins and dienes under relatively mild conditions of temperature and pressure [82,83]. Titanium compounds and aluminum alkyls are used most frequently in commercial polyolefin processes. Both titanium and vanadium compounds are used in conjunction with aluminum alkyls in catalyst systems for synthetic rubber/elastomers.

Early Ziegler–Natta catalysts were produced by direct reaction of $TiCl_4$ with aluminum alkyls and were very inefficient. Later, titanium in a pre-reduced became commercially available by reduction of $TiCl_4$ with aluminum, but catalyst activities remained relatively low. For this reason, catalyst residues in the resultant polymer were high and had to be removed by costly post-reactor treatment to prevent color formation in the polymer and corrosion in extruders of resin processors. Consequently, research was quickly directed toward increasing catalyst activity such that residues would be low enough to obviate removal. Unlike ethylene polymerization, polymerization of propylene may proceed by pathways involving different regioselectivities and stereoselectivities [84]. Development of catalysts wherein control of stereoregularity is achieved while maintaining or increasing activity became a primary objective of polypropylene catalyst research. Polyolefin catalyst research focused primarily on the transition metal portion of the Ziegler–Natta catalyst, rather than the metal alkyl, and resulted in catalysts of much higher activity. Such catalysts are widely used in commercial polyolefin processes today, and many involve use of a support for the catalyst. Suitable supports include silica, alumina, and various magnesium compounds such as $Mg(OR)_2$, HOMgCl, $MgCl_2$, and $Mg(OH)_2$ [81,84–86].

Another large volume application of aluminum alkyls involves their use in Ziegler growth reactions to produce long chain aluminum alkyls which serve as precursors to α-olefins [87] and α-alcohols [88]. Because of the ready availability of ethylene, higher quality of α-olefin products and lower energy requirements, the Ziegler process is supplanting other methods of olefin manufacture, such as wax-cracking and dehydrogenation of paraffins [87]. For α-alcohol production, the long chain aluminum alkyls are first oxidized and subsequently hydrolyzed with water or dilute acid. The resultant linear alcohols are used in production of biodegradable detergents.

16.4.9 Synthesis of Al Compounds

Geate care should be taken in handling neat organaluminums because they are pyrophoric: that is, they ignite spontaneously upon contact with air. Use safety glasses, gloves, and an apron.

(1) Diisobutylaluminum Hydride, $AlH(i\text{-}Bu)_2$ [89]

$$Al(i\text{-}Bu)_3 \longrightarrow AlH(i\text{-}Bu)_2 + CH_2{=}CHMe_2 \tag{95}$$

Diisobutylaluminum hydride was prepared by the thermal elimination of isobutylene from the commercially available triisobutylaluminum (Ethyl Corp.) in accordance with the general procedure of Ziegler and co-workers [90]. A convenient distillation apparatus proved to be a 500 mL, round-bottomed flask bearing one stoppered neck and one neck fused to a Vigreux column. The distillation column, in turn, bore a thermometer joint at its top and was itself fused to a one-piece condenser and fraction distributor. Two-necked flasks, each bearing a three-way stopcock, were connected to the four-way fraction collector to receive the distillate. The condenser was cooled with kerosene which was circulated through copper coils immersed in a cold bath. Since aluminum alkyls can ignite spontaneously in air and can also hydrolyze explosively, this method of cooling is advised to prevent serious accident if the condencer breaks.

In a typical run, triisobutylaluminum (250 mL) was placed in the distillation flask, and the liquid was slowly heated under 1 atom of nitrogen in an oil bath to 160–180 °C. Magnetic stirring of the oil bath permitted a smooth evolution of gas. The isobutylene evolution was monitored by means of a mineral bubbler inserted in the nitrogen line and the olefin could be condensed by passing the evolving gas stream through a dry ice–acetone trap. After 12 h the gas evolution became very slow and hence the slightly turbid liquid was cooled. The system was placed under a reduced pressure of 0.03 mm and gradually heated. From 46–96 °C about 50 mL of triisobutylaluminum was collected, while between 96–107 °C (0.1mm) the diisobutylaluminum hydride distilled over (bath temperature of 125–130 °C). Redistillation gave bp 112–114° (0.3 mm). The main fraction of about 100 mL had a pycnometrically determined density, d_4-0.793. Although the deposition of aluminum and further gas evolution (increase in pressure to 0.1–0.3 mm) signaled some decomposition, satisfactory distillation occurred if the bath temperature was kept below 130 °C. Anal Calcd for $C_8H_{19}Al$: Al, 18.97. Found: Al, 19.08.

(2) Diphenyl(phenylethynyl)aluminum, $AlPh_2(CCPh)$[90]

$$AlPh_3 + HCCPh \longrightarrow AlPh_2(CCPh) + Ph\text{-}H \quad (96)$$

A 250 mL two-necked flask with magnetic stirring is charged with triphenylaluminum (7.4 g, 29 mmol) in the dry box, and then dry benzene (100 mL; from sodium) is distilled into the flask. Then freshly distilled phenylacetylene (3.2 g, 31 mmol) is introduced into the mixture with a syringe. The yellow solution is stirred for 8 h at 40–50 °C. Under a nitrogen flow, the condenser is replaced with a glass frit connected to a nitrogen-filled two-necked flask. The solution is filtered into the receiving flask, and the filtrate freed of benzene *in vacuo* for 1 h. The receiver containing the alkynylaluminum is diluted with a 3 : 2 (v/v) mixture of dry benzene and cyclopentane. The product in the flask is recrystallized twice to give pure diphenyl(phenylethynyl)aluminum, mp 142–144 °C (turning red at 141°C). The yield of the once crystallized product is 85%, and of the highly pure product, 5.2 g (63% yield).

The infrared spectrum shows a strong band at (mineral oil) 2110 cm^{-1}. The NMR spectrum shows absorptions at (δ, benzene-d_6) 8.37 (4H, q, *ortho* H of Ph_2Al), 7.55 (6H, t, *meta* and *para* H of Ph_2Al), and 6.92–7.3 (5H, m) ppm.

Unsymmetric alkynylaluminum derivatives can be prepared by (*a*) the reaction of sodium acetylides with dialkyl- or diaryluminum halides [92–94]; (*b*) the reaction of terminal acetylenes with aluminum hydrides in donor solvents [95,96] and (*c*) the reaction of terminal acetylenes with triorganoaluminums, especially with trimethyl- or triphenylaluminum [91,97]. Where applicable, the last method gives an unsolvated product under mild, simple conditions.

(3) Diethylaluminum Cyanide, $AlEt_2(CN)$[98]

This preparation should be carried out in a good hood.

$$AlEt_3 + HCN \longrightarrow AlEt_2(CN) + Et\text{-}H \quad (97)$$

To a magnetically stirred solution of triethylaluminum (15.7 g, 0.137 mol) in benzene (40 mL) was added dropwise under ice-cooling to a solution of hydrogen cyanide (3.70 g, 0.137 mol) in benzene (35 mL). Vigorous gas evolution occurred during the addition and ceased when the addition was completed, yielding 3.22 L of ethane (calcd: 3.40 L). Benzene was evaporated and the residue was distilled to give 14.1 g (93%) of diethylaluminum cyanide boiling at *ca* 150 °C/0.07 mmHg (bath temperature 200–220 °C) as a highly viscous oil.

(4) Diethylaluminum Dimethylamide, $AlEt_2(NMe_2)$[99]

$$AlEt_2Cl + LiNMe_2 \longrightarrow AlEt_2(NMe_2) + LiCl \quad (98)$$

Diethylaluminium chloride (10.8 g, 0.09 mol) in petroleum ether (10 mL) was added dropwisely to lithium dimethylamide (0.09 mol) previously prepared at –5 °C from n-butyllithium and dimethylamine in petroleum ether. After removal of lithium chloride and solvent, diethylaluminium dimethylamide, bp 67–69°C (0.12 mm), 8.00 g (68% yield), was obtained. (Found: Al, 20.50; active ethyl groups, 2.05; mol. wt. cryoscop. in benzene, 257. $C_6H_{16}AlN$ calcd: Al, 20.93%; active ethyl groups, 2.00 per mole; mol. wt. dimer, 258.); IR 1055 (s), 665 (s), 560 (w), 520 (w) cm^{-1}.)

(5) Methylaluminum Bis(2,6-di-*tert*-butyl-4-methylphenoxide), AlMe $\{O\text{-}C_6H_2(CMe_3)_2Me\}$[100]

$$AlMe_3 + 2\ HO\text{-}C_6H_2(CMe_3)_2Me \longrightarrow AlMe\{O\text{-}C_6H_2(CMe_3)_2Me\}e\ 2\ Me\text{-}H \quad (99)$$

To a solution of 2,6-di-*tert*-butyl-4-methylphenol (13.22 g, 60 mmol) in degassed hexane (40 mL) was added a 2 M hexane solution of Me_3Al (15 mL, 30 mmol) at room temperature. A white precipitate appeared immediately. After 1 h, this mixture was heated until the precipitate redissolved in hexane. The resulting solution was allowed to stand for 3 h, yielding colorless crystals which were filtered in an argon box. Since the crystals contain some impurities such as 2,6-di-*tert*-butyl-4- methylphenol and inorganic aluminum salts, they were further recrystallized from hexane (45 mL) at –20 °C to give essentially pure methylaluminum bis(2,6-di-*tert*-butyl-4-methylphenoxide) (MAD) (7.83 g, 54% yield).

Properties: ^{1}H NMR ($CDCl_3$) δ 7.04 (4H, s, C_6H_2), 2.28 (6H, s, CH_3), 1.53 (36H, s, $C(CH_3)_3$), –0.35 (3H, s, Al—CH_3); ^{13}C NMR ($CDCl_3$) δ 152.02, 138.19, 127.71, 125.94, 34.94, 31.56, 21.40, –9.09 (Al—CH_3).

(6) Diethylaluminum Ethanethiolate, $AlEt_2(SEt)$[99]

$$A]lEt_3 + EtSH \longrightarrow AlEt_2(SEt) + Et\text{-}H \quad (100)$$

A solution of ethanethiol (6.2 g, 0.1 mol) in cyclohexane (20 mL) was carefully added dropwise to a solution of triethylaluminum (11.4 g, 0.1 mol) in cyclohexane (20 mL) at –78 °C. The reaction temeprature was allowed to warm to 40 °C gradually with evolution of ethane gas. Distillation of the crude products under reduced pressure gave the title compound (12.4 g) in 85% yield: bp 84–85 °C (0.05 mm).

Properties: IR 1255 (s), 980 (vs) cm^{-1}; ^{1}H NMR (C_6H_6) δ 2.30 (2H, q, SCH_2), 0.97 (3H, t, $SCCH_3$), 0.25 (4H, q, $AlCH_2$), 0.10 (6H, $AlCCH_3$).

(7) Lithium Butyl(diisobutyl)aluminum Hydride, $AlH(Bu)Li(i\text{-}Bu)_2$[101]

$$AlH(i\text{-}Bu)_2 + BuLi \longrightarrow AlH(Bu)Li(i\text{-}Bu)_2 \quad (101)$$

In a 250 mL flask with a magnetic stirring bar and a rubber septum under nitrogen was placed a 2.8 M hexane solution of diisobutylaluminum hydride (15.0 mL, 42 mmol). THF (42.8 mL) was added and the flask immersed in an ice bath. A 1.6 M hexane solution of *n*-butyllithium (26.2 mL, 42 mmol) was slowly added to the flask with stirring, and the resulting solution was stirred for an additional 30 min to give a solution of the *ate* complex generated from DIBAH and *n*-butyllithium (0.50 M) in THF-hexane. A suspension of the ate complex from DIBAH and *n*-butyllithium in toluene–hexane was prepared in the same manner by substituting toluene for THF. The hydride concentration of the reagent was determined by GLC analysis of reduction products in the reduction of 4-*tert*-butylcyclohexanone. Thus, to a stirred solution of 4-*tert*-butylcyclohexanone (310 mg, 2.0 mmol) in THF (4 mL) was added the solution of the ate complex from DIBAH and *n*-butyllithium (0.5 M, an expected value, 2.0 mL) in THF–hexane or toluene–hexane at 0 °C under nitrogen. The reaction mixture was stirred for 30 min at 0 °C and treated with saturated NaCl solution. The organic layer was separated, dried over anhydrous $MgSO_4$, and subjected to GLC. The hydride concentration determined by GLC analysis was corresponded to the expected value within a 3% error range in all cases.

(8) Triethylgallium, $GaEt_3$[102]

$$2\,GaBr_3 + 3\,Et_2Zn \longrightarrow 2\,GaEt_3 + 3ZnBr_2 \tag{102}$$

Diethylzinc was prepared by treating one molar equivalent of anhydrous zinc chloride with two molar equivalents of triethylaluminum. A turbid solution was formed with little heat evolution. Fractional distillation of the mixture through a Vigreux column gave a 77% yield of colorless diethylzinc boiling at 24–25°C under 17 mm of pressure.

In a three-necked flask equipped with a nitrogen inlet, pressure-equalized dropping funnel and a copper-coil reflex condenser and provided with magnetic stirring were placed dry pentane (100 mL) and gallium(III) bromide (29.3 g, 0.095 mol). Over a 90 min period, diethylzinc (18.2 g, 0.148 mol) was slowly introduced. The mixture deposited a precipitate, presumably zinc bromide, with a marked heat evolution. After a 2 h reflux period, the reaction suspension was centrifuged and the separated liquid phase was freed of pentane. Distillation of the residue yielded 10 g (66%) of triethylgallium, bp 43–44 °C (16 mm). A hydrolyzed sample of this product still showed the presence of trace amounts of zinc.

$$2\,GaBr_3 + n\,Et_3Al \longrightarrow 2\,GaEt_3\,(n = 1 \text{ or } 3) \tag{103}$$

(a) In a 1 : 1 molar ratio. Admixture of gallium(III) bromide (13.5 g, 0.044 mol) in dry hexane (50 mL) with triethylaluminum (5.7 g, 0.05 mol) resulted in the exothermic formation of a turbid solution. After 1 h at reflux the hexane was drawn off under reduced pressure and the residue subjected to vacuum distillation. Under 256 mm only about 0.5 mL of distillate boiling from 57–63 °C was obtained. Lowering the pressure to 84 mm. and raising the bath temperature to 190 °C gave no further distillate.

(b) In a 1 : 3 molar ratio. The residue from the distillation attempted above was dissolved in dry hexane (50 mL), and an additional triethylaluminum (11.4 g, 0.10 mol) was introduced. Only a slight heat evolution was noted. After a 2 h reflux period the hexane was removed and the residue was fractionally distilled at 280 mm. of pressure. The colorless triethylgallium was collected at 106–108 °C, 5.6 g (82%).

$$GaCl_3 + 3\,Et_3Al \longrightarrow GaEt_3 \tag{104}$$

(c) Without potassium chloride: Attempts to substitute the cheaper, more conveniently prepared chloride for the gallium(III) bromide led to markedly lower yields even when a 1 : 3 molar ratio was employed. Thus, treatment of gallium(III) chloride (0.10 mol) partially dissolved in dry pentane (100 mL) with triethylaluminum (42 mL, 0.30 mol) over a 30-min. period resulted in the exothermic formation of a clear solution. The solution was heated at the reflux temperature for 1 h and the solvent then evaporated. Distillation of the residue gave a 42% yield of triethylgallium, bp 108–110 °C (300 mm). In a check run of the same size the yield of the gallium alkyl was 48%.

(d) With potassium chloride. To the distillation residue of the preceding experiment were added finely ground potassium chloride (22.5 g, 0.3 mol) (previously dried at 400 °C). The mixture was heated at 125 °C for 1 h with frequent shaking; two liquid phases formed: distillation yielded a further 42% yield of triethylgallium, bp 117–118°C (320 mm), for a total yield of 42 + 42% = 84%.

(9) Tris(cyclopentadienyl)indium, $In(C_5H_5)_3$[103]

$$MeLi + 3\ C_5H_6 \longrightarrow Li(C_5H_5) + Me\text{-}H \qquad (105)$$

$$InCl_3 + 3\ Li(C_5H_5) \longrightarrow In(C_5H_5)_3 + 3LiCl \qquad (106)$$

Methyllithium (1.15 g, 52.2 mmole) in ether (25 mL) was added to freshly distilled cyclopentadiene (4.2 g, 63.7 mmol) in ether (30 mL) at 0 °C, and the mixture was refluxed for 1 h. Indium trichloride (3.60 g, 16.3 mmol) was added to the resultant heavy white precipitate of cyclopentadienyllithium, and the mixture again refluxed (4 h), after which ether was removed under vacuum and the resultant solid dried (0.01 mmHg, 2 h). Benzene (90 mL) was added to this solid and the mixture heated to reflux. The hot solution was decanted from undissolved solid, and on cooling bright yellow crystals of tris(cyclopentadienyl)indium(III) (1.72 g) deposited. Further crystallization raised the final total yield to 65% (3.28 g)

Properties: IR 3060 (vw), 1610 (w), 1354 (s), 1074 (s), 1056 (m), 1044 (s), 988 (s), 972 (s), 905 (s), 884 (s), 855 (s), 834 (s), 819 (m), 799 (m), 785 (s), 748 (s), 739 (s), 622 (m), 602 (s), 339 (s), 316 (m), 282 (m) cm^{-1}; ^{1}H NMR ($CDCl_3$) δ 5.93. Anal. Calcd for $C_{15}H_{15}In$: C, 58.1; H, 4.87; In, 37.0. Found: C, 57.6; H, 4.85; In, 37.4.

References

[1] K. Ziegler, *Organometallic Chemistry*, H. Zeiss, Reinhold, New York, p. 194 (1960).
[2] G. Wilke, *Coordination Polymerization*, J. C. W. Chien, Academic Press, New York (1975).
[3] T. Mole, E. A. Jeffery, *Organoaluminum Compounds*, Elsevier, Amsterdam (1972).
[4] H. Reinheckel, K. Haage, D. Jahnke, *Organometal. Chem. Rev., A*, **4**, 47 (1969).
[5] H. Lehmkuhl, K. Ziegler, H. G. Gellert, *Houben-Weyl, Methoden der Organischen Chemie*, 4th Ed., Vol. XIII, Part 4. Thieme, Stuttgart (1970).
[6] G. Bruno, *The Use of Aliminum Alkalys in Organic Synthesis*, Ethyl Corporation, Baton Rouge LA, USA (1970,1973 and 1980).
[7] E. Negishi, *J. Organometal. Chem. Libr.*, **1**, 93 (1982).
[8] H. Yamamoto and H. Nozaki, *Angew. Chem. Int. Ed. Engl.*, **17**, 169 (1978).
[9] E. Negishi, *Organometallics in Organic Synthesis*, Vol. 1, p. 286, Wiley, New York (1980).

[10] J. J. Eisch, *Comprehensive Organometallic Chemistry*, G. Wilkinson, F. G. A. Stone and E. W. Abel, Vol. 1, p. 555. Pergamon Press, Oxford (1982).
[11] J. R. Zietz, Jr., G. C. Robinson, K. L. Leindsay, *Comprehensive Organometallic Chemistry*, G. Wilkinson, F. G. A. Stone and E. W. Abel, Vol. 7, p. 365. Pergamon Press, Oxford (1982).
[12] G. Zweifel, J. A. Miller, *Org. React.*, **32**, 375 (1984).
[13] K. Maruoka, H. Yamamoto, *Angew. Chem. Int. Ed. Engl.*, **24**, 668 (1985).
[14] K. Maruoka, T. Miyazaki, M. Ando, Y. Matsumura, S. Sakane, K. Hattori, H. Yamamoto, *J. Am. Chem. Soc.*, **105**, 2831 (1983).
[15] F. Sato, S. Sato, M. Sato, *J. Organometal. Chem.*, **122**, C25 (1976); **131**, C26 (1977).
[16] F. Sato, S. Sato, H. Kodema, M. Sato, *J. Organometal. Chem.*, **142**, 71 (1977).
[17] K. Isagawa, K. Tatsumi, Y. Otsuji, *Chem. Lett.*, 1117 (1977).
[18] E. C. Ashby, J. J. Lin, *J. Org. Chem.*, **43**, 2567 (1978).
[19] F. Sato, T. Okikawa, T. Sato, M. Sato, *Chem. Lett.*, 167 (1979).
[20] F. Sato, H. Kodama, Y. Tomuro, M. Sato, *Chem. Lett.*, 623 (1979).
[21] E. C. Ashby, S. A. Noding, *J. Org. Chem.*, **45**, 1035 (1980).
[22) E. Negishi, T. Yoshida, *Tetrahedron Lett.*, **21**, 1501 (1981).
[23] E. C. Ashby, S. A. Noding, *J. Org. Chem.*, **44**, 4364 (1979).
[24] J. F. Lemarechal, M. Ephritikhine, G. Folcher, *J. Organometal. Chem.*, **309**, C1 (1986).
[25] K. Fischer, K. Jonas, A. Mollbach, G. Wilke, *Z. Naturforsch. Sect.*, *B* **39**, 1011 (1984).
[26] K. Maruoka, H. Sano, K. Shinoda, S. Nakai, H. Yamamoto, *J. Am. Chem. Soc.*, **108**, 6036 (1986).
[27] K. Maruoka, K. Shinoda, H. Yamamoto, *Syn. Commun.*, 18, 1029 (1988).
[28] E. Negishi, M. D. Jensen, D. Y. Kondakov, S. Wang, *J. Am. Chem. Soc.*, **116**, 8404 (1994).
[29] E. Negishi, D. E. Van Horn, T. Yoshida, *J. Am. Chem. Soc.*, **107**, 6639 (1985).
[30] H. Hayami, K. Oshima, H. Nozaki, *Tetrahedron Lett.*, **25**, 4433 (1984).
[31] E. Negishi, *Pure & Appl. Chem.*, **53**, 2333 (1981).
[32] S. Sakane, K. Maruoka, H. Yamamoto, *J. Chem. Soc. Jpn.*, 324 (1985).
[33] B. B. Snider, C. P. Cartaya-Marin, *J. Org. Chem.*, **49**, 153 (1984).
[34] Y. Naruse, J. Ukai, N. Ikeda, H. Yamamoto, *Chem. Lett.*, 1451 (1985).
[35] N. Tsuboniwa, S. Matsubara, Y. Morizawa, K. Oshima, H. Nozaki, *Tetrahedron Lett.*, **25**, 2569 (1984).
[36] S. H. Pine, R. Zahler, D. A. Evans, R. H. Grubbs, *J. Am. Chem. Soc.*, **102**, 3270 (1980).
[37] L. Clawson, S. L. Buchwald, R. H. Grubbs, *Tetrahedron Lett.*, **25**, 5733 (1984).
[38] K. B. Strarowieyski, S. Pasynkiewics, M. Skowronska-Ptasinska, *J. Organometal. Chem.*, **90**, C43 (1975).
[39] K. Maruoka, T. Itoh, H. Yamamoto, *J. Am. Chem. Soc.*, **107**, 4573 (1985).
[40] K. Maruoka, M. Sakurai, H. Yamamoto, *Tetrahedron Lett.*, **26**, 3853 (1985).
[41] K. Maruoka, T. Itoh, T. Shirasaka, H. Yamamoto, *J. Am. Chem. Soc.*, **110**, 310 (1988).
[42] K. Maruoka, A. B. Concepcion, N. Hirayama, H. Yamamoto, *J. Am. Chem. Soc.*, **112**, 7422 (1990).
[43] K. Maruoka, A. B. Concepcion, N. Murase, M. Oishi, N. Hirayama, H. Yamamoto, *J. Am. Chem. Soc.*, **115**, 3943 (1993).
[44] K. Maruoka, S. Saito, A. B. Concepcion, H. Yamamoto, *J. Am. Chem. Soc.*, **115**, 1183 (1993).
[45] K. Maruoka, H. Imoto, S. Saito, H. Yamamoto, *J. Am. Chem. Soc.*, **116**, 4131 (1994).
[46] K. Maruoka, I. Shimada, H. Imoto, H. Yamamoto, *Synlett.*, 519 (1994).
[47] K. Takai, M. Sato, K. Oshima, H. Nozaki, *Bull. Chem. Soc. Jpn.*, **57**, 108 (1984).
[48] W. R Leonard, T. Livinghouse, *Tetrahedron Lett.*, **26**, 6431 (1985).
[49] S. Kiyooka, H. Kuroda, Y. Shimasaki, *Tetrahedron Lett.*, **27**, 3009 (1986).
[50] T. Tsuda, T. Hayashi, H. Satomi, T. Kawamoto, T. Saegusa, *J. Org. Chem.*, **51**, 537 (1986).
[51] T. Mukaiyama, K. Tomimori, T. Oriyama, *Chem. Lett.*, 1359 (1985).
[52] D. A. Evans, K. T. Chapman, J. Bisaha, *J. Am. Chem. Soc.*, **106**, 4261 (1984).
[53] D. A. Evans, K. T. Chapman, J. Bisaha, *Tetrahedron Lett.*, **25**, 4071 (1984).
[54] K. Furuta, K. Iwanaga, H. Yamamoto, *Tetrahedron Lett.*, **27**, 4507 (1986).
[55] J. W. Herndon, *J. Org. Chem.*, **51**, 2853 (1986).

[56] S. G. Davies, I. M. Dordor, P. Warner, *J. Chem. Soc., Chem. Commun.*, 956 (1984).
[57] S. G. Davies, I. M. Doedor-Hedgecock, P. Warner, R. H. Jones, K. Prout, *J. Organometal. Chem.*, **285**, 213 (1985).
[58] G. Tsuchihashi, K. Tomooka, K. Suzuki, *Tetrahedron Lett.*, **25**, 4253 (1984).
[59] K. Suzuki, E. Katayama, G. Tsuchihashi, *Tetrahedron Lett.*, **24**, 4997 (1994); **25**, 1817 (1985).
[60] K. Suzuki, K. Tomooka, M. Shimazaki, G. Tsuchihashi, *Tetrahedron Lett.*, **26**, 4781 (1985).
[61] K. Suzuki, T. Ohkuma, M. Miyazaki, G. Tsuchihashi, *Tetrahedron Lett.*, **27**, 373 (1986).
[62] H. Honda, E. Morita, G. Tsuchihashi, *Chem. Lett.*, 277 (1986).
[63] K. Suzuki, E. Katayama, T. Matsumoto, G. Tsuchihashi, *Tetrahedron Lett.*, **25**, 3715 (1984).
[64] K. Tanino, T. Hatanaka, I. Kuwajima, *Chem. Lett.*, 385 (1987).
[65] B. M. Trost, J. W. Herndon, *J. Am. Chem. Soc.*, **106**, 6835 (1984).
[66] K. Takai, I. Mori, K. Oshima, H. Nozaki, *Bull. Chem. Soc., Jpn.*, **57**, 446 (1986).
[67] K. Maruoka, S. Saito, H. Yamamoto, *J. Am. Chem. Soc.*, **117**, 6153 (1995).
[68] R. S. Matthews, D. J. Eickhoff, *J. Org. Chem.*, **50**, 3923 (1985).
[69] T. Suzuki, H. Saimoto, h. Tomioka, K. Oshima, H. Nozaki, *Tetrahedron Lett.*, **23**, 3597 (1982).
[70] W. R. Roush, M. A. Adam, S. M. Peseckis, *Tetrahedron Lett.*, **24**, 1377 (1983).
[71] K. Maruoka, H. Sato, H. Yamamoto, *Chem. Lett.*, 599 (1985).
[72] A. Mori, J. Fujiwara, K. Maruoka, H. Yamamoto, *J. Organometal. Chem.*, **285**, 83 (1985).
[73] K. Ishihara, A. Mori, I. Arai, H. Yamamoto, *Tetrahedron Lett.*, **26**, 983 (1986).
[74] J. Fujiwara, Y. Fukutani, M. Hasegawa, K. Maruoka, H. Yamamoto, *J. Am. Chem. Soc.*, **106**, 5004 (1984).
[75] K. Maruoka, S. Nakai, M. Sakurai, H. Yamamoto, *Synthesis*, 130 (1986).
[76] Y. Fukutani, K. Maruoka, H. Yamamoto, *Tetrahedron Lett.*, **25**, 5911 (1984).
[77] K. Maruoka, Y. Fukutani, H. Yamamoto, *J. Org. Chem.*, **50**, 4412 (1985).
[78] D. B. Miller, *Tetrahedron Lett.*, 989 (1964).
[79] Y. Wakita, T. Yasunaga, M. Kojima, *J. Organometal. Chem.*, **288**, 261 (1985).
[80] E. Paschke, Kirk–Othmer *Encyclopedia of Chemical Technology*, John Wiley, New York, Third Edition, Vol. 16, p. 434 (1981).
[81] G. Crespi, L. Luciani, Kirk–Othmer, *Encyclopedia of Chemical Technology*, John Wiley, New York, Third Edition, Vol. 16, p. 453 (1981).
[82] J. Boor, Jr., *Ziegler–Natta Catalysts and Polymerization*, Academic Press, New York, 1979.
[83] I. Pasquon, U. Giannini, *Catalysis Science and Technology 6*, Ed. J. R. Anderson and M. Boudart, Springer-Verlag, New York, Chapter 2, 1984.
[84] P. Pino, R. Mulhaupt, *Angew. Chem. Int. Ed. Engl.*, **19**, 857 (1980).
[85] F. J. Karol, *Encyclopedia of Polymer Science and Technology, Supplement 1*, p. 120 (1976).
[86] P. Galli, L. Luciani, G. Cecchin, *Angew. Makromol. Chem.*, **94**, 63 (1981).
[87] D. G. Demianiw, *Kirk–Othmer Encyclopedia of Chemical Technology*, John Wiley, New York, Third Edition, Vol. 16, p. 480 (1981).
[88] M. F. Gauthreaux, W. T. Davis, E. D. Travis, *Kirk-Othmer Encyclopedia of Chemical Technology*, John Wiley, New York, Third Edition, Vol. 1, p. 740 (1978).
[89] J. J. Eisch, W. C. Kaska, *J. Am. Chem. Soc.*, **88**, 2213 (1966).
[90] K. Ziegler, H. G. Gellert, H. Lehmkuhl, W. Pfohl, K. Zosel, *Ann.*, **629**, 11 (1960).
[91] J. J. Eisch, W. C. Kaska, *J. Organometal. Chem.*, **2**, 184 (1964).
[92] G. Wilke, H. Muller, *Justus Liebigs Ann. Chem.*, **629**, 222 (1960).
[93] T. Mole, J. R. Surtees, *Aust. J. Chem.*, **17**, 1229 (1964).
[94] H. Demarne, P. Cadiot, *Bull. Soc. Chim. Fr.*, 216 (1968).
[95] P. Binger, *Angew. Chem.*, **75**, 918 (1963).
[96] J. R. Surtees, *Aust. J. Chem.*, **18**, 14 (1965).
[97] T. Mole, J. R. Surtees, *Chem. and Ind.*, 1727 (1963).
[98] W. Nagata, M. Yoshioka, *Tetrahedron Lett.*, 1913 (1966).

[99] T. Hirabayashi, H. Imaeda, K. Itoh, S. Sakai, Y. Ishii, *J. Organometal. Chem.*, **19**, 299 (1969).
[100] K. Maruoka, S. Nagahara, H. Yamamoto, *J. Am. Chem. Soc.*, **112**, 6155 (1990).
[101] S. Kim, K. H. Ahn, *J. Org. Chem.*, **49**, 1717 (1984).
[102] J. J. Eisch *J. Am. Chem. Soc.*, **84**, 3605 (1962).
[103] J. S. Poland, D. G. Tuck, *J. Organometal. Chem.*, **42**, 307 (1972).

17 Group 14 (Si, Sn, Ge) Metal Compounds

T. Takeda, *Tokyo University of Agriculture and Technology*

17.1 Introduction

Organosilicon, organogermanium, and organotin compounds generally have stable tetrahedral four-coordinate structures similar to those of carbon compounds. Polysilanes, polygermanes, and polystannanes, which correspond to acyclic and cyclic alkanes, are known. Unlike the corresponding carbon chemistry, however, compounds containing M═M or M═C (M=Si, Ge, Sn) bonds are unstable and cannot be isolated unless they are kinetically stabilized with bulky substituents. A variety of compounds with M—X bonds (X=halogens, O, N, S and other chalcogens, pseudohalogens, etc.) are thermally stable and readily available in a pure form.

The relative reactivities of organometallics containing group 14 elements are closely associated with the electronegativity and dissociation energy of the M—C bonds of the elements. For example, some organotin compounds such as allylstannanes are easily decomposed to form radical species whereas homolysis of the carbon–silicon bond of allylsilanes is rarely achieved, which reflects the dissociation energy of the respective M—C bonds.

Since compounds containing these elements are thermally stable and generally not susceptible to oxygen or moisture in the air, they can be synthesized by reactions similar to those employed for the preparation of ordinary organic compounds, and handled by all the usual chromatographic and spectroscopic techniques, with some exceptions. However, there is a clear tendency for the stability of the organic compounds of group 14 elements to decrease with increasing atomic number of the elements; organotin compounds are much less stable than the corresponding silicon or germanium compounds.

The preparative methods are common to most of these organometallics. The following reactions for formation of M—C bonds are representative.

(1) The reaction of metal halides or alkoxides with carbanions including alkyllithiums or Grignard reagents (eq (1))

(2) The reaction of anionic species of group 14 elements with carbon electrophiles such as alkyl halides, acetates, and carbonyl compounds (eq (2))

Synthesis of Organometallic Compounds: A Practical Guide. Edited by S. Komiya

$$R_3MX \quad + \quad R'M' \longrightarrow R_3MR' \tag{1}$$

M = Si, Ge, Sn   M′ = Li, MgX′
X = halogen, R″O

$$R_3MM' \quad + \quad R'X \longrightarrow R_3MR' \tag{2}$$

M = Si, Ge, Sn   X = halogen, AcO, etc
M′ = Li, Na, MgX, Cu, etc

(3) Hydrometallation of carbon–carbon multiple bonds (eq (3))

$$R^1R^2C{=}CR^3R^4 \quad + \quad R^5_3MH \xrightarrow[\text{or catalyst}]{\text{initiator}} R^1R^2C(H){-}CR^3R^4(MR^5_3) \tag{3}$$

(4) The reaction of organic halides with metallic group 14 elements (eq (4))

$$M \quad + \quad 2RX \longrightarrow R_2MX_2 \tag{4}$$

M = Si, Ge, Sn

Since the stability of divalent compounds of group 14 elements increases with increasing atomic number of the elements, certain divalent inorganic as well as organic germanium and tin compounds are commercially available and the oxidative addition of organic halides or dienes to such compounds is an alternative method for the synthesis of tetravalent organometallics of germanium and tin, as shown later in this chapter.

In addition, numerous reagents and methods have also been investigated for the preparation of these compounds. The stereoselective preparations of metal enolates, vinylmetals, and allylmetals of silicon and tin are of special interest because they are employed as useful reagents in organic synthesis.

17.2 Organosilicon Compounds

17.2.1 General Aspects [1]

Organosilicon compounds have become important materials with the increase of range of industrial use together with the development as synthetic reagents since the 1970's. Now more than a thousand organosilicon compounds, ranging from simple haloorganosilanes to rather complex molecules, are commercially available and there may be no difficulty in obtaining suitable starting materials for the synthesis of desired organosilicon compounds.

Among the organometallic compounds of group 14 elements, organosilicon compounds have been the most widely used in organic synthesis. This is attributable to their sufficient preservability as synthetic reagents as well as characteristic reactivities summarized below.

17.2.1.1 *Enhanced Nucleophilic Substitution at Silicon*

Silicon is more electropositive than most of the elements which constitute organosilicon compounds such as hydrogen, carbon, oxygen, nitrogen, or halogens. The positively charged silicon atom is more susceptible to attack of nucleophiles than the corresponding carbon. Silicon forms 5- and 6-coordinated complexes. The formation of such higher valent silicon intermediates or transition state in the nucleophilic substitution at silicon contributes to decrease its activation energy (eq (5)) [2]. The hypercoordinate organosilicon species also play an important role in certain synthetic reactions. For instance, it is believed that the oxidative cleavage of silicon–carbon bond of organosilicon compounds with more than one heteroatom substituent is also enhanced by the formation of 5-coordinated intermediate (eq (6)) [3].

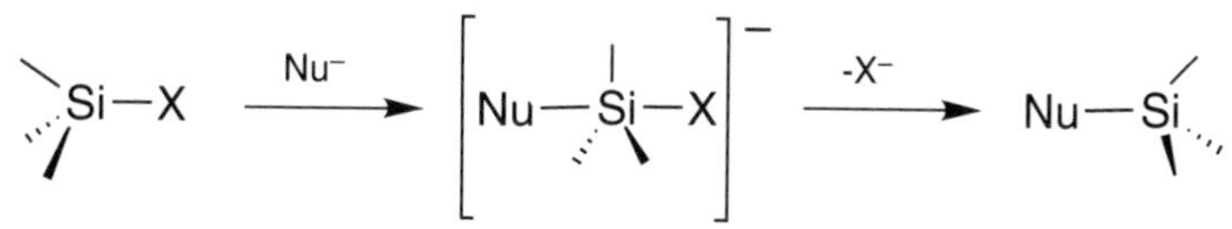

$$R_3Si{-}X \xrightarrow{Nu^-} [Nu{-}SiR_3{-}X]^- \xrightarrow{-X^-} Nu{-}SiR_3 \qquad (5)$$

$$X_2SiRR' \xrightarrow{F^-} [X_2FSiRR']^- \xrightarrow{H_2O_2} [X_2FSiRR'(O{-}OH)] \xrightarrow{-H_2O_2} XFSi(OR)R' \xrightarrow{H_2O} ROH \qquad (6)$$

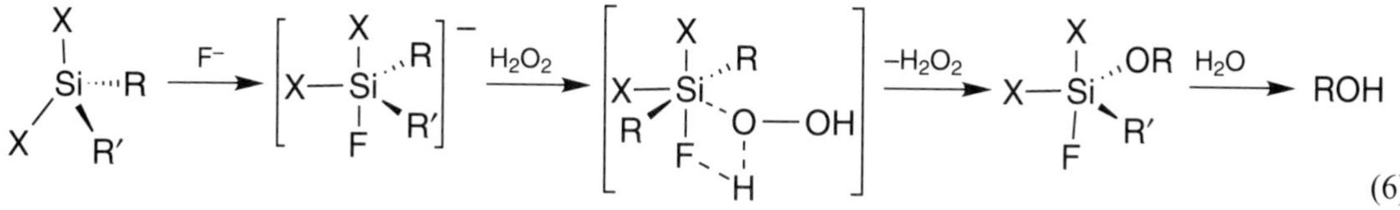

17.2.1.2 *Affinity of Silicon for Heteroatoms*

Silicon forms strong bonds with oxygen (531 kJ / mol), chlorine (471 kJ / mol), and fluorine (808 kJ / mol). In various synthetic reactions, such as the cleavage of silicon–carbon bond by nucleophilic attack of fluoride ion (eq (7)) [4] and the rearrangement of organosilyl group from carbon to oxygen (eq (8)) [5], the formation of stable silicon–heteroatom bonds would serve as the main driving force to effect the reaction.

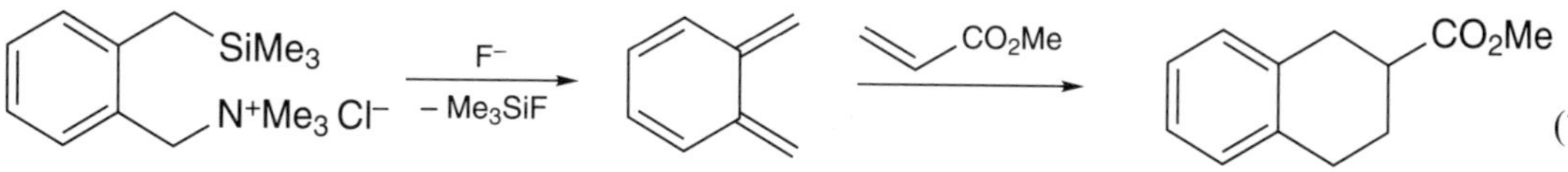

$$o\text{-}(Me_3SiCH_2)C_6H_4CH_2N^+Me_3\,Cl^- \xrightarrow[-\,Me_3SiF]{F^-} \text{o-xylylene} \xrightarrow{CH_2{=}CHCO_2Me} \text{tetrahydronaphthalene-}CO_2Me \qquad (7)$$

$$Pr(OH)C(SiMe_3)CH{=}CH_2 \xrightarrow{BuLi} Pr(OLi)C(SiMe_3)CH{=}CH_2 \xrightarrow{o} Pr(Me_3SiO)C{=}CHCH_2Li \xrightarrow{MeI} Pr(Me_3SiO)C{=}CHPr \qquad (8)$$

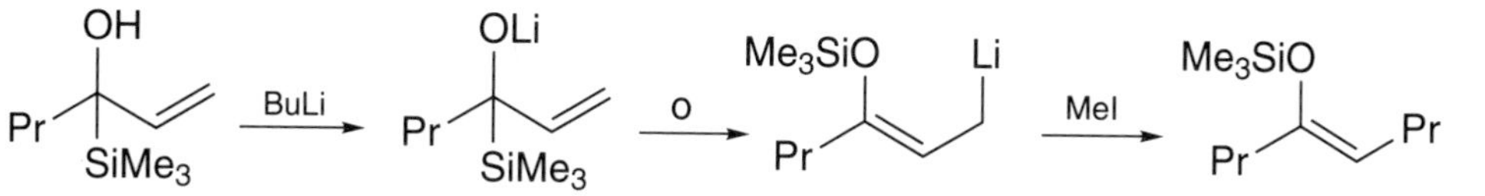

17.2.1.3 *Stabilization of α-Carbanion and β-Carbocation by Silicon*

Carbanions α to silicon are stabilized by the overlap of the filled 2p orbital on carbon with the vacant antibonding orbital of carbon–silicon bond ((σ^*–p)π conjugation) (Scheme 1A). For example, the α-carbanions of allylsilanes or similar compounds are easily generated by base removal of a proton (eq (9)) [6].

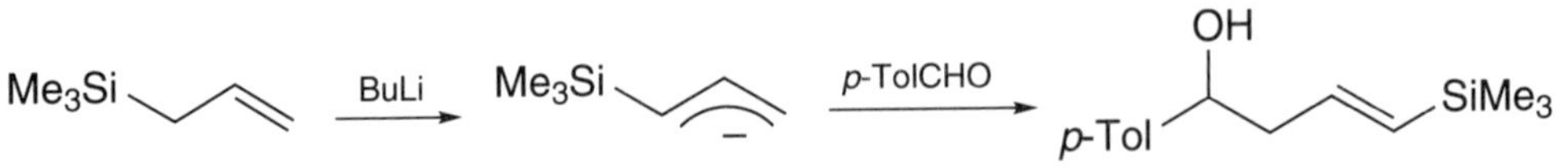

$$Me_3Si\text{-}CH_2CH{=}CH_2 \xrightarrow{BuLi} Me_3Si\text{-allyl}^- \xrightarrow{p\text{-TolCHO}} p\text{-Tol-CH(OH)-CH}_2\text{CH=CH-SiMe}_3 \quad (9)$$

On the other hand, the organosilyl group is capable of stabilizing the carbocation β to silicon. This effect is believed to be attributable to the electron-donating property of the σ-bond between silicon and α-carbon atoms ((σ–p)π conjugation) (Scheme 1B). The reactions of allylsilanes or vinylsilanes with electrophiles are facilitated with this stabilization of intermediate carbocations (eq (10)).

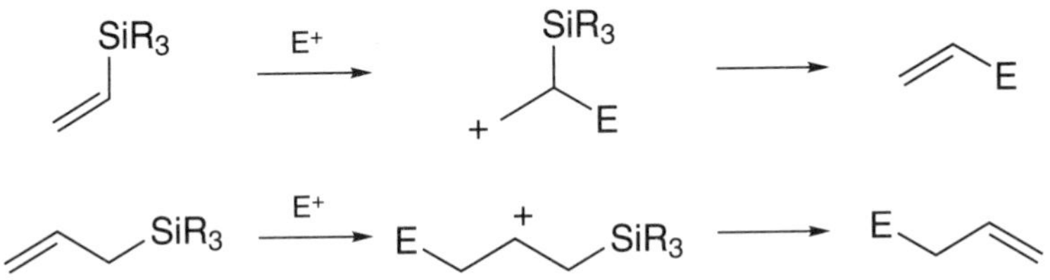

$$(10)$$

All these characteristics of organosilicon compounds are more or less common to other organometallics of group 14 elements, and it is instructive to study their reactions.

17.2.2 Use of Organosilicon Compounds in Organic Synthesis

Organosilicon compounds are employed as carbon nucleophiles, reagents for the transformation of functional groups, reducing agents, and protective groups in organic synthesis. The followings summarize the principal synthetic applications of typical organosilanes. For further details, there are numerous monographs and reviews on the synthetic application of these compounds [7].

17.2.2.1 Allylsilanes

The reactions of trialkylallylsilanes with electrophiles proceed with allylic rearrangement. For example, the treatment with protic acids, such as hydrochloric acid or boron trifluoride /acetic acid, gives the corresponding olefins (eq (11)) [8].

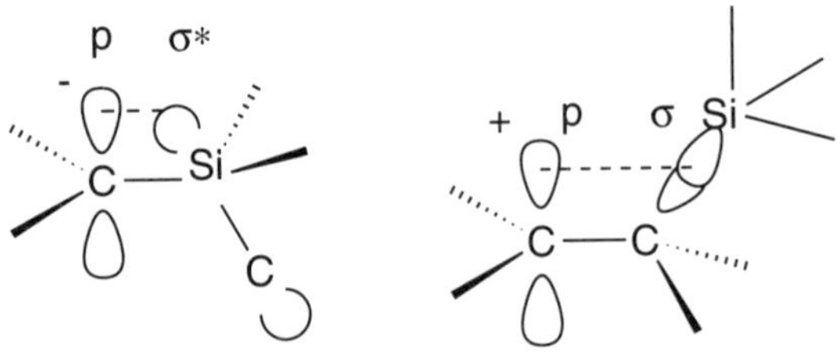

1-A 1-B

Scheme 17.1

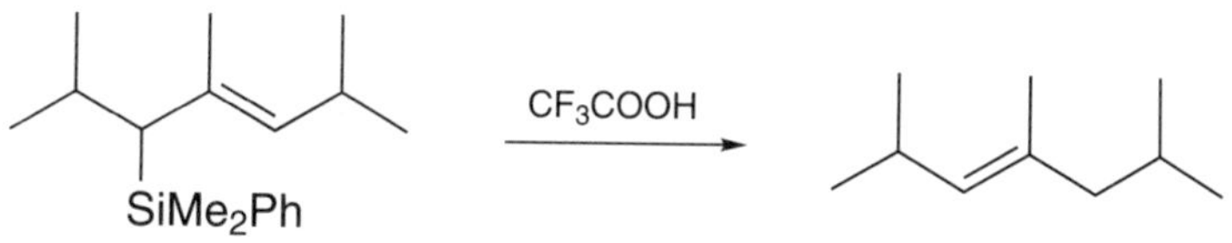

(11)

In the presence of titanium(IV) chloride or aluminum chloride, carbonyl compounds such as aldehydes, ketones, acetals, and acid halides are allylated with allylic silanes (eq (12)) [9]. The conjugate addition to enones is also reported (eq (13)) [10].

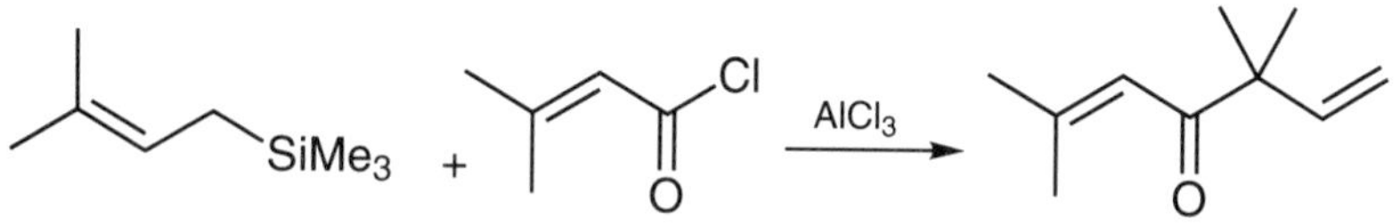

(12)

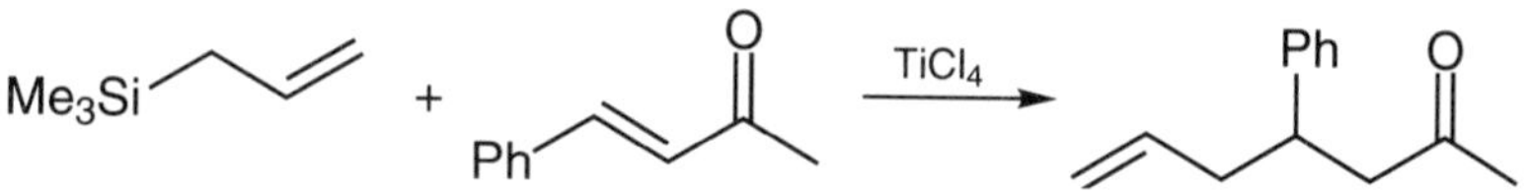

(13)

With fluoride ion allyltrifluorosilanes form 5-coordinated allylsilicates which are highly nucleophilic to react with aldehydes, ketones, and imines with high regio- and diastereoselectivity (eq (14)) [11].

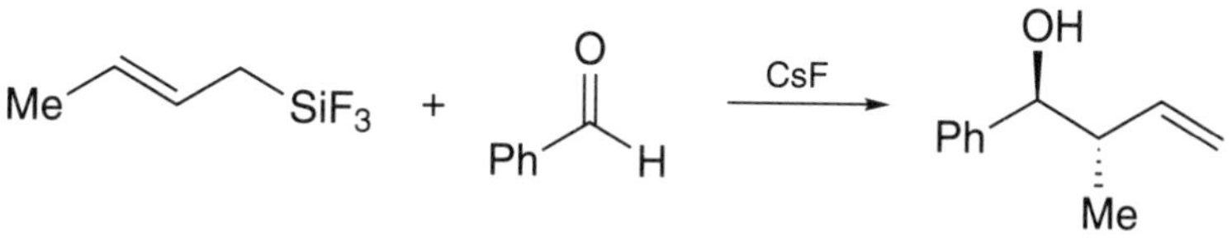

(14)

17.2.2.2 Alkenyl, Alkynyl, and Arylsilanes

Similarly to allylic silanes, alkenylsilanes react with a variety of electrophiles. The *anti* addition of halogens to alkenylsilanes followed by the *anti* elimination of halosilanes yields alkenyl halides with inversion of configuration (eq (15)) [12].

(15)

Although the aluminum chloride or tin(IV) chloride-promoted reactions of alkenylsilanes with acid halides or highly reactive alkyl halides such as chloromethyl alkyl ethers or the corresponding sulfides afford the Friedel–Crafts type products, it is noted that these reactions are not always stereospecific (eq (16)) [13]. The similar reactions of alkynylsilanes with electrophiles have been studied (eq (17)) [14].

The acid-catalyzed isomerization of α-silylepoxides formed by the oxidation of alkenylsilanes affords the corresponding carbonyl compounds, in which the

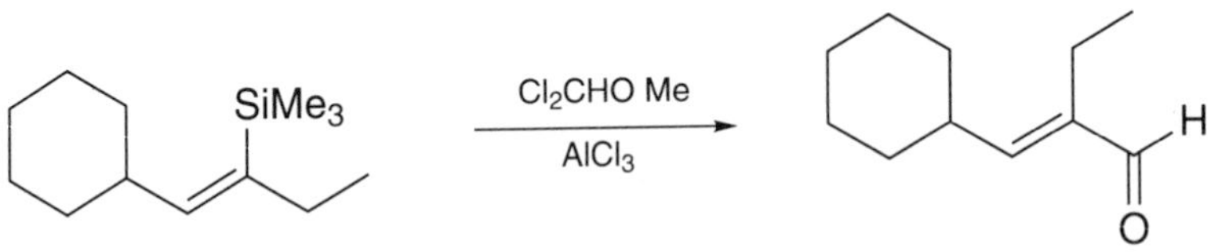

(16)

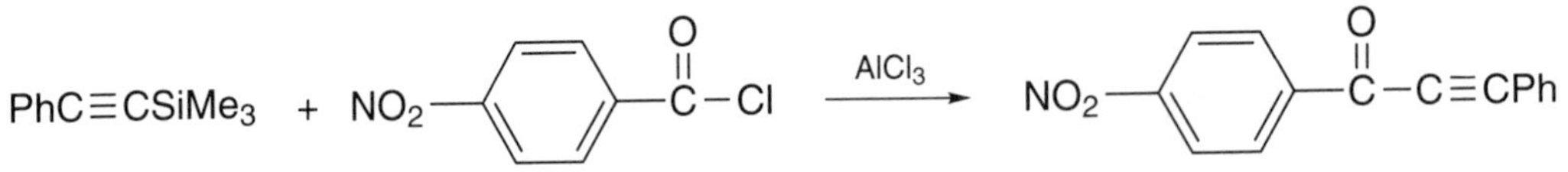

$$PhC{\equiv}CSiMe_3 + NO_2C_6H_4COCl \xrightarrow{AlCl_3} NO_2C_6H_4CO{-}C{\equiv}CPh \quad (17)$$

trimethylsilyl group substituted carbon is selectively converted to carbonyl group due to the enhanced cleavage of the oxygen–carbon bond β to silicon (eq (18)) [15].

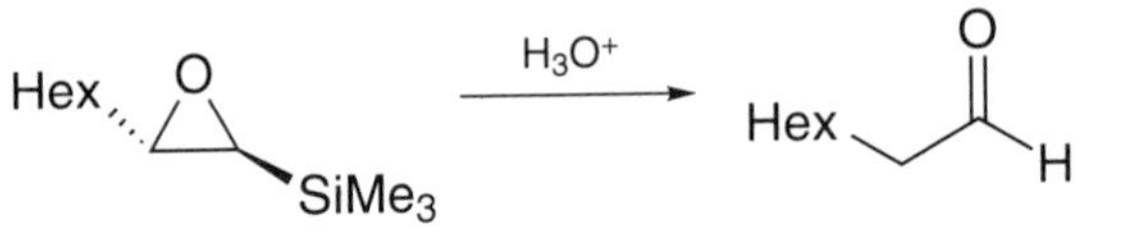

(18)

Although alkenylsilanes undergo palladium-catalyzed coupling reactions with aryl iodides (eq (19)) [16], the similar reactions of 5-coordinated silicates proceeds under milder conditions to give better yields (eq (20)) [17].

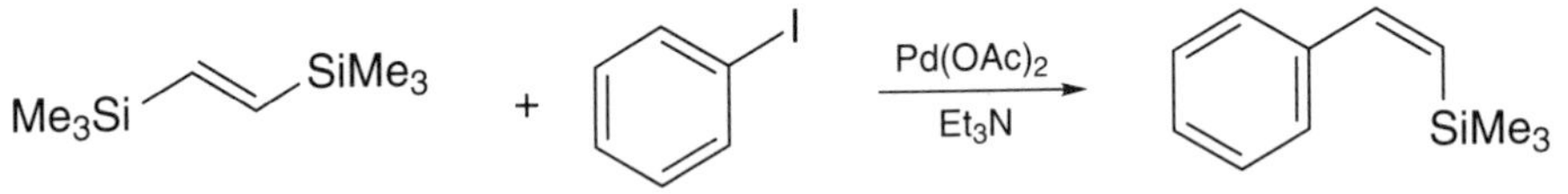

(19)

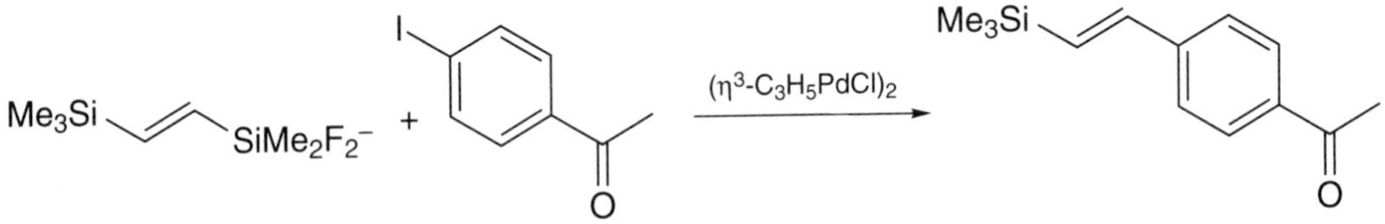

(20)

Arylsilanes also react with various electrophiles to give the corresponding substitution products (eq (21)) [18]. These reactions are superior to conventional electrophilic aromatic substitution in that the displacement occurs at the carbon attached to silicon and is not affected by directing groups on an aromatic ring.

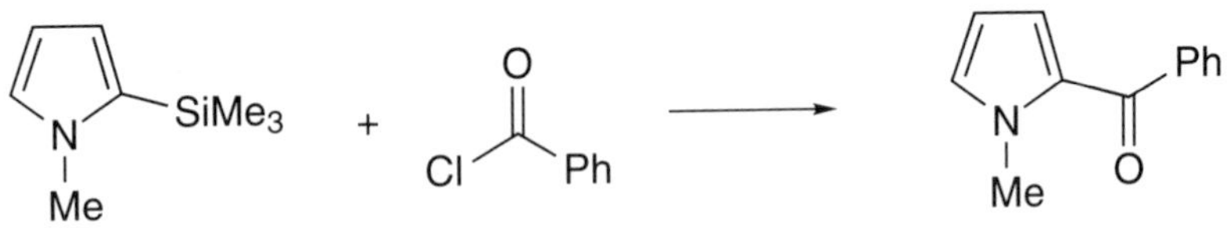

(21)

17.2.2.3 Enol Silyl Ethers and Related Compounds

As the accompanying equations illustrate, the reactions of enol silyl ethers are regarded as those of either electron-rich olefins or stable metal enolates. The treatments of enol silyl ethers with halogens, NBS, sulfenyl halides, and nitronium ions afford the corresponding α-halo, alkylthio, and nitro ketones, respectively (eq (22)) [19].

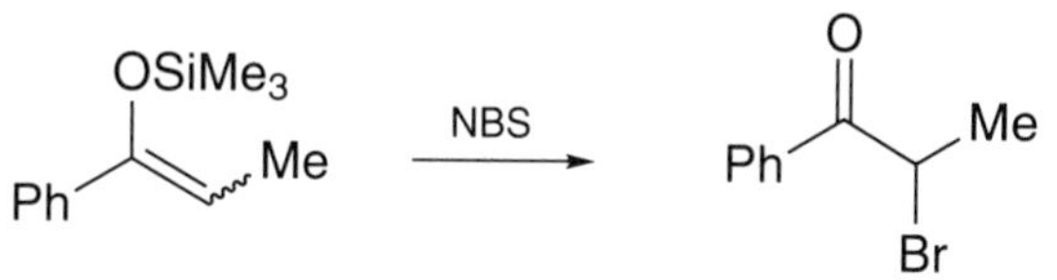

(22)

The titanium(IV) chloride-promoted reactions of enol silyl ethers with aldehydes, ketones, and acetals, known as Mukaiyama reaction, are useful as aldol type reactions which proceed under acidic conditions (eq (23)) [20]. Enol silyl ethers also undergo the Michael type reactions with enones or β,γ-unsaturated acetals (eq (24)) [21]. Under similar reaction conditions, enol silyl ethers are alkylated with reactive alkyl halides such as tertiary halides or chloromethyl sulfides (eq (25)) [22], and acylated with acid halides to give 1,3-diketones (eq (26)) [23].

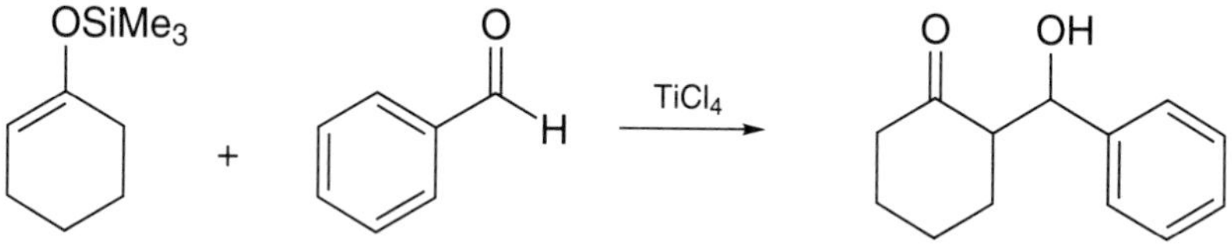

(23)

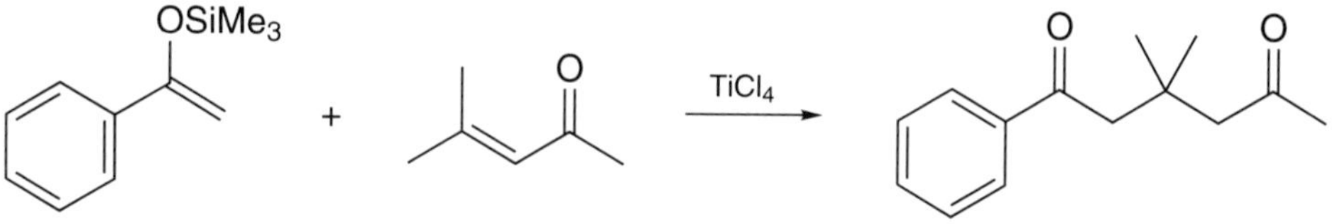

(24)

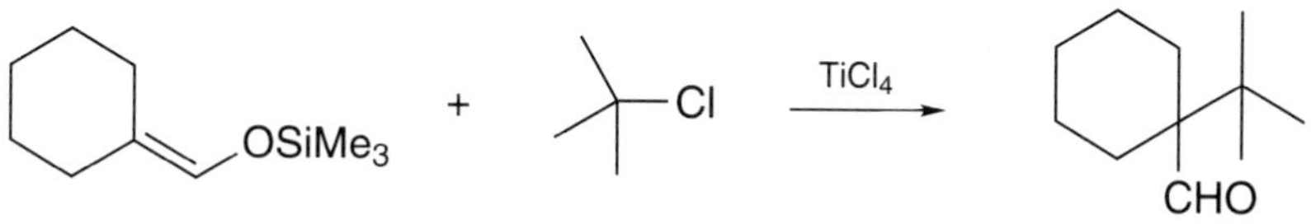

(25)

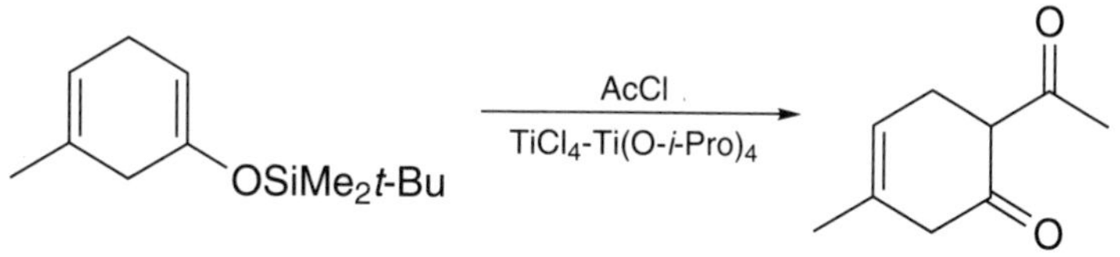

(26)

The active intermediates of these reactions are believed to be titanium enolates formed by the transmetallation with titanium(IV) chloride. Alkylation of enol silyl ethers is also effected by use of benzyltrimethyl ammonium fluoride, in which quaternary ammonium enolates are produced as intermediates (eq (27)) [24].

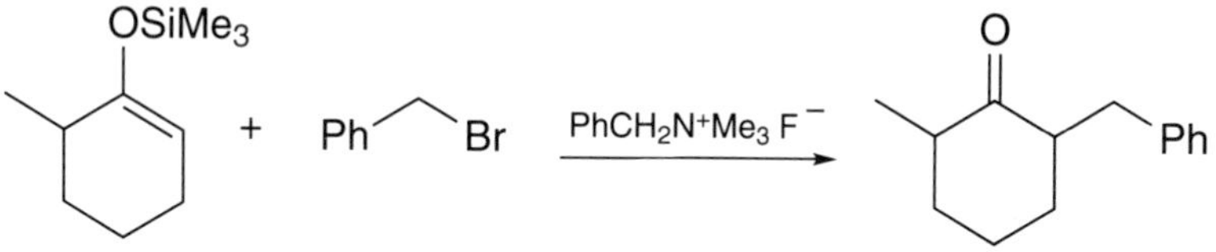

(27)

Enol silyl ethers are also synthetically important as precursors of lithium enolates. The treatment of enol silyl ethers with methyllithium affords the corresponding lithium enolates regiospecifically (eq (28)) [25].

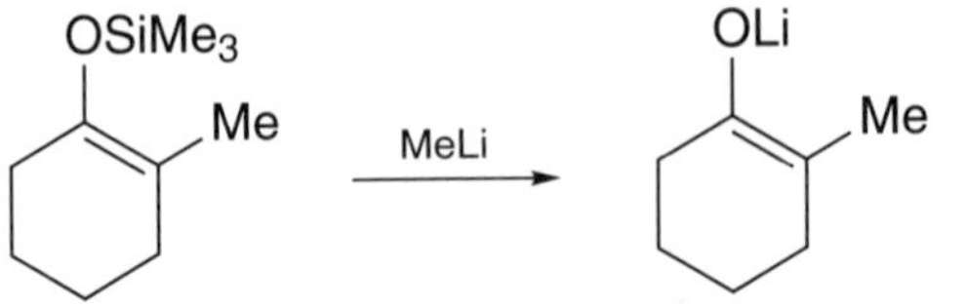

(28)

A variety of silyl enol ethers and silyl dienol ethers are employed in cycloadditions as electron-rich olefins. Regioselective Diels–Alder reactions using these functionalized dienes are used in the syntheses of various natural products (eq (29)) [26].

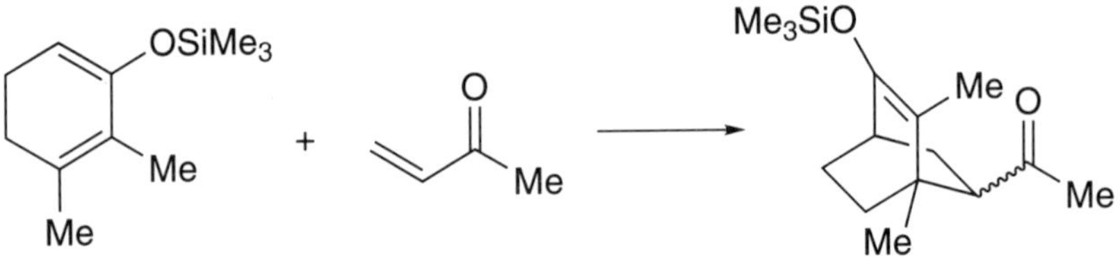

(29)

The oxidation of enol silyl ethers with various oxidizing agents constitutes a convenient methods for the preparation of α,β-unsaturated ketones (DDQ, $Pd(OAc)_2$, trityl BF_4^-), α-hydroxy ketones (MCPBA, LTA, OsO_4), and 1,4-diketones (Ag_2O) (eq (30)) [27].

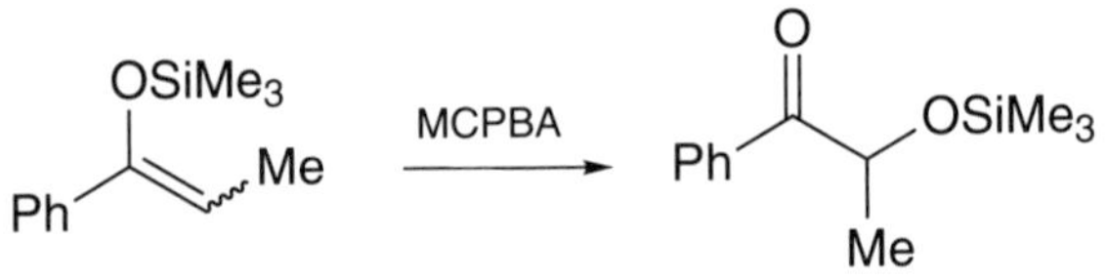

(30)

17.2.2.4 *Trialkylsilanes*

As can be presumed from electronegativity of silicon and hydrogen, silanes serve as hydride donors. Triethylsilane hydrogenates multiple bonds of aldehydes, ketones, and olefins in the presence of trifluoroacetic acid or other catalysts (eq (31)) [28].

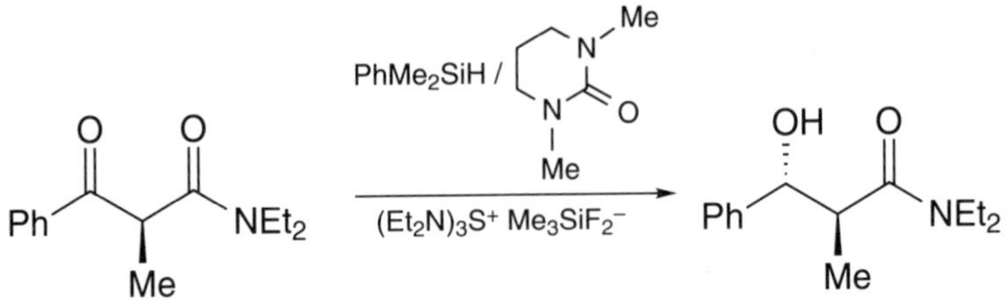

(31)

Homogeneous catalytic hydrosilation of polar multiple bonds such as C=O and C=N bonds is a useful synthetic tool as a better substitute for catalytic hydrogenation. Chiral rhodium complexes are employed for the asymmetric reduction of ketones (eq (32)) [29].

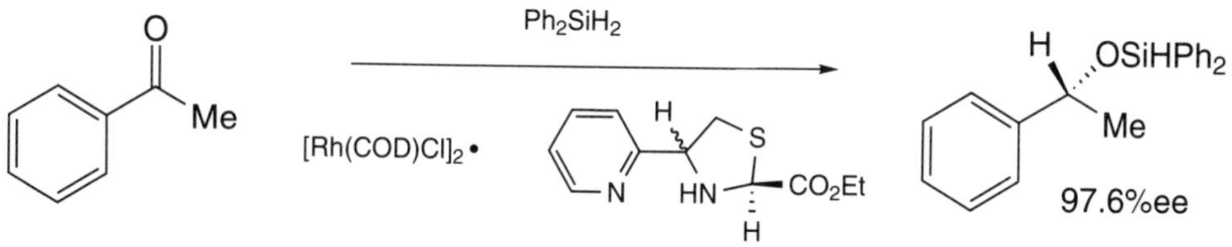

(32)

17.2.2.5 α-Silylcarbanions

Addition of α-silylcarbanions to ketones and subsequent elimination of a silanoate or silanol afford the corresponding olefins (Peterson reaction) (eq (33)) [30]. The use of α-silylcarbanions in olefination of carbonyl compounds offers some advantages over the Wittig reaction that the method can be applied to highly enolizable ketones and both *cis* and *trans* isomers are stereoselectively formed depending on the reaction conditions employed for the elimination step (eq (34)) [31].

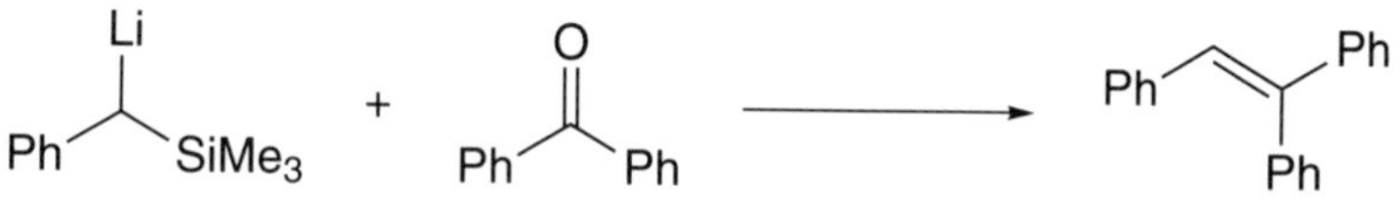

(33)

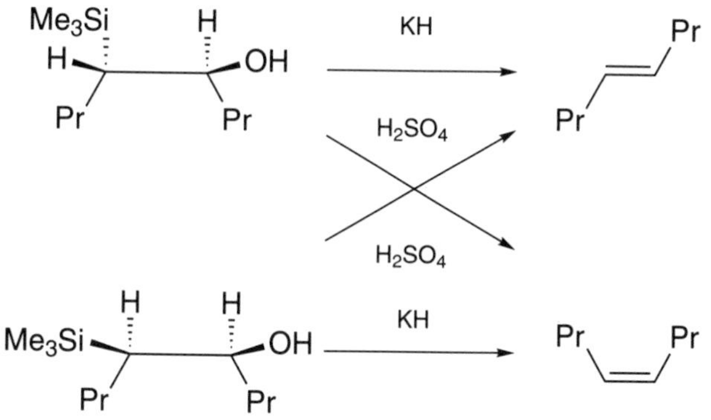

(34)

17.2.2.6 Organosilicon Compounds Containing Heteroatom Substituents and Related Compounds

By the silation of acidic hydrogens of conventional synthetic reagents affords various organosilicon compounds, which are employed for interconversion of functional groups. In certain reactions, diazomethane can be substituted for trimethylsilyldiazomethane which is thermally stable and can be handled safely (eq (35)) [32].

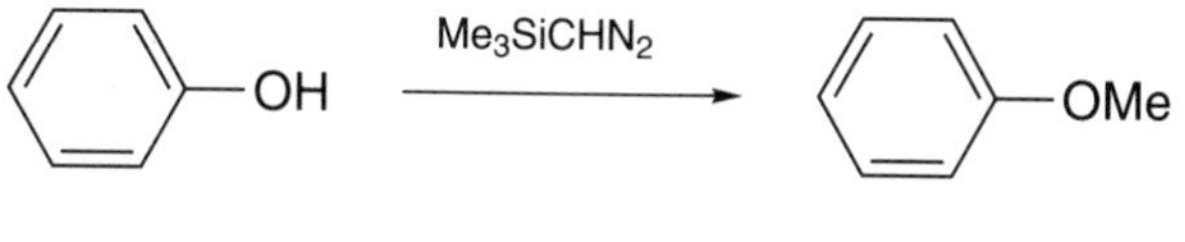

(35)

Trimethylsilyl cyanide reacts with ketones and aldehydes in the presence of a Lewis acid catalyst to afford trimethylsilyl ethers of cyanohydrins (eq (36)) [33].

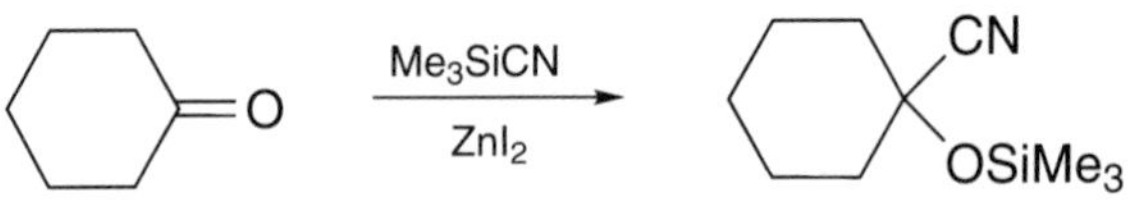

(36)

Trimethylsilyl sulfides are useful reagents for the synthesis of various organosulfur compounds such as thioacetals (eq (37)) [34] and thiolesters (eq (38)) [35].

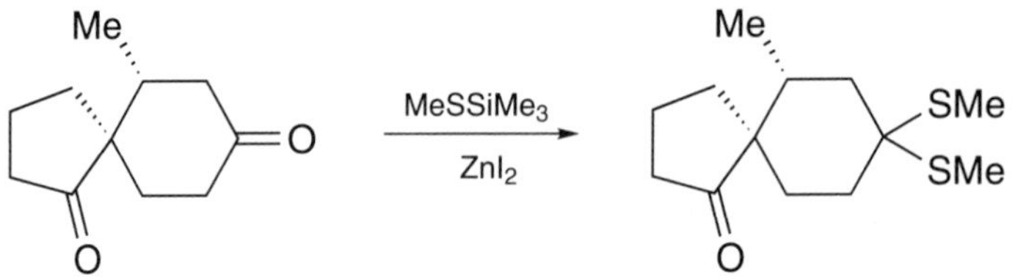

(37)

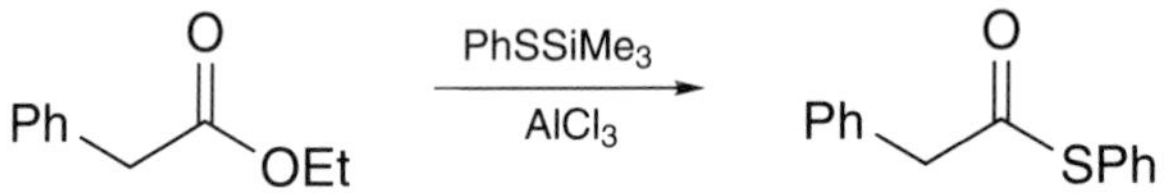

(38)

The C—O bonds of ethers, acetals, and esters are cleaved with iodotrimethylsilane (eq (39)) [36]. Various halosilanes, silylamines, and N,O-bis(trimethylsilyl)acetamide are used not only for the protection of alcohols, amines, and carboxylic acids but also for the conversion of them to the volatilizable compounds for GC analysis (eq (40)) [37]. Silyltriflates act as silating agents as well as Lewis acids (eq (41)) [38].

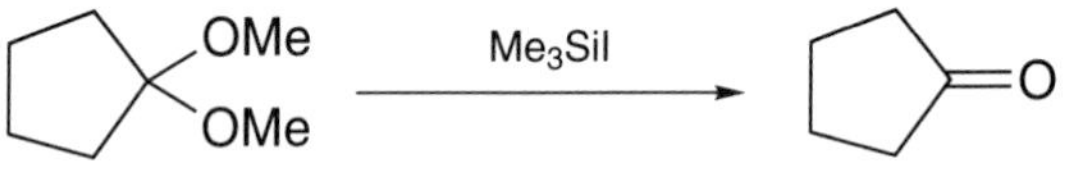

(39)

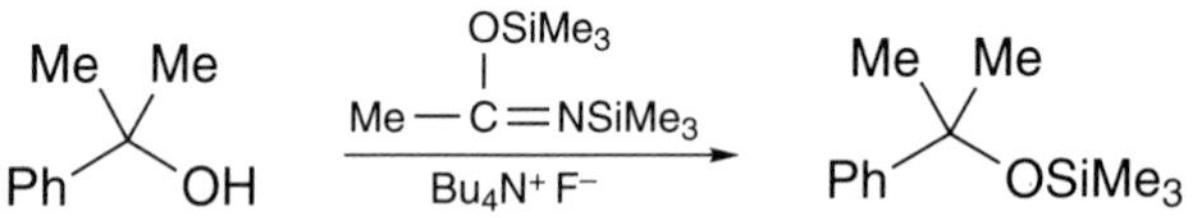

(40)

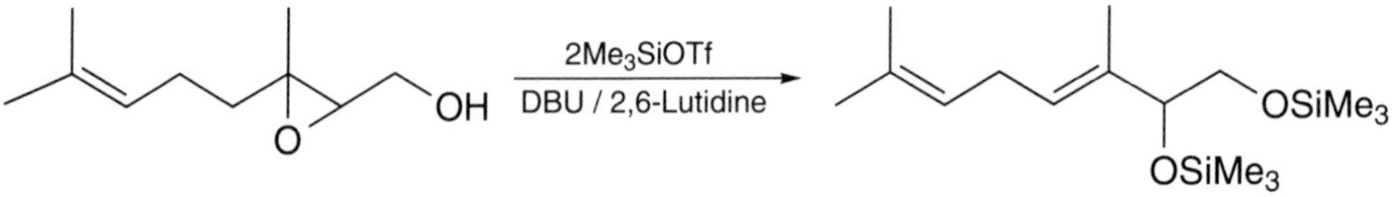

(41)

17.2.3 Preparation of Organosilicon Compounds

17.2.3.1 Haloorganosilanes

As the reaction of halosilanes with organometallic compounds generally proceeds stepwise, haloalkylsilanes with desirable number of alkyl substituents can be synthesized by the slow addition of suitable equivalents of Grignard reagents to polyhalosilanes (inverse addition method) or addition of alkyl halides to a mixture of polyhalosilanes and magnesium metal where the Grignard reagents react with polyhalosilanes immediately as they are formed (*in situ* method) (eq (42)) [39].

$$MeSiCl_3 + PhMgCl \longrightarrow MePhSiCl_2 \quad (42)$$

An alternative method for the preparation of haloalkylsilanes is the treatment of hydrosilanes, disilanes, allylsilanes, or arylsilanes. Iodosilanes are also prepared by the treatment of siloxanes with iodine and aluminum powder (eq (43)) [40].

$$Me_3SiOSiMe_3 + I_2 \xrightarrow{Al} 2\ Me_3SiI \quad (43)$$

Fluorosilanes are obtained by the treatment of corresponding chlorosilanes with metal fluoride such as zinc fluoride, copper(II) fluoride, or antimony(III) fluoride (eq (44)) [11].

$$MeCH{=}CHCH_2SiCl_3 \xrightarrow{SbF_3} MeCH{=}CHCH_2SiF_3 \quad (44)$$

17.2.3.2 Tetraorganosilicon Compounds

Symmetrical tetraorganosilanes are synthesized by the reaction of silicon(IV) chloride with excess Grignard reagents or alkyllithiums (eq (45)) [41]. Similarly the combinations of suitable haloorganosilanes and organometallic compounds afford various unsymmetrical tetraorganosilanes (eq (46)) [42].

$$SiCl_4 + 4PhLi \longrightarrow Ph_4Si \quad (45)$$

$$Me_3SiCl + BuC{\equiv}CLi \longrightarrow BuC{\equiv}CSiMe_3 \quad (46)$$

Since the preparation of anionic species or organosilicon compounds is largely restricted by its reaction conditions, the synthetic utility of silyl anions for the synthesis of organosilicon compounds is limited. However silyllithiums possessing one or more phenyl groups on a silicon atom are readily prepared by the reductive metallation of the corresponding halosilanes with lithium. Allylsilanes are synthesized by the reaction of allyl acetates with the cuprate prepared from such a reagent (eq (47)) [43].

$$Me_2C{=}CHCH(OAc)CH_3 \xrightarrow{(PhMe_2Si)_2CuCNLi_2} Me_2C{=}CHCH(SiMe_2Ph)CH_3 \quad (47)$$

17.2.3.3 Enol Silyl Ethers and Related Compounds

Enol silyl ether is one of the most useful organosilicon reagents, and various methods for the preparation from a variety of precursors have been investigated. The most widely used method is silation of enols or enolates of ketones or aldehydes with trialkylchlorosilanes. The reaction of ketones with triethylamine and chlorotrimethylsilane in DMF affords the thermodynamic equilibrium mixtures of enol silyl ethers (eq (48)) [44]. The use of silyl triflates instead of chlorosilanes generally shortens the reaction time and permits the preparation of some enol silyl ethers which are difficult with halosilanes (eq (49)) [45]. Trialkylsilyl triflates are also employed for the syntheses of enol silyl ethers of esters and S-alkyl thiol esters (eq (50)) [46].

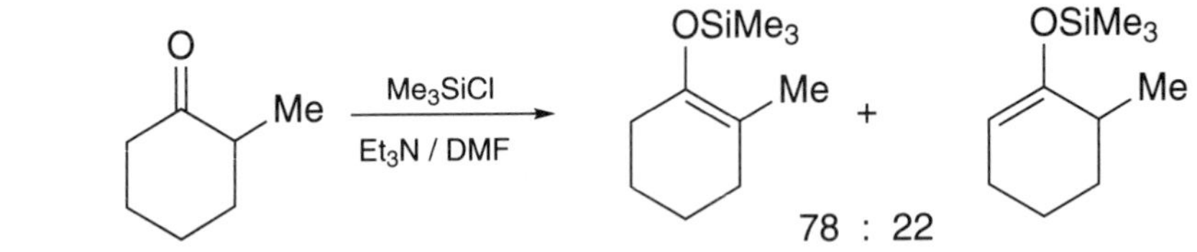

O
Me
Me_3SiCl
Et_3N / DMF
$OSiMe_3$
Me
+
$OSiMe_3$
Me
78 : 22

(48)

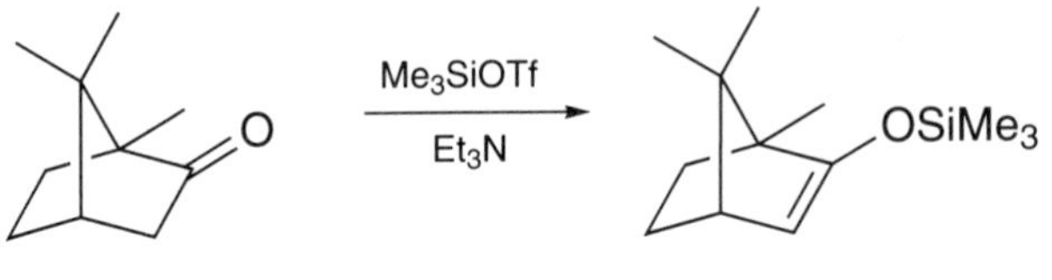

O
Me_3SiOTf
Et_3N
$OSiMe_3$

(49)

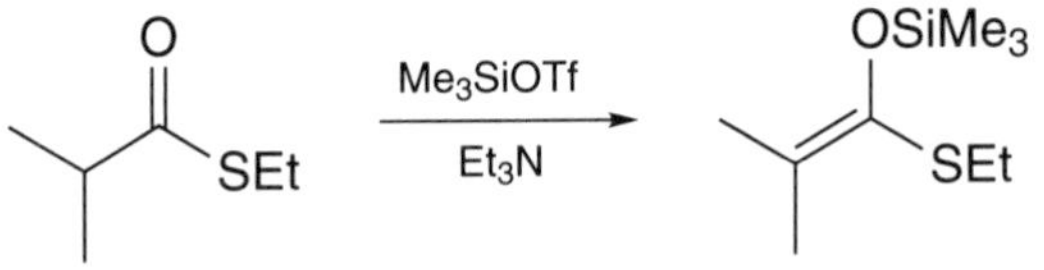

O
SEt
Me_3SiOTf
Et_3N
$OSiMe_3$
SEt

(50)

Enolates prepared by the treatment of ketones with strong bases such as lithium diisopropylamide are trapped with trialkylchlorosilanes to give the kinetically controlled enol silyl ethers (eq (51)) [44].

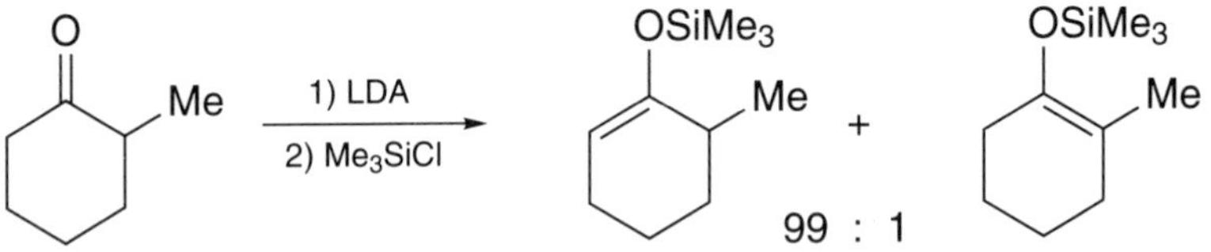

O
Me
1) LDA
2) Me_3SiCl
$OSiMe_3$
Me
+
$OSiMe_3$
Me
99 : 1

(51)

Not only O-silation but also C-silation would take place in the similar reactions of carboxylic esters. The proportion of O- and C-silations is dependent on the structures of esters and alkyl substituents of trialkylhalosilanes (eq (52)) [47].

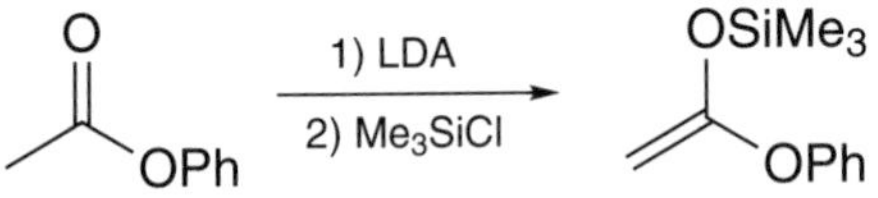

O
OPh
1) LDA
2) Me_3SiCl
$OSiMe_3$
OPh

(52)

Various other preparative methods, such as the trap of enolates formed by dissolving metal reduction of α,β-unsaturated ketones or α-halo ketones (eq (53)) [48], the 1,4-hydrosilation of α,β-unsaturated ketones (eq (54)) [49], and the addition of organometallic compounds to α-silyl ketones followed by the Brook rearrangement (eq (55)) [50], have been investigated.

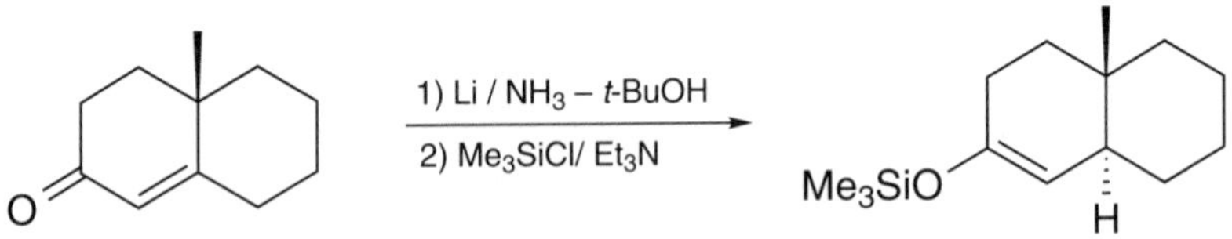

O
1) Li / NH_3 – t-BuOH
2) Me_3SiCl/ Et_3N
Me_3SiO
H

(53)

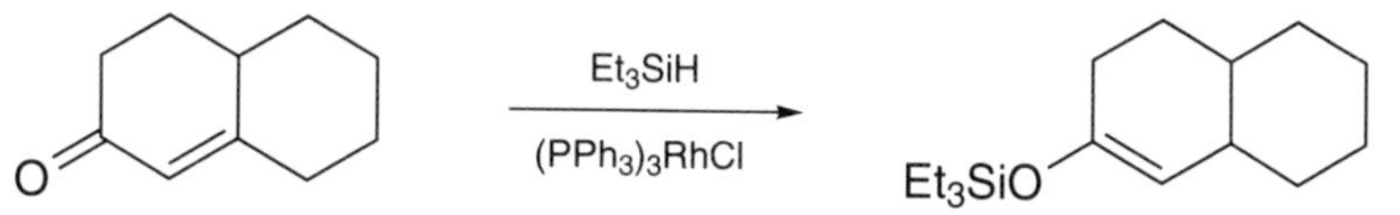

$$\xrightarrow[(PPh_3)_3RhCl]{Et_3SiH} Et_3SiO \qquad (54)$$

$$+ \ Li\diagdown\diagup SO_2Ph \longrightarrow OSiMe_3 \qquad (55)$$

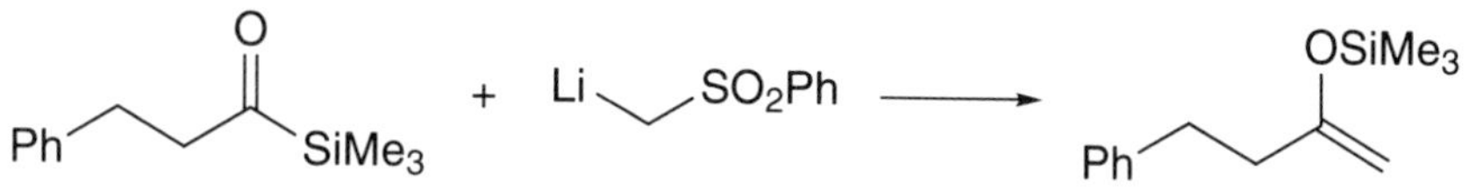

17.2.3.4 Organosilicon Compounds with Silicon-Heteroatom Bonds

Since trialkylsilanols are generally unstable and tend to form siloxanes by the intermolecular dehydration (eq (56)) [51], their preparation was performed by the careful hydrolysis of halosilanes in a neutral medium (eq (57)) [52].

$$2Me_3SiCl \xrightarrow[Me_2NPh]{H_2O} 2Me_3SiOH \longrightarrow Me_3SiOSiMe_3 \qquad (56)$$

$$Et_3SiCl \xrightarrow{NaOH} Et_3SiOH \qquad (57)$$

Preparation of alkoxysilanes is carried out by the treatment of alcohols with trialkylhalosilanes and tertiary amine (eq (58)) [53]. Trialkylsilyltriflates are also employed for the silation, in which N,N-dimethylaminopyridine is frequently used as a catalyst.

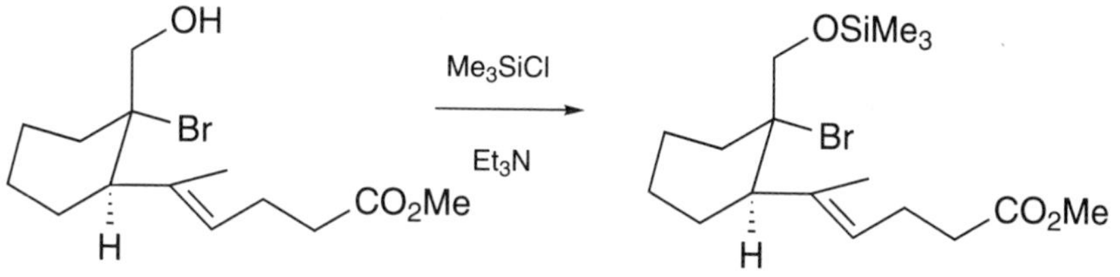

$$\xrightarrow[Et_3N]{Me_3SiCl} \qquad (58)$$

In a similar manner, trialkylsilyltriflates and acyloxysilanes are synthesized by the reaction of trialkylchlorosilanes with trifluoromethane sulfonic acid and carboxylic acids, respectively (eq (59)) [54]. Silyltriflates are also prepared by the dehydrogenation of trialkylsilanes with trifluoromethanesulfonic acid or dealkylation of tetraalkylsilanes with trifluoromethanesulfonic acid (eq (60)) [55].

$$t\text{-}BuMe_2SiCl + CF_3SO_3H \longrightarrow t\text{-}BuMe_2SiOSO_2CF_3 + HCl \qquad (59)$$

$$Me_4Si + CF_3SO_3H \longrightarrow Me_3SiOSO_2CF_3 + CH_4 \qquad (60)$$

Trialkylsilylamines and hexaalkyldisilazanes are obtained by the treatment of the corresponding amines or metal amides with trialkylchlorosilanes (eq (61)) [56]. The reactions of trialkylchlorosilanes with magnesium or lead thiolates afford trialkylsilyl sulfides (eq (62)) [57].

$$Me_3SiCl + Et_2NH \longrightarrow Me_3SiNEt_2 \tag{61}$$

$$2Me_3SiCl + (EtS)_2Pb \longrightarrow 2EtSSiMe_3 + PbCl_2 \tag{62}$$

17.2.3.5 Some Other Synthetically Useful Organosilicon Compounds

Although trialkylsilanes are obtained by the lithium aluminum hydride-reduction of trialkylhalosilanes, they are conveniently prepared by the alkylation of trichlorosilane with Grignard reagents (eq (63)) [58].

$$Cl_3SiH + 3\ EtMgBr \longrightarrow Et_3SiH \tag{63}$$

Pseudohalogenoids of organosilicons such as trialkylsilyl cyanides, trialkylsilyl perchlorates, and azidotrialkylsilanes are synthesized by the treatments of trialkylchlorosilanes with lithium cyanide (eq (64)) [59], silver perchlorate (eq (65)) [60], and sodium azide (eq (66)) [61], respectively.

$$Me_3SiCl + LiCN \longrightarrow Me_3SiCN \tag{64}$$

$$Ph_3SiCl + AgClO_4 \longrightarrow Ph_3SiClO_4 \tag{65}$$

$$Me_3SiCl + NaN_3 \longrightarrow Me_3SiN_3 \tag{66}$$

α-Silylcarbanions are prepared by reduction of halomethylsilanes with lithium or magnesium (eq (67)) [62], addition of organometallics to alkenylsilanes, or α-metallation of some organosilanes (eq (68)) [63].

$$Me_3SiCH_2Cl \xrightarrow{Mg} Me_3SiCH_2MgCl \tag{67}$$

$$(Me_3Si)_2CH_2 \xrightarrow[THF\text{–}HMPA]{t\text{-}BuLi} (Me_3Si)_2CHLi \tag{68}$$

17.3 Synthesis of Organosilicon Compounds

(1) Triethylsilyl Iodide [64]

$$Et_3SiH + I_2 \longrightarrow Et_3SiI \tag{69}$$

Triethylsilane (0.19 mol) is added dropwise to iodine (0.19 mol) in a three-neck flask fitted with a dropping funnel, nitrogen inlet, and a reflux condenser with a $CaCl_2$ tube. Gentle reflux is maintained by the heat of reaction during the addition (15 min), and hydrogen iodide is evolved. After addition is completed, the reaction mixture is refluxed for 15 min with nitrogen bubbling to remove hydrogen iodide. Further triethylsilane (*ca.* 1g) is added to remove the remaining iodine, and refluxing with passage of nitrogen is continued for 15 min. After addition of a little magnesium powder, the mixture is fractionated under nitrogen with protection from light to give triethylsilyl iodide (85%).

Properties: bp 190.5 °C / 744 mmHg.

With the exception of fluorosilanes, haloorganosilanes are readily hydrolyzed by action of moisture in the air similarly to other heteroatom substituted silanes. So it is desirable to handle and store them under nitrogen or argon. If such compounds are stored in glass vessels, attached glass joints should be greased so as not to become tight by adherence of the siloxane formed by the hydrolysis of silanes. Siloxanes adhered to the glass surface are easily removed by soaking them in a warm aqueous solution of sodium hydroxide or a commercial alkaline cleaning solution.

(2) Tetraallylsilane [65]

$$SiCl_4 + 4\ CH_2{=}CHCH_2MgBr \longrightarrow (CH_2{=}CHCH_2)_4Si \qquad (70)$$

Tetrachlorosilane (0.176 mol) is added to an ethereal (600 ml) solution of allylmagnesium bromide (0.9 mol) at a rate sufficient to maintain gentle reflux. After the addition is completed, the slurry is refluxed for 2 h. The mixture is then cooled to –20 °C and a 10% solution of NH_4Cl (200 ml) is slowly added dropwise with efficient stirring over a period of 2 h. The organic layer is separated and the aqueous layer is extracted with ether (3 × 100 ml). The combined organic layers are dried ($MgSO_4$) and fractionated to give tetraallylsilane (88%).

Properties: bp 86.5 °C / 10 mmHg. ^1H NMR (CCl_4): $\delta = 1.58$ (d, $J = 8$ Hz, 8H), 4.87 (m, 8H), 5.74 (m, 4H).

(3) (1-Methoxyvinyl)trimethylsilane [66]

$$CH_2{=}C(OMe)H \xrightarrow{t\text{-BuLi}} CH_2{=}C(OMe)Li \xrightarrow{Me_3SiCl} CH_2{=}C(OMe)SiMe_3 \qquad (71)$$

Methyl vinyl ether (1.24 mol) in THF (450 ml) cooled to –78 °C is transferred to a flask using a double-ended needle. *t*-Butyllithium (1.0 mol, 2.0 M) is added dropwise over *ca*. 90 min so that the reaction mixture is maintained at a temperature below –70 °C. The mixture is warmed slowly to 0 °C over 3 h and subsequently recooled to –78 °C. Chlorotrimethylsilane (0.78 mol) is added dropwise, and the mixture is warmed slowly to room temperature. After being stirred for an additional 1 h, the mixture is quenched with saturated NH_4Cl. The organic layer is separated, washed with water (12 × 250 ml), dried (K_2CO_3), and distilled to give (1-methoxyvinyl)trimethylsilane (84%).

Properties: bp 102–104 °C / 760 mmHg. ^1H NMR (CCl_4): $\delta = 0.09$ (s, 9H), 3.50 (s, 3H), 4.28 (d, 1H, $J = 2.0$ Hz), 4.50 (d, 1H, $J = 2.0$ Hz). IR (neat) 1590 cm^{-1}.

(4) 1-Methyl-2-(trimethylsiloxy)cyclohex-1-ene [67]

$$\text{2-methylcyclohexanone} \xrightarrow[\text{Et}_3\text{N}]{\text{Me}_3\text{SiCl}} \text{1-methyl-2-(OSiMe}_3\text{)cyclohex-1-ene} \quad (72)$$

Chlorotrimethylsilane (150 mmol) is added dropwise with stirring to a solution of 2-methylcyclohexanone (125 mmol) in triethylamine (300 mmol) and DMF (50 ml). The mixture is heated at 130 °C for 90 h under nitrogen. After cooling, the mixture is diluted with ether (200 ml) and washed with saturated $NaHCO_3$ (200 ml). The aqueous layer is extracted with ether (3 × 200 ml), and the combined extracts are washed *rapidly* with 0.5 M HCl (250 ml), saturated $NaHCO_3$ (2 × 200 ml), and water (200 ml). The organic layer is dried ($MgSO_4$), evaporated in vacuo, and distilled to give a mixture of 1-methyl-2-(trimethylsiloxy)cyclohex-1-ene and 6-methyl-1-trimethylsiloxycyclohex-1-ene in ratio of 88 : 12 (83%).

Properties: bp 82–84 °C / 16 mmHg. ^{1}H NMR ($CDCl_3$): δ = 0.17 (s, 9H), 1.45–1.70 (m, 7H), 1.80–2.15 (m, 4H). IR (neat) 1685, 1250 cm^{-1}.

The preparation can be performed at ambient temperature by use of trimethylsilyl iodide prepared *in situ* from trimethylsilyl chloride and sodium iodide [68].

(5) Triethylsilane [58]

$$HSiCl_3 + 3\ EtMgBr \longrightarrow Et_3SiH \quad (73)$$

A cold ethereal (1200 ml) solution of trichlorosilane (3.0 mol) is added to ethylmagnesium bromide (12.6 mol) over 6 h with cooling and vigorous stirring. The mixture is stirred for 8 h at room temperature and then refluxed for 5 h. Ether is removed from the reaction mixture through an efficient fractionating column and the residue is heated on a steam bath for 10 h. With cooling, the solid residue is hydrolyzed with water (180 mL), followed by concentrated HCl (372 mL). The aqueous layer is separated and extracted with ether (2 × 500 mL). The extracts and product are combined, washed with water, and dried (K_2CO_3). Fractional distillation gives triethylsilane (77.5%).

Properties: bp 107 °C / 733 mmHg.

(6) *t*-(Butyldimethylsiloxy)cyclohexane [69]

$$\text{C}_6\text{H}_{11}\text{–OH} \xrightarrow[\text{pyridine}]{t\text{-BuMe}_2\text{SiOClO}_3} \text{C}_6\text{H}_{11}\text{–OSiMe}_2\text{Bu-}t \quad (74)$$

To a MeCN (5 ml) solution of *t*-butyldimethylsilyl perchlorate (4.0 mmol) is added pyridine (4.8 mmol) slowly. Then cyclohexanol (4.0 mmol) is added to the stirred solution. After being stirred for 1.5 h, the reaction mixture is poured into a small separatory funnel containing pentane (15 mL) and saturated $NaHCO_3$ (15 mL). The pentane layer is extracted several times until the smell of pyridine can no longer be detected in the aqueous phase. Evaporation of the pentane solution affords (*t*-butyldimethylsiloxy)cyclohexane (>95% pure by GC, 99%).

Properties: bp 78 °C/5 mmHg [70]. ^{1}H NMR (CCl_4): δ = 0.02 (s, 6H), 0.88 (s, 9H), 1.47 (m, 10H), 3.58 (br s, 1H).

t-Butyldimethylsilation is also performed by stirring a mixture of alcohol, *t*-butyldimethyl silyl chloride, imidazole (1:1.2:2.5 molar ratio), and dimethylformamide (2 ml/g of the alcohol) [70] or treatment of an alcohol with N-*t*-butyldimethylsilylimidazole in the presence of tetrabutylammonium fluoride [71].

(7) Triisopropylsilyl Triflate [54]

$$[(CH_3)_2CH]_3SiH + CF_3SO_3H \longrightarrow [(CH_3)_2CH]_3SiOSO_2CF_3 \qquad (75)$$

Trifluoromethanesulfonic acid (0.226 mol) is added dropwise to triisopropylsilane (0.242 mol), stirred under an inert atmosphere and cooled to 0 °C in a flask fitted with dropping funnel, gas inlet, exit tubes and thermometer. After the addition is completed, stirring is continued at 22 °C for 16 h. Triisopropylsilyl triflate (97%) is isolated by the distillation through a 30 cm vacuum jacketed Vigreux column.

Properties: bp 83–87 °C / 1.7 mmHg. ^{1}H NMR ($CDCl_3$): δ =1.05–1.6 (m).

(8) Trimethyl(methylthio)silane [72]

$$Me_3SiCl + MeSMgI \longrightarrow Me_3SiSMe \qquad (76)$$

Methanethiol (1.04 mol) is added dropwise to methylmagnesium iodide prepared from methyl iodide (1.0 mol) and magnesium (1.13 g. atom) in ether (900 ml) at 0 °C. After the addition is completed, the reaction mixture is warmed to room temperature and chlorotrimethylsilane (0.5 mol) is added. A slightly exothermic reaction occurs and the reaction mixture is refluxed gently for 4 h. All volatile material is then removed under high vacuum and condensed at –78 °C. Finally the reaction flask is slightly warmed to complete the transfer. The distillation through an efficient fractionating column gives trimethylmethylthiosilane (78%).

Properties: bp 110–111 °C.

(9) Triethylsilyl Cyanide [73]

$$Me_3SiCN \xrightarrow{BuLi} LiCN \xrightarrow{Et_3SiCl} Et_3SiCN \qquad (77)$$

Butyllithium (0.32 mol, 1.6 M) is transferred under nitrogen into a flask containing dry toluene (300 ml). With cooling (0 °C) and stirring, trimethylsilyl cyanide (0.36 mol) is added dropwise over a period of 10 min. After the addition is completed, the reaction mixture is stirred for an additional 15 min. Chlorotriethylsilane (0.30 mol) is added to the LiCN slurry in one lot. The reaction mixture is heated to reflux overnight and then filtered by a coarse fritted disk under nitrogen. The solid is washed with two small portions of cyclohexane. The filtrate is condensed under reduced pressure and distilled to give triethylsilyl cyanide (76%).

Properties: bp 180–183 °C.

Since trialkylsilyl cyanides are highly toxic, preparation of these compounds must be performed in a well ventilated food. The preparation of anhydrous LiCN is also conveniently performed by the treatment of acetone cyanohydrin with lithium hydride [59].

17.4 Organogermanium Compounds

17.4.1 General Aspects [74]

The reactivities of organogermanium compounds are believed to be intermediate between those of the corresponding organosilicon and tin compounds, and hence the synthetic utility of organogermanium compounds has remained almost unexplored. This is also because germanium and its derivatives are far more expensive than the corresponding silicon and tin compounds. Since commercially available organogermanium compounds are limited, organogermanium compounds are inevitably synthesized from inorganic germanium compounds, such as metallic germanium, germanium(IV) oxide, germanium(IV) halides, germanium(II) iodide, which can be purchased from companies of laboratory chemicals as well as of materials for semiconductors; certain tetraorganogermanes and haloorganogermanes are commercially available, but they are very high in price. Organogermanium compounds have several advantages over the corresponding organotin compounds that they are more stable and less toxic, and accordingly they are synthesized and handled without special care.

17.4.2 Use of Organogermanium Compounds for Organic Synthesis

Although germanium–carbon bonds of allyl, alkenyl, and arylgermanium compounds are cleaved by action of electrophiles such as bromine or trifluoroacetic acid to afford the corresponding halides (eq (78)) [75] or hydrocarbons (eq (79)) [76], like organosilicon compounds, the synthetic utility of these reactions are largely limited by the unavailability of starting materials.

$$Ph_4Ge + Br_2 \longrightarrow PhBr + Ph_3GeBr \quad (78)$$

$$MeO-C_6H_4-C{\equiv}CGeEt_3 \xrightarrow{MeOH / H^+} MeO-C_6H_4-C{\equiv}CH \quad (79)$$

Recently the chemistry of organogermanium compounds is becoming of considerable importance in organic synthesis. The following equations illustrate some examples in which organogermanium compounds are employed as synthetic intermediates (eq (80) [77], (81) [78], (82) [79], (83) [80], (84) [81]).

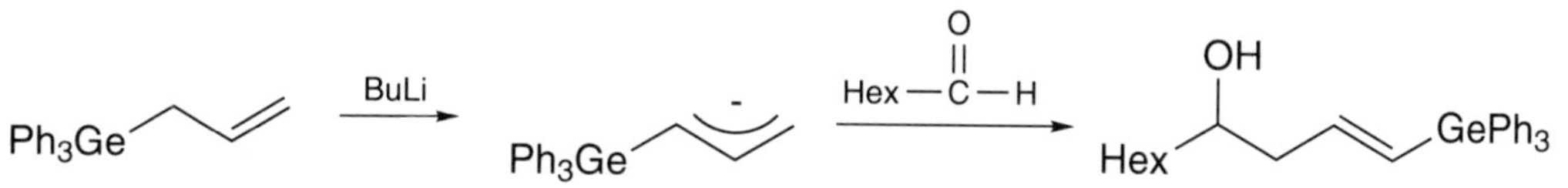

$$Ph_3Ge \quad \xrightarrow{BuLi} \quad Ph_3Ge^{-} \quad \xrightarrow{Hex-C(=O)-H} \quad Hex-CH(OH)\ldots GePh_3 \tag{80}$$

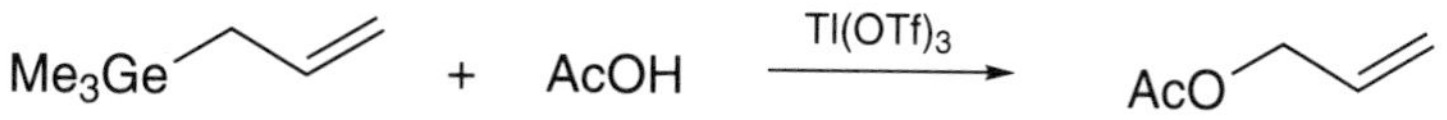

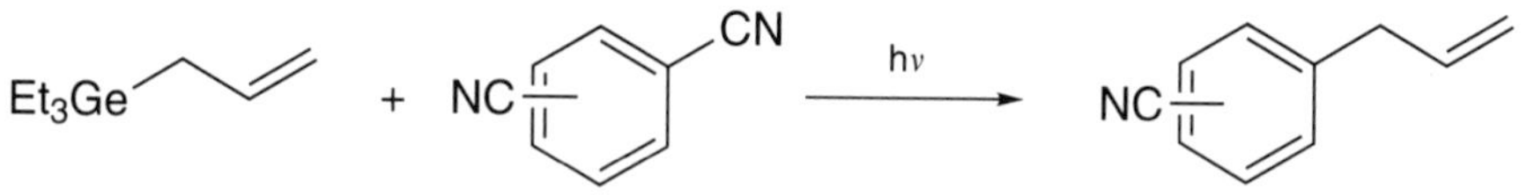

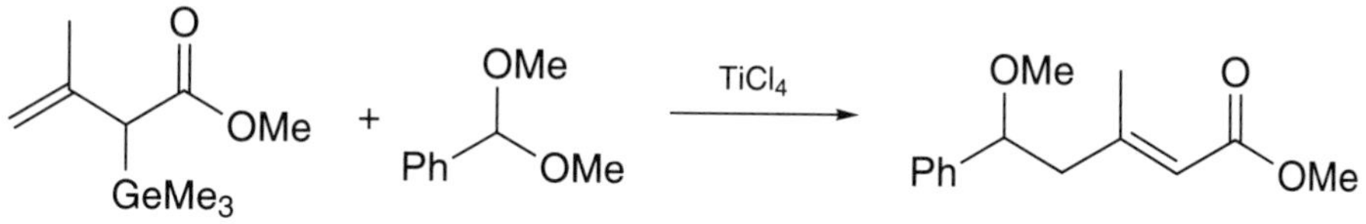

$$Me_3Ge \quad + \quad AcOH \quad \xrightarrow{Tl(OTf)_3} \quad AcO \tag{81}$$

$$Et_3Ge \quad + \quad NC \quad CN \quad \xrightarrow{h\nu} \quad NC \tag{82}$$

$$O \quad OMe \quad GeMe_3 \quad + \quad OMe \quad Ph \quad OMe \quad \xrightarrow{TiCl_4} \quad OMe \quad Ph \quad O \quad OMe \tag{83}$$

$$Me_3Ge \quad O \quad OCMe_3 \quad \xrightarrow[2)\ PhCHO]{1)\ LDA} \quad Me_3Ge \quad Ph \quad OCMe_3 \quad OLi \quad O \quad \longrightarrow \quad Ph \quad O \quad OCMe_3 \tag{84}$$

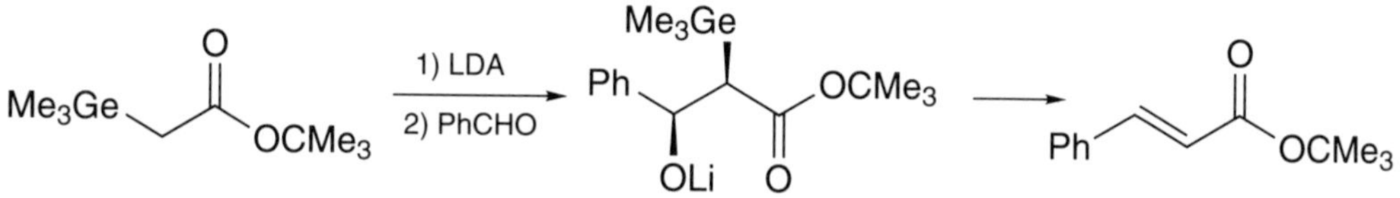

17.4.3 Preparation of Organogermanium Compounds

17.4.3.1 Haloorganogermanium Compounds

The reactions of germanium metal with alkyl halides afford dialkyldihalogermanes (eq (85)) [82]. Alkyltrihalogermanes are obtained by the similar oxidative addition of alkyl halides to germanium(II) iodide (eq (86)) [83].

$$Ge \quad + \quad 2MeCl \quad \xrightarrow{Cu} \quad Me_2GeCl_2 \tag{85}$$

$$GeI_2 \quad + \quad MeI \quad \longrightarrow \quad MeGeI_3 \tag{86}$$

Most trialkylhalogermanes are conveniently prepared by the selective cleavage of one alkyl group of tetraorganogermanes with halogens, hydrogen halides, or acetyl chloride in the presence of aluminum chloride (eq (87)) [84].

$$Me_4Ge \quad + \quad Br_2 \quad \xrightarrow{AlCl_3} \quad Me_3GeBr \quad + \quad MeBr \tag{87}$$

The preparation of trialkyliodogermanes is also performed by the treatment of hexaalkylgermoxanes with hydrogen iodide (eq (88)) [85].

$$Et_3GeOGeEt_3 \quad + \quad 2HI \quad \longrightarrow \quad 2Et_3GeI \tag{88}$$

Since the partial alkylation of germanium(IV) halides dose not proceed stepwise, this reaction cannot be used for the preparation of haloorganogermanes except for those possessing bulky substituents. Although the disproportionation of germanium compounds occurs similarly to that of organotin compounds described later, this reaction is unsuitable as a preparative method for haloorganogermanes because it generally requires very high reaction temperature.

17.4.3.2 *Tetraorganogermanium compounds*

Symmetrical tetraalkylgermanes are synthesized by the treatment of germanium(IV) chloride or bromide with excess Grignard reagents or alkyllithiums (Eq (89)) [86].

$$GeCl_4 + 4CH_2{=}CHMgBr \longrightarrow (CH_2{=}CH)_4Ge \qquad (89)$$

The reactions of organometallic reagents with haloorganogermanes provide general methods for the preparation of unsymmetrically substituted tetraorganogermanes (eq (90)) [87]. Using various carbanions, such as enolates, this method is applied to the preparation of organogermanium compounds with various functional groups (eq (91)) [88].

$$2Me_3GeCl + BrMgC{\equiv}CMgBr \longrightarrow Me_3GeC{\equiv}CGeMe_3 \qquad (90)$$

$$Me_3SiCH_2C(O)OEt \xrightarrow[\text{2) } Me_3GeCl]{\text{1) LDA}} Me_3SiCH(GeMe_3)C(O)OEt \qquad (91)$$

Trialkylgermyl anions are also employed for the synthesis of organogermanium compounds (eq (92)) [89]. As is often observed in the reactions of silyl anions, unexpected substitution products are sometimes formed in the reaction of germyl anions with alkyl halides by an electron-transfer process depending on the structure of alkyl halides. The reaction of bis(triethylgermyl)cuprate with allylic acetates proceeds to give allylgermanes with high regioselectivity (eq (93)) [90]. Acylgermanes are obtained by the reaction of acid chlorides with germyl anions (eq (94)) [91].

$$Ph_3GeLi + PhCH_2Cl \longrightarrow Ph_3GeCH_2Ph \qquad (92)$$

$$\text{1-vinylcyclohexyl acetate (OAc)} \xrightarrow{(Et_3Ge)_2CuLi} \text{cyclohexylidene}{=}CHCH_2GeEt_3 \qquad (93)$$

$$Ph_3GeLi + PhC(O)Cl \longrightarrow Ph_3GeC(O)Ph \qquad (94)$$

The reaction of germanium(II) iodide with conjugated dienes and the following treatment with Grignard reagents constitutes the convenient method for the preparation of cyclic tetraorganogermanes (eq (95)) [92].

$$GeI_2 + \text{butadiene} \longrightarrow I_2Ge(C_4H_6) \xrightarrow{2MeMgI} Me_2Ge(C_4H_6) \quad (95)$$

17.4.3.3 Triorganogermanium Compounds

Triorganogermanium compounds are prepared using triorganohalogermanes as starting materials. The reduction of trialkylhalogermanes with lithium aluminum hydride affords the corresponding trialkylgermanes (eq (96)) [93]. Various organogermanium compounds with heteroatom substituents are also obtained by the reaction of trialkylhalogermanes with metal alkoxides (eq (97)) [94], amides (eq (98)) [84], thiolates (eq (99)) [84], and so on.

$$Ph_3GeBr \xrightarrow{LiAlH_4} Ph_3GeH \quad (96)$$

$$Me_2GeCl_2 + 2NaOMe \longrightarrow Me_2Ge(OMe)_2 \quad (97)$$

$$Me_3GeBr + Et_2NLi \longrightarrow Me_3GeNEt_2 \quad (98)$$

$$Me_3GeBr + EtSNa \longrightarrow Me_3GeSEt \quad (99)$$

Preparation of trialkylgermylmetals is performed by the reduction of trialkylhalogermanes with alkali metals (eq (100)) [89] or the treatment of trialkylgermanes with appropriate bases (eq (101)) [95].

$$Ph_3GeCl + 2\,Li \longrightarrow Ph_3GeLi \quad (100)$$

$$Ph_3GeH + BuLi \longrightarrow Ph_3GeLi \quad (101)$$

17.5 Synthesis of Organogermanium Compounds

(1) Chlorotriethylgermane [96]

$$Et_4Ge + AcCl \xrightarrow{AlCl_3} Et_3GeCl \quad (102)$$

To a flask charged with finely powdered aluminum chloride (33 mmol) is added tetraethylgermane (30 mmol) at 0 °C under argon. Acetyl chloride (33 mmol) is then added dropwise to the resulting slurry with stirring. An exothermic reaction occurs and the reaction mixture becomes homogeneous at the end of addition. After being stirred overnight at room temperature, chlorotriethylgermane (93%) is isolated by distillation of the reaction mixture under reduced pressure.

Properties: bp 42 °C / 4 mmHg (175.9 °C [85]). ^1H NMR ($CDCl_3$): $\delta = 1.13–1.15$ (m).

(2) (*E*)-3,7-Dimethyl-1-triethylgermyl-2,6-octadiene [90]

$$\text{geranyl–OAc} + (Et_3Ge)_2CuLi \longrightarrow \text{geranyl–GeEt}_3 \qquad (103)$$

To a THF (1 ml) solution of triethylgermane (1.0 mmol) and N,N,N′,N′-tetramethylethylenediamine (1.1 mmol) is added a pentane solution of *t*-butyllithium (1.1 mmol) at 0 °C. After being stirred for 30 min, the resultant solution of triethylgermyllithium is added to a suspension of CuI (0.5 mmol) in THF (2 mL) at 0 °C, and the reaction mixture is stirred for 10 min. After cooling (–23 °C), geranyl acetate (0.5 mmol) in THF (1 ml) is added. The mixture is stirred for 30 min and then quenched by addition of a phosphate buffer solution (pH 7). The organic material is extracted with ether. The etherial layer is washed with 3.5% aqueous NH_3 and water successively and dried (Na_2SO_4). After evaporation of solvent, (*E*)-3,7-dimethyl-1-triethylgermyl-2,6-octadiene (80%) is isolated by TLC (silica gel, hexane).

Properties: bp 100 °C (air–bath temperature)/3 mmHg. ^{1}H NMR ($CDCl_3$): $\delta = 0.72$ (6H, q, $J = 7.7$ Hz), 1.01 (9H, t, $J = 7.7$Hz), 1.57 (3H, s), 1.60 (3H, s), 1.52–1.63 (2H, m), 1.67 (3H, d, $J = 1.0$ Hz), 1.89–2.16 (4H, m), 5.01–5.14 (1H, m), 5.22 (1H, t, $J = 8.6$ Hz). ^{13}C NMR ($CDCl_3$): $\delta = 4.24$, 8.97, 13.17, 15.68, 17.67, 25.69, 26.97, 40.05, 121.53, 124.76, 131.03, 131.36.

17.6 Organotin Compounds

17.6.1 General Aspects [97]

Despite the long history of organotin chemistry started with the synthesis of diethyltin diiodide by Frankland in 1849, their use in organic synthesis has scarcely investigated before 1970. As with the development of organosilicon compounds as synthetic reagents, application of organotin compounds has also become of interest in this field.

The principal reactivities of organotin compounds are similar to those of organosilicon compounds. In general, organotin compounds are more reactive than the organosilicon analogues. Moreover, they show remarkable properties in certain reactions such as tin–lithium transmetallation. Other characteristic reactions of organotin compounds are those which involve the homolitic cleavage of tin–carbon, tin–hydrogen, or tin–heteroatom bonds. The representative examples are noted later.

Some organotin compounds are unstable compared with the corresponding organosilicons. For example, α-hydroxystannanes and certain vinylstannanes are readily hydrolyzed during chromatography and allylstannanes are apt to be oxidized by exposure to air for a long period. However, in practice, organotin compounds can be handled without any special experimental technique which are generally required in the preparation of other organometallics.

Since organotin compounds are more toxic than organosilicon and germanium compounds, experiments using these compounds should be performed in a well ventilated food with appropriate chemical resistant gloves. In general the toxicity decreases with increasing the number of heteroatom substituents on tin atom. The toxicity of organotin compounds is also dependent of the structures of alkyl substituents. Since methyl and ethyl derivatives are highly toxic, the use of less toxic butyl derivatives is strongly recommended unless the experiment has a special purpose to use such low boiling compounds.

Organotin compounds have been used not only in organic synthesis but also in industries as stabilizers of poly(vinyl chloride), antifungal agents, agricultural chemicals, and so on. Therefore various fundamental organotin compounds including haloorganostannanes are commercially available.

17.6.2 Use of Organotin Compounds in Organic Synthesis

Synthetically valuable organotin compounds are classified into the following categories; carbon nucleophiles in carbon–carbon bond forming reactions, precursors of organolithiums in transmetallation reactions, donors of alkyl groups in transition metal-catalyzed carbon–carbon bond forming reactions, reagents for the generation of tin-centered or carbon radicals, and reagents for functional group transformation. Reactions of representative organotin compounds follow. For details of these reactions, see recently published monographs [98].

17.6.2.1 Allyl, Alkenyl, Alkynyl and Arylstannanes

The allylation of aldehydes with allylstannanes is effected with Lewis acids such as boron trifluoride etherate or titanium(IV) chloride (eq (104)) [99], in which diastereoselectivity is dependent on the catalyst employed; the most striking feature is that *syn*-homoallyl alcohols are stereoselectively produced using boron trifluoride etherate. Although this allylation occurs without a catalyst, high reaction temperature must be used.

Me⁀SnBu$_3$ + PhCHO $\xrightarrow{BF_3 \cdot OEt_2}$ (product: OH, Ph, Me) (104)

In contrast, allylstannanes act as electrophiles when they were treated with appropriate higher valent metal salts or oxidized by electrochemical process (eq (105)) [100]. In the presence of tin(IV) chloride, allylstannanes react with nucleophiles such as enol silyl ethers or allylsilanes (eq (106)) [101] as well as electrophilic aldehydes.

SnBu$_3$ + (OMe-benzene) $\xrightarrow{(CF_3CO_2)_3Tl}$ (OMe product) (105)

SnBu$_3$ / Ph + SiMe$_3$ $\xrightarrow{SnCl_4}$ Ph (106)

The reactions of allyl, alkenyl, and arylstannanes with allyl, alkenyl, and aryl halides or triflates proceed in the presence of palladium catalysts to give the cross-coupling products (eq (107)) [102]. In the reaction of alkenylstannanes, their configuration is generally retained. If these reactions are carried out in an atmosphere of carbon monoxide, the corresponding ketones are obtained (eq (108)) [103].

OTf + SnBu$_3$ $\xrightarrow[\text{LiCl}]{\text{Pd(Ph}_3\text{P)}_4}$ (107)

I + SnBu$_3$ + CO $\xrightarrow{\text{PdCl}_2\text{(PPh}_3)_2}$ (108)

Ketones are also synthesized by the aluminum chloride or palladium-catalyzed acylation of organotin compounds (eq (109)) [104].

$$Ph_4Sn + PhCOCl \xrightarrow{Pd(PPh_3)_4} PhC(O)Ph \quad (109)$$

One of the most synthetically valuable reactions of tetraorganostannanes is the allylation which includes homolytic cleavage of allyl–tin bonds. The radical chain reactions of allylstannanes with alkyl halides and sulfides effected with radical initiators or photoirradiation afford the corresponding substitution products (eq (110)) [105].

PhO, S, O, O, O, OBn, O $\xrightarrow[h\nu]{\text{SnBu}_4}$ O, O, OBn, O (110)

Organolithiums, such as allyl, vinyl, and α-alkoxyalkyllithiums, are easily synthesized by the metal–metal exchange of organotin compounds with appropriate alkyllithiums in high yields (eq (111)) [106].

Bu$_3$Sn, H, H, OEt $\xrightarrow{\text{BuLi}}$ Li, H, H, OEt (111)

17.6.2.2 *Trialkyltin hydrides*

Trialkyltin hydrides reduce carbon–halogen, carbon–sulfur, and carbon–selenium bonds when they are heated with radical initiators (eq (112)) [107]. The carbon–oxygen bond in thiocarbonates and carbon–nitrogen bonds in organonitro compounds are also hydrogenolyzed chemoselectively. By this reduction, organic halides possessing sp^2 carbon–halogen bonds, such as vinyl, aryl (eq (113)) [108], and also acid halides (eq (114)) [109] are reduced to afford the corresponding hydrogenolyzed compounds.

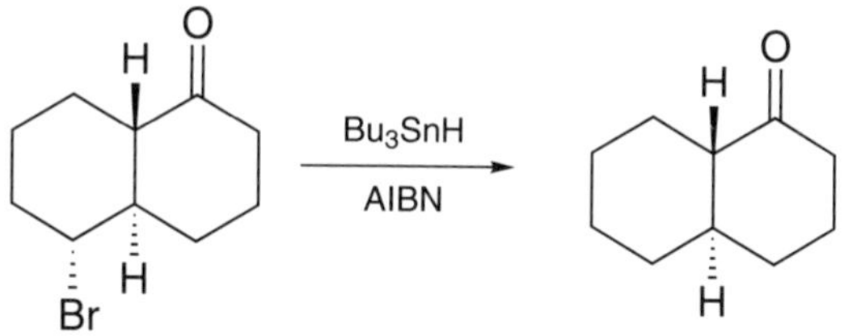

Bu_3SnH / AIBN (112)

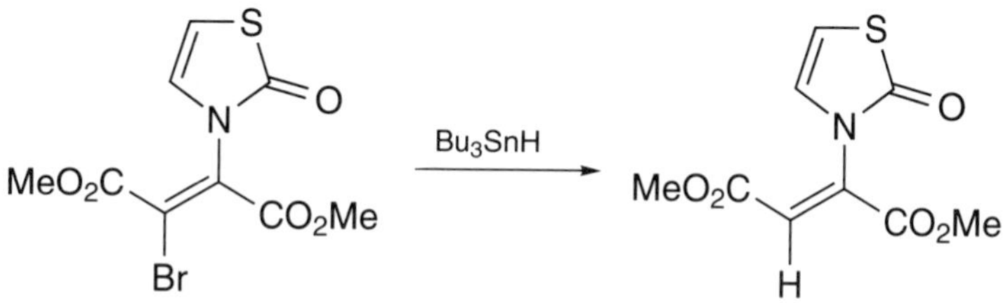

Bu_3SnH (113)

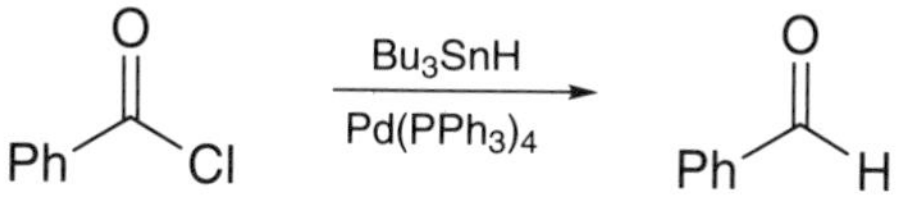

Bu_3SnH / $Pd(PPh_3)_4$ (114)

Since these reductions of carbon–heteroatom bonds proceed via carbon radical intermediates, the cyclized products are obtained by their intramolecular reaction with carbon–carbon multiple bonds when an olefin or acetylene moiety is present in their structures (eq (115)) [110].

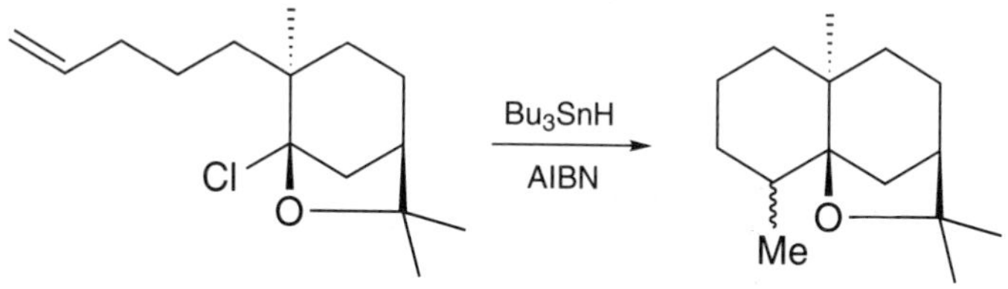

Bu_3SnH / AIBN (115)

Intermolecular trapping of intermediate radicals with acrylonitrile or acrylic acid esters also occurs (eq (116)) [111].

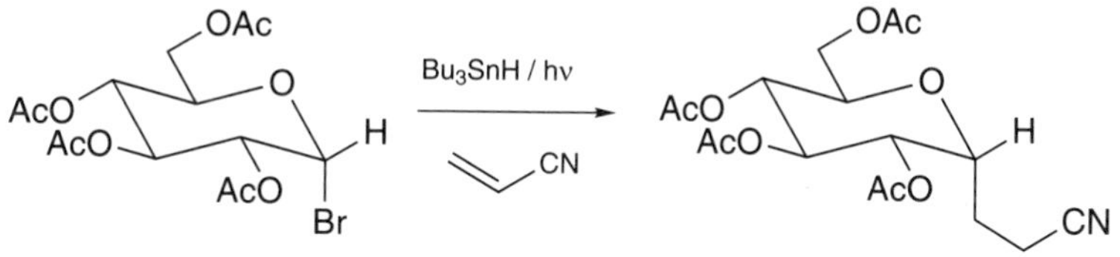

Bu_3SnH / hv (116)

17.6.2.3 The Organotin Compounds with Tin-Heteroatom Bonds

Organotin alkoxides are employed for the etherification and esterification of alcohols, the latter of which is also applied to the macrolide synthesis (eq (117)) [112].

$$\text{methyl glucoside} \xrightarrow{(Bu_3Sn)_2O} \text{6-OSnBu}_3\text{ derivative} \xrightarrow{PhCOCl} \text{6-OCOPh derivative} \quad (117)$$

Aminations of alkyl halides and carboxylic esters are performed with organotin amides, which are also employed for the preparation of enamines (eq (118)) [113].

$$MeC(O)Me \xrightarrow{Bu_2NSnBu_3} CH_2{=}C(NBu_2)Me \quad (118)$$

Organotin compounds with tin–sulfur bonds are useful for the transformations of alkyl halides to sulfides (eq (119)) [114], esters to thiolesters (eq (120)) [115], and acetals to monothioacetals (eq (121)) [116].

$$CH_2{=}CHCH(Cl)C_4H_9 \xrightarrow{MeSn(SMe)_3} MeSCH_2CH{=}CHC_4H_9 \quad (119)$$

$$PhCH_2COCl \xrightarrow{Bu_3SnSPh} PhCH_2COSPh \quad (120)$$

$$Hex(Me)C(OMe)_2 \xrightarrow{Bu_3SnSPh} Hex(Me)C(OMe)(SPh) \quad (121)$$

Without any catalyst, organotin enolates react with allylic halides to afford the monoallylation products (eq (122)) [117]. Alkenyl halides, aryl halides, and allylic acetates can be used for the alkylation of tin enolates when the reactions are catalyzed

$$\text{OSnPh}_3\text{ enol ether (t-BuMe}_2\text{SiO, Pent, t-BuMe}_2\text{SiO)} \xrightarrow{I\text{-CH}_2CH{=}CH(CH_2)_3CO_2Me} \text{alkylated cyclopentanone (CO}_2\text{Me, Pent, t-BuMe}_2\text{SiO)} \quad (122)$$

with transition metal complexes (eq (123)) [118]. The tin(II) enolates prepared by the treatment of ketones with tin(II) triflate are highly reactive and their reaction with aldehydes afford *syn*-aldols (eq (124)) [119].

$$\text{2-methyl-1-(OSnBu}_3\text{)cyclohexene} + \text{CH}_2\text{=C(OEt)CH}_2\text{OAc} \xrightarrow{Pd(PPh_3)_4} \text{2-methyl-2-(2-ethoxyallyl)cyclohexanone} \quad (123)$$

$$PhCOCH_2CH_3 \xrightarrow[\text{N-ethylpiperidine}]{Sn(OTf)_2} Ph(OSnOTf)C{=}CHMe \xrightarrow{PhCHO} PhCOCH(Me)CH(OH)Ph \quad (124)$$

17.6.3 Preparation of Organotin Compounds

17.6.3.1 Haloorganotin Compounds

Direct reactions of tin metal with relatively reactive alkyl halides are widely used methods for the preparation of dialkyldihalostannanes (eq (125)) [120]. Using catalysts such as other metals, metal halides, or Lewis bases, this method can be also applied to unreactive alkyl halides (eq (126)) [121].

$$Sn + 2\,PhCH_2Cl \longrightarrow (PhCH_2)_2SnCl_2 \quad (125)$$

$$Sn + 2\,CH_2{=}CHCH_2Br \xrightarrow{Et_3N\,/\,HgCl_2} (CH_2{=}CHCH_2)_2SnBr_2 \quad (126)$$

Alkyltrihalostannanes are obtained by the Sb(III)-catalyzed reaction of tin(II) halides with alkyl halides at elevated temperatures (eq (127)) [122].

$$SnBr_2 + n\text{-}C_{18}H_{37}Br \xrightarrow{Et_3Sb} n\text{-}C_{18}H_{37}SnBr_3 \quad (127)$$

The most general procedure for the preparation of haloorganostannanes is the disproportionation of tetraalkylstannanes with tin(IV) halides. By a change in the molar ratio of starting materials, trialkylhalostannanes, dialkyldihalostannanes, and alkyltrihalostannanes are selectively formed (eq (128)) [123].

$$3\,Ph_4Sn + SnCl_4 \longrightarrow 4\,Ph_3SnCl \quad (128)$$

The alternative way to trialkylbromostannanes or trialkyliodostannanes is the cleavage of tin–carbon bonds of tetraalkylstannanes with bromine or iodine (eq (129)) [124].

$$Me_4Sn + Br_2 \longrightarrow Me_3SnBr \quad (129)$$

17.6.3.2 Tetraorganotin Compounds

Symmetrical tetraorganotin compounds are obtained by the treatment of tin(IV) chloride with Grignard reagents or alkyllithiums (eq (130)) [125].

$$SnCl_4 + 4\ CH_2{=}CHMgBr \longrightarrow (CH_2{=}CH)_4Sn \quad (130)$$

The treatment of organometallic compounds with appropriate haloorganostannanes affords unsymmetrical tetraorganotin compounds (eq (131)) [126]. By the use of Reformatsky reagents and metal enolates, α-trialkylstannyl esters and ketones are obtained, respectively.

Alternatively these compounds are prepared by the treatment of enol acetates with trialkylalkoxystannanes or addition of alkoxystannanes to ketenes (eq (132)) [127]. The ratio of O- and C-enolates is dependent on the substituents on enol acetates.

NNHTos, Et, Me $\xrightarrow{BuLi}$ Li, Et, Me $\xrightarrow{Me_3SnCl}$ $SnMe_3$, Et, Me (131)

OAc, Me $\xrightarrow{Bu_3SnOMe}$ $OSnBu_3$, Me + O, Me, $SnBu_3$ (132)

75 : 25

The treatment of terminal acetylenes with trialkylstannylamides afford the corresponding alkynylstannanes (eq (133)) [128].

H $\xrightarrow{Me_3SnNMe_2}$ Me_3Sn (133)

Tetraorganotin compounds are also synthesized by the hydrostannation of carbon–carbon multiple bonds effected with radical initiators (eq (134)) [129]. Although these reactions also proceed without initiator, such reactions sometimes give mixtures of regioisomers depending on the substituents and functional groups attached to carbon–carbon multiple bonds (eq (135)) [130].

CN $\xrightarrow[AIBN]{Bu_3SnH}$ $SnBu_3$ CN (134)

CN $\xrightarrow{Bu_3SnH}$ Bu_3Sn CN (135)

Since the various trialkylstannylmetals are available, their reactions with electrophiles provide a convenient methods for the preparation of organotin compounds. Substitutions of alkyl halides, tosylates, acetates, and sulfides with suitable stannylmetals afford a variety of tetraorganotin compounds (eq (136)) [131].

$$\text{PhCH=CHCH}_2\text{OAc} \xrightarrow[\text{Pd(PPh}_3)_4]{\text{Et}_2\text{AlSnBu}_3} \text{PhCH=CHCH}_2\text{SnBu}_3 \qquad (136)$$

The additions of trialkylstannyl anions to aldehydes, ketones, enones, and alkynes also occur to afford tetraorganostannanes with various functional groups. α-Alkoxyalkylstannanes prepared by the reactions of trialkylstannyllithium with carbonyl compounds followed by the protection of their hydroxyl groups are the useful precursors of α-alkoxyalkyllithiums (eq (137)) [132].

Hex–CHO —[1) Bu_3SnLi; 2) EtOCH(Me)Cl]→ Hex–CH($SnBu_3$)–O–CH(Me)–OEt —[1) BuLi; 2) cyclohexanone]→ 1-[1-(1-ethoxyethoxy)heptyl]cyclohexanol (HO, Hex, Me) (137)

17.6.3.3 Some Other Synthetically Useful Organotin Compounds

Trialkyltin hydrides are prepared by the reduction of trialkylhalostannanes with lithium aluminum hydride or the reaction of poly(methylhydrosiloxane) with bis(trialkyltin)oxides (eq (138)) [133]. Starting materials for the latter preparation are obtained by the hydrolysis of trialkylchlorostannanes with sodium hydroxide followed by the further treatment of the resulting trialkylhydroxystannanes with sodium metal.

$$\text{Bu}_3\text{SnOSnBu}_3 + \left(\text{Si(Me)(H)–O}\right)_n \longrightarrow 2\ \text{Bu}_3\text{SnH} \qquad (138)$$

Various organotin compounds with tin–heteroatom bonds are synthesized by the reactions of halostannanes with the corresponding heteroatom nucleophiles. For example, trialkyltin alkoxides and trialkyltin amides are prepared by the treatments of halostannanes with sodium alkoxides (eq (139)) [134] and lithium amides (eq (140)) [135], respectively. Tin alkoxides are also formed by the reaction of alcohols with tin amides (eq (141)) [136].

$$\text{Bu}_3\text{SnCl} + \text{NaOMe} \longrightarrow \text{Bu}_3\text{SnOMe} \qquad (139)$$

$$\text{Me}_3\text{SnCl} + \text{Me}_2\text{NLi} \longrightarrow \text{Me}_3\text{SnNMe}_3 \qquad (140)$$

$$\text{Bu}_3\text{SnNMe}_3 + \text{MeOH} \longrightarrow \text{Bu}_3\text{SnOMe} \qquad (141)$$

Tin hydroxides and oxides are also employed for the syntheses of heteroatom-substituted organotin compounds; the treatment of trifluoromethanesulfonic anhydride with bis(trialkyltin)oxides affords trialkyltin triflates (eq (142)) [137] and the reaction of trialkylchlorostannanes with thiols proceeds via trialkylhydroxystannanes to give trialkylstannyl sulfides (eq (143)) [138].

$$Bu_3SnOSnBu_3 + (CF_3SO_2)_2O \longrightarrow 2\ Bu_3SnO_3SCF_3 \quad (142)$$

$$Me_3SnCl \xrightarrow{NaOH} Me_3SnOH \xrightarrow{MeSH} Me_3SnSMe \quad (143)$$

17.7 Synthesis of Organotin Compounds

(1) Tetramethylstannane [139]

$$SnCl_4 + 4\ CH_3MgI \longrightarrow (CH_3)_4Sn \quad (144)$$

Tetrachlorostannane (0.19–0.29 mol) is added dropwise to a butyl ether solution of methymagnesium iodide prepared from methyl iodide (1.59 mol) and magnesium (2.06 g. atom) at room temperature. Only gentle refluxing should occur during this step which requires 2–2.5 h to complete. The reaction mixture is heated under steady reflux (85–95 °C) for 1h and then allows to stand for several hours. The crude product is distilled from the reaction mixture, and a mixture of tetramethylstannane and butyl ether, distilling at 85–95 °C, is obtained. Tetramethylstannane (85–91%) is isolated by the fractional distillation of the mixture.

Properties: bp 76.6 °C / 748 mmHg.

(2) 1-Hexynyltrimethylstannane [140]

$$\text{1-hexyne (C≡C–H)} \xrightarrow[\text{2) } Me_3SnCl]{\text{1) BuLi}} \text{1-hexynyl–}SnMe_3 \quad (145)$$

To an ethereal (10 mL) solution of 1-hexyne (10.5 mmol) is added dropwise butyllithium (10.5 mmol, 1.74 M) at –78 °C. The reaction mixture is stirred for 30 min at –78 °C followed by warming to 0 °C and stirring for 30 min. Chlorotrimethylstannane (10 mmol) in dry ether (25 mL) is added dropwise to the mixture at –78 °C, and the mixture is slowly warmed to room temperature over a 12 h period. After removal of ether by fractional distillation, all volatiles are transferred at 0.35 mmHg with an oil bath temperature of 45–48 °C. The distillate is collected at –78 °C. 1-Hexynyltrimethylstannane (92.5%) is isolated by distillation.

Properties: bp 26 °C / 0.35 mmHg. ^1H NMR ($CDCl_3$): $\delta = 0.22$ (9H, s), 0.87 (3H, t, $J =$ 6.9Hz), 1.34–1.47 (4H, m), 2.20 (2H, t, $J = 6.9$ Hz). ^{13}C NMR ($CDCl_3$): $\delta = -7.85$, 13.53, 19.78, 21.93, 31.20, 81.78, 111.02. IR (neat) 2144 cm^{-1}.

1-Hexynyltrimethylstannane is also prepared by the heating a mixture of 1-hexyne (10.9 mmol) and (diethylamino)trimethylstannane (10.9 mmol) at 50 °C for 12 h in 89.2% yield [140].

(3) Ethyl (*E*)-3-(tributylstannyl)-2-pentenoate [141]

$$\text{EtC}\equiv\text{CCO}_2\text{Et} \xrightarrow{Bu_3SnCu\cdot MeS\bullet LiBr} (E)\text{-}Bu_3Sn(Et)C{=}CHCO_2Et \quad (146)$$

Butyllithium (26 mmol, 1.63 M) is added to a cold (–20 °C) stirred solution of hexabutyldistannane (26 mmol) in THF (200 mL) under argon. The solution is stirred at –20 °C for 20 min and then cooled to –78 °C. CuBr–SMe_2 complex (26 mmol) is added and the mixture is stirred for 30 min. To the mixture is added a THF (10 mL) solution of ethyl 2-pentynoate (20 mmol) and stirring is continued for 3 h. Ethanol (1 mL), saturated NH_4Cl (20 mL), and ether (200 mL) are added successively and the mixture is warmed to room temperature. The layers are separated and the aqueous layer is extracted with ether. The combined extracts are washed (saturated NH_4Cl) and dried ($MgSO_4$). Removal of the solvent affords an oil which is chromatographed on silica gel (ether : petroleum ether = 1: 50). Ethyl (*E*)-3-(tributylstannyl)-2-pentenoate (83%) is isolated by distillation (air-bath temperature 130 °C / 0.3 mmHg) of the material obtained from the appropriate fractions.

Properties: ^1H NMR ($CDCl_3$): $\delta = 0.7–1.8$ (33H, diffuse m), 2.88 (2H, d of q, $J = 7.5$, 1.2 Hz, with satellite peaks indicating $J_{Sn-H} \approx 56$ Hz), 4.16 (2H, q, $J = 7$ Hz), 5.92 (1H, t, $J = 1.2$ Hz with satellite peaks indicating $J_{Sn-H} = 65$ Hz). IR (neat) 1710, 1585, 1180, 870 cm^{-1}.

(4) (Benzyloxymethyl)tributylstannane [132]

$$ICH_2ZnI + Bu_3SnCl \longrightarrow Bu_3SnCH_2I \xrightarrow{PhCH_2ONa} Bu_3SnCH_2OCH_2Ph \quad (147)$$

A THF (100 mL) solution of iodomethylzinc iodide (0.15 mol) [142] is treated with tributyltin chloride (0.1 mol) under nitrogen. After stirring for 18 h, the reaction mixture is poured into petroleum ether (300 mL) and washed with water (2 × 200 mL). After drying (Na_2SO_4), the solvent is removed under reduced pressure to give tributyl(iodomethyl)stannane (96%). Sodium hydride (50% dispersion, 0.10 mol) is washed with pentane (three times) and suspended in THF (250 mL) under nitrogen. Benzyl alcohol (0.085 mol) was added dropwise with stirring. After stirring at room temperature for *ca.* 1 h, tributyl(iodomethyl)stannane (0.065 mol) was added. The mixture is stirred for 48 h and then treated with a little methanol to destroy any excess sodium hydride. The mixture is poured into petroleum ether (1 L), washed with water (2 × 250 mL), and dried (Na_2SO_4). After removal of solvent under reduced pressure, the residue is distilled through a 20 inch Vigreux column to give (benzyloxymethyl)-tributylstannane (81%).

Properties: bp 140–144 °C / 0.03 mmHg. ^{1}H NMR (CCl_4): δ = 4.40 (2H, s), 7.15 (5H, s). IR (neat) 1455, 1375, 1085, 1065 cm^{-1}.

(5) Triphenyltin hydride [143]

$$Ph_3SnCl \xrightarrow{LiAlH_4} Ph_3SnH \qquad (148)$$

Lithium aluminum hydride (40.9 mmol) and triphenylchlorostannane (100 mmol) are added to ether (150 mL) in a flask cooled with an ice–water bath. The mixture is stirred at the bath temperature for 15 min, and then at room temperature for 3 h. It is slowly hydrolyzed with water (100 mL) with cooling by the ice–water bath. The ether layer is washed with ice–water (2 × 100 mL) and dried ($MgSO_4$). After removal of ether, triphenyltin hydride (77–85%) is obtained by rapid distillation using an oil-bath preheated to 200 °C. Since the hydride is thermally unstable, low distillation pressures are desirable.

Properties: bp 162–168 °C / 0.5 mmHg.

It should be noted that trialkyltin hydrides, especially when they are contaminated with certain impurities, gradually decompose with evolution of hydrogen even if they are stored at low temperatures.

(6) Tributylmethoxystannane [144]

$$Bu_3SnOSnBu_3 + MeOC(=O)OMe \longrightarrow 2\ Bu_3SnOMe \qquad (149)$$

Bis(tributyltin) oxide (10 mmol) and dimethyl carbonate (15 mmol) are refluxed in a bath at 130–135 °C. A slow stream of nitrogen is passed through the liquid and then through $Ca(OH)_2$solution. The evolution of CO_2 begins after *ca.* 15 min and is complete after 40 min. Distillation gives tributylmethoxystannane (93%).

Properties: bp 90 °C / 0.1 mmHg. ^{1}H NMR : δ = 3.6 (OCH_3). IR (neat) 1075 cm^{-1}.

References

[1] a) Y. Apeloig, *The Chemistry of Organic Silicon Compounds*; S. Patai, Z. Rappoport, Eds, Vol. 1, p. 57, John Wiley, Chichester (1989).
(b) S. Pawlenko, *Organosilicon Chemistry*; Walter de Gruyter: Berlin (1986).
(c) D. R. M. Walton, *Organosilicon Compounds, Dictionary of Organometallic Compounds*, Vol. 2, Chapman and Hall, London (1984).
(d) D. A. Armitage, *Comprehensive Organometallic Chemistry*; G. Wilkinson, F. G. A. Stone, E. W. Abel, Eds, Vol. 2, p. 1, Pergamon, Oxford (1982).

[2] A. R. Bassindale, P. G. Taylor, *The Chemistry of Organic Silicon Compounds*; S. Patai, Z. Rappoport, Eds, Vol. 1, p. 839, John Wiley & Sons, Chichester (1989).

[3] K. Tamao, T. Hayashi, Y. Ito, *Nippon Kagaku Kaishi*, 509 (1990); K. Tamao, K. Kobayashi, Y. Ito, *J. Am. Chem. Soc.*, **111**, 6478 (1989).

[4] Y. Ito, M. Nakatsuka, T. Saegusa, *J. Am. Chem. Soc.*, **103**, 476 (1981).
[5] I. Kuwajima, M. Kato, *J. Chem. Soc., Chem. Commum.*, 708 (1979).
[6] D. Ayalon-Chass, E. Ehlinger, P. Magnus, *J. Chem. Soc., Chem. Commum.*, 772 (1977).
[7] (a) G. L. Larson, *The Chemistry of Organic Silicon Compounds*, S. Patai, Z. Rappoport, Eds, Vol. 1, p. 763, John Wiley, Chichester (1989).
(b) E. W. Colvin, *Silicon Reagents in Organic Synthesis*, Academic Press, London (1988).
(c) W. P. Weber, *Silicon Reagents for Organic Synthesis*, Springer-Verlag, Berlin (1983).
(d) P. D. Magnus, T. Sarkar, S. Djuric, *Comprehensive Organometallic Chemistry*; G. Wilkinson, F. G. A. Stone, E. W. Abel, Eds, Vol. 7, p. 515, Pergamon, Oxford (1982).
(e) E. W. Colvin, *Silicon in Organic Synthesis*; Butterworths, London (1981).
[8] I. Fleming, D. Higgins, *J. Chem. Soc., Perkin Trans.*, **1**, 3327 (1992).
[9] J. -P. Pillot, J. Dunogues, R. Calas, *Tetrahedron Lett.*, 1871 (1976).
[10] H. Sakurai, A. Hosomi, J. Hayashi, *Org. Synth. Coll.*, **VII**, 443 (1990).
[11] M. Kira, T. Hino, H. Sakurai, *Tetrahedron Lett.*, **30**, 1099 (1989).
[12] A. W. P. Jarvie, A. Holt, J. Thompson, *J. Chem. Soc., B* 852 (1969).
[13] T. H. Chan, P. W. K. Lau, W. Mychajlowskij, *Tetrahedron Lett.*, 3317 (1977).
[14] L. Birkofer, A. Ritter, H. Uhlenbrauck, *Chem. Ber.*, **96**, 3280 (1963).
[15] G. Stork, E. Colvin, *J. Am. Chem. Soc.*, **93**, 2080 (1971).
[16] K. Karabelas, A. Hallberg, *J. Org. Chem.*, **54**, 1773 (1989).
[17] Y. Hatanaka, T. Hiyama, *J. Org. Chem.*, **54**, 268 (1989).
[18] J. R. Pratt, F. H. Pinkerton, S. F. Thames, *J. Organomet. Chem.*, **38**, 29 (1972).
[19] E. Nakamura, H. Murofushi, M. Shimizu, I. Kuwajima, *J. Am. Chem. Soc.*, **98**, 2346 (1976).
[20] T. Mukaiyama, K. Banno, K. Narasaka, *J. Am. Chem. Soc.*, **96**, 7503 (1974).
[21] K. Narasaka, K. Soai, Y. Aikawa, T. Mukaiyama, *Bull. Chem. Soc. Jpn.*, **49**, 779 (1976).
[22] T. H. Chan, I. Paterson, J. Pinsonnault, *Tetrahedron Lett.*, 4183 (1977).
[23] R. E. Donaldson, P. L. Fuchs, *J. Org. Chem.*, **42**, 2032 (1977).
[24] I. Kuwajima, E. Nakamura, M. Shimizu, *J. Am. Chem. Soc.*, **104**, 1025 (1982).
[25] G. Stork, P. F. Hudrlik, *J. Am. Chem. Soc.*, **90**, 4464 (1968).
[26] M. E. Jung, C. A. McCombs, Y. Takeda, Y.-G. Pan, *J. Am. Chem. Soc.*, **103**, 6677 (1981).
[27] A. Hassner, R. H. Reuss, H. W. Pinnjck, *J. Org. Chem.*, **40**, 3427 (1975).
[28] M. Fujita, T. Hiyama, *J. Am. Chem. Soc.*, **107**, 8294 (1985).
[29] (a) H. Brunner, R. Becker, G. Riepl, *Organometallics*, **3**, 1354 (1984).
(b) H. Brunner, A. Kurzinger, *J. Organomet. Chem.*, **346**, 413 (1988).
[30] D. J. Peterson, *J. Org. Chem.*, **33**, 780 (1968).
[31] P. F. Hudrlik, D. Peterson, R. J. Rona, *J. Org. Chem.*, **40**, 2263 (1975).
[32] T. Aoyama, S. Terasawa, K. Sudo, T. Shioiri, *Chem. Pharm. Bull.*, **32**, 3759 (1984).
[33] D. A. Evans, L. K. Truesdale, G. L. Carroll, *J. Chem. Soc., Chem. Commum.*, 55 (1973).
[34] D. A. Evans, L. K. Truesdale, K. G. Grimm, S. L. Nesbitt, *J. Am. Chem. Soc.*, **99**, 5009 (1977).
[35] T. Mukaiyama, T. Takeda, K. Atsumi, *Chem. Lett.*, 187 (1974).
[36] M. E. Jung, W. A. Andrus, P. L. Ornstein, *Tetrahedron Lett.*, 4175 (1977).
[37] D. A. Johnson, *Carbohydr. Res.*, **237**, 313 (1992).
[38] S. Murata, M. Suzuki, R. Noyori, *J. Am. Chem. Soc.*, **101**, 2738 (1979).
[39] S. D. Rosenberg, J. J. Wallburn, H. E. Ramsden, *J. Org. Chem.*, **22**, 1606 (1957).
[40] M. E. Jung, M. A. Lyster, *Org. Synth. Coll.*, **VI**, 353 (1988).
[41] J. S. Peake, W. H. Nebergall, Y. H. Chen, *J. Am. Chem. Soc.*, **74**, 1526 (1952).
[42] G. Zweifel, W. Lewis, *J. Org. Chem.*, **43**, 2739 (1978).
[43] I. Fleming, D. Higgins, N. J. Lawrence, A. P. Thomas, *J. Chem. Soc., Perkin Trans., I*, 3331 (1992).
[44] H. O. House, L. J. Czuba, M. Gall, H. D. Olmstead, *J. Org. Chem.*, **34**, 2324 (1969).
[45] G. Simchen, W. Kober, *Synthesis*, 259 (1976).
[46] G. Simchen, W. West, *Synthesis*, 247 (1977).
[47] N. Slougui, G. Rousseau, *Synth. Commum.*, **17**, 1 (1987).
[48] G. Stork, J. Singh, *J. Am. Chem. Soc.*, **96**, 6181 (1974).
[49] I. Ojima, M. Nihonyanagi, T. Kogure, M. Kumagai, S. Horiuchi, K. Nakatsugawa, K. *J. Organomet. Chem.*, **94**, 449 (1975).

[50] H. J. Reich, R. C. Holtan, C. Bolm, *J. Am. Chem. Soc.*, **112**, 5609 (1990).
[51] M. E. Jung, M. A. Lyster, *Org. Synth. Coll.*, **VI**, 353 (1988).
[52] L. H. Sommer, E. W. Pietrusza, F. C. Whitmore, *J. Am. Chem. Soc.*, **68**, 2282 (1946).
[53] E. J. Corey, B. B. Snider, *J. Am. Chem. Soc.*, **94**, 2549 (1972).
[54] E. J. Corey, H. Cho, C. Rucker, D. H. Hua, *Tetrahedron Lett.*, **22**, 3455 (1981).
[55] M. Demuth, G. Mikhail, *Synthesis*, 827 (1982).
[56] R. O. Sauer, R. H. Hasek, *J. Am. Chem. Soc.*, **68**, 241 (1946).
[57] E. W. Abel, *J. Chem. Soc.*, 4406 (1960).
[58] F. C. Whitmore, E. W. Pietrusza, L. H. Sommer, *J. Am. Chem. Soc.*, **69**, 2108 (1947).
[59] Livinghouse, *Org. Synth. Coll.*, **VII**, 517 (1990).
[60] G. K. S. Prakash, S. Keyaniyan, R. Aniszfelt, L. Heiliger, G. A. Olah, R. C. Stevens, H. -K. Choi, R. Bau, *J. Am. Chem. Soc.*, **109**, 5123 (1987).
[61] L. Birkofer, P. Wegner, *Org. Synth. Coll.*, **VI**, 1030 (1988).
[62] F. C. Whitmore, L. H. Sommer, J. Gold, R. E. Van Strien, *J. Am. Chem. Soc.*, **69**, 1551 (1947).
[63] B.-T. Gröbel, D. Seebach, *Chem. Ber.*, **110**, 852 (1977).
[64] C. Eaborn, *J. Chem. Soc.*, 2755 (1949).
[65] E. W. Abel, R. J. Rowley, *J. Organomet. Chem.*, **84**, 199 (1975).
[66] J. A. Soderquist, G. J. -H. Hsu, *Organometallics*, **1**, 830 (1982).
[67] I. Fleming, I. Paterson, *Synthesis*, 736 (1979).
[68] P. Cazeau, F. Moulines, O. Laporte, F. Duboudin, *J. Organomet. Chem.*, **201**, C9 (1980).
[69] T. J. Barton, C. R. Tully, *J. Org. Chem.*, **43**, 3649 (1978).
[70] R. F. Cunico, L. Bedell, *J. Org. Chem.*, **45**, 4797 (1980).
[71] Y. Tanabe, M. Murakami, K. Kitaichi, Y. Yoshida, *Tetrahedron Lett.*, **35**, 8409 (1994).
[72] K. A. Hooton, A. L. Allred, *Inorganic Chem.*, **4**, 671 (1965).
[73] K. Mai, G. Patil, *J. Org. Chem.*, **51**, 3545 (1986).
[74] (a) P. Rivière, M. Rivière-Baudet, J. Satgè, *Comprehensive Organometallic Chemistry*; G. Wilkinson, F. G. A. Stone, E. W. Abel, Eds. Pergamon, Oxford, Vol. 2, p 399 (1982). (b) M. Lesbre, J. Mazerolles, J. Satgè, *The Organic Compounds of Germanium*; John Wiley, London (1971).
[75] C. A. Kraus, L. S. Foster, *J. Am. Chem. Soc.*, **49**, 457 (1927).
[76] R. W. Bott, C. Eaborn, D. R. M. Walton, *J. Organomet. Chem.*, **1**, 420 (1964).
[77] K. Wakamatsu, K. Oshima, K. Utimoto, *Chem. Lett.*, 2029 (1987).
[78] M. Ochiai, E. Fujita, M. Arimoto, H. Yamaguchi, *Chem. Pharm. Bull.*, **32**, 5027 (1984).
[79] K. Mizuno, K. Nakanishi, Y. Otsuji, *Chem. Lett.*, 1833 (1988).
[80] Y. Yamamoto, S. Hatsuya, J. Yamada, *J. Org. Chem.*, **55**, 3118 (1990).
[81] S. Inoue, Y. Sato, *J. Org. Chem.*, **56**, 347 (1991).
[82] M. Schmidt, I. Ruidisch, *Z. Anorg. Allg. Chem.*, **311**, 331 (1961).
[83] E. A. Flood, K. L. Godfrey, L. S. Foster, *Inorg. Synth.*, **3**, 64 (1950).
[84] E. W. Abel, D. A. Armitage, D. B. Brady, *J. Organomet. Chem.*, **5**, 130 (1966).
[85] C. A. Kraus, E. A. Flood, *J. Am. Chem. Soc.*, **54**, 1635 (1932).
[86] D. Seyferth, *J. Am. Chem. Soc.*, **79**, 2738 (1957).
[87] D. Seyferth, D. L. White, *J. Organomet. Chem.*, **32**, 317 (1971).
[88] S. Inoue, Y. Sato, T. Suzuki, *Organometallics*, **7**, 739 (1988).
[89] C. Tamborski, F. E. Ford, W. L. Lehn, G. L. Moore, E. J. Soloski, *J. Org. Chem.*, **27**, 619 (1962).
[90] J. Yamaguchi, Y. Tamada, T. Takeda, *Bull. Chem. Soc. Jpn.*, **66**, 607 (1993).
[91] D. A. Nicholson, A. L. Allred, *Inorg. Chem.*, **4**, 1747 (1965).
[92] G. Manuel, P. Mazerolles, *Organometallic Syntheses*, R. B. King, J. J. Eisch, Eds. Elsevier, Amsterdam, Vol. 3, p 552 (1986).
[93] O. H. Johnson, W. H. Nebergall, D. M. Harris, *Inorg. Synth.*, **5**, 76 (1957).
[94] R. West, H. R. Hunt, Jr. R. O. Whipple, *J. Am. Chem. Soc.*, **76**, 310 (1954).
[95] H. Gilman, C. W. Gerow, *J. Am. Chem. Soc.*, **78**, 5435 (1956).
[96] T. Takeda, Y. Tamada, N. Nomura, unpublished results; the preparation was carried out according to the method reported by H. Sakurai, K. Tominaga, T. Watanabe, M. Kumada, *Tetrahedron Lett.*, 5493 (1966)

[97] (a) P. G. Harrison Ed *Chemistry of Tin*, Blackies & Son, Glasgow (1989).
(b) I. Omae, *Organotin Chemistry, Journal of Organometallic Chemistry Library 21*, Elsevier, Amsterdam (1989).
(c) A. G. Davies, P. J. Smith, *Comprehensive Organometallic Chemistry*; G. Wilkinson, F. G. A. Stone, E. W. Abel, Eds. Pergamon, Oxford, Vol. 2, p. 519 (1982).
[98] (a) H. Nozaki, *Organometallics in Synthesis*, M. Schlosser, Ed. John Wiley, Chichester, p. 535 (1994).
(b) J. L. Wardel, *Chemistry of Tin* , P. G. Harrison, Ed. Blackies & Son, Glasgow, p. 315 (1989).
(c) M. Pereyre, J.-P. Quintard, A. Rahm, *Tin in Organic Synthesis*, Butterworth, Guildford (1987).
[99] (a) Y. Yamamoto, H. Yatagai, Y. Naruta, K. Maruyama, *J. Am. Chem. Soc.*, **102**, 7107 (1980).
(b) Y. Yamamoto, *Acc. Chem. Res.*, **20**, 243 (1987).
[100] M. Ochiai, E. Fujita, M. Arimoto, H. Yamaguchi, *Chem. Pharm. Bull.*, **31**, 86 (1983).
[101] T. Takeda, Y. Takagi, H. Takano, T. Fujiwara, *Tetrahedron Lett.*, **33**, 5381 (1992).
[102] W. Scott, J. K. Stille, *J. Am. Chem. Soc.*, **108**, 3033 (1986).
[103] W. Goure, M. Wright, P. Daris, S. Labodie, J. K. Stille, *J. Am. Chem. Soc.*, **106**, 6417 (1984).
[104] M. Kosugi, Y. Shimizu, T. Migita, *Chem. Lett.*, 1423 (1977).
[105] G. E. Keck, D. F. Kachensky, E. J. Enholm, *J. Org. Chem.*, **50**, 4317 (1985).
[106] R. H. Wollenberg, K. F. Albizati, R. Peries, *J. Am. Chem. Soc.*, **99**, 7365 (1977).
[107] (a) D. P. G. Hamon, K. P. Richards, *Aust. J. Chem.*, **36**, 2243 (1983).
(b) W. P. Neumann, *Synthesis* 665 (1987).
[108] A. Medici, M. Fogagnolo, P. Pedrini, A. Dondoni, *J. Org. Chem.*, **47**, 3844 (1982).
[109] P. Four, F. Guibe, *J. Org. Chem.*, **46**, 4439 (1981).
[110] G. Büchi, H. Wüest, *J. Org. Chem.*, **44**, 546 (1979).
[111] B. Giese, J. Dupuis, M. Nix, *Org. Synth.*, **65**, 236 (1987).
[112] Y. Tsuda, M. E. Haque, K. Yoshimoto, *Chem. Pharm. Bull.*, **31**, 1612 (1983).
[113] J. C. Pommier, A. Roubineau, *J. Organomet. Chem.*, **17**, 25 (1969).
[114] H. E. Katz, W. H. Starnes, Jr., *J. Org. Chem.*, **49**, 2758 (1984).
[115] D. N. Harpp, T. Aida,T. H. Chan, *Tetrahedron Lett.*, 2853 (1979).
[116] T. Sato, T. Kobayashi, T. Gojo, E. Yoshida, J. Otera, H. Nozaki, *Chem. Lett.*, 1661 (1987).
[117] M. Suzuki, A. Yanagisawa, R. Noyori, *J. Am. Chem. Soc.*, **107**, 3348 (1985).
[118] B. M. Trost, E. Keinan,*Tetrahedron Lett.*, **21**, 2591 (1980).
[119] (a) T. Mukaiyama, R. W. Stevens, N. Iwasawa, *Chem. Lett.*, 353 (1982).
(b) T. Mukaiyama, *Pure Appl. Chem.*, **58**, 505 (1986).
[120] K. Sisido,Y. Takeda, Z. Kinugawa, *J. Am. Chem. Soc.*, **83**, 538 (1961).
[121] K. Sisido, Y. Takeda, *J. Org. Chem.*, **26**, 2301 (1961).
[122] E. J. Bulten, *J. Organomet. Chem.*, **97**, 167 (1975).
[123] H. Gilman, S. D. Rosenberg, *J. Am. Chem. Soc.*, **74**, 5580 (1952).
[124] C. A. Kraus, W. V. Sessions, *J. Am. Chem. Soc.*, **47**, 2361 (1925).
[125] S. D. Rosenberg, A. J. Gibbons, Jr., H. E. Ramsden, *J. Am. Chem. Soc.*, **79**, 2137 (1957).
[126] W. Barth, L. A. Paquett, *J. Org. Chem.*, **50**, 2438 (1985).
[127] M. Pereyre, B. Bellegarde, J. Mendelsohn, J. Valade, *J. Organomet. Chem.*, **11**, 97 (1968).
[128] J. K. Stille, J. H. Simpson, *J. Am. Chem. Soc.*, **109**, 2138 (1987).
[129] A. J. Leusink, J. G. Noltes, *Tetrahedron Lett.*, 335 (1966).
[130] A. J. Leusink, J. W. Marsman, H. A. Budding, *Recl. Trav. Chim. Pays-Bas*, **84**, 689 (1965).
[131] B. M. Trost, J. W. Herndon, *J. Am. Chem. Soc.*, **106**, 6835 (1984).
[132] W. C. Still, *J. Am. Chem. Soc.*, **100**, 1481 (1978).
[133] K. Hayashi, J. Iyoda, I. Shiihara, *J. Organomet. Chem.*, **10**, 81 (1967).
[134] D. L. Alleston, A. G. Davies, *J. Chem. Soc.*, 2050 (1962).
[135] K. Jones, M. F. Lappert, *J. Chem. Soc.*, 1944 (1965).
[136] K. Jones, M. F. Lappert, *J. Organomet. Chem.*, **3**, 295 (1965).
[137] E. J. Corey, T. M. Eckrich, *Tetrahedron Lett.*, **25**, 2419 (1984).
[138] E. W. Abel, D. B. Brady, *J. Chem. Soc.*, 1192 (1965).

[139] W. F. Edgell, C. H. Ward, *J. Am. Chem. Soc.*, **76**, 1169 (1954).
[140] J. K. Stille, J. H. Simpson, *J. Am. Chem. Soc.*, **109**, 2138 (1987).
[141] E. Piers, J. M. Chong, K. Gustafson, R. J. Andersen, *Can. J. Chem.*, **62**, 1 (1984).
[142] D. Seyferth, S. B. Andrews, *J. Organomet. Chem.*, **30**, 151 (1971).
[143] H. G. Kuivila, O. F. Beumel, Jr., *J. Am. Chem. Soc.*, **83**, 1246 (1961).
[144] A. G. Davies, D. C. Kleinschmidt, P. R. Palan, S. C. Vasishtha, *J. Chem. Soc.*, **(C)** 3972 (1971).

Index

Index compiled by Geoffrey C. Jones

Periodic Table of the Elements

Group →; Period →; Atomic number; Atomic weight

Period	1	2	3	4	5	6	7	8	9	10	11	12	13	14	15	16	17	18
1	1 H 1.0079																	2 He 4.0026
2	3 Li 6.941	4 Be 9.0122											5 B 10.811	6 C 12.011	7 N 14.0067	8 O 15.9994	9 F 18.9984	10 Ne 20.179
3	11 Na 22.9898	12 Mg 24.305											13 Al 26.9815	14 Si 28.0855	15 P 30.9738	16 S 32.066	17 Cl 35.453	18 Ar 39.948
4	19 K 39.0983	20 Ca 40.078	21 Sc 44.9559	22 Ti 47.88	23 V 50.9415	24 Cr 51.996	25 Mn 54.9380	26 Fe 55.847	27 Co 58.9332	28 Ni 58.69	29 Cu 63.546	30 Zn 65.39	31 Ga 69.723	32 Ge 72.59	33 As 74.9216	34 Se 78.96	35 Br 79.904	36 Kr 83.80
5	37 Rb 85.4678	38 Sr 87.62	39 Y 88.9059	40 Zr 91.224	41 Nb 92.9064	42 Mo 95.94	43 Tc 98.9062	44 Ru 101.07	45 Rh 102.9055	46 Pd 106.42	47 Ag 107.8682	48 Cd 112.41	49 In 114.82	50 Sn 118.710	51 Sb 121.75	52 Te 127.60	53 I 126.9045	54 Xe 131.29
6	55 Cs 132.9054	56 Ba 137.33	57 * La 138.9055	72 Hf 178.49	73 Ta 180.9479	74 W 183.85	75 Re 186.207	76 Os 190.2	77 Ir 192.22	78 Pt 195.08	79 Au 196.9665	80 Hg 200.59	81 Tl 204.383	82 Pb 207.2	83 Bi 208.9804	84 Po (209)	85 At (210)	86 Rn (222)
7	87 Fr (223)	88 Ra 226.0254	89 ** Ac 227.0278	104 Rf (261)	105 Ha (262)	106 Sg (266)	107 Ns (261)	108 Hs (265)	109 Mt (266)									

Zintl border →

* LANTHANIDES	58 Ce 140.12	59 Pr 140.9077	60 Nd 144.24	61 Pm (147)	62 Sm 150.36	63 Eu 151.96	64 Gd 157.25	65 Tb 158.9254	66 Dy 162.50	67 Ho 164.9304	68 Er 167.26	69 Tm 168.9342	70 Yb 173.04	71 Lu 174.967
** ACTINIDES	90 Th 232.0381	91 Pa 231.0359	92 U 238.0289	93 Np 237.0482	94 Pu (244)	95 Am (243)	96 Cm (247)	97 Bk (247)	98 Cf (251)	99 Es (252)	100 Fm (257)	101 Md (260)	102 No (259)	103 Lr (262)

The symbols given for elements 104–109 are endorsed by the Nomenclature Committee of the American Chemical Society but have not yet been officially approved by the International Union of Pure and Applied Chemistry.